THIRD EDITION

Math and Dosage Calculations for Health Care

Kathryn A. Booth, RN-BSN, MS, RMA, RPT, CPhT
Total Care Programming, Inc.
Palm Coast, Florida

James E. Whaley, MS, RPh
Baker College
Owosso, Michigan

Boston Burr Ridge, IL Dubuque, IA New York San Francisco St. Louis
Bangkok Bogotá Caracas Kuala Lumpur Lisbon London Madrid Mexico City
Milan Montreal New Delhi Santiago Seoul Singapore Sydney Taipei Toronto

The McGraw·Hill Companies

MATH AND DOSAGE CALCULATIONS FOR HEALTH CARE, THIRD EDITION

Published by McGraw-Hill, a business unit of The McGraw-Hill Companies, Inc., 1221 Avenue of the Americas, New York, NY, 10020.

This book is printed on acid-free paper.

1 2 3 4 5 6 7 8 9 0 VNH/VNH 0 9

ISBN 978-0-07-337417-8

MHID 0-07-337417-2

Vice president/Editor in chief: *Elizabeth Haefele*
Vice president/Director of marketing: *John E. Biernat*
Senior Sponsoring Editor: *Debbie Fitzgerald*
Developmental Editor: *Connie Kuhl*
Executive Marketing Manager: *Roxan Kinsey*
Lead media producer: *Damian Moshak*
Media producer: *Marc Mattson*
Director, Editing/Design/Production: *Jess Ann Kosic*
Project manager: *Christine M. Demma*
Senior production supervisor: *Janean A. Utley*
Senior designer: *Srdjan Savanovic*
Senior photo research coordinator: *Lori Hancock*
Media project manager: *Mark A. S. Dierker*
Cover & interior design: *Srdjan Savanovic*
Typeface: *10/12 Palatino*
Compositor: *Macmillan Publishing Solutions*
Printer: *R. R. Donnelley*
Cover Credit: *©Veer/Stockbyte*

Photo Credits: Front Matter: © Corbis and Getty RF; Chapter 4: Figure 4 4: © Total Care Programming, Inc.; 4.5: Courtesy of Apothecary Products, Inc.; 4.9, 4.11, 4.12, 4.14, 4.15, 4.17-4.20: © Total Care Programming, Inc.; Chapter 8: Figures 8.1, 8.2, 8.6a, b, 8.7, 8.8: © Total Care Programming, Inc.; Chapter 9: Figure 9.14: © Corbis RF; Chapter 10: Figure 10.4, 10 5a-10.10: © Total Care Programming, Inc.; Chapter 11: Figure 11.1: © The McGraw-Hill Companies, Inc. /Jill Braaten, photographer; 11.2: © Creatis/Punchstock RF; 11.3: © Total Care Programming, Inc.

The credits section for this book begins on page 550 and is considered an extension of the copyright page.

Library of Congress Cataloging-in-Publication Data

Booth, Kathryn A., 1957-
Math and dosage calculations for health care / Kathryn A. Booth, James E. Whaley. -- 3rd ed.
p. ; cm.
Includes index.
ISBN-13: 978-0-07-337417-8 (alk. paper)
ISBN-10: 0-07-337417-2 (alk. paper)
1. Pharmaceutical arithmetic--Problems, exercises, etc. I. Whaley, James E. (James Earl)
II. Title.
[DNLM: 1. Drug Dosage Calculations--Problems and Exercises. 2. Pharmaceutical Preparations--administration & dosage--Problems and Exercises. 3. Mathematics—Problems and Exercises. QV 18.2 B725m 2010]
RS57.H334 2010
615'.1401513--dc22

www.mhhe.com

About the Authors

Kathryn A. Booth, RN-BSN, MS, RMA, RPT, CPhT, is an author, consultant, vice president, and owner of Total Care Programming, a multimedia software development company. Her background includes a bachelor's degree in nursing and a master's degree in education. Her 29 years of teaching, nursing, and health care experience span five states. She has authored and developed multimedia CD-ROM software and health occupations educational textbooks and educational materials for Total Care Programming, Inc., Glencoe/McGraw-Hill, Mosby Lifeline, and McGraw-Hill Higher Education. Her most recent textbook is *Medical Assisting: Administrative and Clinical Procedures*, third edition, by McGraw-Hill Higher Education. Mrs. Booth has presented at numerous state and national conferences since 1994. Her current focus is to develop additional health care materials that will assist health care educators and promote the health care profession.

James E. Whaley, RPh, MS, is currently an associate professor of health sciences and chemistry at Baker College of Owosso and coordinator of the Pharmacy Technician program, which is offered at five Baker College campuses. He routinely teaches anatomy and physiology, pathophysiology, a variety of pharmacy technician courses, and general chemistry. Mr. Whaley has taught at Baker College of Owosso, MI, since 1995 and was the first recipient of the College's prestigious Teacher of the Year award. He has been selected as a member of *Who's Who Among College Teachers* numerous times. Prior to coming to Baker, Mr. Whaley was twice ranked in the top 10 percent of instructors at the University of Illinois, where he was awarded a Fellowship in Cellular and Molecular Biology from the National Institutes of Health. Mr. Whaley worked as a retail pharmacist before beginning his career as an educator and has been a registered pharmacist since 1981.

Dedication

To the potential health care employees using this book: may your career goals be achieved and the health care workforce be increased from your accomplishment.

To my husband, Jim, for his enduring encouragement, love, friendship, and patience.

To my first granddaughter, Kaylyn, who has renewed my spirit and hope for the future.

K. Booth

To my brother, Tom, who taught me the power of perseverance. You are missed.

J. Whaley

Brief Contents

Contents

Preface

To the Instructor

Math and Dosage Calculations for Health Care is organized into 12 chapters, along with a pretest and a comprehensive evaluation, or posttest.

The **Pretest** gives the students an opportunity to review the basic math skills they will need for the book. The content of the Pretest parallels the content of Chapters 1 and 2.

Chapter 1, Fractions and Decimals, provides a comprehensive review of fractions and decimals. These are the basic building blocks for all that follows.

Chapter 2, Percents, Ratios, and Proportions, continues the math review by introducing percents, ratios, and proportions. The concept of solution strengths for both dry and liquid medication is also introduced. In both Chapters 1 and 2, word problems draw on medical settings.

Chapter 3, Systems of Weights and Measures, reviews weights and measures. It introduces the metric system in detail as well as the household and apothecary systems. Time and temperature conversions are included. Special attention is given to conversion factors, and the procedures for ratio proportion, fraction proportion, and dimensional analysis methods are introduced as building blocks for later chapters. These methods of dosage calculation are introduced and sustained throughout the rest of the textbook.

Chapter 4, Equipment for Dosage Measurement, covers the equipment used to measure and administer medications. It looks first at equipment used for oral administration, then at hypodermic syringes, and finally at other means of delivering medication, including transdermal patches.

Chapter 5, Drug Orders, begins with the Rights of Medication Administration and then shows the various ways in which drug orders may be written. It emphasizes safety and shows how drug orders can easily be misread.

Chapter 6, Drug Labels and Package Inserts, teaches the students how to find a wide range of information common to all drug labels and then specifically to oral and parenteral medications. A discussion of dosage strength is provided as a building block for later calculations.

Chapter 7, Methods of Dosage Calculations, teaches the techniques for calculating doses. Building on information from earlier chapters, students are taught how to calculate the amount to administer, using all four methods of dosage calculations including ratio proportion, fraction proportion, dimensional analysis, and formula. Students should be encouraged to explore the four methods and determine which they understand the best and want to continue to use as their method of choice for later chapters.

Chapter 8, Oral Dosages, discusses tablets and capsules in depth and gives information about breaking or crushing them. Liquid oral medications are also discussed. The methods of dosage calculations introduced in Chapter 7 are applied and color-coded throughout the chapter.

Chapter 9, Parenteral Dosages, applies techniques learned in Chapter 7 to calculations of parenteral dosages, emphasizing injectable medications. The chapter concludes with a look at other parenteral medication routes such as inhalants and rectal and transdermal systems.

Chapter 10, Intravenous Dosages, presents information and calculations unique to administering intravenous medications. After introducing IV solutions and equipment, we turn our attention to calculating flow rates for electronic and manual regulation. Students also learn calculations for adjusting the flow rate, infusion times, and infusion volumes.

Chapter 11, Calculations for Special Populations, includes drug orders based upon body weight. It introduces body surface area (BSA) calculations. Discussions of special concerns for pediatric and geriatric patients with regard to daily maintenance fluid needs are included. Creatinine clearance, the relationship between ideal and actual body weights, and polypharmacy are also presented.

Chapter 12, Specialized Calculations, includes additional information and calculations related to insulin, heparin, critical care IV, and solutions, solids, and compounds. Students are familiarized with insulin syringes, heparin administration, titrations, and alligations.

A **Comprehensive Evaluation**, or posttest, is available after Chapter 12 for the student's review.

Features of the Textbook

The third edition of *Math and Dosage Calculations for Health Care* keeps all of the well-liked features from the second edition to make the content user-friendly.

- **Error Alert!** boxes point out common errors and focus on avoiding them and calculating correctly.
- **Rules** state the important formulas and facts. **Examples** that follow illustrate the rules.
- **Patient Education** boxes teach students clear and accurate communication with their patients.
- **Tables** concisely summarize key information including the most up-to-date "Do Not Use" and "Undesirable" abbreviations according to the Joint Commission on Accreditation of Healthcare Organizations (JCAHO).
- **Critical Thinking Applications** sections encourage students to problem-solve and use higher-level thinking to answer questions related to actual health care practice.
- **Case Studies** provide questions to be solved based on real-world practice situations.
- **Internet Activity** boxes direct the student to factual content and resources and stimulate the use of the Internet as a professional tool.
- **Four methods of dosage calculations Fraction Proportion, Ratio Proportion, Dimensional Analysis, and Formula** are introduced together and used throughout the textbook. Each method is identified by a different color for easy reference.
- Pocket-size dosage calculation **reference cards.**
- Chapter **introduction, key terms**, and **quotations** are used as an advance organizer or anticipatory set for each chapter.
- Includes more than **250** full-color, up-to-date **drug labels** currently used by the health care profession. (Some of the labels have been enlarged or made smaller for clarity and design purposes.) Labels provide realistic learning.
- Additional **Review and Practice** problems for every section.
- **CD-ROM references** throughout that direct the student to exercises and provide for independent review, reinforcement, and evaluation.
- **Learning Link feature** refers the student to an earlier chapter for a quick review when concepts are repeated.
- Up-to-date **Glossary** with **pronunciations on the student CD-ROM.**

What's New to the Third Edition

- A **Chapter Homework** sheet has been added to each chapter. These homework sheets provide at least one of every type of problem introduced in the chapter. The answers are NOT provided in the back of the book so these can be used as an introduction, a review, or even a chapter quiz.
- **Critical Thinking on the Job** scenarios help the student apply math and dosage calculations to the health care profession. Students must read the scenario and answer critical thinking questions to determine what they would do to solve each situation.

Feature List

TABLES

PATIENT EDUCATION

CRITICAL THINKING ON THE JOB

ERROR ALERTS!

Resources and Ancillaries to Make Your Job Easier

Given the importance of performing accurate math and dosage calculations and the potential lack of time available to teach the subject, the *Math and Dosage Calculations for Health Care* program includes multiple resources and ancillaries for your convenience.

1. Comprehensive student CD-ROM that directly and completely corresponds with the student textbook. The textbook and CD-ROM can be used for classroom study or independent self-study. Students can review with multiple drill and practice question sections that include graphics and animation to attract attention, increase interest, and improve comprehension. After completing a chapter in the book and the exercises on the CD-ROM, students can print their scores to document their proficiency. This comprehensive CD-ROM along with the student text could be used as part of a distance education course for independent student study, or as a hybrid course for dosage calculations.
2. Effective Instruction CD-ROM includes Easy Test generator PowerPoint® presentations for each chapter, plus a figure browser. EZTest allows instructors to create their own tests from the database of test questions. The PowerPoint presentations can be used for classroom instruction or for independent study. These presentations include "Apply Your Knowledge" questions to promote discussion and interactivity. The figure browser provides over 50 labels and other graphics to use for your own tests, worksheets, or

presentations. Electronic *Instructor's Manual* with additional activities and tests for each chapter that include the solutions. Plus detailed solutions to all questions including the NEW homework sheet answers are provided in the student self-study text as well as chapter lesson plans and sample course syllabi.

Alternative chapter evaluation tests and pretest and post test are included. Instructors are also able to modify the interactive question on the student CD-ROM as well as manage the course activities.

3. Online Learning Center provides fluid and valuable resources for you as well as correlated student and password-protected instructor resources. These include case studies; math concept review section; Power Points; Career Opportunities; games such as Concentration, crossword puzzles, flashcards; drug company links; and additional chapter exercises plus alternative pre- and posttests.

ERROR ALERTS!

Chapter #	Name	Page #
1	Reducing a fraction does not automatically mean you have simplified it to its lowest terms.	11
1	Avoid reducing too many terms.	26
1	Write division problems carefully to avoid mistakes.	29
2	Do not forget the units of measurement.	62
3	Remember: The larger the unit, the smaller the quantity. The smaller the unit, the larger the quantity.	85
3	Do not confuse grains and grams.	86
3	Converting quantities for medications.	86
3	Do not confuse the symbols for drams and ounces.	87
4	The utensil you use must provide the calibration you need to accurately measure the dose.	113
4	Pay close attention to the calibration of any syringe you use.	121
5	Never guess what the prescriber meant.	150
5	Always be certain that you are dispensing the correct medication.	151
6	Give the right medication by the right route.	180
6	Consider the age and health needs of your patient when you administer a drug.	183
7	Always cancel units with fraction proportion correctly.	210
7	Always cancel units with ratio proportion correctly.	212
7	Canceling units for dimensional analysis.	215
9	Select the correct instructions for the strength and route ordered.	330
11	Converting ounces carefully.	412
11	Calculating doses for medications strongly bound to lean body tissue.	427
12	When two types of insulin are combined, measure the correct amount of each.	448

Resources and Ancillaries to Make Your Job Easier

Given the importance of performing accurate math and dosage calculations and the potential lack of time available to teach the subject, the *Math and Dosage Calculations for Health Care* program includes multiple resources and ancillaries for your convenience.

1. Comprehensive student CD-ROM that directly and completely corresponds with the student textbook. The textbook and CD-ROM can be used for classroom study or independent self-study. Students can review with multiple drill and practice question sections that include graphics and animation to attract attention, increase interest, and improve comprehension. After completing a chapter in the book and the exercises on the CD-ROM, students can print their scores to document their proficiency. This comprehensive CD-ROM along with the student text could be used as part of a distance education course for independent student study, or as a hybrid course for dosage calculations.
2. Effective Instruction CD-ROM includes Easy Test generator PowerPoint® presentations for each chapter, plus a figure browser. EZTest allows instructors to create their own tests from the database of test questions. The PowerPoint presentations can be used for classroom instruction or for independent study. These presentations include "Apply Your Knowledge" questions to promote discussion and interactivity. The figure browser provides over 50 labels and other graphics to use for your own tests, worksheets, or

presentations. Electronic *Instructor's Manual* with additional activities and tests for each chapter that include the solutions. Plus detailed solutions to all questions including the NEW homework sheet answers are provided in the student self-study text as well as chapter lesson plans and sample course syllabi.

Alternative chapter evaluation tests and pretest and post test are included. Instructors are also able to modify the interactive question on the student CD-ROM as well as manage the course activities.

3. Online Learning Center provides fluid and valuable resources for you as well as correlated student and password-protected instructor resources. These include case studies; math concept review section; Power Points; Career Opportunities; games such as Concentration, crossword puzzles, flashcards; drug company links; and additional chapter exercises plus alternative pre- and posttests.

Acknowledgments

We would like to graciously acknowledge the contributors and reviewers of *Math and Dosage Calculations for Health Care*. Their time, effort, and expert opinions are essential to the outcome of this book and ancillaries. A special thanks to the following for their consultation and assistance.

Suzanne K. Bolt, RN, DpN, BA, BSN, MAEd
Director of Nursing Education
Virginia Career Institute, Virginia Beach, VA

Lynn M. Egler, RMA, AHI, CPhT
Dorsey Schools, Madison Heights, MI

Susan Hurley Findley RN, MSN
Houston, TX

Lenore Huskey
Everest University, Jacksonville, FL

Glen Miller, EdD
Borough of Manhattan Community College
New York, NY

Philip R. Rushing B.S. Pharm
Tahlequah, OK

Reviewers for all Editions

Linda S. Albrecht
Richland College

Cathy Kelley-Arney, CMA, MLTC (ASCP), AS
National College of Business and Technology

Michelle Bach
Kansas City Kansas Community College

Sylvia Banzone
Judy Barcelon
Santa Rosa Junior College

Mary Barko, CMA, MAED/AEDL
Ohio Institute of Health Careers

Ellon Barlow
Fayetteville Technical Community College

Gail A. Chester
Community College of Southern Nevada

Barbara M. Dahl, CMA
Whatcom Community College

James M. Downey
Career Education Institute

Betty Earp, B.S.N., R.N.
Mesa Community College

Dr. Mary Ann Fritz
South Florida Community College

Joe Gallegos
Salt Lake Community College

Deb Getting, BSN
Northwest Iowa Community College

Margaret M. Gingrich, BSN, MSN
Harrisburg Area Community College

Michelle Goosby
Lurleen B. Wallace Community College/Macarthur Campus

Kris A. Hardy, CMA
Brevard Community College

Bernadette A. Hebb, RN, BSN
Florida Metropolitan University

Elizabeth Keene, CMA, BSN
Lansdale School of Business

Rita Michelson, LPN
Middlesex Community College

Peggie Newton, BSN, RN
Health Institute of Louisville

Brigitte Niedzwiecki, RN, BSN, MSN
Chippewa Valley Technical College

Lori Moseley, RN
Hill College

Bonnie Normile
Carnegie Technical Institute

Warren Ruud
Santa Rosa Junior College

Mary E. Stassi, RN
St. Charles County Community College

Susan E. Swords RN, BSN
Bryant and Stratton College

Cynthia Thompson
Davenport University

Donald C. Turner, PhD
Pamlico Community College

Jeremy Watson
National College

Wanda M. Webb, MSN
Hickory House Nursing Home

Janet Westhoff
Mott Community College

Leesa Gray Whicker
Central Piedmont Community College

Holly Ann Williamson, BSN
Lorenzo Walker Institute of Technology

Mindy Wray, BS, CMA, RMA
ECPI College of Technology

To the Student

Math and Dosage Calculations

Accurate math and dosage calculations play a large role in ensuring that medication errors do not occur. Medication errors affect the health and safety of patients and are of great concern for the medical community. As a health care professional, you must carefully perform math and dosage calculations to prevent these errors. This *Math and Dosage Calculations for Health Care* textbook and student CD-ROM are designed to help you do just that.

This course teaches you the skills and techniques to calculate the amount of medication that a patient should receive or you should administer. You will also learn to interpret physician's orders, medication records, and drug labels. This textbook is written for any student who plans to enter the health care field and will be working with math and medications. Right now you may be pursuing a career as a medical assistant, nurse, pharmacy technician, or other health care professional. No matter what profession you have chosen, you must always be alert for medication errors.

This textbook and CD-ROM will guide you through all the steps to find the correct amount of medication for patients. The early chapters provide detailed review of the basic arithmetic and algebra skills that are needed. The later chapters provide options for calculating dosages of medications administered by different routes and to different-age patients.

Since your math backgrounds may be different, this book provides various approaches for calculating dosages, so you can build upon what you already know. You may pick the dosage calculations technique that you understand the best and feel most comfortable with performing. These four methods are introduced: fraction proportion, ratio proportion, dimensional analysis, and formula. Once you have determined the method you prefer, read the color-coded material throughout the book.

How Can I Succeed in this Class, a First Step Toward My Goals?

If you're reading this, you're on the right track.

You are the same today that you are going to be 5 years from now except for two things: the people with whom you associate and the books you read. —Charles Jones

Right now, you're probably leafing through this book, feeling just a little overwhelmed. You're trying to juggle several other classes (which probably are equally intimidating), possibly a job, and, on top of it all, a life.

It's true—you are what you put into your studies. You have a lot of time and money invested in your education. Don't blow it now by only putting in half of the effort that this class requires. Succeeding in this class (and life) requires

- Making a commitment—of time and perseverance.
- Knowing and motivating yourself.
- Getting organized.
- Managing your time.

This special introduction has been designed specifically to help you focus. It's here to help you learn how to manage your time and your studies to succeed. It will help you learn how to be effective in these areas as well as offer guidance in:

- Getting the most out of your lecture.
- Thinking through—and applying—the material.
- Getting the most out of your textbook and CD-ROM.
- Finding extra help when you need it.

Making a Commitment—of Time and Perseverance

Learning—and mastering—takes time. And patience. Nothing worthwhile comes easily. Be committed to your studies, and you will reap the benefits in the long run.

Consider this: Your math and dosage calculations course is building the foundation for your future—a future in your chosen health care profession. Sloppy and hurried craftsmanship now will only lead to ruin later.

Study Tip: A good rule of thumb is to allow 2 hours of study time for every hour you spend in class.

Knowing and Motivating Yourself

What type of a learner are you? When are you most productive? Know yourself and your limits, and work within them. Know how to motivate yourself to give your all to your studies and achieve your goals. Quite bluntly, you are the one who benefits most from your success. If you lack self-motivation and drive, you are the first person who suffers.

Know yourself—there are many types of learners, and there is no right or wrong way of learning. Which category do you fall into?

1. *Visual learner*. You respond best to "seeing" processes and information. Particularly focus on the text's rules, examples, figures, and tables.
2. *Auditory learner*. You work best by listening to—and possibly tape-recording—the lecture and by talking information through with a study partner. Don't miss any lectures. Be sure to listen to the key terms on the audio Glossary and review all the rules on the CD-ROM.
3. *Tactile / kinesthetic learner*. You learn best by being "hands on." You'll benefit by applying what you've learned during lab time. Think of how to apply your critical thinking skills. Work through the Critical Thinking on the Job sections, and be sure to complete all the practice problems and games on the CD-ROM.

Identify your own personal preferences for learning, and seek out the resources that will best help you with your studies. Also, learn by recognizing your weaknesses and try to compensate or work to improve them.

Getting Organized

It's simple, yet it's fundamental. It seems the more organized you are, the more easily things come. Take the time before your course begins to look around and analyze your life and your study habits. Get organized now, and you'll find you have a little more time—and a lot less stress.

Find a calendar system that works for you. The best kind is one that you can take with you everywhere. To be truly organized, you should integrate all aspects of your life into this one calendar—school, work, and leisure. Some people also find it helpful to have an additional monthly calendar posted by their desks for "at a glance" dates and to have a visual of what's to come. If you do this, be sure you are consistently synchronizing both calendars so as not to miss anything. More tips for organizing your calendar can be found in the next section, "Managing Your Time."

By the same token, keep everything for your course or courses in one place—and at your fingertips. A three-ring binder works well because it allows you to add or organize handouts and notes from class in any order you prefer. Incorporating your own custom tabs helps you flip to exactly what you need at a moment's notice.

Find your space. Find a place that helps you be organized and focused. If it's your desk in your dorm room or in your home, keep it clean. Clutter adds confusion and stress, and it wastes time. Or perhaps your "space" is at the library. If that's the case, keep a backpack or bag that's fully stocked with what you might need—your text, binder or notes, pens, highlighters, Post-Its, and phone numbers of study partners (*Hint:* a good place to keep phone numbers is in your "one place for everything calendar").

Helpful Hint. Add extra "padding" into your deadlines to yourself. If you have a report due on Friday, set a goal for yourself to have it done on Wednesday. Then take time on Thursday to look over your project again, with a fresh eye. Make any corrections or enhancements and have it ready to turn in on Friday.

Managing Your Time

Managing your time is the single most important thing you can do to help yourself. And it's probably one of the most difficult tasks to successfully master.

You are taking this course because you want to succeed in life. You are preparing for a career. You are expected to work much harder and to learn much more than you ever have before. To be successful, you need to invest in your education with a commitment of time.

How time slips away. We tend to let an enormous amount of time slip away from us, mainly in three ways:

1. *Procrastination*—putting off chores simply because we don't feel in the mood to do them right away.
2. *Distraction*—getting sidetracked by the endless variety of other things that seem easier or more fun to do, often not realizing how much time they eat up.
3. *Underestimating the value of small bits of time*—thinking it's not worth doing any work because we have something else to do or somewhere else to be in 20 minutes or so.

We all lead busy lives. But we all choose how to spend our time. Choose wisely and make the most of every minute you have by implementing these tips.

Know yourself and when you'll be able to study most efficiently. When are you most productive? Are you a late nighter? Or an early bird? Plan to study when you are most alert and can have uninterrupted segments. This could include a quick 5-minute review before class or a 1-hour problem-solving study session with a friend.

Create a set study time for yourself daily. Having a set schedule for yourself helps you commit to studying and helps you plan instead of cram. Find—and use—a planner that is small enough to take with you—everywhere. This can be a $2.50 paper calendar or a more expensive electronic version. They all work on the same premise—*organize all your activities in one place.*

Make sure you log your projects and homework deadlines in your personal calendar.

Less is more. Schedule study time using shorter, focused blocks with small breaks. Doing this offers two benefits:

1. You will be less fatigued and will gain more from your effort.
2. Studying will seem less overwhelming, and you will be less likely to procrastinate.

Plan time for leisure, friends, exercise, and sleep. Studying should be your main focus, but you need to balance your time—and your life. Try to complete tasks ahead of schedule. This will give you a chance to carefully review your work before you hand it in (instead of at 1 a.m. when you are half awake). You'll feel less stressed in the end.

Prioritize! In your calendar or planner, highlight or number key projects; do them first, and then cross them off when you've completed them. Give yourself a pat on the back for getting them done! Review your calendar and reprioritize daily.

Try to resist distractions by setting and sticking to a designated study time (remember your commitment and perseverance!). Distractions may include friends and surfing the Internet.

Multitask when possible. You may find a lot of extra time you didn't think you had. Review material or organize your term paper in your head while walking to class or doing laundry, or during "mental downtime." (*Note:* Mental downtime does *not* mean in the middle of a lecture.)

Getting the Most Out of Lectures

Believe it or not instructors want you to succeed. They put a lot of effort into helping you learn and into preparing their lectures. Attending class is one of the simplest, most valuable things you can do to help yourself. But it doesn't end there—getting the most out of your lectures means being organized. Here's how:

Prepare before you go to class. Really! You'll be amazed at how much more comprehensible the material will be when you preview the chapter before you go to class. Don't feel overwhelmed by this already. One tip may help you: Plan to arrive at class 5 to 15 minutes before the lecture. Bring your text with you, and skim the chapter before the lecture begins. At the very least, this will give you an overview of what may be discussed. Complete the homework sheet provided at the end of each chapter to see what you already know. Then, review the sections of the chapter where you had the most difficulty.

Be a good listener. Most people think they are good listeners, but few really are. Are you?

Obvious, but important, points to remember include these:

1. You can't listen if you are talking.
2. You aren't listening if you are daydreaming.
3. Listening and comprehending are two different things. If you don't understand something your instructor is saying, ask a question or make a note and visit the instructor after hours. Don't feel dumb or intimidated; you probably aren't the only person who "doesn't get it."

Take good notes

1. Use a standard-size notebook or, better yet, a three-ring binder with looseleaf notepaper. The binder will allow you to organize and integrate your notes and handouts, integrate easy-to-reference tabs, etc.
2. Use a standard black or blue ink pen to take your initial notes. You can annotate later, using a pencil, which can be erased if need be.
3. Start a new page with each lecture or note-taking session (yes, you can and should also take notes from your textbook).
4. Label each page with the date and a heading for each day.
5. Focus on main points, and try to use an outline format to take notes to capture key ideas and organize subpoints.
6. Review and edit your notes shortly after class—at least within 24 hours—to make sure they make sense and that you've recorded core thoughts. You may also want to compare your notes with those of a study partner later, to make sure neither of you missed anything.

Get a study partner. Having a study partner has so many benefits. First, he or she can help you keep your commitment to this class. By having set study dates, you can combine study and social time and maybe even make it fun! In addition, you now have two sets of eyes and ears and two minds to help digest the information from lectures and from the text. Talk through concepts, compare notes, and quiz each other.

An obvious note: Don't take advantage of your study partner by skipping class or skipping study dates. You obviously won't have a study partner—or a friend—much longer if it's not a mutually beneficial arrangement!

Helpful Hint. Take your text to the lecture and keep it open to the topics being discussed. You can take brief notes in your textbook margin or reference textbook pages in your notebook to help you study later.

Getting the Most Out of Your Textbook

McGraw-Hill and the authors of this book, Kathryn Booth and James Whaley, have invested time, research, and talents to help you succeed as well. Our goal is to make learning easier—for you.

Here's how: The textbook has 12 chapters. Each opens with:

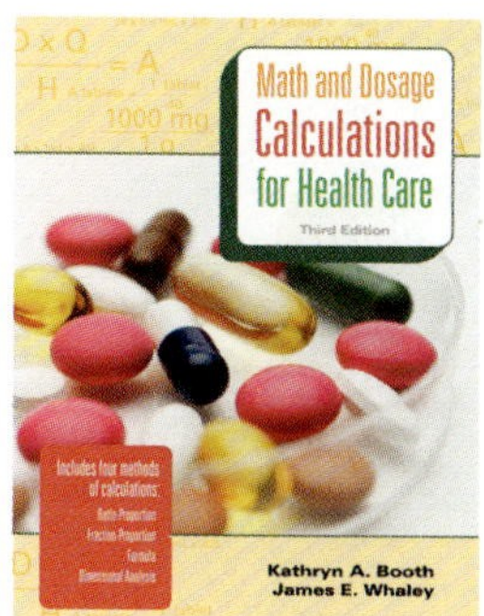

1. *Objectives*, so you understand the key points you should master.
2. *Motivational quote*, so you can think about and get started with the chapter.
3. *Key Terms*, so you will know the terms necessary to perform math and dosage calculations.
4. *Introduction*, for you to understand why each chapter is important.

The text introduces techniques for math and dosage calculations by providing *Rules* to guide you. These rules introduce an important concept that must be followed while you are learning. After you read and study each rule and the examples that follow, you should open the student CD-ROM found in the back of the book and view the rule and complete the practice problems. Once you have mastered a rule, continue in the textbook.

Once you have completed a certain number of rules, you will come to the end of a section. Each section includes multiple Review and Practice problems to ensure your mastery of the content in that section. Check your answers against those in the back of the book.

Complete the chapter Check Up in the textbook for a final review, and take the chapter test on the CD-ROM for a final evaluation of your ability.

The Learning System

Chapter openers feature **Objectives, Key Terms, Introduction,** and an insightful quote to introduce you to the chapter and help prepare you for the information presented. Take time to read and review these sections before you begin.

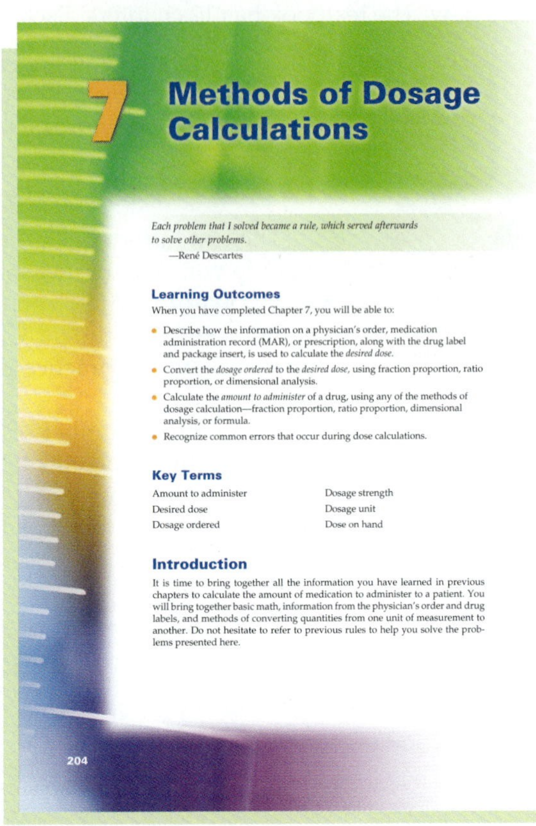

7 Methods of Dosage Calculations

Each problem that I solved became a rule, which served afterwards to solve other problems.
—René Descartes

Learning Outcomes

When you have completed Chapter 7, you will be able to:

- Describe how the information on a physician's order, medication administration record (MAR), or prescription, along with the drug label and package insert, is used to calculate the *desired dose.*
- Convert the *dosage ordered* to the *desired dose,* using fraction proportion, ratio proportion, or dimensional analysis.
- Calculate the *amount to administer* of a drug, using any of the methods of dosage calculation—fraction proportion, ratio proportion, dimensional analysis, or formula.
- Recognize common errors that occur during dose calculations.

Key Terms

Amount to administer	Dosage strength
Desired dose	Dosage unit
Dosage ordered	Dose on hand

Introduction

It is time to bring together all the information you have learned in previous chapters to calculate the amount of medication to administer to a patient. You will bring together basic math, information from the physician's order and drug labels, and methods of converting quantities from one unit of measurement to another. Do not hesitate to refer to previous rules to help you solve the problems presented here.

204

"The text is easy to read and the explanations are simple so that the students with different reading and comprehension levels can be comfortable."

Judy Barcelon, Santa Rosa Junior College

Rule 2-2 To convert a decimal into a percent, multiply the decimal by 100. Then add the percent symbol.

Example 1 Convert 1.42 to a percent.

$1.42 \times 100\% = 142.00\% = 142\%$

You can write this as $1.42 = 1.42.\% = 142\%$.

Example 2 Convert 0.02 to a percent.

$0.02 \times 100\% = 2.00\% = 2\%$

You can write this as $0.02 = 0.02.\% = 2\%$.

Example 3 When you move the decimal point to the right, you may need to insert zeros. Convert 0.8 to a percent.

$0.8 \times 100\% = 80.0\% = 80\%$

You can write this as $0.8 = 0.80 = 0.80.\% = 80\%$.

Rules state important formulas and facts. They provide you with guidelines for calculating dosages. The examples illustrate the rules.

TABLE 5-4 "Do Not Use" Abbreviations

Abbreviation	Potential Problem	Preferred Term
U (for unit)	Mistaken as zero, four, or cc.	Write *unit.*
IU (for international unit)	Mistaken as IV (intravenous) or 10.	Write *international unit.*
Q.D., Q.O.D. (Latin abbreviations for once daily and every other day)	Mistaken for each other. The period after the Q can be mistaken for an I and the O can be mistaken for I.	Write *daily* and *every other day.*
Trailing zero (X.0 mg) [*Note:* Prohibited only for medication-related notations] Lack of leading zero (.X mg)	Decimal point is missed.	Never write a zero by itself after a decimal point (X mg), and always use a zero before a decimal point (0.X mg).
MS, MSO_4, $MgSO_4$	Confused for one another. Can mean morphine sulfate or magnesium sulfate.	Write *morphine sulfate* or *magnesium sulfate.*

Tables summarize key information.

GO TO . . . Open the CD-ROM that accompanies your textbook, and select Chapter 5, Practice 5-3. Review the animation and example problems, then complete the practice problems. Continue to the next section of the book once you have mastered the rule presented. ■

CD-ROM references direct you to CD exercises that help reinforce the concepts previously learned.

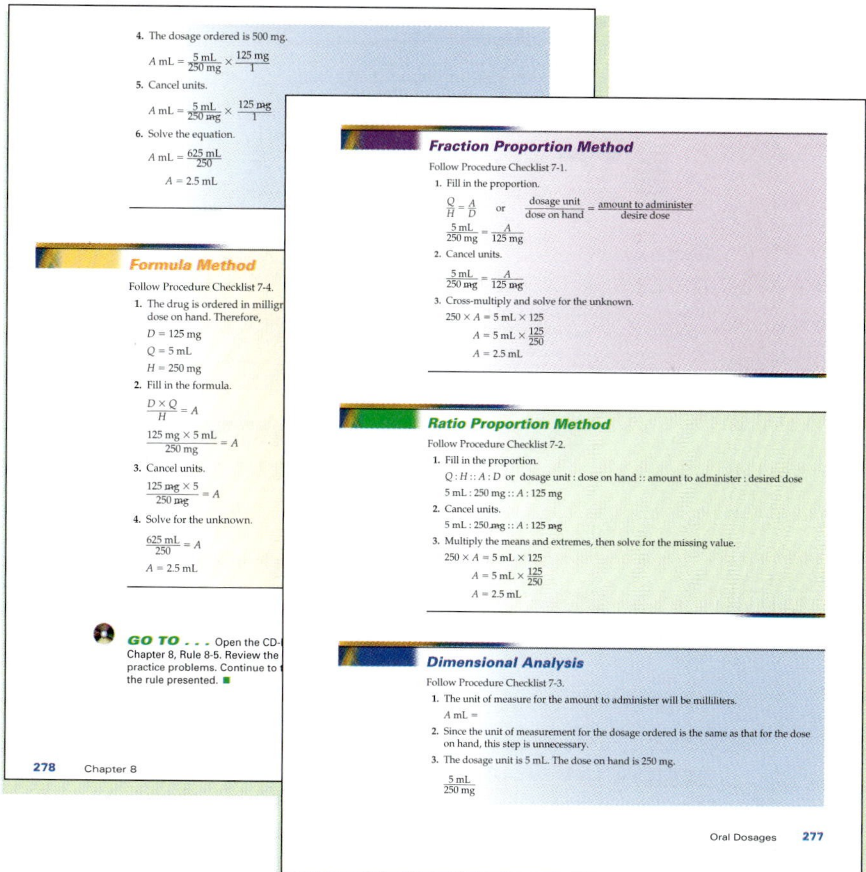

4. The dosage ordered is 500 mg.

$A \text{ mL} = \frac{5 \text{ mL}}{250 \text{ mg}} \times \frac{125 \text{ mg}}{1}$

5. Cancel units.

$A \text{ mL} = \frac{5 \text{ mL}}{250 \text{ mg}} \times \frac{125 \text{ mg}}{1}$

6. Solve the equation.

$A \text{ mL} = \frac{625 \text{ mL}}{250}$

$A = 2.5 \text{ mL}$

Formula Method

Follow Procedure Checklist 7-4.

1. The drug is ordered in milligr... dose on hand. Therefore,

$D = 125$ mg

$Q = 5$ mL

$H = 250$ mg

2. Fill in the formula.

$\frac{D \times Q}{H} = A$

$\frac{125 \text{ mg} \times 5 \text{ mL}}{250 \text{ mg}} = A$

3. Cancel units.

$\frac{125 \text{ mg} \times 5}{250 \text{ mg}} = A$

4. Solve for the unknown.

$\frac{625 \text{ mL}}{250} = A$

$A = 2.5$ mL

GO TO . . . Open the CD-... Chapter 8, Rule 8-5. Review the ... practice problems. Continue to ... the rule presented.

278 Chapter 8

Fraction Proportion Method

Follow Procedure Checklist 7-1.

1. Fill in the proportion.

$\frac{Q}{H} = \frac{A}{D}$ or $\frac{\text{dosage unit}}{\text{dose on hand}} = \frac{\text{amount to administer}}{\text{desire dose}}$

$\frac{5 \text{ mL}}{250 \text{ mg}} = \frac{A}{125 \text{ mg}}$

2. Cancel units.

$\frac{5 \text{ mL}}{250 \text{ mg}} = \frac{A}{125 \text{ mg}}$

3. Cross-multiply and solve for the unknown.

$250 \times A = 5 \text{ mL} \times 125$

$A = 5 \text{ mL} \times \frac{125}{250}$

$A = 2.5$ mL

Ratio Proportion Method

Follow Procedure Checklist 7-2.

1. Fill in the proportion.

$Q : H :: A : D$ or dosage unit : dose on hand :: amount to administer : desired dose

5 mL : 250 mg :: A : 125 mg

2. Cancel units.

5 mL : 250 mg :: A : 125 mg

3. Multiply the means and extremes, then solve for the missing value.

$250 \times A = 5 \text{ mL} \times 125$

$A = 5 \text{ mL} \times \frac{125}{250}$

$A = 2.5$ mL

Dimensional Analysis

Follow Procedure Checklist 7-3.

1. The unit of measure for the amount to administer will be milliliters.

A mL =

2. Since the unit of measurement for the dosage ordered is the same as that for the dose on hand, this step is unnecessary.

3. The dosage unit is 5 mL. The dose on hand is 250 mg.

$\frac{5 \text{ mL}}{250 \text{ mg}}$

Oral Dosages 277

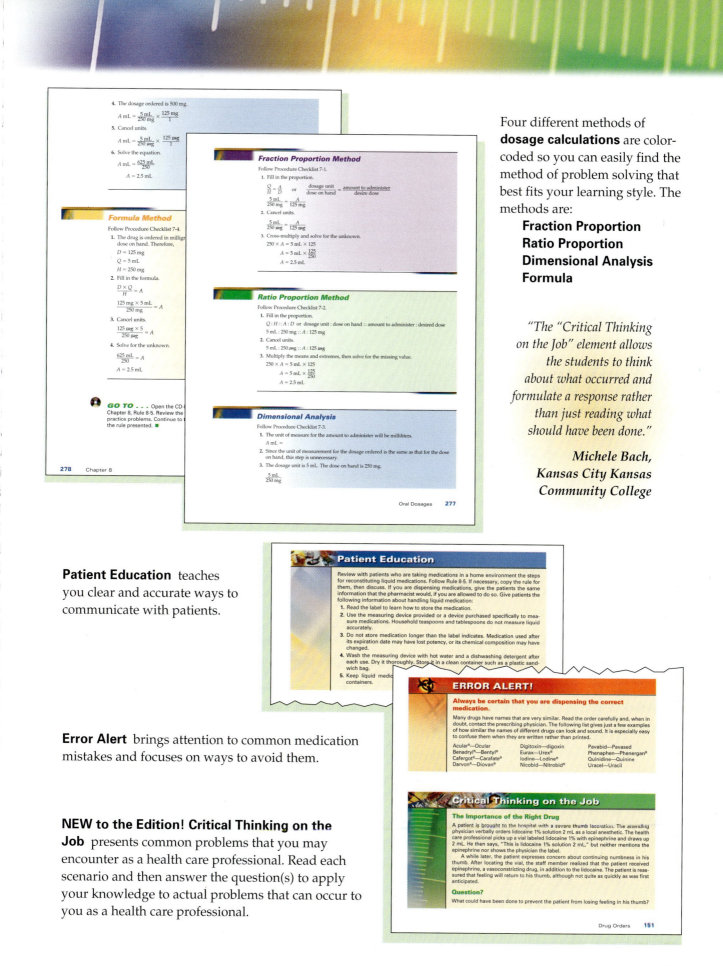

Four different methods of **dosage calculations** are color-coded so you can easily find the method of problem solving that best fits your learning style. The methods are:

Fraction Proportion
Ratio Proportion
Dimensional Analysis
Formula

"The "Critical Thinking on the Job" element allows the students to think about what occurred and formulate a response rather than just reading what should have been done."

Michele Bach, Kansas City Kansas Community College

Patient Education teaches you clear and accurate ways to communicate with patients.

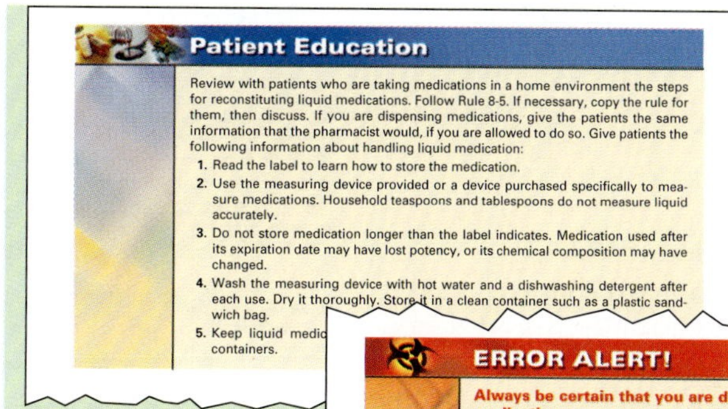

Patient Education

Review with patients who are taking medications in a home environment the steps for reconstituting liquid medications. Follow Rule 8-5. If necessary, copy the rule for them, then discuss. If you are dispensing medications, give the patients the same information that the pharmacist would, if you are allowed to do so. Give patients the following information about handling liquid medication:

1. Read the label to learn how to store the medication.
2. Use the measuring device provided or a device purchased specifically to measure medications. Household teaspoons and tablespoons do not measure liquid accurately.
3. Do not store medication longer than the label indicates. Medication used after its expiration date may have lost potency, or its chemical composition may have changed.
4. Wash the measuring device with hot water and a dishwashing detergent after each use. Dry it thoroughly. Store it in a clean container such as a plastic sandwich bag.
5. Keep liquid medic... containers.

Error Alert brings attention to common medication mistakes and focuses on ways to avoid them.

ERROR ALERT!

Always be certain that you are dispensing the correct medication.

Many drugs have names that are very similar. Read the order carefully and, when in doubt, contact the prescribing physician. The following list gives just a few examples of how similar the names of different drugs can look and sound. It is especially easy to confuse them when they are written rather than printed.

Acular®—Ocular	Digitoxin—digoxin	Pavabid—Pavased
Benadryl®—Bentyl®	Eurax—Urex®	Phenaphen—Phenergan®
Cafergot®—Carafate®	Iodine—Lodine®	Quinidine—Quinine
Darvon®—Diovan®	Nicobid—Nitrobid®	Uracel—Uracil

NEW to the Edition! Critical Thinking on the Job presents common problems that you may encounter as a health care professional. Read each scenario and then answer the question(s) to apply your knowledge to actual problems that can occur to you as a health care professional.

Critical Thinking on the Job

The Importance of the Right Drug

A patient is brought to the hospital with a severe thumb laceration. The attending physician verbally orders lidocaine 1% solution 2 mL as a local anesthetic. The health care professional picks up a vial labeled lidocaine 1% with epinephrine and draws up 2 mL. He then says, "This is lidocaine 1% solution 2 mL," but neither mentions the epinephrine nor shows the physician the label.

A while later, the patient expresses concern about continuing numbness in his thumb. After locating the vial, the staff member realized that the patient received epinephrine, a vasoconstricting drug, in addition to the lidocaine. The patient is reassured that feeling will return to his thumb, although not quite as quickly as was first anticipated.

Question?

What could have been done to prevent the patient from losing feeling in his thumb?

Drug Orders 151

Critical Thinking Applications require you to go beyond simple calculations.

Over **250 up-to-date full-color drug labels** are presented along with common calculation problems that you will see on the job.

Case Study exercises connect you to real-life situations.

Internet Activities help you learn how to use the Internet for research and information.

Review and Practice exercises follow each section of every chapter, giving you the opportunity to apply new concepts.

NEW to this Edition! Homework problems are provided for each chapter. Every type of problem that is introduced in the chapter is represented in this homework assignment sheet.

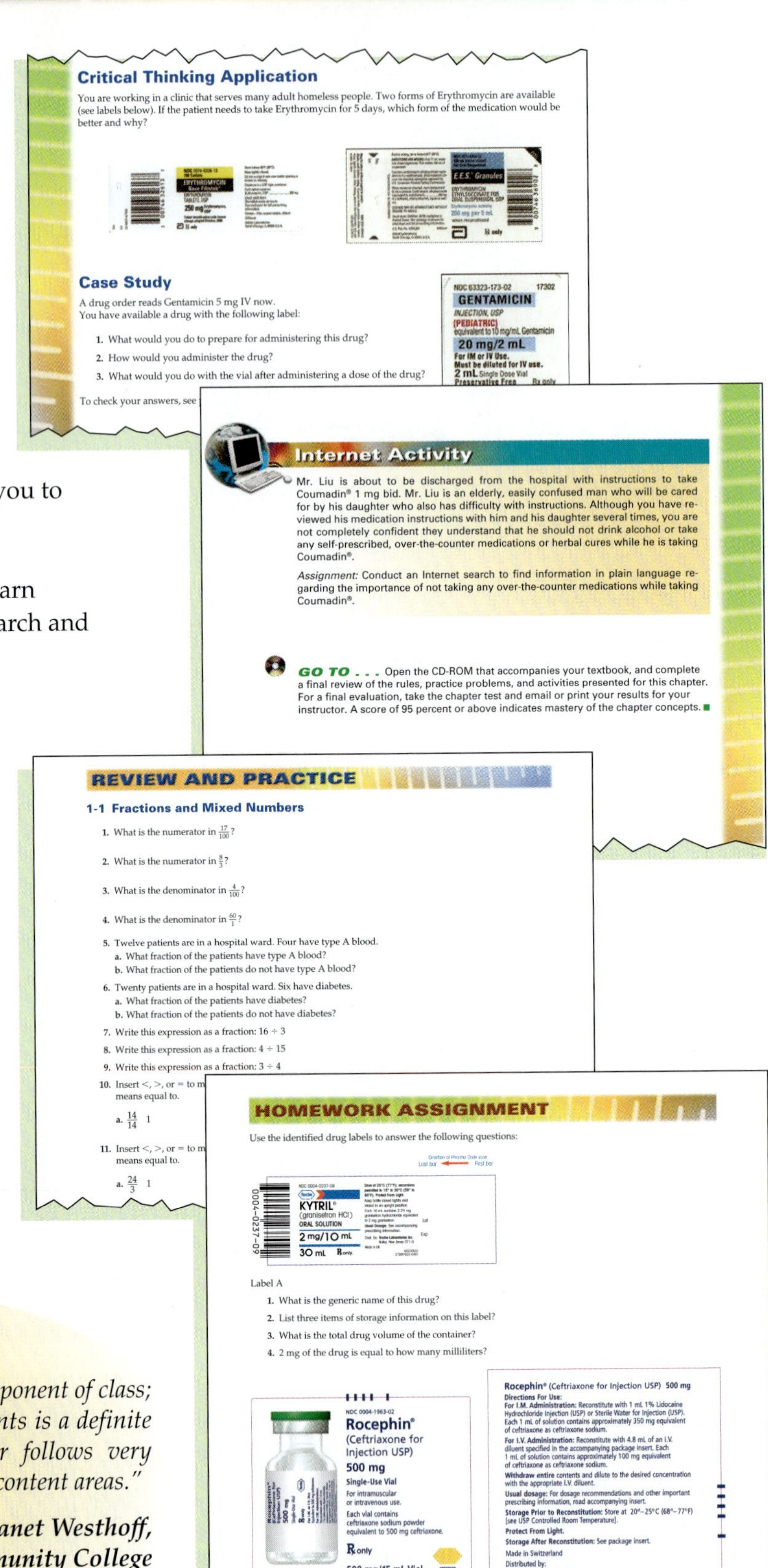

"Homework and practice is a critical component of class; having homework available for the students is a definite plus! Also, homework for each chapter follows very closely and is comprehensive sampling of content areas."

Janet Westhoff, RN, MSN, Mott Community College

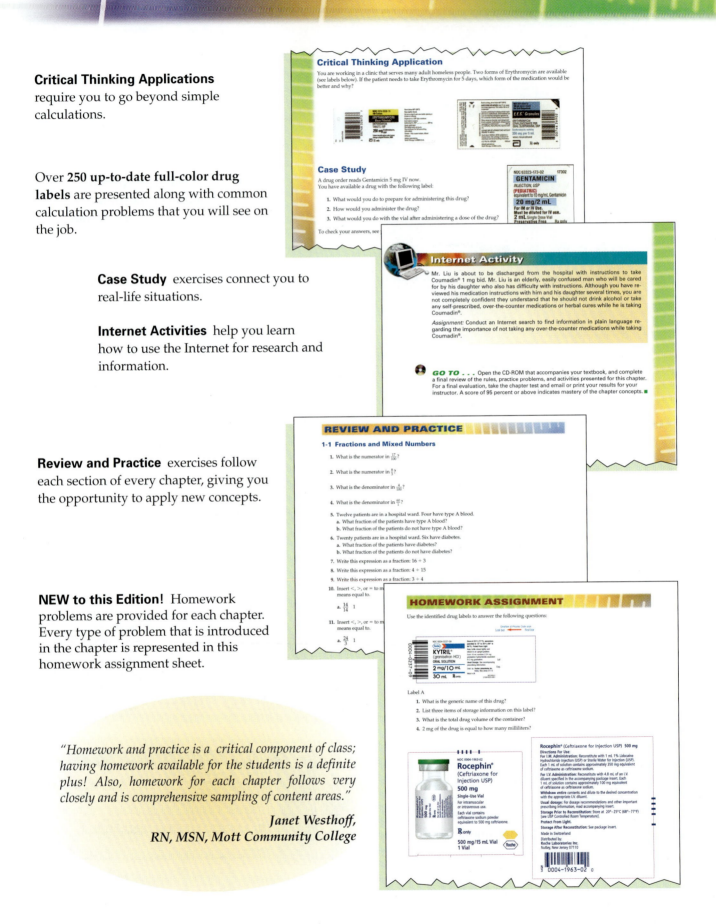

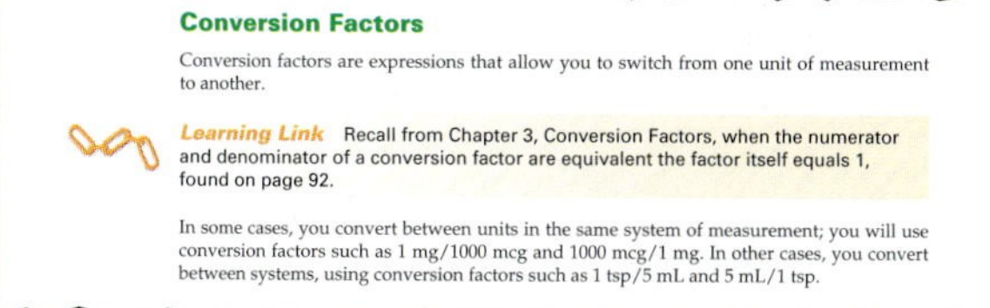

Conversion Factors

Conversion factors are expressions that allow you to switch from one unit of measurement to another.

Learning Link Recall from Chapter 3, Conversion Factors, when the numerator and denominator of a conversion factor are equivalent the factor itself equals 1, found on page 92.

In some cases, you convert between units in the same system of measurement; you will use conversion factors such as 1 mg/1000 mcg and 1000 mcg/1 mg. In other cases, you convert between systems, using conversion factors such as 1 tsp/5 mL and 5 mL/1 tsp.

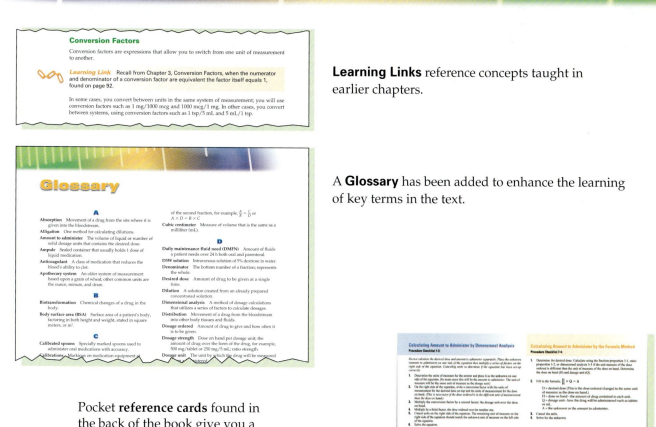

Learning Links reference concepts taught in earlier chapters.

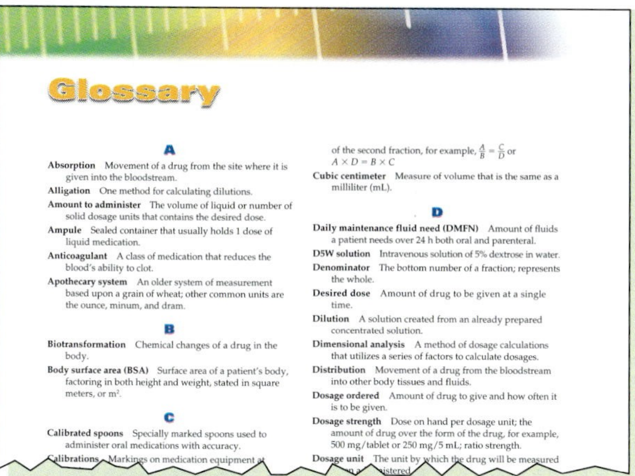

Glossary

A

Absorption Movement of a drug from the site where it is given into the bloodstream.

Alligation One method for calculating dilutions.

Amount to administer The volume of liquid or number of solid dosage units that contains the desired dose.

Ampule Sealed container that usually holds 1 dose of liquid medication.

Anticoagulant A class of medication that reduces the blood's ability to clot.

Apothecary system An older system of measurement based upon a grain of wheat; other common units are the ounce, minum, and dram.

B

Biotransformation Chemical changes of a drug in the body.

Body surface area (BSA) Surface area of a patient's body, factoring in both height and weight, stated in square meters, or m².

C

Calibrated spoons Specially marked spoons used to administer oral medications with accuracy.

Calibrations Markings on medication equipment at

of the second fraction, for example, $\frac{A}{B} = \frac{C}{D}$ or $A \times D = B \times C$

Cubic centimeter Measure of volume that is the same as a milliliter (mL).

D

Daily maintenance fluid need (DMFN) Amount of fluids a patient needs over 24 h both oral and parenteral.

D5W solution Intravenous solution of 5% dextrose in water.

Denominator The bottom number of a fraction; represents the whole.

Desired dose Amount of drug to be given at a single time.

Dilution A solution created from an already prepared concentrated solution.

Dimensional analysis A method of dosage calculations that utilizes a series of factors to calculate dosages.

Distribution Movement of a drug from the bloodstream into other body tissues and fluids.

Dosage ordered Amount of drug to give and how often it is to be given.

Dosage strength Dose on hand per dosage unit; the amount of drug over the form of the drug, for example, 500 mg/tablet or 250 mg/5 mL; ratio strength.

Dosage unit The unit by which the drug will be measured ... istered.

A **Glossary** has been added to enhance the learning of key terms in the text.

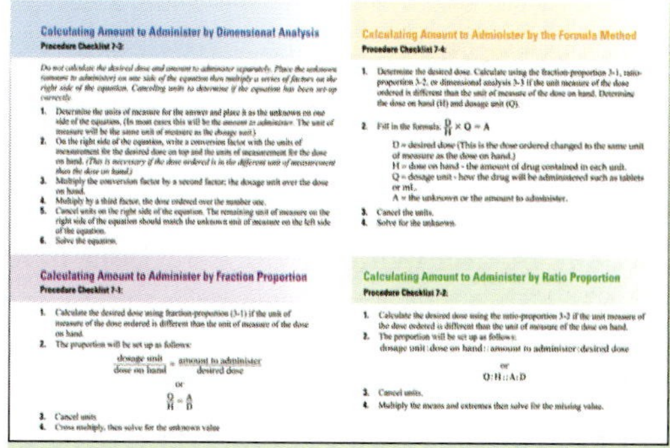

Calculating Amount to Administer by Dimensional Analysis

Calculating Amount to Administer by the Formula Method

Calculating Amount to Administer by Fraction Proportion

Calculating Amount to Administer by Ratio Proportion

Pocket **reference cards** found in the back of the book give you a convenient tool that displays the methods of calculations.

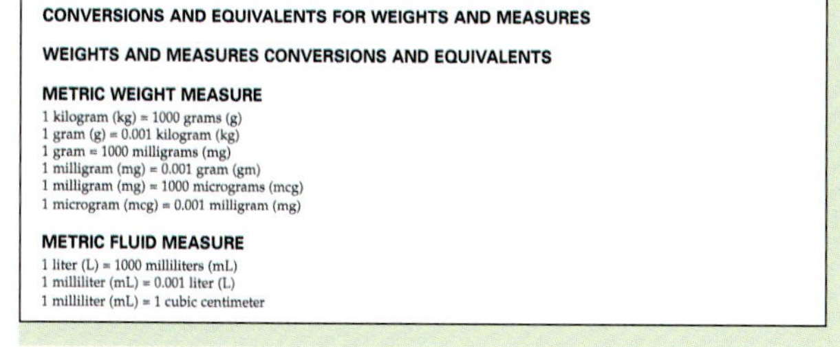

CONVERSIONS AND EQUIVALENTS FOR WEIGHTS AND MEASURES

WEIGHTS AND MEASURES CONVERSIONS AND EQUIVALENTS

METRIC WEIGHT MEASURE

1 kilogram (kg) = 1000 grams (g)
1 gram (g) = 0.001 kilogram (kg)
1 gram = 1000 milligrams (mg)
1 milligram (mg) = 0.001 gram (gm)
1 milligram (mg) = 1000 micrograms (mcg)
1 microgram (mcg) = 0.001 milligram (mg)

METRIC FLUID MEASURE

1 liter (L) = 1000 milliliters (mL)
1 milliliter (mL) = 0.001 liter (L)
1 milliliter (mL) = 1 cubic centimeter

Conversion table printed on inside back cover shows a quick and handy reference to common mathematical equivalents.

Pretest

The following test covers basic mathematical concepts that you will need to understand and calculate dosages. This test will help you determine which concepts you need to review before continuing. You should already be able to perform basic operations—addition, subtraction, multiplication, and division—with whole numbers. The test covers fractions, decimals, percents, ratios, and proportions.

Take 90 minutes to answer the following 50 questions. Then check your answers on page 494. Review the questions you answered incorrectly to learn more about any basic math weaknesses. Then, as needed, review that content in Chapters 1 and 2.

1. Convert $\frac{14}{3}$ to a mixed number.

2. Convert $3\frac{7}{8}$ to a fraction.

Find the missing numerator in the following equations.

3. $\frac{2}{7} = \frac{?}{21}$ 4. $1\frac{1}{8} = \frac{?}{16}$

5. Reduce $\frac{40}{100}$ to lowest terms.

6. Which fraction has the greater value, $\frac{3}{8}$ or $\frac{2}{6}$?

Calculate the following. Reduce fractions to lowest terms and rewrite any fractions as mixed numbers.

7. $\frac{4}{5} + \frac{3}{8}$ **8.** $1\frac{1}{3} + \frac{5}{7}$ **9.** $\frac{7}{10} - \frac{1}{4}$ **10.** $8\frac{1}{4} - 2\frac{1}{3}$

11. $\frac{3}{5} \times \frac{1}{9}$ **12.** $3\frac{1}{5} \times 4\frac{3}{8}$ **13.** $\frac{2}{3} \div \frac{4}{5}$ **14.** $5\frac{1}{4} \div 2\frac{5}{8}$

15. Which number has the lesser value, 1.01 or 1.009?

16. Round 14.42 to the nearest whole number.

17. Round 6.05 to the nearest tenth.

18. Round 19.197 to the nearest hundredth.

19. Convert $3\frac{4}{5}$ to a decimal number. If necessary, round to the nearest hundredth.

20. Convert 0.045 to a fraction or a mixed number. Reduce to lowest terms.

Calculate the following.

21. $7.289 + 8.011$ **22.** $0.012 + 0.9 + 4.2$ **23.** $19.1 - 4.4$

24. $100.03 - 0.6$ **25.** 0.07×3.2 **26.** $0.4 \div 0.02$

27. Convert 0.8 percent to a decimal number.

28. Convert 0.99 to a percent.

29. Convert 260 percent to a fraction or mixed number.

30. Convert $1\frac{1}{8}$ to a percent.

31. Convert 7 : 12 to a fraction.

32. Convert $\frac{10}{50}$ to a ratio. Reduce to lowest terms.

33. Convert 1 : 12 to a decimal. Round to the nearest hundredth, if necessary.

34. Convert 0.4 to a ratio. Reduce to lowest terms.

35. Convert 3 : 8 to a percent. Round to the nearest percent, if necessary.

36. Convert 0.5 percent to a ratio. Reduce to lowest terms.

Find the missing value in the following proportions.

37. 8 : 16 : : ? : 8

38. $\frac{5}{9} = \frac{?}{27}$

39. 8 : 12 : : ? : 9

40. $\frac{2}{7} = \frac{?}{28}$

41. A health care professional is instructed to give a patient $1\frac{1}{2}$ teaspoons of cough syrup 4 times a day. How many teaspoons of cough syrup will the nurse give each day?

42. A health care professional tries to keep the equivalent of 12 bottles of a medication on hand. The hospital's first floor has $1\frac{1}{2}$ bottles, the second floor has $1\frac{3}{4}$ bottles, the third floor has $3\frac{1}{4}$ bottles, and the supply closet has 3 bottles. Is there enough medication on hand? If not, how much should the technician order?

43. A bottle contains 75 milliliters (mL) of a liquid medication. Since the bottle was opened, one patient has received 3 doses of 2.5 mL. A second patient has received 4 doses of 2.2 mL. How much medication remains in the bottle?

44. A tablet contains 0.125 milligram (mg) of medication. A patient receives 3 tablets a day for 5 days. How many milligrams of medication does the patient receive over the 5 days?

45. An IV bag contained 1000 mL of a liquid. The liquid was administered to a patient, and now there is 400 mL left in the bag after 3 hours. How much IV fluid did the patient receive each hour?

46. The patient is taking 0.5 mg of medication 4 times a day. How many milligrams would the patient receive after $1\frac{1}{2}$ days?

47. The patient took 0.88 microgram (mcg) every morning and 1.2 mcg each evening for 4 days. What was the total amount of medication taken?

48. Write a ratio that represents that 500 mL of solution contains 5 mg of drug.

49. Write a ratio that represents that every tablet in a bottle contains 25 mg of drug.

50. Write a ratio that represents that 3 mL of solution contains 125 mg of drug.

1 Fractions and Decimals

He who is ashamed of asking is ashamed of learning.
—Danish proverb

Learning Outcomes

When you have completed Chapter 1, you will be able to:

- Compare the values of fractions in various formats.
- Accurately add, subtract, multiply, and divide fractions.
- Convert fractions to mixed numbers and decimals.
- Recognize the format of decimals and measure their relative values.
- Accurately add, subtract, multiply, and divide decimals.
- Round decimals to the nearest tenth, hundredth, or thousandth.

Key Terms

Complex fraction
Denominator
Equivalent fractions
Least common denominator
Mixed number
Numerator
Prime number

Introduction

Basic math skills, such as working with fractions and decimals, are the building blocks for accurate dosage calculations. To prepare yourself mathematically, you must be confident in your math skills; so do not be afraid to ask for help while you are learning these important concepts. Remember that a minor mistake in basic math can mean major errors in the patient's medication.

Fractions and Mixed Numbers

Fractions measure a portion or part of a whole amount. They are written in two ways: as common fractions or as decimals. In medical settings, you must often convert from one type of fraction to another.

Common Fractions

A common fraction represents equal parts of a whole. It consists of two numbers and a fraction bar, and it is written in the form:

$$\frac{\text{Numerator}}{\text{Denominator}}$$

The **denominator**—the bottom part of the fraction—represents the whole. It can *never* equal zero. Suppose the whole is 1 yard (yd). You could express the denominator in many ways: as 1 (the yard as a whole), 3 (the yard as 3 feet), or 36 (the yard as 36 inches).

The **numerator**—the top part of the fraction—represents parts of the whole. If you buy 2 feet (ft) of fabric out of a yard, you can express the numerator as 2 with the denominator as 3 feet (ft). The fraction $\frac{2}{3}$ represents how much of a yard of fabric you buy, or 2 of 3 ft. If you buy less fabric, say, 9 inches (in.), you can express the numerator as 9 with the denominator as 36 inches (in.). The fraction $\frac{9}{36}$ represents how much of a yard of fabric you buy, or 9 of 36 inches (in). The numerator and denominator should always be expressed in the same unit of measure—inches/inches or feet/feet.

Suppose you are working with a medicine tablet that is scored (marked) for division in two parts, and you must administer one part of that tablet each day. The denominator represents the whole tablet. The numerator represents one part, the amount that you administer each day. The fraction, or part, of the tablet that you must administer each day is written as:

$$\frac{\text{Numerator}}{\text{Denominator}} = \frac{1 \text{ part}}{2 \text{ parts}} = \frac{1}{2}$$

This number is read *one-half*. The denominator is 2, since two parts make up the whole. If you administer 1 part each day, you administer $\frac{1}{2}$ of the tablet.

If you have trouble remembering which number is the numerator and which is the denominator, note that the words *denominator* and *down* begin with the letter d. The <u>d</u>enominator is <u>d</u>own, under the fraction bar.

The fraction $\frac{2}{3}$, read as *two-thirds*, means two parts out of the three parts that make up the whole. The fraction bar also means *divided by*. Thus, $\frac{2}{3}$ can be read as "two divided by three," or $2 \div 3$ This definition is important when you change fractions to decimals.

Sometimes fractions show the relationship between part of a group and the whole group. For example, in a group of 15 patients with hyperthyroidism, 9 patients respond well to a medication. The other 6 patients show no change. The number of patients in the full group, 15, is the whole, or the denominator. You write the fraction of patients who respond well to the medication as:

$$\frac{\text{Part}}{\text{Whole}} = \frac{\text{respond well}}{\text{whole group}} = \frac{9}{15}$$

Similarly, the fraction of patients who show no change is:

$$\frac{\text{Part}}{\text{Whole}} = \frac{\text{show no change}}{\text{whole group}} = \frac{6}{15}$$

Rule 1-1 When the denominator is 1, the fraction equals the number in the numerator.

Example $\frac{4}{1} = 4 \quad \frac{100}{1} = 100$

Check these equations by treating each fraction as a division problem.

$4 \div 1 = 4 \quad 100 \div 1 = 100$

GO TO . . . Open the CD-ROM that accompanies your textbook, and select Chapter 1, Rule 1-1. Review the animation and example problems; then complete the practice problems. Continue to the next section of the book once you have mastered the rule presented. ■

Mixed Numbers

Fractions with a value greater than 1 are more properly written as mixed numbers. A **mixed number** combines a whole number with a fraction. Examples include $2\frac{2}{3}$ (two and two-thirds), $1\frac{7}{8}$ (one and seven-eighths) and $12\frac{31}{32}$ (twelve and thirty-one thirty-seconds).

Rule 1-2

1. If the numerator of the fraction is less than the denominator, the fraction has a value less than ($<$) 1
2. If the numerator of the fraction is equal to the denominator, the fraction has a value equal to ($=$) 1
3. If the numerator of the fraction is greater than the denominator, the fraction has a value greater than ($>$) 1 Fractions with a value greater than one may be written as a mixed number.

GO TO . . . Open the CD-ROM that accompanies your textbook, and select Chapter 1, Rule 1-2. Review the animation and example problems; then complete the practice problems. Continue to the next section of the book once you have mastered the rule presented. ■

Example 1

a. The fraction $\frac{8}{9}$ is less than 1 because the numerator (8) is less than the denominator (9). This can be written $\frac{8}{9} < 1$

b. The fraction $\frac{3}{4}$ is less than 1 because the numerator (3) is less than the denominator (4). This can be written $\frac{3}{4} < 1$

Example 2

a. $\frac{12}{12}$ is equal to 1 because the numerator (12) is equal to the denominator (12). This can be written $\frac{12}{12} = 1$

b. $\frac{32}{32}$ is equal to 1 because the numerator (32) is equal to the denominator (32). This can be written $\frac{32}{32} = 1$

Example 3 **a.** $\frac{8}{5}$ is greater than 1 because the numerator (8) is greater than the denominator (5). This can be written $\frac{8}{5} > 1$

b. $\frac{11}{4}$ is greater than 1 because the numerator (11) is greater than the denominator (4). This can be written $\frac{11}{4} > 1$

Rule 1-3 To convert a fraction to a mixed number:

1. Divide the numerator by the denominator. The result will be a whole number plus a remainder.
2. Write the remainder as the numerator over the original denominator.
3. Combine the whole number and the fractional remainder. This mixed number equals the original fraction.

Reminder: This rule is only applied when the numerator is greater than the denominator.

Example 1 Convert $\frac{11}{4}$ to a mixed number.

1. Divide the numerator by the denominator:

   $11 \div 4 = 2$ R3 (R3 means a *remainder* of 3)

   The result is the whole number 2 with a remainder of 3

2. Write the remainder as the numerator over the original denominator of 4

   $$\frac{\text{Remainder}}{\text{Denominator}} = \frac{3}{4}$$

3. Combine the whole number and the fractional remainder:

   $$2 + \frac{3}{4} = 2\frac{3}{4}$$

   The mixed number $2\frac{3}{4}$ equals the original fraction $\frac{11}{4}$.

Example 2 Convert $\frac{23}{7}$ to a mixed number.

1. $23 \div 7 = 3$ R2

   The result is the whole number 3 with a remainder of 2

2. $\frac{\text{Remainder}}{\text{Denominator}} = \frac{2}{7}$

3. $3 + \frac{2}{7} = 3\frac{2}{7}$

   The fraction $\frac{23}{7}$ equals the mixed number $3\frac{2}{7}$.

GO TO . . . Open the CD-ROM that accompanies your textbook, and select Chapter 1, Rule 1-3. Review the animation and example problems; then complete the practice problems. Continue to the next section of the book once you have mastered the rule presented. ■

You can also convert mixed numbers to fractions. This is often necessary before you use the number in a calculation.

Rule 1-4 To convert a mixed number to a fraction:

1. Multiply the whole number by the denominator of the fraction.
2. Add the product from step 1 to the numerator of the fraction.
3. Write the sum from step 2 over the original denominator. The result is a fraction equal to the original mixed number.

You can also use this equation for converting a mixed number to a fraction:

$$\text{Whole number}\left(\frac{\text{numerator}}{\text{denominator}}\right) = \frac{(\text{whole number} \times \text{denominator}) + \text{numerator}}{\text{denominator}}$$

Example 1 Convert $5\frac{1}{3}$ to a fraction.

The whole number is 5. The denominator of the fraction is 3. The numerator of the fraction is 1

1. Multiply the whole number by the denominator of the fraction.

   $5 \times 3 = 15$

2. Add the product from step 1 to the numerator of the fraction.

   $15 + 1 = 16$

3. Write the sum from step 2 over the original denominator.

   $\frac{16}{3}$

Thus, $5\frac{1}{3} = \frac{16}{3}$.

Example 2 Convert $10\frac{7}{8}$ to a fraction.

The whole number is 10 The denominator is 8 The numerator is 7

1. $10 \times 8 = 80$
2. $80 + 7 = 87$
3. $\frac{87}{8}$

Thus, $10\frac{7}{8} = \frac{87}{8}$.

GO TO . . . Open the CD-ROM that accompanies your textbook, and select Chapter 1, Rule 1-4. Review the animation and example problems; then complete the practice problems. Continue to the next section of the book once you have mastered the rule presented. ■

REVIEW AND PRACTICE

1-1 Fractions and Mixed Numbers

1. What is the numerator in $\frac{17}{100}$?

2. What is the numerator in $\frac{8}{3}$?

3. What is the denominator in $\frac{4}{100}$?

4. What is the denominator in $\frac{60}{1}$?

5. Twelve patients are in a hospital ward. Four have type A blood.
   a. What fraction of the patients have type A blood?
   b. What fraction of the patients do not have type A blood?

6. Twenty patients are in a hospital ward. Six have diabetes.
   a. What fraction of the patients have diabetes?
   b. What fraction of the patients do not have diabetes?

7. Write this expression as a fraction: $16 \div 3$

8. Write this expression as a fraction: $4 \div 15$

9. Write this expression as a fraction: $3 \div 4$

10. Insert $<$, $>$, or $=$ to make a true statement, where $<$ means less than, $>$ means greater than, and $=$ means equal to.

    a. $\frac{14}{14}$ 1 b. $\frac{24}{32}$ 1 c. $\frac{125}{100}$ 1

11. Insert $<$, $>$, or $=$ to make a true statement, where $<$ means less than, $>$ means greater than, and $=$ means equal to.

    a. $\frac{24}{3}$ 1 b. $\frac{75}{100}$ 1 c. $\frac{18}{18}$ 1

Convert the following fractions to mixed or whole numbers.

12. $\frac{43}{6}$
13. $\frac{17}{3}$
14. $\frac{100}{20}$
15. $\frac{50}{50}$
16. $\frac{8}{5}$
17. $\frac{167}{25}$
18. $\frac{16}{12}$

Convert the following mixed numbers to fractions.

19. $2\frac{16}{17}$
20. $8\frac{8}{9}$
21. $1\frac{1}{10}$
22. $4\frac{1}{8}$
23. $103\frac{2}{3}$
24. $6\frac{7}{8}$
25. $8\frac{1}{5}$

To check your answers, see page 494.

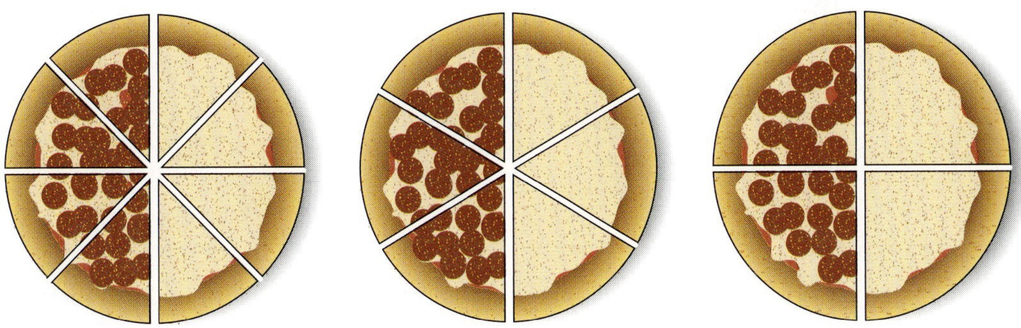

Figure 1-1 Equivalent fractions.

Equivalent Fractions

Two fractions may have the same value even when they are written differently. These are known as **equivalent fractions**. Suppose you and a friend are sharing a pizza equally, dividing it in half. If you cut the pizza into eight slices, you will each get four pieces, or $\frac{4}{8}$ of the whole pizza. If you cut the pizza into six slices, you will each get three pieces, or $\frac{3}{6}$. And if you cut the pizza into four slices, you will each get two slices, or $\frac{2}{4}$. Whether you get $\frac{4}{8}$, $\frac{3}{6}$, or $\frac{2}{4}$ of the pizza, you still have the same amount: one-half or $\frac{1}{2}$ of the pizza (see Figure 1-1). Thus, $\frac{4}{8} = \frac{3}{6} = \frac{2}{4} = \frac{1}{2}$. These four fractions are equivalent fractions.

Converting Fractions. Equivalent fractions help you compare measurements more easily. They also help you add and subtract fractions that have different denominators.

Rule 1-5 To find an equivalent fraction, multiply or divide both the numerator and denominator by the same number. *Exception:* The numerator and denominator cannot be multiplied or divided by zero.

Example 1 Find equivalent fractions for $\frac{2}{4}$.

$$\frac{2 \times 2}{4 \times 2} = \frac{4}{8} \qquad \frac{2 \times 3}{4 \times 3} = \frac{6}{12} \qquad \frac{2 \div 2}{4 \div 2} = \frac{1}{2} \qquad \frac{2 \times 10}{4 \times 10} = \frac{20}{40}$$

Thus, $\frac{2}{4} = \frac{4}{8} = \frac{6}{12} = \frac{1}{2} = \frac{20}{40}$ These are equivalent fractions.

Example 2 Find equivalent fractions for $\frac{4}{7}$.

$$\frac{4 \times 3}{7 \times 3} = \frac{12}{21} \qquad \frac{4 \times 5}{7 \times 5} = \frac{20}{35} \qquad \frac{4 \times 10}{7 \times 10} = \frac{40}{70} \qquad \frac{4 \times 100}{7 \times 100} = \frac{400}{700}$$

Thus, $\frac{4}{7} = \frac{12}{21} = \frac{20}{35} = \frac{40}{70} = \frac{400}{700}$. These are equivalent fractions.

Example 3 Find some equivalent fractions for 4

To find equivalent fractions for a whole number, first write the whole number as a fraction. Then proceed as before.

$$4 = \frac{4}{1}$$

$$\frac{4 \times 2}{1 \times 2} = \frac{8}{2} \qquad \frac{4 \times 3}{1 \times 3} = \frac{12}{3} \qquad \frac{4 \times 4}{1 \times 4} = \frac{16}{4} \qquad \frac{4 \times 5}{1 \times 5} = \frac{20}{5}$$

Thus, $4 = \frac{4}{1} = \frac{8}{2} = \frac{12}{3} = \frac{16}{4} = \frac{20}{5}$. These are some equivalent fractions.

Example 4 Find some equivalent fractions for $1\frac{4}{6}$.

To find equivalent fractions for a mixed number, first convert the mixed number to a fraction.

1. $1 \times 6 = 6$
2. $6 + 4 = 10$
3. $1\frac{4}{6} = \frac{10}{6}$

Now follow the same steps used in Examples 1 through 3.

$$\frac{10 \times 2}{6 \times 2} = \frac{20}{12} \qquad \frac{10 \times 3}{6 \times 3} = \frac{30}{18} \qquad \frac{10 \div 2}{6 \div 2} = \frac{5}{3} \qquad \frac{10 \times 10}{6 \times 10} = \frac{100}{60}$$

Thus, $1\frac{4}{6} = \frac{10}{6} = \frac{20}{12} = \frac{30}{18} = \frac{5}{3} = \frac{100}{60}$.

GO TO . . . Open the CD-ROM that accompanies your textbook, and select Chapter 1, Rule 1-5. Review the animation and example problems; then complete the practice problems. Continue to the next section of the book once you have mastered the rule presented. ■

Finding Missing Numerators. Suppose you want to convert a fraction into an equivalent one with a specific denominator. To convert $\frac{1}{5}$ to tenths, find the missing value in $\frac{1}{5} = \frac{?}{10}$. The ? stands for the number you want to find. In this case, compare the denominators. The new denominator, 10, is larger than the original denominator, 5. First divide the larger denominator by the smaller one. Here, $10 \div 5 = 2$. Multiplying the numerator and denominator by the result, or 2, will lead to an equivalent fraction with 10 as the denominator.

$$\frac{1}{5} = \frac{1 \times 2}{5 \times 2} = \frac{2}{10} = \frac{?}{10} \qquad \text{and} \qquad ? = 2$$

Rule 1-6 To find the missing numerator in an equivalent fraction:

1. If the denominator of the equivalent fraction is larger than the original denominator, then:
   a. Divide the larger denominator by the smaller one. (The answer to a division problem is the *quotient*.)
   b. Multiply the original numerator by the quotient from step a.

2. If the denominator of the equivalent fraction is smaller than the original denominator, then:
   a. Divide the larger denominator by the smaller one.
   b. Divide the original numerator by the quotient from step a.

Example 1 $\frac{2}{3} = \frac{?}{12}$

a. Divide the larger denominator by the smaller one.

$12 \div 3 = 4$

b. Multiply the original numerator by the quotient from step a.

$2 \times 4 = 8$

Thus, $\frac{2}{3} = \frac{2 \times 4}{3 \times 4} = \frac{8}{12}$ and $? = 8$

Example 2 $\frac{28}{60} = \frac{?}{15}$

a. Here the original denominator is the larger one. Divide the larger denominator by the smaller one.

$60 \div 15 = 4$

b. Divide the original numerator by the quotient from step a.

$28 \div 4 = 7$

Thus, $\frac{28}{60} = \frac{28 \div 4}{60 \div 4} = \frac{7}{15}$ and $? = 7$

Example 3 $2\frac{1}{2} = \frac{?}{6}$

First convert the mixed number into a fraction.

$$2\frac{1}{2} = \frac{(2 \times 2) + 1}{2} = \frac{5}{2}$$

The equation is now $\frac{5}{2} = \frac{?}{6}$.

a. $6 \div 2 = 3$

b. $5 \times 3 = 15$

Thus, $2\frac{1}{2} = \frac{15}{6}$ and $? = 15$

GO TO . . . Open the CD-ROM that accompanies your textbook, and select Chapter 1, Rule 1-6. Review the animation and example problems; then complete the practice problems. Continue to the next section of the book once you have mastered the rule presented. ■

You can use a similar method to find a missing denominator in an equivalent fraction. In Chapter 2, you will learn another way to find the missing numerator by cross-multiplying.

REVIEW AND PRACTICE

1-2 Equivalent Fractions

Find three equivalent fractions for each of the following.

1. $\frac{4}{5}$	**2.** $\frac{1}{10}$	**3.** $\frac{4}{2}$	**4.** $\frac{15}{9}$	**5.** 9
6. 24	**7.** $2\frac{1}{3}$	**8.** $3\frac{6}{9}$	**9.** $\frac{7}{12}$	**10.** $4\frac{1}{4}$

Find the missing numerator in the following equations:

11. $\frac{3}{8} = \frac{?}{16}$	**12.** $\frac{1}{3} = \frac{?}{27}$	**13.** $\frac{16}{24} = \frac{?}{6}$	**14.** $\frac{18}{15} = \frac{?}{5}$	**15.** $3 = \frac{?}{4}$
16. $5 = \frac{?}{12}$	**17.** $1\frac{5}{16} = \frac{?}{160}$	**18.** $4\frac{2}{8} = \frac{?}{4}$	**19.** $\frac{8}{12} = \frac{?}{24}$	**20.** $\frac{32}{16} = \frac{?}{4}$

To check your answers, see page 495.

Simplifying Fractions to Lowest Terms

The last section showed how to find equivalent fractions by multiplying or dividing the numerator and denominator by the same number. When you divide, you simplify (or reduce) a fraction. Reduced or simplified equivalent fractions may be easier to use when you are performing a calculation. It is considered proper form to express your final answer in a fraction that is reduced to its lowest terms.

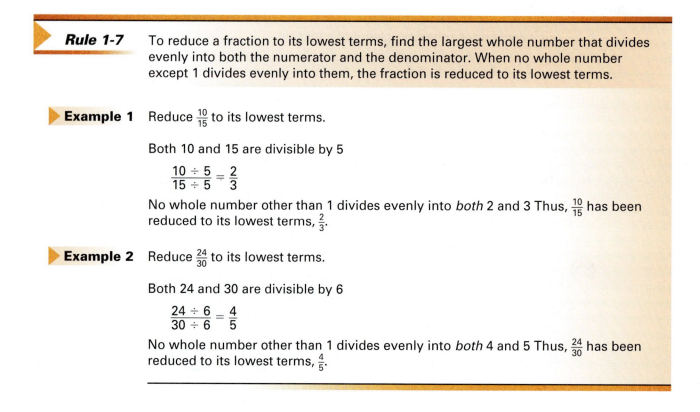

Rule 1-7 To reduce a fraction to its lowest terms, find the largest whole number that divides evenly into both the numerator and the denominator. When no whole number except 1 divides evenly into them, the fraction is reduced to its lowest terms.

Example 1 Reduce $\frac{10}{15}$ to its lowest terms.

Both 10 and 15 are divisible by 5

$$\frac{10 \div 5}{15 \div 5} = \frac{2}{3}$$

No whole number other than 1 divides evenly into *both* 2 and 3 Thus, $\frac{10}{15}$ has been reduced to its lowest terms, $\frac{2}{3}$.

Example 2 Reduce $\frac{24}{30}$ to its lowest terms.

Both 24 and 30 are divisible by 6

$$\frac{24 \div 6}{30 \div 6} = \frac{4}{5}$$

No whole number other than 1 divides evenly into *both* 4 and 5 Thus, $\frac{24}{30}$ has been reduced to its lowest terms, $\frac{4}{5}$.

GO TO . . . Open the CD-ROM that accompanies your textbook, and select Chapter 1, Rule 1-7. Review the animation and example problems; then complete the practice problems. Continue to the next section of the book once you have mastered the rule presented. ■

ERROR ALERT!

Reducing a fraction does not automatically mean you have simplified it to its lowest terms.

More than one number may divide evenly into both the numerator and the denominator. For example, both 18 and 42 are even numbers. To reduce $\frac{18}{42}$, you can divide by 2, so that:

$$\frac{18 \div 2}{42 \div 2} = \frac{9}{21}$$

You are not done, though. Both 9 and 21 are divisible by 3, so that:

$$\frac{18}{42} = \frac{9}{21} = \frac{9 \div 3}{21 \div 3} = \frac{3}{7}$$

Some fractions are easy to reduce. Looking at $\frac{2}{4}$, you can guess that 2 divides evenly into both the numerator and the denominator, so that $\frac{2 \div 2}{4 \div 2} = \frac{1}{2}$. In other cases, you may have to use several steps. See Table 1-1 for numbers divisible by 2, 3, 4, 5, 6, 8, 9, or 10

TABLE 1-1 Is a Number Divisible by 2, 3, 4, 5, 6, 8, 9, or 10?

Number	Hint	Example
2	Even numbers (numbers ending with 2, 4, 6, 8, or 0) are divisible by 2	112; 734; 2936; 10,118; 356, 920
3	If the sum of the digits of a number is divisible by 3, then the number is divisible by 3	37,887 The sum of the digits is 3 + 7 + 8 + 8 + 7 = 33; 33 is divisible by 3
4	If the last two digits of a number are divisible by 4, the entire number is divisible by 4	126,936 The last two digits form a number, 36, that is divisible by 4
5	Any number that ends with 5 or 0 is divisible by 5	735 12,290
6	Combine the rules for 2 and 3 If a number is even *and* the sum of its digits is divisible by 3 then the number is divisible by 6	582 The number is even. The sum of its digits, 5 + 8 + 2 = 15, is divisible by 3
8	If the last three digits are divisible by 8, then the entire number is divisible by 8	42,376 Here, 376 is divisible by 8
9	If the sum of the digits is a multiple of 9, the number is divisible by 9	42,705 4 + 2 + 7 + 0 + 5 = 18, which is divisible by 9
10	If a number ends with 0, then the number is divisible by 10	640

Prime numbers are whole numbers other than 1 that can be evenly divided only by themselves and 1 The first 10 prime numbers are 2, 3, 5, 7, 11, 13, 17, 19, 23, and 29 If either the numerator or the denominator is a prime number, and if the other term is not divisible by that prime number, then the fraction is in lowest terms. For example, $\frac{17}{24}$ is in lowest terms. However, you can simplify $\frac{17}{34}$ to $\frac{1}{2}$, dividing both the numerator and denominator by 17

REVIEW AND PRACTICE

1-3 Simplifying Fractions to Lowest Terms

Reduce the following fractions to their lowest terms.

1. $\frac{10}{12}$	**2.** $\frac{3}{6}$	**3.** $\frac{27}{81}$	**4.** $\frac{11}{22}$	**5.** $\frac{10}{100}$
6. $\frac{55}{100}$	**7.** $\frac{4}{5}$	**8.** $\frac{6}{17}$	**9.** $\frac{21}{27}$	**10.** $\frac{35}{50}$
11. $\frac{48}{90}$	**12.** $\frac{49}{84}$	**13.** $\frac{10}{28}$	**14.** $\frac{5}{8}$	**15.** $\frac{33}{99}$
16. $\frac{25}{35}$	**17.** $\frac{18}{48}$	**18.** $\frac{49}{77}$	**19.** $\frac{8}{32}$	**20.** $\frac{24}{44}$

To check your answers, see page 495.

Finding Common Denominators

A *common denominator* is any number that is a common multiple of all the denominators in a group of fractions. The **least common denominator** (LCD) is the smallest of these numbers. Before you can add and subtract fractions with different denominators, you must first convert them to equivalent fractions with a common denominator. To compare fractions with different denominators, you must also convert them to equivalent fractions with a common denominator.

Rule 1-8 To find the least common denominator (LCD) of a group of fractions:

1. List the multiples of each denominator.
2. Compare the lists. Any numbers that appear on all lists are common denominators.
3. The smallest number that appears on all the lists is the LCD.

Once you have found the LCD, you can convert each fraction to an equivalent fraction with the LCD as the denominator.

Example 1 Find the least common denominator of $\frac{1}{3}$ and $\frac{1}{2}$. Then convert each to an equivalent fraction with the LCD.

1. The number 3 divides evenly into 3, <u>6</u>, 9, <u>12</u>, 15, <u>18</u>, and 21 The number 2 divides evenly into 2, 4, <u>6</u>, 8, 10, <u>12</u>, 14, 16, and <u>18</u>
2. The numbers 6, 12, and 18 are common denominators.
3. The smallest number that appears on both lists is 6 It is the least common denominator and is divisible by both 3 and 2

   Now convert $\frac{1}{3}$ and $\frac{1}{2}$ so that $\frac{1}{3} = \frac{?}{6}$, and $\frac{1}{2} = \frac{?}{6}$.

4. To convert $\frac{1}{3}$ to the equivalent fraction $\frac{?}{6}$.

   a. $6 \div 3 = 2$

   b. $\frac{1}{3} = \frac{1 \times 2}{3 \times 2} = \frac{2}{6}$

5. To convert $\frac{1}{2}$ to the equivalent fraction $\frac{?}{6}$.

   $6 \div 2 = 3$

   $\frac{1}{2} = \frac{1 \times 3}{2 \times 3} = \frac{3}{6}$

The least common denominator is 6 Using this denominator the equivalent fractions are $\frac{1}{3} = \frac{2}{6}$ and $\frac{1}{2} = \frac{3}{6}$.

Example 2 Find the least common denominator of $\frac{1}{4}$, $\frac{1}{6}$, and $\frac{1}{8}$. Then convert each to an equivalent fraction with the LCD.

1. The number 4 divides evenly into 4, 8, 12, 16, 20, and 24

   The number 6 divides evenly into 6, 12, 18, and 24

   The number 8 divides evenly into 8, 16, and 24
2. The number 24 is a common denominator.
3. In this case, 24 is the LCD.
4. $\frac{1}{4} = \frac{?}{24}$

   $24 \div 4 = 6$

   $\frac{1}{4} = \frac{1 \times 6}{4 \times 6} = \frac{6}{24}$

5. $\frac{1}{6} = \frac{?}{24}$

   $24 \div 6 = 4$

   $\frac{1}{6} = \frac{1 \times 4}{6 \times 4} = \frac{4}{24}$

6. $\frac{1}{8} = \frac{?}{24}$

   $24 \div 8 = 3$

   $\frac{1}{8} = \frac{1 \times 3}{8 \times 3} = \frac{3}{24}$

The least common denominator is 24 Using this denominator, we see the equivalent fractions are:

$\frac{1}{4} = \frac{6}{24}$

$\frac{1}{6} = \frac{4}{24}$

$\frac{1}{8} = \frac{3}{24}$

GO TO . . . Open the CD-ROM that accompanies your textbook, and select Chapter 1, Rule 1-8. Review the animation and example problems; then complete the practice problems. Continue to the next section of the book once you have mastered the rule presented. ■

You may find it difficult to find common denominators of fractions with large denominators. However, you can simply multiply the individual denominators to find a common denominator.

Rule 1-9 To convert fractions with large denominators to equivalent fractions with a common denominator:

1. List the denominators of all the fractions.
2. Multiply the denominators. The product is a common denominator. Convert each fraction to an equivalent one with the common denominator.

Example 1 Convert $\frac{1}{7}$ and $\frac{1}{19}$ to equivalent fractions with a common denominator.

1. The denominators are 7 and 19
2. Multiply 7 × 19 The common denominator is 133

$$\frac{1}{7} = \frac{1 \times 19}{7 \times 19} = \frac{19}{133} \quad \text{and} \quad \frac{1}{19} = \frac{1 \times 7}{19 \times 7} = \frac{7}{133}$$

The equivalent fractions are $\frac{19}{133}$ and $\frac{7}{133}$.

Example 2 Convert $\frac{2}{37}$ and $\frac{7}{90}$ to equivalent fractions with a common denominator.

1. The denominators are 37 and 90
2. 37 × 90 = 3330

$$\frac{2 \times 90}{37 \times 90} = \frac{180}{3330} \quad \text{and} \quad \frac{7 \times 37}{90 \times 37} = \frac{259}{3330}$$

The equivalent fractions are $\frac{180}{3330}$ and $\frac{259}{3330}$.

GO TO . . . Open the CD-ROM that accompanies your textbook, and select Chapter 1, Rule 1-9. Review the animation and example problems; then complete the practice problems. Continue to the next section of the book once you have mastered the rule presented. ■

REVIEW AND PRACTICE

1-4 Finding Common Denominators

For each set of fractions, find the least common denominator. Then convert each fraction to an equivalent fraction with the LCD.

1. $\frac{1}{3}$ and $\frac{1}{7}$ **2.** $\frac{1}{5}$ and $\frac{1}{8}$ **3.** $\frac{1}{25}$ and $\frac{1}{40}$ **4.** $\frac{1}{24}$ and $\frac{1}{36}$ **5.** $\frac{1}{2}$ and $\frac{1}{12}$

6. $\frac{1}{6}$ and $\frac{1}{18}$ **7.** $\frac{5}{6}$ and $\frac{4}{7}$ **8.** $\frac{3}{4}$ and $\frac{5}{8}$ **9.** $\frac{1}{9}$ and $\frac{1}{36}$ **10.** $\frac{5}{24}$ and $\frac{9}{96}$

11. $\frac{4}{5}$ and $\frac{9}{11}$ 12. $\frac{5}{6}$ and $\frac{7}{12}$ 13. $\frac{11}{30}$ and $\frac{21}{80}$ 14. $\frac{5}{48}$ and $\frac{7}{72}$ 15. $\frac{1}{2}, \frac{1}{3},$ and $\frac{1}{4}$

16. $\frac{1}{6}, \frac{4}{9},$ and $\frac{13}{24}$ 17. $\frac{2}{3}, \frac{4}{9},$ and $\frac{7}{15}$ 18. $\frac{1}{4}, \frac{5}{6},$ and $\frac{7}{16}$ 19. $\frac{1}{5}, \frac{3}{10},$ and $\frac{7}{20}$ 20. $\frac{1}{6}, \frac{5}{24},$ and $\frac{9}{40}$

To check your answers, see page 495.

Comparing Fractions

Suppose a home patient is to take $\frac{3}{4}$ tablespoon (tbs) of medication with lunch. You learn that the patient took $\frac{2}{3}$ tbs. Did the patient take too little, too much, or just the right amount?

To determine this answer, you must be able to compare fractions by comparing the numerators of the equivalent fractions. For this example the equivalent fractions are:

$$\frac{3}{4} = \frac{9}{12} \quad \frac{2}{3} = \frac{8}{12}$$

$\frac{9}{12}$ is more than $\frac{8}{12}$, so the patient did not take enough.

Rule 1-10 To compare fractions:

1. Write all fractions as equivalent fractions with a common denominator.
2. Write the fractions in order by the size of the numerator. The fraction with the largest numerator is the largest in the group.
3. Restate the comparisons with the original fractions.

Example 1 Order from smallest to largest: $\frac{1}{5}, \frac{4}{5},$ and $\frac{3}{10}$.

1. Write the fractions as equivalent fractions with a common denominator. The least common denominator of $\frac{1}{5}, \frac{4}{5},$ and $\frac{3}{10}$ is 10

   $$\frac{1}{5} = \frac{?}{10} = \frac{1 \times 2}{5 \times 2} = \frac{2}{10}$$

   $$\frac{4}{5} = \frac{?}{10} = \frac{4 \times 2}{5 \times 2} = \frac{8}{10}$$

   $$\frac{3}{10} = \frac{3}{10}$$

   If you have difficulty with this step, review "Equivalent Fractions" and "Finding Common Denominators" in this chapter.

2. Order the fractions by the size of their numerators, in this case, from smallest to largest.

   $$\frac{2}{10} \quad \frac{3}{10} \quad \frac{8}{10}$$

   Insert the proper comparison signs.

   $$\frac{2}{10} < \frac{3}{10} < \frac{8}{10}$$

3. Restate with the original equivalent fractions.

$$\frac{1}{5} < \frac{3}{10} < \frac{4}{5}$$

Example 2 Order from largest to smallest: $1\frac{7}{8}, \frac{6}{3}, 2, \frac{2}{8}$.

First, convert all whole and mixed numbers to fractions:

$$1\frac{7}{8} = \frac{15}{8}$$

$$2 = \frac{2}{1}$$

If you have difficulty with this step, review "Fractions and Mixed Numbers" in this chapter.

1. Write all fractions as equivalent fractions with a common denominator. The LCD for this set of fractions is 24

$$1\frac{7}{8} = \frac{15}{8} = \frac{?}{24} = \frac{15 \times 3}{8 \times 3} = \frac{45}{24}$$

$$\frac{6}{3} = \frac{?}{24} = \frac{6 \times 8}{3 \times 8} = \frac{48}{24}$$

$$2 = \frac{2}{1} = \frac{?}{24} = \frac{2 \times 24}{1 \times 24} = \frac{48}{24}$$

$$\frac{2}{8} = \frac{?}{24} = \frac{2 \times 3}{8 \times 3} = \frac{6}{24}$$

2. Write the fractions in order by the size of their numerator.

$$\frac{48}{24} \quad \frac{48}{24} \quad \frac{45}{24} \quad \frac{6}{24}$$

Insert the proper comparison signs.

$$\frac{48}{24} = \frac{48}{24} > \frac{45}{24} > \frac{6}{24}$$

3. Restate with the original equivalent fractions.

$$2 = \frac{6}{3} > 1\frac{7}{8} > \frac{2}{8}$$

GO TO . . . Open the CD-ROM that accompanies your textbook, and select Chapter 1, Rule 1-10. Review the animation and example problems; then complete the practice problems. Continue to the next section of the book once you have mastered the rule presented. ■

REVIEW AND PRACTICE

1-5 Comparing Fractions

Insert >, <, or = to make a true statement.

1. $\frac{1}{5} \quad \frac{3}{5}$ 2. $\frac{7}{9} \quad \frac{4}{9}$ 3. $\frac{2}{8} \quad \frac{1}{4}$ 4. $\frac{7}{10} \quad \frac{7}{20}$ 5. $\frac{3}{24} \quad \frac{1}{8}$

6. $\frac{11}{3}$ $\frac{13}{9}$ 7. 1 $\frac{2}{3}$ 8. $\frac{9}{4}$ 2 9. $\frac{3}{12}$ $\frac{13}{36}$ 10. $1\frac{1}{12}$ $1\frac{5}{12}$

11. $3\frac{3}{5}$ $3\frac{2}{5}$ 12. $1\frac{3}{4}$ $1\frac{7}{8}$ 13. $2\frac{1}{10}$ $2\frac{1}{8}$ 14. $3\frac{1}{5}$ $\frac{25}{8}$ 15. $\frac{9}{5}$ $1\frac{8}{10}$

Place in order from largest to smallest.

16. $\frac{3}{4}, \frac{2}{5}, \frac{5}{6}, \frac{4}{7}$ 17. $\frac{1}{3}, \frac{4}{7}, \frac{5}{9}, \frac{1}{2}$ 18. $1\frac{3}{16}, \frac{9}{8}, \frac{5}{2}, 2\frac{1}{10}$ 19. $2\frac{1}{2}, \frac{5}{3}, \frac{12}{9}, 1\frac{5}{6}$ 20. $\frac{1}{2}, \frac{3}{5}, \frac{6}{7}, \frac{5}{8}$

21. A home patient is supposed to take $\frac{2}{3}$ tbs of medication with lunch. You learn that the patient took $\frac{1}{2}$ tbs. Did the patient take too little, too much, or just the right amount?

22. You want to prepare 150 units of a solution that consists of 25 units of medication mixed with 125 units of water. You find an already prepared solution that has 1 unit of medication for every 5 units of water. Can you use 150 units of the already prepared solution? Explain your answer.

23. Of 12 patients in the north wing, 8 have high blood pressure. Of 15 patients in the east wing, 9 have high blood pressure. Which wing has a larger portion of patients with high blood pressure?

24. You give George his medication once every 8 hours (h). You give Martha the same medication 4 times a day. Who receives medication more often? (In this problem, a day means 24 h.)

25. You have $1\frac{3}{4}$ h until your shift starts. Your friend has $1\frac{7}{12}$ h. Who has more time before his or her shift starts?

To check your answers, see pages 495–496.

Adding Fractions

Suppose you gave a patient $\frac{1}{2}$ tbs of medication with breakfast, $\frac{3}{8}$ tbs with lunch, $\frac{3}{4}$ tbs with dinner, and another $\frac{3}{8}$ tbs at bedtime. To determine the total amount of medication you have given the patient, you must add the fractions. (This problem will be solved in the Review and Practice section, question 24.)

Rule 1-11 To add fractions:

1. Rewrite any mixed numbers as fractions.
2. Write equivalent fractions with common denominators. The LCD will be the denominator of your answer.
3. Add the numerators. The sum will be the numerator of your answer.

Reminder: Answers should be reported in the proper form. If the answer has a value greater than 1, convert it to a mixed number. If the fraction in the answer can be reduced to lower terms, do so.

Example 1 Add $\frac{2}{6} + \frac{3}{6}$

1. There are no mixed numbers.
2. The fractions already have a common denominator. The denominator of the answer is 6
3. Add the numerators: $2 + 3 = 5$ The numerator of the answer is 5

The answer is $\frac{5}{6}$. It is already in the proper form.

Example 2 Add $3\frac{1}{4} + 2\frac{1}{2}$.

1. Rewrite any mixed numbers as fractions.

   $3\frac{1}{4} = \frac{13}{4} \quad 2\frac{1}{2} = \frac{5}{2}$

2. Write equivalent fractions with common denominators.
   The LCD of the fractions is 4
   The denominator of the answer is 4

   You don't need to change $\frac{13}{4}$.

   An equivalent fraction for $\frac{5}{2}$ is $\frac{10}{4}$.

3. Add the numerators.

   $13 + 10 = 23$

   The numerator of the answer is 23

The answer is $\frac{23}{4}$ The proper form for this answer is $5\frac{3}{4}$.

Example 3 Add $1\frac{7}{8} + \frac{2}{3}$.

1. Rewrite any mixed numbers as fractions.

   $1\frac{7}{8} = \frac{15}{8}$. $\frac{2}{3}$ is not a mixed number.

2. Write equivalent fractions with common denominators.
   The LCD of the fractions is 24
   The denominator of the answer is 24

   An equivalent fraction for $\frac{15}{8}$ is $\frac{45}{24}$.

   An equivalent fraction for $\frac{2}{3}$ is $\frac{16}{24}$.

3. Add the numerators.

   $45 + 16 = 61$

   The numerator of the answer is 61

The answer is $\frac{61}{24}$ The proper form for this answer is $2\frac{13}{24}$.

Example 4 Add $1\frac{2}{5} + \frac{1}{2}$.

1. Rewrite any mixed numbers as fractions:

   $1\frac{2}{5} = \frac{7}{5}$ $\frac{1}{2}$ is not a mixed number.

2. Write equivalent fractions with common denominators.
   The LCD of the fractions is 10
   The denominator of the answer is 10

An equivalent fraction for $\frac{7}{5}$ is $\frac{14}{10}$.

An equivalent fraction for $\frac{1}{2}$ is $\frac{5}{10}$.

3. Add the numerators.

$14 + 5 = 19$

The numerator of the answer is 19

The answer is $\frac{19}{10}$ The proper form for this answer is $1\frac{9}{10}$.

GO TO . . . Open the CD-ROM that accompanies your textbook, and select Chapter 1, Rule 1-11. Review the animation and example problems; then complete the practice problems. Continue to the next section of the book once you have mastered the rule presented. ■

REVIEW AND PRACTICE

1-6 Adding Fractions

Find the following sums. (Rewrite answers in the proper form.)

1. $\frac{1}{8} + \frac{3}{8}$
2. $\frac{1}{7} + \frac{3}{7}$
3. $\frac{1}{7} + \frac{2}{14}$
4. $\frac{2}{5} + \frac{4}{15}$
5. $\frac{1}{6} + \frac{3}{8}$
6. $\frac{4}{10} + \frac{2}{25}$
7. $\frac{5}{8} + \frac{7}{12}$
8. $\frac{5}{6} + \frac{7}{9}$
9. $2 + \frac{4}{5}$
10. $\frac{8}{11} + 3$
11. $1\frac{1}{2} + \frac{1}{3}$
12. $2\frac{3}{8} + \frac{1}{5}$
13. $\frac{7}{9} + \frac{9}{12}$
14. $1\frac{2}{5} + 4\frac{3}{7}$
15. $2\frac{1}{8} + 1\frac{1}{2}$
16. $\frac{1}{2} + \frac{1}{5} + \frac{1}{8}$
17. $\frac{1}{3} + \frac{1}{4} + \frac{1}{5}$
18. $\frac{3}{4} + \frac{3}{8} + \frac{7}{12}$
19. $\frac{1}{2} + \frac{2}{3} + \frac{3}{5}$
20. $\frac{1}{4} + \frac{2}{5} + \frac{3}{10}$
21. The patient's chart indicates that he weighed 158 pounds (lb) at the end of April. He then gained $\frac{3}{4}$ lb in May and $1\frac{1}{2}$ lb in June. What did he weigh at the end of June?
22. $2\frac{1}{2}$ ounces (oz), $3\frac{3}{4}$ oz, and 5 oz were used from a bottle of solution in the office laboratory. What is the total amount of solution used?
23. Since breakfast, Kelly drank $1\frac{1}{4}$ cups of water, $\frac{2}{3}$ cup of juice, and $\frac{3}{4}$ cup of milk. How much liquid has Kelly had since breakfast?
24. During the day, you gave one of your patients $\frac{1}{2}$ tbs of medication with breakfast, $\frac{3}{8}$ tbs with lunch, $\frac{3}{4}$ tbs with dinner, and $\frac{3}{8}$ tbs at bedtime. What is the total amount of medication you gave your patient?
25. You are observing your patient's sleep pattern over the past 24 h. She slept $7\frac{1}{2}$ h at night, $1\frac{3}{4}$ h after breakfast, and $2\frac{1}{4}$ h after lunch. She also had a $\frac{1}{4}$ h nap before lunch and a $\frac{1}{4}$ h nap after dinner. How many hours did she sleep?

To check your answers, see page 496.

Subtracting Fractions

Subtracting fractions is similar to adding fractions.

Rule 1-12 To subtract fractions:

1. Rewrite any mixed numbers as fractions.
2. Write equivalent fractions with common denominators. The LCD will be the denominator of your answer.
3. Subtract the numerators. The difference will be the numerator of your answer.

Reminder: Answers should be reported in the proper form. If the answer has a value greater than 1, convert it to a mixed number. If the fraction in the answer can be reduced to lower terms, do so.

Example 1 Subtract $\frac{2}{6} - \frac{3}{12}$.

1. There are no mixed numbers.
2. Write equivalent fractions with common denominators.
   The LCD of the fractions is 12
   The denominator of the answer is 12

   An equivalent fraction for $\frac{2}{6}$ is $\frac{4}{12}$.

   You don't need to change $\frac{3}{12}$.
3. Subtract the numerators.

   $4 - 3 = 1$

   The numerator of the answer is 1

   The answer is $\frac{1}{12}$. It is already in the proper form.

Example 2 Subtract $3\frac{1}{4} - 2\frac{1}{2}$.

1. Rewrite any mixed numbers as fractions.

   $3\frac{1}{4} = \frac{13}{4}$ $\quad$ $2\frac{1}{2} = \frac{5}{2}$
2. Write equivalent fractions with common denominators.
   The LCD of the fractions is 4
   The denominator of the answer is 4

   You don't need to change $\frac{13}{4}$.

   An equivalent fraction for $\frac{5}{2}$ is $\frac{10}{4}$.
3. Subtract the numerators.

   $13 - 10 = 3$

   The numerator of the answer is 3

   The answer is $\frac{3}{4}$. It is already in the proper form.

Example 3 Subtract $9\frac{3}{4} - 2\frac{3}{8}$.

1. Rewrite any mixed numbers as fractions.

   $9\frac{3}{4} = \frac{39}{4}$ $\quad 2\frac{3}{8} = \frac{19}{8}$

2. Write equivalent fractions with common denominators.
   The LCD of the fractions is 8
   The denominator of the answer is 8

   An equivalent fraction for $\frac{39}{4}$ is $\frac{78}{8}$.

   You don't need to change $\frac{19}{8}$.

3. Subtract the numerators.

   $78 - 19 = 59$

   The numerator of the answer is 59

   The answer is $\frac{59}{8}$ Written as a mixed number, this is $7\frac{3}{8}$.

Example 4 Subtract $6 - 1\frac{1}{2}$.

1. Rewrite any mixed numbers as fractions.
   6 is not a mixed number.

   $1\frac{1}{2} = \frac{3}{2}$

2. Write equivalent fractions with common denominators.
   The LCD of the fractions is 2
   The denominator of the answer is 2

   An equivalent fraction for 6 is $\frac{12}{2}$.

   You don't need to change $\frac{3}{2}$.

3. Subtract the numerators.

   $12 - 3 = 9$

   The numerator of the answer is 9

   The answer is $\frac{9}{2}$ The proper form for the answer is $4\frac{1}{2}$.

GO TO . . . Open the CD-ROM that accompanies your textbook, and select Chapter 1, Rule 1-12. Review the animation and example problems; then complete the practice problems. Continue to the next section of the book once you have mastered the rule presented. ■

REVIEW AND PRACTICE

1-7 Subtracting Fractions

Find the following differences.

1. $7\frac{7}{15} - 4\frac{4}{15}$
2. $\frac{7}{25} - \frac{2}{25}$
3. $\frac{11}{3} - \frac{2}{6}$
4. $\frac{4}{7} - \frac{3}{21}$
5. $\frac{5}{6} - \frac{4}{9}$
6. $\frac{3}{4} - \frac{1}{6}$
7. $1\frac{7}{8} - \frac{1}{4}$
8. $2\frac{5}{8} - \frac{1}{2}$
9. $6\frac{1}{3} - \frac{5}{6}$
10. $4\frac{1}{2} - \frac{3}{4}$
11. $14\frac{9}{10} - 3\frac{1}{3}$
12. $6\frac{6}{7} - 2\frac{3}{5}$
13. $24\frac{1}{8} - 3\frac{3}{16}$
14. $8\frac{7}{10} - 3\frac{3}{4}$
15. $6 - \frac{2}{3}$
16. $7 - \frac{3}{7}$
17. $5 - \frac{3}{5}$
18. $\frac{2}{3} - \frac{1}{2}$
19. $10\frac{1}{5} - \frac{7}{15}$
20. $7\frac{2}{3} - \frac{7}{12}$
21. You give a patient $\frac{3}{4}$ cup (c) of juice, but he only drinks $\frac{3}{8}$ c. How much juice remains?
22. You give a patient $1\frac{1}{4}$ c of water to drink before supper. When you bring in the meal, you see that $\frac{5}{8}$ c remains in the glass. How much water did the patient drink?
23. At the beginning of the day, you have $6\frac{1}{2}$ bottles of a medication on hand. At the end of the day, $2\frac{3}{4}$ bottles remain. How much of the medication was used during the day?
24. Brenda weighed $153\frac{1}{2}$ lb when she began a diet. The first month she lost $2\frac{3}{4}$ lb. The second month she lost $4\frac{1}{2}$ lb. The third month she lost $2\frac{1}{2}$ lb. What does she weigh now? (*Hint*: You can subtract each month separately, or you can calculate her total weight loss first.)
25. The patient's temperature is $101\frac{1}{2}$ degrees. The patient's normal temperature is $98\frac{3}{4}$ degrees. How many degrees above normal is this patient's temperature?

To check your answers, see page 496.

Multiplying Fractions

Unlike adding and subtracting fractions, multiplying fractions does not need a common denominator. Think about what $\frac{2}{3} \times \frac{1}{2}$ means. This problem could be read as "two-thirds times one-half" or "two-thirds of one-half." In Figure 1-2a, a pizza is divided into six slices. Half of the pizza (three slices) has pepperoni. When you look for $\frac{2}{3}$ of $\frac{1}{2}$ of the pizza, you are looking for two-thirds of the pepperoni half.

In Figure 1-2b, two-thirds of the pepperoni half also has mushrooms. The mushroom slices represent $\frac{2}{3}$ of $\frac{1}{2}$ of the pizza, or $\frac{2}{3} \times \frac{1}{2}$. They also represent $\frac{2}{6}$ of the entire pizza. Thus, $\frac{2}{3} \times \frac{1}{2} = \frac{2}{6} = \frac{1}{3}$.

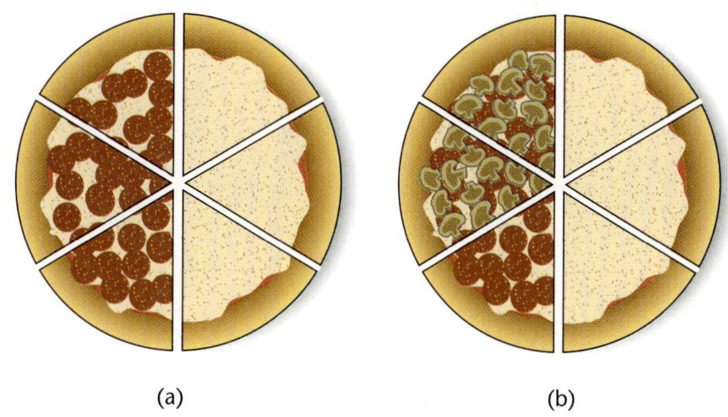

Figure 1-2 Multiplying fractions.

Rule 1-13 To multiply fractions:

1. Convert any mixed or whole numbers to improper fractions.
2. Multiply the numerators. Then multiply the denominators.
3. Reduce the product to its lowest terms.

Example 1 Multiply $\frac{1}{6} \times \frac{3}{4}$.

1. There are no mixed or whole numbers.
2. The product of the numerators is $1 \times 3 = 3$ The product of the denominators is $6 \times 4 = 24$ Thus, $\frac{1}{6} \times \frac{3}{4} = \frac{3}{24}$.
3. $\frac{3}{24}$ reduces to $\frac{1}{8}$

Example 2 Multiply $\frac{1}{2} \times \frac{7}{3} \times \frac{4}{9}$.

The only difference from Example 1 is that now you are multiplying three numerators and three denominators.

$$\frac{1}{2} \times \frac{7}{3} \times \frac{4}{9} = \frac{1 \times 7 \times 4}{2 \times 3 \times 9} = \frac{28}{54}$$

$\frac{28}{54}$ reduces to $\frac{14}{27}$

Example 3 Multiply $1\frac{4}{7} \times 2\frac{3}{5}$.

1. First convert the mixed numbers to improper fractions.

$$1\frac{4}{7} = \frac{11}{7} \quad \text{and} \quad 2\frac{3}{5} = \frac{13}{5}$$

2. Now multiply the numerators and denominators.

$$1\frac{4}{7} \times 2\frac{3}{5} = \frac{11}{7} \times \frac{13}{5} = \frac{11 \times 13}{7 \times 5} = \frac{143}{35}$$

3. $\frac{143}{35}$ converts to $4\frac{3}{35}$

Example 4 Multiply $3 \times \frac{2}{3}$

First convert 3 to the improper fraction $\frac{3}{1}$ Now solve.

$$3 \times \frac{2}{3} = \frac{3}{1} \times \frac{2}{3} = \frac{3 \times 2}{1 \times 3} = \frac{6}{3}$$

$\frac{6}{3}$ reduces to 2

GO TO . . . Open the CD-ROM that accompanies your textbook, and select Chapter 1, Rule 1-13. Review the animation and example problems; then complete the practice problems. Continue to the next section of the book once you have mastered the rule presented. ■

Reducing terms provides a shortcut that makes multiplying fractions easier. It lets you work with smaller numbers, decreasing the potential for arithmetic errors. If you divide both the numerator and the denominator of a fraction by the same number, you have not changed the fraction's value. You already use this rule to reduce a fraction. This rule also applies to two or more fractions that are being multiplied.

To multiply $\frac{8}{21} \times \frac{7}{16}$, you could multiply the numerators and multiply the denominators.

$$\frac{8}{21} \times \frac{7}{16} = \frac{8 \times 7}{21 \times 16} = \frac{56}{336}$$

$\frac{56}{336}$ reduces to $\frac{1}{6}$, although that may not be immediately clear to you. By reducing terms before multiplying, however, you work with smaller numbers.

Rule 1-14 To reduce terms when you are multiplying fractions, divide both a numerator and a denominator by the same number. You may reduce terms only if a numerator and denominator can both be divided evenly.

Example 1 Reduce terms to solve $\frac{8}{21} \times \frac{7}{16}$.

Both the numerator 8 and the denominator 16 can be divided evenly by 8 You can now write the problem as:

$$\frac{\overset{1}{\cancel{8}}}{21} \times \frac{7}{\underset{2}{\cancel{16}}} \quad \text{which is equivalent to} \quad \frac{1}{21} \times \frac{7}{2}$$

The slash marks indicate that 8 and 16 were reduced. In this case, they were divided by 8, reducing 8 and 16 to 1 and 2, respectively. Both the numerator 7 and the denominator 21 are divisible by 7 After you reduce again, you can rewrite the problem as:

$$\frac{\overset{1}{\cancel{8}}}{\underset{3}{\cancel{21}}} \times \frac{\overset{1}{\cancel{7}}}{\underset{2}{\cancel{16}}} \quad \text{which is equivalent to} \quad \frac{1}{3} \times \frac{1}{2}$$

Now when you solve, the answer will already be in lowest terms.

$$\frac{8}{21} \times \frac{7}{16} = \frac{\overset{1}{\cancel{8}}}{\underset{3}{\cancel{21}}} \times \frac{\overset{1}{\cancel{7}}}{\underset{2}{\cancel{16}}} = \frac{1}{3} \times \frac{1}{2} = \frac{1}{6}$$

Example 2 $\frac{27}{36} \times \frac{4}{5}$

In this problem, one of the fractions has not been reduced to lowest terms. Both 27 and 36 are divisible by 9

$$\frac{\overset{3}{\cancel{27}}}{\underset{4}{\cancel{36}}} \times \frac{4}{5} \quad \text{becomes} \quad \frac{3}{4} \times \frac{4}{5}$$

You can also reduce the numerator 4 and what had begun as the denominator 36. The problem now becomes:

$$\frac{\overset{3}{\cancel{27}}}{\underset{\underset{1}{\cancel{4}}}{\cancel{36}}} \times \frac{\overset{1}{\cancel{4}}}{5} = \frac{3}{\underset{1}{\cancel{4}}} \times \frac{\overset{1}{\cancel{4}}}{5} = \frac{3}{1} \times \frac{1}{5} = \frac{3}{5}$$

The answer $\frac{3}{5}$ is already reduced to lowest terms.

Example 3 $2\frac{1}{2} \times \frac{8}{15} \times \frac{45}{4}$

First convert the mixed number $2\frac{1}{2}$ to the improper fraction $\frac{5}{2}$ Now the problem becomes:

$$\frac{5}{2} \times \frac{8}{15} \times \frac{45}{4}$$

Both the numerator 45 and the denominator 15 are divisible by 15 Both the numerator 8 and the denominator 2 are divisible by 2

$$\frac{5}{\underset{1}{\cancel{2}}} \times \frac{\overset{4}{\cancel{8}}}{\underset{1}{\cancel{15}}} \times \frac{\overset{3}{\cancel{45}}}{4} = \frac{5}{1} \times \frac{4}{1} \times \frac{3}{4}$$

You have another opportunity to reduce both a numerator and a denominator by 4. The problem now becomes:

$$\frac{5}{\underset{1}{\cancel{2}}} \times \frac{\overset{4}{\cancel{8}}}{\underset{1}{\cancel{15}}} \times \frac{\overset{3}{\cancel{45}}}{4} = \frac{5}{1} \times \frac{\overset{1}{\cancel{4}}}{1} \times \frac{3}{\underset{1}{\cancel{4}}} = \frac{5}{1} \times \frac{1}{1} \times \frac{3}{1} = \frac{15}{1} = 15$$

If you are not sure what numbers will divide evenly into both the numerator and the denominator, review Table 1-1.

GO TO . . . Open the CD-ROM that accompanies your textbook, and select Chapter 1, Rule 1-14. Review the animation and example problems; then complete the practice problems. Continue to the next section of the book once you have mastered the rule presented. ■

ERROR ALERT!

Avoid reducing too many terms.

A term that you plan to reduce may be a factor in more than one numerator or more than one denominator. Each time you reduce a term, you must reduce it from *one numerator* ***and*** *one denominator.*

Suppose you are multiplying $\frac{7}{12} \times \frac{8}{20}$. You can reduce 4 from the numerator 8 You can also reduce either of the denominators (12 or 20) by 4, but not both. Either of the following is correct.

$$\frac{7}{\overset{}{\cancel{12}}_{3}} \times \frac{\overset{2}{\cancel{8}}}{20} \quad \text{or} \quad \frac{7}{12} \times \frac{\overset{2}{\cancel{8}}}{\underset{5}{\cancel{20}}}$$

However, *you cannot reduce by 4 as follows* in this problem:

$$\frac{7}{\underset{3}{\cancel{12}}} \times \frac{\overset{2}{\cancel{8}}}{\underset{5}{\cancel{20}}}$$

If the problem were $\frac{5}{12} \times \frac{9}{24}$, where the numerator can be reduced by 3 twice, then you could reduce 3 twice in the denominators. Thus,

$$\frac{5}{12} \times \frac{9}{24} = \frac{5}{\underset{4}{\cancel{12}}} \times \frac{\overset{\overset{1}{\cancel{3}}}{\cancel{9}}}{\underset{8}{\cancel{24}}} = \frac{5}{4} \times \frac{1}{8} = \frac{5}{32}$$

REVIEW AND PRACTICE

1-8 Multiplying Fractions

Find the following products. (Rewrite answers in the proper form.)

1. $\frac{1}{6} \times \frac{1}{8}$
2. $\frac{2}{7} \times \frac{3}{5}$
3. $\frac{1}{2} \times \frac{6}{8}$
4. $\frac{6}{9} \times \frac{1}{6}$
5. $\frac{3}{8} \times \frac{4}{9}$
6. $\frac{5}{12} \times \frac{6}{15}$
7. $\frac{10}{14} \times \frac{7}{5}$
8. $\frac{5}{3} \times \frac{9}{10}$
9. $\frac{9}{8} \times \frac{8}{2}$
10. $\frac{4}{3} \times \frac{15}{8}$
11. $1\frac{7}{8} \times \frac{4}{5}$
12. $3\frac{1}{3} \times \frac{9}{15}$
13. $3\frac{6}{8} \times 5\frac{2}{9}$
14. $1\frac{5}{6} \times 7\frac{4}{5}$
15. $\frac{7}{16} \times \frac{4}{3} \times \frac{1}{2}$
16. $\frac{5}{7} \times \frac{3}{10} \times \frac{3}{4}$
17. $\frac{11}{32} \times \frac{4}{22} \times 12$
18. $5 \times \frac{7}{15} \times \frac{3}{14}$
19. $\frac{12}{25} \times \frac{8}{9} \times \frac{15}{16}$
20. $\frac{49}{20} \times \frac{12}{7} \times \frac{5}{21}$
21. A bottle of liquid medication contains 24 doses. If the hospital has a supply of $9\frac{3}{4}$ bottles of the medication, how many doses are available?
22. A tablet contains $\frac{1}{4}$ grain (gr) of a medication. If you give a patient $1\frac{1}{2}$ tablets 3 times per day, how many grains of the medication are you giving to the patient each day?
23. A patient is supposed to take $\frac{1}{3}$ teaspoon (tsp) of medicine 4 times per day. However, the patient misunderstood the directions and took $\frac{1}{4}$ tsp of medicine 3 times per day.
    - **a.** How much medicine should the patient have taken per day?
    - **b.** How much medicine did the patient take per day?
    - **c.** What is the difference per day between the two amounts?

24. For 4 days, you give a patient $1\frac{1}{2}$ oz of a medication 5 times per day. How much medication did you give the patient over the 4 days?

25. One tablet contains 500 milligrams (mg) of medication. How many milligrams are in $3\frac{1}{2}$ tablets?

To check your answers, see pages 496–497.

Dividing Fractions

You have now learned most of the steps needed to divide fractions. Suppose you have $\frac{3}{4}$ bottle of liquid medication available. The regular dose you would give a patient is $\frac{1}{16}$ bottle, and you want to know how many doses remain in the bottle. You solve this problem by dividing fractions.

You want to solve $\frac{3}{4} \div \frac{1}{16}$, where $\frac{3}{4}$ is the dividend, $\frac{1}{16}$ is the divisor, and your answer is the quotient. The problem is read, "three-quarters divided by one-sixteenth," where you are finding out how many times $\frac{1}{16}$ goes into $\frac{3}{4}$.

To solve this problem, multiply the dividend $\frac{3}{4}$ by the reciprocal of the divisor $\frac{1}{16}$ You find the *reciprocal* of a fraction by inverting it—flipping it so that the numerator becomes the denominator and the denominator becomes the numerator. The reciprocal of $\frac{1}{16}$ is $\frac{16}{1}$ Thus,

$$\frac{3}{4} \div \frac{1}{16} = \frac{3}{4} \times \frac{16}{1}$$

You now solve this as a multiplication problem.

$$\frac{3}{\underset{1}{\cancel{4}}} \times \frac{\overset{4}{\cancel{16}}}{1} = \frac{3}{1} \times \frac{4}{1} = \frac{12}{1} = 12$$

The bottle has 12 doses remaining.

Rule 1-15 To divide fractions:

1. Convert any mixed or whole numbers to improper fractions.
2. Invert (flip) the divisor to find its reciprocal.
3. Multiply the dividend by the reciprocal of the divisor and reduce.

Example 1 Divide $\frac{1}{2} \div \frac{1}{4}$.

1. The problem has no mixed or whole numbers.
2. Invert (flip) the divisor $\frac{1}{4}$ to find its reciprocal $\frac{4}{1}$.
3. Multiply the dividend by the reciprocal of the divisor.

$$\frac{1}{2} \div \frac{1}{4} = \frac{1}{2} \times \frac{4}{1} = \frac{1}{\underset{1}{\cancel{2}}} \times \frac{\overset{2}{\cancel{4}}}{1} = \frac{2}{1} = 2$$

Example 2 Divide $1\frac{1}{2} \div \frac{1}{4}$.

1. Convert the mixed number to a fraction.

$$1\frac{1}{2} = \frac{3}{2}$$

2. Invert (flip) the divisor $\frac{1}{4}$ to find its reciprocal $\frac{4}{1}$.
3. Multiply the dividend by the reciprocal of the divisor.

$$1\frac{1}{2} \div \frac{1}{4} = \frac{3}{2} \div \frac{1}{4} = \frac{3}{\cancel{2}_1} \times \frac{\cancel{4}^2}{1} = \frac{3}{1} \times \frac{2}{1} = \frac{6}{1} = 6$$

You may have to solve a complex fraction, in which the numerator and the denominator are themselves fractions. The main fraction bar will often be wider and darker than the fraction bars within the numerator and the denominator. You can simply rewrite a complex fraction as an ordinary division problem and proceed.

Example 3 Simplify $\frac{\frac{7}{10}}{\frac{3}{5}}$.

Here $\frac{7}{10}$ is the numerator and $\frac{3}{5}$ is the denominator. Rewrite the complex fraction as a regular division problem, then solve.

$$\frac{\frac{7}{10}}{\frac{3}{5}} = \frac{7}{10} \div \frac{3}{5} = \frac{7}{10} \times \frac{5}{3} = \frac{7}{\cancel{10}_2} \times \frac{\cancel{5}^1}{3} = \frac{7}{2} \times \frac{1}{3} = \frac{7}{6} = 1\frac{1}{6}$$

Example 4 Divide $20\frac{1}{6} \div 4$

1. Convert $20\frac{1}{6}$ to $\frac{121}{6}$. Change the whole number 4 to $\frac{4}{1}$.

2. $\frac{121}{6} \div \frac{4}{1}$

   Flip (invert) the divisor $\frac{4}{1}$ to find its reciprocal $\frac{1}{4}$.

3. Multiply $\frac{121}{6} \times \frac{1}{4} = \frac{121}{24}$. Reduce answer to proper form $5\frac{1}{24}$.

GO TO . . . Open the CD-ROM that accompanies your textbook, and select Chapter 1, Rule 1-15. Review the animation and example problems; then complete the practice problems. Continue to the next section of the book once you have mastered the rule presented. ■

ERROR ALERT!

Write division problems carefully to avoid mistakes.

1. Convert whole numbers to fractions, especially if you use complex fractions. You cannot tell if $\frac{3}{\frac{4}{8}}$ is $3 \div \frac{8}{4}$ or $\frac{3}{8} \div 4$

2. Be sure to find the reciprocal of the correct fraction. You want the reciprocal of the divisor when you convert the problem from division to multiplication.

   $\frac{2}{3} \div \frac{4}{5} = \frac{2}{3} \times \frac{5}{4}$ not $\frac{3}{2} \times \frac{4}{5}$

REVIEW AND PRACTICE

1-9 Dividing Fractions

Find the following quotients. (Rewrite the answers in proper form.)

1. $\frac{4}{9} \div \frac{5}{7}$
2. $\frac{3}{11} \div \frac{4}{5}$
3. $\frac{3}{8} \div \frac{1}{2}$
4. $\frac{1}{6} \div \frac{3}{4}$
5. $\frac{3}{5} \div \frac{2}{8}$
6. $\frac{6}{9} \div \frac{5}{11}$
7. $\frac{9}{10} \div \frac{3}{5}$
8. $\frac{7}{12} \div \frac{21}{36}$
9. $1\frac{3}{4} \div \frac{2}{3}$
10. $\frac{7}{8} \div 1\frac{3}{4}$
11. $4\frac{2}{9} \div 2\frac{3}{8}$
12. $3\frac{1}{2} \div 1\frac{1}{4}$
13. $1\frac{7}{8} \div 9$
14. $6 \div \frac{5}{8}$
15. $\frac{\frac{9}{12}}{\frac{4}{6}}$
16. $\frac{\frac{2}{9}}{\frac{1}{8}}$
17. $\frac{2}{7} \div \frac{1}{4}$
18. $\frac{7}{10} \div \frac{3}{5}$
19. $5 \div \frac{4}{5}$
20. $\frac{\frac{4}{5}}{\frac{3}{8}}$
21. A bottle of pills has 40 tablets scored so that each tablet can be divided into four pieces. If a typical dose is $\frac{1}{4}$ tablet, how many doses does the bottle contain?
22. A health care professional administered doses of $2\frac{1}{2}$ milliliters (mL) of medication from a bottle that contained 150 mL. How many $2\frac{1}{2}$ mL doses was the health care professional able to give from the bottle?
23. A patient is told to drink the equivalent of 8 glasses of water each day. How many times must the patient drink $\frac{1}{2}$ glass of water to reach the daily goal?
24. A health care professional opens a case that has a total of 84 oz of medication. If each vial in the case holds $1\frac{3}{4}$ oz, how many vials are in the case?
25. How many $\frac{2}{3}$ h periods of time are in $8\frac{1}{2}$ h?

To check your answers, see page 497.

Decimals

The decimal system provides another way to represent whole numbers and their fractional parts. Health care practitioners use decimals in their daily work. The metric system, which is decimal-based, is used in dosage calculations, instrument calibrations, and general charting work. You must be able to work with decimals and convert fractions and mixed numbers to decimals.

Working with Decimals

In the decimal system, the location of a digit relative to the decimal point determines its value. The decimal point separates the whole number from the decimal fraction.

Writing Decimals. Each position in a decimal number has a place value. You already know values to the left of a decimal point. The places to the right of a decimal point represent fractions.

The number 1542.567 is read "one thousand five hundred forty-two *and* five hundred sixty-seven thousandths." Note that when you are writing decimal numbers using words or speaking decimal numbers, the word *and* replaces the decimal point. All numbers to the right of the decimal point are read and written as whole numbers with the place value of the decimal number written last. See Table 1-2.

TABLE 1-2 Decimal Place Values

The number 1542.567 can be represented as follows:

Whole Number				Decimal Point	Decimal Fraction		
Thousands	Hundreds	Tens	Ones	.	Tenths	Hundredths	Thousandths
1	5	4	2	.	5	6	7

Rule 1-16 To write a decimal number:

1. Write the whole-number part to the left of the decimal point.
2. Write the decimal fraction part to the right of the decimal point. Decimal fractions are equivalent to fractions that have denominators of 10, 100, 1000, and so forth.
3. Use zero as a placeholder to the right of the decimal point just as you use zero for whole numbers. The decimal number 1.203 represents 1 ones, 2 tenths, 0 hundredths, and 3 thousandths.

Example 1

Decimal	Description	Mixed Number
12.5	Twelve and five tenths	$12\frac{5}{10}$
206.34	Two hundred six and thirty-four hundredths	$206\frac{34}{100}$
0.33	Thirty-three hundredths	$\frac{33}{100}$
1.125	One and one hundred twenty-five thousandths	$1\frac{125}{1000}$

Example 2 Write $3\frac{4}{10}$ in decimal form.

$3\frac{4}{10}$ is 3 ones and 4 tenths. In decimal form, $3\frac{4}{10} = 3.4$

Example 3 Write $20\frac{7}{100}$ in decimal form.

$20\frac{7}{100}$ is 2 tens, 0 ones, 0 tenths, and 7 hundredths. In decimal form, $20\frac{7}{100} = 20.07$

GO TO . . . Open the CD-ROM that accompanies your textbook, and select Chapter 1, Rule 1-16. Review the animation and example problems; then complete the practice problems. Continue to the next section of the book once you have mastered the rule presented. ■

Rule 1-17 Always write a zero to the left of the decimal point when the decimal number has no whole-number part. Using the zero makes the decimal point more noticeable. Never place a trailing zero after the decimal point when working with medication dosages. Using zeros correctly helps to prevent medication errors.

Example Write the following fractions in decimal form.

a. $\frac{4}{10}$

$\frac{4}{10} = 0.4$ Do *not* write .4

b. $\frac{25}{1000}$

$\frac{25}{1000} = 0.025$ Do *not* write .025

c. $\frac{100}{25}$

$\frac{100}{25} = 4$ Do *not* write 4.0

GO TO . . . Open the CD-ROM that accompanies your textbook, and select Chapter 1, Rule 1-17. Review the animation and example problems; then complete the practice problems. Continue to the next section of the book once you have mastered the rule presented. ■

Comparing Decimals. The more places a number is to the right of the decimal point, the smaller its value. For example, 0.3 is $\frac{3}{10}$, or three-tenths; 0.03 is $\frac{3}{100}$, or three-hundredths; and 0.003 is $\frac{3}{1000}$, or three-thousandths. Think of it like this: $\frac{3}{10}$ is similar to three dimes and $\frac{3}{100}$ is similar to three cents. If a dollar is divided further into 1000 parts, $\frac{3}{1000}$ (three thousandths) is very small.

Rule 1-18 To compare the values of a group of decimal numbers:

1. Look first at the whole-number part. The decimal number with the greatest whole number is the greatest decimal number.

2. If the whole numbers of two decimals are equal, compare the digits in the tenths place. The tenths place is the first place to the right of the decimal point.
3. If the tenths places are equal, move to the right and compare the hundredths place digits.
4. Continue moving to the right, comparing digits until one is greater than the other. This will be the larger number. Zeros added to the right of the last nonzero digit after the decimal point do not change the value of the number.

Example 1 Which is larger, 2.1 or 2.3?

The whole number 2 is the same in both numbers. Move one space to the right of the decimal. Compare the tenths digits. Because $3 > 1$,

$2.3 > 2.1$

Example 2 Which is larger, 0.3 or 0.05?

There is no whole number. Move one space to the right of the decimal. Compare the tenths digits. Because $3 > 0$,

$0.3 > 0.05$

Example 3 Which is larger, 0.121 or 0.13?

There is no whole number. Move one space to the right of the decimal. Compare the tenths digits. These digits are equal. Move one space to the right and compare the hundredths digits. Because $3 > 2$,

$0.13 > 0.121$

GO TO . . . Open the CD-ROM that accompanies your textbook, and select Chapter 1, Rule 1-18. Review the animation and example problems; then complete the practice problems. Continue to the next section of the book once you have mastered the rule presented. ■

REVIEW AND PRACTICE

1-10 Working with Decimals

Write the following fractions in decimal form.

1. $\frac{2}{10}$ **2.** $\frac{17}{100}$ **3.** $6\frac{5}{10}$ **4.** $7\frac{19}{100}$

5. $\frac{3}{1000}$ **6.** $\frac{23}{1000}$ **7.** $5\frac{67}{1000}$ **8.** $7\frac{151}{1000}$

Place $>$ or $<$ between each pair of decimals to make a true statement.

9. 4.27 4.02 **10.** 12.25 12.18 **11.** 0.4 0.6 **12.** 2.22 2.20

13. 0.0170 0.0172 **14.** 0.3001 0.2998 **15.** 5.41 5.34 **16.** 34.58 34.85

17. 0.7 0.9 **18.** 0.67 0.53 **19.** 0.0542 0.0524 **20.** 0.6891 0.8619

To check your answers, see page 497.

Rounding Decimals

In health care settings, you will usually round decimals to the nearest tenth or hundredth, especially if you use a calculator. The answer you get may contain many more decimal places than you need, and you must round the answer. For example, 10 ÷ 6 = 1.66666666666 . . . . In some circumstances you will be asked to round to the nearest tenth, which is 1.7. In other situations your may need to round to the nearest hundredth, which is 1.67.

Rule 1-19 To round decimals:

1. Underline the place value to which you want to round.
2. Look at the digit to the right of this target place value. If this digit is 4 or less, do not change the digit in the target place value. If this digit is 5 or more, round the digit in the target place value up one unit.
3. Drop all digits to the right of the target place value.

Example 1 Round 2.42 to the nearest tenth.

1. Underline the tenths place (the target place value): $2.\underline{4}2$
2. The digit to the right of the tenths place is 2. Do not change the digit in the tenths place.
3. Drop the digits to the right of the tenths place. The number 2.42 rounded to the nearest tenth equals 2.4

Example 2 Round 0.035 to the nearest hundredth.

1. $0.0\underline{3}5$
2. The digit to the right of the hundredths place is 5 Round the digit in the hundredths place up one unit: $0.0\underline{4}$
3. The number 0.035 rounded to the nearest hundredth equals 0.04

Example 3 Round 3.99 to the nearest tenth.

1. $3.\underline{9}9$
2. The digit to the right of the tenths place is 9 Round the digit in the tenths place up one unit. When 9 is rounded up, it becomes 10 Place the 0 in the tenths place, and carry the 1 to the ones place. When 1 is added to the ones place, 3 becomes 4. The rounded number becomes $4.\underline{0}0$
3. The number 3.99 rounded to the nearest tenth equals 4.0

GO TO . . . Open the CD-ROM that accompanies your textbook, and select Chapter 1, Rule 1-19. Review the animation and example problems; then complete the practice problems. Continue to the next section of the book once you have mastered the rule presented. ■

Critical Thinking on the Job

Rounding Errors with 9

A health care professional is calculating how much medication to inject. The patient should receive 4.95 mL, but the syringe being used is calibrated (marked) in tenths. He must round the calculation to the nearest tenth.

The health care professional looks at the 5 in the hundredths place and rounds the tenths place up from 9 to 0. However, the health care professional neglects to carry the unit to the ones place and draws 4.0 mL of medication into the syringe.

Question?

What mistake was made and how could it have been avoided?

REVIEW AND PRACTICE

1-11 Rounding Decimals

Round to the nearest tenth.

1. 14.34 **2.** 3.45 **3.** 0.86 **4.** 0.19

5. 1.007 **6.** 0.2083 **7.** 152.68

Round to the nearest hundredth.

8. 9.293 **9.** 55.168 **10.** 4.0060 **11.** 2.2081

12. 5.5195 **13.** 11.999 **14.** 767.4562

Round to the nearest whole number.

15. 11.493 **16.** 19.98 **17.** 2.099 **18.** 50.505

19. You are preparing a syringe to inject 3.75 mL of medication. The syringe is calibrated in tenths. How much medication should you draw into the syringe?

20. A health care professional is preparing a syringe for injection. The calculations indicate that 0.38 mL should be given to the patient. The syringe is calibrated in hundredths. How much medication should the health care professional draw into the syringe?

To check your answers, see page 497.

Converting Fractions into Decimals

Conversions between fractions and decimals is important in health care settings. When you convert fractions, the equivalent decimals are greater than 1. When you convert fractions to decimals, think of the fractions as division problems. You can write $\frac{1}{4}$ as $1 \div 4$ Reducing fractions first (if possible) often makes the division easier.

Rule 1-20 To convert a fraction to a decimal, divide the numerator by the denominator.

Example 1 Convert $\frac{3}{4}$ to a decimal.

Divide the numerator by the denominator.

$$\begin{array}{r} 0.75 \\ 4\overline{)3.00} \\ \underline{2\,8} \\ 20 \\ \underline{20} \end{array} \qquad \frac{3}{4} = 0.75$$

Example 2 Convert $\frac{2}{3}$ to a decimal.

$$\begin{array}{r} 0.666 \\ 3\overline{)2.000} \\ \underline{1\,8} \\ 20 \\ \underline{18} \\ 20 \\ \underline{18} \\ 2.\,.\,. \end{array}$$

Sometimes the decimal repeats rather than terminates, as with $\frac{2}{3}$. In such cases, you round, for example, to the nearest hundredth.

$$\frac{2}{3} \cong 0.67$$

Example 3 Convert $\frac{8}{5}$ to a decimal.

$$\begin{array}{r} 1.6 \\ 5\overline{)8.0} \\ \underline{5} \\ 3.0 \\ \underline{3.0} \end{array} \qquad \frac{8}{5} = 1.6$$

Example 4 Convert $1\frac{7}{8}$ to a decimal.

$$\begin{array}{r} 0.875 \\ 8\overline{)7.000} \\ \underline{6\,4} \\ 60 \\ \underline{56} \\ 40 \\ \underline{40} \end{array} \qquad 1\frac{7}{8} = 1 + \frac{7}{8} = 1 + 0.875 = 1.875$$

GO TO . . . Open the CD-ROM that accompanies your textbook, and select Chapter 1, Rule 1-20. Review the animation and example problems; then complete the practice problems. Continue to the next section of the book once you have mastered the rule presented. ■

REVIEW AND PRACTICE

1-12 Converting Fractions into Decimals

Convert the following numbers into decimals. Where necessary, round to the nearest thousandth.

1. $\frac{2}{5}$	2. $\frac{7}{20}$	3. $\frac{9}{12}$	4. $\frac{12}{24}$	5. $\frac{1}{3}$
6. $\frac{4}{9}$	7. $\frac{15}{27}$	8. $\frac{21}{36}$	9. $\frac{12}{8}$	10. $\frac{11}{5}$
11. $\frac{7}{3}$	12. $\frac{9}{8}$	13. $1\frac{4}{5}$	14. $2\frac{1}{10}$	15. $6\frac{3}{4}$
16. $3\frac{1}{2}$	17. $\frac{7}{8}$	18. $\frac{9}{45}$	19. $\frac{20}{7}$	20. $7\frac{2}{3}$

To check your answers, see pages 497–498.

Converting Decimals into Fractions

Sometimes you need to convert a decimal to a fraction, especially when you use a calculator that provides decimals, but you need a fraction. When you work with decimals, treat the number to the left of the decimal point as a whole number and the number to the right of the decimal point as a fraction. For example, 12.5 is twelve and five tenths, or $12\frac{5}{10}$ The place value of the digit farthest to the right of the decimal point is the denominator. For 12.5, this place value is the tenths place. The denominator is 10 The numerator is 5, the number to the right of the decimal point.

Rule 1-21 To convert a decimal to a fraction or mixed number:

1. Write the number to the left of the decimal point as the whole number.
2. Write the number to the right of the decimal point as the numerator of the fraction.
3. Use the place value of the digit farthest to the right of the decimal point as the denominator.
4. Reduce the fraction part to its lowest terms.

Example 1 Convert 3.75 to a mixed number.

1. Write the number to the left of the decimal point, 3, as the whole number.
2. Write the number to the right of the decimal point, 75, as the numerator of the fraction.
3. The digit farthest to the right of the decimal point, 5, is in the hundredths place.

   Thus, the denominator is 100 The mixed number is $3\frac{75}{100}$.

4. Reduce to lowest terms: $3\frac{75}{100} = 3\frac{3}{4}$.

Example 2 Convert 0.010 to a fraction.

1. The number to the left of the decimal point is 0, so 0.010 has no whole number.
2. The number 010 is to the right of the decimal point. Because 010 = 10, write 10 as the numerator of the fraction.
3. The digit farthest to the right of the decimal point is 0, in the thousandths place. The denominator is 1000 The fraction is $\frac{10}{1000}$.
4. $\frac{10}{1000} = \frac{1}{100}$

GO TO . . . Open the CD-ROM that accompanies your textbook, and select Chapter 1, Rule 1-21. Review the animation and example problems; then complete the practice problems. Continue to the next section of the book once you have mastered the rule presented. ■

REVIEW AND PRACTICE

1-13 Converting Decimals into Fractions

Convert the following decimals to fractions or mixed numbers. Reduce the answer to its lowest terms.

1. 1.2	**2.** 98.6	**3.** 0.3	**4.** 0.442	**5.** 5.03
6. 0.301	**7.** 100.04	**8.** 206.070	**9.** 10.68	**10.** 7.44

To check your answers, see page 498.

Adding and Subtracting Decimals

When you add or subtract decimals, you align them by their place value, just as you do to add or subtract whole numbers.

Rule 1-22 To add or subtract decimals:

1. Write the problem vertically, as you would with whole numbers. Align the decimal points.
2. Add or subtract, starting from the right. Include the decimal point in your answer.

Example 1 Add 2.47 + 0.39

1. Write the problem vertically. Align the decimal points.

$$\begin{array}{r} 2.47 \\ +\ 0.39 \\ \hline \end{array}$$

2. Add.

$$\begin{array}{r} 2.47 \\ +\ 0.39 \\ \hline 2.86 \end{array}$$

Example 2 Subtract 52.04 − 14.31

$$\begin{array}{r} 52.04 \\ -\ 14.31 \\ \hline 37.73 \end{array}$$

Align the decimal points.

When the decimals have an unequal number of places, add zeros to the end of the decimal fraction so that all numbers are the same length past the decimal point. Writing zeros after the last digit to the right of the decimal point does not change the number's value. Including these zeros helps prevent errors in calculations.

Example 3 Add 14.3 + 1.56 + 9 + 0.352

Align the numbers. (Rewrite 9 as 9.0 to help you align it properly.) Fill in zeros so that all decimal fractions are of equal length. Then add.

$$\begin{array}{r} 14.3 \\ 1.56 \\ 9.0 \\ +\ 0.352 \\ \hline \end{array} \qquad \begin{array}{r} 14.300 \\ 1.560 \\ 9.000 \\ +\ 0.352 \\ \hline 25.212 \end{array}$$

Example 4 Subtract 7.3 − 1.005

Align the numbers. Fill in zeros so that all decimal fractions are of equal length. Then subtract.

$$\begin{array}{r} 7.3 \\ -\ 1.005 \\ \hline \end{array} \qquad \begin{array}{r} 7.300 \\ -\ 1.005 \\ \hline 6.295 \end{array}$$

Example 5 Subtract 10 − 0.75

$$\begin{array}{r} 10.00 \\ -\ 0.75 \\ \hline 9.25 \end{array}$$

GO TO . . . Open the CD-ROM that accompanies your textbook, and select Chapter 1, Rule 1-22. Review the animation and example problems; then complete the practice problems. Continue to the next section of the book once you have mastered the rule presented. ■

REVIEW AND PRACTICE

1-14 Adding and Subtracting Decimals

Add or subtract the following pairs of numbers.

1. 7.58 + 3.24 **2.** 143.05 + 22.07 **3.** 13.561 + 0.099 **4.** 24.102 + 2.410 **5.** 2.01 + 0.5

6. 2.30 + 0.005 **7.** 0.075 + 0.73 **8.** 4 + 0.025 **9.** 31.64 − 17.39 **10.** 16.250 − 1.625

11. 5.66 − 0.09 **12.** 14.7 − 0.9 **13.** 1.22 − 0.4 **14.** 12.2 − 0.972 **15.** 8 − 0.076

16. 12 − 0.02 **17.** 8.67 + 0.93 **18.** 121.04 + 56.75 **19.** 70.22 − 4.23 **20.** 526.10 − 7.41

21. Steve's temperature on Wednesday morning was 101.4 degrees Fahrenheit (101.4°F). By Thursday afternoon, it was 99.5°F. By how many degrees had his temperature changed?

22. While waiting to see her father, Helene ate at the hospital cafeteria, where she spent $1.30 for a soda, $2.65 for a bowl of soup, and $3.50 for a garden salad. How much did Helene spend?

23. You are supposed to administer 9 grams (g) of a medication. You give the patient one tablet with 4.5 grams (g) and a second tablet with 2.25 grams (g). How much more medication should you administer?

24. A bottle of liquid medication contains 50 mL. The following amounts are given to patients from the bottle: 2.5 mL, 3.1 mL, 1.75 mL, 3 mL, and 2.25 mL. How much medication remains in the bottle?

25. Your patient weighed 70.57 kilograms (kg) 2 months ago. His weight then increased by 2.3 kg one month and 1.75 kg the next month. What is this patient's current weight in kilograms?

To check your answers, see page 498.

Multiplying Decimals

Multiplying decimals is similar to multiplying whole numbers, except you must determine where to place the decimal point.

Rule 1-23 To multiply decimals:

1. First multiply without considering the decimal points, as if the numbers were whole numbers.
2. Count the total number of places to the right of the decimal points in *both* numbers.
3. To place the decimal point in the answer, start at its right end. Move the decimal point to the left the same number of places as the answer from step 2.

Example 1 Multiply 3.42 × 2.5

When you are multiplying, the decimal points *do not* need to line up as they do for adding and subtracting.

1. First multiply without considering the decimal points.

```
  3.42
 × 2.5
 -----
  1710
  684
 -----
  8550
```

2. Count the total number of decimal places (to the right of the decimal point) in the numbers. The number 3.42 has two decimal places; 2.5 has one decimal place. The numbers have a total of three decimal places.
3. Place the decimal point in the answer. Start at the right of the answer 8550 Move the decimal point three places to the left: 8.550 *After* placing the decimal point, you can drop the final zero so that the answer is 8.55

Example 2 Multiply 0.001 × 0.02

1. Multiply.

```
  0.001
 × 0.02
 ------
      2
```

2. The number 0.001 has three decimal places, and 0.02 has two decimal places. The numbers have a total of five decimal places.
3. Start to the right of the answer 2 Move the decimal point five places to the left. Insert zeros to the left of 2 in order to correctly place the decimal point. The correct answer is 0.00002

GO TO . . . Open the CD-ROM that accompanies your textbook, and select Chapter 1, Rule 1-23. Review the animation and example problems; then complete the practice problems. Continue to the next section of the book once you have mastered the rule presented. ■

REVIEW AND PRACTICE

1-15 Multiplying Decimals

Multiply the following numbers.

1. 7.4 × 8.2
2. 8.21 × 1.1
3. 4.2 × 0.3
4. 3.04 × 0.04
5. 0.55 × 0.5
6. 0.027 × 0.4
7. 0.003 × 0.02
8. 0.25 × 0.75
9. 1.03 × 14
10. 12 × 0.09
11. 0.004 × 15.5
12. 0.004 × 40.01
13. 5.2 × 3.1
14. A patient is given 7.5 mL of liquid medication 5 times per day. How many milliliters does she receive per day?
15. A small syringe is used to give a patient 0.28 mL of medication 4 times per day for 4 days. How much medication does he receive over the 4 days?
16. A tablet has a strength of 0.25 milligram (mg) of medication. You give the patient $1\frac{1}{2}$ tablets 3 times per day. How many milligrams of medication do you give the patient each day? (*Hint:* Convert $1\frac{1}{2}$ to decimal form first.)
17. A tablet has a strength of 0.4 mg of medication. If you give the patient $\frac{1}{4}$ tablet twice a day, how many milligrams of medication does the patient receive per day?
18. 72.4 × 0.24
19. 90 × 30.01
20. A case of isopropyl alcohol contains 12 bottles. Each bottle contains 15.95 oz. How many ounces of alcohol are in each case?

To check your answers, see page 498.

Dividing Decimals

The key to dividing decimals correctly is to place the decimal point properly. Recall that the dividend is the number that will be divided. If the divisor is a decimal, you want to convert the problem to one in which the divisor is a whole number.

Rule 1-24 To divide decimals:

1. Write the problem as a fraction. (The dividend is the numerator, and the divisor is the denominator.)
2. Move the decimal point to the right the same number of places in both the numerator and the denominator until the denominator is a whole number. Insert zeros in the numerator as necessary.
3. Complete the division as you would with whole numbers. Align the decimal point of the quotient with the decimal point of the numerator, if needed.

Example 1 Divide 0.8 ÷ 0.02

1. Write the problem as a fraction: $\frac{0.8}{0.02}$.

2. Move the decimal point two places to the right in both the numerator and the denominator. The denominator is now a whole number.

   $$\frac{0.8}{0.02} = \frac{8}{0.2} = \frac{80}{2}$$

   (This step is equivalent to multiplying $\frac{0.8}{0.02} \times \frac{100}{100}$.)

3. Complete the division.

   $$\begin{array}{r} 40 \\ 2\overline{)80} \\ \underline{80} \end{array}$$ so $0.8 \div 0.02 = 40$

Example 2 Divide 0.066 ÷ 0.11

1. $\frac{0.066}{0.11}$

2. $\frac{0.066}{0.11} = \frac{6.6}{11}$

   Move the decimal point two places to the right so that the denominator (divisor) is a whole number.

3. $$\begin{array}{r} 0.6 \\ 11\overline{)6.6} \\ \underline{6.6} \end{array}$$

Align the decimal point of the quotient with the decimal point of the numerator (dividend). Here, $0.066 \div 0.11 = 0.6$

GO TO . . . Open the CD-ROM that accompanies your textbook, and select Chapter 1, Rule 1-24. Review the animation and example problems; then complete the practice problems. Continue to the next section of the book once you have mastered the rule presented. ■

Critical Thinking on the Job

Placing Decimals Correctly

A practitioner was instructed to give 0.25 gram (g) of medication for every 1.0 kilogram (kg) of body weight. A baby she was treating weighed 6.25 kg. She set up this calculation:

```
  6.25
× 0.25
------
  3125
 1250
------
156.25
```

Questions?

What mistake did the health care professional make? What could have happened if the mistake was not corrected?

REVIEW AND PRACTICE

1-16 Dividing Decimals

Divide the following numbers. When, necessary, round to the nearest thousandth.

1. 3.2 ÷ 1.6	**2.** 48.6 ÷ 1.8	**3.** 24.5 ÷ 0.2	**4.** 0.004 ÷ 0.002
5. 1.25 ÷ 0.5	**6.** 0.32 ÷ 0.8	**7.** 0.05 ÷ 4	**8.** 12.6 ÷ 4
9. 40 ÷ 0.8	**10.** 0.44 ÷ 4.4	**11.** 29.05 ÷ 100	**12.** 3.48 ÷ 1000
13. 39.666 ÷ 0.03	**14.** 54.54 ÷ 0.009	**15.** 59.48 ÷ 66.93	**16.** 84.3 ÷ 68.48

17. A bottle holds 60 mL of medication. If the average dose is 0.75 mL, how many doses does the bottle hold?

18. A bottle contains 32 oz of medication. If the average dose is 0.4 oz, how many doses does the bottle contain?

19. A patient received a total of 2.25 g of a medication. If the patient received the total over a 3-day period and was given 3 doses per day, what was the strength of each dose?

20. A patient weighs 197.5 lb. The patient's goal is to weigh 152.5 lb a year from now. How much weight should the patient lose per month to be successful?

To check your answers, see pages 498–499.

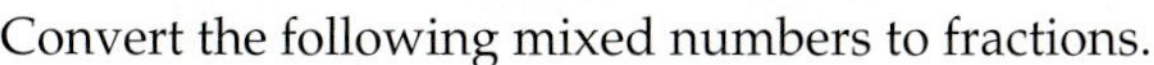

HOMEWORK ASSIGNMENT

Convert the following mixed numbers to fractions.

1. $3\frac{1}{5}$
2. $5\frac{2}{3}$
3. $8\frac{5}{8}$

Reduce the following fractions to their lowest terms. Convert improper fractions to mixed numbers when necessary.

4. $\frac{14}{42}$
5. $\frac{12}{45}$
6. $\frac{42}{8}$

Find the least common denominator for each set of fractions, and then use the common denominator to write equivalent fractions.

7. $\frac{3}{4}$ and $\frac{5}{6}$
8. $\frac{3}{5}, \frac{2}{3},$ and $\frac{3}{4}$

Place >, <, or = between the following pairs of fractions to make a true statement.

9. $\frac{4}{9} \quad \frac{2}{5}$
10. $\frac{14}{25} \quad \frac{3}{5}$

Perform the following calculations. Give the answer in the proper form.

11. $\frac{3}{2} + \frac{1}{3}$
12. $\frac{5}{8} \times \frac{4}{9}$
13. $3\frac{2}{5} - 1\frac{1}{5}$
14. $\frac{5}{7} + \frac{11}{14} + \frac{16}{21}$
15. $\frac{3}{8} \div \frac{2}{3}$
16. $\frac{5}{9} \times \frac{3}{10}$
17. $\frac{4}{9} - \frac{1}{3}$
18. $\frac{7}{15} \div 1\frac{5}{9}$

Place >, <, or = between the following pairs of decimals to make a true statement.

19. 0.017   0.09
20. 3.092   3.27

Round to the nearest hundredth.

21. 1.1834
22. 17.526

Round to the nearest tenth.

23. 6.158
24. 0.135

Round to the nearest whole number.

25. 12.185
26. 1.518

Convert the following fractions to decimals.

27. $2\frac{1}{8}$
28. $\frac{12}{5}$

Convert the following decimals to fractions. Reduce the answers to lowest terms.

29. 1.35
30. 0.025

Perform the following calculations.

31. 4.25×1.2
32. $1.86 \div 0.3$
33. $3.26 + 0.015$
34. 0.325×2.8
35. $12.05 - 7.6$
36. $10.5 \div 1.5$
37. $0.321 + 0.0075$
38. $3.65 - 0.125$
39. During the day, a patient drank the following quantities of fluid: $1\frac{1}{2}$ cups, 2.4 cups, $1\frac{3}{4}$ cups, and 1.2 cups.
    a. Convert each of the measurements into fractions. Add the fractions to find the total volume of fluid.
    b. Convert each of the measurements into decimals. Add the decimals to find the total volume of fluid.

Chapter 1 Review

Check Up

Convert the following mixed numbers to fractions.

1. $2\frac{3}{8}$ **2.** $1\frac{2}{7}$ **3.** $9\frac{9}{10}$ **4.** $12\frac{11}{12}$

Reduce the following fractions to their lowest terms. Convert improper fractions to mixed numbers when necessary.

5. $\frac{12}{36}$ **6.** $\frac{39}{48}$ **7.** $\frac{45}{9}$ **8.** $\frac{58}{8}$

Find the least common denominator. Then write an equivalent fraction for each.

9. $\frac{3}{10}$ and $\frac{4}{5}$ **10.** $\frac{5}{6}$ and $\frac{4}{9}$ **11.** $\frac{3}{8}, \frac{3}{4}$, and $\frac{1}{6}$ **12.** $\frac{7}{10}, \frac{1}{4}$, and $\frac{2}{3}$

Place >, <, or = between the following pairs of fractions to make a true statement.

13. $\frac{3}{10}$ $\frac{3}{16}$ **14.** $\frac{3}{2}$ $\frac{8}{5}$ **15.** $1\frac{2}{3}$ $1\frac{16}{24}$ **16.** $\frac{4}{25}$ $\frac{16}{75}$

Perform the following calculations. Give the answer in the proper form.

17. $\frac{9}{4} + \frac{2}{3}$ **18.** $\frac{3}{5} + 1\frac{2}{5}$ **19.** $\frac{2}{10} + \frac{1}{100} + \frac{4}{50}$ **20.** $6 + \frac{5}{8} + \frac{1}{3} + \frac{5}{12}$

21. $\frac{11}{9} - \frac{1}{3}$ **22.** $\frac{4}{5} - \frac{3}{4}$ **23.** $3\frac{1}{4} - 1\frac{7}{8}$ **24.** $3 - \frac{2}{7}$

25. $\frac{5}{6} \times \frac{2}{3}$ **26.** $\frac{7}{9} \times \frac{3}{14}$ **27.** $2\frac{2}{5} \times \frac{10}{3}$ **28.** $\frac{3}{8} \times 11$

29. $\frac{1}{7} \div \frac{3}{4}$ **30.** $\frac{12}{13} \div \frac{3}{52}$ **31.** $2\frac{5}{8} \div \frac{1}{6}$ **32.** $\frac{1}{3} \div 1\frac{1}{4}$

Place >, <, or = between the following pairs of decimals to make a true statement.

33. 5.7 5.09 **34.** 0.04 0.004 **35.** 6.3 6.300 **36.** 9.033 9.303

Round to the nearest hundredth.

37. 0.229 **38.** 7.091 **39.** 46.001 **40.** 9.885

Round to the nearest tenth.

41. 4.34 **42.** 3.65 **43.** 6.991 **44.** 0.073

Round to the nearest whole number.

45. 8.96 **46.** 20.6 **47.** 0.931 **48.** 12.449

Convert the following fractions to decimals.

49. $\frac{7}{14}$ **50.** $\frac{5}{8}$ **51.** $2\frac{3}{5}$ **52.** $\frac{32}{4}$

Convert the following decimals to fractions. Reduce the answers to lowest terms.

53. 0.82 **54.** 0.65 **55.** 3.5 **56.** 1.001

Perform the following calculations.

57. $7.23 + 12.38$
58. $4.59 + 0.2$
59. $0.031 + 0.99$
60. $12 + 0.004 + 1.7$
61. $7.49 - 0.38$
62. $4.28 - 3.39$
63. $0.852 - 0.61$
64. $14.01 - 0.788$
65. 2.3×4.9
66. 0.33×0.002
67. 5×0.999
68. 12.01×1.005
69. $38.85 \div 2.1$
70. $4.875 \div 3.25$
71. $2.2 \div 0.11$
72. $1.4 \div 0.07$
73. A medical unit has 18 patients. Eight have type O blood. Five have type A blood. Two have type AB blood. Three have type B blood. Write the fractions that describe the portions of the medical unit patients that have each blood type.
74. A patient is supposed to receive $\frac{1}{2}$ cup (c) of medication 3 times per day. Instead, the patient receives $\frac{1}{3}$ c twice per day. During the day, how much medicine does the patient receive? How does that amount compare with the amount ordered?
75. During the day, Brian drank $\frac{3}{4}$ c of water 7 times, 1 c of milk 2 times, and $\frac{1}{2}$ c of juice 3 times. How much liquid did Brian consume?
76. A bottle contains 48 mL of liquid medication.
    a. If the average dose is $\frac{3}{4}$ mL, how many doses does the bottle contain?
    b. If the average dose is 1.2 mL, how many doses does the bottle contain?

Critical Thinking Applications

A health care professional is asked to arrange a set of instruments on a tray in order from smallest to largest on the basis of the instruments' diameters. The diameters are marked $\frac{1}{4}$, $\frac{7}{16}$, $\frac{1}{2}$, $\frac{1}{8}$, $\frac{1}{16}$, $\frac{3}{16}$, and $\frac{5}{16}$. How should the health care professional arrange the instruments? Look at the pattern of increase in these measurements. Are any instruments missing in the sequence? If so, which ones?

Case Study

A health care professional is tracking the weight of a patient who is retaining fluids because of congestive heart failure. On day 3, the patient is given a diuretic. Here is a summary of the weight changes that occurred. In the column marked "Change," write the amount of weight change since the previous measurement. Use a plus (+) sign to indicate weight gained and a minus sign (−) to indicate weight lost. Day 2 has been completed as an example.

Time	Weight in lbs	Change
Day 1, 8:00 a.m.	$142\frac{1}{2}$	n/a
Day 2, 8:00 a.m.	144	$+1\frac{1}{2}$
Day 3, 8:00 a.m.	$145\frac{3}{4}$	$+1\frac{3}{4}$
Day 3, 8:00 a.m.	Patient receives diuretic Lasix (furosemide) 40 mg	
Day 3, 2:00 p.m.	$144\frac{3}{4}$	
Day 3, 4:00 p.m.	lasix 40 mg	
Day 4, 8:00 a.m.	$142\frac{3}{4}$	
Day 4, 8:00 a.m.	lasix 20 mg	
Day 4, 4:00 p.m.	$140\frac{1}{2}$	

To check your answers, see page 500.

Internet Activity

Search the Web for more practice problems with fractions and decimals. You might search for "Dosage Calculations Practice Problems," "Math and Dosage Practice Problems," or "Fractions and Decimals Practice."

GO TO . . . Open the CD-ROM that accompanies your textbook, and complete a final review of the rules, practice problems, and activities presented for this chapter. For a final evaluation, take the chapter test and email or print your results for your instructor. A score of 95 percent or above indicates mastery of the chapter concepts. ■

2 Percents, Ratios, and Proportions

If what you're working for really matters, you'll give it all you've got.
—Nido Qubein

Learning Outcomes

When you have completed Chapter 2, you will be able to:

- Describe the relationship among percents, ratios, decimals, and fractions.
- Calculate equivalent measurements, using percents, ratios, decimals, and fractions.
- Indicate solution strengths by using percents and ratios.
- Explain the concept of proportion.
- Calculate missing values in proportions by using ratios (means and extremes) and fractions (cross-multiplying).

Key Terms

Cross-multiplying
Dosage strength
Fraction proportion
Means and extremes
Percent
Percent strength
Proportion
Ratio
Ratio proportion
Ratio strength
Solute
Solution
Solvent (diluent)

Introduction

As a health care employee, you may be responsible for preparing solutions and solids in varying amounts. To do this, you must have a keen understanding of percents, ratios, and proportions. Additionally the process of finding the missing value in a ratio or fraction proportion problem (included in this chapter) is the necessary first step in many dosage calculations. So in preparation for your health career, give Chapter 2 all you've got!

Percents

Percents, like decimals and fractions, provide a way to express the relationship of parts to a whole. Indicated by the symbol %, **percent** literally means "per 100" or "divided by 100" The whole is always 100 units, just as a test is worth

TABLE 2-1 Comparing Decimals, Fractions, and Percents

Words	Decimal	Fraction	Percent
Eight hundredths	0.08	$\frac{8}{100}$	8%
Twenty-three hundredths	0.23	$\frac{23}{100}$	23%
Seven-tenths	0.7	$\frac{7}{10}$	70%
One	1.0	$\frac{1}{1}$	100%
One and five-tenths or one and one-half	1.5	$1\frac{5}{10}$ or $1\frac{1}{2}$	150%

100 points. A grade of 75% means that 75 per 100 points are answered correctly. Table 2-1 shows the same number expressed as a decimal, a fraction, and a percent. A number less than 1 is expressed as less than 100 percent. A number greater than 1 is expressed as greater than 100 percent. Any expression of 1 (for instance, 1.0 or $\frac{5}{5}$) equals 100 percent.

Working with Percents

Converting between percents and decimals requires dividing and multiplying by 100. Converting a percent to a decimal is similar to dividing a number by 100—you move the decimal point two places to the left. If the percent is a fraction or a mixed number, first convert it to a decimal (see Chapter 1 for review). Then divide by 100, moving the decimal point two places to the left.

Rule 2-1 To convert a percent to a decimal, remove the percent symbol. Then divide the remaining number by 100.

Example 1 Convert 42 percent to a decimal.

42% = 42.% = .42. = 0.42

Insert the zero before the decimal point for clarity.

Example 2 Convert 175 percent to a decimal.

175% = 175.% = 1.75. = 1.75

Example 3 When you move the decimal point to the left, you may need to insert zeros. Convert 0.3 percent to a decimal.

0.3% = 000.3% = 0.00.3 = 0.003

Example 4 Convert $25\frac{1}{2}$ percent to a decimal.

$25\frac{1}{2}\% = 25\frac{5}{10}\% = 25.5\% = .25.5 = 0.255$

Add a zero in front of the decimal point for clarity.

Example 5 Convert $\frac{3}{4}$ percent to a decimal.

First convert $\frac{3}{4}$ to a decimal.

$$\frac{3}{4} = 4\overline{)3.00}^{\,0.75} = 0.75$$

$$\frac{3}{4}\% = 0.75\% = 000.75\% = 0.00.75 = 0.0075$$

Insert zeros when necessary.

GO TO . . . Open the CD-ROM that accompanies your textbook, and select Chapter 2, Rule 2-1. Review the animation and example problems; then complete the practice problems. Continue to the next section of the book, once you have mastered the rule presented. ■

Converting a decimal to a percent is similar to multiplying a number by 100—you move the decimal point two places to the right. Because 100% = 1.00, multiplying a number by 100 percent does not change its value.

Rule 2-2 To convert a decimal into a percent, multiply the decimal by 100. Then add the percent symbol.

Example 1 Convert 1.42 to a percent.

$$1.42 \times 100\% = 142.00\% = 142\%$$

You can write this as 1.42 = 1.42.% = 142%.

Example 2 Convert 0.02 to a percent.

$$0.02 \times 100\% = 2.00\% = 2\%$$

You can write this as 0.02 = 0.02.% = 2%.

Example 3 When you move the decimal point to the right, you may need to insert zeros. Convert 0.8 to a percent.

$$0.8 \times 100\% = 80.0\% = 80\%$$

You can write this as 0.8 = 0.80 = 0.80.% = 80%.

Remember that when you multiply a number by 100, you move the decimal point two places to the right. When you divide a number by 100, move the decimal point two places to the left.

GO TO . . . Open the CD-ROM that accompanies your textbook, and select Chapter 2, Rule 2-2. Review the animation and example problems; then complete the practice problems. Continue to the next section of the book, once you have mastered the rule presented. ■

Because percent means "per 100" or "divided by 100," you can easily convert percents to equivalent fractions.

Rule 2-3 To convert a percent to an equivalent fraction, write the value of the percent as the numerator and 100 as the denominator. Then reduce the fraction to its lowest terms.

Example 1 Convert 8 percent to an equivalent fraction.

$$8\% = \frac{8}{100} = \frac{\overset{2}{\cancel{8}}}{\underset{25}{\cancel{100}}} = \frac{2}{25}$$

Example 2 Convert 130 percent to an equivalent mixed number.

$$130\% = \frac{130}{100} = \frac{\overset{13}{\cancel{130}}}{\underset{10}{\cancel{100}}} = \frac{13}{10} = 1\frac{3}{10}$$

Example 3 Change 0.6 percent to an equivalent fraction.

$$0.6\% = \frac{0.6}{100}$$

To reduce a fraction, it must not include a decimal point. To eliminate the decimal point, multiply it by $\frac{10}{10}$. (Remember that a fraction with the same number in the numerator and denominator is equal to 1).

$$\frac{0.6}{100} \times \frac{10}{10} = \frac{6}{1000} = \frac{3}{500}$$

Example 4 Change $\frac{3}{4}$ percent to an equivalent fraction.

$$\frac{3}{4}\% = \frac{3}{4} \div 100 = \frac{3}{4} \times \frac{1}{100} = \frac{3}{400}$$

For a review of division with fractions, see Chapter 1.

GO TO . . . Open the CD-ROM that accompanies your textbook, and select Chapter 2, Rule 2-3. Review the animation and example problems; then complete the practice problems. Continue to the next section of the book, once you have mastered the rule presented. ■

Rule 2-4 To convert a fraction to a percent, first convert the fraction to a decimal. Round the decimal to the nearest hundredth. Then follow the rule for converting a decimal to a percent.

Example 1 Convert $\frac{1}{2}$ to a percent.

First convert $\frac{1}{2}$ to a decimal.

$$\frac{1}{2} = 1 \div 2 = 0.5$$

Now convert the decimal to a percent.

$$\frac{1}{2} = 0.5 = 0.5 \times 100\% = 50\%$$

You can write this as 0.5 = 0.50 = 0.50.% = 50%.

Example 2 Convert $\frac{2}{3}$ to a percent.

Convert $\frac{2}{3}$ to a decimal. Round to the nearest hundredth.

$$\frac{2}{3} = 2 \div 3 = 0.666 = 0.67$$

Now convert to a percent.

$$\frac{2}{3} = 0.67 = 0.67 \times 100\% = 67\%$$

You can write this as 0.67 = 0.67.% = 67%.

Example 3 Convert $1\frac{3}{4}$ to a percent.

$$1\frac{3}{4} = \frac{7}{4} = 1.75 = 1.75 \times 100\% = 175\%$$

You can write this as $1\frac{3}{4}$ = 1.75.% = 175%.

An alternative method for converting a fraction to a percent is to multiply the fraction by 100 percent.

Example 1 Convert $\frac{1}{2}$ to a percent.

$$\frac{1}{2} \times 100\%$$

$$\frac{1}{2} \times \frac{100\%}{1} = \frac{100}{2} = 50\%$$

Example 2 Convert $\frac{2}{3}$ to a percent.

$$\frac{2}{3} \times \frac{100\%}{1} = \frac{200}{3} = 66.66 \text{ rounded to } 67\%.$$

Example 3 Convert $1\frac{3}{4}$ to a percent.

Change $1\frac{3}{4}$ to the fraction $\frac{7}{4}$.

$$\frac{7}{4} \times \frac{100\%}{1} = \frac{700}{4} = 175\%$$

GO TO . . . Open the CD-ROM that accompanies your textbook, and select Chapter 2, Rule 2-4. Review the animation and example problems; then complete the practice problems. Continue to the next section of the book, once you have mastered the rule presented. ■

REVIEW AND PRACTICE

2-1 Working with Percents

Convert the following percents to decimals. Round to the nearest thousandth.

1. 14%	**2.** 30%	**3.** 2%	**4.** 9%	**5.** 103%
6. 300%	**7.** 0.021%	**8.** 0.4%	**9.** $42\frac{1}{2}\%$	**10.** $3\frac{4}{5}\%$
11. 4.5%	**12.** 250.75%	**13.** $23\frac{2}{3}\%$	**14.** $1\frac{5}{6}\%$	**15.** $14\frac{1}{2}\%$

Convert the following decimals to percents.

16. 4.04	**17.** 2.3	**18.** 0.7	**19.** 0.33	**20.** 0.06
21. 0.013	**22.** 15	**23.** 32	**24.** 121	

Convert the following percents to fractions. Reduce the answers to their lowest terms.

25. 22%	**26.** 4%	**27.** 158%	**28.** 300%	**29.** 0.1%
30. 0.8%	**31.** $\frac{9}{10}\%$	**32.** $1\frac{2}{5}\%$	**33.** 0.3%	

Convert the following fractions to percents. Round to the nearest percent.

34. $\frac{6}{8}$	**35.** $\frac{4}{5}$	**36.** $\frac{1}{6}$	**37.** $\frac{5}{9}$	**38.** $1\frac{1}{10}$
39. $2\frac{1}{4}$	**40.** $\frac{175}{100}$	**41.** $\frac{40}{100}$	**42.** $5\frac{2}{3}$	

To check your answers, see page 500.

Percent Strength of Mixtures

Percents are commonly used to indicate the concentration of ingredients in mixtures such as solutions, lotions, creams, and ointments. Mixtures can be divided into two different categories: fluid and solid or semisolid. *Fluid* mixtures are those that flow. One example is a solution. These terms are important to understanding solutions: solute, solvent or diluent, and solution.

- The **solute** is the drug or substance being dissolved.
- The **solvent** or **diluent** is the liquid with which the solute is combined.
- The **solution** is the combined mixture.

Suppose you dissolve 1 teaspoon (tsp) of salt into 8 oz of freshwater. The salt is the solute. The freshwater is the solvent or diluent. The salt water is the solution. Other fluid mixtures include suspensions and lotions.

Solid or semisolid mixtures are creams and ointments. While these mixtures may contain many different components, it is the amount of the medication or medications in the mixture that is important when you are calculating doses.

Rule 2-5 For fluid mixtures prepared with a dry medication, the **percent strength** represents the number of grams of the medication contained in 100 mL of the mixture.

For fluid mixtures prepared with a liquid medication, the percent strength represents the number of milliliters of medication contained in 100 mL of the mixture.

Example 1 Determine the amount of hydrocortisone powder in 300 mL of 2% hydrocortisone lotion.

Each percent represents 1 gram (g) of hydrocortisone per 100 mL of lotion. A 2% hydrocortisone lotion will contain 2 g of hydrocortisone powder in every 100 mL. Therefore, 300 mL of the lotion will contain 3 times as much, or 6 g, of hydrocortisone powder.

Example 2 Determine the amount of ethanol in 500 mL of a 30% ethanol solution. (Ethanol is a liquid.)

The percent strength (30%) tells us that every 100 mL of the solution contains 30 mL of ethanol. So 500 mL of the solution would contain 5 × 30 mL, or 150 mL, of ethanol.

GO TO . . . Open the CD-ROM that accompanies your textbook, and select Chapter 2, Rule 2-5. Review the animation and example problems; then complete the practice problems. Continue to the next section of the book, once you have mastered the rule presented. ■

Since percent means "per 100," the percent strength of a solution is always calculated on the amount of a drug per 100 mL. Solutions with higher percent strengths contain more drug in each 100 mL and, therefore, are stronger solutions. For example, a 10% solution has twice as much drug (10 g per 100 mL) as a 5% solution (5 g per 100 mL). See Table 2-2 to compare solution strengths.

TABLE 2-2 Comparing Solution Strengths

Amount of Solution	Solution Strength	Dry Drug	Liquid Drug
50 mL	1%	0.5 g	0.5 mL
100 mL	1%	1 g	1 mL
200 mL	1%	2 g	2 mL
50 mL	2%	1 g	1 mL
100 mL	2%	2 g	2 mL
200 mL	2%	4 g	4 mL
100 mL	5%	5 g	5 mL
50 mL	10%	5 g	5 mL
100 mL	10%	10 g	10 mL
200 mL	10%	20 g	20 mL

Rule 2-6 For solid or semisolid mixtures prepared with a dry medication, the percent strength represents the number of grams of the medication contained in 100 g of the mixture.

For solid or semisolid mixtures prepared with a liquid medication, the percent strength represents the number of milliliters of medication contained in 100 g of the mixture.

Example Determine the amount of hydrocortisone powder in 50 g of a 1% hydrocortisone ointment.

Each percent represents 1 g of hydrocortisone per 100 g of ointment. A 1% hydrocortisone ointment will contain 1 g of hydrocortisone powder in every 100 g. Therefore, 50 g of the ointment will contain $\frac{1}{2}$ as much, or 0.5 g of hydrocortisone powder.

GO TO . . . Open the CD-ROM that accompanies your textbook, and select Chapter 2, Rule 2-6. Review the animation and example problems; then complete the practice problems. Continue to the next section of the book, once you have mastered the rule presented. ■

Critical Thinking on the Job

Confusing Percent Strength with Percent Conversions

A health care professional is asked to make a 2% drug solution using a dry drug. Before mixing the solution, the professional converts 2% to a decimal. Since 2% = 0.02, the professional adds 0.02 g of drug to 100 mL of solution.

Questions?

What math mistake did the health care professional make? How many grams of drug dissolved in 100 mL will equal 2%?

REVIEW AND PRACTICE

2-2 Percent Strength of Mixtures

Note: 100 mL = 100 cc. The abbreviation *cc* represents cubic centimeters and *mL* represents milliliters. Measurements and abbreviations such as mL and cc will be discussed in Chapter 3.

1. How many grams of drug are in 100 mL of a 5% solution?
2. How many grams of drug are in 100 mL of a 6% solution?
3. How many grams of drugs are in 100 mL of a 10% solution?
4. An intravenous (IV) bag contains 200 cc of dextrose 5% (a 5% drug solution). How many grams of dextrose are in the bag?
5. How many grams of dextrose will a patient receive from a 100 mL bag of dextrose 10% (a 10% solution of dextrose)?
6. How many grams of dextrose will a patient receive from 20 mL of a 5% solution?
7. A full 1000 cc bag of a 10% drug solution contains how many grams of the drug?
8. A 1000 cc bag of a 5% drug solution is now half empty. How many grams of the drug remain in the bag?

9. The drug label states that 50 mL of pure drug is contained in 100 mL of solution. What is the strength of the solution in percent?

10. Your patient pulled out his IV when it was $\frac{3}{4}$ empty. He had a 1000 mL bag of 10% drug solution that is now $\frac{1}{4}$ full. How many grams of drug did the patient receive?

To check your answers, see page 501.

Ratios

Ratios also express the relationship of a part to the whole. They may relate a quantity of liquid drug to a quantity of solution, as with intravenous drugs. Ratios can also be used to calculate dosages of dry medication such as tablets.

Ratios can be written as fractions but are usually written in the form *A* : *B*. The colon tells you to compare *A* to *B*, and *A* : *B* is read "*A* to *B*." For example, a 1 : 100 drug solution ("one to one hundred") describes a solution that has 1 part drug to every 100 parts of solution. The ratio 1 : 100 is equivalent to the fraction $\frac{1}{100}$, the decimal 0.01, and the percent 1%.

When ratios describe dry medications, the whole unit is often 1, as in 1 tablet. Thus, 25 milligrams (mg) of drug to one tablet would be written as the ratio 25 : 1 This concept of ratio is important when you calculate dosage.

Working with Ratios

You may use only whole numbers when you write a ratio. Correct ratios include 8 : 1, 2 : 5, and 1 : 100 Incorrect ratios include 2.5 : 10, 1 : 4.5, and $3\frac{1}{2}$: 100

Ratios should almost always be expressed in lowest terms. Just as $\frac{4}{100}$ reduces to $\frac{1}{25}$, the ratio 4 : 100 should be written 1 : 25 Similarly, you reduce 2 : 10 to 1 : 5 and 10 : 12 to 5 : 6

Rule 2-7 Reduce a ratio as you would a fraction. Find the largest whole number that divides evenly into both values *A* and *B*.

Example 1 Reduce 2 : 12 to its lowest terms.

Both values 2 and 12 are divisible by 2

$2 \div 2 = 1 \quad 12 \div 2 = 6$

Thus, 2 : 12 is written 1 : 6

Example 2 Reduce 10 : 15 to its lowest terms.

Both values 10 and 15 are divisible by 5

$10 \div 5 = 2 \quad 15 \div 5 = 3$

$10 : 15 = 2 : 3$

GO TO . . . Open the CD-ROM that accompanies your textbook, and select Chapter 2, Rule 2-7. Review the animation and example problems; then complete the practice problems. Continue to the next section of the book, once you have mastered the rule presented. ■

Because a ratio relates two quantities, value A and value B, ratios can be written as fractions. Within this textbook for simplicity when two numbers are expressed as $A : B$, this is a ratio. When two numbers are expressed as $\frac{A}{B}$, this is a fraction.

Rule 2-8 To convert a ratio to a fraction, write value A (the first number) as the numerator and value B (the second number) as the denominator, so that $A : B = \frac{A}{B}$.

Example Convert the following ratios to fractions.

a. $1 : 2 = \frac{1}{2}$

b. $4 : 5 = \frac{4}{5}$

c. $1 : 100 = \frac{1}{100}$

d. $7 : 3 = \frac{7}{3}$

e. $8 : 5 = \frac{8}{5}$

GO TO . . . Open the CD-ROM that accompanies your textbook, and select Chapter 2, Rule 2-8. Review the animation and example problems; then complete the practice problems. Continue to the next section of the book, once you have mastered the rule presented. ■

Rule 2-9 To convert a fraction to a ratio, write the numerator as the first value A and the denominator as the second value B.

$$\frac{A}{B} = A : B$$

Convert a mixed number to a ratio by first writing the mixed number as a fraction.

Example Convert the following to ratios.

a. $\frac{7}{12} = 7 : 12$

b. $\frac{3}{10} = 3 : 10$

c. $\frac{3}{2} = 3 : 2$

d. $\frac{47}{12} = 47 : 12$

e. $3\frac{1}{3} = \frac{10}{3} = 10 : 3$

f. $2\frac{1}{2} = \frac{5}{2} = 5 : 2$

GO TO . . . Open the CD-ROM that accompanies your textbook, and select Chapter 2, Rule 2-9. Review the animation and example problems; then complete the practice problems. Continue to the next section of the book, once you have mastered the rule presented. ■

When you convert between ratios and fractions, you do not have to perform any calculations. You simply rearrange the presentation of the numbers.

Rule 2-10 To convert a ratio to a decimal:

1. Write the ratio as a fraction.
2. Convert the fraction to a decimal. (See Chapter 1.)

Example 1 Convert 1 : 10 to a decimal.

1. Write the ratio as a fraction.

$1:10 = \frac{1}{10}$

2. Convert the fraction to a decimal.

$\frac{1}{10} = 1 \div 10 = 0.1$

Thus, $1:10 = \frac{1}{10} = 0.1$

Example 2 Convert 3 : 2 to a decimal.

1. $3:2 = \frac{3}{2}$
2. $\frac{3}{2} = 3 \div 2 = 1.5$

Thus, $3:2 = \frac{3}{2} = 1.5$

GO TO . . . Open the CD-ROM that accompanies your textbook, and select Chapter 2, Rule 2-10. Review the animation and example problems; then complete the practice problems. Continue to the next section of the book, once you have mastered the rule presented. ■

Rule 2-11 To convert a decimal to a ratio:

1. Write the decimal as a fraction. (See Chapter 1.)
2. Reduce the fraction to lowest terms.
3. Restate the fraction as a ratio by writing the numerator as value *A* and the denominator as value *B*; in the form *A* : *B*.

Example 1 Convert 0.8 to a ratio.

1. Write the decimal as a fraction.

$0.8 = \frac{8}{10}$

2. Reduce the fraction to lowest terms.

$$\frac{8}{10} = \frac{\overset{4}{\cancel{8}}}{\underset{5}{\cancel{10}}} = \frac{4}{5}$$

3. Restate the number as a ratio.

$$\frac{4}{5} = 4:5$$

Thus, $0.8 = \frac{8}{10} = \frac{4}{5} = 4:5$

Example 2 Convert 0.05 to a ratio.

1. $0.05 = \frac{5}{100}$

2. $\frac{5}{100} = \frac{\overset{1}{\cancel{5}}}{\underset{20}{\cancel{100}}} = \frac{1}{20}$

3. $\frac{1}{20} = 1:20$

Thus, $0.05 = \frac{1}{20} = 1:20$

Example 3 Convert 2.5 to a ratio.

1. $2.5 = 2\frac{5}{10} = \frac{25}{10}$

2. $\frac{25}{10} = \frac{\overset{5}{\cancel{25}}}{\underset{2}{\cancel{10}}} = \frac{5}{2}$

3. $\frac{5}{2} = 5:2$

Thus, $2.5 = \frac{5}{2} = 5:2$

GO TO . . . Open the CD-ROM that accompanies your textbook, and select Chapter 2, Rule 2-11. Review the animation and example problems; then complete the practice problems. Continue to the next section of the book, once you have mastered the rule presented. ■

Recall that you can write a percent as a fraction with the denominator of 100 This step helps you to convert a ratio to a percent and a percent to a ratio.

Rule 2-12 To convert a ratio to a percent:

1. Convert the ratio to a decimal.
2. Write the decimal as a percent by multiplying the decimal by 100 and adding the percent symbol.

Example 1 Convert 1 : 50 to a percent.

1. Convert the ratio to a decimal.

   $1 : 50 = \frac{1}{50} = 0.02$

2. Multiply 0.02 by 100 and add the percent symbol.

   $0.02 \times 100\% = 2\%$

   $1 : 50 = \frac{1}{50} = 0.02 = 2\%$

Example 2 Convert 2 : 3 to a percent.

1. $2 : 3 = \frac{2}{3} = 0.67$
2. $0.67 \times 100\% = 67\%$

   $2 : 3 = \frac{2}{3} = 0.67 = 67\%$

Example 3 Convert 5 : 2 to a percent.

1. $5 : 2 = \frac{5}{2} = 2.5 = 2.50$
2. $2.50 \times 100\% = 250\%$

   $5 : 2 = \frac{5}{2} = 2.50 = 250\%$

GO TO . . . Open the CD-ROM that accompanies your textbook, and select Chapter 2, Rule 2-12. Review the animation and example problems; then complete the practice problems. Continue to the next section of the book, once you have mastered the rule presented. ■

Rule 2-13 To convert a percent to a ratio:

1. Write the percent as a fraction.
2. Reduce the fraction to lowest terms.
3. Write the fraction as a ratio by writing the numerator as value *A* and the denominator as value *B*, in the form *A* : *B*.

Example 1 Convert 25 percent to a ratio.

1. Write the percent as a fraction.

   $25\% = \frac{25}{100}$

2. Reduce the fraction.

   $\frac{25}{100} = \frac{1}{4}$

3. Restate the fraction as a ratio. Write the numerator as value A and the donominator as value B.

$\frac{1}{4} = 1 : 4$

Thus, $25\% = \frac{1}{4} = 1 : 4$

Example 2 Convert 450 percent to a ratio.

1. $450\% = \frac{450}{100}$
2. $\frac{450}{100} = \frac{45}{10} = \frac{9}{2}$
3. $\frac{9}{2} = 9 : 2$

Thus, $450\% = \frac{9}{2} = 9 : 2$

Example 3 Convert 0.3 percent to a ratio.

1. Write the percent as a fraction. In this case, rewrite the fraction without decimal points.

   $0.3\% = \frac{0.3}{100} = \frac{3}{1000}$
2. $\frac{3}{1000}$ is reduced to lowest terms.
3. $\frac{3}{1000} = 3 : 1000$

Thus, $0.3\% = \frac{3}{1000} = 3 : 1000$

GO TO . . . Open the CD-ROM that accompanies your textbook, and select Chapter 2, Rule 2-13. Review the animation and example problems; then complete the practice problems. Continue to the next section of the book, once you have mastered the rule presented. ■

REVIEW AND PRACTICE

2-3 Working with Ratios

Convert the following ratios to fractions or mixed numbers.

1. 3 : 4
2. 4 : 9
3. 5 : 3
4. 10 : 1
5. 1 : 20
6. 1 : 250
7. 4 : 12

Convert the following fractions to ratios.

8. $\frac{2}{3}$
9. $\frac{6}{7}$
10. $\frac{5}{4}$
11. $\frac{7}{3}$
12. $1\frac{7}{8}$
13. $3\frac{1}{3}$
14. $\frac{6}{10}$
15. $\frac{18}{27}$
16. $\frac{1}{50}$
17. $\frac{1}{75}$
18. $5\frac{4}{5}$

Convert the following ratios to decimals. Round to the nearest hundredth, if necessary.

19. 1 : 4
20. 1 : 8
21. 3 : 4
22. 2 : 5
23. 50 : 1
24. 25 : 2
25. 8 : 3
26. 5 : 6
27. 5 : 75

Convert the following decimals to ratios.

28. 0.9
29. 0.3
30. 0.01
31. 0.45
32. 6
33. 2.4
34. 8
35. 9.8

Convert the following ratios to percents. If necessary, round to the nearest percent.

36. 1 : 4
37. 1 : 25
38. 2 : 9
39. 7 : 17
40. 20 : 1
41. 15 : 2
42. 3 : 8

Convert the following percents to ratios.

43. 14 %
44. 65 %
45. 400 %
46. 175 %
47. 0.6 %
48. 0.18 %
49. 84 %
50. 0.57 %

To check your answers, see page 501.

Ratio Strengths

Ratio strengths can be used to express the amount of drug in a solution or the amount of drug in a solid dosage form such as a tablet or capsule. This relationship represents the **dosage strength** of the medication. The first number of the ratio represents the amount of drug, while the second number represents the amount of solution or the number of tablets or capsules. For example, a 1 mg : 5 mL dosage strength represents 1 mg of drug in every 5 mL of solution. A 250 mg : 1 tablet dosage strength means that each tablet contains 250 mg of drug.

Example 1 Write the ratio strength to describe 50 mL of solution containing 3 g of a drug.

The first number in the ratio strength always represents the amount of drug, which is 3 g. The second number represents the amount of solution, 50 mL. The ratio is 3 g : 50 mL.

Example 2 Write the ratio strength to describe 25 mg of a drug dissolved in 150 mL of solution.

The amount of the drug is 25 mg, while the amount of solution is 150 mL. The ratio is 25 mg : 150 mL.
Recall from the previous section that ratios should always be reduced to lowest terms. Since both numbers in the ratio can be divided by 25, the ratio needs to be reduced. The ratio should be written 1 mg : 6 mL.

Example 3 Write the ratio strength to describe 3 tablets containing 75 mg of a drug.

Remember that the amount of drug always goes first in the ratio. The amount of the drug is 75 mg. The ratio is 75 mg : 3 tablets, which is then reduced to 25 mg : 1 tablet.

ERROR ALERT!

Do not forget the units of measurement.

Including units in the dosage strength will help you to avoid some common errors. Consider the case in which we have two solutions of a drug. One of the solutions contains 1 g of drug in 50 mL; the other contains 1 mg of drug in 50 mL. While both of these solutions have ratio strengths of 1 : 50, they are obviously different from each other. To distinguish between them, the first solution could be written as 1 g : 50 mL while the second is written 1 mg : 50 mL.

Critical Thinking on the Job

Reversing Terms in a Ratio Strength

A health care professional is preparing 1000 mL of a 1 g : 10 mL solution of dextrose. He mixes 10,000 g of dextrose in 1000 mL of solution. The health care professional has mistakenly reversed the terms and prepared a 10 g : 1 mL solution.

Questions?

How should the health care professional correctly set up the problem? What should have alerted the professional to a problem?

REVIEW AND PRACTICE

2-4 Ratio Strengths

Write a ratio to describe the dosage strength in each of the following. Reduce the ratio to its lowest terms.

1. 100 mL of solution contains 5 g of drug.
2. 500 mL of solution contain 25 g of dextrose.
3. Each capsule contains 5 mg of drug.
4. 40 mg of drug is in every tablet.
5. 20 mg of drug is in 100 mL of solution.
6. 150 mg of drug is in 1500 mL of solution.
7. 250 mL of solution contains 50 g of drug.
8. 500 mg of drug in each capsule.
9. 500 mg of drug in 1000 mL of solution.
10. Two tablets contain 500 mg of drug.
11. A 1000 mL solution contains 100 mL of drug.
12. 50 mL of solution contains 10 mL of drug.
13. 5 (grains) gr of drug is in 1 tablet.

14. Three tablets contain a total of 750 mg of drug.
15. Three tablets contain 90 mg of drug.
16. Two tablets contain 20 mg of drug.
17. 75 mL of solution contains 15 mL of drug.
18. Each 500 mL of solution contains 250 mL of drug.
19. 10 gr of drug is in 2 tablets.
20. Four tablets contain a total of 1 g of drug.

To check your answers, see page 501.

Proportion

A **proportion** is a mathematical statement that two ratios are equal. Because ratios are often written as fractions, a proportion is also a statement that two fractions are equal.

Writing Proportions

You have learned that 2 : 3 is read "two to three." A double colon in a proportion means *as*. The proportion 2 : 3 : : 4 : 6 reads "two is to three as four is to six." This **ratio proportion** states that the relationship of 2 to 3 is the same as the relationship of 4 to 6 By now, you know that 2 divided by 3 is the same as 4 divided by 6 *When you write proportions, do not reduce the ratios to their lowest terms.*

You can write proportions by replacing the double colon with an equal sign. Thus, 2 : 3 : : 4 : 6 is the same as 2 : 3 = 4 : 6. You can also write a proportion with fractions. For example, 2 : 3 : : 4 : 6 can be written:

$$\frac{2}{3} = \frac{4}{6}$$

This format is referred to later as a **fraction proportion.**

Rule 2-14 To write a ratio proportion as a fraction proportion:

1. Change the double colon to an equal sign.
2. Convert both ratios to fractions.

Example 1 Write 3 : 4 : : 9 : 12 as a fraction proportion.

1. Change the double colon to an equal sign.

   $3 : 4 : : 9 : 12 \rightarrow 3 : 4 = 9 : 12$

2. Convert both ratios to fractions. Here 3 : 4 becomes $\frac{3}{4}$ and 9 : 12 becomes $\frac{9}{12}$, so that:

   $3 : 4 = 9 : 12 \rightarrow \frac{3}{4} = \frac{9}{12}$

Example 2 Write 5 : 10 : : 50 : 100 as a fraction proportion.

1. $5 : 10 : : 50 : 100 \rightarrow 5 : 10 = 50 : 100$
2. $5 : 10 = 50 : 100 \rightarrow \frac{5}{10} = \frac{50}{100}$

Example 3 Write 8 : 6 : : 4 : 3 as a fraction proportion.

1. 8 : 6 : : 4 : 3 → 8 : 6 = 4 : 3
2. 8 : 6 = 4 : 3 → $\frac{8}{6} = \frac{4}{3}$

GO TO . . . Open the CD-ROM that accompanies your textbook, and select Chapter 2, Rule 2-14. Review the animation and example problems; then complete the practice problems. Continue to the next section of the book, once you have mastered the rule presented. ■

Rule 2-15 To write a fraction proportion as a ratio proportion:

1. Convert each fraction to a ratio.
2. Change the equal sign to a double colon.

Example 1 Write $\frac{5}{6} = \frac{10}{12}$ as a ratio proportion.

1. Convert each fraction to a ratio.

   $\frac{5}{6}$ = 5 : 6 and $\frac{10}{12}$ = 10 : 12 so 5 : 6 = 10 : 12

2. Change the equal sign to a double colon.

   5 : 6 = 10 : 12 → 5 : 6 : : 10 : 12

Example 2 Write $\frac{3}{8} = \frac{9}{24}$ as a ratio proportion.

1. $\frac{3}{8}$ = 3 : 8 and $\frac{9}{24}$ = 9 : 24, so that 3 : 8 = 9 : 24
2. 3 : 8 = 9 : 24 → 3 : 8 : : 9 : 24

Example 3 Write $\frac{10}{2} = \frac{5}{1}$ as a ratio proportion.

1. $\frac{10}{2}$ = 10 : 2 and $\frac{5}{1}$ = 5 : 1 so that 10 : 2 = 5 : 1
2. 10 : 2 = 5 : 1 → 10 : 2 : : 5 : 1

GO TO . . . Open the CD-ROM that accompanies your textbook, and select Chapter 2, Rule 2-15. Review the animation and example problems; then complete the practice problems. Continue to the next section of the book, once you have mastered the rule presented. ■

REVIEW AND PRACTICE

2-5 Writing Proportions

Write the following ratio proportions as fraction proportions.

1. 4 : 5 : : 8 : 10
2. 5 : 12 : : 10 : 24
3. 1 : 10 : : 100 : 1000
4. 2 : 3 : : 20 : 30
5. 50 : 25 : : 10 : 5
6. 6 : 4 : : 18 : 12
7. 5 : 24 : : 10 : 48
8. 75 : 100 : : 150 : 200
9. 4 : 16 : : 16 : 64
10. 125 : 100 : : 375 : 300

Write the following fraction proportions as ratio proportions.

11. $\frac{3}{4} = \frac{75}{100}$
12. $\frac{1}{5} = \frac{3}{15}$
13. $\frac{8}{4} = \frac{2}{1}$
14. $\frac{8}{7} = \frac{24}{21}$
15. $\frac{18}{16} = \frac{9}{8}$
16. $\frac{10}{1} = \frac{40}{4}$
17. $\frac{5}{7} = \frac{15}{21}$
18. $\frac{45}{5} = \frac{9}{1}$
19. $\frac{1}{100} = \frac{100}{10{,}000}$
20. $\frac{36}{12} = \frac{72}{24}$

To check your answers, see page 502.

Means and Extremes

You often work with proportions to calculate dosages. When you know three of four values of a proportion, you will find the missing value. In this section, you will learn to find the missing value in a ratio proportion. Then you will learn to find the missing value in a fraction proportion. Both methods lead to the same answer. Which method you select is a matter of personal preference.

It is not enough to learn to find the missing value. You must also learn to set up the proportion correctly. If you set up the proportion incorrectly, you could give the wrong amount of medication, with serious consequences for the patient. Use critical thinking skills to select the appropriate information and set up the proportion. In later chapters, you will learn to read physician's orders and drug labels, the sources for the information that goes into the proportion. The remainder of this chapter focuses on finding missing values.

When you set up a ratio proportion in the form *A* : *B* : : *C* : *D*, the values *A* and *D* are the **extremes**. The values *B* and *C* are the **means.** If you have trouble remembering which is which, think, "Extremes are on the ends. Means are in the middle."

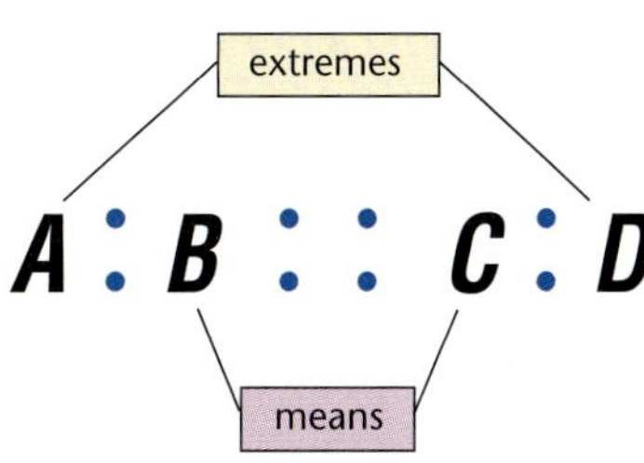

In any true ratio proportion, the product of the means always equals the product of the extremes.

Rule 2-16 To determine if a ratio proportion is true:

1. Multiply the means.
2. Multiply the extremes.
3. Compare the product of the means with the product of the extremes. If the products are equal, the proportion is true.

Example 1 Determine if 1 : 2 : : 3 : 6 is a true proportion.

1. Multiply the means: $2 \times 3 = 6$
2. Multiply the extremes: $1 \times 6 = 6$
3. Compare the products of the means and the extremes.

   $6 = 6$

The statement 1 : 2 : : 3 : 6 is a true proportion.

Example 2 Determine if 100 : 40 : : 50 : 20 is a true proportion.

1. $40 \times 50 = 2000$
2. $100 \times 20 = 2000$
3. $2000 = 2000$

The proportion 100 : 40 : 50 : 20 is true.

Example 3 Determine if 100 : 20 : : 5 : 4 is a true proportion.

1. $20 \times 5 = 100$
2. $100 \times 4 = 400$
3. $400 \neq 100$

The proportion 10 : 20 : : 5 : 4 is not a true proportion.

GO TO . . . Open the CD-ROM that accompanies your textbook, and select Chapter 2, Rule 2-16. Review the animation and example problems; then complete the practice problems. Continue to the next section of the book, once you have mastered the rule presented. ■

By definition, both sides of an equation are equal. If you perform the same calculation on both sides of an equation, the two sides will still be equal. For instance, consider the equation:

$$4 \times 2 = 8$$

You can add 3 to both sides or subtract 5 from both sides of the equation, and the resulting equations are still true.

$$(4 \times 2) + 3 = 8 + 3 \qquad (4 \times 2) - 5 = 8 - 5$$
$$11 = 11 \qquad\qquad 3 = 3$$

You can also multiply or divide both sides by the same nonzero number and the sides remain equal.

$$(4 \times 2) \times 6 = 8 \times 6 \qquad (4 \times 2) \div 4 = 8 \div 4$$
$$48 = 48 \qquad\qquad 2 = 2$$

Now you can use the means and extremes to help you find a missing value in a proportion. Suppose you have the proportion.

$$2 : 4 : : ? : 12$$

where ? represents the missing value. The product of the means equals the product of the extremes.

$$4 \times ? = 2 \times 12$$
$$4 \times ? = 24$$

To find the value of ?, you must write the equation so that ? stands alone on one side of the equal sign. Here you simply divide both sides by the number before ?, or 4

$$4 \times ? = 24$$

$$\frac{4 \times ?}{4} = \frac{24}{4}$$

$$? = 6$$

Check that the proportion is now true:

$$2 : 4 :: 6 : 12$$

$$4 \times 6 = 24 \quad \text{and} \quad 2 \times 12 = 24$$

Because $4 \times 6 = 2 \times 12$, the proportion is true. *Remember*, taking the time to check your work will help you avoid errors.

Rule 2-17 To find the missing value in a ratio proportion:

1. Write an equation setting the product of the means equal to the product of the extremes.
2. Solve the equation for the missing value.
3. Restate the proportion, inserting the missing value.
4. Check your work. Determine if the ratio proportion is true.

Example Find the missing value in 25 : 5 : : 50 : ?

1. Write an equation setting the product of the means equal to the product of the extremes.

   $$5 \times 50 = 25 \times ?$$

   $$250 = 25 \times ?$$

2. Solve the equation. Here, divide both sides by 25

   $$\frac{250}{25} = \frac{25 \times ?}{25}$$

   $$10 = ?$$

3. Restate the proportion, inserting the missing value.

   25 : 5 : : 50 : 10

4. Check your work.

   $$5 \times 50 = 25 \times 10$$

   $$250 = 250$$

The missing value is 10

GO TO . . . Open the CD-ROM that accompanies your textbook, and select Chapter 2, Rule 2-17. Review the animation and example problems; then complete the practice problems. Continue to the next section of the book, once you have mastered the rule presented. ■

Canceling Units in Proportions. Remember that it is important to include units when you are writing ratios. Since proportions show the relationship between two ratios, it is also important to include units in proportions. Including units, and learning how to cancel like units, will help you to determine the correct units for the answer when you solve problems using proportions. For example, you have a solution containing 200 mg of drug in 5 mL, and you are asked to determine how many milliliters of a solution contain 500 mg of drug. You can solve the problem by using the following two ratios.

200 mg : 5 mL

500 mg : ?

If the units of the first parts of the two ratios are the same, they can be dropped or canceled. Likewise, if the units from the second part of the two ratios are the same, they can be canceled. In this case, the units for the first part of each ratio are milligrams. Canceling the units leaves us with the following proportion.

200 : 5 mL : : 500 : ?

The product of the means equals the product of the extremes.

$5 \text{ mL} \times 500 = 200 \times ?$

We then divide both sides of the equation by 200 so that ? stands alone.

$$\frac{5 \text{ mL} \times 500}{200} = \frac{200 \times ?}{200}$$

$12.5 \text{ mL} = ?$

There are 500 mg of the drug in 12.5 mL of the solution. Critically thinking about this answer, we realize there is 200 mg for every 5 mL, and we need 500 mg of medication, which is $2\frac{1}{2}$ times the 200 mg. Our solution of 12.5 mL is $2\frac{1}{2}$ times the 5 mL of solution.

Rule 2-18 If the units in the first part of the ratios in a proportion are the same, the units can be canceled. If the units in the second part of the ratios in a proportion are the same, the units can be canceled.

Example 1 If 100 mL of solution contains 20 mg of drug, how many milligrams of the drug will be in 500 mL of the solution?

Start by setting up the ratios.

20 mg : 100 mL and ? : 500 mL

Compare the units used in the two ratios to see if any can be canceled. In this case, the units for the second part of both ratios are milliliters. These can be canceled when we set up the proportion.

20 mg : 100 : : ? : 500

Now solve for ?, the missing value.

1. $100 \times ? = 20 \text{ mg} \times 500$
2. $\frac{100 \times ?}{100} = \frac{20 \text{ mg} \times 500}{100}$
3. $? = 100 \text{ mg}$

The second solution will contain 100 mg of drug in 500 mL of solution.

Example 2 15 g of drug is dissolved in 300 mL of solution. If you need 45 g of the drug, how many milliliters of the solution are needed?

Set up the ratios.

15 g : 300 mL and 45 g : ?

Cancel units and set up the proportion.

15 : 300 mL : : 45 : ?

Solve for the missing value.

1. 300 mL × 45 = 15 × ?
2. $\frac{300 \text{ mL} \times 45}{15} = \frac{15 \times ?}{15}$
3. 900 mL = ?

You will need 900 mL of the solution to have 45 g of drug.

GO TO . . . Open the CD-ROM that accompanies your textbook, and select Chapter 2, Rule 2-18. Review the animation and example problems; then complete the practice problems. Continue to the next section of the book, once you have mastered the rule presented. ■

REVIEW AND PRACTICE

2-6 Means and Extremes

Determine whether the following proportions are true.

1. 6 : 12 : : 12 : 24 **2.** 3 : 8 : : 9 : 32 **3.** 5 : 75 : : 15 : 250

4. 8 : 100 : : 20 : 250 **5.** 6 : 18 : : 18 : 54

Use the means and extremes to find the missing values.

6. 10 : ? : : 5 : 8 **7.** 10 : 4 : : 20 : ? **8.** 4 : 25 : : 16 : ?

9. ? : 15 : : 100 : 75 **10.** 21 : 27 : : ? : 45 **11.** 100 : ? : : 50 : 2

12. 3 : 12 : : ? : 36 **13.** 33 : 39 : : 55 : ? **14.** ? : 24 : : 5 : 30

15. 18 : ? : : 27 : 6

16. If 1 tablet contains 30 mg of drug, how many milligrams of drug do 5 tablets contain?

17. If 10 g of drug is in 250 mL of solution, how many grams of drug are in 1000 mL of solution?

18. If 3 tablets contain 45 mg of drug, how many milligrams of drug are in 1 tablet?

19. If 60 mg of drug is in 500 mL of solution, how many milliliters of solution contain 36 mg of drug?

20. If 80 mg of drug is in 480 mL of solution, how many milliliters of solution contain 60 mg of drug?

To check your answers, see page 502.

Cross-Multiplying

Earlier in this chapter, you learned that a proportion can be written with ratios or with fractions. For example, the proportion 2 : 4 : : 5 : 10 can be written as $\frac{2}{4} = \frac{5}{10}$. More generally,

$A : B : : C : D$ can be written $\frac{A}{B} = \frac{C}{D}$

To determine whether a proportion is true, compare the products of the extremes (*A* and *D*) with the product of the means (*B* and *C*). When a proportion is written with fractions, use a similar method, **cross-multiplying,** to determine if it is true. The numerator of the first fraction, *A*, and the denominator of the second fraction, *D*, are the extremes of the ratio proportion. The denominator of the first fraction, *B*, and the numerator of the second fraction, *C*, are the means.

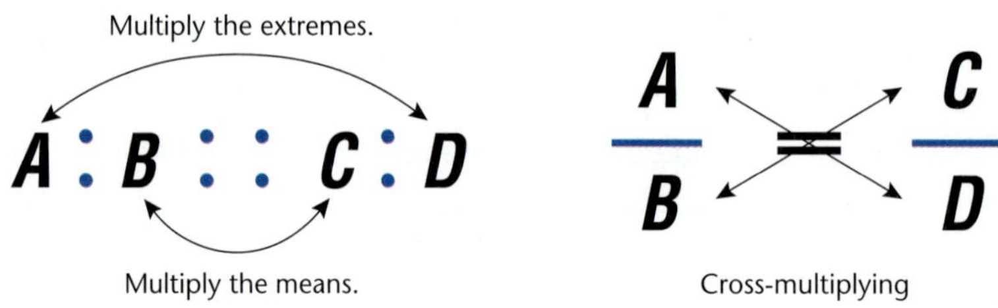

Cross-multiplying

Rule 2-19 To determine if a fraction proportion is true:

1. Cross-multiply. Multiply the numerator of the first fraction with the denominator of the second fraction. Then multiply the denominator of the first fraction with the numerator of the second fraction.
2. Compare the products. The products must be equal.

Example 1 Determine if $\frac{2}{5} = \frac{10}{25}$ is a true proportion.

1. Cross-multiply.

   $\frac{2}{5} \times \frac{10}{25} \rightarrow 2 \times 25 = 5 \times 10$
2. Compare the products on both sides of the equal sign.

   $50 = 50$

   $\frac{2}{5} = \frac{10}{25}$ is a true proportion.

Example 2 Determine if $\frac{100}{1000} = \frac{500}{5000}$ is a true proportion.

1. $\frac{100}{1000} \times \frac{500}{5000} \rightarrow 100 \times 5000 = 1000 \times 500$
2. $500{,}000 = 500{,}000$

   $\frac{100}{1000} = \frac{500}{5000}$ is a true proportion.

Example 3 Determine if $\frac{5}{8} = \frac{40}{72}$ is a true proportion.

1. $\frac{5}{8} \times \frac{40}{72} \rightarrow 5 \times 72 = 8 \times 40$
2. $360 \neq 320$

The proportion $\frac{5}{8} = \frac{40}{72}$ is not true.

GO TO . . . Open the CD-ROM that accompanies your textbook, and select Chapter 2, Rule 2-19. Review the animation and example problems; then complete the practice problems. Continue to the next section of the book, once you have mastered the rule presented. ■

In the previous section, you learned to use means and extremes to find a missing value in a ratio proportion. As you might expect, you can cross-multiply to find the missing value in a fraction proportion.

Rule 2-20 To find the missing value in a fraction proportion:

1. Cross-multiply. Write an equation setting the products equal to each other.
2. Solve the equation to find the missing value.
3. Restate the proportion, inserting the missing value.
4. Check your work. Determine if the fraction proportion is true.

Example 1 Find the missing value in $\frac{3}{5} = \frac{6}{?}$.

1. Cross-multiply.

$$\frac{3}{5} \times \frac{6}{?} \rightarrow 3 \times ? = 5 \times 6$$

$$3 \times ? = 30$$

2. Solve the equation. Here, divide both sides by 3

$$\frac{3 \times ?}{3} = \frac{30}{3}$$

$$? = 10$$

3. Restate the proportion, inserting the missing value.

$$\frac{3}{5} = \frac{6}{10}$$

4. Check your work by cross-multiplying.

$$3 \times 10 = 5 \times 6$$

$$30 = 30$$

The missing value is 10.

Example 2 Find the missing value in $\frac{25}{5} = \frac{50}{?}$.

1. $\frac{25}{5} \times \frac{50}{?} \rightarrow 25 \times ? = 5 \times 50$

$$25 \times ? = 250$$

2. $\frac{25 \times ?}{25} = \frac{250}{25}$

$$? = 10$$

3. $\frac{25}{5} = \frac{50}{10}$
4. $25 \times 10 = 5 \times 50$

$$250 = 250$$

The missing value is 10

GO TO . . . Open the CD-ROM that accompanies your textbook, and select Chapter 2, Rule 2-20. Review the animation and example problems; then complete the practice problems. Continue to the next section of the book, once you have mastered the rule presented. ■

Canceling Units in Fraction Proportions. Just as we were able to cancel units in ratio proportions, we can cancel them in fraction proportions. Now, however, we need to compare the units used in the top and bottom of the two fractions in the proportion. For example, you have a solution containing 200 mg of drug in 5 mL, and you are asked to determine how many milliliters of a solution contain 500 mg of drug. You can solve the problem by using the following two fractions. Must make sure that the labels are in like positions. In this case, mg is placed in the numerator in both fractions.

$$\frac{200 \text{ mg}}{5 \text{ mL}} \qquad \frac{500 \text{ mg}}{?}$$

Because the units for both numerators of the fraction are milligrams, they can be canceled. Canceling the units leaves us with the following proportion.

$$\frac{200}{5 \text{ mL}} = \frac{500}{?}$$

The missing value can now be found by cross-multiplying and solving the equation as before.

Rule 2-21 If the units of the numerator of the two fractions are the same, they can be dropped or canceled before you set up a proportion. Likewise, if the units from the denominator of the two fractions are the same, they can be canceled.

Example 1 If 100 mL of solution contains 20 mg of drug, how many milligrams of the drug will be in 500 mL of the solution?

Start by setting up the fractions.

$$\frac{20 \text{ mg}}{100 \text{ mL}} \quad \text{and} \quad \frac{?}{500 \text{ mL}}$$

Compare the units used in the two fractions to see if any can be canceled. In this case, the units for the denominators of both fractions are milliliters. These can be canceled when you set up the proportion.

$$\frac{20 \text{ mg}}{100} = \frac{?}{500}$$

Now solve for ?, the missing value.

1. $100 \times ? = 20 \text{ mg} \times 500$
2. $\frac{100 \times ?}{100} = \frac{20 \text{ mg} \times 500}{100}$
3. $? = 100 \text{ mg}$

The second solution will contain 100 mg of drug in 500 mL of solution.

Example 2 15 grams of drug is dissolved in 300 mL of solution. If you need 45 g of the drug, how many milliliters of the solution are needed?

Set up the fractions.

$$\frac{15 \text{ g}}{300 \text{ mL}} \quad \text{and} \quad \frac{45 \text{ g}}{?}$$

Cancel units and set up the proportion.

$$\frac{15}{300 \text{ mL}} = \frac{45}{?}$$

Solve for the missing value.

1. $300 \text{ mL} \times 45 = 15 \times ?$
2. $\frac{300 \text{ mL} \times 45}{15} = \frac{15 \times ?}{15}$
3. $900 \text{ mL} = ?$

You will need 900 mL of the solution to have 45 g of drug.

GO TO . . . Open the CD-ROM that accompanies your textbook, and select Chapter 2, Rule 2-21. Review the animation and example problems; then complete the practice problems. Continue to the next section of the book, once you have mastered the rule presented. ■

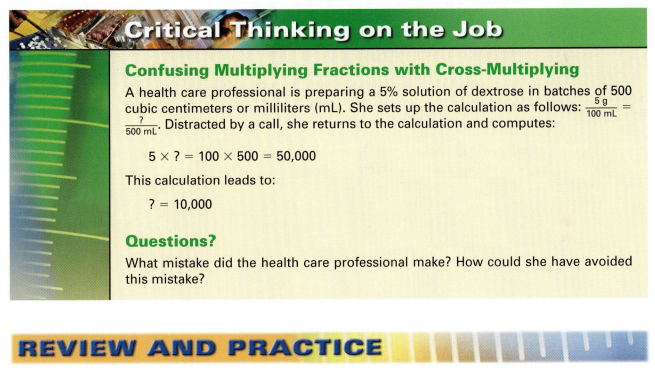

Critical Thinking on the Job

Confusing Multiplying Fractions with Cross-Multiplying

A health care professional is preparing a 5% solution of dextrose in batches of 500 cubic centimeters or milliliters (mL). She sets up the calculation as follows: $\frac{5 \text{ g}}{100 \text{ mL}} = \frac{?}{500 \text{ mL}}$. Distracted by a call, she returns to the calculation and computes:

$$5 \times ? = 100 \times 500 = 50{,}000$$

This calculation leads to:

$$? = 10{,}000$$

Questions?

What mistake did the health care professional make? How could she have avoided this mistake?

REVIEW AND PRACTICE

2-7 Cross-Multiplying

Determine if the following proportions are true.

1. $\frac{7}{16} - \frac{28}{48}$
2. $\frac{6}{9} - \frac{24}{36}$
3. $\frac{100}{250} = \frac{150}{375}$
4. $\frac{50}{125} = \frac{125}{300}$
5. $\frac{60}{96} = \frac{80}{108}$

Cross-multiply to find the missing value.

6. $\frac{3}{15} = \frac{?}{5}$
7. $\frac{2}{?} = \frac{8}{100}$
8. $\frac{?}{20} = \frac{120}{100}$
9. $\frac{50}{75} = \frac{100}{?}$
10. $\frac{10}{3} = \frac{?}{60}$

11. $\frac{?}{4} = \frac{4}{16}$ 12. $\frac{25}{?} = \frac{75}{3}$ 13. $\frac{2}{3} = \frac{6}{?}$ 14. $\frac{?}{9} = \frac{18}{27}$ 15. $\frac{25}{125} = \frac{?}{150}$

16. A patient must take 3 tablets per day for 14 days. How many tablets should the pharmacy supply to fill this order?

17. If 15 mL of solution contains 75 mg of drug, how many milligrams of drug are in 60 mL of solution?

18. A health care professional is instructed to administer 600 mL of a solution every 8 h. How many hours will be needed to administer 1800 mL of the solution?

19. Two tablets contain a total of 50 mg of drug. How many milligrams of drug are in 10 tablets?

20. If 250 mL of solution contains 90 mg of drug, there is 450 mg of drug in how many milliliters of solution?

To check your answers, see page 502.

Critical Thinking on the Job

Setting Up the Correct Proportion

A physician's order calls for a patient to receive 250 mg of amoxicillin oral suspension 3 times a day. Amoxicillin oral suspension is a dry medication that is mixed with water before being given to the patient. Each 5 mL of suspension contains 125 mg of drug. The health care professional needs to calculate how many milliliters he will need to give the patient for each dose. He sets up the proportion as 5 mL/125 mg = 250 mg/?.

Questions?

What mistake did the health care professional make? How should it be corrected? What dose should the patient receive?

HOMEWORK ASSIGNMENT

In each row of the table below, use the information to calculate the equivalent values. When necessary, round decimals to the nearest hundredth. Round percents to the nearest tenth of a percent. Do not reduce ratios or fractions.

	Fraction	Decimal	Ratio	Percent
1.		0.625		
2.	$\frac{5}{2}$			
3.			3 : 5	
4.				35%
5.	$\frac{1}{6}$			
6.		11.4		
7.				18.5%
8.			6 : 20	

Find the missing value:

9. 2 : ? : : 6 : 15

10. ? : 8 : : 3 : 12

11. 5 : 8 : : ? : 40

12. $\frac{1}{6} = \frac{?}{18}$

13. $\frac{4}{12} = \frac{2}{?}$

14. $\frac{15}{?} = \frac{6}{8}$

15. A solution contains 200 mg of drug in 5 mL.
   a. What is the ratio strength of the solution?
   b. How much drug is in 2 mL of the solution?

16. How many grams of drug are in 75 mL of a 4% solution?

17. A solution contains 20 g of drug in 400 mL of solution. What is the strength of the solution in percent?

Chapter 2 Review

Check Up

In each row of the table below, use the information to calculate the equivalent values. For instance, in row 1, convert the ratio 2 : 3 to a fraction, a decimal, and a percent. Where necessary, round decimals to the nearest hundredth. Round percents to the nearest percent. Do not reduce ratios and fractions.

	Fraction	Decimal	Ratio	Percent
1.			2 : 3	
2.	$\frac{5}{4}$			
3.				28%
4.		0.03		
5.	$\frac{40}{8}$			
6.			4 : 12	
7.	$\frac{9}{27}$			
8.		1.4		
9.				0.5%
10.			3 : 50	
11.				25%
12.		6		
13.	$\frac{1}{9}$			
14.				150%
15.			6 : 97	
16.		12.8		

Find the missing value.

17. 1 : 10 : : 4 : ?

18. 3 : 27 : : ? : 9

19. ? : 6 : : 8 : 12

20. 5 : ? : : 10 : 50

21. $\frac{4}{8} = \frac{24}{?}$

22. $\frac{?}{14} = \frac{5}{70}$

23. $\frac{3}{?} = \frac{30}{20}$

24. $\frac{1}{25} = \frac{?}{125}$

25. If 1 tablet contains 25 mg of drug, how many milligrams of drug are in 3 tablets?
26. If 100 mL of drug is in 600 mL of solution, how many milliliters of drug are in 1800 mL of solution?
27. A solution is 5% dextrose. How many grams (g) of dextrose are in 500 cc of solution?
28. A solution contains 1 g of drug for every 50 mL of solution. How much solution would you need to give a patient to administer 3 g of drug?
29. 10 mL of a liquid medication contains 250 mg of drug. How many milliliters contain 50 mg of drug?
30. If 30 g of drug is in 100 mL of solution, what is the ratio strength of the solution?

To check your answers, see page 503.

Critical Thinking Applications

A health care professional has just finished preparing 250 mL of a 6% solution (using a dry drug) when he learns that the physician wants an 8% solution, not a 6% solution. How many additional grams of dry drug should the health care professional add to the 6% solution he has just prepared? (Assume that adding the dry drug does not affect the total volume of solution.)

Case Study

A physician's order calls for Ceclor® oral suspension 750 mg to be given daily for 14 days. The daily dosage is to be divided into 3 equal doses per day. Ceclor® oral suspension is a drug that is mixed with solvent before it is administered.

1. How many milligrams of drug should be given in each dose?
2. If 5 mL of solution contains 125 mg of drug, how many milliliters should be given in each dose?

To check your answers, see page 504.

Internet Activity

To obtain a better understanding of solutions, lotions, creams, suspensions, and ointments, search for these terms on the Internet. Find their definitions and at least three example medications of each type. You can start with an online dictionary such as found on www.refdesk.com and then search websites such as www.fda.gov or www.rxlist.com for example medications.

Go To . . . Open the CD-ROM that accompanies your textbook, and complete a final review of the rules, practice problems, and activities presented for this chapter. For a final evaluation, take the chapter test and email or print your results for your instructor. A score of 95 percent or above indicates mastery of the chapter concepts. ■

3 Systems of Weights and Measures

Practice is the best instruction of them all.

—Publilius Syrus

Learning Outcomes

When you have completed Chapter 3, you will be able to:

- List the fundamental units of the metric system for length, weight, and volume.
- Summarize metric notation.
- Calculate equivalent measurements within the metric system.
- Identify the most frequently used equivalent measurements among metric, household, and apothecaries' measurements.
- Recognize the symbols for dram, ounce, grain, and drop.
- Convert measurements within and among the metric, household, and apothecary systems of measurement.
- Calculate temperature and time conversions.

Key Terms

Centi (c)
Dram (ʒ)
Grain (gr)
Gram (g)
International Unit (IU)
Kilo (k)
Liter (L)
Meter (m)
Micro (mc)
Milli (m)
Milliequivalents (mEq)
Minim (♏)
Ounce (℥)
Unit

Introduction

A large number of medications taken by patients are measured in grams or milligrams. These are basic units of the metric system. Additionally, understanding and converting systems of weights and measures are traditional skills required by health care employees. Practice these concepts until you are certain you understand.

Metric System

The metric system is the most widely used system of measurement in the world today. The system, which was defined in 1792, gets its name from the **meter**, the basic unit of length. A meter is approximately 3 inches (8.56 cm) longer than a yard. See Figure 3-1.

Units of measurement in the metric system are sometimes referred to as SI units, an abbreviation for International System of Units. This system was established in 1960 to make units of measurement for the metric system standard throughout the world. Table 3-1 lists the basic metric units for length, weight, and volume.

Notice that meter and gram are abbreviated with lowercase letters, but **liter** is abbreviated with an uppercase L. Using the uppercase L minimizes the chance of confusing the lowercase letter L (l) with the digit 1. You will use length mostly when expressing measurements such as patient height, infant head circumference, and lesion or wound size. However, you will use weight and volume frequently when you calculate dosages. Most dosages and drug strengths are expressed using the metric system.

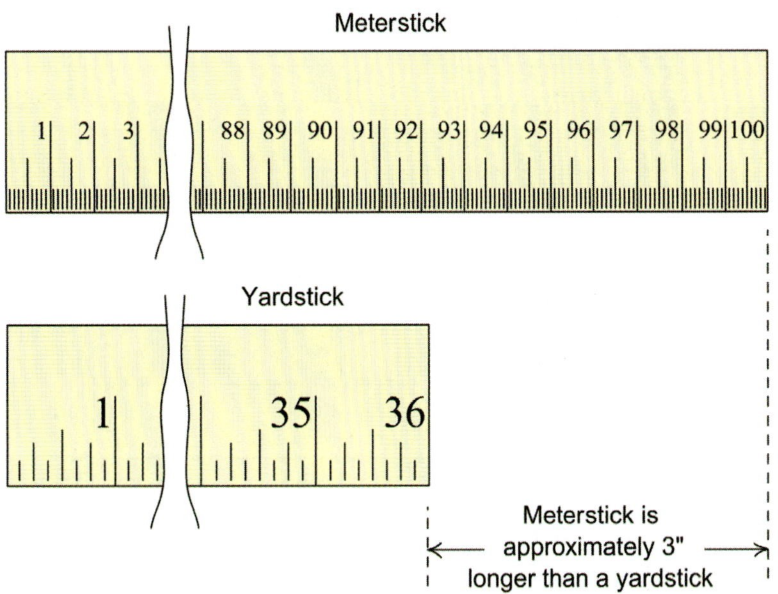

Figure 3-1 Meter versus yard.

TABLE 3-1 Basic Units of Metric Measurement

Type of Measure	Basic Unit	Abbreviation
Length	meter	m
Weight (or mass)	gram	g (or gm)
Volume	liter	L

Understanding Metric Notation

Like the decimal system, the metric system is based on multiples of 10. The greater confidence you have working with decimals, the more comfortable you will be working with metric units. (To review decimals, see Chapter 1.)

A prefix before the basic unit indicates relative size. For example, **kilo-** indicates that you multiply the basic unit by 1000. A kilometer is 1000 meters, a kilogram is 1000 grams, and a kiloliter is 1000 liters. When you divide 1 meter into 1000 equal lengths, each length is 1 millimeter. The prefix **milli-** means one-thousandth. A millimeter is one-thousandth of a meter, a milliliter is one-thousandth of a liter, and a milligram is one-thousandth of a gram. Table 3-2 lists several common prefixes, their abbreviations, and their value relative to the basic unit.

As a health care provider, you will most often use the metric prefixes *kilo-*, **centi-**, *milli-*, and **micro-**. Table 3-3 shows the most commonly used combinations of prefixes and basic units.

TABLE 3-2 Common Metric System Prefixes

Prefix	Length	Value
kilo- (k)	kilometer (km)	1 km = 1000 m
(basic unit)	meter (m)	1 m
centi- (c)	centimeter (cm)	1 cm = $\frac{1}{100}$ m = 0.01 m
milli- (m)	millimeter (mm)	1 mm = $\frac{1}{1000}$ m = 0.001 m
micro- (mc)	micrometer (mcm)	1 mcm = $\frac{1}{1{,}000{,}000}$ m = 0.000001 m

TABLE 3-3 Combining Prefixes and Units

Prefix	Length (meter)	Weight (Mass) (gram)	Volume (liter)
kilo- (× 1000)	kilometer km	kilogram kg	kiloliter kL
centi- (÷ 100)	centimeter cm	centigram cg	centiliter cL
milli- (÷ 1000)	millimeter mm	milligram mg	milliliter mL
micro- (÷ 1,000,000)	micrometer mcm	microgram mcg	microliter mcL

Rule 3-1 Use Arabic numerals, with decimals to represent any fractions.

Example Write 1.25 g to represent $1\frac{1}{4}$g.

Rule 3-2 If the quantity is less than 1, include a 0 before the decimal point. Delete any other zeros that are not necessary.

Example Do not write .750; instead, write 0.75, adding a zero before the decimal point and deleting the unnecessary zero at the end.

Rule 3-3 Write the unit after the quantity with a space between them.

Example Write 30 mg, not mg 30.

Rule 3-4 Use lowercase letters for metric abbreviations. However, use uppercase L to represent liter.

Example Write mg, not MG. Write mL, not ml. While ml is technically correct, you will avoid errors if you use an uppercase L.

Considering Rules 3-1 to 3-4, we will determine the correct metric notation for six and two-eighths milliliters. First, $6\frac{2}{8}$ must be converted to decimals.

Learning Link Recall from Chapter 1, Rules 1-5 and 1-20, that $\frac{2}{8}$ can be reduced to $\frac{1}{4}$ and $\frac{1}{4} = 0.25$ found on pages 7 and 35.

$$6\frac{2}{8} = 6.25$$

Next, write the unit after the quantity, leaving a space between them and using the abbreviation mL for milliliters. So:

6.25 mL is correct metric notation for six and two-eighths milliliters.

Now consider how you would write one-half milligram.

$$\frac{1}{2} = 0.50$$

Place a zero in front of the decimal point, and delete any unnecessary zeros.

0.5

Place the unit after the quantity with a space between them.

0.5 mg is correct metric notation for one-half milligram.

GO TO . . . Open the CD-ROM that accompanies your textbook, and select Chapter 3, Rules 3-1 to 3-4. Review the animation and example problems, then complete the practice problems. Continue to the next section of the book once you have mastered the rules presented. ■

REVIEW AND PRACTICE

3-1 Understanding Metric Notation

In Exercises 1–10, select the correct metric notation.

1. Two and one-half kilograms
   a. 2.5 Kg **b.** 2.05 kg **c.** $2\frac{1}{2}$kg **d.** 2.5 kg
2. Seven-tenths of a milliliter
   a. $\frac{7}{10}$mL **b.** .7mL **c.** ml 0.7 **d.** 0.7 mL
3. Four-hundredths of a gram
   a. 400 G **b.** 0.4g **c.** 0.04 g **d.** .04 g
4. Thirty-one millimeters
   a. 31 mm **b.** 0.031 mm **c.** 31.0 mlm **d.** 31 mm
5. Eight liters
   a. 8.0 l **b.** 8 L **c.** 8.0L **d.** 0.8 l
6. One hundred twenty-five micrograms
   a. 125 mg **b.** 0.125 mcg **c.** 125 mcg **d.** 125 mg
7. Seventy-eight centimeters
   a. 78 ctm **b.** 78.0 Cm **c.** 0.78 cm **d.** 78 cm
8. Two hundred fifty microliters
   a. mcL 250 **b.** 250 mcL **c.** 25.0 mcL **d.** 250 mL
9. Nine and one-quarter milligrams
   a. $9\frac{1}{4}$mg **b.** 9.25 mg **c.** 9.25 mg **d.** 9.25 mgm
10. Four-tenths of a liter
    a. 0.4 L **b.** $\frac{4}{10}$L **c.** 0.40 L **d.** 0.40 l

In Exercises 11–20, write the indicated amounts.

11. Four and one-half milliliters ______
12. Sixty-two hundredths of a gram ______
13. Three-quarters of a milliliter ______
14. Seven-tenths of a meter ______
15. Twelve liters ______
16. Nine-twelfths of a kilogram ______
17. One hundred fifty-seven kilometers ______
18. Seven and three-quarters centimeters ______
19. Ninety-three micrograms ______
20. Eight-hundredths of a milligram ______

To check your answers, see page 504.

Converting Within the Metric System

Recall from Chapters 1 and 2 that when you multiply a decimal number by 100, you move the decimal point two places to the right and get a larger number. When you divide a decimal number by 100, you move the decimal point two places to the left and get a smaller number.

Converting one metric unit of measurement to another is similar to multiplying and dividing decimal numbers. For example, if you travel 1 kilometer, you travel 1000 meters. When you convert from the larger unit of measurement (kilometer) to the smaller unit (meter), the quantity of units increases. Therefore, you multiply, moving the decimal point to the right.

If you convert from meters to kilometers, the quantity of units decreases. If you travel 1000 meters, you travel 1 kilometer. When you convert from a smaller unit (meter) to a larger unit (kilometer), the quantity of units decreases. Therefore, you divide, moving the decimal point to the left.

When you calculate dosages, you will work most often with four metric units of weight and three metric units of volume. The four units of weight are kilogram (kg), gram (g), milligram (mg), and microgram (mcg). Two of the units of volume are liter (L) and milliliter (mL). The third is cubic centimeter (cc), which is equivalent to milliliter (mL). Although the abbreviation cc for cubic centimeter may be seen in practice, it should not be used. Instead use the abbreviation mL for milliliter.

Rule 3-5 When you convert a quantity from one unit of metric measurement to another:

1. Move the decimal point to the right if you convert from a larger to a smaller unit.
2. Move the decimal point to the left if you convert from a smaller to a larger unit.

Figure 3-2 will help you determine both the direction and the number of places to move the decimal point when you convert between units of metric measurement. For example, milliliter is three places to the right of liter, the basic unit. To convert a quantity from liters (larger) to milliliters (smaller), move the decimal point three places to the right. Similarly, to convert a quantity from grams (smaller) to kilograms (larger), move the decimal point three places to the left.

Example 1 Convert 4 L to milliliters (mL).

A milliliter (mL) is smaller than a liter (L); a quantity will have more milliliters than liters. Using Figure 3-2, you can see that milliliter is three places to the right of liter. Move the decimal point for 4 (or 4.0) three places to the right to find the number of milliliters. Add zeros as necessary to help you with your calculation.

4 L = 4.000 L = 4000 mL

Prefix	kilo-	hecto-	deka-	base unit	deci-	centi-	milli-	decimilli-	centimilli-	micro-
Meaning	1000	100	10	1	$\frac{1}{10}$	$\frac{1}{100}$	$\frac{1}{1000}$	$\frac{1}{10,000}$	$\frac{1}{100,000}$	$\frac{1}{1,000,000}$
Abbreviation	km kg kL			meter gram liter			mm mg mL			mcm mcg mcL
Decimal Value	1000	100	10	1.0	0.1	0.01	0.001	0.0001	0.00001	0.000001

Figure 3-2 Metric system place values.

Example 2 How many meters (m) are in 75 mm?

A meter (m) is larger than a millimeter (mm); a quantity will have fewer meters than millimeters. Using Figure 3-2, you can see that meter is three places to the left of millimeter. Write 75 as 75.0, and move the decimal point in 75 three places to the left. Add zeros as necessary.

75 mm = 75.0 mm = 0.075 m

Example 3 Convert 4.5 mcg to milligrams (mg).

You are converting from a smaller unit to a larger one. Use Figure 3-2 or divide 4.5 by 1000, moving the decimal point three places to the left.

4.5 mcg ÷ 1000 = 0004.5 mcg ÷ 1000 = 0.0045 mg

Example 4 Convert 62 kg to grams (g).

You are converting from a larger unit to a smaller one. Use Figure 3-2 or multiply 62 by 1000, moving the decimal point three places to the right.

62 kg × 1000 = 62.000 kg × 1000 = 62,000 g

Example 5 Convert 300 mg to grams (g).

You are converting from a smaller unit to a larger one. Use Figure 3-2 or divide 300 by 1000, moving the decimal point three places to the left.

300 mg ÷ 1000 = 0300.0 mg ÷ 1000 = 0.3 g

GO TO . . . Open the CD-ROM that accompanies your textbook, and select Chapter 3, Rule 3-5. Review the animation and example problems, then complete the practice problems. Continue to the next section of the book once you have mastered the rule presented. ■

The four units of weight, or mass, are related to each other by a factor of 1000. A kilogram is 1000 times larger than a gram. Thus, 1 kg = 1000 g. In turn, a gram is 1000 times larger than a milligram, which is 1000 times larger than a microgram. The same relationship is true for liters and milliliters; a liter is 1000 times larger than a milliliter. Table 3-4 lists four of the most commonly used equivalent measurements. Because they are so important to dosage calculations, you should memorize them.

TABLE 3-4 Equivalent Metric Measurements

1 kg = 1000 g	1 mg = 1000 mcg
1 g = 1000 mg	1 L = 1000 mL

ERROR ALERT!

Remember: The larger the unit, the smaller the quantity. The smaller the unit, the larger the quantity.

You may be tempted to multiply when you convert from a smaller unit to a larger unit, thinking that you are increasing in size. If you find yourself confused, think about conversions you have made all your life.

For example, a dollar bill is a larger unit of money than a quarter, which is a larger unit of money than a penny. When you write their relationship, look at how the quantity changes:

1 dollar bill = 4 quarters = 100 pennies

When you convert from the larger unit to the smaller one, the quantity increases. Writing the money relationship as:

100 pennies = 4 quarters = 1 dollar bill

shows you that as the unit increases in size, the quantity decreases. You see the same relationship with units of time and in the metric system:

1 hour = 60 minutes = 3600 seconds

1 g = 1000 mg = 1,000,000 mcg

Critical Thinking on the Job

Placing the Decimal Point Correctly

A child suffering from congestive heart failure is rushed into an emergency room. The physician orders 0.05 mg of Lanoxin® for the child. The health care professional quickly calculates that 0.05 mg = 500 mcg. Lanoxin® is available for injection in quantities of 500 mcg. The nurse hands the syringe to the doctor.

Fortunately, the doctor catches the error before the Lanoxin® is administered. The child should be given 50 mcg, not 500 mcg of Lanoxin®. As it turns out, Lanoxin® is available as an elixir in doses of 50 mcg. This quantity should be administered. The larger dose of 500 mcg could be fatal to the child.

Questions?

How can the attending health care professional ensure that she converted the quantity correctly. How should the problem have been solved? Why is it important in this situation?

REVIEW AND PRACTICE

3-2 Converting Within the Metric System

In Exercises 1–20. complete the conversions.

1. 7 g = _______ mg **2.** 1200 mg = _______ g **3.** 23 g = _______ kg

4. 8 kg = _______ g **5.** 8.01 L = _______ mL **6.** 100 mL = _______ L

7. 3.6 m = _______ mm **8.** 5233 mm = _______ m **9.** 500 m = _______ km

10. 3.25 km = ________ m

11. 0.25 mg = ________ mcg

12. 462 mg = ________ mcg

13. 250 mcg = ________ mg

14. 75 mcg = ________ mg

15. 0.06 g = ________ mcg

16. 0.5 g = ________ mcg

17. 8000 mcg = ________ g

18. 20,000 mcg = ________ g

19. 562 mm = ________ cm

20. 4.32 cm = ________ m

To check your answers, see page 504.

Other Systems of Measurement

The trend in health care is clearly toward using the metric system of measurement. However, two other systems, the apothecary and household systems, are still in use. Plus, two other measures—milliequivalents and units—are used for some medications.

Apothecary System

The apothecary system is an old system of measurement. Used first by apothecaries (early pharmacists), it traveled across Europe from Greece and Rome to France and England. Eventually, it crossed the Atlantic to colonial America. The household system familiar to most Americans evolved from the apothecary system. Certain medications, especially older ones such as aspirin and morphine, sometimes show apothecary units. The use of apothecary units is being phased out since it is less familiar and can be confused with metric units. Metric units are preferred for calculations in most cases.

Units of Measure. The basic unit of weight in the apothecary system is the grain (gr). Originally, the grain was defined as the weight of a single grain of wheat, hence its name.

ERROR ALERT!

Do not confuse grains and grams.

Because they have names and abbreviations that are similar, **grains** (gr) and **grams** (g) are easily confused. A grain is a measure in the apothecary system; a gram is a measure in the metric system. If you are not sure whether an order refers to grains or grams, check with the physician or pharmacist. For most conversions, 1 grain equals either 60 or 65 milligrams (mg), which means:

1 gr = 60 mg = 0.06 g

or 1 gr = 65 mg = 0.065 g

In either case, 1 grain is significantly smaller than 1 gram. Medications that are measured in grains do not all use the same conversion. However, typically their labels list the metric units as well.

ERROR ALERT!

Converting quantities for medications.

When you convert quantities from one unit of measure to another within the metric system, pay close attention to the decimal point. For example, when going from milligrams (mg) to micrograms (mcg), the quantity should be multiplied by 1000; the decimal should move three places to the right. If you move the decimal the wrong direction a dangerous error can occur.

Three units of volume in the apothecary system are the **minim ♏**, the **dram ʒ**, and the **ounce ℥**. The apothecary ounce has become part of the common system of measures used in the United States. There are 8 ounces (oz) to 1 cup (c) in our commonly used household system of measures. The minim is seldom used these days, although many syringes continue to have marks that indicate minims. The dram symbol ʒ is most frequently used as an abbreviation for a teaspoonful, which is nearly the same volume.

ERROR ALERT!

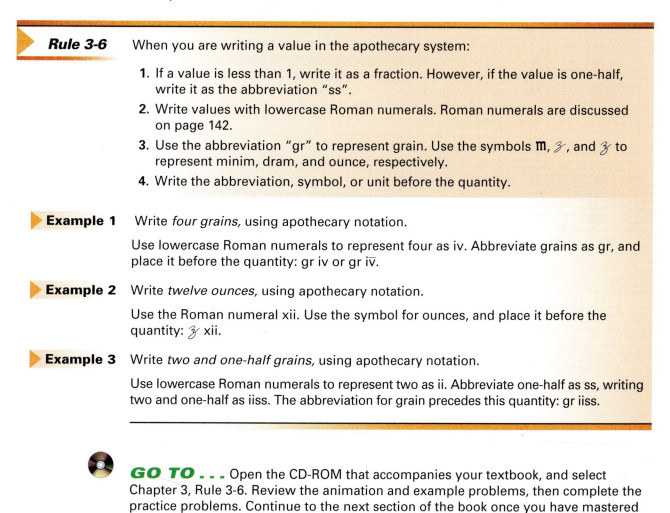

Do not confuse the symbols for drams and ounces.

The symbols for drams and ounces appear similar. When they are typed, the symbol for dram ʒ looks similar to the numeral *3*; the symbol for ounce ℥ has an extra line at the top. When they are handwritten, the symbol for dram ʒ can look like the written letter *z*. Again, the symbol for ounce ℥ has an extra line at the top. As always, if you are in doubt about which symbol is used, check with the physician who wrote the order.

Apothecary Notation. The system of apothecary notation has special rules that combine fractions, Roman and Arabic numerals, symbols, and abbreviations. Even the order in which information is written differs from the order most familiar to you. Recall that Roman numerals may be written with a bar above them.

Rule 3-6 When you are writing a value in the apothecary system:

1. If a value is less than 1, write it as a fraction. However, if the value is one-half, write it as the abbreviation "ss".
2. Write values with lowercase Roman numerals. Roman numerals are discussed on page 142.
3. Use the abbreviation "gr" to represent grain. Use the symbols ♏, ʒ, and ℥ to represent minim, dram, and ounce, respectively.
4. Write the abbreviation, symbol, or unit before the quantity.

Example 1 Write *four grains,* using apothecary notation.

Use lowercase Roman numerals to represent four as iv. Abbreviate grains as gr, and place it before the quantity: gr iv or gr i$\bar{\text{v}}$.

Example 2 Write *twelve ounces,* using apothecary notation.

Use the Roman numeral xii. Use the symbol for ounces, and place it before the quantity: ℥ xii.

Example 3 Write *two and one-half grains,* using apothecary notation.

Use lowercase Roman numerals to represent two as ii. Abbreviate one-half as ss, writing two and one-half as iiss. The abbreviation for grain precedes this quantity: gr iiss.

GO TO . . . Open the CD-ROM that accompanies your textbook, and select Chapter 3, Rule 3-6. Review the animation and example problems, then complete the practice problems. Continue to the next section of the book once you have mastered the rule presented. ■

TABLE 3-5 Abbreviations for Household Measures

Unit of Measurement	Abbreviation
Drop	gtt
Teaspoon	tsp, t, or ʒ*
Tablespoon	tbsp, tbs or T
Ounce	oz or ℥*
Cup	cup or c
Pint	pt
Quart	qt
Gallon	gal

*When you refer to the apothecary symbols for teaspoon and ounce, Roman numerals are used.

Household System

The household system of measurement is still commonly used today. Patients who take medication at home are more likely to use everyday household measures than metric ones. Many over-the-counter medications provide instructions for patients relying on household measures. For instance, a patient will be told to take two teaspoons of a cough syrup.

While the household system is the most familiar one to patients, in practice it is the least accurate. For instance, patients who take a teaspoon of a syrup will often use everyday spoons that vary in size, rather than baking or other calibrated spoons. Instructions for over-the-counter medications can even invite inaccuracies. A patient may be told to mix a rounded teaspoon of powder with a quantity of water. The interpretation of *rounded* will vary from patient to patient.

Units of Measure. Basic units of volume in the household system, in increasing size, include the drop, teaspoon, tablespoon, ounce, cup, pint, quart, and gallon. Of these, the four smallest measures are most commonly used for medications.

When one is specifically discussing medications, the word *ounce* generally implies volume; it represents fluid ounce. In other contexts, *ounce* may represent a unit of weight, as does *pound*.

Household Notation. As with the metric system, household notation places the quantity in Arabic numerals before the abbreviation for the unit. Table 3-5 summarizes the standard abbreviations.

Example 1 Write *six drops,* using household notation.

Write the quantity with Arabic numerals before the abbreviation for the unit: 6 gtt.

Example 2 Write *twelve ounces,* using household notation.

Write the quantity with Arabic numerals before the abbreviation for the unit: 12 oz.

Apothecary and Household Equivalents

Units of measurement found in both the apothecary and the household systems are equal: an apothecary ounce equals a household ounce. Unlike the metric system, neither the apothecary system nor the household system is based on multiples of 10. You must become familiar with their equivalent measures (see Table 3-6). In practice, the size of a drop depends on the dropper and the liquid. A drop by itself is not a reliably accurate form of measurement.

TABLE 3-6 Apothecary and Household Equivalent Measures

Drop	1 drop	=	1 minim
Teaspoon	1 teaspoon	=	60 drops
Tablespoon	1 tablespoon	=	3 teaspoons
Ounce	1 ounce	=	2 tablespoons
Cup	1 cup	=	8 ounces

Example 1 How many teaspoons of solution are contained in 1 ounce (oz) of solution?

From Table 3-6, you can see that 1 ounce contains 2 tablespoons. In turn, each tablespoon contains 3 teaspoons. Therefore,

$$1 \text{ oz} = 2 \times 1 \text{ tbsp} = 2 \times 3 \text{ tsp} = 6 \text{ tsp}$$

One ounce of solution contains six teaspoons of solution.

Example 2 How many tablespoons are in $\frac{1}{2}$ cup of solution?

Convert 1 cup to ounces, then ounces to tablespoons. From Table 3-6, you know that 1 cup = 8 oz and 1 oz = 2 tbsp:

$$\frac{1}{2} \text{ cup} = \frac{1}{2} \times 1 \text{ cup} = \frac{1}{2} \times 8 \text{ oz} = 4 \text{ oz}$$

$$= 4 \times 1 \text{ oz} = 4 \times 2 \text{ tbsp} = 8 \text{ tbsp}$$

One-half cup of solution contains eight tablespoons of solution.

Milliequivalents and Units

Some drugs are measured in **milliequivalents** (mEq). A unit of measure based on the chemical combining power of the substance, one milliequivalant is defined as $\frac{1}{1000}$ of an equivalent weight of a chemical. Electrolytes, such as sodium and potassium, are often measured in milliequivalents. Sodium bicarbonate and potassium chloride are examples of drugs that are prescribed in milliequivalents. You do not need to learn to convert from milliequivalent to another system of measurement.

Medications such as insulin, heparin, and penicillin are measured in *USP units*. (See Chapter 6 to learn more about USP.) A **unit** is the amount of a medication required to produce a certain effect. *The size of a unit varies for each drug*. Some medications, such as vitamins, are measured in standardized units called **international units** (IU). These IUs represent the amount of medication needed to produce a certain effect, but they are standardized by international agreement. As with milliequivalent, you do not need to convert from units to other measures. Medications that are ordered in units will also be labeled in units.

REVIEW AND PRACTICE

3-3 Other Systems of Measurement

In Exercises 1–10, write the symbols or abbreviations.

1. Minim **2.** Dram **3.** Grain **4.** Ounce **5.** Drop

6. Teaspoon **7.** Tablespoon **8.** Pint **9.** Milliequivalent **10.** Unit

In Exercises 11–25, write the amounts, using either apothecary or household notation, as appropriate. (Some exercises may require you to write the amount using both notations.)

11. Seven grains
12. Five drams
13. Three ounces
14. Eight ounces
15. Fourteen grains
16. Seventeen grains
17. One-half teaspoon
18. One-half tablespoon
19. One-half grain
20. One-half ounce
21. Two and one-half ounces
22. Five and one-half ounces
23. Two thousand units
24. Forty milliequivalents
25. Forty-nine units

To check your answers, see page 505.

Converting Among Metric, Apothecary, and Household Systems

When you calculate dosages, you must often convert among the metric, apothecary, and household systems of measurement. To do so, you will need to know how the measure of a quantity in one system compares with its measure in another system. For example, you learned the relationships between milliliter and liter and between teaspoon and tablespoon. To convert between systems, you may also need to know the relationship between milliliter and teaspoon. When you convert between systems, you lose a certain amount of exactness, especially when you round numbers. Thus, two measures are often approximately the same, though not exactly the same.

Equivalent Volume Measurements

A standard equivalent measure is that 1 tsp = 5 mL.

1 tsp = 5 mL

You can now determine most relationships between household or apothecary systems and metric systems. For instance, because 1 tbsp = 3 tsp,

1 tbsp = 3 tsp = 3 × 1 tsp = 3 × 5 mL = 15 mL

Therefore, 1 tbsp = 15 mL. Furthermore, because 1 oz = 2 tbsp,

1 oz = 2 tbsp = 2 × 1 tbsp = 2 × 15 mL = 30 mL

Therefore, 1 oz = 30 mL.

1 oz = 30 mL

Earlier you learned that 1 oz = 8 dr. Therefore,

1 oz = 8 dr = 2 tbsp = 30 mL

The importance of this relationship will become clear in Chapter 4, when you learn about medicine cups.

Table 3-7 summarizes several important volume relationships.

TABLE 3-7 Approximate Equivalent Measures for Volume

Metric	Household	Apothecary
5 mL	1 tsp	1 dram*
15 mL	1 tbsp	3 drams*
30 mL	2 tbsp = 1 oz	1 oz = 8 drams*
240 mL	8 oz = 1 c	8 oz
480 mL	2 c = 1 pt	16 oz
960 mL	2 pt = 1 qt	32 oz

*The dram (dr) is used today to represent 1 tsp or 5 mL. Actually the dram has an exact volume of 3.7 mL rounded to 4 mL when performing conversions.

Equivalent Weight Measurements

Earlier you learned that 1 grain is equivalent to either 60 or 65 milligrams:

gr i = 60 mg *or* gr i = 65 mg

The relationship between grains and milligrams or grams is actually more complex. The conversion varies from 60 mg to 66.7 mg per grain.

In cases where you use 60 mg as the equivalent measure, one way to remember the relationship between grains and milligrams is to think of a clock (see Figure 3-3). If each "minute" is one milligram, then an entire hour is one grain. This image of the clock may help you when you need to find the equivalent of $\frac{1}{2}$ grain (gr ss) or $\frac{1}{4}$ grain $\left(\text{gr}\frac{1}{4}\right)$. Each half of an "hour" or grain is 30 "minutes" or milligrams. Similarly, each quarter of an "hour" or grain is 15 "minutes" or milligrams.

Table 3-8 summarizes important weight equivalent measures, including the relationship between kilograms and pounds (lb).

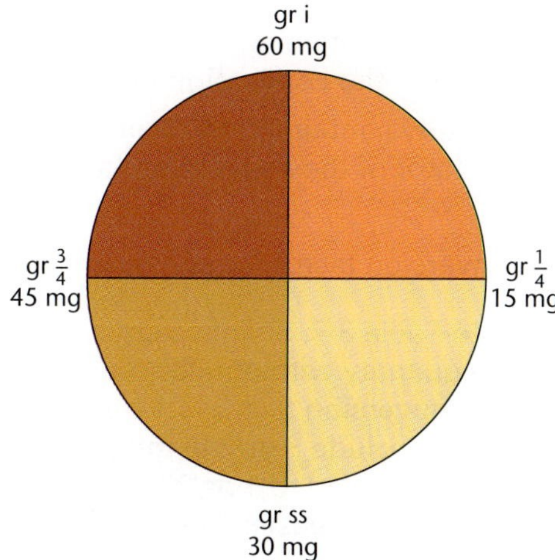

Figure 3-3 Use a round image to help you remember the comparison between 60 mg and 1 gr.

TABLE 3-8 Approximate Equivalent Measures for Weight

Metric	Apothecary
60 mg or 65 mg	gr i (1 grain)
30 mg	gr ss ($\frac{1}{2}$ grain)
15 mg	gr $\frac{1}{4}$
1 mg	gr $\frac{1}{60}$
1 g (1000 mg)	gr xv (15 grains)
0.5 g	gr viiss ($7\frac{1}{2}$ grains)
1 kg	2.2 lb

Conversion Factors

A *conversion factor* is a fraction made of two quantities that are equal to each other but are expressed in different units. For example, Table 3-8 tells us that 1 kg and 2.2 lb are equal to each other. Forming a fraction from these two quantities gives us a conversion factor. Two different conversion factors can be formed from these quantities, 1 kg/2.2 lb and 2.2 lb/1 kg. The first of these is used to convert from pounds to kilograms, while the second is used to convert from kilograms to pounds.

Rule 3-7 Using Conversion Factors

When you are writing conversion factors:

1. The two quantities in the conversion factor must be equal to each other.
2. The quantity containing the units that you wish to convert *to* goes in the numerator of the conversion factor.
3. The quantity containing the units that you are converting *from* goes in the denominator of the conversion factor.

Example Write a conversion factor for converting from milliliters to ounces.

According to Table 3-7, 30 mL is equivalent to 1 oz. Since we wish to convert to ounces, the quantity with ounces must be the numerator of the conversion factor. The correct conversion factor is 1 oz/30 mL. (*Note*: Other correct conversion factors from Table 3-7 include 8 oz/240 mL, 16 oz/480 mL, and 32 oz/960 mL. Each of these factors can be reduced to 1 oz/30 mL, but using the reduced conversion factor will make the conversion easier to solve.)

GO TO . . . Open the CD-ROM that accompanies your textbook, and select Chapter 3, Rule 3-7. Review the animation and example problems, then complete the practice problems. Continue to the next section of the book once you have mastered the rule presented. ■

Fraction Proportion Method

Cross-multiplying can be used to convert from one unit to another if you know a conversion factor. Recall from Chapter 2, Rule 2-20, you can solve proportions for an unknown value by cross-multiplying, found on page 71. One of the fractions in your proportion is the conversion factor itself. The other fraction contains the unknown value in the numerator and the value that you wish to convert in the denominator.

Procedure Checklist 3-1

Converting by the Fraction Proportion Method

1. Write a conversion factor with the units needed in the answer in the numerator and the units you are converting from in the denominator.
2. Write a fraction with the unknown, or ?, in the numerator and the number that you need to convert in the denominator.
3. Set the two fractions up as a proportion.
4. Cancel units.
5. Cross-multiply, then solve for the unknown value.

Example 1 Convert 66 lb to kilograms.

Follow Procedure Checklist 3-1.

1. Table 3-8 tells us that 1 kg is equal to 2.2 lb. Since we are converting to kilograms, kilograms must appear in the numerator of our conversion factor. Our conversion factor is 1 kg/2.2 lb.
2. The other fraction for our proportion has the unknown ? for a numerator and 66 lb as the denominator: $\frac{?}{66\text{ lb}}$.
3. Setting the two fractions into a proportion gives us the following equation:

$$\frac{?}{66\text{ lb}} = \frac{1\text{ kg}}{2.2\text{ lb}}$$

4. Cancel units.

$$\frac{?}{66} = \frac{1\text{ kg}}{2.2}$$

5. Solve for the unknown by cross-multiplying.

$$2.2 \times ? = 1\text{ kg} \times 66$$

$$\frac{2.2 \times ?}{2.2} = \frac{1\text{ kg} \times 66}{2.2}$$

$$? = 30\text{ kg}$$

Example 2 A patient needs to take 10 mL of a medication, but is going to be measuring the medication with a teaspoon. How many teaspoons should he use?

Follow Procedure Checklist 3-1.

1. Table 3-7 tells us that 5 mL is equal to 1 tsp. Since we are converting to teaspoons, our conversion factor is $\frac{1 \text{ tsp}}{5 \text{ mL}}$.
2. The other fraction for our proportion is $\frac{?}{10 \text{ mL}}$.
3. Setting the two fractions into a proportion gives us the following equation:

$$\frac{?}{10 \text{ mL}} = \frac{1 \text{ tsp}}{5 \text{ mL}}$$

4. Cancel units.

$$\frac{?}{10} = \frac{1 \text{ tsp}}{5}$$

5. Solve for the unknown.

$$5 \times ? = 1 \text{ tsp} \times 10$$

$$\frac{5 \times ?}{5} = \frac{1 \text{ tsp} \times 10}{5}$$

$$? = 2 \text{ tsp}$$

Example 3 A recipe from your aunt in Italy calls for 60 mL of olive oil. How many tablespoons of olive oil do you use?

Follow Procedure Checklist 3-1.

1. 30 mL is equal to 2 tablespoons (T). Our conversion factor is $\frac{2 \text{ T}}{30 \text{ mL}}$.
2. The other fraction for our proportion is $\frac{?}{60 \text{ mL}}$.
3. Setting the two fractions into a proportion gives us the following equation:

$$\frac{?}{60 \text{ mL}} = \frac{2 \text{ T}}{30 \text{ mL}}$$

4. Cancel units.

$$\frac{?}{60} = \frac{2 \text{ T}}{30}$$

5. Solve for the unknown.

$$30 \times ? = 2 \text{ T} \times 60$$

$$\frac{30 \times ?}{30} = \frac{2 \text{ T} \times 60}{30}$$

$$? = 4 \text{ T}$$

GO TO . . . Open the CD-ROM that accompanies your textbook, and select Chapter 3, Procedure Checklist 3-1. Review the animation and example problems, then complete the practice problems. Continue to the next section of the book once you have mastered the procedure checklist presented. ■

USING CONVERSION FACTORS

Ratio Proportion Method

By writing a conversion factor as a ratio, you can use means and extremes to convert a quantity from one unit of measure to another. Recall from Chapter 2, Rule 2-17, you can use means and extremes to find a missing value in ratio proportion, found on page 67.

Procedure Checklist 3-2

Converting by the Ratio Proportion Method

1. Write the conversion factor as a ratio $A : B$ so that A has the units needed in the answer.
2. Write a second ratio $C : D$ so that C is the missing value and D is the number that is being converted.
3. Write the proportion in the form $A : B :: C : D$.
4. Cancel units.
5. Solve the proportion by multiplying the means and extremes.

Example 1 Convert 66 lb to kilograms.

Follow Procedure Checklist 3-2.

1. Since we are converting to kilograms, our conversion ratio will have kilograms as the first part. Since 1 kg = 2.2 lb, our first ratio is 1 kg : 2.2 lb.
2. The second ratio is ? : 66 lb.
3. Our proportion is:

   1 kg : 2.2 lb : : ? : 66 lb
4. Cancel units.

   1 kg : 2.2 : : ? : 66
5. Solve for the missing value.

   $2.2 \times ? = 1 \text{ kg} \times 66$

   $$\frac{2.2 \times ?}{2.2} = \frac{1 \text{ kg} \times 66}{2.2}$$

   $? = 30 \text{ kg}$

Example 2 A patient needs to take 10 mL of a medication, but is going to be measuring the medication with a teaspoon. How many teaspoons should he use?

Follow Procedure Checklist 3-2.

1. We are converting to teaspoons, so they must appear in the first part of the conversion ratio. Our conversion ratio is 1 tsp : 5 mL.
2. The second ratio is ? : 10 mL.
3. Our proportion is:

   1 tsp : 5 mL : : ? : 10 mL
4. Cancel units.

   1 tsp : 5 : : ? : 10

5. Solve for the missing value.

$5 \times ? = 1 \text{ tsp} \times 10$

$\frac{5 \times ?}{5} = \frac{1 \text{ tsp} \times 10}{5}$

$? = 2 \text{ tsp}$

Example 3 A recipe from your aunt in Italy calls for 60 mL of olive oil. How many tablespoons of olive oil do you use?

Follow Procedure Checklist 3-2.

1. Our conversion factor is 2 T : 30 mL.
2. The second ratio is ? : 60 mL.
3. Our proportion is:

   2 T : 30 mL : : ? : 60 mL
4. Cancel units.

   2 T : 30 : : ? : 60
5. Solve for the missing value.

$30 \times ? = 2 \text{ T} \times 60$

$\frac{30 \times ?}{30} = \frac{2 \text{ T} \times 60}{30}$

$? = 4 \text{ T}$

GO TO . . . Open the CD-ROM that accompanies your textbook, and select Chapter 3, Procedure Checklist 3-2. Review the animation and example problems, then complete the practice problems. Continue to the next section of the book once you have mastered the procedure checklist presented. ■

USING CONVERSION FACTORS

Dimensional Analysis

The *dimensional analysis* (DA) method for using conversion factors is a modification of the fraction proportion and ratio proportion methods. Sometimes the DA method is also known as the *factor method* or *factor analysis*. When you are using DA, the unknown value ? stands alone on one side of the equation. The conversion factor multiplied by the number being converted is placed on the other side of the equation.

Procedure Checklist 3-3

Converting by Dimensional Analysis

1. Determine the unit of measure for the answer, and place it as the unknown on one side of the equation.
2. On the other side of the equation, write a conversion factor with the units of measure for the answer on top and the units you are converting from on the bottom.
3. Multiply the numerator of the conversion factor by the number that is being converted divided by 1.
4. Cancel the units on the right side of the equation. The remaining unit of measure on the right side of the equation should match the unknown unit of measure on the left side of the equation.
5. Solve the equation.

Example 1 Convert 66 lb to kilograms.

Follow Procedure Checklist 3-3.

1. The unit of measure for the answer is kilograms.

   ? kg =

2. Our conversion factor is $\frac{1\text{ kg}}{2.2\text{ lb}}$. Place it on the other side of an equal sign.

   $$? \text{ kg} = \frac{1\text{ kg}}{2.2\text{ lb}}$$

3. Multiplying the numerator of the conversion factor by 66 lb gives us the following equation.

   $$? \text{ kg} = \frac{1\text{ kg}}{2.2\text{ lb}} \times \frac{66\text{ lb}}{1}$$

4. Cancel units. The remaining unit on both sides is kilograms.

   $$? \text{ kg} = \frac{1\text{ kg}}{2.2\,\cancel{\text{lb}}} \times \frac{66\,\cancel{\text{lb}}}{1}$$

5. Solve for the unknown.

   ? = 30 kg

Example 2 A patient needs to take 10 mL of a medication, but is going to be measuring the medication with a teaspoon. How many teaspoons should he use?

Follow Procedure Checklist 3-3.

1. The unit of measure for the answer is teaspoons.

   ? tsp

2. Our conversion factor is $\frac{1\text{ tsp}}{5\text{ mL}}$. Place it on the other side of an equal sign.

   $$? \text{ tsp} = \frac{1\text{ tsp}}{5\text{ mL}}$$

3. Multiplying the numerator of the conversion factor by 10 mL gives us the following equation.

   $$? \text{ tsp} = \frac{1\text{ tsp}}{5\text{ mL}} \times \frac{10\text{ mL}}{1}$$

4. Cancel units. The remaining units indicate we are solving for teaspoons.

   $$? \text{ tsp} = \frac{1\text{ tsp}}{5\,\cancel{\text{mL}}} \times \frac{10\,\cancel{\text{mL}}}{1}$$

5. Solve for the unknown.

   ? = 2 tsp

Example 3 A recipe from your aunt in Italy calls for 60 mL of olive oil. How many tablespoons of olive oil do you use?

Follow Procedure Checklist 3-3.

1. The unit of measure for the answer is tablespoons.

   ? tbsp

2. Our conversion factor is $\frac{2 \text{ tbsp}}{30 \text{ mL}}$. Place it on the other side of an equal sign.

$$? \text{ tbsp} = \frac{2 \text{ tbsp}}{30 \text{ mL}}$$

3. Multiplying the numerator of the conversion factor by 60 mL gives us the following equation.

$$? \text{ tbsp} = \frac{2 \text{ tbsp}}{30 \text{ mL}} \times \frac{60 \text{ mL}}{1}$$

4. Cancel units. The remaining units indicate we are solving for tablespoons.

$$? \text{ tbsp} = \frac{2 \text{ tbsp}}{30 \cancel{\text{mL}}} \times \frac{60 \cancel{\text{mL}}}{1}$$

5. Solve for the unknown.

$$? = 4 \text{ tbsp}$$

GO TO . . . Open the CD-ROM that accompanies your textbook, and select Chapter 3, Procedure Checklist 3-3. Review the animation and example problems, then complete the practice problems. Continue to the next section of the book once you have mastered the procedure checklist presented. ■

Critical Thinking on the Job

Selecting the Correct Conversion Factor

Greg is teaching a patient how much liquid medication to take. The physician has ordered 30 mL of Milk of Magnesia, but the patient will be using teaspoons to measure her medication. Using a conversion chart, Greg confuses 1 tbsp with 1 tsp, and he reads 1 tsp = 15 mL. Using the incorrect information, he calculates the dose as follows:

$$\frac{?}{30 \text{ mL}} = \frac{1 \text{ tsp}}{15 \text{ mL}}$$

$$? \times 15 = 1 \text{ tsp} \times 30$$

$$? = 2 \text{ tsp}$$

Greg tells the patient to take 2 tsp of Milk of Magnesia. This amount is only one-third of the amount that the physician ordered; the patient does not get the relief desired.

Questions?

What mistake did Greg make? How could the error have been avoided? What is the correct dose the patient should receive?

REVIEW AND PRACTICE

3-4 Converting Among Metric, Apothecary, and Household Systems

In Exercises 1–11, convert the measures from one system of measurement to another. When necessary, round to the nearest tenth.

1. 4 dr = _______ tsp
2. 125 mL = _______ tsp
3. 120 mL = _______ tsp
4. 240 mL = _______ oz
5. 15 mg = gr _______
6. gr 15 = _______ mg
7. 10 mg = gr _______
8. 2.5 g = gr _______
9. 42 kg = _______ lb
10. 44 lb = _______ kg
11. 6 oz = _______ mL

12. During the total course of his treatment, a patient will receive 720 mL of medication. How many pints will he receive?

13. If an order calls for the patient to receive 2 tsp of cough syrup, how many milliliters of syrup should the patient receive?

14. A patient weighs 65 kg. How many pounds does she weigh?

15. A patient weighs 187 lb. How many kilograms does he weigh?

16. A physician orders 8 dr of liquid medication. How many tablespoons should the patient take?

17. A patient drinks 4 c of liquid during the morning. How many milliliters did the patient drink?

18. An order is for gr iii of medication. How many milligrams should the patient be given?

19. A physician orders that a patient be given 10 mg of medication 3 times per day. How many grains of medication should the patient be given per day?

20. An order calls for 1.5 tbsp of medicated mouthwash. How many milliliters of medicated mouthwash should the patient receive?

To check your answers, see page 505.

Temperature

Both the Fahrenheit (F) and Celsius (C) temperature scales are used in health care settings. The Celsius temperature scale is also known as the Centigrade (C) temperature scale. If you examine the two thermometers in Figure 3-4, you will notice that the Fahrenheit scale sets the temperature at which water freezes at 32 degrees, or 32°F. It also measures the temperature at which water boils as 212°F. On the Celsius scale, water freezes at 0°C and boils at 100°C. In Fahrenheit, average body temperature is 98.6°F. In Celsius, average body temperature is 37°C.

As a health care worker, you may need to convert between these two temperature scales. The following formula can be used for converting between the two systems:

$$5F - 160 = 9C$$

In this formula, F represents the temperature in degrees Fahrenheit and C represents the temperature in degrees Celsius.

You may also use these formulas to convert between temperature scales.

From Fahrenheit to Celsius use:

$$\frac{°\text{F} - 32}{1.8} = °\text{C}$$

From Celsius to Fahrenheit use:

$$(1.8 \times °\text{C}) + 32 = °\text{F}$$

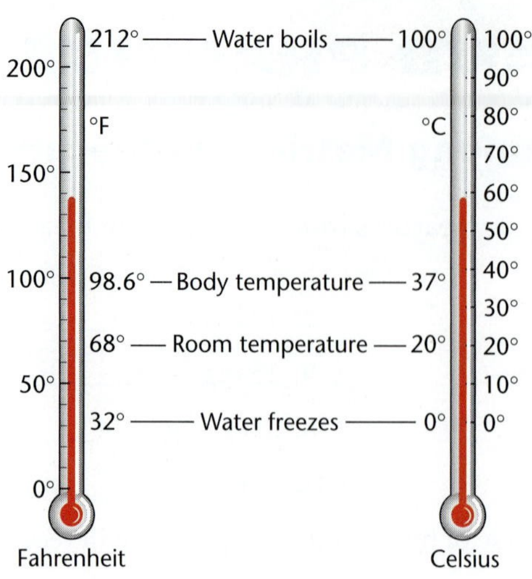

Figure 3-4 Fahrenheit and Celsius scales.

Rule 3-8 Converting Between Temperature Systems

Use any of the following formulas to convert a temperature from Fahrenheit to Celsius.

$5F - 160 = 9C$ $\qquad \frac{°F - 32}{1.8} = °C$ $\qquad (1.8 \times °C) + 32 = °F$

Example 1 Convert 98.6°F to degrees Celsius.

Substituting 98.6 for *F* in the first formula gives us:

$$(5 \times 98.6) - 160 = 9C \quad \text{(Multiply before subtracting.)}$$
$$493 - 160 = 9C$$
$$333 = 9C \quad \left(\text{Since } \tfrac{9}{9} \text{ equals 1, divide both sides by 9 to solve for } C.\right)$$
$$\frac{333}{9} = \frac{9C}{9}$$
$$37 = C$$

Thus 98.6°F = 37°C; both measures represent normal body temperature.

Example 2 Convert 100°C to degrees Fahrenheit.
Substituting 100 for *C* in the first formula gives us:

$$5F - 160 = 9 \times 100$$
$$5F - 160 = 900 \quad \text{(Add 160 to both sides.)}$$
$$5F = 900 + 160$$
$$\frac{5F}{5} = \frac{1060}{5} \quad \left(\text{Since } \tfrac{5}{5} \text{ equals 1, divide both sides by 5 to solve for } F.\right)$$
$$F = 212$$

So 100°C = 212°F; both measures represent the boiling point of water.

Example 3 Convert 98.6°F to degrees Celsius.

Substituting 98.6 for °F in the second formula gives:

$$\frac{98.6 - 32}{1.8} = \frac{66.6}{1.8} = 37$$

So 98.6°F = 37°C; both measures represent normal body temperature.

Example 4 Convert 37°C to degrees Fahrenheit.

Substituting 37 for °C in the third formula gives:

$$(1.8 \times 37) + 32 = 66.6 + 32 = 98.6$$

Thus 37°C = 98.6°F.

GO TO . . . Open the CD-ROM that accompanies your textbook, and select Chapter 3, Rule 3-8. Review the animation and example problems, then complete the practice problems. Continue to the next section of the book once you have mastered the rule presented. ■

REVIEW AND PRACTICE

3-5 Temperature

In Exercises 1–10, convert the temperatures. Round to the nearest tenth, when necessary.

1. 34°C = ________°F **2.** 41°C = ________°F **3.** 95°F = ________°C

4. 102°F = ________°C **5.** 45.3°F = ________°C **6.** 212°F = ________°C

7. 25°C = ________°F **8.** 100°C = ________°F **9.** 59°F = ________°C

10. 67°C = ________°F

To check your answers, see page 505.

Time

Many health care facilities use the clock with 24 hours (h), known as military or international time. A traditional 12-h clock is a source of errors in administering medication. On the 12-h clock, each time occurs twice a day. For instance, the hour 10:00 is recorded as both 10:00 a.m. and 10:00 p.m. The abbreviation "a.m." means *ante meridian* or before noon; "p.m." means *post meridian* or after noon. If these abbreviations are not clearly marked, the patient could receive medication at the wrong time.

The 24-h clock (international time) bypasses this opportunity for error. Each time occurs only once per day. In international time, 10:00 a.m. is written as 1000, whereas 10:00 p.m. is written as 2200. (See Figure 3-5.)

Figure 3-5 International time is based on a 24-h clock.

When you write the time using a 12-h clock, you separate the hour from the minutes by a colon. You write a single digit for hours 1 through 9 You then add a.m. or p.m. to indicate before or after noon. When you write the time using a 24-h clock, you use a four-digit number with no colon. The first two digits represent the hour; the last two digits, the minutes.

Rule 3-9 When you are using a 24-h clock for international time:

1. Write 00 as the first two digits to represent the first hour after midnight.
2. Write 01, 02, 03, . . . , 09 as the first two digits to represent the hours 1:00 a.m. through 9:00 a.m.
3. Add 12 to the first two digits to represent the hours 1:00 p.m. through 11:00 p.m., so that 13, 14, 15, . . . , 23 represent these hours.
4. Write midnight as either 2400 (international) or 0000 (military time).

Example 1 Convert 9:00 a.m. to international time.

Remove the colon and the abbreviation a.m. Write the hour 9 with two digits, starting with zero.

9:00 a.m. = 0900

Example 2 Convert 12:19 a.m. to international time.

Remove the colon and the abbreviation a.m. Because this time occurs in the first hour after midnight, use 00 for the hour.

12:19 a.m. = 0019

Example 3 Convert 4:28 p.m. to international time.

Remove the colon and the abbreviation p.m. Because this time is after noon, add 12 to the hour.

4:28 p.m. = 1628

Example 4 Convert 1139 to traditional time.

Insert a colon to separate the hour from the minutes. Because this time occurs before noon, add a.m. following the time.

1139 = 11:39 a.m.

Example 5 Convert 1515 to traditional time.

Insert a colon to separate the hour from the minutes. Subtract 12 from the hour, and add the abbreviation p.m.

1515 = 3:15 p.m.

GO TO . . . Open the CD-ROM that accompanies your textbook, and select Chapter 3, Rule 3-9. Review the animation and example problems, then complete the practice problems. Continue to the next section of the book once you have mastered the rule presented. ■

Rule 3-10 To state the time using international time:

1. Say *zero* if the first digit is a zero.
2. Say *zero zero* if the first two digits are both zero.
3. If the minutes are represented by 00, then say *hundred* after you say the hour.

Example 1 State the time 0900.

Say *zero nine* for the hours and *hundred* for the minutes. Thus, 0900 is stated as *zero nine hundred.*

Example 2 State the time 1139.

Say *eleven* for the hours and *thirty-nine* for the minutes. Thus, 1139 is stated as *eleven thirty-nine.*

Example 3 State the time 0023.

Say *zero zero* for the hours and *twenty-three* for the minutes. Thus, 0023 is stated *zero zero twenty-three.*

GO TO . . . Open the CD-ROM that accompanies your textbook, and select Chapter 3, Rule 3-10. Review the animation and example problems, then complete the practice problems. Continue to the next section of the book once you have mastered the rule presented. ■

REVIEW AND PRACTICE

3-6 Time

In Exercises 1–10, convert the times to international time.

1. 2:35 a.m.	**2.** 7:57 a.m.	**3.** 12:08 a.m.	**4.** 12:55 a.m.	**5.** 1:49 p.m.
6. 3:14 p.m.	**7.** 11:54 p.m.	**8.** 10:19 p.m.	**9.** 6:59 p.m.	**10.** 4:26 a.m.

In Exercises 11–20, convert the times to traditional time.

11. 0011	**12.** 0036	**13.** 0325	**14.** 0849	**15.** 1313
16. 1527	**17.** 2145	**18.** 2359	**19.** 2037	**20.** 1818

To check your answers, see page 506.

Workspace

HOMEWORK ASSIGNMENT

Perform the following conversions

1. 3.5 g = ________ mg
2. 30 mg = ________ gr
3. 41°F = ________°C
4. 0.25 mg = ________ mcg
5. 110 lb = ________ kg
6. 3 tsp = ________ mL
7. 180 mL = ________ oz
8. 1.2 mm = ________ cm
9. 5 gr = ________ mg
10. 36°C = ________°F

Convert the following times to international time.

11. 5:30 p.m.
12. 11:15 a.m.

Convert the following times to traditional time.

13. 0730
14. 1234
15. Bill has lost 14.6 pounds. Fred has lost 6.2 kilograms. Who has lost more weight?

Chapter 3 Review

Check Up

In Exercises 1–8, write the indicated amounts, using numerals and abbreviations.

1. Twenty-five and one-half kilograms
2. Forty-five hundredths of a centimeter
3. Forty micrograms
4. Three-quarters of a liter
5. Nine-tenths of a milligram
6. One and one-half grains
7. Three hundred seventy-five thousandths of a gram
8. Twelve milliliters

In Exercises 9–30, calculate the conversions.

9. 0.06 g = ________ mg
10. 125 mcg = ________ mg
11. 0.004 km = ________ m
12. 0.75 cm = ________ mm
13. 965 mL = ________ L
14. 0.008 L = ________ mL
15. 0.32 kg = ________ g
16. 0.05 mg = ________ mcg
17. 988 m = ________ km
18. 1725 cm = ________ km
19. 368 mg = ________ g
20. 247 g = ________ kg
21. 8 g = gr ________
22. gr iiss = ________ mg
23. 90 mL = ________ tbsp
24. 5 tsp = ________ mL
25. 8 dr = ________ mL
26. 1200 mL = ________oz
27. 540 mg = gr ________
28. gr $\frac{3}{4}$ = ________ mg
29. 178.2 lb = ________ kg
30. 47 kg = ________ lb
31. An order is placed for gr v of medication. If the medication is supplied in milligrams, how many milligrams should be given? (For this example, assume gr i = 65 mg.)
32. If a patient weighs 44 lb, how many kilograms does she weigh?
33. A physician orders $\frac{1}{2}$ oz of medication for a patient. How many milliliters of medication should the patient be given?
34. The maximum dose of a medication is 3 tbsp. What is the maximum number of milliliters that the patient should be given?
35. A physician tells a patient to drink 2400 mL of fluid per day. How many quarts of liquid should this patient drink?
36. Several months ago, a patient weighed 95 kg. When he comes in for his next appointment, he tells you he has lost 11 lb. If he is correct, how many kilograms should he weigh?

Convert the following temperatures to Celsius. Round to the nearest tenth, when necessary.

37. 97.6°F **38.** 72°F **39.** 57.4°F **40.** 82.8°F

Convert the following temperatures to Fahrenheit. Round to the nearest tenth, when necessary.

41. 24°C **42.** 43.8°C **43.** 15.6°C **44.** 8.8°C

Convert the following times to international time.

45. 3:21 a.m. **46.** 4:42 p.m. **47.** 10:47 p.m. **48.** 11:20 a.m.

Convert the following times to traditional time.

49. 0029 **50.** 1417 **51.** 2053 **52.** 0912

Critical Thinking Applications

A patient suffering from a cold is given a prescription for 15 mL of cough suppressant every 8 h for 10 days. The patient will be using a household measuring device to measure the dose of medication.

1. Which household device should the patient use?
2. How much medication should the patient take, given the device you recommend?
3. If the medication is supplied in half-pint bottles, how many bottles will the patient need during the 10 days?

Case Study

The state health department requires that certain medications be stored between 36°F and 41°F. The refrigerator in the medication room has a Celsius thermometer. What temperature range is appropriate, using the Celsius thermometer?

To check your answers, see page 506.

Internet Activity

Find a reliable metric conversion chart on the Internet, and use it to convert the following.

1. Your weight in pounds to your weight in kilograms.
2. A temperature of 98.2°F to degrees Celsius.
3. A 500-mg dose of medication to grams.
4. A 2-tbsp dose of medication to ounces.

Go To . . . Open the CD-ROM that accompanies your textbook, and complete a final review of the rules, practice problems, and activities presented for this chapter. For a final evaluation, take the chapter test and email or print your results for your instructor. A score of 95 percent or above indicates mastery of the chapter concepts. ■

4 Equipment for Dosage Measurement

Nothing will work unless you do.

—Maya Angelou

Learning Outcomes

When you have completed Chapter 4, you will be able to:

- Identify equipment used to administer medication.
- Indicate the appropriate equipment for delivering various types of medicine.
- Measure medications by using the calibrations on the equipment.
- Describe the method of administration appropriate for each piece of measuring equipment.

Key Terms

Ampule
Calibrated spoons
Calibrations
Cartridges
Eccentric
Hypodermic syringes
Instillations
Jejunostomy tube
Leading ring
Meniscus
Metered dose inhaler
Nasogastric
Parenteral
PEG tube
Trailing ring
Transdermal
Vial

Introduction

To prepare the correct dosage of medications, you must know the equipment you will be using. You will be required to accurately select and read this equipment. This chapter will introduce you to common equipment used to prepare dosages and administer medications.

Oral Administration

Many medications are available in liquid form and can be administered orally. Several types of equipment are used to measure and administer oral liquid medications. These include medicine cups, droppers, calibrated spoons, and oral syringes.

Each measuring device has a series of **calibrations**, or marks numbered at varying intervals. Calibrations enable you to measure the amount of liquid in the device. When you choose a measuring device, compare its calibrations with the desired dose of medication. They may represent different units of measurement. If your equipment does not match the order, then you will have to convert the order to the unit of measurement you will use to administer the medication. For example, a patient is required to take 10 mL of a medication. You have available a container that is marked in teaspoons only. In this case you will have to obtain a different container that is marked in milliliters or convert the amount of medication (10 mL) to teaspoons.

Medicine Cups

Medicine cups are used to measure oral liquid medications and administer them to patients. Cups provide a measured dose that is easy for most patients to swallow. Usually, medicine cups are plastic and measure up to 1 fluid ounce, or its equivalent. To make dose calculation easier, most cups are typically marked with metric, household, and apothecary systems of measurement. Thus, cups include units such as tablespoons (tbs), teaspoons (tsp), milliliters (mL), drams (dr), and ounces (oz).

Use the two views of a cup shown in Figure 4-1 to compare the different calibrations. In Figure 4-1A, milliliters are displayed in units of 5. Teaspoons and tablespoons are marked in units of 1 or $\frac{1}{2}$ On the other side of the cup in Figure 4-1B, ounces are displayed in units of $\frac{1}{8}$ or $\frac{1}{4}$, and drams are marked in units of 1 or 2. You can see that 5 mL is equivalent to 1 tsp. The slight curve in the surface of a liquid is the **meniscus.** The quantity of liquid is measured at the bottom of the meniscus, not by the higher levels at the edges. (Refer to Figure 4-2.)

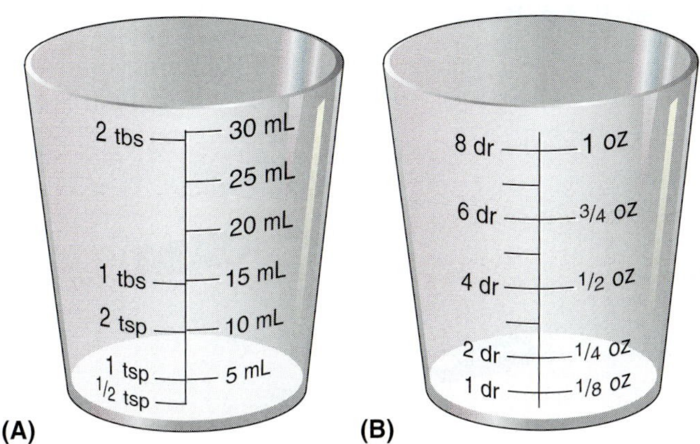

Figure 4-1 Two views of a medicine cup.

Rule 4-1 Do not use medicine cups for doses less than 5 mL, even if the cup has calibrations smaller than 5 mL. Instead, use a dropper, calibrated spoon, or oral syringe to ensure accuracy.

GO TO . . . Open the CD-ROM that accompanies your textbook, and, select Chapter 4, Rule 4-1. Review the animation and example problems, then complete the practice problems. Continue to the next section of the book once you have mastered the rule presented. ■

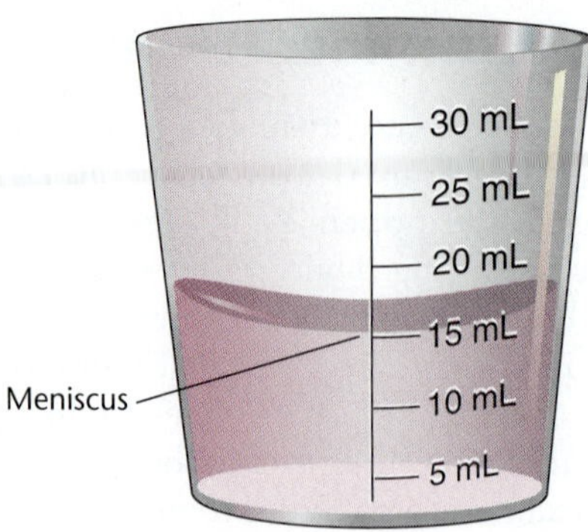

Figure 4-2 Fluid in a medicine cup should be measured at the meniscus.

Droppers

Droppers help you measure and administer small amounts of oral liquid medication. You may also use them to deliver certain liquid medications to the eyes, ears, and nose. Droppers are especially helpful with oral pediatric doses. A product that requires a dropper is often packaged with one calibrated for the specified dose. The indicated units of measurement (calibrations) are usually milliliters (mL), cubic centimeters, drops, or even teaspoons. (Recall that 1 mL = 1 cc.) See Figure 4-3 for various types of droppers.

Droppers have different-size openings. The diameter of the opening affects the size of the individual drops. For example, 3 drops from a dropper with a large opening provides more medication than 3 drops from one with a smaller opening. So do not interchange droppers that are packaged with medications. However, separate calibrated droppers can be reused if properly cleaned between uses.

Calibrated Spoons

In some cases, you can deliver small amounts of medication by using **calibrated spoons** (see Figure 4-4). Spoons are often used with pediatric or elderly patients. They come in many sizes, calibrated to a variety of doses.

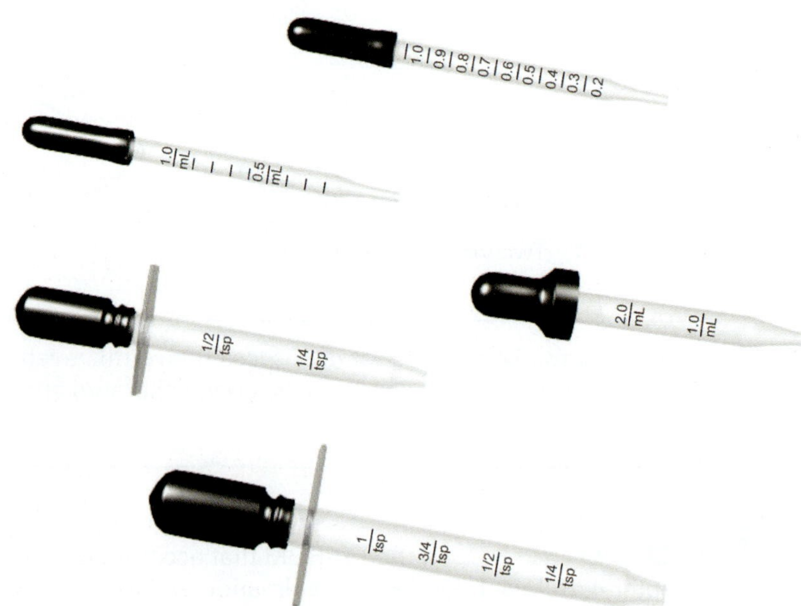

Figure 4-3 Droppers come in various sizes with different calibrations.

You can use the spoons to administer medication directly into the mouth. You can also use them to measure medication into food or a beverage for a child or elderly patient. Children who are used to being fed from a household spoon may accept medication if it comes from a calibrated spoon rather than from a dropper or a medicine cup. You can also use spoons for thick liquids that cannot be easily delivered through the small openings of a dropper.

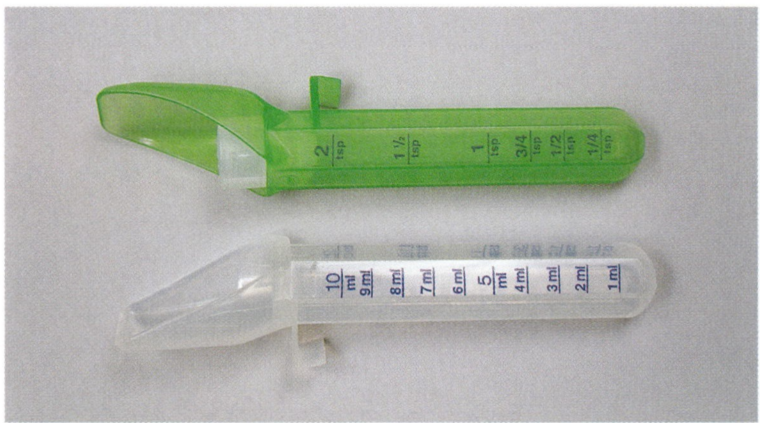

Figure 4-4 Calibrated spoons.

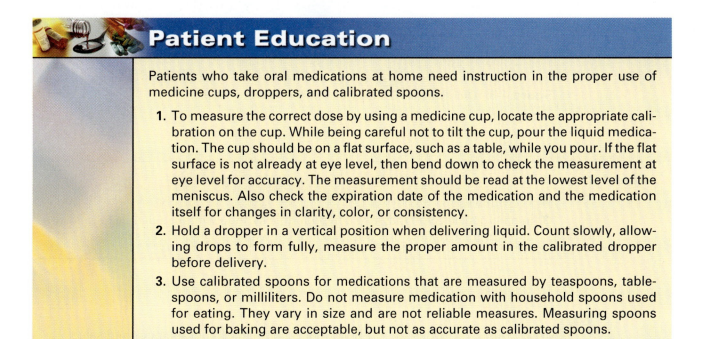

Patient Education

Patients who take oral medications at home need instruction in the proper use of medicine cups, droppers, and calibrated spoons.

1. To measure the correct dose by using a medicine cup, locate the appropriate calibration on the cup. While being careful not to tilt the cup, pour the liquid medication. The cup should be on a flat surface, such as a table, while you pour. If the flat surface is not already at eye level, then bend down to check the measurement at eye level for accuracy. The measurement should be read at the lowest level of the meniscus. Also check the expiration date of the medication and the medication itself for changes in clarity, color, or consistency.
2. Hold a dropper in a vertical position when delivering liquid. Count slowly, allowing drops to form fully, measure the proper amount in the calibrated dropper before delivery.
3. Use calibrated spoons for medications that are measured by teaspoons, tablespoons, or milliliters. Do not measure medication with household spoons used for eating. They vary in size and are not reliable measures. Measuring spoons used for baking are acceptable, but not as accurate as calibrated spoons.

Oral Syringes

For small quantities of liquid, especially less than 5 mL, oral syringes provide accurate readings. Generic oral syringes are often calibrated for milliliters and teaspoons, with additional calibrations between these numbers (see Figure 4-5). Oral syringes are designed with safety features to keep them from being confused with hypodermic syringes. Oral syringes often have **eccentric,** or off-center, tips that have a different shape and size than the tips of hypodermic syringes. Oral syringes may be tinted, whereas hypodermic syringes are clear.

Oral syringes are not to be confused with hypodermic syringes. Oral syringes are not sterile. Oral syringes include a small cap that must be removed before administering medication orally to prevent choking. To administer sterile medications, you must use a sterile hypodermic syringe.

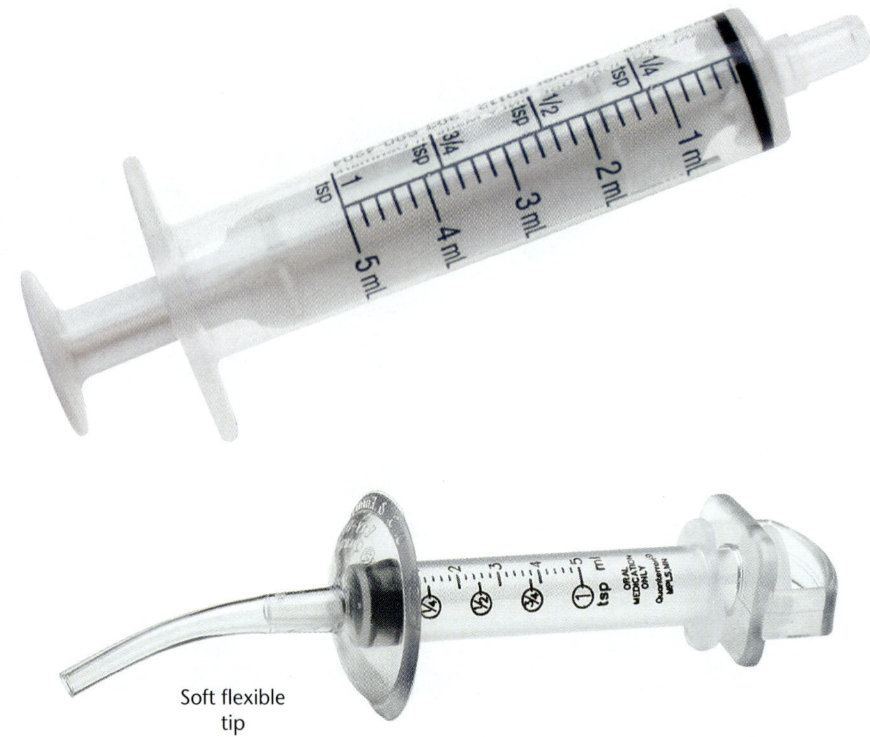

Figure 4-5 Oral syringes. Courtesy of Apothecary Products, Inc.

Rule 4-2

1. Never attach a hypodermic needle to an oral syringe.
2. Never inject an oral dose.
3. In emergencies, you may use a hypodermic syringe without a needle to measure and administer liquid oral doses, but never while its needle is attached.

GO TO . . . Open the CD-ROM that accompanies your textbook and select Chapter 4, Rule 4-2. Review the animation and example problems, then complete the practice problems. Continue to the next section of the book once you have mastered the rule presented. ■

Other Equipment for Oral Medications

Sometimes oral medications, intended for absorption in the stomach or intestines, cannot be delivered orally. The patient may have difficulty swallowing or some condition or trauma that prevents taking the medication orally.

Liquid medications are preferred for **nasogastric** tubes, which delivers medication into the stomach (see Figure 4-6). Sometimes, you can crush a solid medication, adding water to

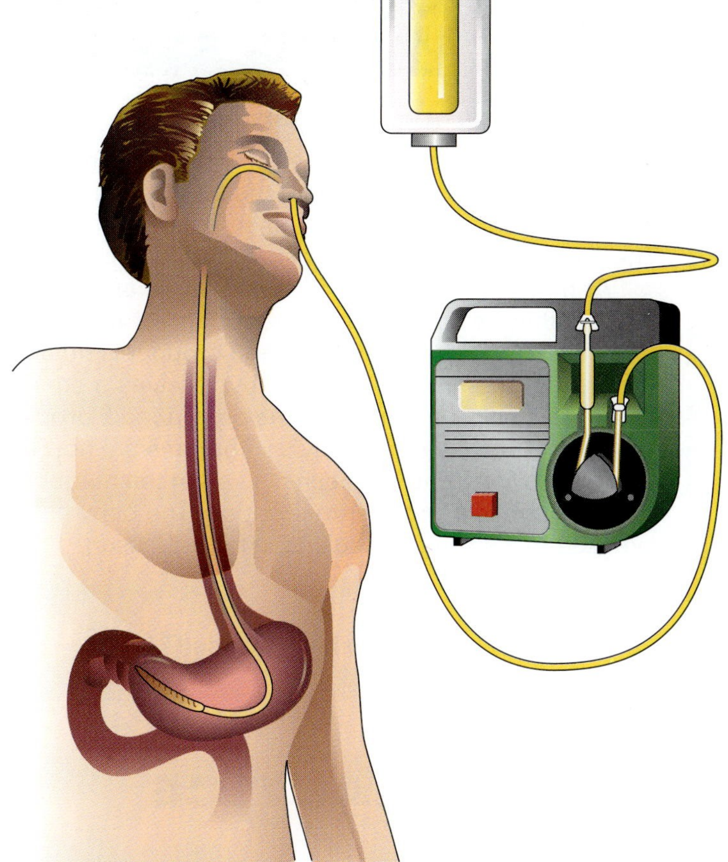

Figure 4-6 A nasogastric tube delivers medication into the stomach.

transport it through the tube. Consult your facility for the appropriate procedure. Many solid oral medications, such as gelcaps and extended-release medications, may *not* be crushed. See Chapter 8 for further discussion about these medications.

Other tubes are used to deliver medications directly to the stomach or intestines. A nasal-small-bowel tube extends from the nose to the small intestines. A **percutaneous (through the skin) endoscopic (PEG) tube** delivers medication and nutrients directly to the stomach. A **jejunostomy tube** delivers medication and nutrients directly to the small intestine.

No matter what equipment is used to administer medication, it must be measured accurately using the equipment's calibrations.

ERROR ALERT!

The utensil you use must provide the calibration you need to accurately measure the dose.

Suppose the volume of the dose is less than 0.5 mL. The calibrations on the utensil must measure increments of less than 0.5 mL. Otherwise, you cannot accurately measure the medication. Using a utensil that is marked in 1-mL increments and estimating the halfway point is not accurate.

Critical Thinking on the Job

Use the Correct Dropper

A baby with a fever is prescribed acetaminophen for home administration. The health care professional tells the baby's father that he will be given a bottle of liquid acetaminophen and a medicine dropper for measuring the prescribed number of drops. The father is told to give the baby 8 drops of medicine at regular intervals.

At home, the father accidentally drops and breaks the dropper. He remembers, though, that he has a dropper from another medication. He uses it instead to measure the acetaminophen. The second dropper is much smaller than the one that came with the acetaminophen. The baby receives a smaller dose than prescribed, even though the father administers the prescribed 8 drops.

Two nights later, the baby's symptoms are not relieved. The father calls the physician, who asks if he has been delivering the number of prescribed drops. The physician prescribes a stronger medication. The baby has suffered needlessly and now is exposed to a stronger medication for no reason.

Question?

How could this error have been avoided?

REVIEW AND PRACTICE

4-1 Oral Administration

In Exercises 1–14, determine if the statement is true or false. If false, explain why it is incorrect.

1. You may use a hypodermic syringe with a needle to measure liquid for oral administration.
2. You may use a medicine cup to measure liquid doses of less than 1 mL for oral administration.
3. Oral and hypodermic syringes are identical in appearance.
4. When you measure liquid in any type of utensil for oral administration, the dose volume must be level with the corresponding calibration line on the device.
5. If the dropper supplied by a drug manufacturer for a specific medication is not available, you may substitute a dropper supplied for another medication, as long as the replacement dropper has never been used.
6. If a patient does not have a calibrated spoon, then any household spoon may be substituted.
7. Measuring utensils are often calibrated with more than one system of measurement.
8. A prescribed dose of liquid oral medication cannot be dispensed reliably without calibrated cups, spoons, oral syringes, or droppers.
9. When you calculate volume and dosage, it is helpful to remember that 1 mL is equal to 1.5 cubic centimeters.
10. If a prescribed dose and the calibrated device used for administering that dose use different systems of measurement, then the device cannot be used.
11. Measure the quantity of liquid in a measuring cup by the higher levels at the edge.
12. Droppers may be used to deliver certain liquid medications to the eyes and ears but not the nose.
13. Measuring spoons used for baking are acceptable for measuring liquid medication.
14. Gelcaps and extended-release medications may be delivered through a nasogastric tube as long as they are crushed and flushed with water.

In Exercises 15–20, convert the dosage ordered to the same units as those marked on calibrated utensils. You may wish to refer to the conversion factors in Chapter 3. There may be more than one correct answer for each conversion.

15. The prescribed dose is 2 tbs. Which of the following is *not* equivalent, as marked on the medicine cup?

a. 30 mL **b.** 1 oz **c.** 8 dr **d.** 15 mL

16. An oral medication comes in a bottle labeled 10 units per cubic centimeter. The dose to be administered is 50 units. Which of the following is a correct dose?

a. 50 mL **b.** 500 mL **c.** 5 mL **d.** 0.5 mL

17. Which of the following statements about calibrated droppers is true?

a. 0.25 mL equals 25 drops.
b. 0.25 mL equals 5 drops.
c. 1 mL equals 10 drops.
d. The number of drops in each milliliter varies per dropper.

18. The dose to be given is 5 mL. The medicine cup is labeled in tablespoons, teaspoons, ounces, and drams. Which of the following is a correct dose?

a. 5 dr **b.** $\frac{1}{3}$ tbs **c.** $\frac{1}{3}$ tsp **d.** $\frac{1}{3}$ oz

19. An oral medication comes in a bottle labeled 200 mg per 5 mL. The dose to be administered is 600 mg. Which of the following is a correct dose?

a. 1 tsp **b.** 2 tsp **c.** 1 tbs **d.** 2 tbs

20. The dose to be given is 15 mL. The dose cup is labeled in tablespoons, teaspoons, and ounces. Which of the following is a correct dose?

a. 3 tsp **b.** 1 tbs **c.** $\frac{1}{2}$ oz **d.** All the above

To check your answers, see page 507.

Hypodermic Syringes

Many medications must be administered **parenterally**, bypassing the digestive tract. (*Parenterally* means "outside the intestines.") While parenteral dosage forms include topical and transdermal medications, inhalers, and sublingual tablets, the term most often refers to injections. The most common injection routes are *intravenous* (IV), *intramuscular* (IM), *intradermal* (ID), and *subcutaneous* (SC). See Chapters 9 and 10 for more information about these methods of injection. Different **hypodermic syringes** are used to administer injections. These syringes are calibrated with different measurements. The type of syringe you use depends on the type and amount of medication to be administered. Remember 1 milliliter (mL) equals 1 cubic centimeter (cc).

Standard Syringes

The 3-mL syringe is one of the most common standard syringes used for parenteral administration. Standard syringes have scales calibrated in milliliters. Syringes with smaller capacities may have divisions of tenths, two-tenths, or even hundredths of a milliliter, allowing for measurement of small doses. The 3-mL syringe in Figure 4-7 is calibrated in tenths; it has 10 calibrations for each milliliter. Calibrations for half and whole milliliters are numbered. Standard syringes may also be marked with a minim scale from the apothecary system. However, the metric system is almost always used.

Any health care worker who uses a syringe must be familiar with its calibrations so that the correct dose is administered. On all syringes, the zero calibration is the edge of the barrel closest to the needle. The barrel is filled with liquid up to the point of the wide ring, known as the **leading ring**, on the tip of the plunger closest to the needle. Liquid in the barrel does not go past this ring. While the leading ring might have a raised middle, measure from the ring itself. Do *not* measure from **the trailing ring**, which is the ring farther from the needle.

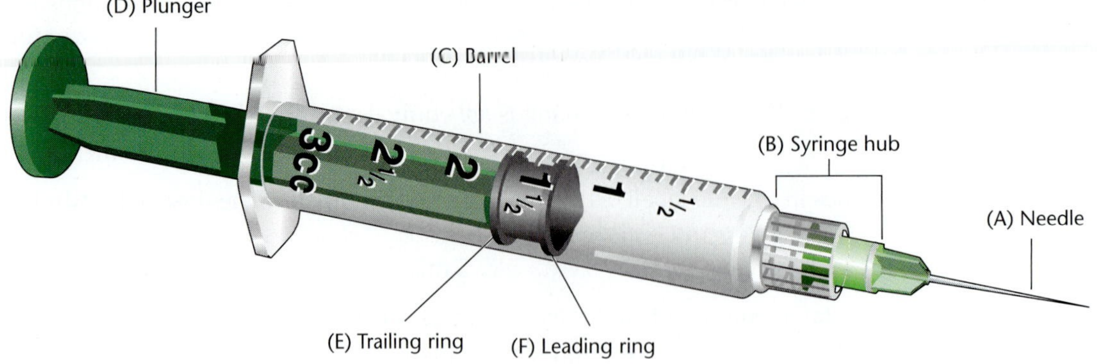

Figure 4-7 The parts of a standard syringe include (A) the needle, (B) the syringe hub, (C) the barrel that contains the liquid, (D) the plunger, (E) the trailing ring, and (F) the plunger tip, also called the leading ring.

Safety Syringes

Safety syringes have the same components as their standard counterparts: a needle, a hub, a barrel, and a plunger. See Figures 4-8 and 4-9. However, their needles are protected by plastic shields. These shields help prevent needlestick injury. Although safety syringes do not guarantee that health care workers will not receive accidental needlesticks, the syringes reduce the chances of such accidents. Safety syringes come in all sizes with various calibrations. They are calibrated in milliliters or units. Smaller-capacity safety syringes are divided into tenths, two-tenths, and hundredths of a milliliter.

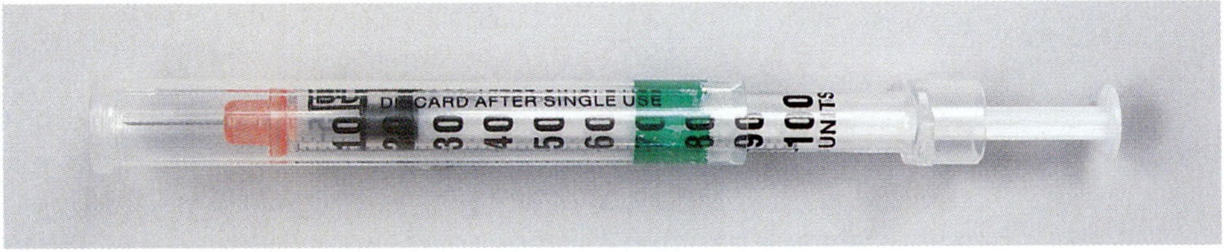

Figure 4-8 This 100-unit insulin safety syringe has a shield that covers the needle, minimizing needlestick injuries.

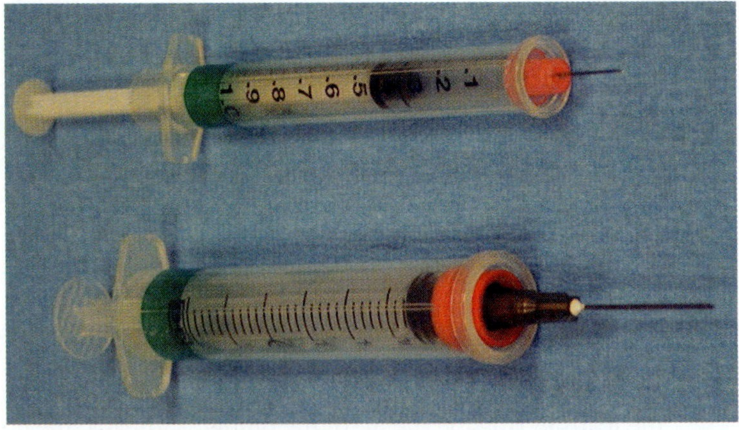

Figure 4-9 All syringes should have a safety mechanism. The syringes pictured here include a protective sheath that is pushed over the needle after use.

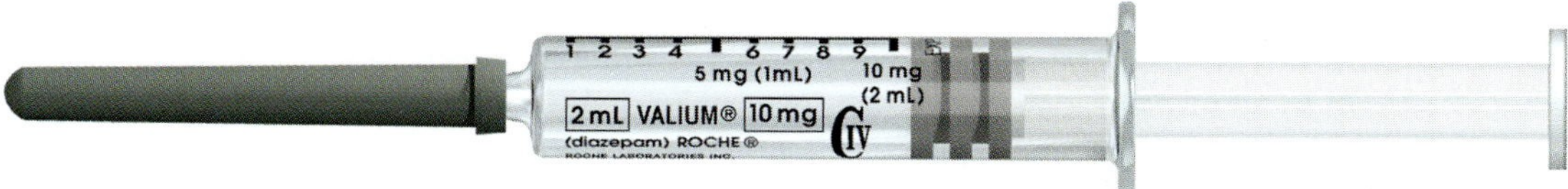

Figure 4-10 Prefilled syringe. Notice the calibration marks indicate milligrams.

Prefilled Syringes

Prefilled syringes are shipped from the manufacturer filled with a single dose of medication. If the patient is given only a portion of the dose, the remainder must be discarded before the medication is administered. Prefilled syringes have the same parts as a standard syringe: a needle, a syringe hub, a barrel, and a plunger (see Figure 4-8). Prefilled syringes are sometimes marked in units other than milliliters (mL). For example, the Valium syringe in Figure 4-10 contains a total of 10 mg of medication in 2 mL of solution. Note that the syringe is marked in units from 1 to 10, and it indicates milligrams of drug rather than milliliters. It is also marked showing one and 2 mL of medication. Since the dose markings are on the syringe, it would not be necessary to convert the dosage to milliliters, which simplifies the use of the product. If an order calls for 6 mg, you discard the excess by pushing the plunger until the leading ring indicates 6. You then inject the remaining contents of the syringe.

Rule 4-3 When you are using a prefilled syringe, always examine the markings to determine whether the syringe is calibrated in milliliters or milligrams, and calculate the dose accordingly.

GO TO . . . Open the CD-ROM that accompanies your textbook and select Chapter 4, Rule 4-3. Review the animation and example problems, then complete the practice problems. Continue to the next section of the book once you have mastered the rule presented. ■

Insulin Syringes

Insulin syringes are used only to measure and administer insulin. Insulin is measured in units. Insulin syringes are unique. They are calibrated in the amount of medication in units rather than calibrated by volume (milliliters). Whether for adults or children, insulin doses are smaller than many other doses. In turn, insulin syringes are calibrated in smaller increments. The most common form of insulin is U-100; it contains 100 units of insulin per 1 mL.

Rule 4-4 Never use any type of syringe other than an insulin syringe to measure and administer insulin.

Figure 4-11 shows a standard U-100 insulin syringe. It contains up to 100 units (1 mL) of insulin. The larger numbers mark increments of 10 units. The smaller calibrations indicate every 2 units. Many 100-unit syringes have 2 scales–one on the right showing the even number unit line and one on the left showing the odd number unit lines. Not all insulin syringes are marked with these increments. Figure 4-12 shows a syringe that holds up to 50 units of insulin. The larger numbers show increments of 5 units, and the smaller calibrations show increments of 1 unit. This syringe is often used to measure and administer pediatric or adult doses of insulin that are less than 50 units. (See Figure 4-13 for a comparison.)

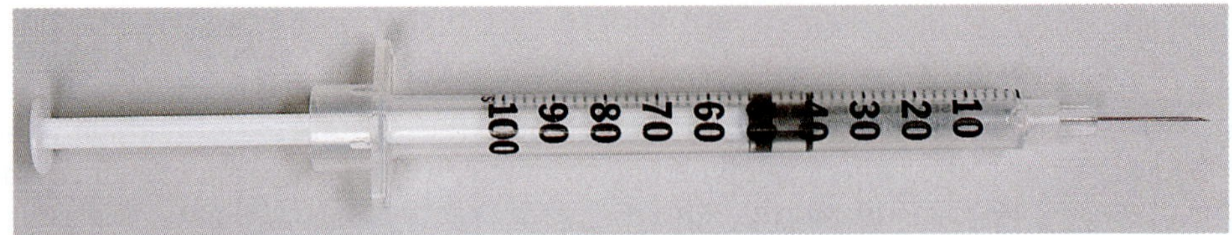

Figure 4-11 A standard insulin syringe can hold up to 100 units of insulin. (Picture for comparison only. All syringes should include a safe-needle device to prevent accidental needle sticks.)

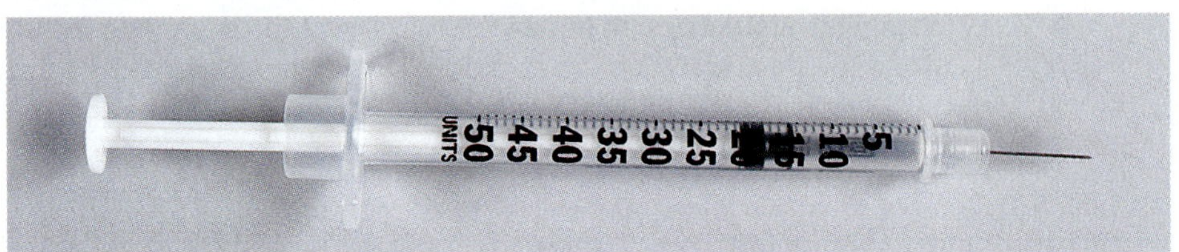

Figure 4-12 This insulin syringe can hold up to 50 units of insulin. (Picture for comparison only. All syringes should include a safe-needle device to prevent accidental needle sticks.)

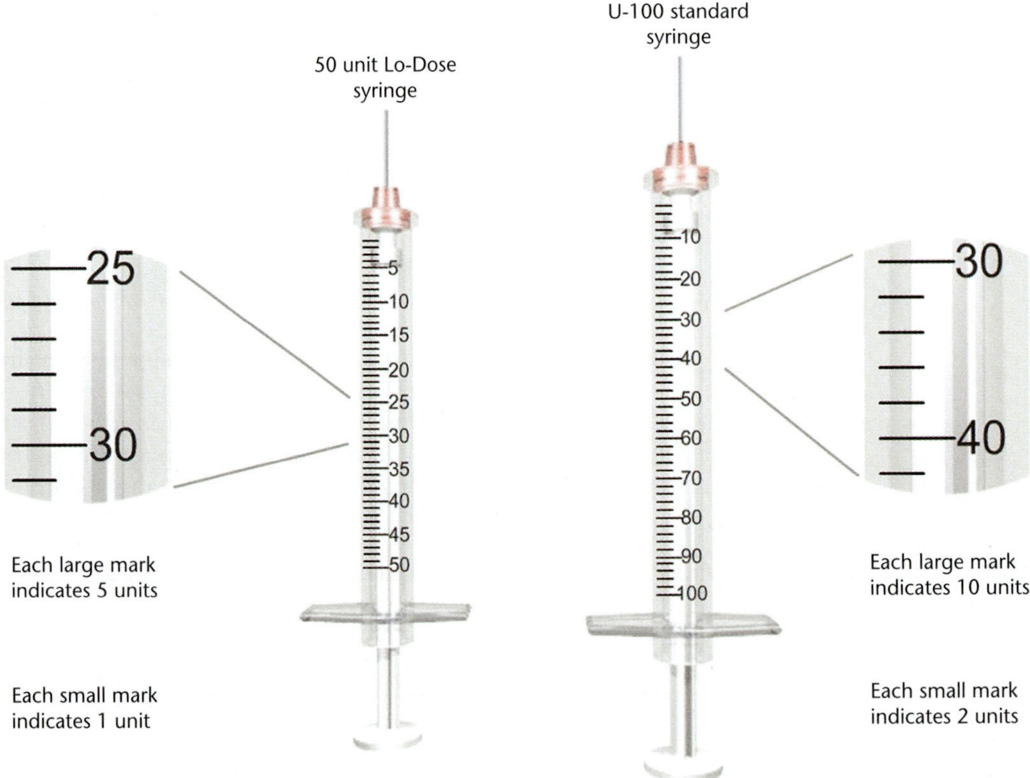

Figure 4-13 When comparing a 50 unit versus standard U-100 insulin syringe, always check the calibrations carefully.

GO TO . . . Open the CD-ROM that accompanies your textbook and select Chapter 4, Rule 4-4. Review the animation and example problems, then complete the practice problems. Continue to the next section of the book once you have mastered the rule presented. ■

Tuberculin Syringes

Tuberculin syringes are used to administer subcutaneous injections as well as the intradermal purified protein derivative (PPD) skin test that determines if a person has been exposed to tuberculosis. More than that, they are simply small syringes used when small doses of medication—less than 1 mL—are administered. Vaccines, heparin, pediatric medicines, and allergen extracts are typically administered with a tuberculin syringe.

Tuberculin syringes usually hold a total volume of 1 mL and are calibrated in hundredths of a milliliter. The numbering is slightly different from that of the other syringes. In Figure 4-14, the marked numbers represent tenths of a milliliter. The first number, located on the tenth calibration, is 0.1 mL. Each smaller calibration represents one-hundredth (0.01) of a milliliter. Some syringes are even smaller. The tuberculin syringe in Figure 4-15 holds a total volume of 0.5 mL.

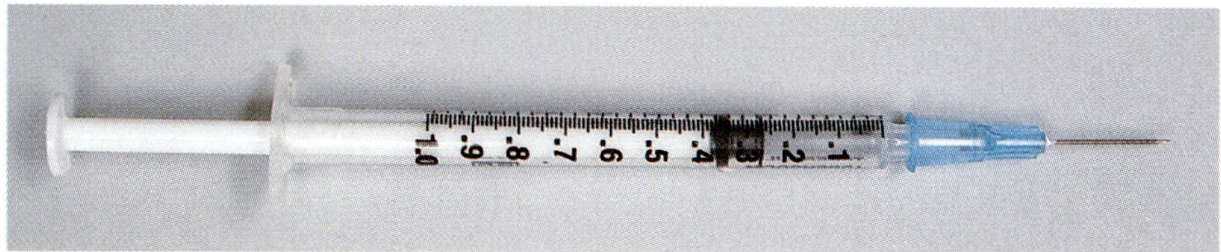

Figure 4-14 A 1-mL tuberculin syringe. (Picture for comparison only. All syringes should include a safe-needle device to prevent accidental needle sticks.)

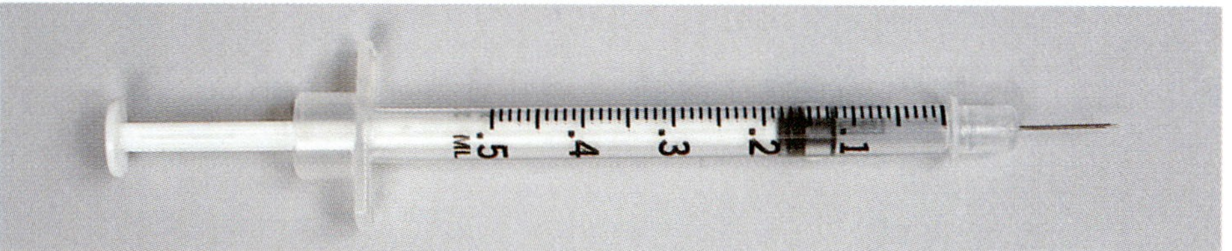

Figure 4-15 A 0.5-mL tuberculin syringe. (Picture for comparison only. All syringes should include a safe-needle device to prevent accidental needle sticks.)

Measuring the correct dose with a tuberculin syringe requires extreme care. The calibrations are close together and marked with a number only at every one-tenth or two-tenths calibration. Be sure the leading ring is aligned with the proper calibration. See Figure 4-16 to become familiar with these calibrations. Sometimes tuberculin syringes are also calibrated with the apothecary scale in minims although these calibrations are not used very often. You must always take great care when reading any syringe to ensure you are reading the correct scale.

Syringes for Established Intravenous Lines

Some syringes are used to administer medication through already established intravenous lines that deliver medication and fluids directly into a patient's veins. Figure 4-17 shows an example of such a syringe.

Adding medication through existing lines has several advantages. Using the injection ports, IV medications can be administered quickly without the patient being punctured repeatedly. Because the syringes do not have needles, accidental needlesticks to patients and

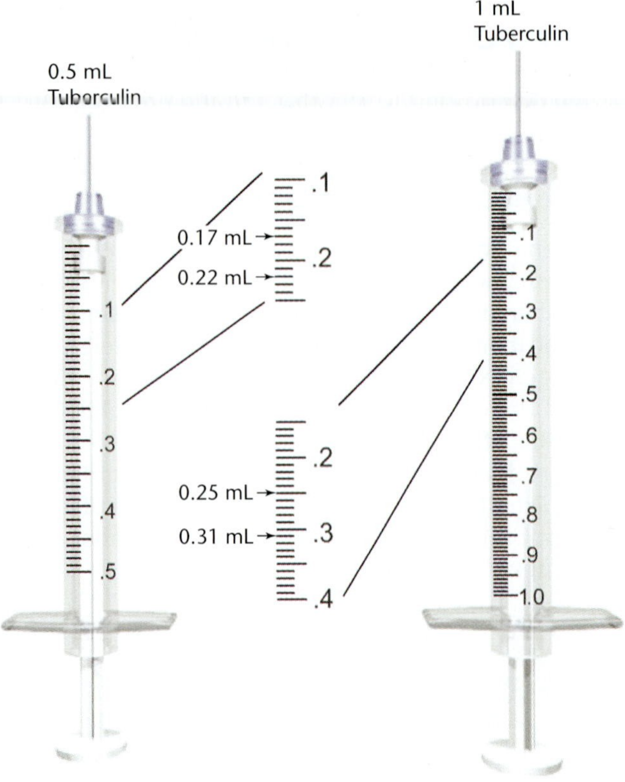

Figure 4-16 Both syringes are calibrated to 0.01 mL. Each number indicates 0.1 mL.

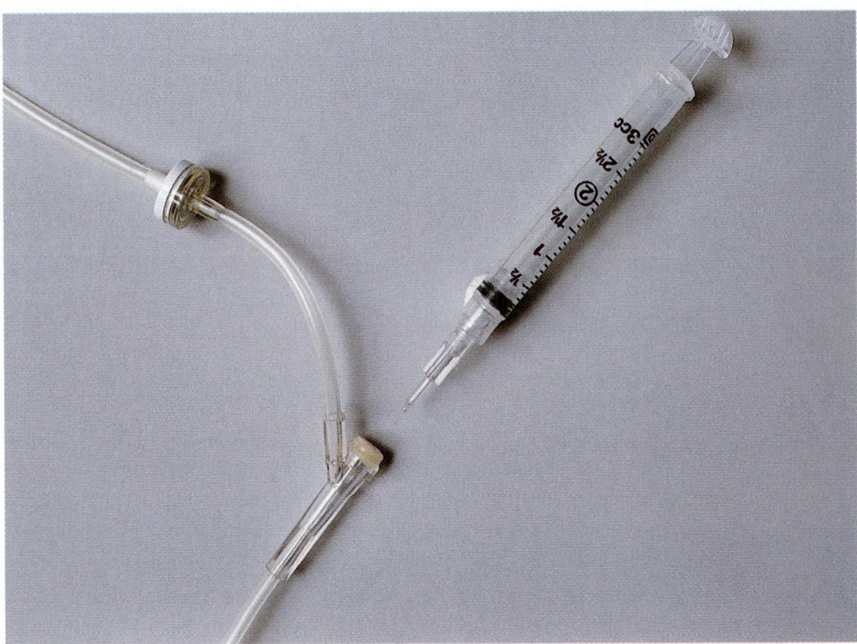

Figure 4-17 A needleless syringe is used to inject medication into existing IV tubing.

health care workers are avoided. An intravenous system with needleless syringes allows more than one drug to be administered at a time, provided that the drugs are compatible. Needleless syringes also enable you to deliver drugs on a periodic basis and to dilute the medication. Although these syringes are needleless, they still have calibrations that must be read carefully.

Large-Capacity Syringes

Not all medications can be delivered in doses of 3 mL or less. Therefore, syringes with 5, 6, 10, and 12 mL or even more are available (see Figure 4-18.) As with other syringes, volume is measured in milliliters, and the number of calibrations between numbered milliliters varies. Because of this you must look carefully at the marks to measure an accurate amount of medication.

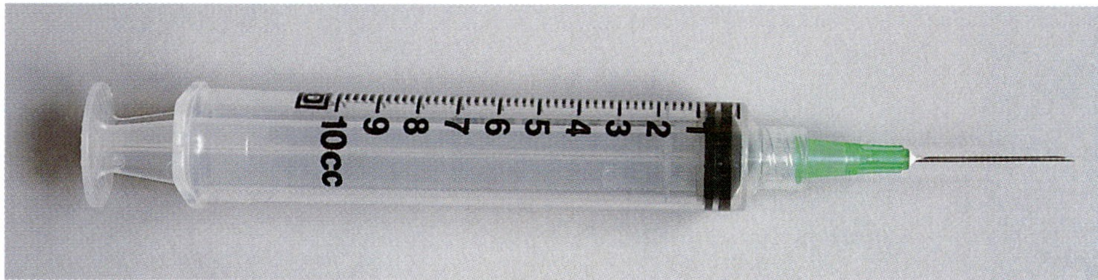

Figure 4-18 A large-capacity syringe.

ERROR ALERT!

Pay close attention to the calibration of any syringe you use.

Perhaps the most important aspect of a syringe is its calibration. If medication is not accurately measured, serious problems can result for the patient. Do not assume the calibration of any syringe you use. Check the marks carefully. After you determine the value of each calibration mark, align the plunger at the appropriate level. **Remember**: No matter what type of syringe you use, *the leading ring, closest to the needle, is the part that must be aligned with the calibration.*

Ampules, Vials, and Cartridges

Parenteral medications may be packaged in ampules, vials, or cartridges (see Figure 4-19). **Cartridges** are prefilled containers shaped like syringe barrels. They generally hold one dose of medication and fit a reusable syringe. Tubex® and Carpujet® are examples of cartridges. **Ampules** are sealed containers and usually hold one dose of liquid medication. You snap

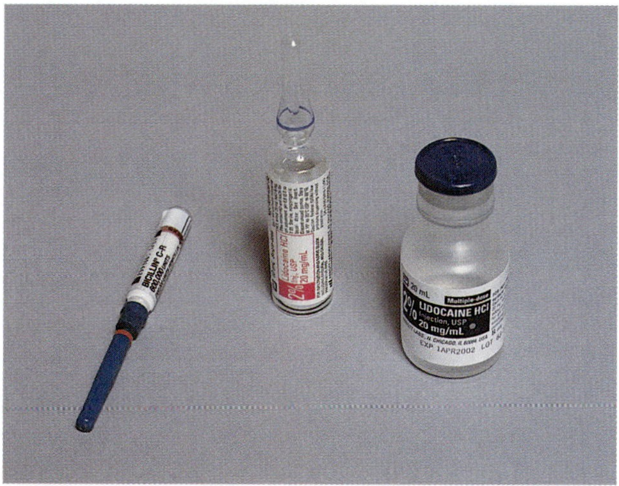

Figure 4-19 A cartridge (left), an ampule (center), and a vial (right).

them open and use any standard syringe to withdraw the medication. **Vials** are containers covered with a rubber stopper, or diaphragm. They may hold more than a single dose of medication, in either liquid or powder form. If powder, then diluent is injected in the vial to reconstitute the medication.

Preparing the Syringe

If you fill a syringe, you must administer it yourself or properly label the contents for the person who will use it.

Rule 4-5 In most circumstances, the person who prepares a syringe for injection should deliver the injection. Exceptions include:

1. Pharmacy technicians who prefill syringes for nurses, medical assistants, or patients.
2. Nurses or medical assistants preparing a syringe for a physician.
3. Health care workers teaching a patient to administer his or her own medication.

This last exception occurs, for instance, when you teach a patient with diabetes how to administer insulin.

GO TO . . . Open the CD-ROM that accompanies your textbook and select Chapter 4, Rule 4-5. Review the animation and example problems, then complete the practice problems. Continue to the next section of the book once you have mastered the rule presented. ■

Needle Gauge and Length

When you administer an injection, you must choose a needle with an appropriate gauge. A needle's *gauge* is its interior diameter. Lower numbered gauges correspond to larger diameters; an 18-gauge needle is wider than a 22-gauge needle. The gauge you use depends on the viscosity (thickness) of the medication as well as the injection site. More viscous drugs and deeper injections require larger needles (those with a lower gauge number), as seen in Table 4-1.

The injection site also determines the length of the needle. You should select a needle that is long enough to reach the area of tissue specified. However, the needle should not be so long that it penetrates beyond that area.

TABLE 4-1 Needle Gauge and Length

Type of Injection	Needle Gauge	Length (inches) for Adults
Intradermal	25 to 26	$\frac{3}{8}, \frac{5}{8}$
Subcutaneous	23 to 27	$\frac{1}{2}, \frac{5}{8}$
Intramuscular	18 to 23	1 to 2

Critical Thinking on the Job

Finishing What You Start

A health care professional prepares to deliver a single dose of Valium® by injection. He has a prefilled syringe containing 2.5 mL, but the prescribed dose is 1.5 mL. Before he can administer the injection, his pager goes off and he rushes to another patient's room. As he does, he asks another health care professional to administer the Valium.

She administers the Valium®, assuming that the prefilled syringe contains the appropriate dose. As a result of receiving more Valium® than was prescribed, the patient's blood pressure drops. The first health care professional does not ask if the correct dose has been administered, and the patient undergoes tests to find the cause of the drop in blood pressure.

Question?

How could this error have been avoided?

REVIEW AND PRACTICE

4-2 Hypodermic Syringes

In Exercises 1–4, provide a brief answer.

1. What is the standard calibration of a 3-mL syringe?
2. What is the standard calibration of a 50-unit insulin syringe?
3. What is the standard calibration of a tuberculin syringe?
4. What is the standard calibration of a large-capacity syringe?

In Exercises 5–13, determine if the statement is true or false. If false, explain why it is incorrect.

5. Any extra medication in a syringe should be discarded before the injection is given.
6. The first calibration on any syringe is always zero (0).
7. When you are measuring a dose in a syringe, read the calibration that aligns with the trailing ring on the plunger.
8. Some prefilled syringes are overfilled with 0.1 to 0.2 mL of medication to allow for air expulsion from the needle.
9. Prefilled syringes and standard syringes do not have the same calibrations.
10. You can use an insulin syringe to measure 6 mL of medication.
11. A patient is punctured each time a syringe is used with an established intravenous line.
12. Safety syringes are a guaranteed way to avoid accidental needlestick injuries.
13. Tuberculin syringes are used to administer the subcutaneous PPD skin test.

In Exercises 14–25, identify the type of syringe and the volume of the dosage it contains. Identify the correct units of measurement.

Example: Refer to the sample syringe below:

Type: tuberculin Volume: 0.3 mL

Sample

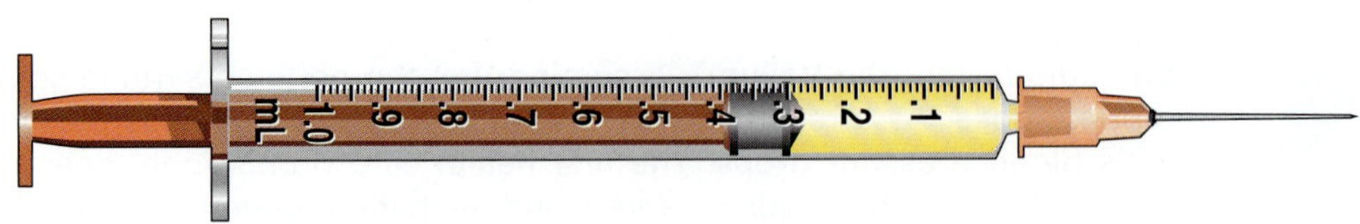

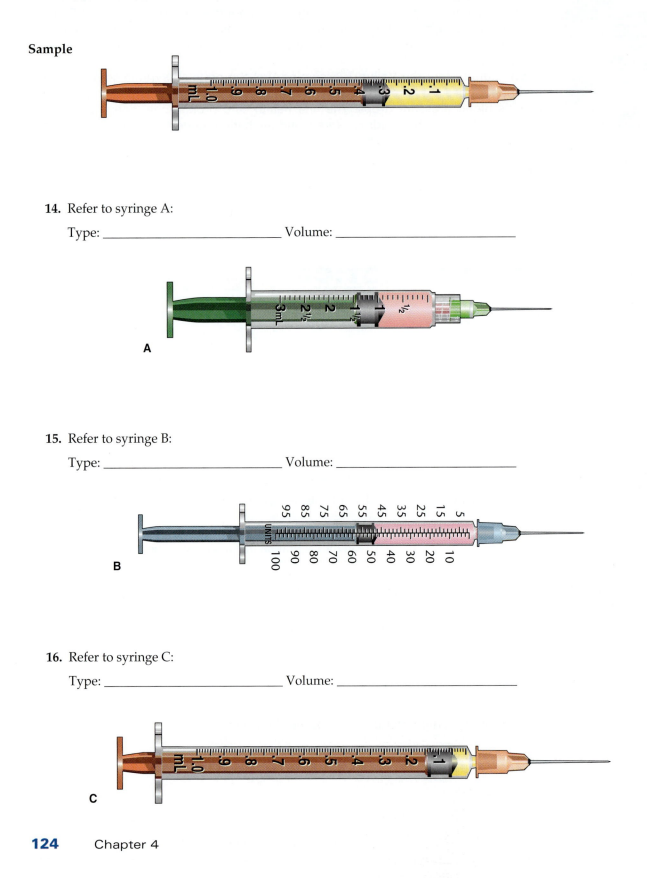

14. Refer to syringe A:

Type: ______________________ Volume: ______________________

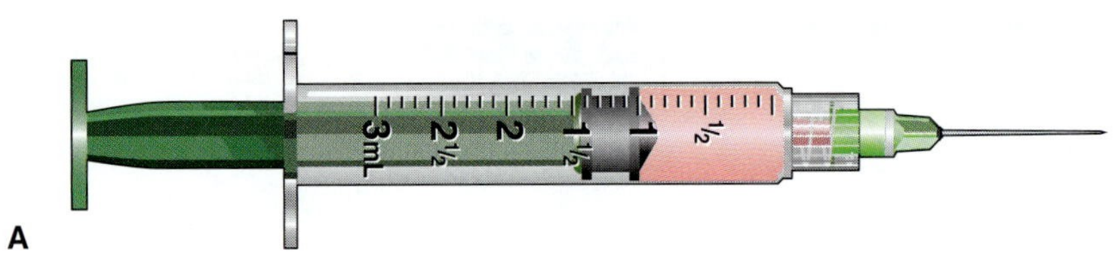

A

15. Refer to syringe B:

Type: ______________________ Volume: ______________________

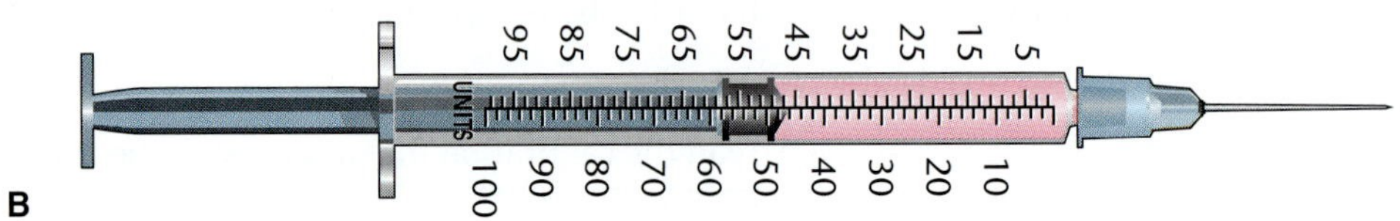

B

16. Refer to syringe C:

Type: ______________________ Volume: ______________________

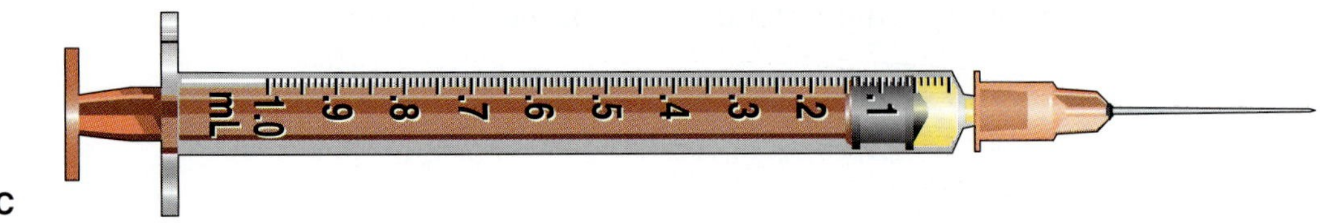

C

17. Refer to syringe D:

Type: ______________________ Volume: ______________________

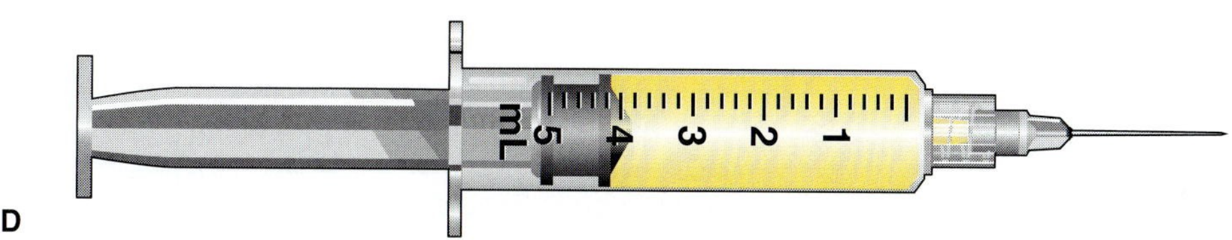

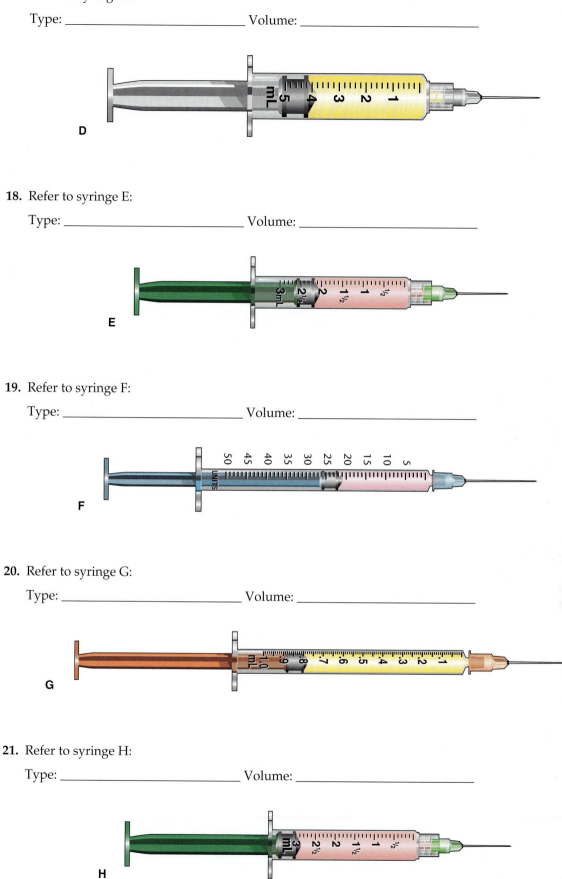

18. Refer to syringe E:

Type: ______________________ Volume: ______________________

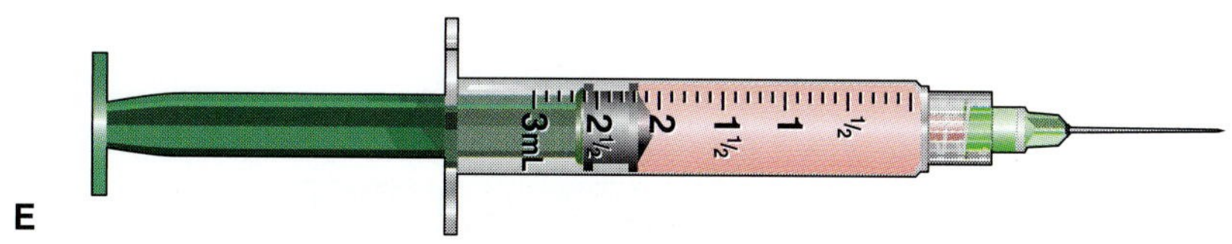

19. Refer to syringe F:

Type: ______________________ Volume: ______________________

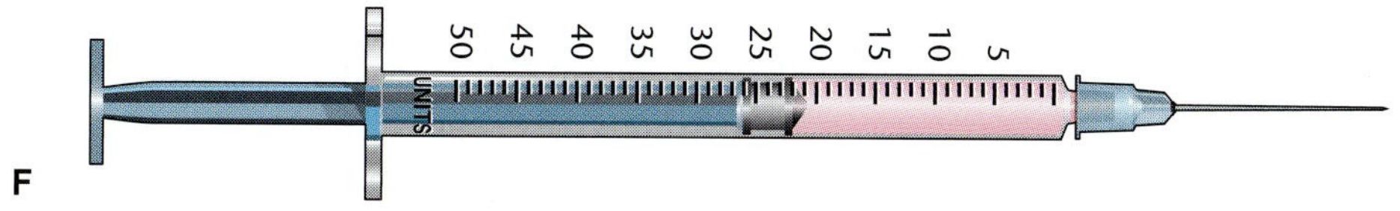

20. Refer to syringe G:

Type: ______________________ Volume: ______________________

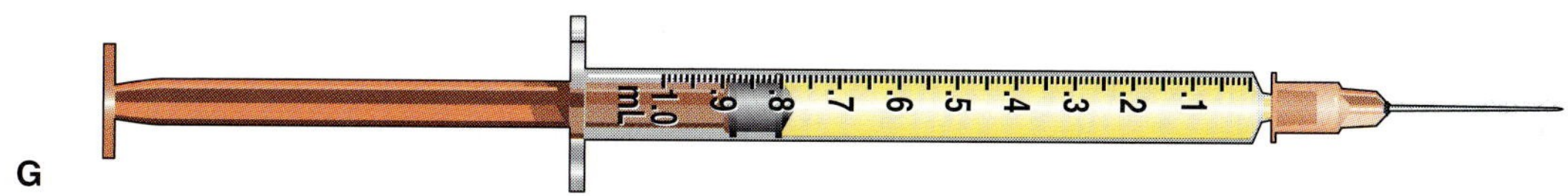

21. Refer to syringe H:

Type: ______________________ Volume: ______________________

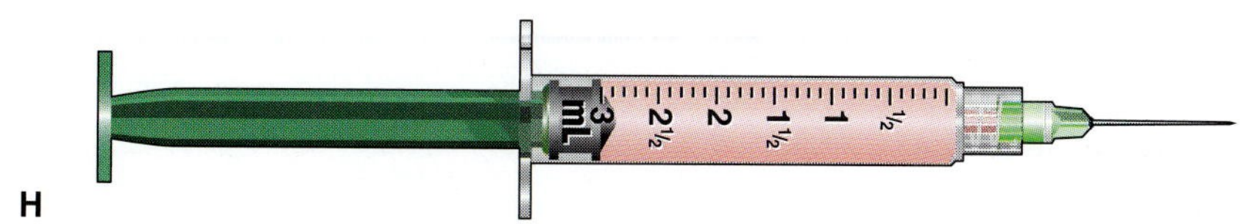

22. Refer to syringe I:

Type: ______________________ Volume: ______________________

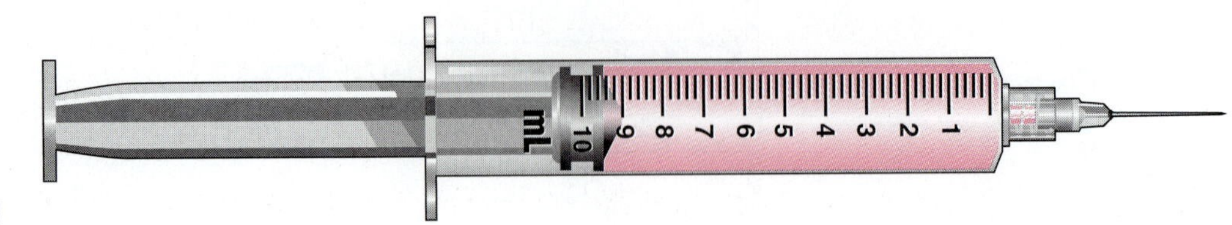

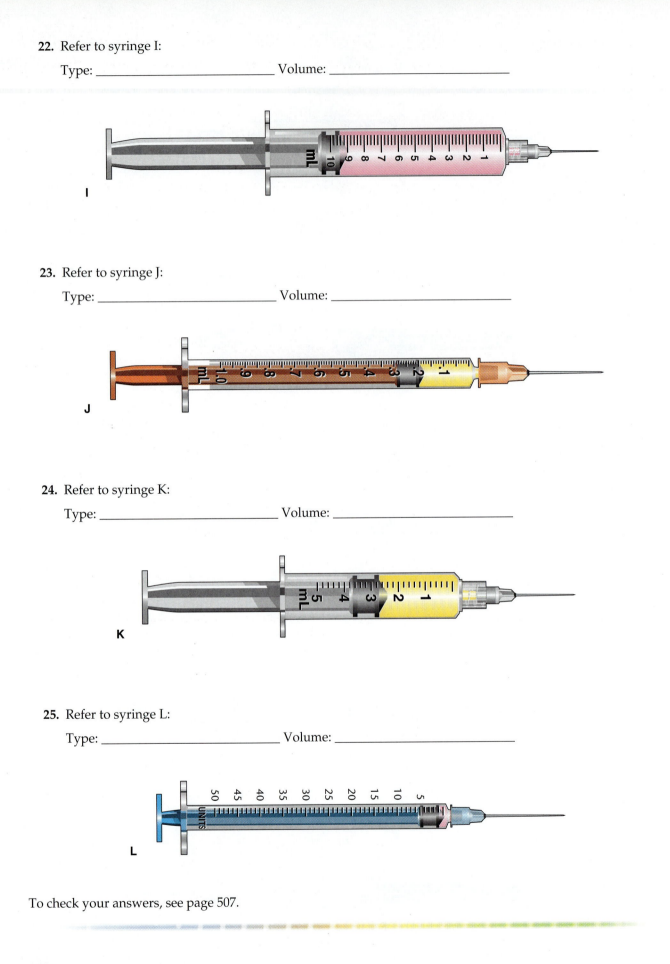

23. Refer to syringe J:

Type: ______________________ Volume: ______________________

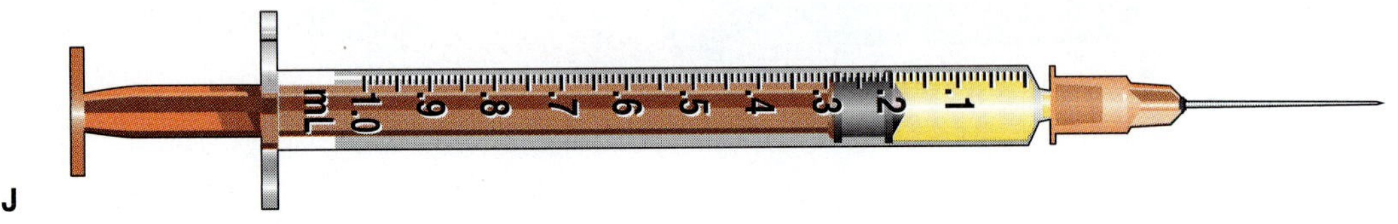

24. Refer to syringe K:

Type: ______________________ Volume: ______________________

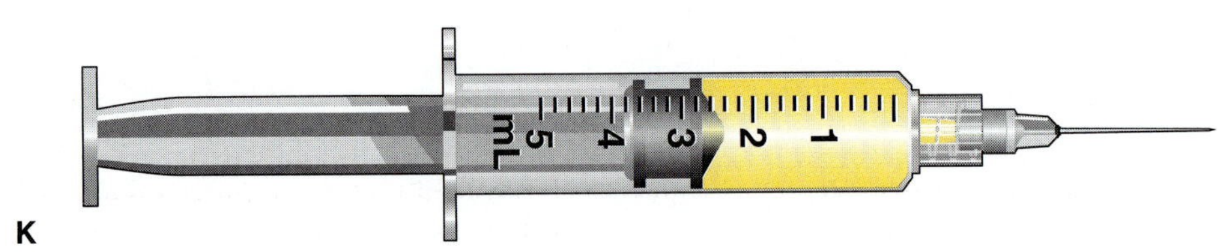

25. Refer to syringe L:

Type: ______________________ Volume: ______________________

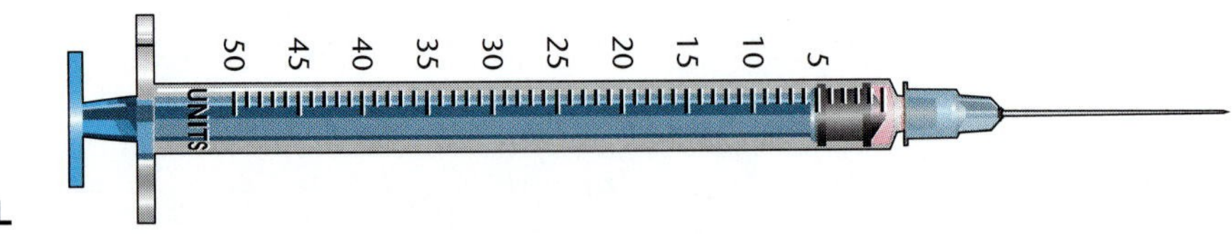

To check your answers, see page 507.

Other Administration Forms—Equipment

You can administer medications in other ways. You need to become familiar with them to correctly measure and administer each dose of medication.

Topical Medications

Topical medications, such as gels, creams, ointments, and pastes, are applied directly to the skin. You administer the drug with a glove, tongue blade, or cotton-tipped applicator. Topical medications are usually given for their therapeutic effect in or on the skin. Follow the instructions that accompany the product to determine how to remove the medication from its container and administer it. Avoid letting any of the medication contact your own skin.

Transdermal Medications

A **transdermal** medication is a form of topical medication in which the medication is absorbed through the skin into the bloodstream. Often you administer a transdermal medication by using a self-adhesive patch (see Figure 4-20). You place the medication-filled patch on clean, dry, hairless skin that has no rash or irritation but has good circulation. The medication is then released continuously through the skin to the entire body for a specified period. The package is marked with the dose rate.

Because patches allow for continuous regular absorption, they maintain consistent levels of the medication in the blood. Patches are beneficial for people who have trouble swallowing oral drugs. Patches bypass the *gastrointestinal* (GI) system, where the medication could cause digestive and other problems.

When you administer a patch, you must document its placement so that other health care providers can readily find it. Mark each new patch with your initials, the date, and the time of administration. Rotate placement sites. Remove older patches and dispose of them safely. Medications commonly administered through the transdermal method include cardiovascular drugs, hormones, and allergy and pain medications.

Drops, Sprays, and Mists

Different types of equipment are used to deliver medication to the nose, ears, eyes, and throat in the form of drops, sprays, and mists. Drops, also called **instillations**, deliver medication to the nose, eyes, and ears. Sprays deliver medication to the nose and throat. Drops and sprays are measured according to the dose prescribed and the manufacturer's instructions. Therefore, be sure to use the equipment that accompanies the drug when you administer these medications.

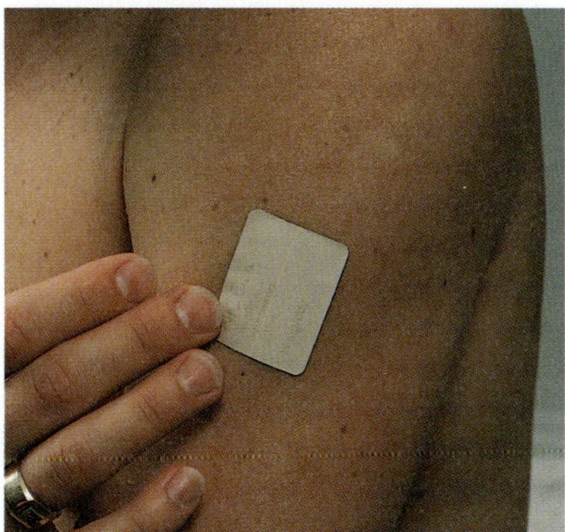

Figure 4-20 A transdermal patch.

Droppers, similar to those for oral medications, are used to administer drops. Plastic squeeze bottles are used for drops and sprays. Atomizers deliver sprays by using a rubber bulb to propel the spray from a medicine container into the nose.

Another way to administer medication is to use a mist that the patient inhales. Vaporizers, or steam inhalers, use boiling water to create a mist from liquid medication. Nebulizers, often used by patients with asthma, and **metered dose inhalers** (MDIs) also help deliver medication to the patient.

Vaginal and Rectal Medications

You administer some medications vaginally or rectally. Suppositories are available for vaginal and rectal administration, and tablets are available for vaginal administration. These medications are usually solid dosage forms administered into the patient's vagina or rectum, depending on the product. Each suppository or vaginal tablet contains a predetermined dose. Some of the vaginal medications are available in oral form, but the vaginal form is preferred for patients who have difficulty with oral medications or who need the medication directly on the site.

Whereas suppositories deliver medication in solid forms, liquid forms of medication can also be administered. Douches deliver liquid vaginal medication through a special nozzle. Enemas deliver liquid rectal medication through a tube inserted into the patient's rectum. Other medications are administered through creams and foams.

GO TO . . . Open the CD-ROM that accompanies your textbook and select Chapter 4, Practice 4-3. Review the animation and example problems, then complete the practice problems. Continue to the next section of the book once you have mastered the rule presented. ■

Critical Thinking on the Job

Document the Use of Patches

An elderly nursing home patient with cancer has chronic pain. She is receiving oral codeine and morphine, but still has debilitating pain between doses. She complains about the problem to the health care professional, who suggests to the physician that the patient may benefit from a more continuous administration of medication.

The doctor prescribes a transdermal Fentanyl patch. The patch contains enough medication for 3 days. The health care professional applies the patch to the right side of the patient's back. Later that night, another professional checks to see whether the patient's patch has been applied. It is difficult to roll the patient on her side to examine her back. Because the patch is not readily visible, the second professional assumes the patch has not been placed and puts another patch on the patient's chest.

The next morning, the first health care professional notes that the patient is barely breathing and cannot be awakened. She quickly finds the patch on her chest. She remembers placing a patch on the back, not the chest. She confirms that two patches are in place, strips off both, and calls the physician. Narcan is administered to reverse the drug's effects; the patient quickly becomes more coherent.

Question?

How could this error have been avoided?

REVIEW AND PRACTICE

4-3 Other Administration Forms—Equipment

Match the medication forms with their descriptions. Medication forms may be used more than once.

a. Topical **b.** Transdermal **c.** Drops

d. Sprays **e.** Mists **f.** Vaginal

g. Rectal

_________ **1.** Can be delivered in the form of a tablet, suppository, or douche.

_________ **2.** Applied directly to the skin.

_________ **3.** Delivered by a nebulizer, vaporizer, or metered-dose inhaler.

_________ **4.** Medication from a patch is absorbed through the skin.

_________ **5.** Can be delivered in the form of a suppository or an enema.

_________ **6.** Must be marked with your initials, date, and time of administration.

_________ **7.** Also known as instillations.

_________ **8.** An atomizer is used to deliver this medication form.

_________ **9.** Are useful when a patient has difficulty or trouble swallowing oral drugs.

_________ **10.** Are usually delivered to the nose, eyes, and ears.

To check your answers, see page 508.

Workspace

HOMEWORK ASSIGNMENT

Determine if the statement is true or false. If false, explain what is incorrect about the statement.

1. Oral dosage forms are also called parenteral.
2. The leading ring is closer to the needle than the trailing ring.
3. Prefilled syringes can be used to give multiple injections as long as they are given to the same patient.
4. Tuberculin syringes have a capacity of 3 mL.
5. A vial can be used to give injections to multiple patients.
6. Syringes are always calibrated in milliliters.

Using the accompanying illustrations of equipment for the remaining questions, mark with a line or shading where you would measure the required dose.

7. 3 drams

8 dr 1 oz
6 dr 3/4 oz
4 dr 1/2 oz
2 dr 1/4 oz
1 dr 1/8 oz

8. 0.7 mL

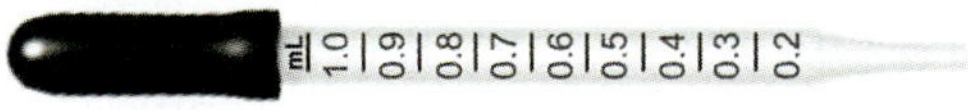

mL 1.0 0.9 0.8 0.7 0.6 0.5 0.4 0.3 0.2

9. 2.4 mL

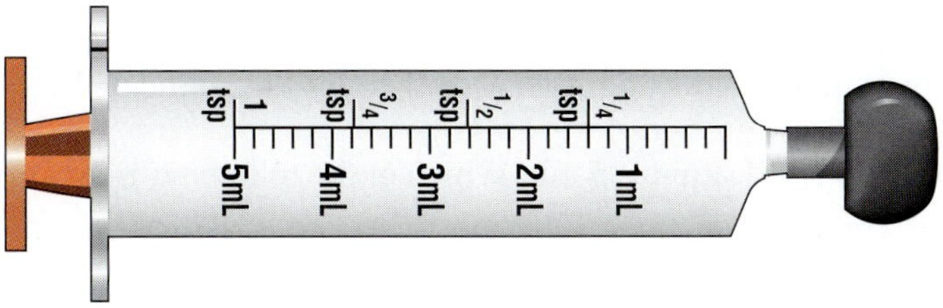

1 tsp 3/4 tsp 1/2 tsp 1/4 tsp
5mL 4mL 3mL 2mL 1mL

10. 2.4 mL

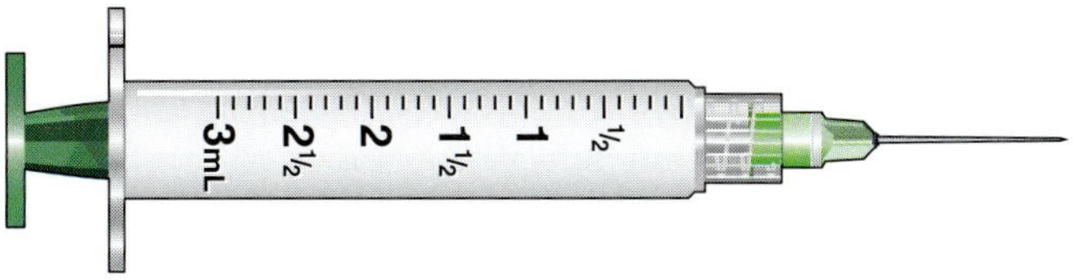

3mL 2 1/2 2 1 1/2 1 1/2

11. 18 units

UNITS 50 45 40 35 30 25 20 15 10 5

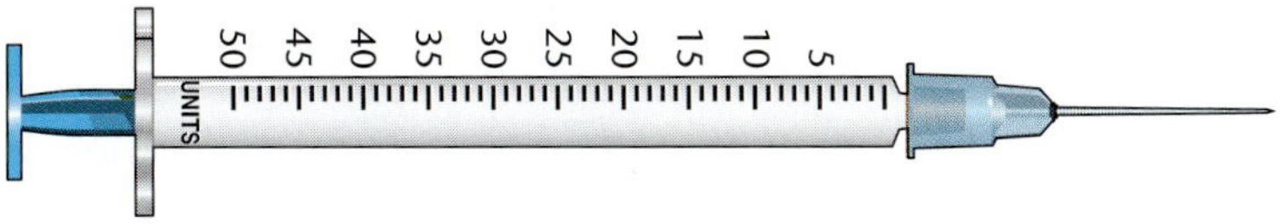

12. 0.38 mL

mL 1.0 .9 .8 .7 .6 .5 .4 .3 .2 .1

Chapter 4 Review

Check Up

In Exercises 1–10, answer the multiple-choice questions. More than one answer may be correct.

1. Which of the following equals 1 oz?

a. 2 tbs **b.** 20 mL **c.** 8 dr **d.** 2 tsp

2. A patient is supposed to receive 15 mL of Mylanta. A measuring cup cannot be found. What is an equivalent dose?

a. 1 tsp **b.** 2 tsp **c.** 3 tsp **d.** 5 tsp

3. The dose of a liquid medication for oral administration is $\frac{3}{4}$ oz. Which is the correct equivalent dose?

a. 20 mL **b.** 25 mL **c.** 7.5 dr **d.** 1.5 tbs

4. The ordered dosage of a liquid medication for oral administration is 2.5 mL. What is the appropriate method of oral administration?

a. $\frac{1}{8}$ oz as measured in a medicine cup
b. 1 dr as measured in a medicine cup
c. $\frac{1}{2}$ tsp as measured in a calibrated spoon
d. 2.5 mL as measured in a calibrated dropper

5. The prescribed dosage of a medication is 5 drops. Which of the following is an appropriate method of administering the dose?

a. 0.5 mL using only the calibrated dropper that accompanies the medicine bottle
b. 5.0 mL using the calibrated dropper that accompanies the medicine bottle
c. 5 drops using any calibrated dropper
d. 5 drops using only the calibrated dropper that accompanies the medicine bottle

6. The ordered oral dosage of a medication is 5 mL. Which of the following is an appropriate method of administering the dose?

a. 1 tsp using a calibrated spoon
b. 5 mL using a syringe for parenteral administration with the needle removed
c. 25 drops using a calibrated dropper
d. 5 tsp using a calibrated spoon

7. The ordered dosage of a medication is 10 mg. The medication is mixed at a strength of 5 mg per milliliter. Using a 2.5-mL prefilled syringe, you should discard how much medication before administration?

a. 0.2 mL **b.** 1 mL **c.** 0.5 mL **d.** 2 mL

8. A tuberculin syringe is being used to administer 0.25 mL of a given medication. Which of the following is the equivalent dose?

a. 2.5 hundredths of a milliliter
b. 25 hundredths of a milliliter
c. 250 hundredths of a milliliter
d. 2.5 drops using a calibrated dropper

9. Which of the following can be used to administer a 7-mL dose via the parenteral route?

a. A large-capacity syringe
b. A standard syringe
c. A tuberculin syringe
d. An insulin syringe

10. Which of the following has the most precise and accurate calibrations?
    a. A 5-mL syringe to an established intravenous line
    b. A 3-mL syringe to an established intravenous line
    c. A 3-mL safety syringe
    d. A 1-mL tuberculin syringe

In Exercises 11–20, determine if the statement is true or false. If false, explain why it is incorrect.

11. It is apparent from the calibrations on the medicine cups that 30 mL is equivalent to 1 oz.
12. If an ordered dose calls for 3 mL, a medicine cup may be used to measure it.
13. A calibrated dropper dispenses a standard drop of 3 mL.
14. The standard syringe can hold more than 1 mL.
15. The standard syringe is calibrated in tenths of a milliliter.
16. Prefilled, single-dose syringes can be used more than once.
17. The standard U-100 insulin syringe can hold up to 100 units or 1 mL.
18. The tuberculin syringe is used to measure doses of drugs larger than 3 mL.
19. The syringes used to deliver medication through already established intravenous systems are hypodermic syringes.
20. You may administer an oral medication by injecting it parenterally.

In Exercises 21–40, use the accompanying illustrations of equipment. For each question, mark with a line or with shading where you would measure the required dose:

21. 30 mL (Refer to medicine cup A.)

22. $\frac{1}{2}$ oz (Refer to medicine cup B.)

23. 1 mL (Refer to calibrated dropper C.)

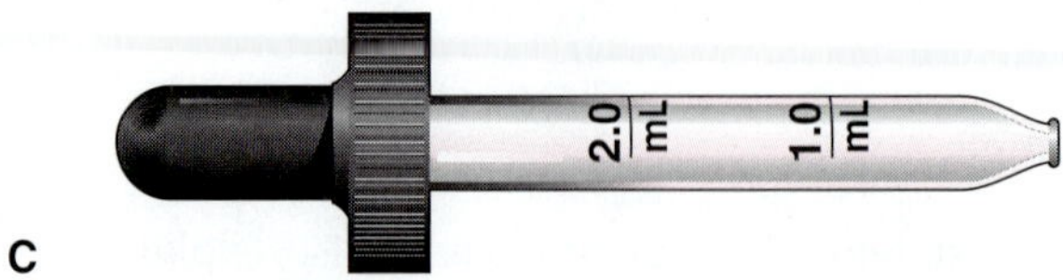

24. 0.6 mL (Refer to calibrated dropper D.)

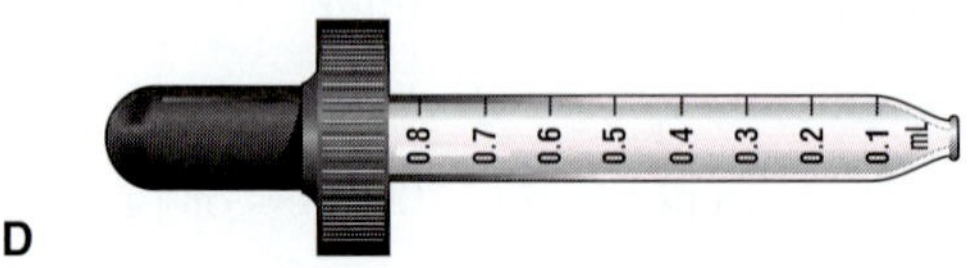

25. $\frac{3}{4}$ tsp (Refer to dropper E.)

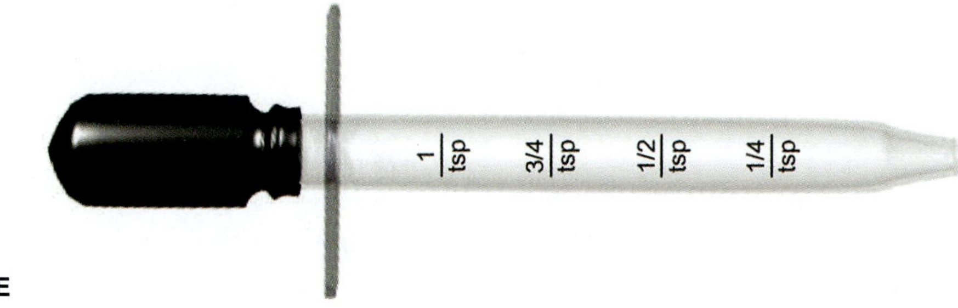

26. 0.5 mL (Refer to dropper F.)

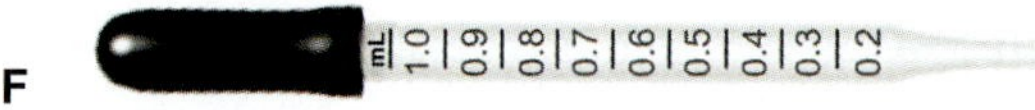

27. $\frac{1}{4}$ tsp (Refer to dropper G.)

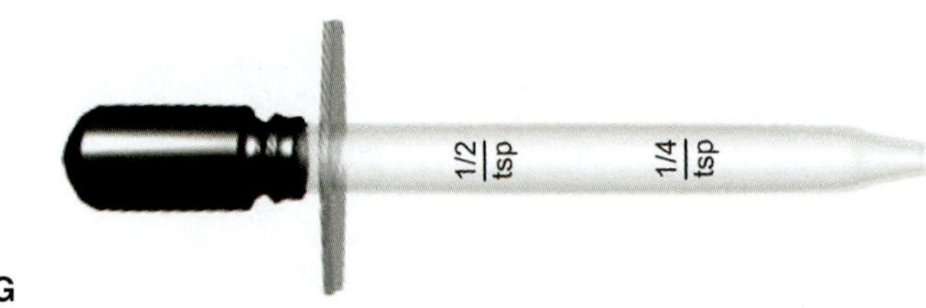

28. 0.32 mL (Refer to syringe H.)

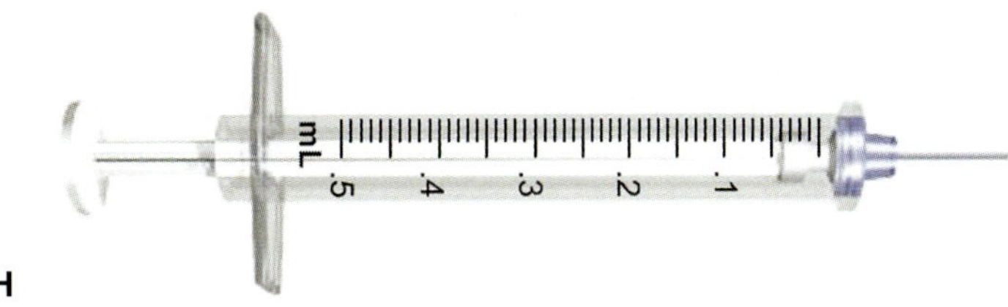

29. 2 mL (Refer to syringe I.)

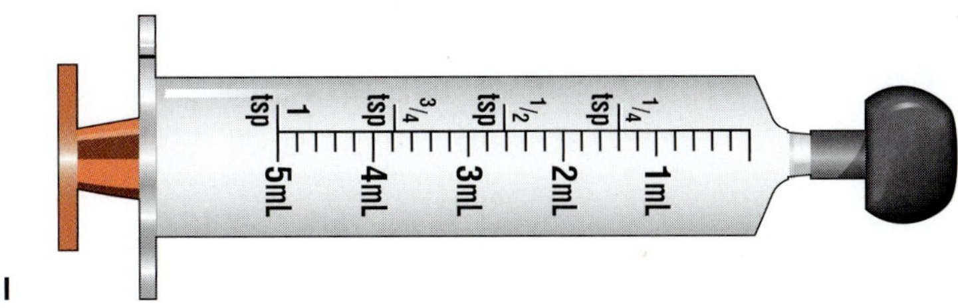

30. $1\frac{1}{2}$ tsp (Refer to syringe J.)

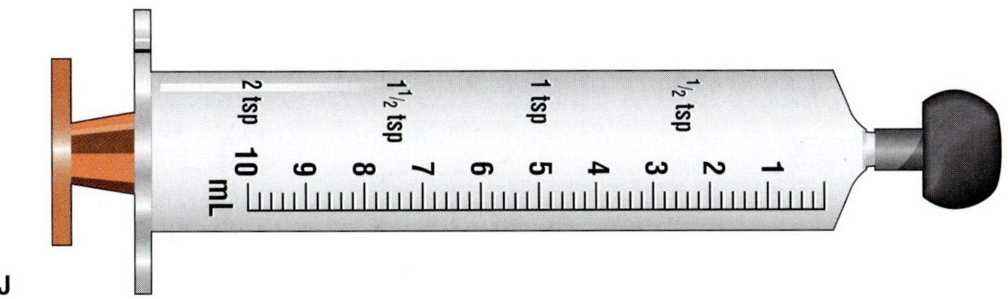

31. 1.5 mL (Refer to syringe K.)

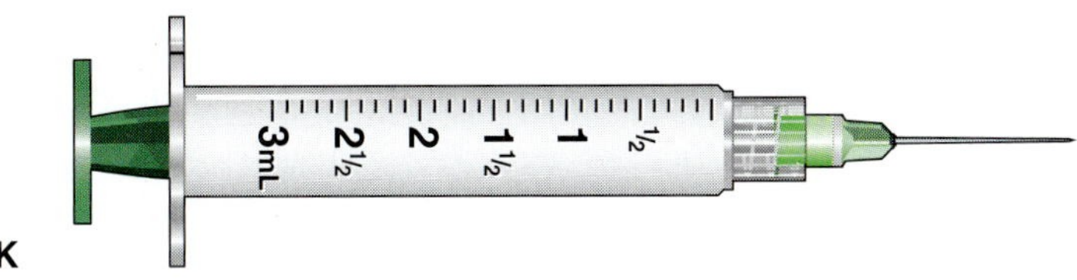

32. 2.3 mL (Refer to syringe L.)

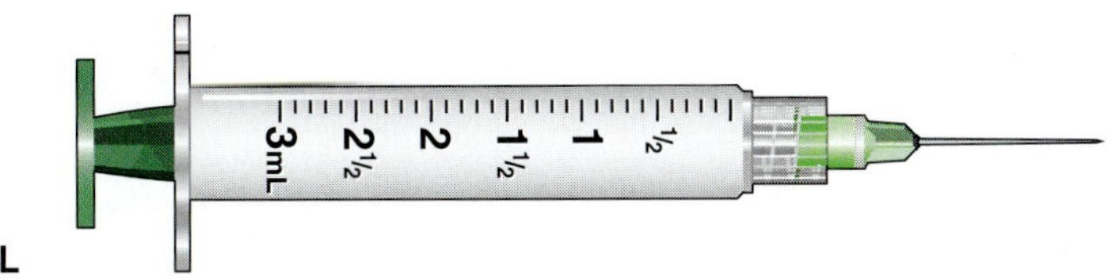

33. 80 units (Refer to syringe M.)

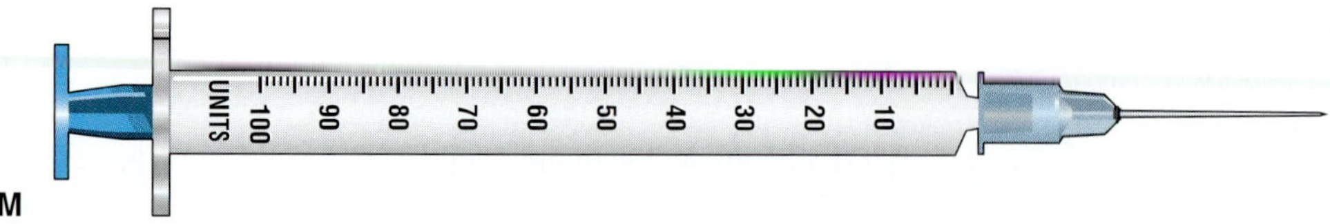

34. 45 units (Refer to syringe N.)

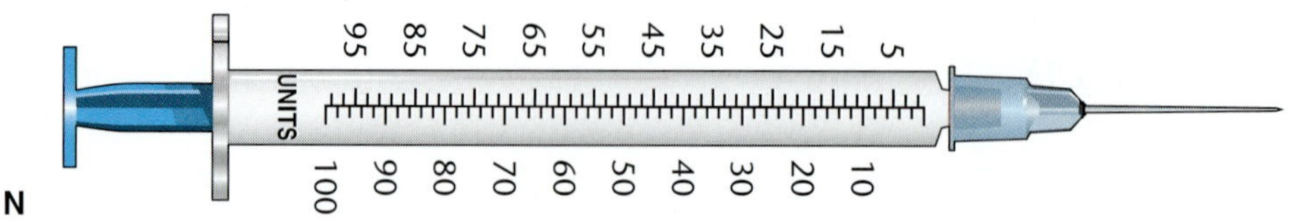

35. 35 units (Refer to syringe O.)

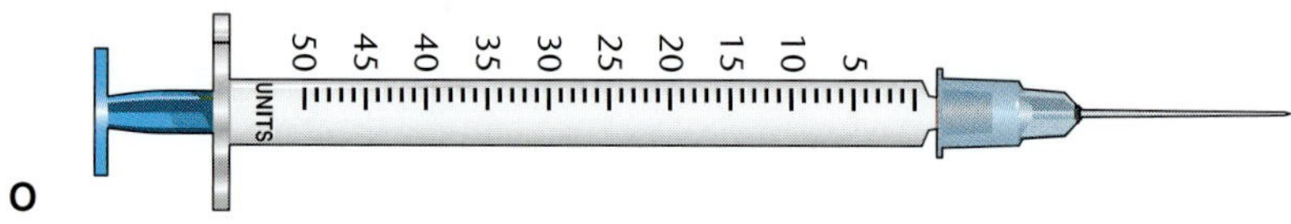

36. 27 units (Refer to syringe P.)

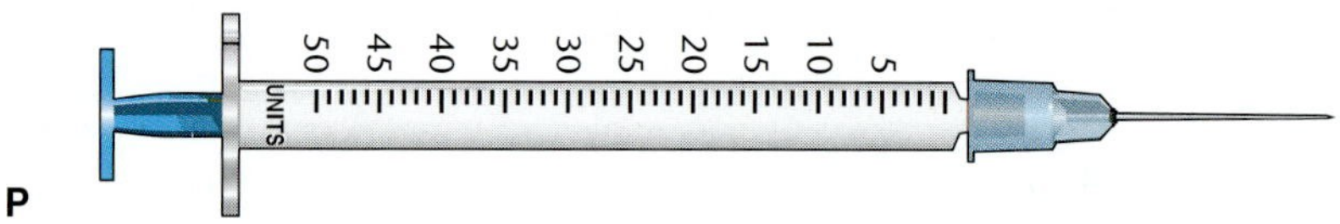

37. 0.5 mL (Refer to syringe Q.)

38. 0.25 mL (Refer to syringe R.)

39. 5 mL (Refer to syringe S.)

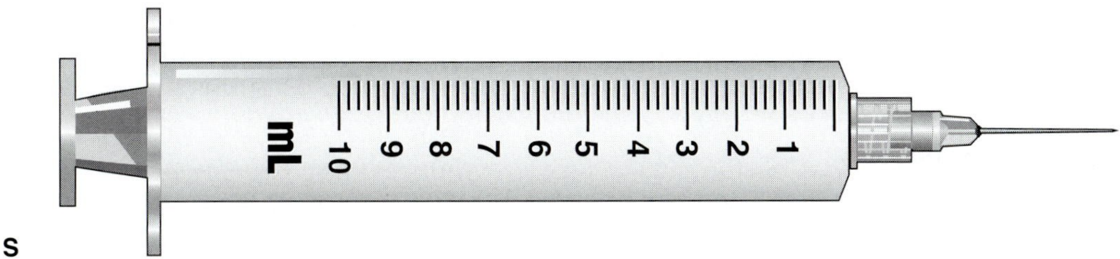

40. 7.2 mL (Refer to syringe T.)

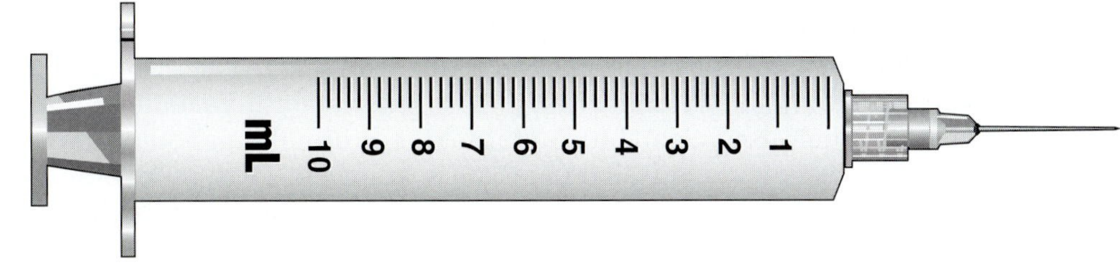

Critical Thinking Applications

What are the best utensils for measuring and administering the doses in each of the following situations? Choose from any of the equipment discussed in this chapter.

1. 1.5 mL, to be delivered parenterally

2. 29 units of insulin, to be delivered parenterally

3. 0.3 mL, to be delivered parenterally

4. Oral liquid dose of 29 mL

5. Oral liquid dose of 1 tbs

6. Parenteral dose of 1.25 mL

7. Oral liquid dose of 2 tbs

8. 8 mL, to be administered parenterally

9. 0.3 mg, to be administered transdermally

10. 0.4 mL, to be administered orally

11. $\frac{3}{4}$ tsp to be administered orally

Case Study

A health care professional must administer 10 drops of an oral medication. In attempting to administer the dose, the health care professional breaks the calibrated dropper. Instructions on the bottle say that 20 drops is equal to 1.5 tsp. What is the dose to be administered, and what is the best utensil for measuring it, now that the dropper is broken?

To check your answers, see page 512.

Internet Activity

In the facility where you work, they want to implement the use of safety syringes and needleless systems for the protection of the health care employees. You have been asked to determine the types of syringes and systems available and to become familiar with their uses. Search the Internet for information about safety syringes and needleless systems. Determine what is available with regard to brand names, use, price, and safety statistics. Prepare a brief report or table comparing the items you have found.

Go To . . . Open the CD-ROM that accompanies your textbook, and complete a final review of the rules, practice problems, and activities presented for this chapter. For a final evaluation, take the chapter test and email or print your results for your instructor. A score of 95 percent or above indicates mastery of the chapter concepts. ■

5 Drug Orders

You must motivate yourself everyday!

—Matthew Stasior

Learning Outcomes

When you have completed Chapter 5, you will be able to:

- Summarize the Rights of Medication Administration.
- Interpret a written drug order.
- Identify on physicians' orders and prescriptions the information needed to dispense medication.
- Locate on medication administration records or electronic medication records the information needed to administer medication.
- Recognize drug orders that do not contain all the necessary information to carry out the orders safely.
- Select appropriate action for confusing, incomplete, or illegible drug orders.

Key Terms

Electronic medication administration record (eMAR)

Medication administration record (MAR)

Physicians' Desk Reference (PDR)

Physician order form

Prescription

Sig

Introduction

To correctly calculate a medication dose, first you must be able to read and understand the drug order. Motivate yourself for accuracy when completing this chapter about drug orders, the patient's rights, medication administration systems, and your responsibilities related to drug orders.

The Rights of Medication Administration

When you carry out a medication order, you can be held responsible if an error occurs, regardless of its source. You must be able to interpret and confirm physicians' orders. To assist you, the medical field has created guidelines, called the Rights of Medication Administration (see Table 5-1). The basic five rights of medication administration include right patient, right drug, right dose, right route, and right time. Additional rights include right reason, right to refuse, right to be educated, right technique, and right documentation. Since about one-third of all medication errors occur during medication administration, these rights must always be followed.

TABLE 5-1 The Rights of Medication Administration

1. Right patient	**6.** Right reason
2. Right drug	**7.** Right to know
3. Right dose	**8.** Right to refuse
4. Right route	**9.** Right technique
5. Right time	**10.** Right documentation

Right Patient. Before you give a medication to a patient you must check two identifiers including this full name and another identifier such as the patient's date of birth, social security number, or medical record number. Check that the name on the medication order is exactly the same as the name of the patient. Verify the full name. Two patients with the same last name may have the same first initials, or even the same first names. Ask the patient to state her or his full name and date of birth or other identifying information. (Follow the policy at the facility where you are employed.) If the patient is unable to do so, ask the parent or caregiver to state the patient's full name. Compare the name to the medication order. In outpatient settings, you may be required to ask for photographic identification, such as a driver's license. In an inpatient setting, check the bed number and the patient's identification number. In many facilities you may be required to scan the bar code on the patient's identification bracelet as well as the medication you will be administering.

Right Drug. To be certain that a patient receives the *right drug,* administer only drugs you have prepared yourself or that are clearly and completely labeled. Check that the drug order has not expired and that the medication is still in date. (See Chapter 6 for more details on drug labels.)

If a patient questions a medication, recheck the original order. Be sure you have the correct drug. Patients are often familiar with their medications. Listening to them may prevent an error. A patient *always* has the right to refuse medication. If this happens, do not return the medication to its container. Instead, follow your facility's guidelines for disposing of it. Record the patient's refusal and notify the physician.

Rule 5-1 Check medication 3 times: when you take it off the shelf, when you prepare it, and when you replace it on the shelf. Check the medication 3 times even if the dose is prepackaged, labeled, and ready to be administered.

GO TO . . . Open the CD-ROM that accompanies your textbook, and select Chapter 5, Rule 5-1. Review the animation and example problems, then complete the practice problems. Continue to the next section of the book once you have mastered the rule presented. ■

Right Dose. The patient must be given the *right dose* of medication. In later chapters, you will learn to convert from the *dosage ordered* by the physician to the *desired dose*—the amount of drug that a patient should receive at any one time. You will also learn to calculate the *amount to administer* to a patient, factoring in the strength of the medication and the equipment you are using.

Use extreme caution when you calculate the desired dose and the amount to administer. Pay special attention to decimal points. They can easily be placed in the wrong location or missed altogether when an order is copied. If you misread a decimal point, then the patient could receive a dose significantly different from the one ordered.

Right Route. You must give patients drugs by the *right route.* A drug intended for one route is often not safe if administered via another route. For example, only drugs labeled *for ophthalmic use* should be instilled (applied with a dropper) into the eye.

Some medications are produced in different versions for different routes. The drug label (see Chapter 6) will indicate the intended route. For example, Compazine® is available as a suppository, a tablet, and an injection. Always check that the route listed on the drug label matches the route ordered by the physician.

Right Time. Give medications at the *right time.* In most cases, to be "on time," you must administer medications within 30 minutes (min) of schedule. The right time may refer to an absolute time, such as 6:00 p.m., or to a relative time, such as "before breakfast."

Some medications, such as insulin, antibiotics, and antidysrhythmic drugs, must be given at specific times because of how they interact with food or the patient's body. Other medications may be spaced over waking hours without changing their effectiveness. The drug order must identify special timing considerations to be followed. If a medication is ordered PRN (whenever necessary), a time interval or condition must usually be specified (e.g., for temperature >101 or blood pressure >120). Before you administer such a PRN drug, check that enough time has passed since the previous dose was given. Otherwise, the patient could receive the medication too soon, leading to severe consequences.

Right Reason. The health care professional who administers the medication should know the reason the drug is given. This should be the right reason. Depending upon health care profession, it may be your responsibility to ensure that a medication is given for the right reason. You may need to check the patient's medication and/or with the prescribing physician to be certain the medication ordered is correct for the patient.

Right to Know. All patients have the right to be educated about the medications they are receiving. This should include the reason, the effect and the side effects of medications. This basic "right to know" is an essential right of all patients. When a medication is administered in a health care facility, this information must be provided to the patient by a health care professional.

Right to Refuse. Every patient has the *right to refuse* a medication. If a patient does refuse a medication, this information should be documented in the patient's medical record.

Right Technique. Be familiar with the *right technique* to administer a medication. For example, both buccal and sublingual medications are applied to the mucous membranes of the mouth. A buccal medication, such as Cepacol lozenges, is placed between the cheek and the gum, whereas a sublingual medication, such as nitroglycerin, is placed under the tongue. If you are not familiar with the correct technique to use, check resources such as the ***Physicians' Desk Reference*** (PDR), a current drug reference book, the facility policy and procedure manual, or a valid Internet resource for more information.

Right Documentation. Be sure that the *right documentation* is completed. If medication administration is not documented it is not considered complete. For example, if you administer

Patient Education

Even though patient education is not listed specifically as one of the rights of medication administration, you should provide patients with basic information regarding their drugs.

1. Explain the purpose of a medication and its side effects.
2. Review the dose, route, frequency, and time that the physician has prescribed.
3. When appropriate, be certain that the patient understands how to self-administer the medication.
4. If the patient is taking liquid oral medications at home, emphasize the importance of using calibrated spoons and measuring cups.

a medication, you must sign the **medication administration record** (MAR) (or similar form) *immediately after* the patient takes the medication. Until a procedure is documented, it is not complete. Document that you have administered a medication *only after* and *immediately after* you have actually administered it. If the patient declines the medication, consumes only part of the dose, or vomits shortly after taking the medication, document that information as well.

REVIEW AND PRACTICE

5-1 The Rights of Drug Administration

Match the rights of drug administration with an example of how that right can be violated.

a. Right patient **b.** Right drug **c.** Right dose **d.** Right route **e.** Right time
f. Right technique **g.** Right documentation **h.** Right reason **i.** Right to know

______ **1.** The medication bottle said *for optic use,* and the medication was instilled into the patient's ears.

______ **2.** A patient has a bleeding disorder and they are to be given a anticoagulant (blood thinner).

______ **3.** James F. Jones received James E. Jones' medication.

______ **4.** The nurse charted a medication on the medication record before the patient had taken the medication.

______ **5.** A patient asks why she is getting a medication and although the nurse knows why, she does not answer and administers the medication anyway.

______ **6.** The dose to be administered was $\frac{1}{2}$ tsp, and the patient received 5 mL.

______ **7.** The medication was ordered at bedtime, and the patient took it at 9 a.m.

______ **8.** The physician ordered Uracel, and the patient received uracil.

______ **9.** The medication was to be given under the tongue, and the patient was told to swallow it.

To check your answers, see page 512.

Roman Numerals

Roman numerals are sometimes used in drug orders including prescriptions and physician's orders. You will therefore need to understand how to change Roman numerals to Arabic numbers to perform dosage calculations.

In the Roman numeral system, letters are used to represent numbers. Recall from Chapter 3 that lowercase letters are frequently used in pharmacy, especially for apothecary measurements. The Roman numerals that you are likely to encounter in a medical setting include ss $\left(\frac{1}{2}\right)$, I (1), V (5), and X (10). These numerals may be written in either uppercase or lowercase, and a line is sometimes written above lowercase symbols. Thus, the number "one" can be written as I, i, or $\bar{\text{i}}$.

The Roman numerals from 1 to 30 are the ones you are most likely to see in physicians' orders. Table 5-2 summarizes these numerals for you. Remember that "ss" could be added to the end of any of these expressions to add the value of $\frac{1}{2}$.

Combining Roman Numerals

Obviously, it is often necessary to write Roman numerals other than $\frac{1}{2}$, 1, 5, or 10. To do so, the letters are combined into a single expression. The expression can be translated to an Arabic number by following two simple steps.

Rule 5-2 When you read a Roman numeral containing more than one letter, follow these two steps:

1. If any letter with a smaller value appears *before* a letter with a larger value, subtract the smaller value from the larger.
2. Add the values of all the letters not affected by step 1 to those that were combined.

Example 1 a. IX = 10 − 1 = 9 b. iv = 5 − 1 = 4

Example 2 a. XIV = 10 + (5 − 1) = 14 b. VII = 5 + 2 = 7

c. ivss = $(5 - 1) + \frac{1}{2} = 4\frac{1}{2}$ d. VI = (5 + 1) = 6

TABLE 5-2 Converting Roman Numerals

Roman Numeral*	Arabic Number	Roman Numeral	Arabic Number	Roman Numeral	Arabic Number
SS, $\overline{ss}$	$\frac{1}{2}$	XI, xi	11	XXII, xxii	22
I, i	1	XII, xii	12	XXIII, xxiii	23
II, ii	2	XIII, xiii	13	XXIV, xxiv	24
III, iii	3	XIV, xiv	14	XXV, xxv	25
IV, iv	4	XV, xv	15	XXVI, xxvi	26
V, v	5	XVI, xvi	16	XXVII, xxvii	27
VI, vi	6	XVII, xvii	17	XXVIII, xxviii	28
VII, vii	7	XVIII, xviii	18	XXIX, xxix	29
VIII, viii	8	XIX, xix	19	XXX, xxx	30
IX, ix	9	XX, xx	20		
X, x	10	XXI, xxi	21		

* Roman numerals written with small letters are also correctly written with a line over the top; for example, iv and $\overline{iv}$ are both correct.

GO TO . . . Open the CD-ROM that accompanies your textbook, and select Chapter 5, Rule 5-2. Review the animation and example problems, then complete the practice problems. Continue to the next section of the book once you have mastered the rule presented. ■

Critical Thinking on the Job

Understand the Order of Roman Numerals

A medication Administration Record lists a drug order of gr ix. (The abbreviation "gr" stands for grains. See Chapter 3.) A health care professional reading the order thinks, "The Roman numeral i equals 1 and x equals 10, so I should give 11 grains of the drug."

Question?

What mistake did the health care professional make?

REVIEW AND PRACTICE

5-2 Roman Numerals

Convert the following Roman numerals to Arabic numbers.

1. VI **2.** XII **3.** IX **4.** XIV **5.** xxiv **6.** xviii

7. vss **8.** ixss **9.** xiss **10.** xxv **11.** xix **12.** viiiss

Write the answers to the following exercises as Arabic numbers.

13. IV + XVII **14.** xii + xiv **15.** VIII + III **16.** V + V

17. XXIII − VII **18.** xvi − ix **19.** XXI − III **20.** XXX − V

To check your answers, see pages 512–513.

Physicians' Orders and Prescriptions

Physicians use abbreviations when writing orders. Memorize the commonly used ones and have available a complete list of those accepted at your facility. Approved abbreviations vary among facilities. You may encounter either uppercase or lowercase letters and orders with or without punctuation marks. You may also notice slight differences in the way that the abbreviations are spelled. See Table 5-3 for some commonly used abbreviations.

TABLE 5-3 Abbreviations Commonly Used in Drug Orders

General Abbreviations			
Abbreviation	**Meaning**	**Abbreviation**	**Meaning**
aq	water	NPO, n.p.o.	nothing by mouth
aq dist	distilled water	$\bar{p}$, p	after
a, $\bar{a}$	before, ante	q, q., $\bar{q}$	every
aa, $\overline{aa}$	of each	qs	quantity sufficient
BP	blood pressure	R	take
c, $\bar{c}$	with	$\bar{s}$	without
d.c., D/C*	discontinue	sig. s	write on label
disp	dispense	ss, $\overline{ss}$	one-half
et	and	sys	systolic
iss, $\overline{iss}$	one and one-half	tbsp, tbs, T	tablespoon
NKA	no known allergies	tsp, t	teaspoon
NKDA	no known drug allergies	ut dict, ud	as directed
Form of Medication			
Abbreviation	**Meaning**	**Abbreviation**	**Meaning**
cap, caps	capsule	ext, ext.	extract
comp	compound	fld., fl	fluid
dil, dil.	dilute	gtt, gtts	drop, drops
EC	enteric-coated	H	hypodermic
elix, elix.	elixir	LA	long-acting

TABLE 5-3 Continued

liq	liquid	syr, syp.	syrup
MDI	metered-dose inhaler	syr	syringe
sol, soln.	solution	tab	tablet
SR	slow-release	tr, tinct, tinc.	tincture
supp, supp.	suppository	ung, oint	ointment
susp, susp.	suspension		

Route (Where to Administer)

Abbreviation	Meaning	Abbreviation	Meaning
ad, A.D., AD*	right ear	ID	intradermal
as, A.S., AS*	left ear	IM, I.M.	intramuscular
au, A.U., AU*	both ears	IV, I.V.	intravenous
GT	gastrostomy tube	IVP	intravenous push

Route (Where to Administer)

Abbreviation	Meaning	Abbreviation	Meaning
IVPB	intravenous piggyback	per	per, by, through
IVSS	intravenous soluset	po, p.o., PO, P.O.	by mouth; orally
KVO, TKO	keep vein open	R, P.R., p.r.	rectally
NG, NGT, ng	nasogastric tube	sc* SC,* s.c.,* sq,* SQ,* sub-q, Sub-q	subcutaneous, beneath the skin
NJ	nasojejunal tube		
od, O.D., OD	right eye	SL, sl	sublingually, under the tongue
os, O.S., OS	left eye	top, TOP	topical, applied to skin surface
ou, O.U., OU	both eyes		

Frequency

Abbreviation	Meaning	Abbreviation	Meaning
a.c., ac, AC, $\overline{ac}$	before meals	qam, q.a.m.	every morning
ad. lib., ad lib	as desired, freely	qpm, o.n., q.n.	every night
b.i.d., bid, BID	twice a day	q.h., qh	every hour
b.i.w.	twice a week	q. _____ hrs, q_____h	every _____ hours
h, hr	hour	qhs, q.h.s.	every night, at bedtime
h.s.,* hs,* HS*	hour of sleep, at bedtime	q.i.d., qid, QID	4 times a day
LOS	length of stay	rep	repeat
min	minute	SOS, s.o.s.	once if necessary, as necessary
non rep	do not repeat		
n, noc, noct	night	stat	immediately
p.c., pc, PC, $\overline{pc}$	after meals	t.i.d., tid, TID	3 times a day
p.r.n., prn, PRN	when necessary, when required	t.i.w.*	3 times a week

*Indicates an "undesirable" abbreviation as established by JCAHO.

The Joint Commission on Accreditation of Healthcare Organizations (JCAHO), a regulating agency for health care facilities, has established a list of "Do not use" and "Undesirable" abbreviations. (See Tables 5-4 and 5-5.) JCAHO has also developed a list of additional abbreviations, acronyms, and symbols for possible future inclusion in the official "Do Not Use" list. As you will note some of these are already considered "undesirable" and should not be used. See Table 5-6. These lists were developed after research indicated that use of the abbreviations increased the number of medication errors. Be certain to check abbreviations carefully when you read drug orders.

Some physicians use lowercase Roman numerals, such as ii, to indicate numbers. You may see these numerals with a line over them, such as ii. Physicians often use this format for apothecary measurements such as grains. They may also put a line over general and frequency abbreviations, such as *a*, *ac*, *c*, *p*, and *s*, when the abbreviations are lowercase.

Always verify that a drug order contains all information needed to carry it out safely and accurately. It should include the full name of the patient, the full name of the drug, the dose, the route, the time and frequency, the signature of the prescribing physician, and the date of the order. A prn order must include the reason for administering the medication. If an order is unclear, talk with the physician before you carry it out.

TABLE 5-4 "Do Not Use" Abbreviations

Abbreviation	Potential Problem	Preferred Term
U (for unit)	Mistaken as zero, four, or cc.	Write *unit.*
IU (for international unit)	Mistaken as IV (intravenous) or 10.	Write *international unit.*
Q.D., Q.O.D. (Latin abbreviations for once daily and every other day)	Mistaken for each other. The period after the Q can be mistaken for an I and the O can be mistaken for I.	Write *daily* and *every other day.*
Trailing zero (X.0 mg) [*Note:* Prohibited only for medication-related notations] Lack of leading zero (.X mg)	Decimal point is missed.	Never write a zero by itself after a decimal point (X mg), and always use a zero before a decimal point (0.X mg).
MS MSO_4 $MgSO_4$	Confused for one another. Can mean morphine sulfate or magnesium sulfate.	Write *morphine sulfate* or *magnesium sulfate.*

Outpatient Settings

For outpatient settings, physicians' orders are given as **prescriptions** that are filled at a pharmacy or through the mail. Prescriptions include all the elements of a physician's order, as well as the physician's name and prescriber number, the quantity to be dispensed, the number of refills permitted, and instructions for the label of the container. These instructions are preceded by the word **sig** (see Figure 5-1).

The patient is Arthur Simons. The drug is Doxycycline. The dose is 100 mg. From the **sig** line, the instructions on the label should read, "Take one capsule twice a day after meals" or one capsule (cap i), twice a day (BID), by mouth (po), after meals (pc). Form, number, route, frequency, and timing are all shown. The quantity (quan) of capsules is 20 The prescription cannot be refilled. The physician's name, prescriber number, and signature are present. This order contains all the required elements.

TABLE 5-5 Undesirable Abbreviations

Abbreviation	Potential Problem	Preferred Term
µg (for microgram)	Mistaken for mg (milligrams), resulting in 1000-fold dosing overdose.	Write *mcg.*
H.S. (for half-strength, or Latin abbreviation for bedtime)	Mistaken for either half-strength or hour of sleep (at bedtime). q.H.S. mistaken for every hour. All can result in a dosing error.	Write out *half-strength* or *at bedtime.*
T.I.W. (for 3 times a week)	Mistaken for 3 times a day or twice weekly, resulting in an overdose.	Write *3 times weekly* or *three times weekly.*
S.C. or S.Q. (for subcutaneous)	Mistaken as SL for sublingual, or "5 every."	Write *Sub-Q, subQ,* or *subcutaneously.*
D/C (for discharge)	Interpreted as discontinue whatever medications follow (typically discharge medications).	Write *discharge.*
c.c. (for cubic centimeter)	Mistaken for U (units) when poorly written.	Write *mL* for milliliters.
A.S., A.D., A.U. (Latin abbreviation for left, right, or both ears)	Mistaken for OS, OD, and OU.	Write *left ear, right ear,* or *both ears.*

TABLE 5-6 Additional Abbreviations, Acronyms, and Symbols (For possible future inclusion in the Official "Do Not Use" List)

Do Not Use	Potential Problem	Use Instead
> (greater than); < (less than)	Misinterpreted as the number 7 (seven) or the letter L. Confused for one another.	Write "greater than" or "less than."
Abbreviations for drug names	Misinterpreted because of similar abbreviations for multiple drugs.	Write drug names in full.
Apothecary units	Unfamiliar to many practitioners. Confused with metric units.	Use metric units.
@	Mistaken for the number 2 (two).	Write "at."
cc	Mistaken for U (units) when poorly written.	Write "mL" or "milliliters."
µg	Mistaken for mg (milligrams), resulting in a thousand-fold overdose.	Write "mcg" or "micrograms."

GO TO . . . Open the CD-ROM that accompanies your textbook, and select Chapter 5, Practice 5-3. Review the animation and example problems, then complete the practice problems. Continue to the next section of the book once you have mastered the rule presented. ■

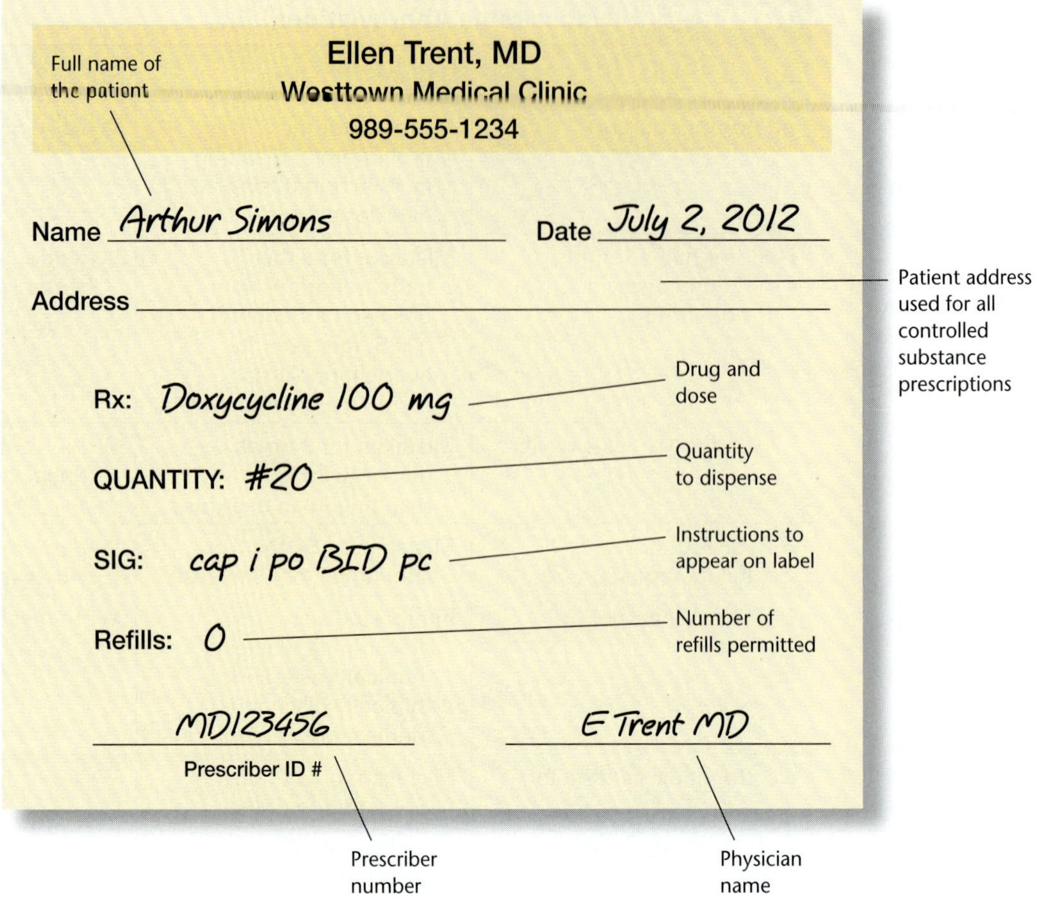

Figure 5-1 A typical outpatient prescription.

In an outpatient setting, sometimes the physician wants the medication given right away in addition to or instead of writing a prescription. In this case, the physician will write the order in the chart or provide it verbally. Verbal orders are discussed later. In either case, you must check the order for completeness, including the seven rights. Once the order is administered, it should be documented in the patient chart on the progress notes. For example, you might document a medication like this:

5/16/06 500 mg penicillin administered IM to left ventrogluteal site at 1000 (your initials)

Inpatient Settings

For inpatient settings, drug orders are usually written on **physicians' order forms**, with space for multiple orders. Orders may also be entered into a computer. The patient's name and the physician's signature appear once on the form. Under *Medication Orders*, the physician writes the components of each medication requested in the following sequence: name of drug, dose, route, frequency, and additional instructions.

The form in Figure 5-2 shows several medication orders; some are correct and others have errors. Order 1 contains all necessary components. The drug is Lasix, the dose is 20 mg, the route is oral (po), and the frequency is once a day (daily).

Order 2, for KCl (potassium chloride) elixir, is not complete. The order lists 1 tsp as the amount, but KCl elixir is available in strengths of 10, 20, 30, or 40 mEq per 15 mL. Each strength provides a different dose per teaspoon.

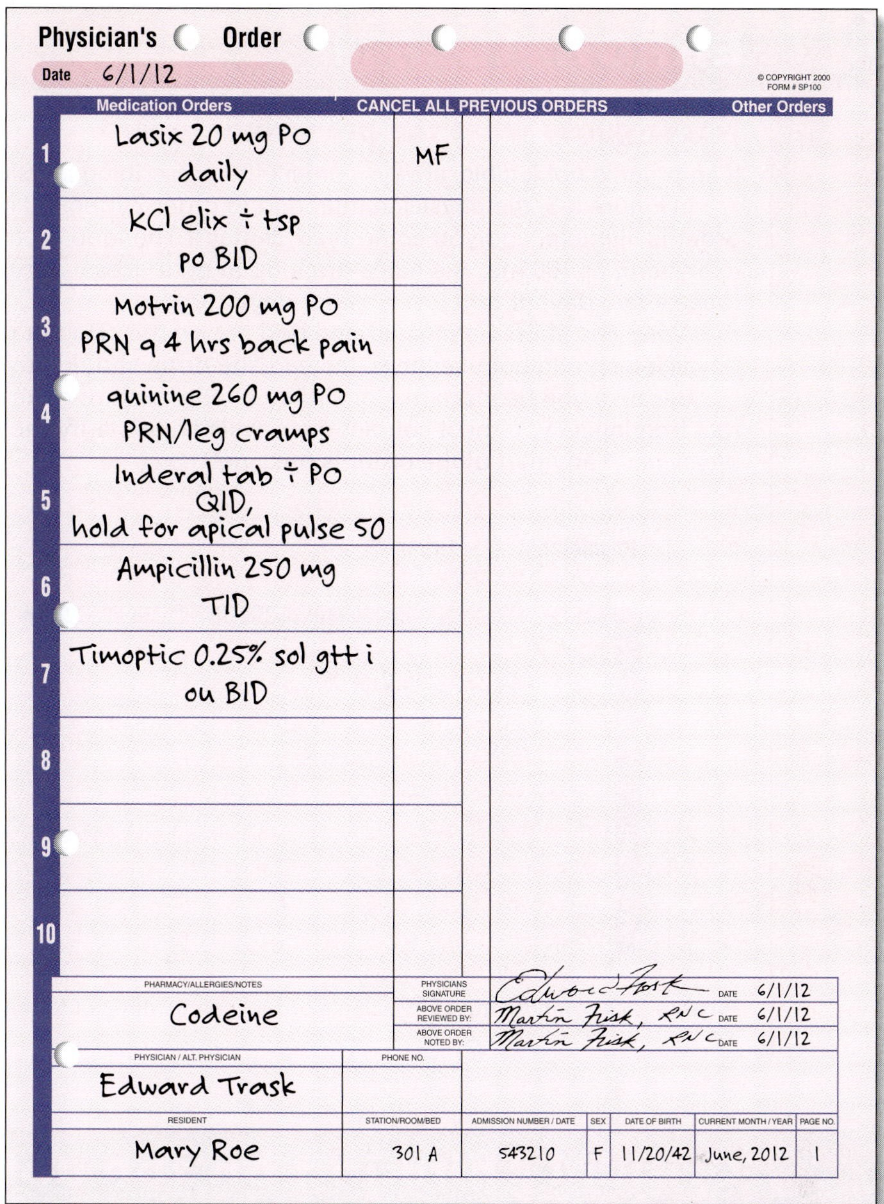

Physician's Order

Date 6/1/12

© COPYRIGHT 2000
FORM # SP100

	Medication Orders		CANCEL ALL PREVIOUS ORDERS	Other Orders
1	Lasix 20 mg PO daily	MF		
2	KCl elix ī tsp po BID			
3	Motrin 200 mg PO PRN q 4 hrs back pain			
4	quinine 260 mg PO PRN/leg cramps			
5	Inderal tab ī PO QID, hold for apical pulse 50			
6	Ampicillin 250 mg TID			
7	Timoptic 0.25% sol gtt i ou BID			
8				
9				
10				

PHARMACY/ALLERGIES/NOTES: Codeine

PHYSICIANS SIGNATURE: Edward Trask DATE 6/1/12

ABOVE ORDER REVIEWED BY: Martin Fisk, RNC DATE 6/1/12

ABOVE ORDER NOTED BY: Martin Fisk, RNC DATE 6/1/12

PHYSICIAN / ALT. PHYSICIAN: Edward Trask

PHONE NO.

RESIDENT	STATION/ROOM/BED	ADMISSION NUMBER / DATE	SEX	DATE OF BIRTH	CURRENT MONTH / YEAR	PAGE NO.
Mary Roe	301 A	543210	F	11/20/42	June, 2012	1

Figure 5-2 Physician's order form.

Order 3 is correct. The drug is Motrin®, the dose is 200 mg, the route is oral (po), and the frequency is whenever necessary (PRN) every 4 hours (q 4 h). For PRN drugs, the physician must specify how often it may be given. Here, Motrin® may not be administered more often than every 4 h.

Order 4 is not complete because it does not include an adequate frequency. The quinine is ordered PRN, when necessary for leg cramps. Yet no minimum time interval is specified.

In order 5, the physician has included instructions for "holding" (not administering) the medication. You need to check if the patient's apical pulse is below 50 before administering Inderal. The order is not complete because the physician has ordered one tablet, not the actual dose. Inderal is available in 10-, 20-, 40-, 60-, and 90-mg tablets.

Order 6 does not specify the route for 250 mg Ampicillin, available in oral, intramuscular, and intravenous forms. Consult the physician to find the intended route.

Unlike order 6, order 7 is complete. The physician specified the number of drops as well as the strength of Timoptic® desired.

ERROR ALERT!

Never guess what the prescriber meant.

If an order is not legible, always contact the prescribing physician to clarify the order.

In Figure 5-3, the physician intended to order Zyrtec® 10 mg po qd. However, the order is illegible, and you could read Zantac®. The loop on the m in mg could be an extra zero, or 100 mg. A small extra loop in qd makes it hard to tell if the order is *qd* (once every day) or qod (every other day).

Note: The abbreviations qd and qod are on the "Do not use" list by JCAHO. The physician should not use them. Instead she or he should have written *daily.*

Because you bear the responsibility of administering the correct dose at the correct frequency, you must contact the physician to clarify the order. Also contact the physician if any part of the order is missing.

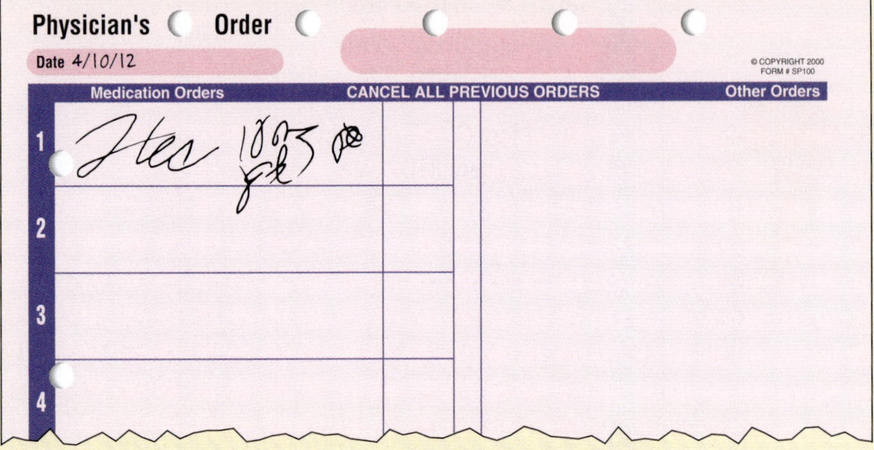

Figure 5-3 Handwritten Physician's orders can be difficult to read.

Critical Thinking on the Job

The Importance of the Right Dose

A physician's order reads Compazine® supp i pr q4h PRN/nausea. The health care professional interprets this order as "administer 1 Compazine® suppository rectally every four hours as necessary for nausea."

He assumes that the patient is an adult and dispenses 25-mg suppositories, the normal adult dose. In turn another health care professional, who does not notice that the dose is not specified in the order, administers the 25-mg suppository to the patient, a 6-year-old boy.

The usual dose of Compazine® for children is a 2.5-mg suppository. The pediatrician who wrote the order did not include the dose, assuming the staff would know this information. The child receives 10 times the normal dose of Compazine®. He has a seizure and develops fever, respiratory distress, severe hypotension, and tachycardia because of drug toxicity. He is admitted to the intensive care unit for treatment.

Question?

What could have been done to prevent this problem?

Verbal Orders

Usually, orders must be written or personally signed by a physician. However, if a physician is not able to write an order that must be carried out quickly, verbal orders may be permitted. State laws govern which personnel may accept such orders and how soon the physician must countersign them.

If you are legally permitted to accept a telephone order, write it carefully and legibly *as* you receive it, *not after* the call. In some cases, you may write the order on the physician's order form, identifying it as a verbal order. *Always read the order back to the physician to verify that you have transcribed it correctly.* If you are not certain of the spelling of the drug name, ask the physician to spell it. Many drugs have names that are pronounced or spelled similarly.

If you must prepare a medication from a verbal order, take special precautions. Double-check that you have prepared the medication accurately. If a physician is present, show the label before you administer the medication or before you hand it to the physician to administer. State the medication and the dose so that the physician can verify that what you have prepared is what was ordered.

ERROR ALERT!

Always be certain that you are dispensing the correct medication.

Many drugs have names that are very similar. Read the order carefully and, when in doubt, contact the prescribing physician. The following list gives just a few examples of how similar the names of different drugs can look and sound. It is especially easy to confuse them when they are written rather than printed.

Acular®—Ocular
Benadryl®—Bentyl®
Cafergot®—Carafate®
Darvon®—Diovan®
Digitoxin—digoxin
Eurax—Urex®
Iodine—Lodine®
Nicobid—Nitrobid®
Pavabid—Pavased
Phenaphen—Phenergan®
Quinidine—Quinine
Uracel—Uracil

Critical Thinking on the Job

The Importance of the Right Drug

A patient is brought to the hospital with a severe thumb laceration. The attending physician verbally orders lidocaine 1% solution 2 mL as a local anesthetic. The health care professional picks up a vial labeled lidocaine 1% with epinephrine and draws up 2 mL. He then says, "This is lidocaine 1% solution 2 mL," but neither mentions the epinephrine nor shows the physician the label.

A while later, the patient expresses concern about continuing numbness in his thumb. After locating the vial, the staff member realized that the patient received epinephrine, a vasoconstricting drug, in addition to the lidocaine. The patient is reassured that feeling will return to his thumb, although not quite as quickly as was first anticipated.

Question?

What could have been done to prevent the patient from losing feeling in his thumb?

REVIEW AND PRACTICE

5-3 Physicians' Orders and Prescriptions

In Exercises 1–5, refer to prescription A.

1. What components, if any, are missing from prescription A?
2. How many Lopressor® tablets should the pharmacy technician dispense?
3. How often should the patient take Lopressor®?
4. What strength tablets should be dispensed?
5. If the patient gets all the refills permitted, how long will the medication covered by this prescription last?

Alan Capsella, MD
Westtown Medical Clinic
989-555-1234

Name Ann Pechin Date July 9, 2012

Address

Rx: Lopressor 50 mg

QUANTITY: #90

SIG: tab i po tid

Refills: 5

MD398475 Alan Capsella MD

Prescriber ID #

Prescription A

In Exercises 6–10, refer to prescription B.

6. What components, if any, are missing from prescription B?
7. How much Amoxil® should the pharmacy technician dispense?
8. How much Amoxil® should the patient take at one time?
9. How many times can this prescription be refilled?
10. How often should the patient take Amoxil®?

Alan Capsella, MD
Westtown Medical Clinic
989-555-1234

Name Mark Ward Date April 10, 2012

Address

Rx: Amoxil - oral susp

QUANTITY: 100 mL

SIG: i tsp po q8h

Refills: 0

MD398475 Alan Capsella MD

Prescriber ID #

Prescription B

In Exercises 11–15, refer to prescription C.

Alan Capsella, MD
Total Care Clinic
989-555-1234

Name B. Talbott Date May 10, 2012

Address

Rx: Norvasc

QUANTITY: #30

SIG: i po

Refills: 2

MD239485 Cynthia Buckwalter MD

Prescriber ID #

Prescription C

11. How many components, if any, are missing from prescription C?
12. How many Norvasc® tablets should the pharmacy technician dispense?
13. How much Norvasc® should the patient take at a time?
14. How many times can this prescription be refilled?
15. If the patient gets all the refills permitted, how long will the medication covered by the prescription last?

In Exercises 16–25, refer to Physician's Order Form 1.

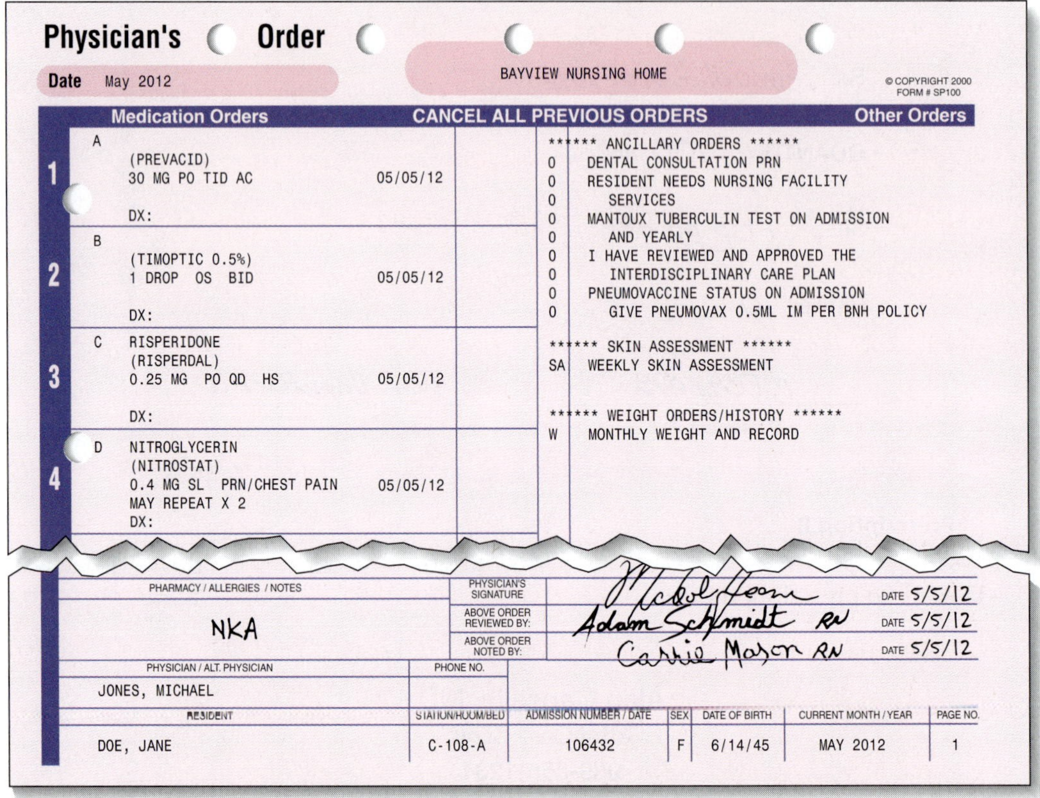

Physician's Order

Date May 2012

BAYVIEW NURSING HOME

© COPYRIGHT 2000 FORM # SP100

Medication Orders — CANCEL ALL PREVIOUS ORDERS — Other Orders

1 A (PREVACID) 30 MG PO TID AC 05/05/12 DX:

2 B (TIMOPTIC 0.5%) 1 DROP OS BID 05/05/12 DX:

3 C RISPERIDONE (RISPERDAL) 0.25 MG PO QD HS 05/05/12 DX:

4 D NITROGLYCERIN (NITROSTAT) 0.4 MG SL PRN/CHEST PAIN 05/05/12 MAY REPEAT X 2 DX:

****** ANCILLARY ORDERS ******
0 DENTAL CONSULTATION PRN
0 RESIDENT NEEDS NURSING FACILITY SERVICES
0 MANTOUX TUBERCULIN TEST ON ADMISSION AND YEARLY
0 I HAVE REVIEWED AND APPROVED THE INTERDISCIPLINARY CARE PLAN
0 PNEUMOVACCINE STATUS ON ADMISSION
0 GIVE PNEUMOVAX 0.5ML IM PER BNH POLICY

****** SKIN ASSESSMENT ******
SA WEEKLY SKIN ASSESSMENT

****** WEIGHT ORDERS/HISTORY ******
W MONTHLY WEIGHT AND RECORD

PHARMACY / ALLERGIES / NOTES: NKA

PHYSICIAN'S SIGNATURE: Michael Jones DATE 5/5/12
ABOVE ORDER REVIEWED BY: Adam Schmidt RN DATE 5/5/12
ABOVE ORDER NOTED BY: Carrie Mason RN DATE 5/5/12

PHYSICIAN / ALT. PHYSICIAN: JONES, MICHAEL — PHONE NO.:

RESIDENT	STATION/ROOM/BED	ADMISSION NUMBER / DATE	SEX	DATE OF BIRTH	CURRENT MONTH / YEAR	PAGE NO.
DOE, JANE	C-108-A	106432	F	6/14/45	MAY 2012	1

Physician's Order Form 1

16. When should Prevacid® be administered?
17. By what route should Timoptic® be given?
18. How often should Risperdal® be administered?
19. How much Nitrostat® should be given in a single dose?
20. By what route should Prevacid® be given?
21. How often should Timoptic® be administered?
22. How much Risperdal® should be given in a single dose?
23. When should Risperdal® be given?
24. By what route should Nitrostat® be given?
25. If medications are delivered once a week to this facility, how many doses of Risperdal® should be dispensed?

To check your answers, see pages 513–514.

Medication Administration Systems

Most facilities have a standard schedule for administering medication (see Table 5-7). To minimize errors, many of them use the 24-h clock (see Chapter 3). The person who verifies the transcription ensures that the times listed are appropriate for the medications. For example, some medications need to be given with food or after a meal; others, on an empty stomach. Times may need to be adjusted to accommodate a patient's meals. A patient may take two or more medications with conflicting schedules. Again, the timing may need to be adjusted.

TABLE 5-7 Sample Times for Medication Administration

Frequency Ordered	Times to Administer
Daily	0800
bid	0800, 2000
tid	0800, 1400, 2000
qid	0800, 1200, 1600, 2000
q 12 hrs	0800, 2000
q 8 hrs	2400, 0800, 1600
q 6 hrs	2400, 0600, 1200, 1800
qhs	2000

Medication Administration Records

Medication Administration Records (MARs) are legal documents that may be handwritten forms or computerized printouts. MARs contain the same information as a physicians' order form and also specifies the actual times to administer the medication. Additionally it provides a place to document that each medication has been given. *Remember:* by law, when you administer medication, you must *immediately* document this step.

Rule 5-3 MARs must include the following information:

1. The full name of the medication, the dose, the route, and the frequency.
2. Times that accurately reflect the frequency specified.
3. The full name and identification number of the patient.
4. The date the order was written. If no start date is listed, then the assumption is that the date of order is the start date. Orders for narcotics and antibiotics should include end dates, according to your facility's policies.
5. Any special instructions or information as required by your facility. This includes, but is not limited to, the patient's diagnosis and weight.

Example 1 Determine whether the MAR in Figure 5-4 is complete.

In Figure 5-4 all three orders are written correctly. In order A, the drug is Nitrobid® 2% cream, the dose is $\frac{1}{2}$ inch, the route is topical, and the frequency is every 6 hr. The scheduled times are 2400 (midnight), 0600 (6:00 a.m.), 1200 (12:00 p.m.), and 1800 (6:00 p.m.).

In order B, the drug is Vasotec®, the dose is 10 mg, the route is oral, and the frequency is twice a day. This order includes a special instruction to hold the medication if the systolic blood pressure is below 100

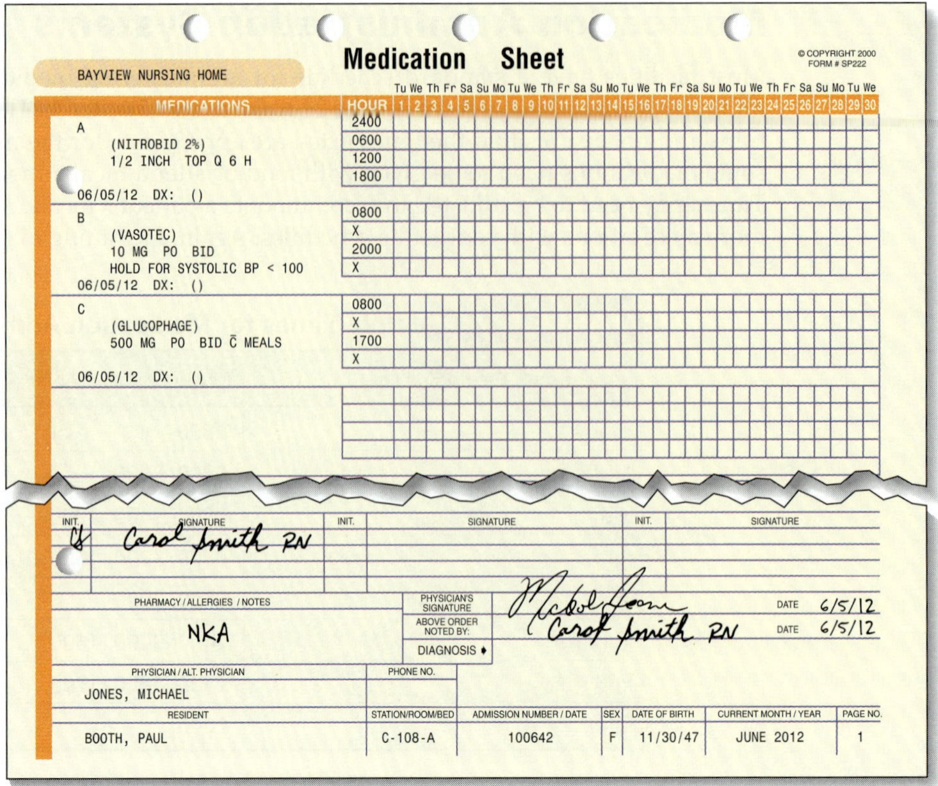

BAYVIEW NURSING HOME

Medication Sheet

© COPYRIGHT 2000
FORM # SP222

MEDICATIONS	HOUR
A (NITROBID 2%) 1/2 INCH TOP Q 6 H 06/05/12 DX: ()	2400 0600 1200 1800
B (VASOTEC) 10 MG PO BID HOLD FOR SYSTOLIC BP < 100 06/05/12 DX: ()	0800 X 2000 X
C (GLUCOPHAGE) 500 MG PO BID c̄ MEALS 06/05/12 DX: ()	0800 X 1700 X

Tu We Th Fr Sa Su Mo Tu We Th Fr Sa Su Mo Tu We Th Fr Sa Su Mo Tu We Th Fr Sa Su Mo Tu We
1 2 3 4 5 6 7 8 9 10 11 12 13 14 15 16 17 18 19 20 21 22 23 24 25 26 27 28 29 30

INIT.	SIGNATURE	INIT.	SIGNATURE	INIT.	SIGNATURE
CS	Carol Smith RN				

PHARMACY / ALLERGIES / NOTES: NKA

PHYSICIAN'S SIGNATURE: Michael Jones — DATE 6/5/12

ABOVE ORDER NOTED BY: Carol Smith RN — DATE 6/5/12

DIAGNOSIS ♦

PHYSICIAN / ALT. PHYSICIAN	PHONE NO.
JONES, MICHAEL	

RESIDENT	STATION/ROOM/BED	ADMISSION NUMBER / DATE	SEX	DATE OF BIRTH	CURRENT MONTH / YEAR	PAGE NO.
BOOTH, PAUL	C-108-A	100642	F	11/30/47	JUNE 2012	1

Figure 5-4 Medication administration record (MAR).

In order C, the drug is Glucophage®, the dose is 500 mg, the frequency is twice a day, and the route is oral. Glucophage® must be administered with meals. Therefore, the times have been adjusted to fit the facility's meal schedule.

Example 2 Determine whether the MAR in Figure 5-5 is complete.

In Figure 5-5, order A is correct. The drug is Synthroid®, the dose is 50 mcg, the route is oral, and the frequency is daily.

Order B may seem to have all required information, but the dose—1 tablet—is not adequate. Erythromycin is available in several strengths. The order does not specify which dosage strength is intended.

Order C contains an error in the times listed. Persantine® 75 mg is to be given q6h, or every 6 h. The times (2400, 0800, 1600) are for every 8 h. If the error is not recognized, then the patient may not receive a therapeutic level of the drug.

Order D does not include a route. Heparin can be administered either subcutaneously or intravenously. This order cannot be carried out as written.

GO TO . . . Open the CD-ROM that accompanies your textbook, and select Chapter 5, Rule 5-3. Review the animation and example problems, then complete the practice problems. Continue to the next section of the book once you have mastered the rule presented. ■

Medication Cards

Medication cards are used infrequently but discussed here in case you encounter them at your place of employment. Each drug order is listed on its own card. The cards for all patients

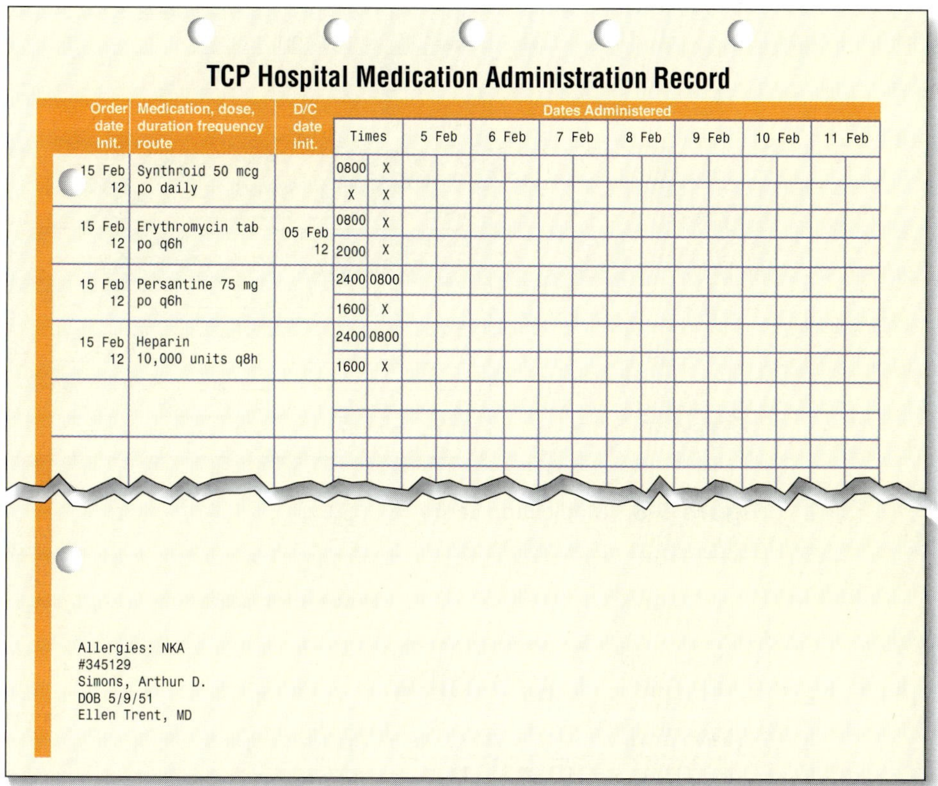

TCP Hospital Medication Administration Record

Order date Init.	Medication, dose, duration frequency route	D/C date Init.	Times		5 Feb	6 Feb	7 Feb	8 Feb	9 Feb	10 Feb	11 Feb
					Dates Administered						
15 Feb 12	Synthroid 50 mcg po daily		0800	X							
			X	X							
15 Feb 12	Erythromycin tab po q6h	05 Feb 12	0800	X							
			2000	X							
15 Feb 12	Persantine 75 mg po q6h		2400	0800							
			1600	X							
15 Feb 12	Heparin 10,000 units q8h		2400	0800							
			1600	X							

Allergies: NKA
#345129
Simons, Arthur D.
DOB 5/9/51
Ellen Trent, MD

Figure 5-5 Medication administration record (MAR).

are stored on a special rack according to the time of the next medication administration. At these specified hours, a health care worker gathers the cards and prepares the medications. After the medications have been administered, the cards are refiled based on the next set of administration times. Most medication cards have a grid representing all the hours of the day. The times that medication is to be given are marked on the grid. A medication card must contain the patient's name and room number, the full name of the drug, the dose, the route, the frequency, any special instructions, the administration times, and the date ordered.

Medication card systems are used infrequently because they have several disadvantages. Because of their small size, the cards are easily misplaced. Each medication is listed separately, increasing the potential for omitting a medication. Also, medication cards generally do not list allergies or diagnoses. By contrast, MARs list all drugs together where health care workers can see the big picture, including orders for medications that should not interact with each other. See Figure 5-6 for a sample medication card.

Computerized Medical Records

To provide the most effective care, health care providers should be aware of a patient's complete medical history, including information about earlier reactions to medications and any previously diagnosed medical conditions. This information is typically recorded on paper, but in many facilities computers are used.

Patient information is often entered directly into a computer either via a hand module or console at the patient's bedside or at a centralized console. Information can be easily updated and is accessible from the nurse's station, pharmacy, physician desk, or other areas. All information is stored on the hard drive or server. Hard copies can be printed at will and used for backup copies.

A computerized medical record may include all orders for medications, administration times, and the patient's reaction to that medication. Lab results and observations, information about allergies, and other records can be obtained immediately. A computerized record can even be used to schedule anticipated procedures or to plan for the patient's eventual discharge. (See Figure 5-7.)

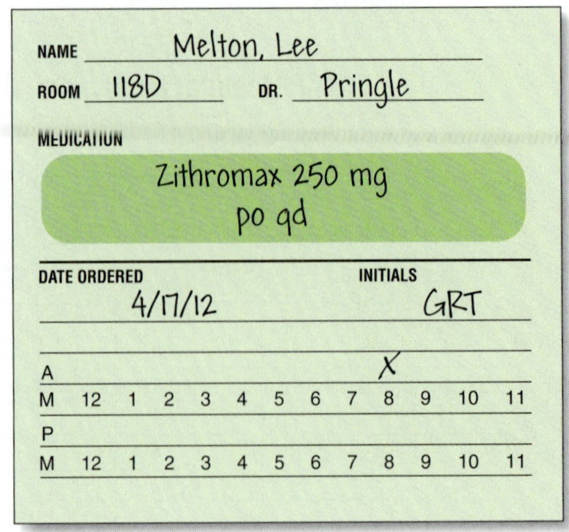

NAME Melton, Lee

ROOM 118D DR. Pringle

MEDICATION

Zithromax 250 mg
po qd

DATE ORDERED 4/17/12 INITIALS GRT

A
M 12 1 2 3 4 5 6 7 8 9 10 11

P
M 12 1 2 3 4 5 6 7 8 9 10 11

Figure 5-6 Medication card A.

```
6K   -0626                        CENTRAL CITY HOSPITAL                         PAGE 001
09/14/12  07:05

=================================
SMITH, JOHN                     M     39
SSN: 123456789    ACCT#: 501330269251
SERV: MULT        6K        067999
MD: JONES, MARY
DX: STAB WOUND IN ABD                                              PATIENT CARE WORKSHEET
SHORT STAY:                                                               TEMPORARY
=================================
WORKSHEET: 09/14 07:15    TO 15:15

 DATA RETRIEVED ACROSS ALL ADMISSIONS:
      09/08/12       DRUG ALLERGY: NO KNOWN ALLERGY
PATIENT INFORMATION:
      09/08          ADMIT DX:STAB WOUND IN ABD
      09/08          PRINCIPAL DX:S/P EX LAP, DUODENAL REPAIR, MESENTERIC VESSEL REPAIR
      09/08          WT: ADM WT: 77.1 KG
ALL CURRENT MEDICAL ORDERS:
  DOCTOR TO NURSE ORDERS:
      09/09   85.00 ACTIVITY, OUT OF BED, AMBULATE IN HALL TID, <09/09/12>
      09/08          --ALL ORDERS ON MULTIPLE TRAUMA PATIENTS TO BE WRITTEN OR COUNTERSIGNED
                     BY TRAUMA SERVICE EXCEPT FOR ICP MANAGEMENT ORDERS
      09/08          --NURSING CARE PER BURN/TRAUMA STANDARD OF CARE
      09/08          NOTIFY MD --FOR TEMP GREATER THAN 101.5
      09/08          --PNEUMATIC STOCKINGS THIGH -- FOR DVT PROPHYLAXIS
      09/08          --NG TO LCS.
      09/08          INCENTIVE SPIROMETER Q1H W/AWAKE
      09/08          JEJUNOSTOMY TUBE --FLUSH W/20CC NS Q6HRS.
      09/09          INCENTIVE SPIROMETER Q1H --COUGH/DEEP BREATH
      09/10          --INCREASE TF BY 20CC Q4HR UNTIL REACH GOAL OF 110/HR DONT CHECK
                     RESIDUALS --DECREASE IVF PROPORTIONAL TO TF
      09/12          --GLYCERIN SWABS TO BEDSIDE
      09/14          --HEP.LOCK IV
DIET:
      09/08          DIET: NPO, <09/08/12>
SCHEDULED MEDICATIONS:
 *    09/08          (ZANTAC) RANITIDINE IVPB 50 MG, IN 50ML 1/2 N/S, Q8H,
                     (09/08/12 13:00-..)  13
F     09/13          (ATIVAN) *LORAZEPAM INJ (2MG/1ML) 0.5 MG, IV, NOW, --FOR AGITATION,
                     (09/13/12 05:24)
 *    09/13          THIAMINE TAB 100MG,, #1, (GIVE 1 TAB) JT, DAILY, (09/13/12 09:00-..) 09
 *    09/13          (FOLATE) FOLIC ACID TAB 1MG,, #1 (GIVE 1 TAB) JT, DAILY
                     (09/13/12 09:00-..) 09
 *    09/13          MULTIVIT THERAPEUTIC LIQ 5ML, JT, DAILY, (09/13/12 09:00-..) 09
F*    09/13          (ATIVAN) *LORAZEPAM INJ (2MG/1ML) 0.5 MG, IV, Q6H, STARTING AT 14:00,
                     START TODAY, (09/13/12 14:00-..) 09 14
 *    09/14          (DEMEROL +**MEPERIDINE INJ (50MG/1ML) 50MG, IM, NOW, (09/14/12 05:10)
PRN MEDICATIONS:
      09/14          (DEMEROL) +** MEPERIDINE 50MG, IM, (GIVE: 50MG TO 75MG), Q3H, PRN,
                     PAIN, <09/14/12 05:10-..>
      09/14          +**MEPERIDINE 75MG, IM, (GIVE: 50MG TO 75MG), Q3H, PRN, PAIN, <09/14/12
                     05:10-..>

      R=TIME TO RENEW    U=UNVERIFIED    F=REFRIGERATE    *=SCHEDULED MED
      APOSTROPHE=BILL ON ORDER    +=MED TEACHING SHEET AVAILABLE
      ++=PROVIDE PT W/DRUG LEAFLET OBTAINED FROM PHARMACY
IV-S: (WITH OR WITHOUT ADDITIVES):
U     09/13          IV SALINE LOCK CONVERT TO
U     09/12          DRUG --KCL20MEQ/MVI5ML/THIAMINE10 OMG/MGSO4 2GM, $25.00. 00999 D5-
                     1/2NS, 1000ML, 100ML/H, DAILY, CONT TIL DC'D (MISC MED), <09/12/12
                     09:14-..>
U     09/12          KCL 20MEQ, PREMIXED/D5-1/2NS 1000ML, IV 40ML/H, CONT TIL DC'D
                     (MISC MED), <09/12/12 12:56-..>
==========================================================================================
SMITH, JOHN                                                         PATIENT CARE WORKSHEET
```

Figure 5-7 A computerized medical record contains all information connected with a patient's hospital stay, including all ordered medications.

A computerized record is sometimes used as the ongoing working document that records medications as they are administered. These are called **electronic Medication Administration Records (eMARs)**. Users of these systems must log in to the computer, entering their names and secure passwords. When they administer medications they must scan the bar code on the medication as well as the patient's identification band. The information is documented directly into the computer after the medication is administered. This system allows the physician to enter the order, the pharmacy to fill the order, and the medication to be administered and tracked electronically to help reduce medication errors. (See Figures 5-8 and 5-9)

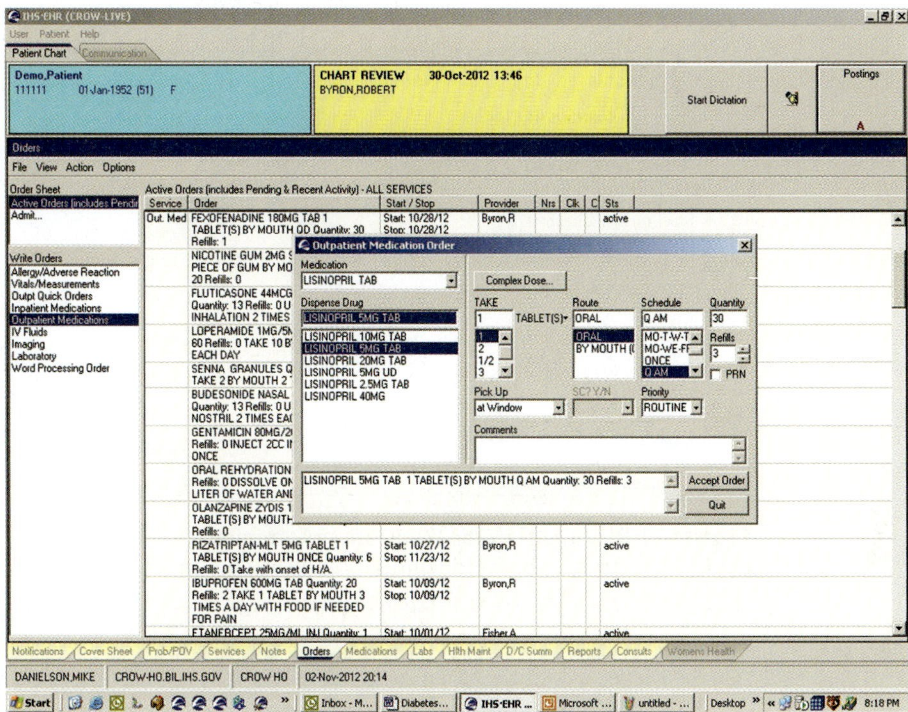

Figure 5-8 For easy access, computerized medication records like this one are available at computer terminals throughout the health care facility.

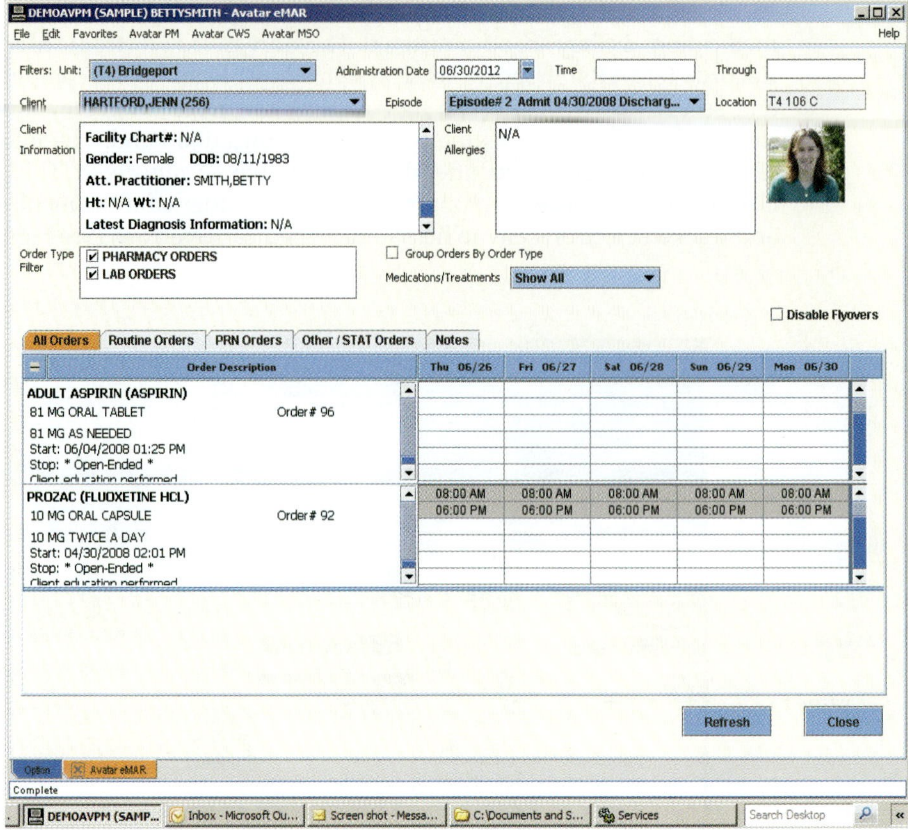

Figure 5-9 An electronic Medication Administration Record (eMAR) like this one allows the health care practitioner to document medication administration electronically.

**Netsmart Technologies (www.ntst.com)*

Critical Thinking on the Job

When in Doubt, Check

A health care professional is preparing medications. An entry in the MAR calls for 600 mg of Lasix® IV, higher than what she usually administers. She checks the *Physicians' Desk Reference.* It indicates that doses this high may be used for congestive heart failure, the patient's diagnosis. Still, she is not comfortable with this level. She checks the original order on the physician's order form and discovers a potential error (see Figure 5-10).

Question?

View the physician order in Figure 5-10 and explain what serious error was prevented and how.

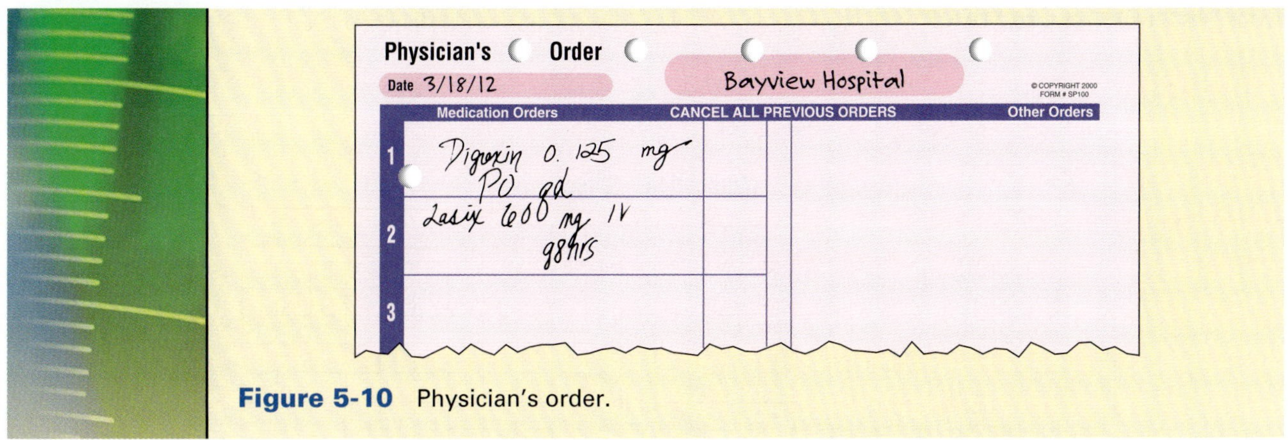

Physician's Order

Date 3/18/12

Bayview Hospital

© COPYRIGHT 2000
FORM # SP100

	Medication Orders	CANCEL ALL PREVIOUS ORDERS		Other Orders
1	Digoxin 0.125 mg PO qd			
2	Lasix 600 mg IV q8hrs			
3				

Figure 5-10 Physician's order.

REVIEW AND PRACTICE

5-4 Medication Administration Systems

Exercises 1–7 refer to MAR 1.

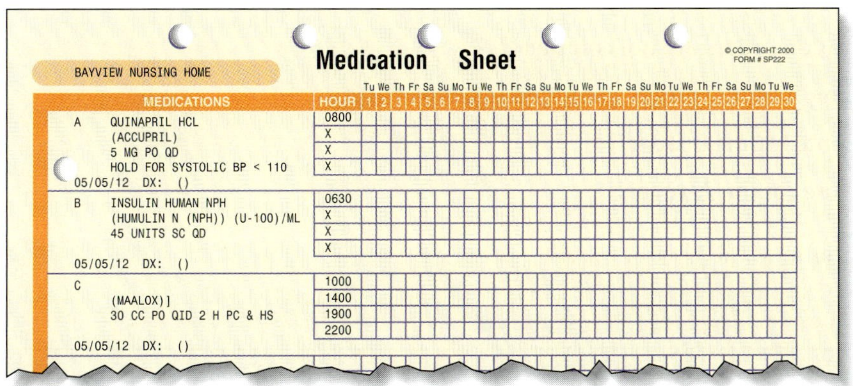

BAYVIEW NURSING HOME

Medication Sheet

© COPYRIGHT 2000
FORM # SP222

MEDICATIONS	HOUR
A QUINAPRIL HCL (ACCUPRIL) 5 MG PO QD HOLD FOR SYSTOLIC BP < 110 05/05/12 DX: ()	0800, X, X, X
B INSULIN HUMAN NPH (HUMULIN N (NPH)) (U-100)/ML 45 UNITS SC QD 05/05/12 DX: ()	0630, X, X, X
C (MAALOX)] 30 CC PO QID 2 H PC & HS 05/05/12 DX: ()	1000, 1400, 1900, 2200

MAR 1

1. What action must you take before administering Accupril®?
2. What dose of Accupril® should this patient receive?
3. At what time is the insulin to be given?
4. By what route is the insulin given?
5. This unit's schedule for QID medications is 0800, 1200, 1700, 2000. Why is Maalox® scheduled for different times?
6. How much Maalox® will this patient receive at 1400?
7. What dose of insulin should this patient receive?

Exercises 8 to 13 refer to MAR 2.

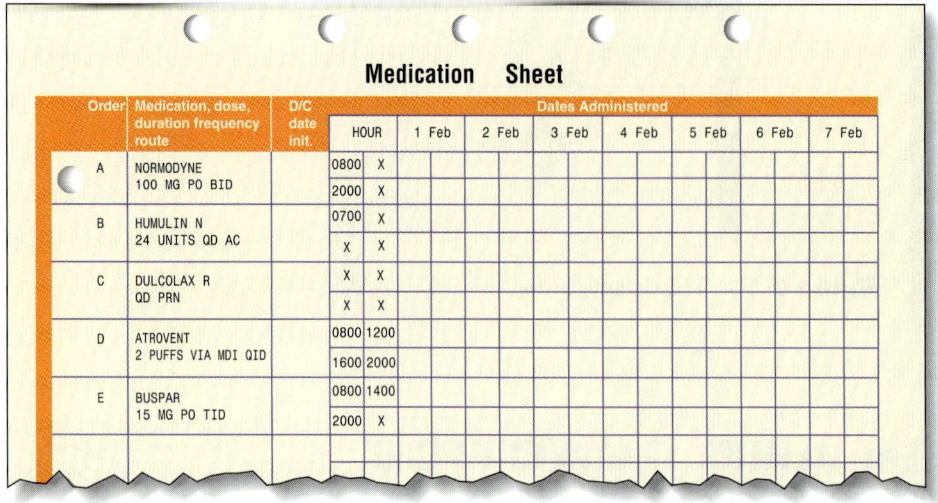

Medication Sheet

Order	Medication, dose, duration frequency route	D/C date init.	HOUR		1 Feb	2 Feb	3 Feb	4 Feb	5 Feb	6 Feb	7 Feb
A	NORMODYNE 100 MG PO BID		0800	X							
			2000	X							
B	HUMULIN N 24 UNITS QD AC		0700	X							
			X	X							
C	DULCOLAX R QD PRN		X	X							
			X	X							
D	ATROVENT 2 PUFFS VIA MDI QID		0800	1200							
			1600	2000							
E	BUSPAR 15 MG PO TID		0800	1400							
			2000	X							

MAR 2

8. Why are no times marked beside the Dulcolax®?
9. Which medications should be given at 8 a.m.?
10. Which medication orders are not complete?
11. What is missing from the orders identified in answer 10?
12. What is the route for the Buspar®?
13. What is the dose for the Atrovent®?

Exercises 14 to 20 refer to MAR 3.

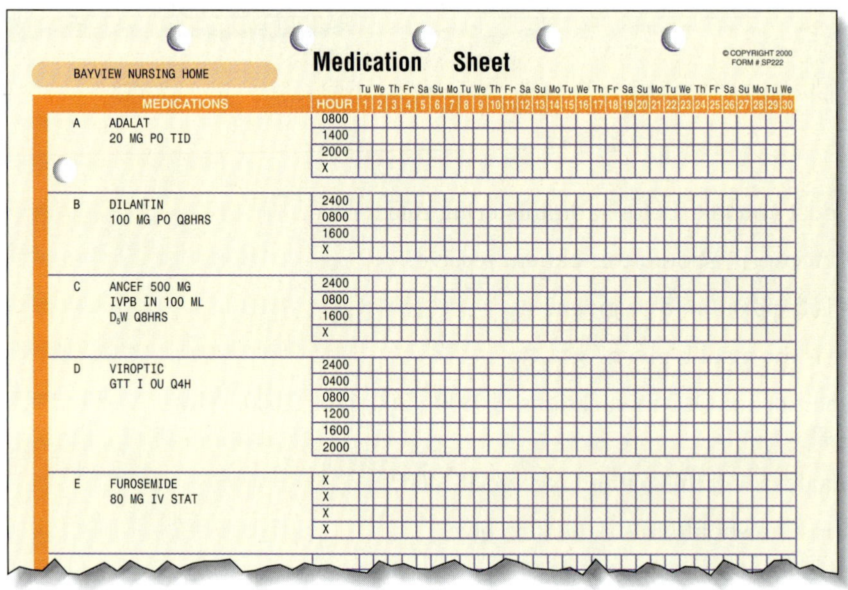

BAYVIEW NURSING HOME

Medication Sheet

© COPYRIGHT 2000
FORM # SP222

	MEDICATIONS	HOUR	Tu 1	We 2	Th 3	Fr 4	Sa 5	Su 6	Mo 7	Tu 8	We 9	Th 10	Fr 11	Sa 12	Su 13	Mo 14	Tu 15	We 16	Th 17	Fr 18	Sa 19	Su 20	Mo 21	Tu 22	We 23	Th 24	Fr 25	Sa 26	Su 27	Mo 28	Tu 29	We 30
A	ADALAT 20 MG PO TID	0800																														
		1400																														
		2000																														
		X																														
B	DILANTIN 100 MG PO Q8HRS	2400																														
		0800																														
		1600																														
		X																														
C	ANCEF 500 MG IVPB IN 100 ML D_5W Q8HRS	2400																														
		0800																														
		1600																														
		X																														
D	VIROPTIC GTT I OU Q4H	2400																														
		0400																														
		0800																														
		1200																														
		1600																														
		2000																														
E	FUROSEMIDE 80 MG IV STAT	X																														
		X																														
		X																														
		X																														

MAR 3

14. When should the Furosemide be given?
15. What is the route of administration of the Viroptic?
16. What is the route of administration of the Ancef®?
17. Which medications will be given at 4 p.m.?
18. If medications were delivered daily, how many doses of medication would be delivered for each of the medications this patient is scheduled to receive?
19. How many times a day does the patient receive Viroptic?
20. What two medications will be administered through an IV?

To check your answers, see page 514.

Medication Reference Materials

When you dispense or administer medications, you are responsible for knowing their effects. Hundreds of drugs exist. New ones are produced and approved all the time. You cannot memorize all the information you might need to know. Therefore, you need to be familiar with drug information sources.

Package inserts provided by the manufacturers with each medication are important reference tools. They describe intended effects, possible side effects, typical doses, dosage forms available, conditions under which the drug should not be used, and special precautions to be taken while using the drug. See Chapter 6 to learn more.

Information from package inserts is also printed in the *Physicians' Desk Reference* (PDR). Other versions feature nonprescription medications and herbal medications. The PDR has information about most currently available prescription drugs. A new volume is produced each year. Many physicians' offices, pharmacies, and health care facilities have the PDR available for employee use.

Many other guides are available for health care professionals, including the United States Pharmacopeia/National Formulary, found in most pharmacies. Most are updated every year or two. Other books have titles suggesting they are for nurses, but they are useful to all health care workers. Their information is similar to that of the PDR, but they often have simpler language.

Internet users can access information about the 200 most commonly prescribed drugs at www.rxlist.com/top200.htm. This site provides information about the most frequently prescribed drugs based on a list published in *American Druggist*. For each drug listed, the site lists appropriate doses of the drug for specific indications, available dosages and dosage forms, descriptions of the pills or liquids, and the drug's effects. Another Internet site, www.druginfonet.com, lists drug information, allows searches by brand or generic names, and provides many other useful features.

Also available are software programs designed to run on a handheld computer, known as a *personal digital assistant*, or PDA. One such program is called Epocrates. Used mostly by physicians, this program is updated regularly and is a handy resource for any health care professional needing the latest medication information.

REVIEW AND PRACTICE

5-5 Medication Reference Materials

Match the medication reference material with its description.

- **a.** PDR
- **b.** Package inserts
- **c.** Rxlist.com
- **d.** Epocrates

_____ **1.** Provides information about one medication.

_____ **2.** Includes a list of the top 200 drugs used.

_____ **3.** Software used on a PDA to reference medications.

_____ **4.** Compilation of package inserts, updated yearly.

To check your answers, see page 514.

HOMEWORK ASSIGNMENT

Mark DeSantis
123 Baker Drive
Owosso, MI 48867
989-555-1234

Name Jeannies Kucharek Date 1/23/2012

Address

Rx: Synthroid 0.1 mg

QUANTITY: #30

SIG: tab i po tid

MD1234567 Mark Desantis, MD

Prescription 1

Referring to prescription 1:

1. What instructions should be printed for the patient?
2. By what route is the medication to be administered?
3. How many refills may the patient be given?

Mark DeSantis
123 Baker Drive
Owosso, MI 48867
989-555-1234

Name Jeannies Kucharek Date 1/23/2012

Address

Rx: Amoxil 250 mg/5ml

SIG: 1 tsp. p.o. q8h until gone

Refills:

MD1234567 Mark Desantis, MD

Prescription 2

Referring to Prescription 2:

4. What information must be obtained before this order can be filled?
5. What instructions should be printed for the patient?

Mark DeSantis
123 Baker Drive
Owosso, MI 48867
989-555-1234

Name *Jeannies Kucharek* Date *1/23/2012*

Address

Rx: *Cortisporin Otic Drops*

QUANTITY: *5 ml*

SIG: *gtts. ii os quid*

Refills: *0*

MD1234567 *Mark Desantis, MD*

Prescription 3

Referring to Prescription 3:

6. What action must be taken before this prescription can be filled?

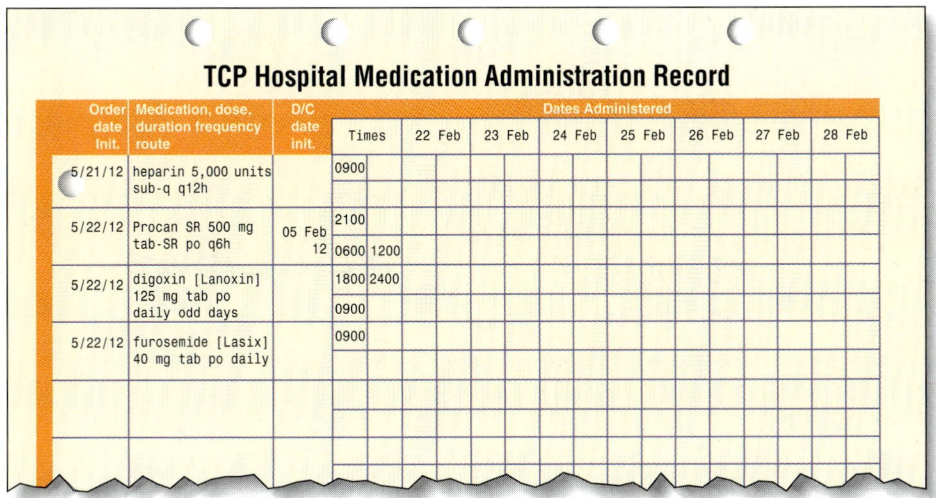

TCP Hospital Medication Administration Record

Order date Init.	Medication, dose, duration frequency route	D/C date init.	Times		Dates Administered 22 Feb	23 Feb	24 Feb	25 Feb	26 Feb	27 Feb	28 Feb
5/21/12	heparin 5,000 units sub-q q12h		0900								
5/22/12	Procan SR 500 mg tab-SR po q6h	05 Feb 12	2100 0600	1200							
5/22/12	digoxin [Lanoxin] 125 mg tab po daily odd days		1800 0900	2400							
5/22/12	furosemide [Lasix] 40 mg tab po daily		0900								

Referring to MAR4,-answer the following questions.

7. By what route will the heparin be given?
8. How many times a day will the patient receive Procan?
9. When should the patient receive his first dose of Lanoxin®?
10. How many oral medications is the patient receiving?
11. If the patient has a dose of Procan at 6 p.m., when should he receive the next dose?

Chapter 5 Review

Check Up

1. What action should you take when you receive this drug order? Dilaudid tab 2 mg po PRN for pain q4h
2. What action should you take when you receive this drug order? Codeine tab 30 mg po qid

In Exercises 3–6, refer to prescription 1.

Alan Capsella, MD
Westtown Medical Clinic
989-555-1234

Name Maria Ortiz Date July 8, 2012

Address

Rx: Timoptic 0.5%

QUANTITY: 5 mL

SIG: gtts ii od QID

Refills: 2

MD398475 Alan Capsella MD

Prescriber ID #

Prescription 1

3. What instructions should be printed for the patient?
4. How many times can this prescription be refilled?
5. By what route should this medication be administered?
6. What information should be on the drug label to verify that this medication is appropriate for the ordered route?

In Exercises 7–12, refer to Physician's Order Form 2.

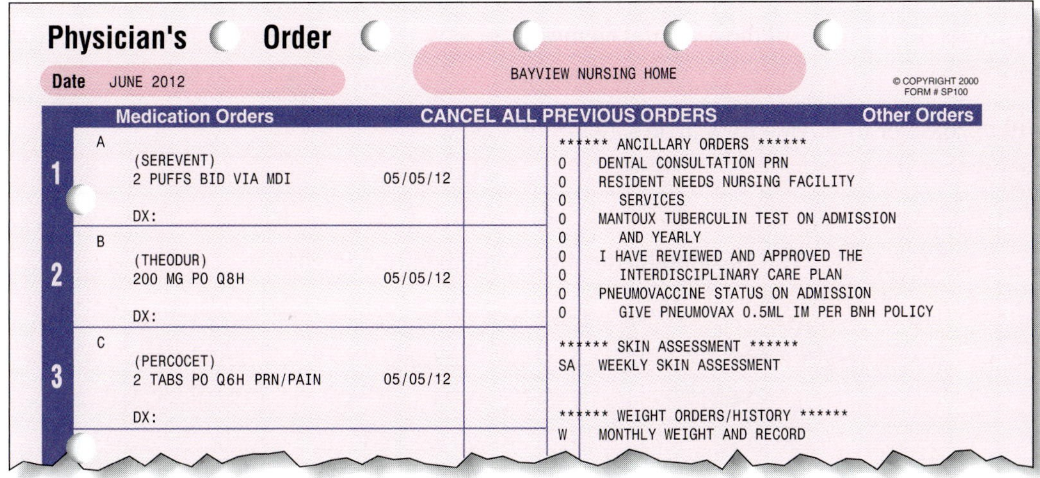

Physician's Order

Date JUNE 2012

BAYVIEW NURSING HOME

© COPYRIGHT 2000 FORM # SP100

Medication Orders — CANCEL ALL PREVIOUS ORDERS — Other Orders

1 A (SEREVENT) 2 PUFFS BID VIA MDI 05/05/12 DX:

2 B (THEODUR) 200 MG PO Q8H 05/05/12 DX:

3 C (PERCOCET) 2 TABS PO Q6H PRN/PAIN 05/05/12 DX:

****** ANCILLARY ORDERS ******
0 DENTAL CONSULTATION PRN
0 RESIDENT NEEDS NURSING FACILITY SERVICES
0 MANTOUX TUBERCULIN TEST ON ADMISSION AND YEARLY
0 I HAVE REVIEWED AND APPROVED THE INTERDISCIPLINARY CARE PLAN
0 PNEUMOVACCINE STATUS ON ADMISSION GIVE PNEUMOVAX 0.5ML IM PER BNH POLICY

****** SKIN ASSESSMENT ******
SA WEEKLY SKIN ASSESSMENT

****** WEIGHT ORDERS/HISTORY ******
W MONTHLY WEIGHT AND RECORD

Physician's Order Form 2

7. How often should Theodur® be given?
8. By what route is Serevent® administered?
9. If Jane Doe receives Percocet® for pain at 4:00 a.m., when will she be permitted to have another dose?
10. If medications are delivered to this unit once daily, how many Percocet® tablets should be dispensed at one time?
11. What dose of Theodur® should Jane Doe be given?
12. How often should Serevent® be given?

In Exercises 13–18, refer to MAR 3 for Arthur Simmons.

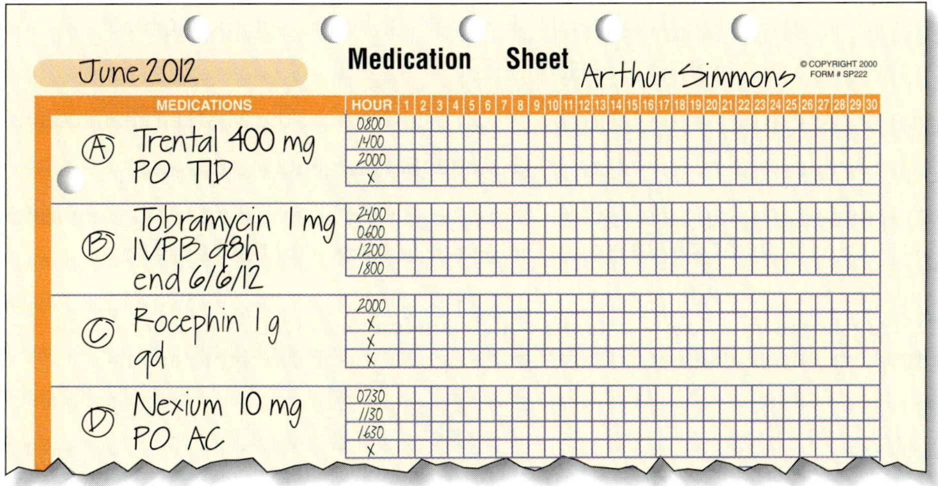

June 2012 — **Medication Sheet** — Arthur Simmons

© COPYRIGHT 2000 FORM # SP222

MEDICATIONS	HOUR	1	2	3	4	5	6	7	8	9	10	11	12	13	14	15	16	17	18	19	20	21	22	23	24	25	26	27	28	29	30
Ⓐ Trental 400 mg PO TID	0800 1400 2000 X																														
Ⓑ Tobramycin 1 mg IVPB q8h end 6/6/12	2400 0600 1200 1800																														
Ⓒ Rocephin 1 g qd	2000 X X X																														
Ⓓ Nexium 10 mg PO AC	0730 1130 1630 X																														

MAR 3

13. By what route is the tobramycin to be given?
14. What dose of Nexium® does Arthur Simmons receive?
15. What medications does Arthur Simmons receive at 8:00 p.m.?

16. How frequently does Arthur Simmons receive Trental®?

17. Should Arthur Simmons be served breakfast at 7:00 a.m., 7:30 a.m., or 8:00 a.m.?

18. Which orders do *not* have all the essential elements for a MAR order? State what is missing from each of them.

In Exercises 19 to 26, refer to MAR 4 for Carrie Kay Smith.

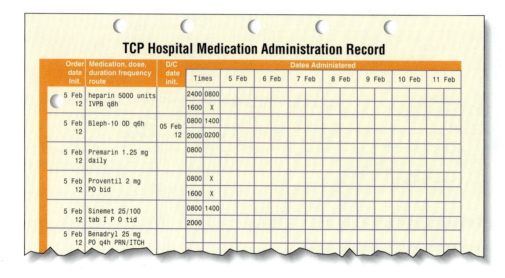

TCP Hospital Medication Administration Record

Order date Init.	Medication, dose, duration frequency route	D/C date init.	Times		5 Feb	6 Feb	7 Feb	8 Feb	9 Feb	10 Feb	11 Feb
5 Feb 12	heparin 5000 units IVPB q8h		2400	0800							
			1600	X							
5 Feb 12	Bleph-10 OD q6h	05 Feb 12	0800	1400							
			2000	0200							
5 Feb 12	Premarin 1.25 mg daily		0800								
5 Feb 12	Proventil 2 mg PO bid		0800	X							
			1600	X							
5 Feb 12	Sinemet 25/100 tab I P O tid		0800	1400							
			2000								
5 Feb 12	Benadryl 25 mg PO q4h PRN/ITCH										

MAR 4

19. What route will the Bleph-10 be given?

20. What dose of Premarin® will be given to Carrie Kay Smith?

21. At what times should Carrie Smith receive her Sinemet®?

22. Which medications will be administered at 1600?

23. If Carrie Smith has a dose of Benadryl® at 0200, when can she have her next dose?

24. How often should Carrie Smith receive Heparin?

25. If the Proventil® were to be given TID instead of BID, how would the MAR be changed?

26. Which two medication orders are not complete, and what elements are missing?

27. At 1:30 p.m., you are preparing medications for patients who have just returned from lunch. Describe some precautions you might take to ensure that you administer the right dose.

28. You have just started your shift on a unit with geriatric and pediatric patients. You are scheduled to administer drugs to several elderly patients with Alzheimer's as well as to several children. What steps should you take to ensure that you administer the right drugs to the right patients?

Critical Thinking Applications

What action should you take with the following drug order?

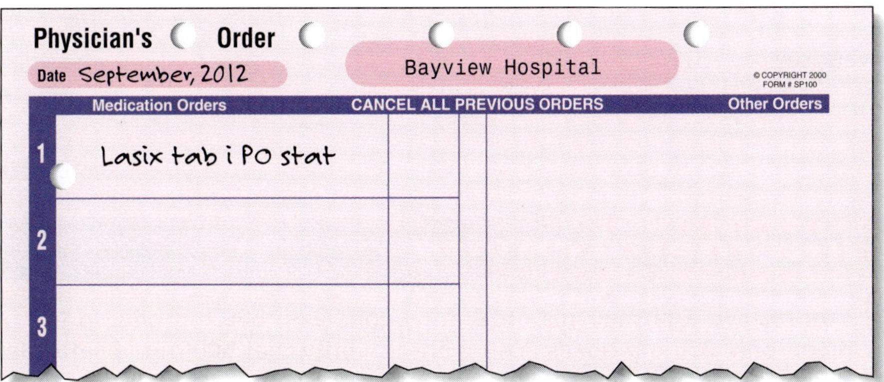

Physician's Order

Date September, 2012

Bayview Hospital

© COPYRIGHT 2000 FORM # SP100

Medication Orders | CANCEL ALL PREVIOUS ORDERS | Other Orders

1 Lasix tab i PO stat

2

3

Case Study

A physician gives a patient the following prescription.

Alan Capsella, MD
Westtown Medical Clinic
989-555-1234

Name Martin Burke Date January 6, 2012

Address 105 North Main, Bolivia, KS 88807

Rx: Xanax 0.5 mg

QUANTITY: 120

SIG: tab i po tid with meals

Refills: 1

MD398475 — Prescriber ID #

Alan Capsella MD

1. One of your jobs is to instruct patients on how to take prescription drugs. What should you tell Martin Burke?

2. If you are filling the prescription, how many tablets will you dispense to the patient? Will you refill the prescription?

To check your answers, see page 515.

Internet Activity

You receive a physician's order for Cephalexin 500 mg q6h. In the past, you have given only 250-mg doses. You want to verify that 500 mg is safe. Use the Top 200 Prescribed Drugs list from the Internet to check the safety of this dose.

Assignment: Type *www.rxlist.com* in the address bar of your Internet search program. Find Cephalexin, then read "Dosage and Administration." Determine whether the ordered dose is safe.

GO TO . . . Open the CD-ROM that accompanies your textbook, and complete a final review of the rules, practice problems, and activities presented for this chapter. For a final evaluation, take the chapter test and email or print your results for your instructor. A score of 95 percent or above indicates mastery of the chapter concepts. ■

6 Drug Labels and Package Inserts

Read in order to live.

—Henry Fielding

Learning Outcomes

When you have completed Chapter 6, you will be able to:

- Identify on a drug label the drug name, form, dosage strength, route, warnings, and manufacturing and storage information.
- Locate directions on drug labels and package inserts for reconstituting and diluting medications.
- Recognize different types of tablets and capsules.
- Distinguish administration routes for medications.
- Locate additional information in a package insert.

Key Terms

Dosage strength
Generic name
ID—intradermal
IM—intramuscular
IV—intravenous
Package insert
Parenteral
Reconstitute
Route
Sub-Q—subcutaneous
Trade name
Transdermal

Introduction

Now that you have learned basic math (Chapters 1 to 3), equipment (Chapter 4), and drug orders (Chapter 5), it is time to learn a little bit about drugs. The drug label and package insert contain the information you need to perform dosage calculations, and they must be read carefully. Make sure you know exactly what is found on a drug label, and do not forget to read the fine print. Very essential information is located there.

Locating Information on Drug Labels and Package Inserts

To prepare and administer drugs, you must understand information that appears on drug labels, including the drug name, form, dosage strength, total amount in the container, route of administration, warnings, storage requirements, and manufacturing information.

Drug Name

Every drug has an official name—its **generic name**. By law, this name must appear on the drug's label. It is also recorded with a national listing of drugs: the *United States Pharmacopeia* (USP) and the *National Formulary* (NF). In Figures 6-1, 6-2, 6-3, and 6-4, each drug's generic name has been identified. If USP appears on the label, it indicates that this drug's name is recorded with the *United States Pharmacopeia*.

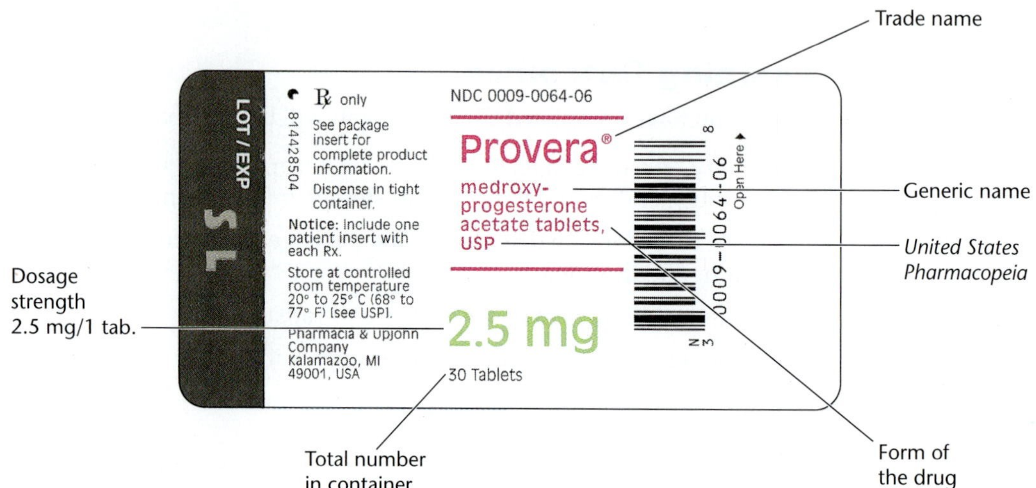

Figure 6-1 A ® beside the trade name indicates it is legally registered with the U.S. Patent and Trademark Office.

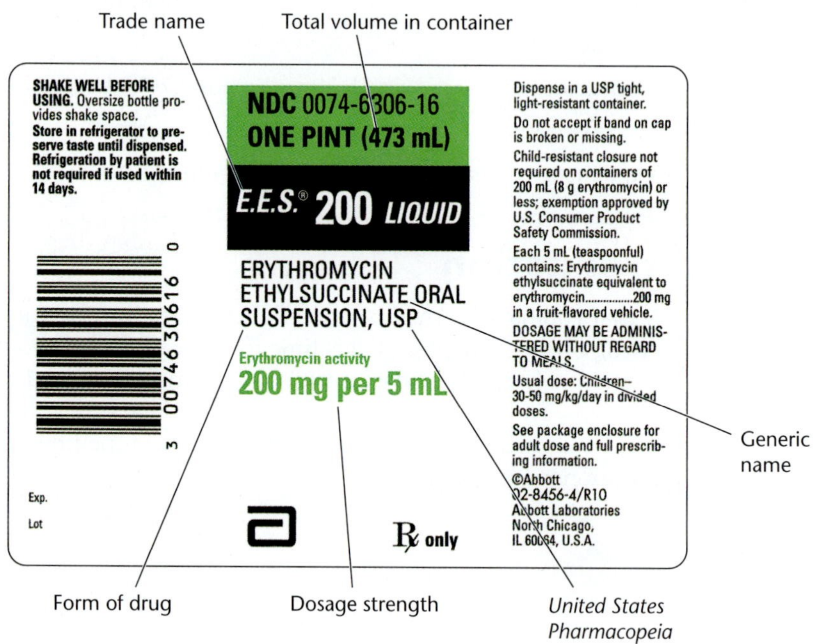

Figure 6-2 E.E.S.® 200 Liquid is the trade name.

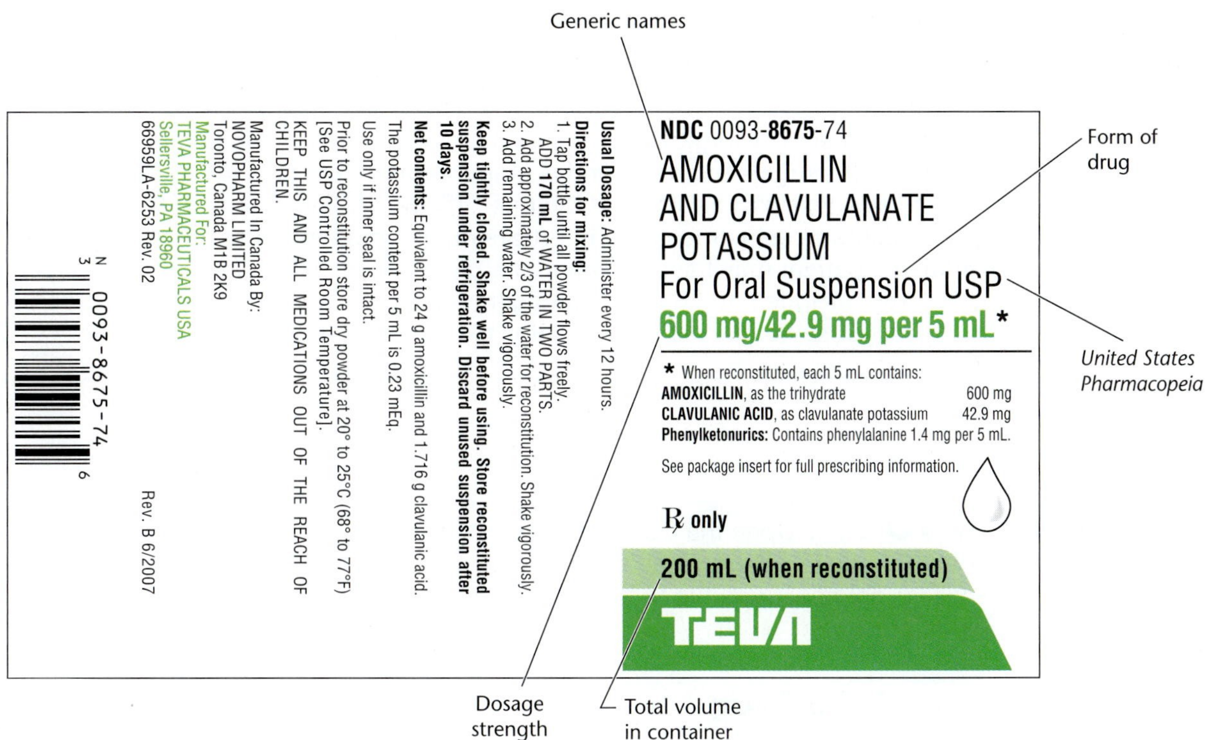

Figure 6-3 Amoxicillin and clavulanate potassium is a combination medication.

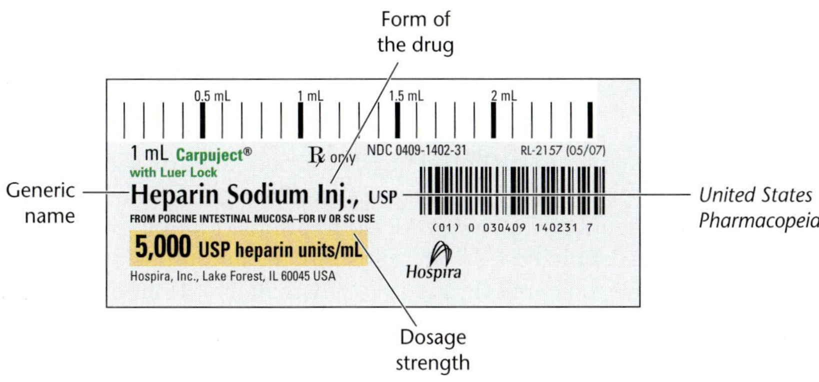

Figure 6-4 Some medications are generic only and have no trade name.

Many drug labels include the **trade**, or *brand*, **name** used to market the drug. In Figures 6-1 and 6-2 note that the trade name is listed before the generic name. However, some drug labels list only a generic name (see Figures 6-3 and 6-4). A trade name is the property of a specific drug company. The registered mark ® indicates the name has been legally registered with the U.S. Patent and Trademark Office. Several companies may manufacture a drug but market it under different trade names.

Physicians can write drug orders using either generic or trade names. Some companies produce drugs under their generic names and market them at a lower cost than that of the trade name equivalents. For example, ibuprofen is sold under its generic name as well as trade names such as Advil® and Motrin®.

Rule 6-1 You must know both the generic and trade names of drugs.

Example Suppose a patient is allergic to Vicodin, a narcotic painkiller. The generic drugs in Vicodin—hydrocodone bitartrate and acetaminophen—are also found in the trade name drugs Anexsia®, Lortab®, and Zydone®. If you administer one of these drugs or any drug containing hydrocodone bitartrate or acetaminophen as an alternative to Vicodin, the patient may have a similar allergic reaction. When you record a patient's drug allergy, include both the trade and generic names. Resources such as the PDR (*Physicians' Desk Reference*) provide information about a drug's ingredients.

GO TO . . . Open the CD-ROM that accompanies your textbook, and select Chapter 6, Rule 6-1. Review the animation and example problems, then complete the practice problems. Continue to the next section of the book once you have mastered the rule presented. ■

Form of the Drug

Manufacturers may offer the same drug in different forms. For example, penicillin is available as a tablet, a capsule, a liquid for oral administration, and an injection. Every label indicates the drug's form. Solid oral medications come in the form of tablets, capsules, gelcaps, and caplets. Liquid forms include oral, injections, inhalants, drops, sprays, and mists. Other forms of medication include ointments, creams, lotions, patches, suppositories, and shampoos. See Figures 6-1, 6-2, 6-3, and 6-4.

Dosage Strength

Drug labels include information about the amount of the drug present. This amount, combined with information about the form of the drug, identifies the drug's **dosage strength.** On the label, the dosage strength is stated as the amount of drug per dosage unit. In most cases, the amount of the drug is listed in grams (g), milligrams (mg), micrograms (mcg), or grains (gr). In certain cases, such as insulin, the amount is listed in units. Certain liquid drugs, such as hydrogen peroxide and glycerin, may list the amount in milliliters (mL).

For solid medications, the dosage strength is the amount of drug present per tablet, capsule, or other form. In Figure 6-1, note the amount of Provera® (2.5 mg) present in a tablet, the form of medication. The dosage strength is 2.5 mg per tablet. *Note: For most solid medications, if the unit is not listed, assume it is one tablet, one capsule, one gelcap, and so forth.*

For liquid medications, the dosage strength is the amount of the drug present in a certain quantity of solution. Recall from Chapter 2 that a solute (the drug) is mixed with a solvent or diluent (such as saline) to create a solution. You need to know both the amount of the drug and the amount of the total solution in which it is contained.

The amount of solution that is considered a dosage unit varies. For example, in Figure 6-2 when E.E.S.® is prepared properly, 200 mg of drug is present in every 5 mL of solution. In this case, 5 mL is the dosage unit. The dosage strength is 200 mg/5 mL. In Figure 6-4, however, 5000 units of Heparin Sodium is present in every 1 mL of solution. Here, 1 mL is the dosage unit. The dosage strength for this medication is 5000 U/1 mL.

Pharmaceutical companies manufacture medications with dosage strengths corresponding to commonly prescribed doses. This practice reduces the risk of medical error by reducing the number of dosage calculations.

Combination Drugs

The generic names and dosage strengths of all components of a combination drug must appear on the label. The label in Figure 6-3 lists the components amoxicillin and clavulanate. It also provides information about the individual drugs' dosage strengths. The line 600 mg/42.9 mg per 5 mL indicates that this medication contains 600 mg of amoxicillin and 42.9 mg of clavulanate in every 5 mL. Combination drugs sometimes have a trade name, which may be used in physician orders. For example, for the drug Lortab® 5/500 (trade name), which includes 5 mg of hydrocodone bitratrate and 500 mg of acetaminophen (generic names), the order might read Lortab® 5/500 1 tab q 4-6h PRN for pain. The order would *not* read hydrocodone bitartrate 5 mg, acetaminophen 500 mg, q 4–6h PRN for pain.

Total Number or Volume in Container

Many oral medications are packaged separately in *unit doses.* These packages may contain a single dosage unit, for example, a single tablet or a vial with 2 mL of solution for injection. If the container holds more than one dosage unit, the total number or volume must be listed on the label. See Figures 6-1, 6-2, 6-3, and 6-4. Prescription and nonprescription medications are often packaged in multiple-dose containers. Figure 6-5 indicates that prescription-strength Prilosec® is available in containers of 30 capsules, each capsule with 10 mg of drug.

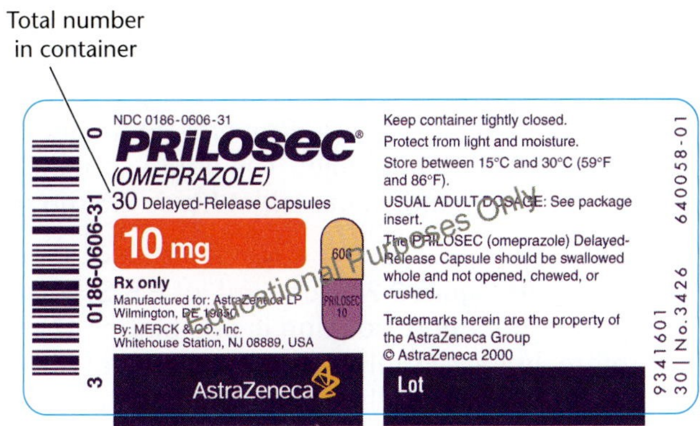

Figure 6-5 Prescription-strength Prilosec® comes in bottles with 30 capsules.

Figure 6-6 shows the label for a dose pack of Azithromycin a prescription medication. The container has 6 tablets taken over 5 days. Two tablets are taken on the first day. Each tablet is packaged as a unit dose. In Figure 6-7, the container of Rocephin® has 15 mL and provides a unit dose of 1 g/15 mL. The term single-use vial on the label. The label's directions indicate that you reconstitute the drug, administer it intravenously or intramuscularly, and discard any unused portion. The container is used once.

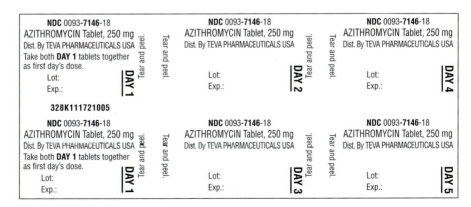

NDC 0093-**7146**-18
AZITHROMYCIN Tablet, 250 mg
Dist. By TEVA PHARMACEUTICALS USA
Take both **DAY 1** tablets together as first day's dose.
Lot:
Exp.:
DAY 1
Tear and peel.
Tear and peel.

NDC 0093-**7146**-18
AZITHROMYCIN Tablet, 250 mg
Dist. By TEVA PHARMACEUTICALS USA
Lot:
Exp.:
DAY 2
Tear and peel.
Tear and peel.

NDC 0093-**7146**-18
AZITHROMYCIN Tablet, 250 mg
Dist. By TEVA PHARMACEUTICALS USA
Lot:
Exp.:
DAY 4

328K111721005

NDC 0093-**7146**-18
AZITHROMYCIN Tablet, 250 mg
Dist. By TEVA PHARMACEUTICALS USA
Take both **DAY 1** tablets together as first day's dose.
Lot:
Exp.:
DAY 1
Tear and peel.
Tear and peel.

NDC 0093-**7146**-18
AZITHROMYCIN Tablet, 250 mg
Dist. By TEVA PHARMACEUTICALS USA
Lot:
Exp.:
DAY 3
Tear and peel.
Tear and peel.

NDC 0093-**7146**-18
AZITHROMYCIN Tablet, 250 mg
Dist. By TEVA PHARMACEUTICALS USA
Lot:
Exp.:
DAY 5

Figure 6-6 A five day dose pack of Azithromycin has one tablet in each package.

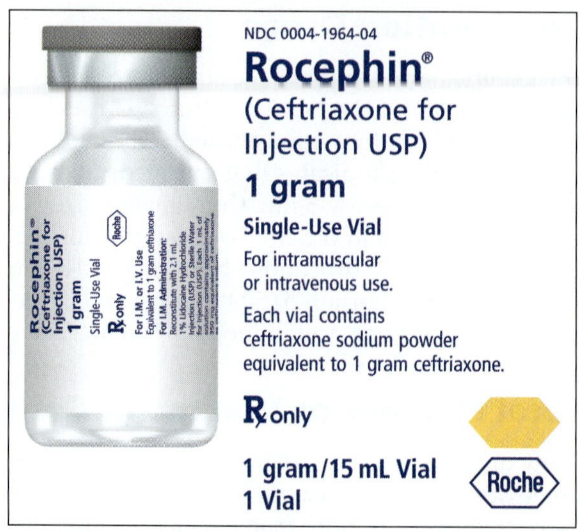

NDC 0004-1964-04
Rocephin®
(Ceftriaxone for Injection USP)
1 gram
Single-Use Vial
For intramuscular or intravenous use.
Each vial contains ceftriaxone sodium powder equivalent to 1 gram ceftriaxone.
℞ only
1 gram/15 mL Vial
1 Vial
Roche

Figure 6-7 Unit-dose package of Rocephin® for IV or IM administration.

Rule 6-2 Do not confuse the total amount of drug in the container with the dosage strength.

Example 1 According to Figure 6-1, the Provera® container holds 30 tablets, with a dosage strength of 2.5 mg per tablet. The entire container holds 30 × 2.5 mg, or 75 mg, of drug whereas an individual tablet holds 2.5 mg of drug.

Example 2 According to Figure 6-2, each 5 mL of E.E.S.® solution contains 200 mg of drug. The entire amount of solution is one pint or 473 mL, not 5 mL. The entire container, therefore, holds 473 mL ÷ 5 mL = 94.6 or approximately 95 unit doses. Even though the entire contents of the container is one pint (473 mL), you will only administer a small portion (5 mL) of it at a time.

GO TO . . . Open the CD-ROM that accompanies your textbook, and select Chapter 6, Rule 6-2. Review the animation and example problems, then complete the practice problems. Continue to the next section of the book once you have mastered the rule presented. ■

Route of Administration

Directions for the **route** of administration may be specified on the label. This information may not be included for oral medications. However, if a tablet or a capsule is not to be swallowed, additional information will be provided. For example, the label for Nitrostat® (Figure 6-8) shows it is administered sublingually, under the tongue. Chewable tablets will be labeled as such. Medications for topical use only will also be marked on the label (Figure 6-9).

Liquid medications may be given orally or injected. Labels will indicate whether an injection is given **intradermally** (ID), **intravenously** (IV), **intramuscularly** (IM), or **subcutaneously** (Sub-Q) (see Figures 6-10 and 6-11). Labels will indicate other routes as well. For example, Aerobid-M is a solution for oral inhalation only (see Figure 6-12).

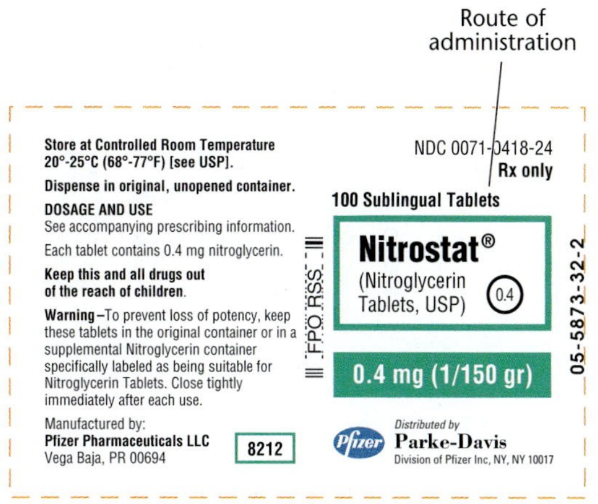

Route of administration
Store at Controlled Room Temperature 20°-25°C (68°-77°F) [see USP].
Dispense in original, unopened container.
DOSAGE AND USE
See accompanying prescribing information.
Each tablet contains 0.4 mg nitroglycerin.
Keep this and all drugs out of the reach of children.
Warning–To prevent loss of potency, keep these tablets in the original container or in a supplemental Nitroglycerin container specifically labeled as being suitable for Nitroglycerin Tablets. Close tightly immediately after each use.
Manufactured by:
Pfizer Pharmaceuticals LLC
Vega Baja, PR 00694
8212
NDC 0071-0418-24
Rx only
100 Sublingual Tablets
Nitrostat®
(Nitroglycerin Tablets, USP)
0.4
0.4 mg (1/150 gr)
Pfizer
Distributed by
Parke-Davis
Division of Pfizer Inc, NY, NY 10017
05-5873-32-2

Figure 6-8 Sublingual tablets should be placed only under the tongue.

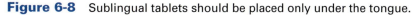

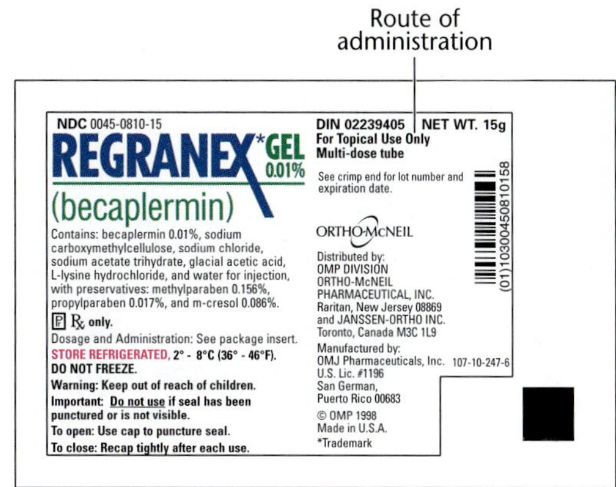

Figure 6-9 Regranex® is a topical gel.

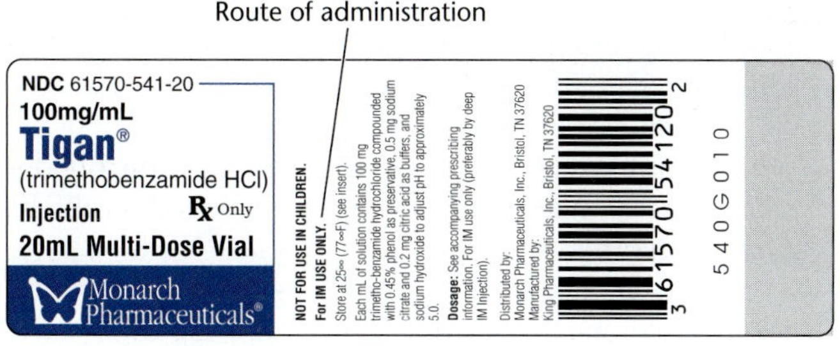

Route of administration
NDC 61570-541-20
100mg/mL
Tigan®
(trimethobenzamide HCl)
Injection
℞ Only
20mL Multi-Dose Vial
Monarch Pharmaceuticals®
NOT FOR USE IN CHILDREN.
For IM USE ONLY.
3 61570 54120 2
540G010

Figure 6-10 For intramuscular (IM) administration only.

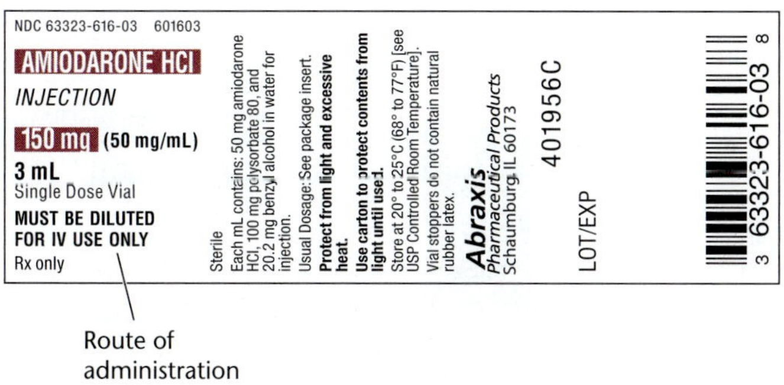

NDC 63323-616-03 601603
AMIODARONE HCl
INJECTION
150 mg (50 mg/mL)
3 mL
Single Dose Vial
MUST BE DILUTED
FOR IV USE ONLY
Rx only
Sterile
Each mL contains: 50 mg amiodarone HCl, 100 mg polysorbate 80, and 20.2 mg benzyl alcohol in water for injection.
Usual Dosage: See package insert.
Protect from light and excessive heat.
Use carton to protect contents from light until used.
Store at 20° to 25°C (68° to 77°F) [see USP Controlled Room Temperature].
Vial stoppers do not contain natural rubber latex.
Abraxis Pharmaceutical Products
Schaumburg, IL 60173
401956C
LOT/EXP
3 63323-616-03 8
Route of administration

Figure 6-11 For intravenous (IV) administration only.

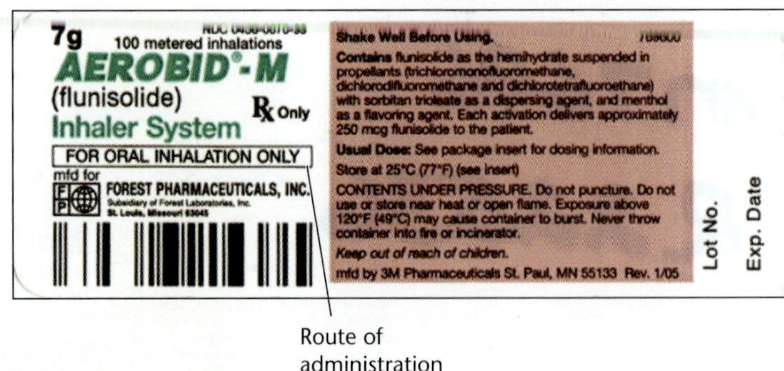
Route of administration

Figure 6-12 For oral inhalation only.

ERROR ALERT!

Give the right medication by the right route.

Cortisporin® is an anti-inflammatory medication that is available in a variety of forms, including an antibacterial suspension for otic (ear) use and an antimicrobial suspension for ophthalmic (eye) use. The usual dosage of the otic suspension is 4 drops instilled 3 to 4 times a day into the affected ear. The usual dosage of the ophthalmic suspension is 1 or 2 drops instilled into the affected eye every 3 or 4 h. If you were to carry out an order for the otic suspension by administering it to the patient's eye, you would not only fail to provide appropriate care for the ear, but also cause considerable irritation to the eye. *The bottom line: do not administer drugs by any route other than intended, as described on the drug label and on the order.* Remember, this is one of the rights of medication administration.

Warnings

Warnings on labels help health care workers administer drugs safely. They include statements such as "It is recommended that drug dispersing should not exceed weekly supply. Dispersing should be contigent upon the results of a WBC count." (See Figure 6-13.) Other labels indicate that the contents are poisonous. Labels may carry warnings for specific groups of patients. For example a label may caution that a medication may not be safe for pregnant women or for children. (See Figure 6-14.) Controlled substances may contain a warning that says "May be habit forming." Other labels describe harmful effects resulting from combinations with other products.

Every facility follows guidelines for disposing of drugs that are not used. The guidelines for medications that carry warnings are especially strict. For example, in some cases, you dispose of medications such as narcotics with a coworker as witness, then provide appropriate documentation.

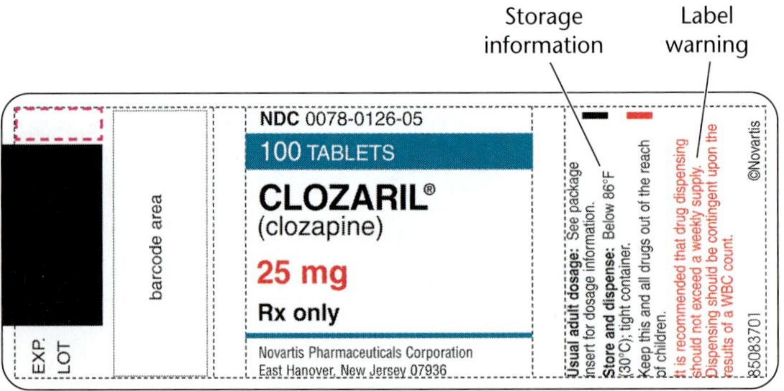
Storage information
Label warning
NDC 0078-0126-05
100 TABLETS
CLOZARIL®
(clozapine)
25 mg
Rx only
Novartis Pharmaceuticals Corporation
East Hanover, New Jersey 07936
barcode area
EXP.
LOT
Usual adult dosage: See package insert for dosage information.
Store and dispense: Below 86°F (30°C); tight container.
Keep this and all drugs out of the reach of children.
It is recommended that drug dispensing should not exceed a weekly supply. Dispensing should be contingent upon the results of a WBC count.
©Novartis
85083701

Figure 6-13 Always read label warnings and storage information.

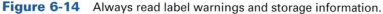

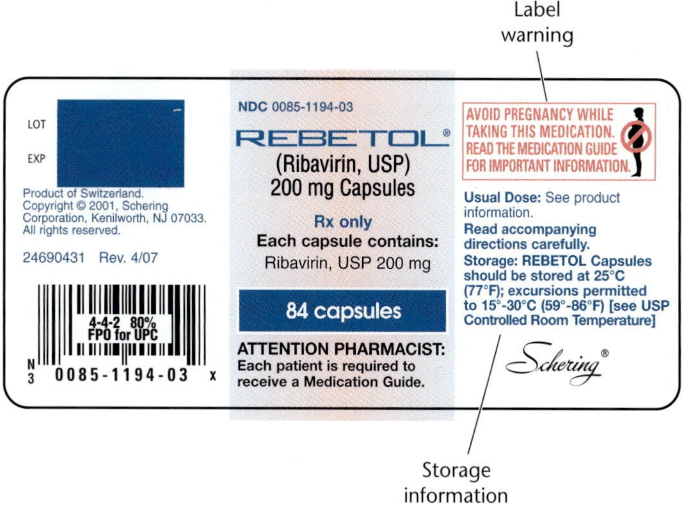

Figure 6-14 Always read label warnings and storage information.

Storage Information

Some drugs must be stored under specific conditions to maintain their potency and effectiveness. Storage information will appear on the drug's label. The label may have information about storage temperature, exposure to light, or the length of time the drug will remain potent after the container has been opened. Storage at the wrong temperature or exposure to light can trigger a chemical reaction that makes the drug unusable. (See Figures 6-13 and 6-14 for storage information.)

Manufacturing Information

Pharmaceutical manufacturers are strictly regulated by the U.S. Food and Drug Administration (FDA). FDA regulations state that every drug label must include the name of the manufacturer; an expiration date, abbreviated EXP, after which the drug may no longer be used; and the lot number (see Figure 6-15).

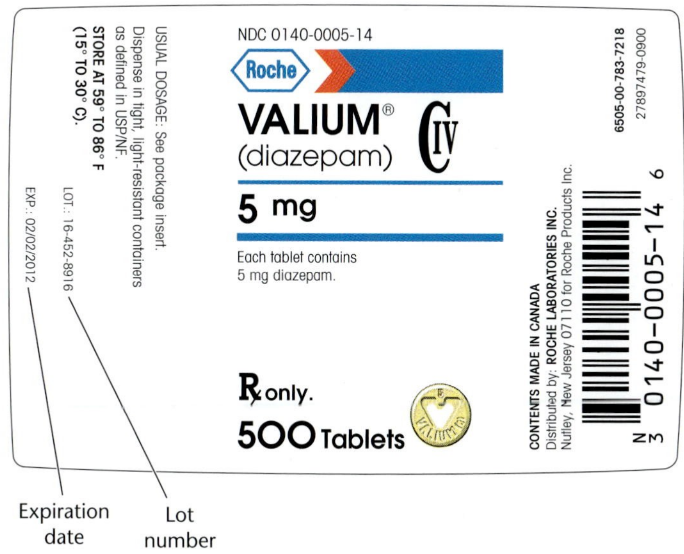

Figure 6-15 The lot number and expiration date are stamped on the drug label during manufacturing.

Medications are produced in batches, known as *lots*. The lot number is a code that indicates when and where a drug was produced. It allows the manufacturer to trace problems linked to a particular batch. If a manufacturer has to remove an entire lot from the market because of contamination, suspected tampering, or unexpected side effects, the lot number helps identify which batch to recall.

Rule 6-3 Never use a drug after the expiration date has passed.

Example Older drugs may become chemically unstable or altered. As a result, they may not provide the correct dosage strength. Worse, they could have an effect different from the intended one. Advise patients to check the expiration dates on all drug labels. If patients have not used a product by the date listed, they should discard it. At an inpatient setting, the medication may need to be returned to the pharmacy, depending on the facility's policy.

GO TO . . . Open the CD-ROM that accompanies your textbook, and select Chapter 6, Rule 6-3. Review the animation and example problems, then complete the practice problems. Continue to the next section of the book once you have mastered the rule presented. ■

Information About Reconstituting Drugs

Some drugs, such as antibiotics, are packaged in powder form. You **reconstitute** the drug (add liquid to the powder) shortly before administering it. Reconstituted medications remain potent for only a short time. The label indicates the time period within which they can be safely administered (see Figures 6-16 and 6-17). Other drugs must be diluted before they are administered; they, too, must be used within a limited time. Directions for reconstituting or diluting a drug appear on the label (see Figures 6-16 and 6-17). Additional information can be found in the package insert, discussed in the next section.

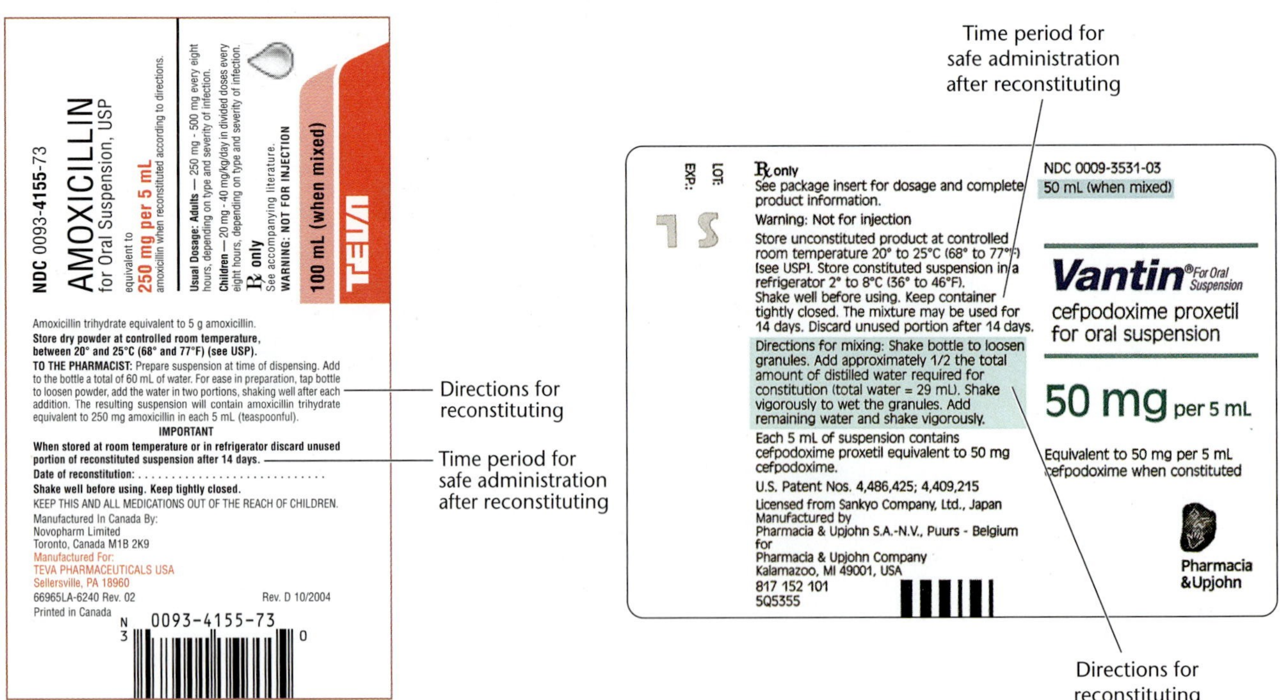

Figure 6-16 To reconstitute a drug, you add liquid to a powder.

Figure 6-17 Follow the information about reconstituting a drug carefully.

ERROR ALERT!

Consider the age and health needs of your patient when you administer a drug.

Suppose the drug order reads Biaxin® 250 mg po b.i.d. Biaxin®, an antibiotic, is available in 250-mg tablets and as an oral suspension with a reconstituted dosage strength of 125 mg/5 mL. (An *oral suspension* is a liquid that contains solid particles of medication. You shake the medication before administering it, suspending the particles.)

It may seem logical to fill the order with one tablet. Yet the age or health of the patient may make a liquid the better choice, especially for children or patients who have difficulty swallowing. If you see a situation in which another form of a drug may work better, consult the physician or pharmacist about changing the form of the drug.

Package Inserts

Package inserts provide complete and authoritative information about a medication. The information in package inserts can also be found in the PDR and other guides. Figure 6-18 shows a portion of a package insert. Table 6-1 summarizes the sections of a package insert.

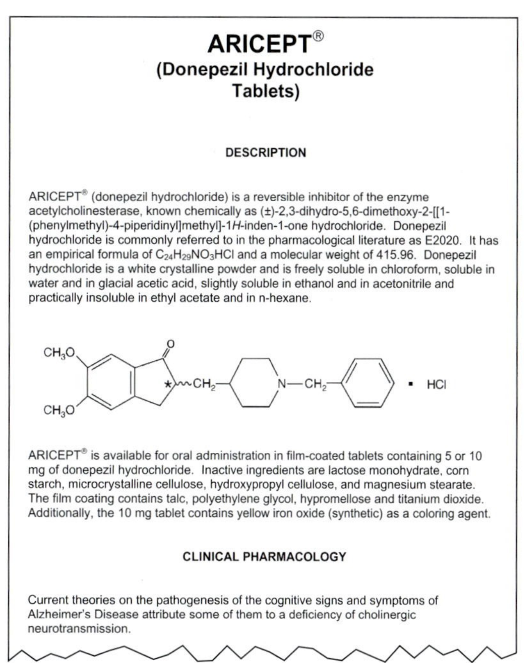

ARICEPT®
(Donepezil Hydrochloride Tablets)

DESCRIPTION

ARICEPT® (donepezil hydrochloride) is a reversible inhibitor of the enzyme acetylcholinesterase, known chemically as (±)-2,3-dihydro-5,6-dimethoxy-2-[[1-(phenylmethyl)-4-piperidinyl]methyl]-1*H*-inden-1-one hydrochloride. Donepezil hydrochloride is commonly referred to in the pharmacological literature as E2020. It has an empirical formula of $C_{24}H_{29}NO_3HCl$ and a molecular weight of 415.96. Donepezil hydrochloride is a white crystalline powder and is freely soluble in chloroform, soluble in water and in glacial acetic acid, slightly soluble in ethanol and in acetonitrile and practically insoluble in ethyl acetate and in n-hexane.

ARICEPT® is available for oral administration in film-coated tablets containing 5 or 10 mg of donepezil hydrochloride. Inactive ingredients are lactose monohydrate, corn starch, microcrystalline cellulose, hydroxypropyl cellulose, and magnesium stearate. The film coating contains talc, polyethylene glycol, hypromellose and titanium dioxide. Additionally, the 10 mg tablet contains yellow iron oxide (synthetic) as a coloring agent.

CLINICAL PHARMACOLOGY

Current theories on the pathogenesis of the cognitive signs and symptoms of Alzheimer's Disease attribute some of them to a deficiency of cholinergic neurotransmission.

Figure 6-18 The top portion of a typical package insert.

TABLE 6-1 Sections of a Package Insert

Section	Description	Example Information
Description	Chemical and physical description of the drug	Aricept® (donepezil hydrochloride) is a reversible inhibitor of the enzyme acetylcholinesterase, known chemically as (±)-2,3-dihydro-5, 6-dimethoxy-2-[[1-(phenylmethyl)-4-piperidinyl]methyl]-1*H*-inden-1-one hydrochloride. 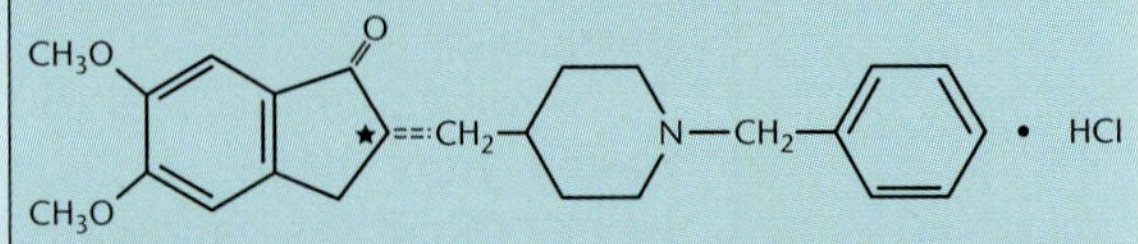
Clinical pharmacology	Description of the actions of the drug	Current theories on the pathogenesis of the cognitive signs and symptoms of Alzheimer's Disease attribute some of the them to a deficiency of cholinergic neurotransmission.
Indications and usage	Medical conditions in which the drug is safe and effective; instructions for use	Aricept® is indicated for the treatment of mild to moderate dementia of the Alzheimer's type.
Contraindications	Conditions and situations under which the drug should not be administered	Aricept® is contraindicated in patients with known hypersensitivity to donepezil hydrochloride or to piperidine derivatives.
Warnings	Information about serious, possibly fatal, side effects	*Gastrointestinal conditions:* Through their primary action, cholinesterase inhibitors may be expected to increase gastric acid secretion due to increased cholinergic activity.
Precautions	Information about drug interactions and other conditions that may cause unwanted side effects	*Use with anticholinergics:* Because of their mechanism of action, cholinesterase inhibitors have the potential to interfere with the activity of anticholinergic medications.
Adverse reactions	Less serious, anticipated side effects that can be caused by the drug	These include nausea, diarrhea, insomnia, vomiting, muscle cramps, fatigue, and anorexia. These adverse events were often of mild intensity and transient, resolving during continued Aricept® treatment without the need for dose modification.
Overdosage	Effects of overdoses and instructions for treatment	As in any case of overdose, general supportive measures should be utilized. Overdosage with cholinesterase inhibitors can result in cholinergic crisis characterized by severe nausea, vomiting, salivation, sweating, bradycardia, hypotension, respiratory depression, collapse, and convulsions.
Dosage and administration	Recommended dosages under various conditions and recommendations for administration routes	The dosages of Aricept® shown to be effective in controlled clinical trials are 5 mg and 10 mg administered once per day.
Preparation for administration	Directions for reconstituting or diluting the drug, if necessary	Aricept® should be taken in the evening, just prior to retiring. Aricept® can be taken with or without food.
Manufacturer supply	Information on dosage strengths and forms of the drug available	Aricept® is supplied as film-coated, round tablets containing either 5 mg or 10 mg of donepezil hydrochloride.

Critical Thinking on the Job

Read Labels Carefully

A health care professional is filling the order Synthroid® 0.05 mg p.o. daily. Synthroid® is available in tablets of 11 different strengths, each in a different color. The health care professional has access to tablets in 0.025-mg (orange), 0.05-mg (white), 0.125-mg (brown), and 0.15-mg (blue) doses.

Looking quickly at two labels (see Figures 6-19 and 6-20), the health care professional sees a Synthroid® label with "5" on it. Without realizing it is for 0.15 mg, he removes a tablet. When he tries to administer it, the patient tells him that her usual pill is white, not blue.

Questions?

What should the health care professional do? What mistake did he make?

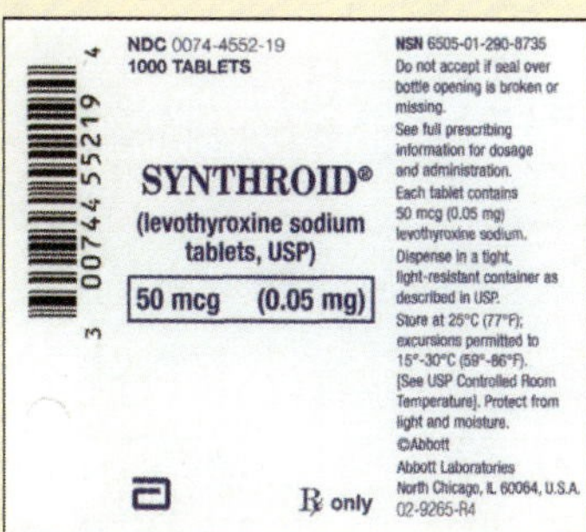
NDC 0074-4552-19
1000 TABLETS
3 00744 55219 4
SYNTHROID®
(levothyroxine sodium tablets, USP)
50 mcg (0.05 mg)
Rx only
NSN 6505-01-290-8735
Do not accept if seal over bottle opening is broken or missing.
See full prescribing information for dosage and administration.
Each tablet contains 50 mcg (0.05 mg) levothyroxine sodium.
Dispense in a tight, light-resistant container as described in USP.
Store at 25°C (77°F); excursions permitted to 15°-30°C (59°-86°F). [See USP Controlled Room Temperature]. Protect from light and moisture.
©Abbott
Abbott Laboratories
North Chicago, IL 60064, U.S.A.
02-9265-R4

Figure 6-19 Read medication labels carefully.

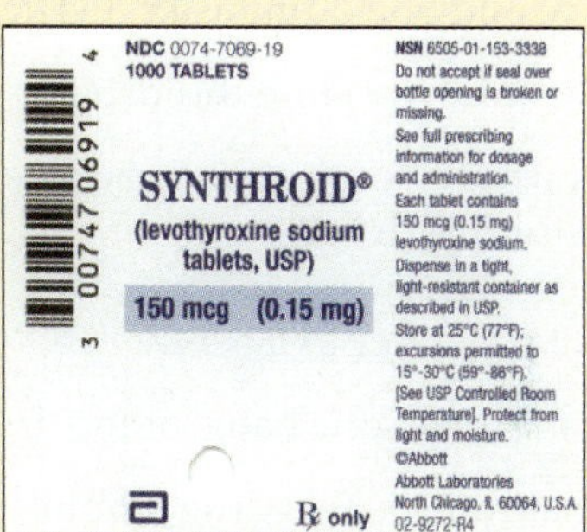
NDC 0074-7069-19
1000 TABLETS
3 00747 06919 4
SYNTHROID®
(levothyroxine sodium tablets, USP)
150 mcg (0.15 mg)
Rx only
NSN 6505-01-153-3338
Do not accept if seal over bottle opening is broken or missing.
See full prescribing information for dosage and administration.
Each tablet contains 150 mcg (0.15 mg) levothyroxine sodium.
Dispense in a tight, light-resistant container as described in USP.
Store at 25°C (77°F); excursions permitted to 15°-30°C (59°-86°F). [See USP Controlled Room Temperature]. Protect from light and moisture.
©Abbott
Abbott Laboratories
North Chicago, IL 60064, U.S.A.
02-9272-R4

Figure 6-20 Two labels may look alike yet are different dosage strengths.

REVIEW AND PRACTICE

6-1 Locating Information on Drug Labels and Package Inserts

In Exercises 1–6, refer to label A.

1. What is the trade name of the drug?
2. What is the generic name of the drug?
3. Does this container hold multiple doses or a unit dose? How do you know?
4. What is the name of the manufacturer?
5. What is the dosage strength?
6. At what temperature should the drug be stored?

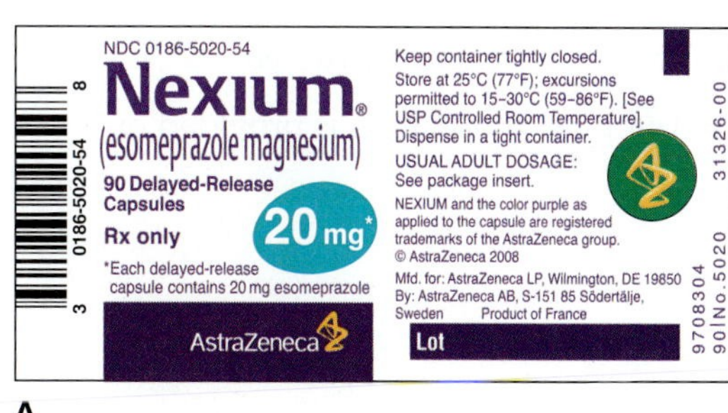
NDC 0186-5020-54
Nexium®
(esomeprazole magnesium)
90 Delayed-Release Capsules
Rx only
20 mg*
*Each delayed-release capsule contains 20 mg esomeprazole
3 0186-5020-54 8
AstraZeneca
Keep container tightly closed.
Store at 25°C (77°F); excursions permitted to 15–30°C (59–86°F). [See USP Controlled Room Temperature].
Dispense in a tight container.
USUAL ADULT DOSAGE:
See package insert.
NEXIUM and the color purple as applied to the capsule are registered trademarks of the AstraZeneca group.
© AstraZeneca 2008
Mfd. for: AstraZeneca LP, Wilmington, DE 19850
By: AstraZeneca AB, S-151 85 Södertälje, Sweden
Product of France
Lot
9708304
90lNo.5020
31326-00

A

In Exercises 7–12, refer to label B.

7. What is the generic name of the drug?
8. What is the trade name of the drug?
9. What is the dosage strength?
10. What type of tablets are in this bottle?
11. What is the lot number?
12. What are the storage requirements for this drug?

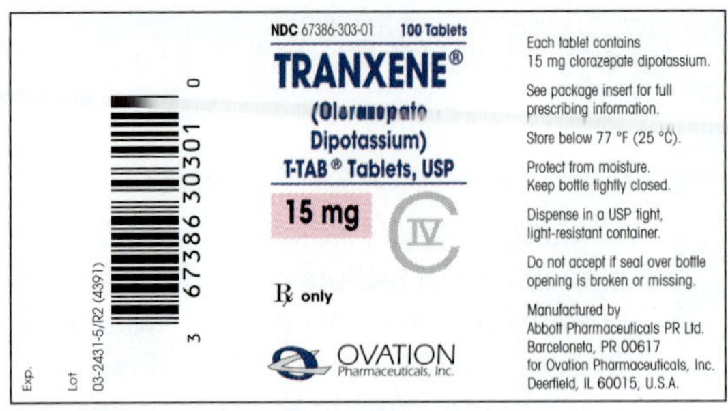

B

In Exercises 13–18, refer to label C.

13. What is the trade name of the drug?
14. What is the name of the manufacturer?
15. What is the dosage strength?
16. How would you administer this drug?
17. How would you store this drug?
18. When would you read the package insert for this drug?

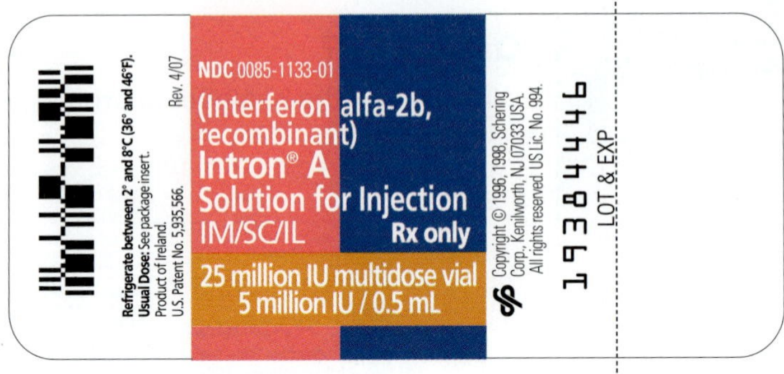

C

In Exercises 19–24, refer to label D.

19. What is the generic name of the drug?
20. By what route is this drug administered?
21. What is the usual dose?
22. What is the dosage strength?
23. If you had a drug order for 250 mg of Omnicef®, how many teaspoons would you administer to the patient?
24. How long would two bottles last if you administered 2 doses of 10 mL daily?

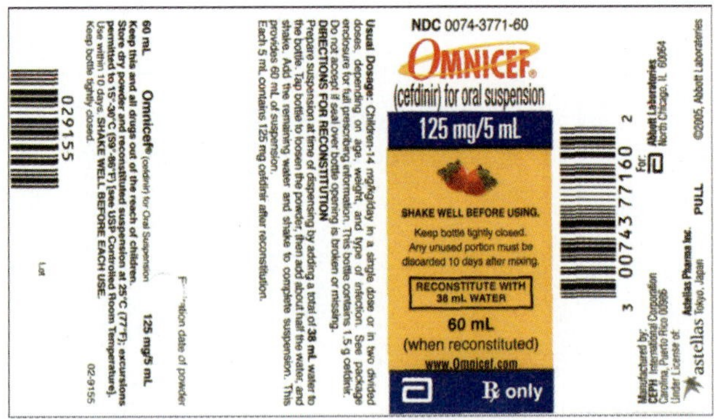

D

In Exercises 25–30, refer to label E.

25. What is the generic name of the drug?
26. What is the trade name of the drug?
27. By what route is this drug administered?
28. What special storage information is provided on the label?

29. How much medication is in one vial?

30. How many doses are in one vial?

In Exercises 31–36, refer to label F.

31. When it is reconstituted, how many milliliters are in the bottle?

32. How would you reconstitute this drug?

33. How much water should initially be added to the powder?

34. When it is reconstituted, what is the dosage strength?

35. How should this medication be stored and when should it be discarded?

36. If the usual dose is 5 mL, how many doses are in this container?

In Exercises 37–43, refer to label G.

37. What is the trade name of the drug?

38. What is the generic name of the drug?

39. What is the dosage strength?

40. How should this medication be stored?

41. Through what route is this drug administered?

42. How many prescriptions could be filled from this bottle if each prescription's quantity was 30 tablets?

43. What is the name of the manufacturer?

In Exercises 44–50, refer to labels A through G.

44. Which of these drugs are tablets?

45. Which of these drugs are given orally?

E

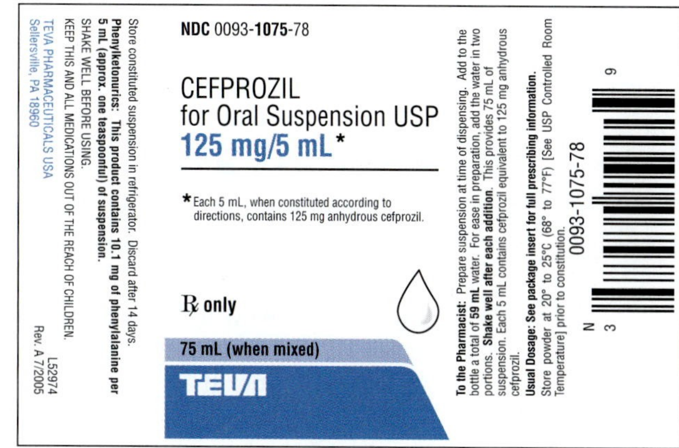

F

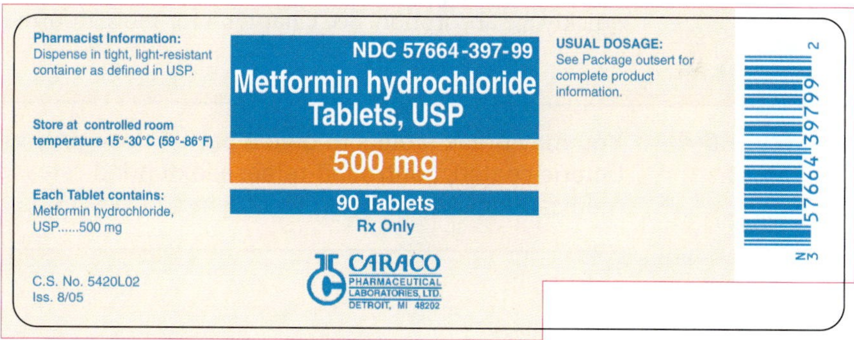

G

46. Which of these is delayed release drug?
47. Which of these drugs must be refrigerated?
48. Which of these drugs would be administered parenterally?
49. Which medication(s) must be reconstituted?
50. Which medication does not include a trade name?

To check your answers, see pages 516–517.

Oral Drugs

Oral medications are available in either solid or liquid form. Tablets are the most common form. They may be scored, chewable, or enteric-coated. Scored tablets can be broken into equal portions so that you can administer a partial dosage, if necessary. Chewable tablets must be chewed to be effective. Enteric-coated tablets must be swallowed whole. Chewing them or dividing them breaks the seal provided by their coating, allowing the drug to be absorbed sooner than intended.

Capsules have a gelatin shell that contains the drug. In most cases, they should be swallowed whole. In some cases, capsules may be opened and mixed with food. *Controlled-release capsules*, also called *sustained-release* or *extended-release capsules*, release the drug over a long time. If these capsules are not swallowed whole, they may release too much of the drug too quickly for absorption. See Chapter 8 for more information about solid oral medications.

Rule 6-4 You may break tablets to give a partial dose *only* when the tablets are scored. Enteric-coated, controlled-release, extended-release, and sustained-release medications should *never* be crushed or broken.

GO TO . . . Open the CD-ROM that accompanies your textbook, and select Chapter 6, Rule 6-4. Review the animation and example problems, then complete the practice problems. Continue to the next section of the book once you have mastered the rule presented. ■

Abbreviations such as SR, CR, and ER listed after the drug name indicate a special drug action. The SR following the brand name means the drug is designed for *sustained release;* CR means that a drug is *controlled release;* and ER or XL following the brand name means the drug has an *extended-release* mechanism (Figures 6-21 and 6-22).

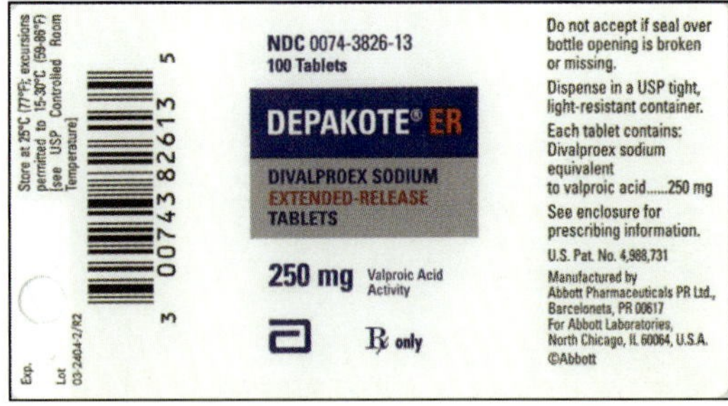

Figure 6-21 Do not break this tablet.

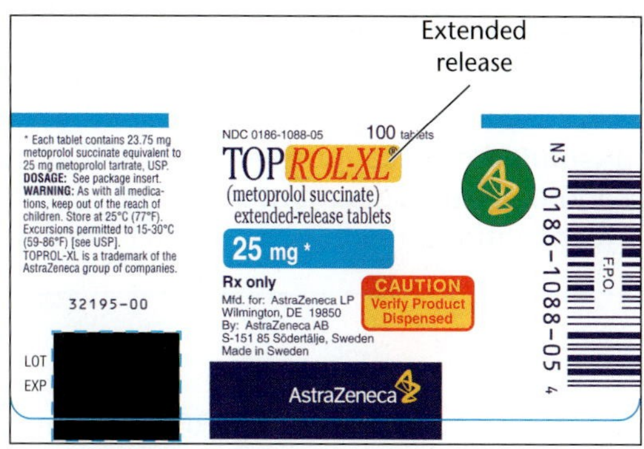

Figure 6-22 XL indicates that this medication has an extended-release mechanism.

Liquid oral medications are described as oral solutions, syrups, elixirs, oral suspensions, and simply liquids (see Figures 6-23 and 6-24). In liquid medications, the dosage strength corresponds to a specific volume of the solution, for example, 250 mg/5 mL.

If a medication needs to be reconstituted, the instructions will be on the label.

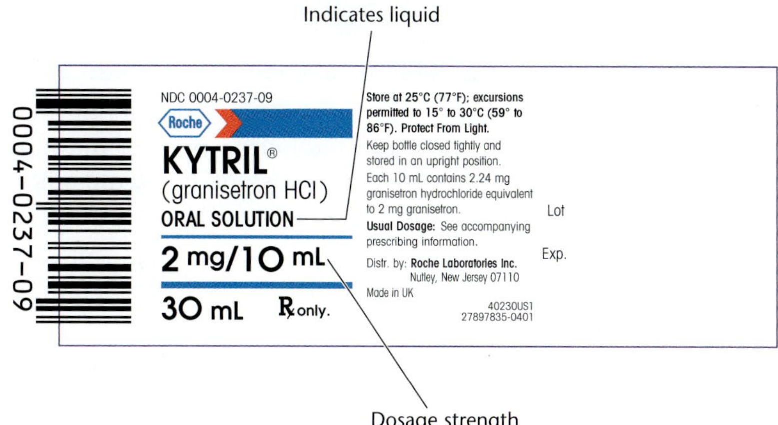

Figure 6-23 For liquid medications the dosage strength corresponds to a number of milligrams per a specific volume of solution.

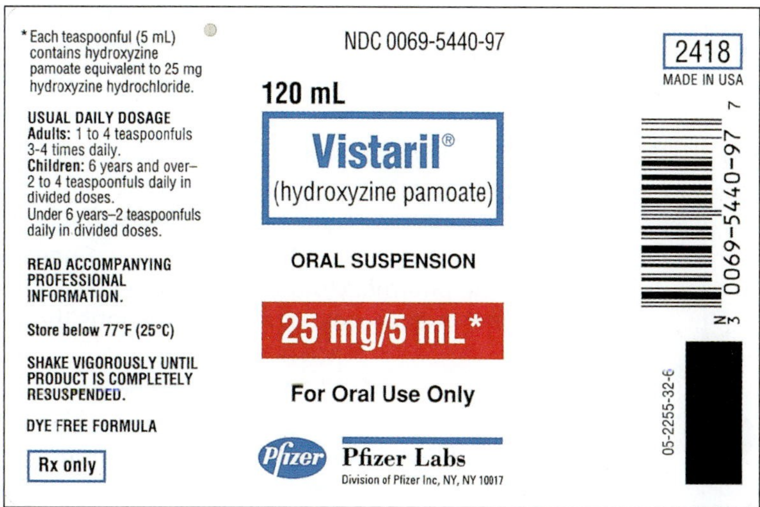

Figure 6-24 The dosage strength is indicated in 25 mg per 5 mL volume for this medication.

Rule 6-5 When you reconstitute a drug that is to be used for more than one dose, you must write your initials as well as the time and date of reconstitution on the label. Reconstituted medications will be usable only for a certain length of time. So the date and time will document when the medication will expire. Your initials document who reconstituted the medication in case a question arises.

GO TO . . . Open the CD-ROM that accompanies your textbook, and select Chapter 6, Rule 6-5. Review the animation and example problems, then complete the practice problems. Continue to the next section of the book once you have mastered the rule presented. ■

Oral liquids may be measured in droppers, calibrated spoons, medicine cups, or oral syringes. Calibrated cups and spoons are available at most pharmacies and sometimes come with the medication. Advise patients who take oral liquid medications at home to use a medicine cup or baking measuring spoon—not a household cup or spoon—if they do not have calibrated cups or spoons.

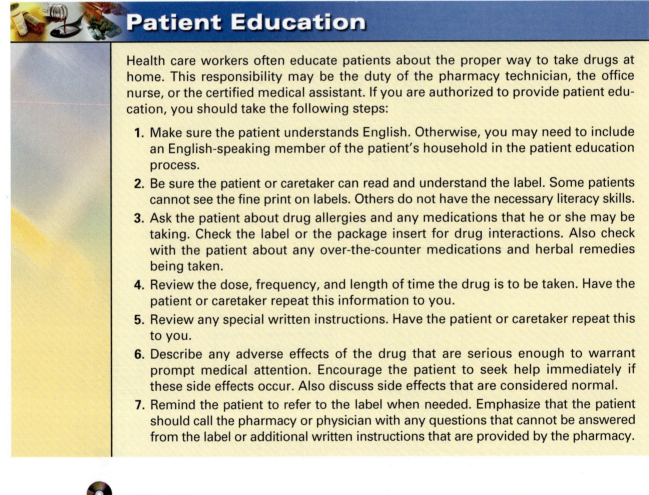

Patient Education

Health care workers often educate patients about the proper way to take drugs at home. This responsibility may be the duty of the pharmacy technician, the office nurse, or the certified medical assistant. If you are authorized to provide patient education, you should take the following steps:

1. Make sure the patient understands English. Otherwise, you may need to include an English-speaking member of the patient's household in the patient education process.
2. Be sure the patient or caretaker can read and understand the label. Some patients cannot see the fine print on labels. Others do not have the necessary literacy skills.
3. Ask the patient about drug allergies and any medications that he or she may be taking. Check the label or the package insert for drug interactions. Also check with the patient about any over-the-counter medications and herbal remedies being taken.
4. Review the dose, frequency, and length of time the drug is to be taken. Have the patient or caretaker repeat this information to you.
5. Review any special written instructions. Have the patient or caretaker repeat this to you.
6. Describe any adverse effects of the drug that are serious enough to warrant prompt medical attention. Encourage the patient to seek help immediately if these side effects occur. Also discuss side effects that are considered normal.
7. Remind the patient to refer to the label when needed. Emphasize that the patient should call the pharmacy or physician with any questions that cannot be answered from the label or additional written instructions that are provided by the pharmacy.

GO TO . . . Open the CD-ROM that accompanies your textbook, and select Chapter 6, Practice 6-3. Review the animation and example problems, then complete the practice problems. Continue to the next section of the book once you have mastered the information presented. ■

REVIEW AND PRACTICE

6-2 Oral Drugs

In Exercises 1–4, refer to label A.

1. What is the NDC number?
2. What is the dosage strength?
3. Can you store this drug on a shelf in the storeroom?
4. How many tablets are in the container?

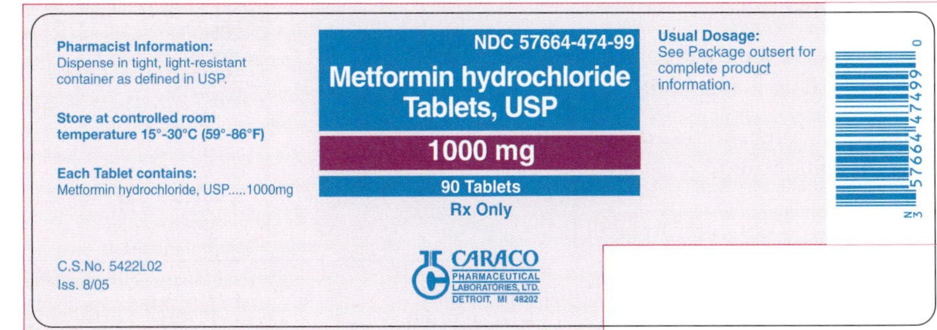

A

In Exercises 5–8, refer to label B.

5. What is the trade name of this drug?
6. Who is the manufacturer?
7. What is the dosage strength?
8. How many tablets are in the container?

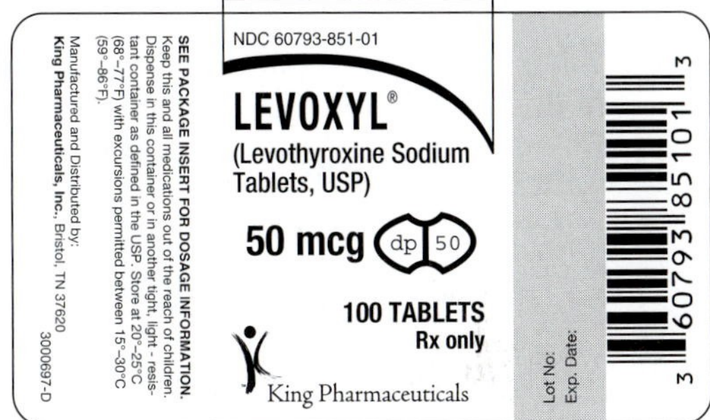

B

In Exercises 9–12, refer to label C.

9. What is the generic name of this drug?
10. How is this medication administered?
11. What is the dosage strength and type?
12. Can these tablets be broken or crushed?

C

In Exercises 13–16, refer to label D.

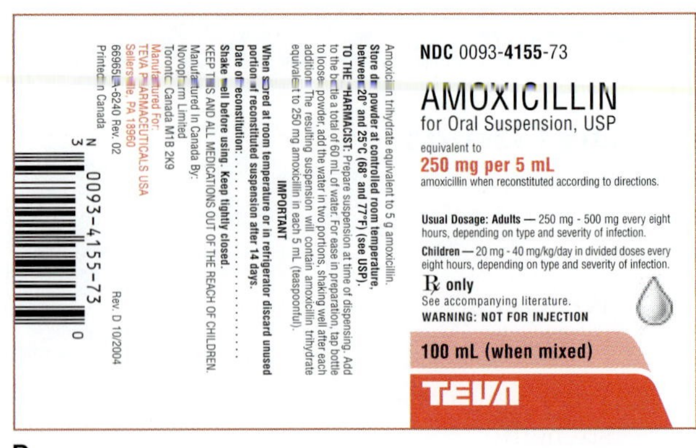

D

13. How many milliliters of water should be used to reconstitute this drug?
14. What is the dosage strength when the drug is reconstituted?
15. What is the total volume in the container when the drug is reconstituted?
16. How long can this drug be stored after it is reconstituted?

In Exercises 17–20, refer to label E.

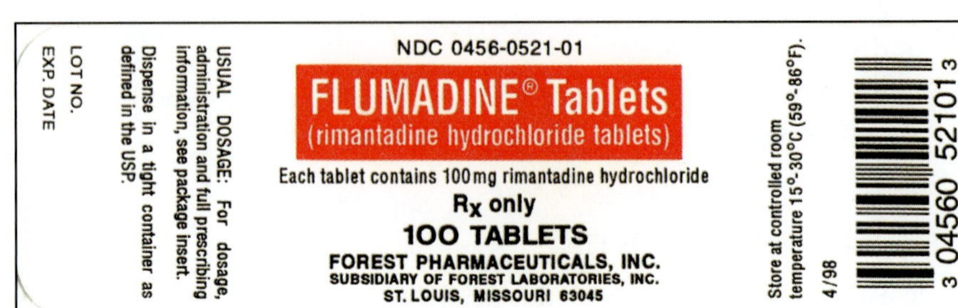

E

17. What is the generic name of this drug?
18. What is the dosage strength?
19. What is the trade name of this medication?
20. How should this drug be stored?

In Exercises 21–25, refer to labels A through E.

21. Which of these medication labels indicate that there are 100 tablets in the bottle?
22. Which of these drugs could be divided to give a partial dose? Why?
23. According to the labels alone, which of these drugs can be safely administered to children?
24. Which of these drugs can be stored at room temperature?
25. Which medications should not be broken or divided?

To check your answers, see page 517.

Parenteral Drugs

Parenteral drugs may be packaged in single-use ampules or vials, single-use prefilled syringes, or multiuse vials. These small containers have small labels that have limited space for providing comprehensive information (Figure 6-25). You must read these labels with extra care. You will often need to review the package insert to obtain complete drug information.

Parenteral drugs can be injected intradermally (ID), intramuscularly (IM), intravenously (IV), or subcutaneously (Sub-Q). Parenteral drugs can also be inhalants and **transdermal** (through the skin) medications or any mode of administration other than through the gastrointestinal tract. The drug label specifies the appropriate route. Tigan® is made for IM intramuscular use only (see Figure 6-10). Primaxin® is a drug that can be administered either intramuscularly or intravenously. Figure 6-11 indicates Amiodarone HCl for IV (intravenous) use only. Betaseron® is made for subcutaneous use only (see Figure 6-25).

On the label of parental medications the dosage strength may be labeled as the amount of drug per dosage unit (Figure 6-26), or in units per mL. (Figures 6-27 and 6-28). In some cases the label is marked with milligrams and micrograms. Dosage strength may also be expressed in milliequivalents per milliliter or as a percent, which is grams per 100 mL. *Remember*, the actual dosage strength is a rate, which is an amount of medication found in an amount of liquid or solid, such as milligram per milliliter or grams per tablet.

Look at the labels for insulin in Figures 6-27 and 6-28. In addition to the standard components, these labels contain information about the origin of the medication and how quickly the insulin takes effect. Insulin can be made from human sources (recombinant DNA origin, or rDNA) or animal sources (beef or pork). Most animal-source insulin is being phased out in the United States. Different types of insulin take effect over different time periods. NPH insulin (Figure 6-27) is an intermediate-acting insulin. Regular insulin (Figure 6-28) is fast acting. See Chapter 12 for more information about insulin.

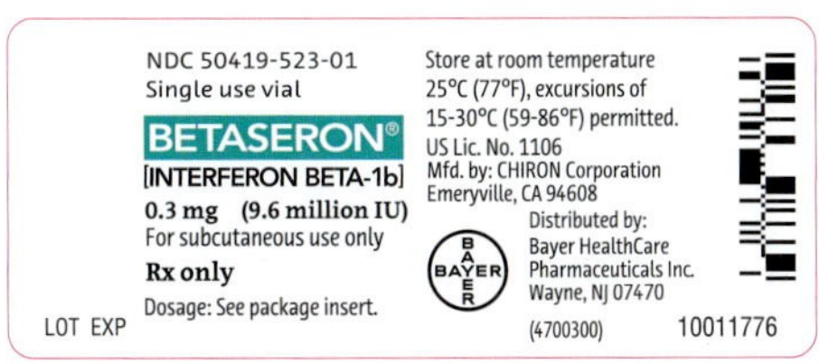

Figure 6-25 This container holds 0.3 mg (9.6 million IU) of Betaseron® for subcutaneous use only. You must read the package insert for the dosage information.

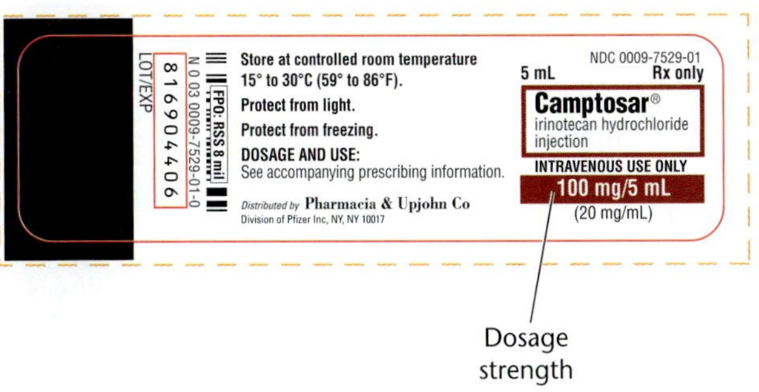

Figure 6-26 This dosage strength is expressed in milligrams per milliliter 100 mg/5 mL.

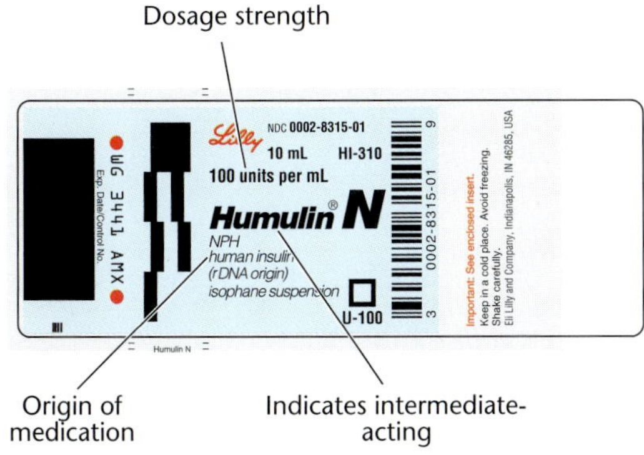

Figure 6-27 Insulin dosage strength is expressed in units per milliliter.

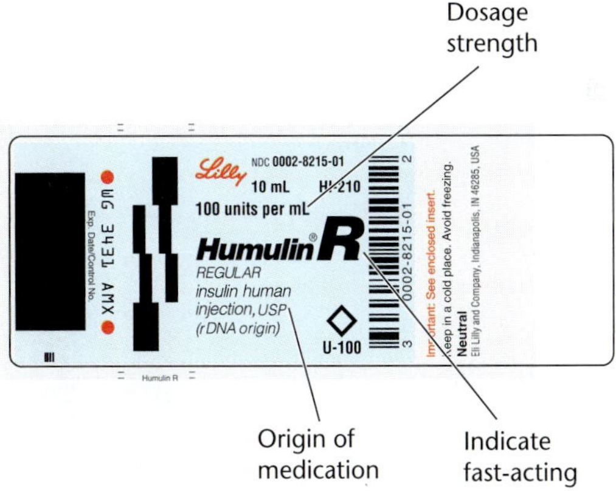

Figure 6-28 Fast-acting Humulin® R insulin has a dosage strength of 100 units/mL.

REVIEW AND PRACTICE

6-3 Parenteral Drugs

In Exercises 1–4, refer to label A.

1. What is the dosage strength?
2. By what route of administration is this drug given?
3. What other instructions does this label provide?
4. How long can this product be stored?

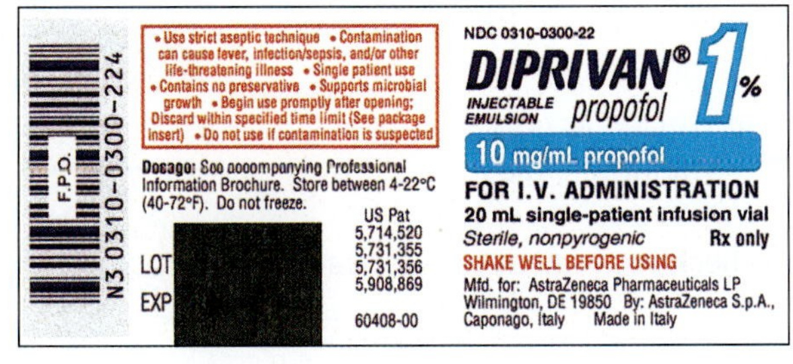

A

In Exercises 5–8, refer to label B.

5. What is the dosage strength of this drug?
6. Who manufactures this medication?
7. What is the generic name of this medication?
8. If you were not familiar with this drug, would you be able to administer it with only the information on the label? Why?

B

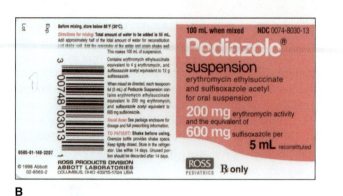

In Exercises 9–12, refer to label C.

9. What are the storage requirements for this medication?
10. What is the dosage strength?
11. By what route of administration is this drug given?
12. If the dose was 1 mcg (2 capsules), how many doses does the container hold?

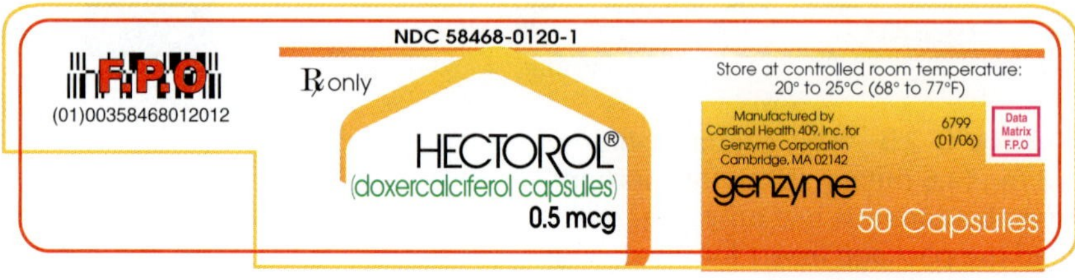

C

In Exercises 13–16, refer to label D.

13. What is the origin of this insulin?
14. What is the dosage strength of the insulin?
15. What is the generic name of this insulin?
16. If the usual dose is 10 units, how many doses are in this container?

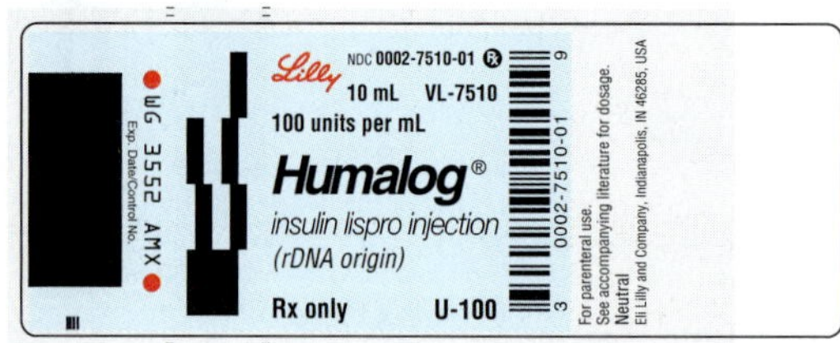

D

In Exercises 17–20, refer to label E.

17. What is the dosage strength?
18. What is the generic name of this medication?
19. How many 100-mg doses does this container hold?
20. What are the storage requirements for this drug?

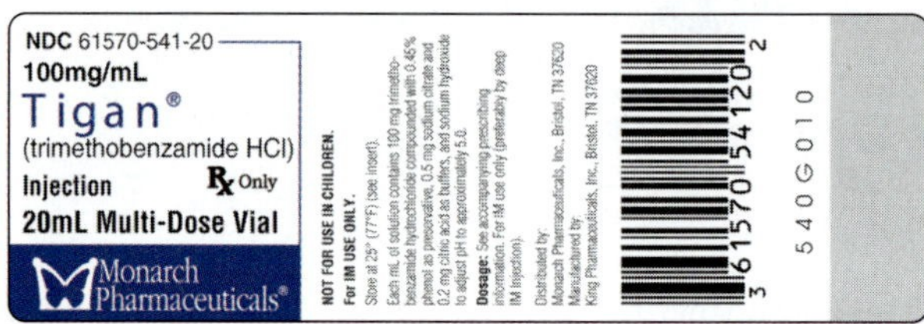

E

To check your answers, see page 518.

Critical Thinking on the Job

Avoid Unnecessary Risks

A drug order calls for the patient to receive Gammagard® S/D. At 0830 the health care professional prepares the medication according to the label. See label below. The health care professional is called away due to an emergency with another one of her patients. She returns at 1100 and the medication still needs to be administered.

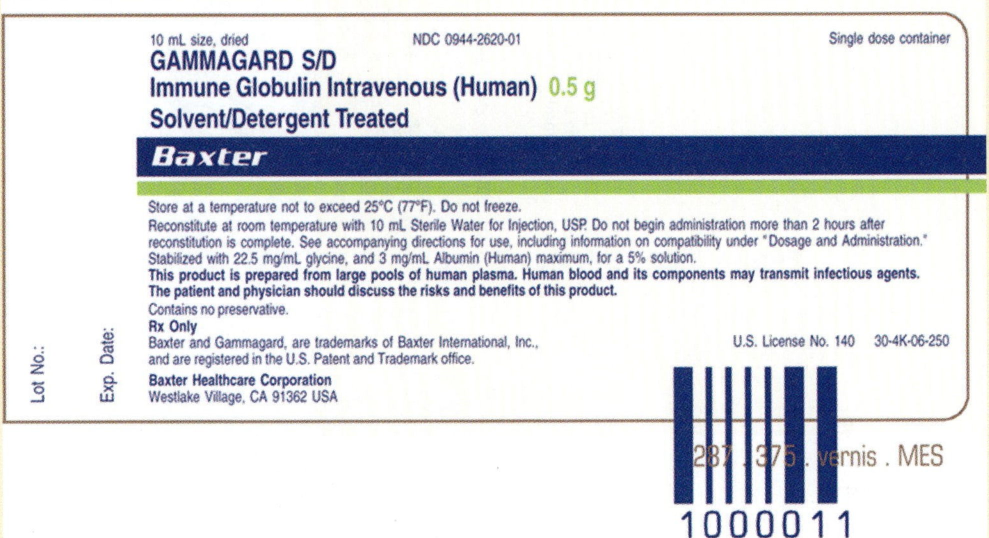

10 mL size, dried
NDC 0944-2620-01
Single dose container
GAMMAGARD S/D
Immune Globulin Intravenous (Human) 0.5 g
Solvent/Detergent Treated
Baxter
Store at a temperature not to exceed 25°C (77°F). Do not freeze.
Reconstitute at room temperature with 10 mL Sterile Water for Injection, USP. Do not begin administration more than 2 hours after reconstitution is complete. See accompanying directions for use, including information on compatibility under "Dosage and Administration."
Stabilized with 22.5 mg/mL glycine, and 3 mg/mL Albumin (Human) maximum, for a 5% solution.
This product is prepared from large pools of human plasma. Human blood and its components may transmit infectious agents.
The patient and physician should discuss the risks and benefits of this product.
Contains no preservative.
Rx Only
Baxter and Gammagard, are trademarks of Baxter International, Inc., and are registered in the U.S. Patent and Trademark office.
Baxter Healthcare Corporation
Westlake Village, CA 91362 USA
U.S. License No. 140
30-4K-06-250
Lot No.:
Exp. Date:
1000011

Question?

What should she do and why?

Drugs Administered by Other Routes

Although many drugs use oral and parenteral routes of administration, other routes exist. They include *sublingual* (under the tongue), *buccal* (between the tongue and cheek), *rectal*, and *vaginal*. Drugs may also be administered as topical ointments (used on the skin); eye or ear drops; patches applied to the skin (transdermal delivery); or nasal, oral, and throat inhalants (see Figures 6-29, 6-30, and 6-31).

The dosage strength is expressed slightly differently on these labels. In Figure 6-29, the dosage strength is the percentage of the lotion that is the active ingredient betamethasone. The total amount of lotion is given in both grams and milliliters. In Figure 6-30, the dosage rate is given as 100 mcg/h; this drug is absorbed over time through the skin. Thus the dosage rate indicates 100 mcg are delivered each hour. In Figure 6-31, the medication is measured in mcg. The delivery system contains 28 metered sprays. The dosage strength is actually 50 mcg per metered spray.

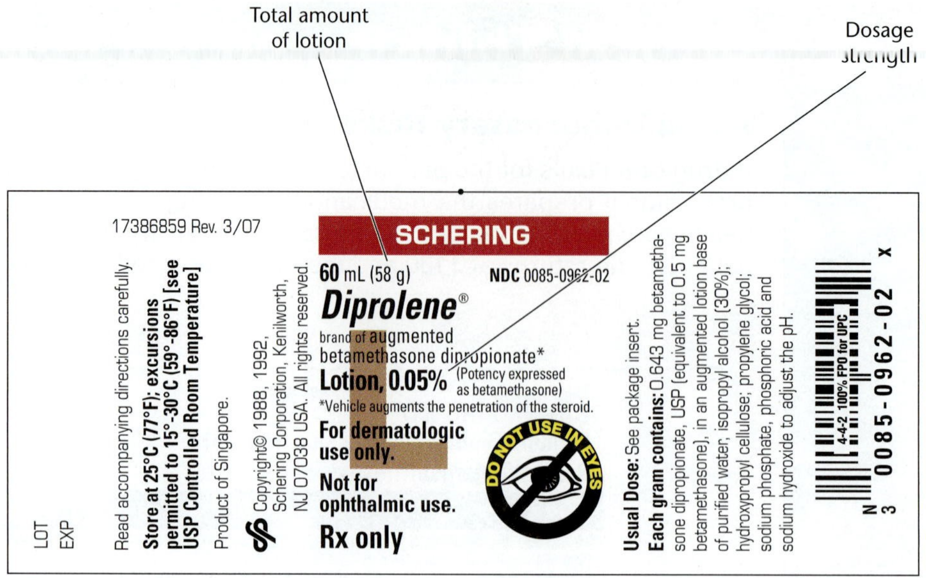

Figure 6-29 This drug is used on the skin, or dermatologic use only.

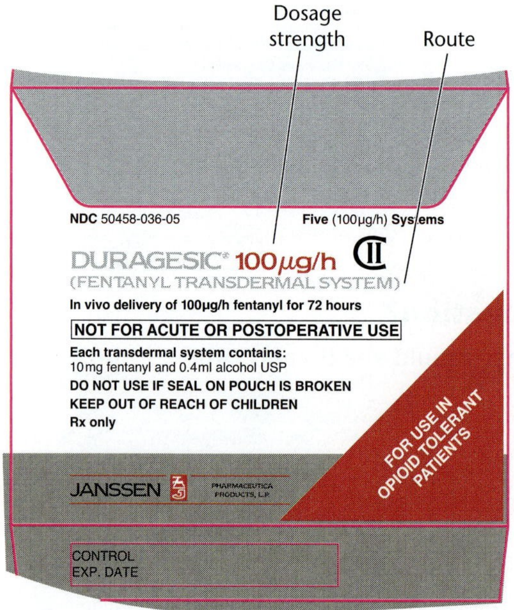

Figure 6-30 Transdermal medications deliver medication through the skin at a dosage rate.

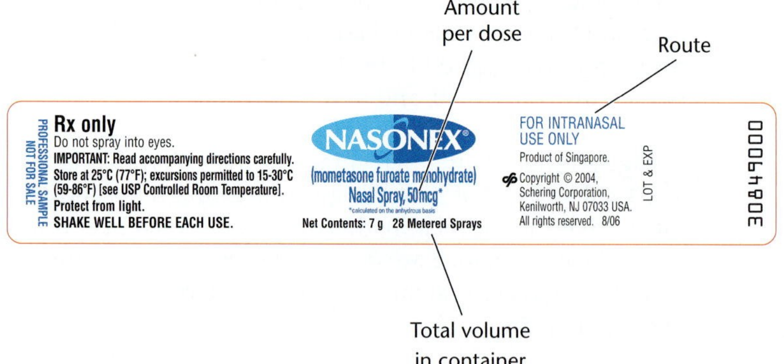

Figure 6-31 Each metered spray delivers 50 mcg of medication.

GO TO . . . Open the CD-ROM that accompanies your textbook, and select Chapter 6, Practice 6-4. Review the animation and example problems, then complete the practice problems. Continue to the next section of the book once you have mastered the rule presented. ■

REVIEW AND PRACTICE

6-4 Drugs Administered by Other Routes

In Exercises 1–4, refer to label A.

1. What is the generic name?

2. What is the dosage strength?

3. What is the usual dosage for adults?

4. How would this medication be stored?

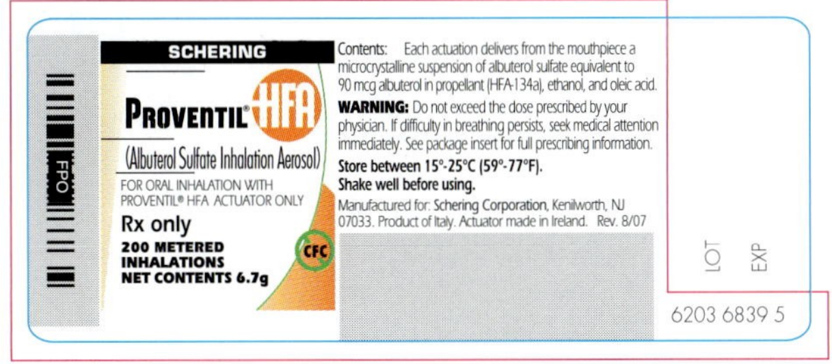

A

In Exercises 5–8, refer to label B.

5. By what route is this drug to be administered?

6. What are the drug's storage requirements?

7. What is the dosage strength?

8. On what ingredient is the dosage strength based?

NDC 0045-0810-15
REGRANEX* GEL 0.01%
(becaplermin)
Contains: becaplermin 0.01%, sodium carboxymethylcellulose, sodium chloride, sodium acetate trihydrate, glacial acetic acid, L-lysine hydrochloride, and water for injection, with preservatives: methylparaben 0.156%, propylparaben 0.017%, and m-cresol 0.086%.
℞ only.
Dosage and Administration: See package insert.
STORE REFRIGERATED, 2° - 8°C (36° - 46°F).
DO NOT FREEZE.
Warning: Keep out of reach of children.
Important: Do not use if seal has been punctured or is not visible.
To open: Use cap to puncture seal.
To close: Recap tightly after each use.
DIN 02239405 NET WT. 15g
For Topical Use Only
Multi-dose tube
See crimp end for lot number and expiration date.
(01)10300450810158
ORTHO-McNEIL
Distributed by:
OMP DIVISION
ORTHO-McNEIL PHARMACEUTICAL, INC.
Raritan, New Jersey 08869
and JANSSEN-ORTHO INC.
Toronto, Canada M3C 1L9
Manufactured by:
OMJ Pharmaceuticals, Inc. 107-10-247-6
U.S. Lic. #1196
San German,
Puerto Rico 00683
© OMP 1998
Made in U.S.A.
*Trademark

B

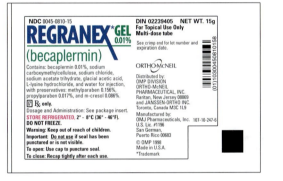

In Exercises 9–12, refer to label C.

9. What is the route of administration?

10. What is the dosage strength?

11. How many metered sprays are in this container?

12. Can this spray be delivered through more than one route?

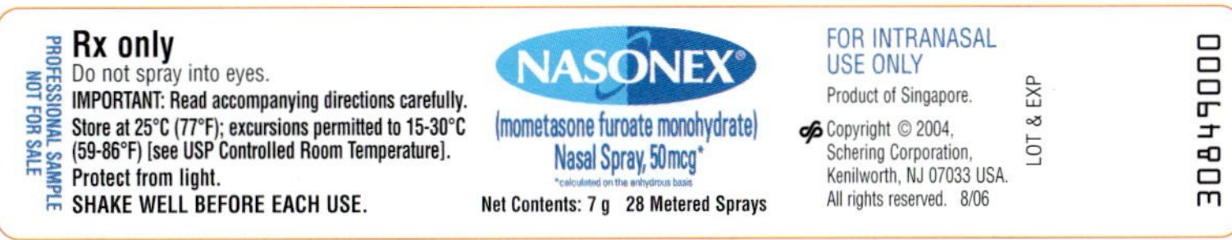

C

In Exercises 13–16, refer to label D.

13. By what route is this drug delivered?
14. What is the dosage strength?
15. What type of patients should use this medication?
16. How many doses are in this box?

In Exercises 17–20, refer to label E.

17. What is the generic name?
18. What is the dosage strength?
19. By what route is this drug to be administered?
20. How many doses are in this container?

To check your answers, see pages 518–519.

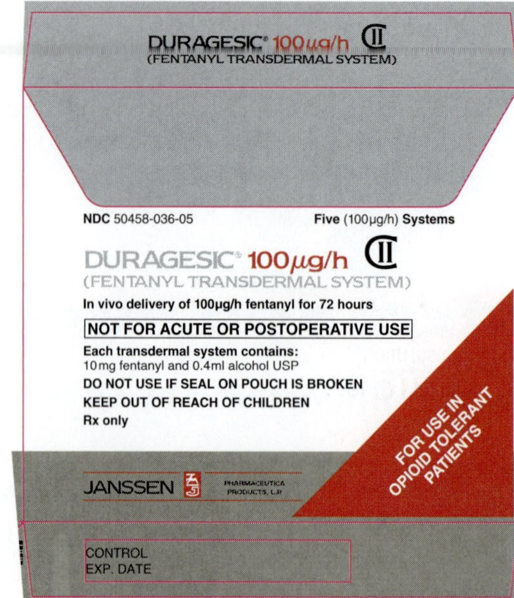

D

E

HOMEWORK ASSIGNMENT

Use the identified drug labels to answer the following questions:

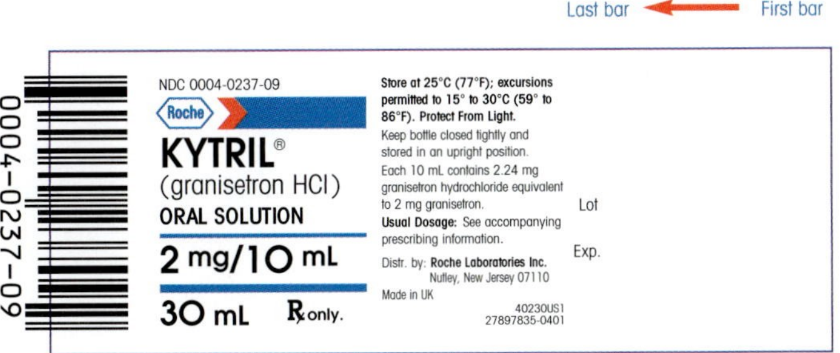

Label A

1. What is the generic name of this drug?
2. List three items of storage information on this label?
3. What is the total drug volume of the container?
4. 2 mg of the drug is equal to how many milliliters?

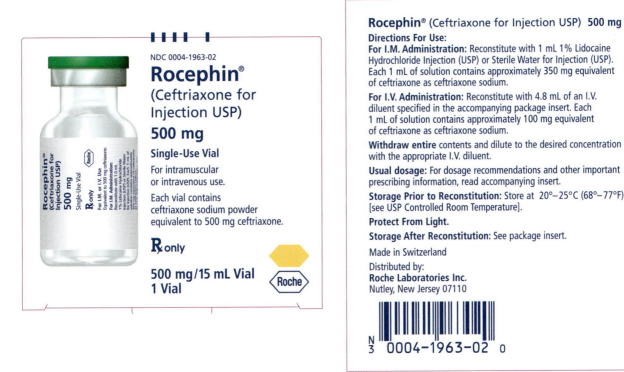

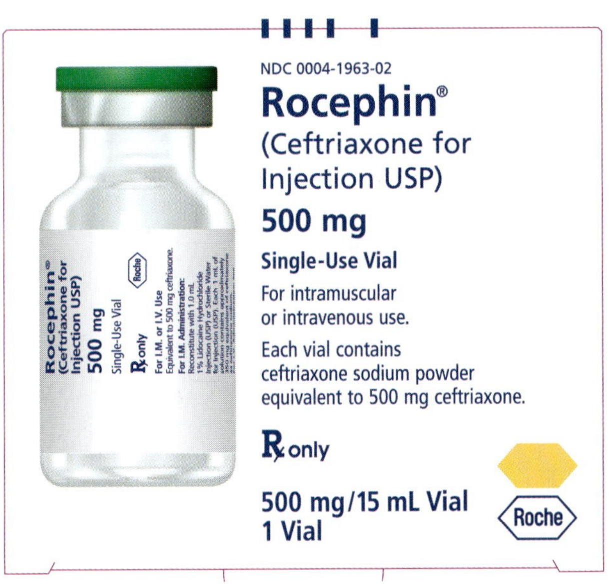

Rocephin® (Ceftriaxone for Injection USP) **500 mg**

Directions For Use:

For I.M. Administration: Reconstitute with 1 mL 1% Lidocaine Hydrochloride Injection (USP) or Sterile Water for Injection (USP). Each 1 mL of solution contains approximately 350 mg equivalent of ceftriaxone as ceftriaxone sodium.

For I.V. Administration: Reconstitute with 4.8 mL of an I.V. diluent specified in the accompanying package insert. Each 1 mL of solution contains approximately 100 mg equivalent of ceftriaxone as ceftriaxone sodium.

Withdraw entire contents and dilute to the desired concentration with the appropriate I.V. diluent.

Usual dosage: For dosage recommendations and other important prescribing information, read accompanying insert.

Storage Prior to Reconstitution: Store at 20°–25°C (68°–77°F) [see USP Controlled Room Temperature].

Protect From Light.

Storage After Reconstitution: See package insert.

Made in Switzerland

Distributed by:
Roche Laboratories Inc.
Nutley, New Jersey 07110

N 3 0004-1963-02 0

Label B

5. What is the brand name of this drug?
6. What two routes of administration are listed on this label?
7. What are the reconstitution directions for an IM injection?
8. How many times may this vial be used?

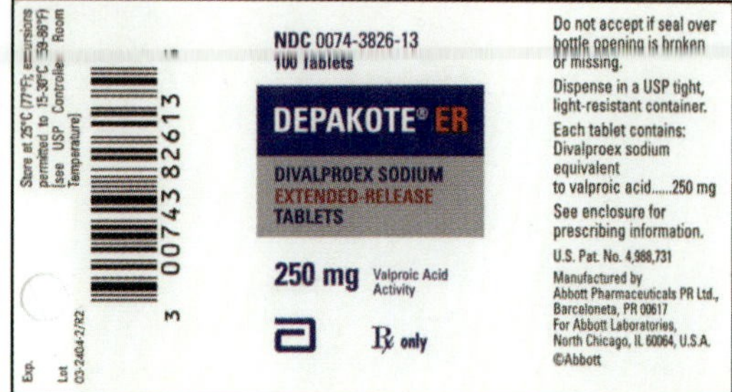

Label C

9. What is the brand name of this drug?
10. What does ER after the drug name mean?
11. How many tablets are in the container?
12. What is name of the manufacturer?

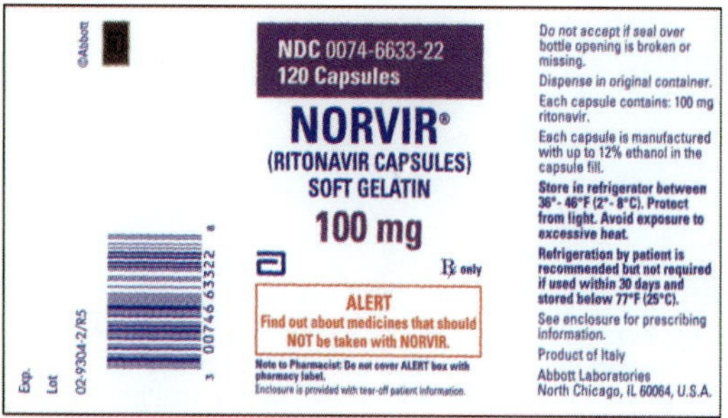

Label D

13. What is the generic name of this drug?
14. If the usual dose is one capsule, how many doses are available in this container?
15. What form of capsule is in this container?
16. How long may the drug be stored without refrigeration?

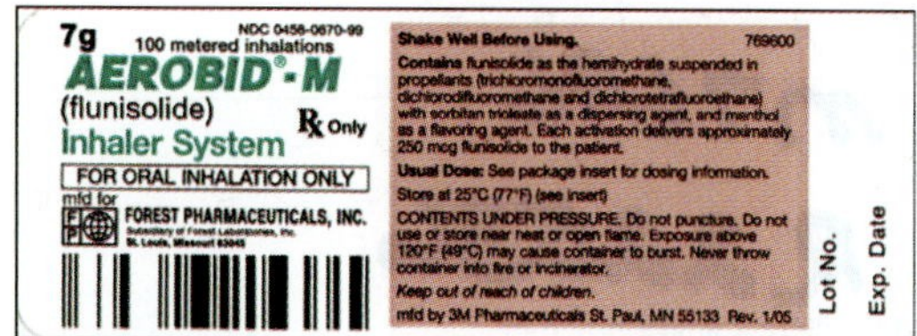

Label E

17. What is the route of administration for this drug?
18. Approximately how much medication is delivered with each inhalation?
19. How many inhalations are in this container?
20. What are the before-use instructions for this drug?

Chapter 6 Review

Check Up

1. Distinguish a drug's trade name from its generic name.
2. Explain why a drug order might be written using only the trade name of a drug.
3. Explain the difference between IM and IV.
4. List the types of tablets that cannot be divided, broken, or crushed for administration, and explain why.
5. Describe when you would use a package insert.
6. Explain the importance of a lot number.

In Exercises 7–10, refer to label A.

7. What is the dosage strength of the drug?
8. How many tablets of medication are in this container?
9. What is the name of the manufacturer of the drug?
10. How is it administered?

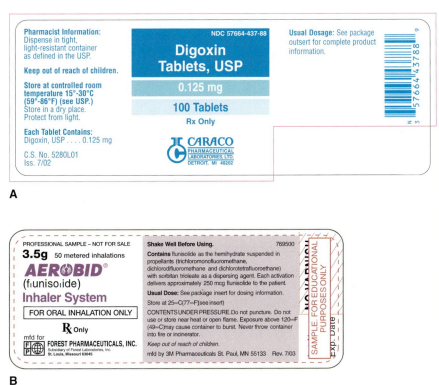

Pharmacist Information:
Dispense in tight, light-resistant container as defined in the USP.
Keep out of reach of children.
Store at controlled room temperature 15°-30°C (59°-86°F) (see USP.)
Store in a dry place.
Protect from light.
Each Tablet Contains:
Digoxin, USP . . . . 0.125 mg
C.S. No. 5280L01
Iss. 7/02
NDC 57664-437-88
Digoxin Tablets, USP
0.125 mg
100 Tablets
Rx Only
CARACO PHARMACEUTICAL LABORATORIES, LTD.
DETROIT, MI 48202
Usual Dosage: See package outsert for complete product information.
N 3 57664 43788 6

A

In Exercises 11–14, refer to label B.

11. What is the generic name of this drug?
12. How is this drug administered?
13. What is the dosage strength?
14. How many doses are in the container?

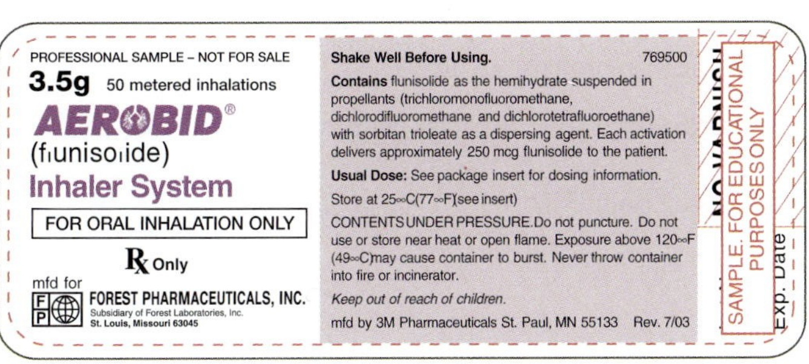

PROFESSIONAL SAMPLE – NOT FOR SALE
3.5g 50 metered inhalations
AEROBID®
(flunisolide)
Inhaler System
FOR ORAL INHALATION ONLY
℞ Only
mfd for
FOREST PHARMACEUTICALS, INC.
St. Louis, Missouri 63045
Shake Well Before Using.
769500
Contains flunisolide as the hemihydrate suspended in propellants (trichloromonofluoromethane, dichlorodifluoromethane and dichlorotetrafluoroethane) with sorbitan trioleate as a dispersing agent. Each activation delivers approximately 250 mcg flunisolide to the patient.
Usual Dose: See package insert for dosing information.
Store at 25∞C(77∞F)[see insert]
CONTENTS UNDER PRESSURE. Do not puncture. Do not use or store near heat or open flame. Exposure above 120∞F (49∞C) may cause container to burst. Never throw container into fire or incinerator.
Keep out of reach of children.
mfd by 3M Pharmaceuticals St. Paul, MN 55133 Rev. 7/03
SAMPLE. FOR EDUCATIONAL PURPOSES ONLY
Exp. Date

B

In Exercises 15–18, refer to label C.

15. How is this drug administered?
16. What is the dosage strength?
17. What is the drug name?
18. How many tablets are in the container?

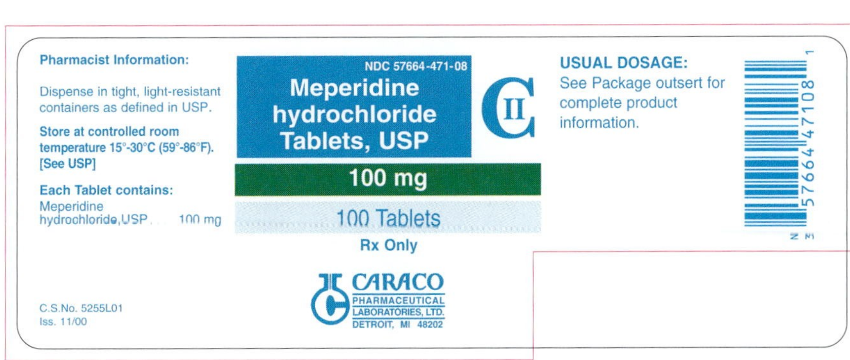

C

In Exercises 19–22, refer to label D.

19. What is the origin of this insulin?
20. What word on the label describes the time frame in which this insulin acts?
21. How is this drug administered?
22. What is the dosage strength?

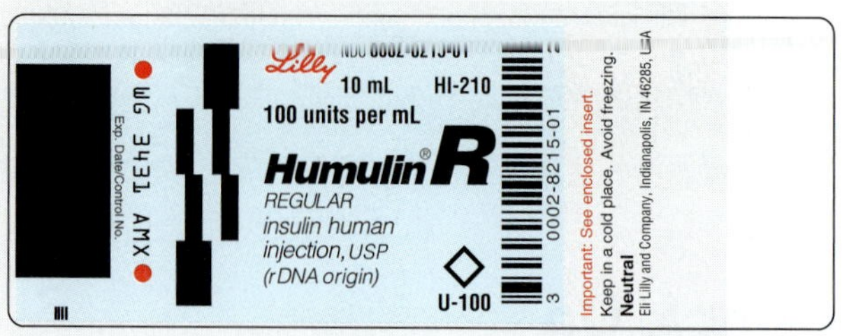

D

Critical Thinking Application

You are working in a clinic that serves many adult homeless people. Two forms of Erythromycin are available (see labels below). If the patient needs to take Erythromycin for 5 days, which form of the medication would be better and why?

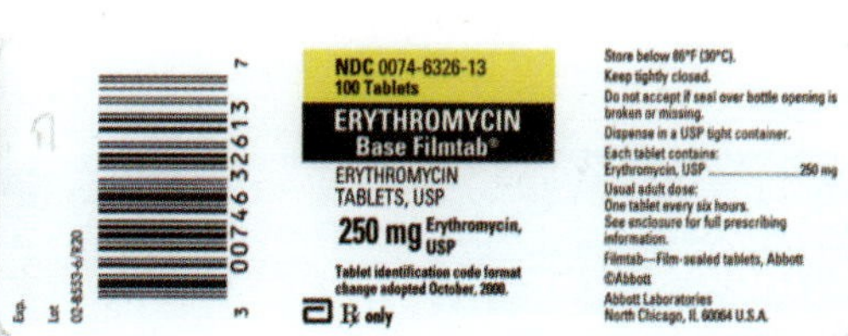

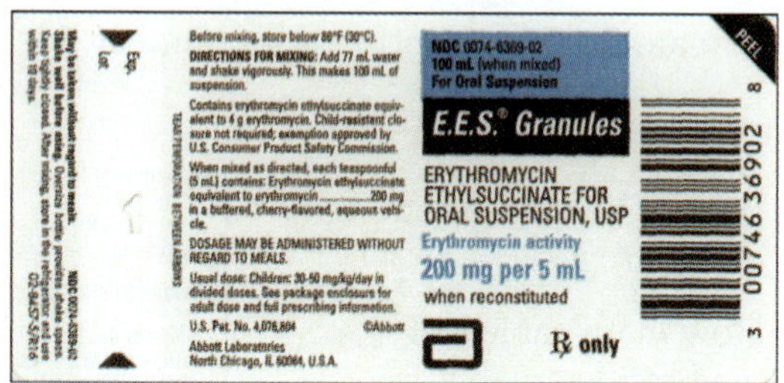

Case Study

A drug order reads Gentamicin 5 mg IV now.
You have available a drug with the following label:

1. What would you do to prepare for administering this drug?
2. How would you administer the drug?
3. What would you do with the vial after administering a dose of the drug?

To check your answers, see page 519.

Internet Activity

Mr. Liu is about to be discharged from the hospital with instructions to take Coumadin® 1 mg bid. Mr. Liu is an elderly, easily confused man who will be cared for by his daughter who also has difficulty with instructions. Although you have reviewed his medication instructions with him and his daughter several times, you are not completely confident they understand that he should not drink alcohol or take any self-prescribed, over-the-counter medications or herbal cures while he is taking Coumadin®.

Assignment: Conduct an Internet search to find information in plain language regarding the importance of not taking any over-the-counter medications while taking Coumadin®.

GO TO . . . Open the CD-ROM that accompanies your textbook, and complete a final review of the rules, practice problems, and activities presented for this chapter. For a final evaluation, take the chapter test and email or print your results for your instructor. A score of 95 percent or above indicates mastery of the chapter concepts. ■

7 Methods of Dosage Calculations

Each problem that I solved became a rule, which served afterwards to solve other problems.

—René Descartes

Learning Outcomes

When you have completed Chapter 7, you will be able to:

- Describe how the information on a physician's order, medication administration record (MAR), or prescription, along with the drug label and package insert, is used to calculate the *desired dose.*
- Convert the *dosage ordered* to the *desired dose,* using fraction proportion, ratio proportion, or dimensional analysis.
- Calculate the *amount to administer* of a drug, using any of the methods of dosage calculation—fraction proportion, ratio proportion, dimensional analysis, or formula.
- Recognize common errors that occur during dose calculations.

Key Terms

Amount to administer

Desired dose

Dosage ordered

Dosage strength

Dosage unit

Dose on hand

Introduction

It is time to bring together all the information you have learned in previous chapters to calculate the amount of medication to administer to a patient. You will bring together basic math, information from the physician's order and drug labels, and methods of converting quantities from one unit of measurement to another. Do not hesitate to refer to previous rules to help you solve the problems presented here.

Dosages and Doses

Before you can calculate in practical terms how many dosage units of a medication to administer to a patient, you must first find the **desired dose**—the amount of the actual drug that the patient is to be given at one time. To determine the desired dose, you must know the following information: the *dosage ordered* and the *dose on hand.*

The **dosage ordered** is the amount of drug the physician has ordered along with the frequency with which it is given. You will find this information on the physician's order or prescription, and the medication administration record (MAR). You may also receive this information verbally from a physician, especially in emergency situations. (Remember to check a verbal order by repeating it as you are writing it down.) *Always check the physician's original order the first time you administer a medication to a patient and if you have any questions about the information on the MAR.*

Dosage strength measures the amount of drug per dosage unit. Many medications are available in different dosage strengths. For example, a medication may be produced in two versions, 75 mg per tablet and 100 mg per tablet. In both of these versions, the **dosage unit** is a tablet.

Learning Link Recall from Chapter 6, dosage strength is a ratio that indicates the amount of medication found in an amount of liquid or solid, found on page 176.

Once you have determined the dosage strength, you are able to determine two other terms—the *dose on hand* and the *dosage unit.* The **dose on hand** is the specific amount of drug that is present in a dosage unit. In this example, the dose on hand for the first tablet is 75 mg. The dose on hand for the second tablet is 100 mg. The dosage unit is one tablet for both. For any medication you have available, you can read the drug label to determine the dose on hand and the dosage unit. Dosage strength, then, is the dose on hand per dosage unit.

Sometimes the *dosage ordered* and the *dose on hand* have the same unit of measurement. In these cases, the desired dose is the same as the dosage ordered and no conversion is necessary. Frequently, however, the dosage ordered and the dose on hand are expressed in different units of measurement. In these cases, you must convert the dosage ordered so that it has the same unit of measurement as the dose on hand. This conversion leads you to the ***desired dose.***

For example, suppose the dosage ordered is written for 0.5 grams of medication. The dose on hand is 1000 mg. To find the desired dose, you would need to convert grams (the dosage ordered) to milligrams (the dose on hand).

See Table 7-1 for a summary of the language of dosage calculations.

Conversion Factors

Conversion factors are expressions that allow you to switch from one unit of measurement to another.

Learning Link Recall from Chapter 3, Conversion Factors, when the numerator and denominator of a conversion factor are equivalent the factor itself equals 1, found on page 92.

In some cases, you convert between units in the same system of measurement; you will use conversion factors such as 1 mg/1000 mcg and 1000 mcg/1 mg. In other cases, you convert between systems, using conversion factors such as 1 tsp/5 mL and 5 mL/1 tsp.

Calculating the Desired Dose

Before you calculate the *amount to administer,* it is first necessary to determine the desired dose. As stated in Table 7-1, the desired dose (the amount of drug given at one time) must have the same unit of measurement as the dose on hand (found on the drug label). Unfortunately, the

TABLE 7-1 The Language of Dosage Calculations

Term	Abbreviation	Definition	Examples Refer to Figure 7-1
Dosage ordered	*O*	The amount of drug to be given to the patient and how often it is to be given *This value will be found on the drug order or prescription.*	40 mg bid (twice a day)
Desired dose	*D*	The amount of drug to be given at a single time *The unit of measurement must be the same as the dose on hand.* *The drug order and the drug label must be consulted.*	This amount must be calculated when dosage ordered is a different unit from the dose on hand. In this problem a calculation will not be necessary.
Dosage unit	*Q*	The units by which the drug will be measured when it is administered *This value will be found on the drug label.*	In Figure 7-1 this is capsules. Other examples include: tablets, 1 mL, 5 mL, drops, or units.
Dose on hand	*H*	The amount of drug contained in each dosage unit *This value will be found on the drug label.*	In Figure 7-1 this is 20 mg.
Amount to administer	*A*	The volume of a liquid or the number of solid dosage units that contains the desired dose *This value is found with a calculation.*	If the desired dose is 40 mg and the dosage strength is 20 mg per capsule, the amount to administer is 2 capsules.
Dosage strength	$\frac{H}{Q}$	Dose on hand per dosage unit *This value can be determined from the drug label.*	In Figure 7-1 this is 20 mg per capsule.

dosage ordered will not always be written in the same units as are found on the drug label. For example, an order may be written in grams while the drug is labeled in milligrams. When this occurs, it is necessary to convert the dosage ordered to a desired dose having the same units as the dose on hand.

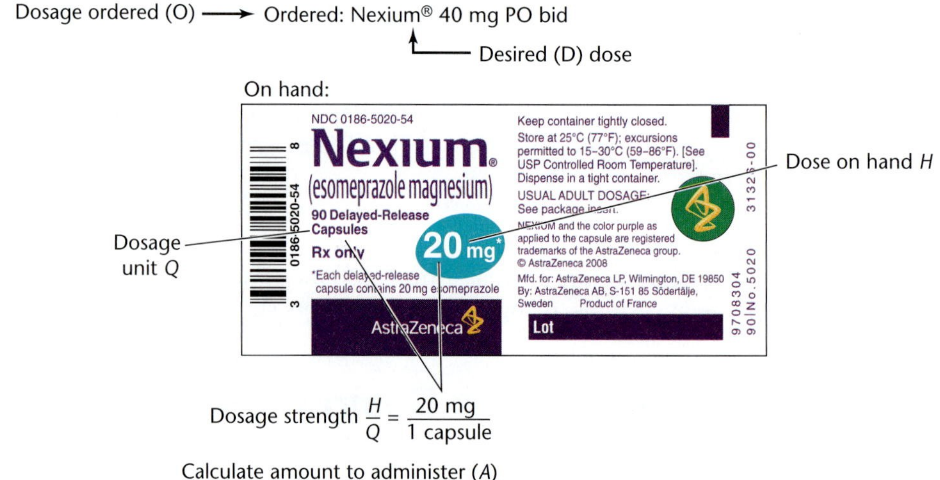

Figure 7-1 The language of dosage calculations.

This section reviews three methods that can be used to calculate the desired dose: the fraction proportion method, the ratio proportion method, and dimensional analysis. Remember, each will give you the same result, and the method you use is a matter of personal preference. Once you identify the method you prefer, follow the color coding of that method. Regardless of the method that you choose, you will want to become familiar with the terms contained in Table 7-1 before you proceed.

Rule 7-1 Calculating the Desired Dose

The unit of measurement for the desired dose must be the same as the unit of measurement of the dose on hand before the amount to administer can be calculated. This is calculated by converting the dosage ordered into the same unit of measurement as the dose on hand; once converted, it becomes the desired dose.

Example Consider that the physician has ordered the patient to receive 0.2 mg of medication. This is the dosage ordered.

The dosage strength is 100 mcg/tablet, making the unit of measure of the dose on hand micrograms (mcg). *Recall the dosage strength is the dose on hand per dosage unit.*

Since the bottle of medication comes in micrograms and the order is for milligrams, you must change the dosage ordered (0.2 mg) to the same unit of measurement as the dose on hand (micrograms) to obtain the desired dose.

GO TO . . . Open the CD-ROM that accompanies your textbook, and select Chapter 7, Rule 7-1. Review the animation and example problems, then complete the practice problems. Continue to the next section of the book once you have mastered the rule presented. ■

Fraction Proportion Method

You can use fraction proportions to convert from one unit of measure to another. Procedure Checklist 3-1 was used. This same procedure checklist can be used to determine the desired dose from the dosage ordered.

Learning Link Recall Chapter 3, Procedure Checklist 3-1, found on page 93.

Procedure Checklist 3-1

Converting by the Fraction Proportion Method

1. Write a conversion factor with the units that you are converting to in the numerator and the units you are converting from in the denominator.
2. Write a fraction with the unknown? in the numerator and the number that you need to convert in the denominator. *(The unknown is the desired dose D. The number you need to convert is the dosage ordered O.)*
3. Set up the two fractions as a proportion.
4. Cancel units.
5. Cross-multiply and then solve for the unknown value.

Example 1 The dosage ordered is 0.2 mg once a day.

The dosage strength is 100 mcg per tablet.

Find the desired dose.

In this case, the drug is measured in milligrams on the drug order and micrograms on the drug label. The units for the desired dose must match those found on the drug label, which means that we must convert 0.2 mg to micrograms.

Follow the steps of Procedure Checklist 3-1.

1. Since we are converting to micrograms, micrograms must appear in the numerator of our conversion factor. Our conversion factor is $\frac{1000 \text{ mcg}}{1 \text{ mg}}$.

Learning Link Recall from Chapter 3, Table 3-4, 1 mg = 1000 mcg, found on page 84.

2. The other fraction for our proportion has the unknown *D* for a numerator. The value that is being converted, 0.2 mg or the dosage ordered, must appear as the denominator. Our conversion factor is $\frac{D}{0.2 \text{ mg}}$.

3. Setting the two fractions into a proportion gives us the following equation:

$$\frac{D}{0.2 \text{ mg}} = \frac{1000 \text{ mcg}}{1 \text{ mg}}$$

4. Cancel units.

$$\frac{D}{0.2 \not{\text{mg}}} = \frac{1000 \text{ mcg}}{1 \not{\text{mg}}}$$

5. Cross-multiply and then solve for the unknown.

$$1 \times D = 1000 \text{ mcg} \times 0.2$$

$$D = 200 \text{ mcg} = \text{desired dose}$$

Example 2 The order reads ASA gr v PO daily.

The drug label indicates 325-mg tablets.

Find the desired dose.

Again, the drug order and the drug label use different units. In this case, we must convert the dosage ordered (5 gr) to milligrams to find the desired dose.

Follow the steps of Procedure Checklist 3-1.

1. Since we are converting to milligrams, our conversion factor is $\frac{65 \text{ mg}}{1 \text{ gr}}$. *Note:* For this medication 65 mg per 1 gr is used.

Learning Link Recall from Chapter 3, Table 3-8, 1 gr = 65 mg, found on page 92.

2. The other fraction for our proportion is $\frac{D}{5 \text{ gr}}$.

3. Setting the two fractions into a proportion gives the following equation:

$$\frac{D}{5 \text{ gr}} = \frac{65 \text{ mg}}{1 \text{ gr}}$$

4. Cancel units.

$$\frac{D}{5 \not{\text{gr}}} = \frac{65 \text{ mg}}{1 \not{\text{gr}}}$$

5. Cross-multiply and then solve for the unknown.

$1 \times D = 65 \text{ mg} \times 5$

$D = 325 \text{ mg} = \text{desired dose}$

Example 3 Ordered: 500 mg q6h

Dosage strength available: 250 mg per tablet

Find the desired dose.

In this case the unit for the dose ordered is the same as that for the dose on hand. No calculation is needed. The drug is measured in milligrams on both the order and the drug label, so the desired dose $D = 500$ mg.

Example 4 Ordered: Nitrostat® 800 mcg sublingually PRN chest pain

On hand: See Figure 7-2.

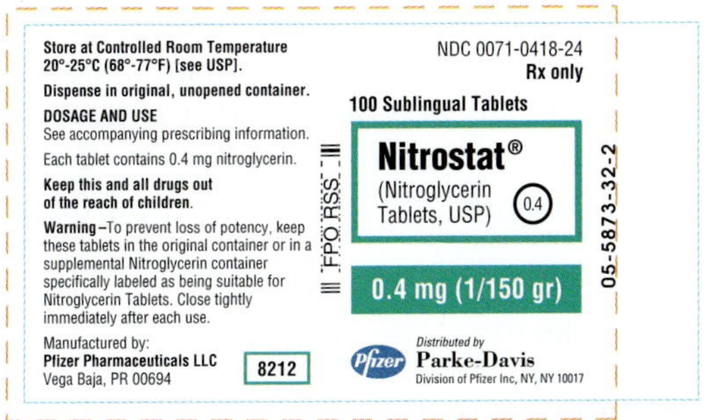

Store at Controlled Room Temperature 20°-25°C (68°-77°F) [see USP].
Dispense in original, unopened container.
DOSAGE AND USE
See accompanying prescribing information.
Each tablet contains 0.4 mg nitroglycerin.
Keep this and all drugs out of the reach of children.
Warning–To prevent loss of potency, keep these tablets in the original container or in a supplemental Nitroglycerin container specifically labeled as being suitable for Nitroglycerin Tablets. Close tightly immediately after each use.
Manufactured by:
Pfizer Pharmaceuticals LLC
Vega Baja, PR 00694
8212
NDC 0071-0418-24
Rx only
100 Sublingual Tablets
Nitrostat®
(Nitroglycerin Tablets, USP)
0.4
0.4 mg (1/150 gr)
Distributed by
Pfizer
Parke-Davis
Division of Pfizer Inc, NY, NY 10017
05-5873-32-2

Figure 7-2

In this case, the drug is measured in micrograms on the drug order and in milligrams on the drug label. The units for the *desired dose* must match those found on the drug label, which means that we will convert 800 mcg to milligrams.

Follow the steps of Procedure Checklist 3-1.

1. Conversion factor 1 mg = 1000 mcg. Since we are converting to milligrams, our first fraction is:

$$\frac{1 \text{ mg}}{1000 \text{ mcg}}$$

2. The other fraction is:

$$\frac{D}{800 \text{ mcg}}$$

3. Set up the fractions.

$$\frac{D}{800 \text{ mcg}} = \frac{1 \text{ mg}}{1000 \text{ mcg}}$$

4. Cancel units.

$$\frac{D}{800 \cancel{\text{mcg}}} = \frac{1 \text{ mg}}{1000 \cancel{\text{mcg}}}$$

5. Cross-multiply and then solve for the unknown.

$D \times 1000 = 800 \times 1 \text{ mg}$

$D \times \frac{1000}{1000} = \frac{800}{1000} \times 1 \text{ mg}$

$D = \frac{800}{1000} = 0.8 \text{ mg} = \text{desired dose}$

ERROR ALERT!

In a fraction proportion, units from the two fractions can be canceled *only* when they are in the same portion of the fraction. Units in the denominator of one fraction cannot be canceled with units found in the numerator of the other.

If you set up a proportion with mismatched units, you will calculate the desired dose incorrectly. In an earlier example, the dosage ordered was in grains, and the dosage strength was measured in milligrams. Suppose that you had used the conversion factor $\frac{1 \text{ gr}}{65 \text{ mg}}$ instead of $\frac{65 \text{ mg}}{1 \text{ gr}}$. Your proportion would then have been:

$$\frac{D}{5 \text{ gr}} = \frac{1 \text{ gr}}{65 \text{ mg}}$$

Here, the units are mismatched and cannot be canceled. You should immediately realize that the conversion factor is incorrect because the units in the denominators of the proportion do not match. If you had not included the units when setting up the proportion, the error may have gone unnoticed. *Always include the units when you perform calculations.*

Ratio Proportion Method

The following checklist (Procedure 3-2) was used in Chapter 3 to convert units by ratio proportions. It can also be followed when you are calculating the desired dose from the dosage ordered.

Procedure Checklist 3-2

Converting by the Ratio Proportion Method

1. Write the conversion factor as a ratio $A : B$ so that A has the units of the value that you are converting (the dosage ordered) and B has the unit of value of the dose on hand.
2. Write a second ratio $C : D$ so that C is the missing value (desired dose) and D is the number that is being converted (the dosage ordered).
3. Write the proportion in the form $A : B :: C : D$. *Note: When you are using the ratio proportion method to calculate the desired dose, C indicates the unknown value (desired dose).*
4. Cancel units.
5. Solve the proportion by multiplying the means and extremes.

Example 1 The dosage ordered is 0.2 mg once a day.

The dosage strength is 100 mcg per tablet.

Find the desired dose.

The drug is measured in milligrams on the drug order and micrograms on the drug label. The units for the desired dose must match those found on the drug label, which means that we must determine how many micrograms are equivalent to 0.2 mg.

Follow the steps of Procedure Checklist 3-2.

1. Since we are converting to micrograms, micrograms must appear at the beginning of our conversion ratio. Our conversion ratio is 1000 mcg : 1 mg.

Learning Link Recall from Chapter 3, Table 3-4, 1 mg = 1000 mcg, found on page 84.

2. The second ratio in our proportion will be *C* : 0.2 mg, with *C* being the unknown value or desired dose and *D* being the dosage ordered or the number that is being converted.
3. Our proportion is:

   1000 mcg : 1 ~~mg~~ : : *C* : 0.2 ~~mg~~

4. Cancel units.

   1000 mcg : 1 : : *C* : 0.2

5. Multiply the means and extremes, and then solve for the missing value.

   $1 \times C = 1000 \text{ mcg} \times 0.2$

   $C = 200 \text{ mcg} = \text{desired dose}$

Example 2 The order reads ASA gr v PO daily.

The drug label indicates 325-mg tablets.

Find the desired dose.

Again, the drug order and the drug label use different units. In this case, we must convert the dosage ordered (5 gr) to milligrams to find the desired dose.

Follow the steps of Procedure Checklist 3-2.

1. Since we are converting to milligrams, our conversion ratio is 65 mg : 1 gr.

Learning Link Recall from Chapter 3, Table 3-8, 1 gr = 65 mg, found on page 92.

Note: For this medication 65 mg per 1 gr is used.

2. Our second ratio will be *C* : 5 gr.
3. Our proportion is:

   65 mg : 1 ~~gr~~ : : *C* : 5 ~~gr~~

4. Cancel units.

   65 mg : 1 : : *C* : 5

5. Multiply the means and extremes, and then solve for the missing value.

   $1 \times C = 65 \text{ mg} \times 5$

   $C = 325 \text{ mg} = \text{desired dose}$

Example 3 Ordered: 500 mg q6h

Dosage strength available: 250 mg per tablet

Find the desired dose.

In this case the unit for the dose ordered is the same as that of the dose on hand. No calculation is needed. The drug is measured in milligrams on both the order and the drug label, so the desired dose $D = 500$ mg.

Example 4 Ordered: Nitrostat 800 mcg sublingually PRN chest pain.

On hand: Nitrostat 0.4 mg sublingual tablets (see Figure 7-2).

In this case, the drug is measured in micrograms on the drug order and in milligrams on the drug label. The units for the desired dose must match those found on the drug label, which means that we will convert 800 mcg to milligrams.

Follow the steps of Procedure Checklist 3-2.

1. Recall from Chapter 3 that 1 mg = 1000 mcg. Since we are converting to milligrams, milligrams must appear in the numerator of our conversion factor. Our conversion factor is $\frac{1\text{ mg}}{1000\text{ mcg}}$.
2. The second ratio of our proportion is C : 800 mcg. Here C is the missing value (desired dose), and D is the number being converted, 800 mcg (dosage ordered).
3. Our proportion is:

   $1\text{ mg} : 1000\text{ mcg} :: C : 800\text{ mcg}$
4. Cancel units.

   $1\text{ mg} : 1000\,\cancel{\text{mcg}} :: C : 800\,\cancel{\text{mcg}}$
5. Multiply the means and extremes, and then solve for the missing value.

   $C \times 1000 = 1\text{ mg} \times 800$

   $C \times 1000 = 800$

   $C = \frac{800}{1000}\text{ mg} = 0.8\text{ mg} = \text{desired dose}$

ERROR ALERT!

In a ratio proportion, units can be canceled *only* when they are found in the same part of each of the ratios.

In an earlier example, the dosage ordered was in milligrams, and the dosage strength was measured in micrograms. Suppose that you had used the conversion ratio 1 mg : 1000 mcg instead of 1000 mcg : 1 mg. Your proportion would then have been:

1 mg : 1000 mcg : : C : 0.2 mg

The common unit in the two ratios is not found in the same part of the ratios—milligrams are found at the beginning of the first and at end of the second ratio. Therefore, the units cannot be canceled. You should immediately realize that the conversion ratio is incorrect because the units cannot be canceled. If you had not included the units when setting up the proportion, the error might have gone unnoticed. *Always include the units when you are performing calculations.*

Dimensional Analysis

Recall from Chapter 3 that the dimensional analysis (DA) method is a modification of the fraction proportion and ratio proportion methods. When we are using DA, the unknown value ? stands alone on one side of an equation. In this case, the unknown is the desired dose. The conversion factor is placed on the other side of the equation, and the number being converted is placed over 1

Procedure Checklist 3-3

Converting Using the Dimensional Analysis Method

1. Determine the unit of measure for the answer, and place it as the unknown on one side of the equation.
2. On the other side of the equation, write a conversion factor with the units of measure for the answer on top and the units you are converting from on the bottom.
3. Multiply the conversion factor by the number that is being converted over 1
4. Cancel units on the right side of the equation. The remaining unit of measure on the right side of the equation should match the unknown unit of measure on the left side of the equation.
5. Solve the equation.

Example 1 The dosage ordered is 0.2 mg once a day.

The dosage strength is 100 mcg per tablet.

Find the desired dose.

The drug is measured in milligrams on the drug order and in micrograms on the drug label. The units for the desired dose must match the units of the dose on hand. We must determine how many micrograms is equivalent to 0.2 mg.

Follow the steps of Procedure Checklist 3-3.

1. The unit of measure for the answer is micrograms. Place this on the left side of the equation.

   $D \text{ mcg} =$

   (D represents the desired dose, which is the unknown.)

2. Since we are converting to micrograms, micrograms must appear in the numerator of our conversion factor. Our conversion factor is $\frac{1000 \text{ mcg}}{1 \text{ mg}}$. This will go on the other side of the equation.

Learning Link Recall from Chapter 3, Table 3-4, 1 mg = 1000 mcg, found on page 84.

3. Multiply the numerator of the conversion factor by the number that is being converted, the dosage ordered over 1.

   $$D \text{ mcg} = \frac{1000 \text{ mcg}}{1 \text{ mg}} \times \frac{0.2 \text{ mg}}{1}$$

4. Cancel units. The remaining unit on both sides is micrograms.

$$D \text{ mcg} = \frac{1000 \text{ mcg}}{1 \cancel{\text{mg}}} \times \frac{0.2 \cancel{\text{mg}}}{1}$$

5. Solve the equation.

$D = 1000 \text{ mcg} \times 0.2 = 200 \text{ mcg} =$ desired code

Example 2 The order reads ASA gr v PO daily.

The drug label indicates 325-mg tablets.

Find the desired dose.

Again, the drug order and the drug label use different units. In this case, we must convert the dosage ordered (5 gr) into milligrams (mg) to find the desired dose.

Follow the steps of Procedure Checklist 3-3.

1. The unit of measure for the answer is milligrams. Place this on the left side of the equation.

   D mg =

   (D represents the desired dose, which is the unknown.)

2. Since we are converting to milligrams, our conversion factor is $\frac{65 \text{ mg}}{1 \text{ gr}}$.

Learning Link Recall from Chapter 3, Table 3-8, 1 gr = 65 mg, found on page 92.

Note: For this medication 65 mg per 1 gr is used.

3. Multiplying the numerator of the conversion factor by the number being converted (dosage ordered over 1) gives us the following equation.

$$D = \frac{65 \text{ mg}}{1 \text{ gr}} \times \frac{5 \text{ gr}}{1}$$

4. Cancel units.

$$D = \frac{65 \text{ mg}}{1 \cancel{\text{gr}}} \times \frac{5 \cancel{\text{gr}}}{1}$$

5. Solve the equation.

$D = 65 \text{ mg} \times 5 = 325 \text{ mg} =$ desired dose

Example 3 Ordered: 500 mg q6h

Dosage strength available: 250 mg per tablet

Find the desired dose.

In this case the unit for the dose ordered is the same as that for the dose on hand. No calculation is needed. The drug is measured in milligrams on both the order and the drug label, so the desired dose $D = 500$ mg.

Example 4 Ordered: Nitrostat 800 mcg sublingually PRN chest pain

On hand: Nitrostat 0.4 mg sublingual tablets

In this case, the drug is measured in micrograms on the drug order and in milligrams on the drug label. The units for the desired dose must match those found on the drug label, which means that we will convert 1500 mcg into milligrams.

Follow the steps of Procedure Checklist 3-3.

1. The unit of measure for the answer is milligrams. Place this on the left side of the equation.

   $D \text{ mg} =$

2. Since we are converting to milligrams, milligrams must appear in the numerator of our conversion factor. Our conversion factor is $\frac{1 \text{ mg}}{1000 \text{ mcg}}$.

3. Multiply the numerator of the conversion factor by the number that is being converted, the dosage ordered over 1

   $$D \text{ mg} = \frac{1 \text{ mg}}{1000 \text{ mcg}} \times \frac{800 \text{ mcg}}{1}$$

4. Cancel units.

   $$D \text{ mg} = \frac{1 \text{ mg}}{1000 \cancel{\text{mcg}}} \times \frac{800 \cancel{\text{mcg}}}{1}$$

5. Solve the equation.

   $$D = \frac{800 \text{ mg}}{1000} = 0.8 \text{ mg} = \text{desired dose}$$

ERROR ALERT!

In dimensional analysis, units can be canceled *only* when they are found in both the numerator and the denominator of the fraction.

In an earlier example, the dosage ordered was in milligrams and the dosage strength was measured in micrograms. Suppose that you had used the conversion factor $\frac{1 \text{ mg}}{1000 \text{ mcg}}$ instead of $\frac{1000 \text{ mcg}}{1 \text{ mg}}$. Your equation would then have been:

$$D = \frac{1 \text{ mg}}{1000 \text{ mcg}} \times \frac{0.2 \text{ mg}}{1}$$

You may cancel units within a fraction only when they are found in both the numerator and the denominator. Here, the common unit (milligrams) is found in the numerator only and cannot be canceled. You should immediately realize that the conversion factor is incorrect because the units cannot be canceled. If you had not included the units when setting up the equation, the error might have gone unnoticed. *Always include the units when you are performing calculations.*

REVIEW AND PRACTICE

7-1 Calculating the Desired Dose

In Exercises 1–20, convert the dosage ordered to the same unit as that of the dose on hand or measuring device. Use conversion tables from Chapter 3 as needed.

1. Ordered: Amoxicillin 0.25 g — Desired dose: _____
   On hand: Amoxicillin 125-mg capsules
2. Ordered: Erythromycin 0.5 g — Desired dose: _____
   On hand: Erythromycin 500-mg tablets
3. Ordered: Phenobarbital gr ss — Desired dose: _____
   On hand: Phenobarbital 15-mg tablets
4. Ordered: Penicillin VK 0.25 g — Desired dose: _____
   On hand: Penicillin VK 500 mg
5. Ordered: Levoxyl® 0.15 mg — Desired dose: _____
   On hand: Levoxyl® 300-mcg tablets
6. Ordered: Duratuss® HD 5 mL — Desired dose: _____
   Available measuring device is marked in teaspoons.
7. Ordered: Robitussin® DM 2 tsp — Desired dose: _____
   Available measuring device is marked in mL.
8. Ordered: Biaxin® 1 g PO daily — Desired dose: _____
   On hand: Refer to label A.

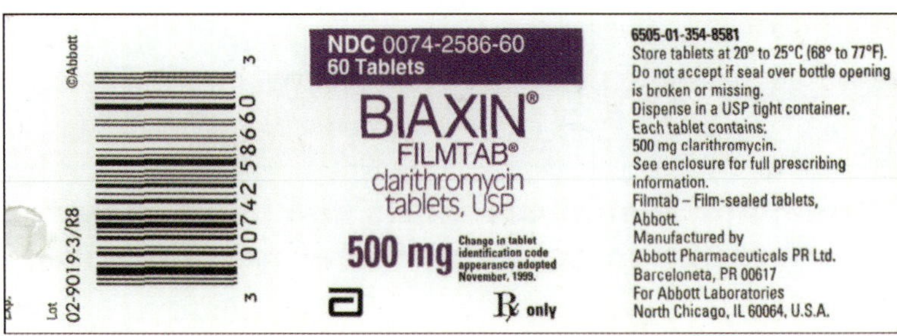

A

9. Ordered: Meloxicam gr $\frac{1}{4}$ — Desired dose: _____
   On hand: Refer to label B.

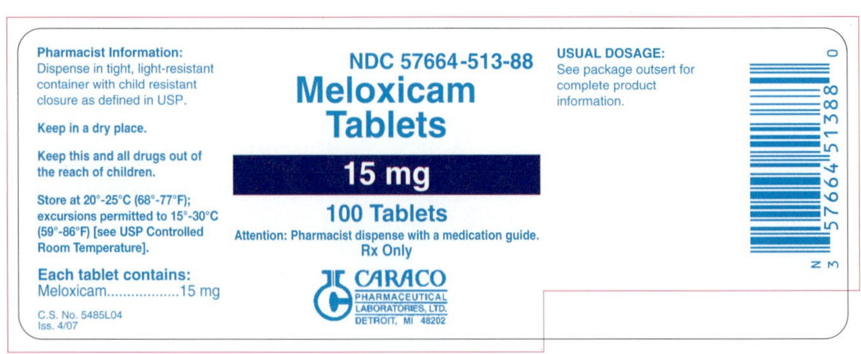

B

10. Ordered: Synthroid® 0.05 mg Desired dose: _____
On hand: Refer to label C.

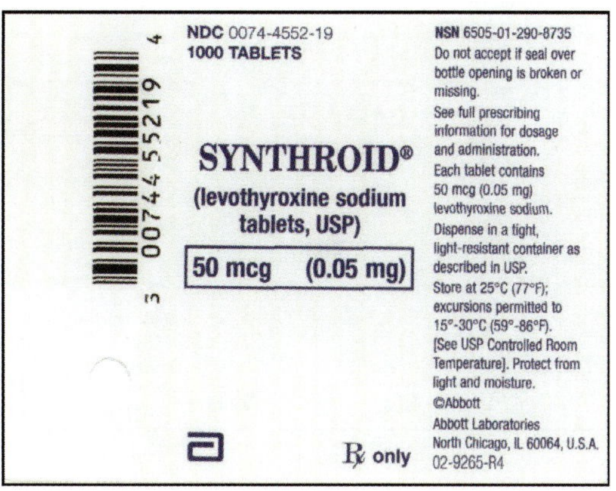

C

11. Ordered: Synthroid® 0.088 mg Desired dose: _____
On hand: Refer to label D.

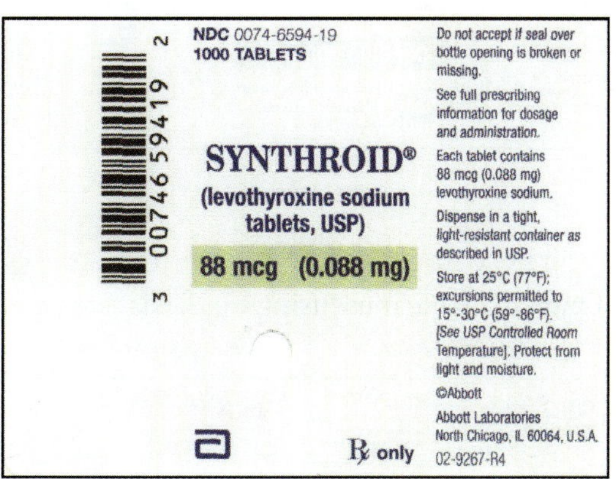

D

12. Ordered: Depakote® 0.5 g Desired dose: _____
On hand: Refer to label E.

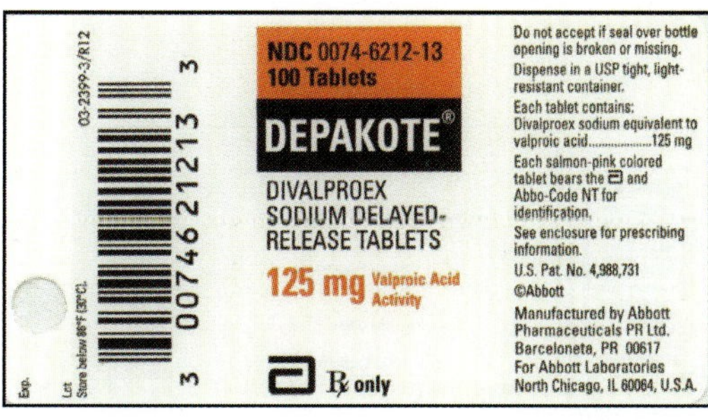

E

13. Ordered: Synthroid® 250 mcg Desired dose: _____
On hand: Refer to label F.

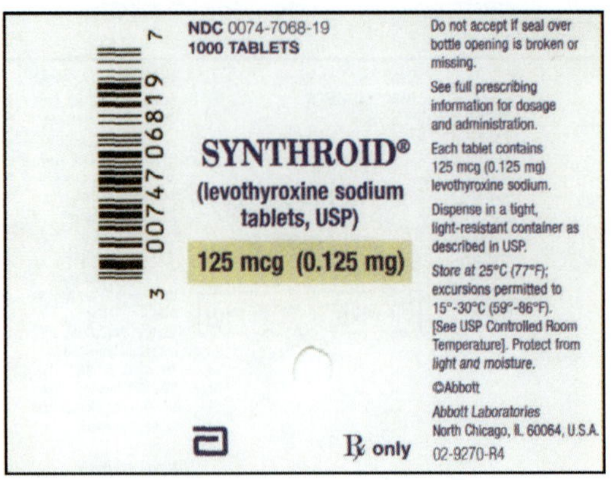

F

14. Ordered: 1½ teaspoon Zithromax® 200 mg/ 5 mL PO q6h Desired dose: _____
On hand: Refer to letter G. (Only available measuring device is marked in mL.)

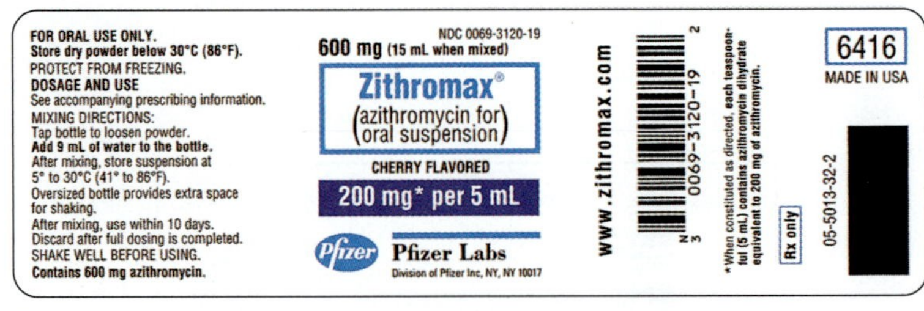

G

15. Ordered: 7½ mL clarithromycin PO q4h Desired dose: _____
On hand: Refer to label H. (Only available measuring device is a teaspoon.)

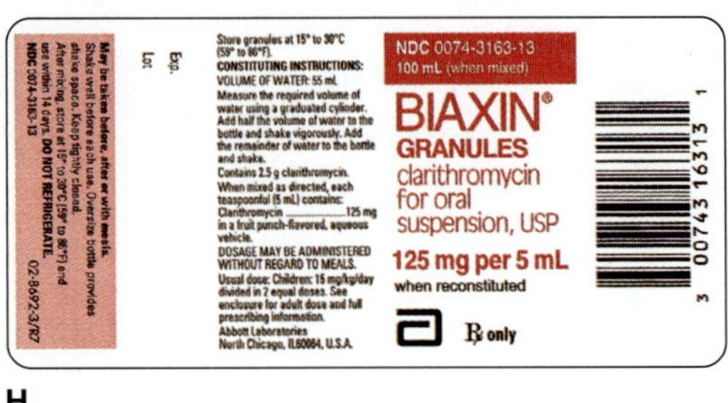

H

16. Ordered: Levothroid® 0.137 mg PO daily Desired dose: _____
On hand: Refer to label I.

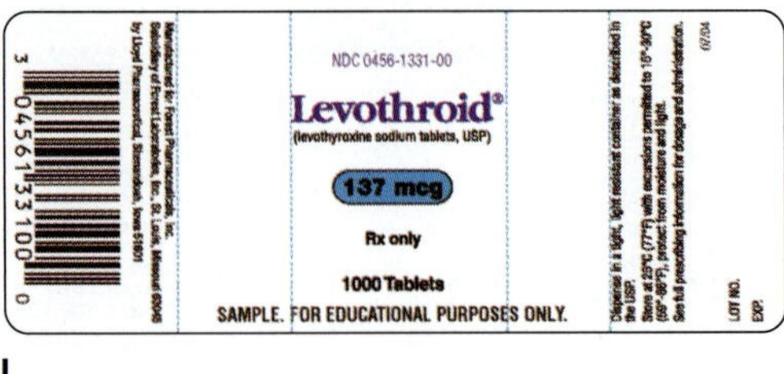

I

17. Ordered: Levothroid® 0.112 mg
On hand: Refer to label J.

Desired dose: _____

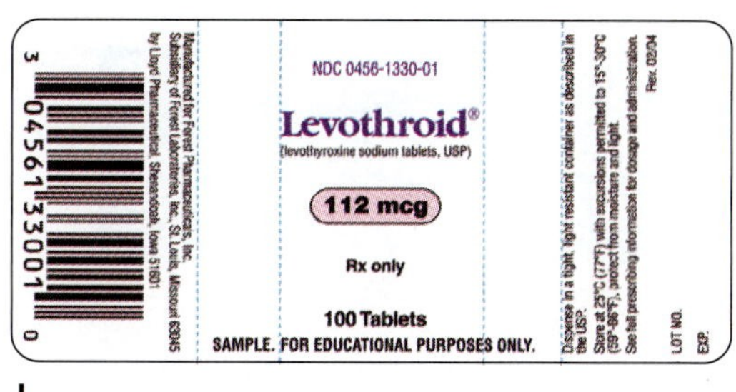

J

18. Ordered: Risperdal® 250 mcg
On hand: Refer to label K.

Desired dose: _____

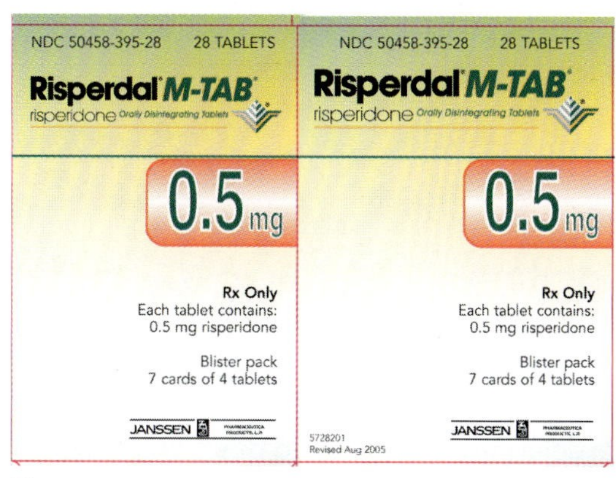

K

19. Ordered: Prandin® 750 mcg
On hand: Refer to label L.

Desired dose: _____

L

20. Ordered: Metformin 1 g Desired dose: _____
On hand: Refer to label M.

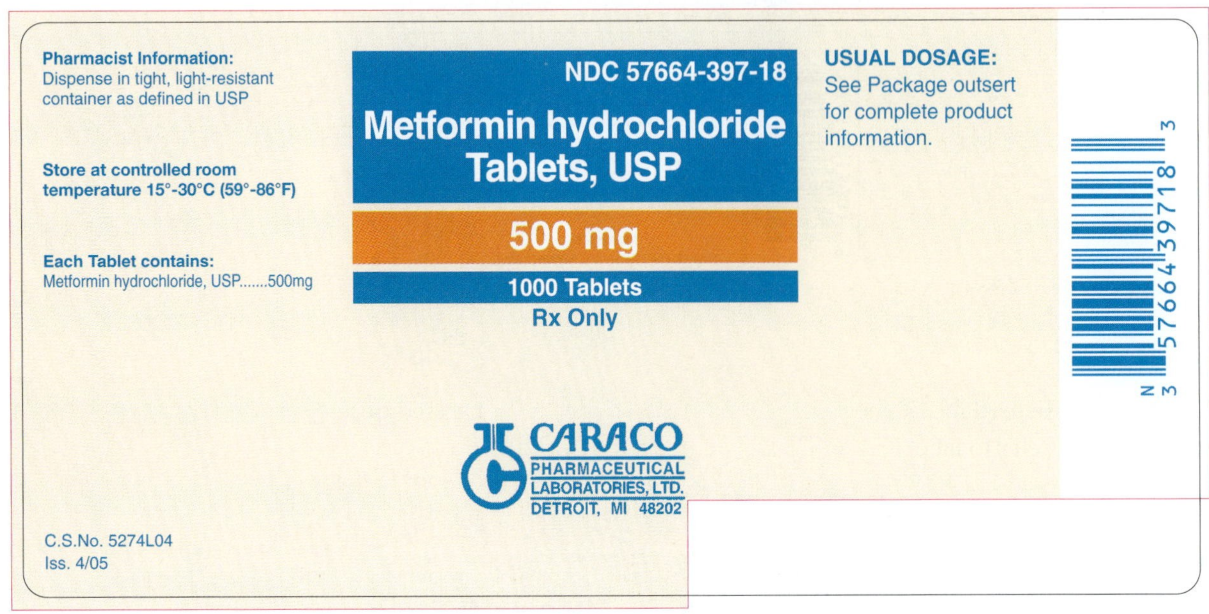

Pharmacist Information:
Dispense in tight, light-resistant container as defined in USP
Store at controlled room temperature 15°-30°C (59°-86°F)
Each Tablet contains:
Metformin hydrochloride, USP.......500mg
NDC 57664-397-18
Metformin hydrochloride Tablets, USP
500 mg
1000 Tablets
Rx Only
USUAL DOSAGE:
See Package outsert for complete product information.
N 3 57664 39718 3
CARACO
PHARMACEUTICAL LABORATORIES, LTD.
DETROIT, MI 48202
C.S.No. 5274L04
Iss. 4/05

M

To check your answers, see page 520.

Calculating the Amount to Administer

Once you have determined the desired dose, you still have one more step that must be completed. While the desired dose tells you how many grams, milligrams, or grains of a drug the patient is to receive, you will need to know how many tablets, capsules, teaspoons, or milliliters of the medication must be given to deliver the desired dose. You must calculate an *amount to administer* to the patient.

In this section, you will be presented with four methods for calculating the *amount to administer*. As with previous calculations, you may choose to use fraction proportion, ratio proportion, or dimensional analysis. We will also introduce the formula method. Again, the method that you choose to use is up to you—each will give you the same result.

Rule 7-2 Calculating the Amount to Administer

To calculate the amount to administer A, the following information must be known:

- The desired dose D, or the amount of drug to be given at a single time. This is the dosage ordered converted to the same units as those of the dose on hand, if necessary.
- The dosage strength or the dose on hand H per dosage unit Q. Recall, the dose on hand H is the amount of drug contained in a dosage unit. The dosage unit Q is the unit by which you will measure the medication—tablets, capsules, milliliters, teaspoons, etc. This is obtained from the medication label.

Example Ordered: Erythromycin 0.5 g PO twice daily.

On hand: See label (Figure 7-3).

In this case the dose on hand is 250 mg and the dosage unit is one capsule. The dose ordered is 0.5 g. Thus, you need to convert the dose ordered to the same unit of measurement as that of the dose on hand, to determine the desired dose. In this case you will be converting 0.5 g to milligrams. After calculating the desired dose, you have all the necessary information to calculate the amount to administer.

- Desired dose = 500 mg
- Dose on hand = 250 mg
- Dosage unit = 1 capsule

Dose ordered (O) [must be (converted to milligrams) to determine desired dose (D)]

Dosage ordered (O): Erythromycin 0.5 g twice daily

On hand:

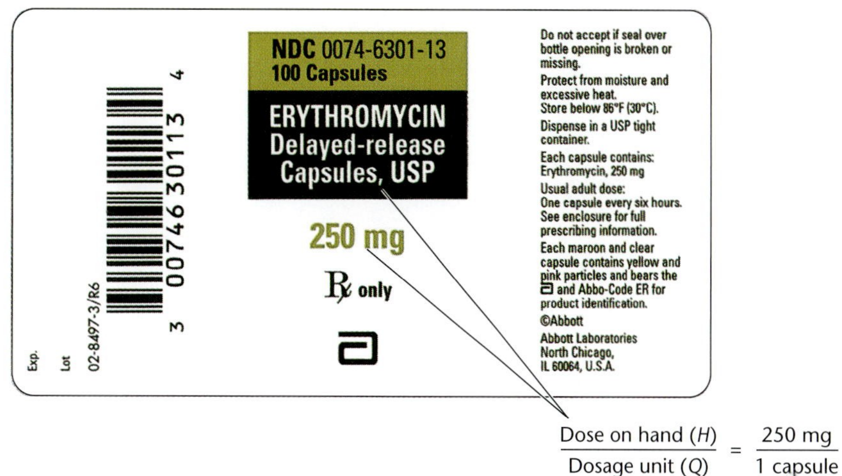

Figure 7-3 Information necessary to calculate the amount to administer. In this case the dose ordered must be converted to milligrams to obtain the desired dose.

GO TO . . . Open the CD-ROM that accompanies your textbook, and select Chapter 7, Rule 7-2. Review the animation and example problems, then complete the practice problems. Continue to the next section of the book once you have mastered the rule presented. ■

Procedure Checklist 7-1

Calculating the Amount to Administer by Fraction Proportion Method

1. The proportion will be set up as follows:

$$\frac{\text{Dosage unit}}{\text{Dose on hand}} = \frac{\text{amount to administer}}{\text{desired dose}} \quad \text{or} \quad \frac{Q}{H} = \frac{A}{D}$$

2. Cancel units.
3. Cross-multiply and then solve for the unknown value.

Example 1 Find the amount to administer.

Ordered: Famvir® 500 mg PO q8h

On hand: See label in Figure 7-4.

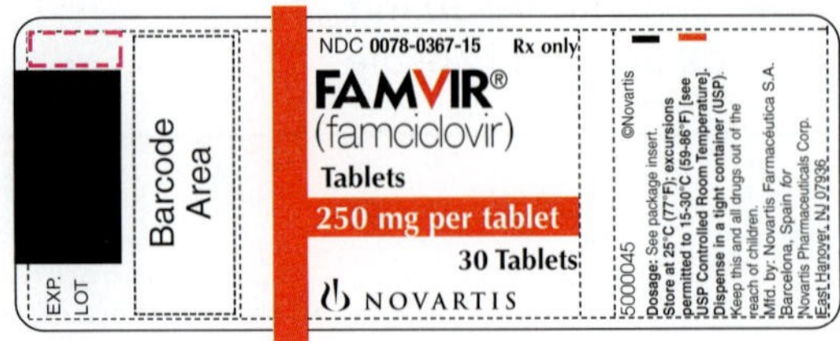

Figure 7-4

The drug is ordered in milligrams, which is the same unit used on the label. Therefore, the dosage ordered is the same as the desired dose (500 mg). By reading the label we find that the dosage unit is 1 tablet and the dose on hand is 250 mg. Therefore,

$D = 500$ mg

$Q = 1$ tablet

$H = 250$ mg

Follow the Procedure Checklist 7-1.

1. Fill in the proportion. (Think: *If 1 tablet equals 250 mg, then how many tablets equals 500 mg?*)

$$\frac{1 \text{ tablet}}{250 \text{ mg}} = \frac{A}{500 \text{ mg}}$$

2. Cancel units.

$$\frac{1 \text{ tablet}}{250 \cancel{\text{mg}}} = \frac{A}{500 \cancel{\text{mg}}}$$

3. Cross-multiply and solve for the unknown.

$250 \times A = 1 \text{ tablet} \times 500$

$A = \frac{500}{250}$ tablets

$A = 2$ tablets $=$ amount to administer

Example 2 Find the amount to administer.

Ordered: Norvir® 200 mg PO now

On hand: See label in Figure 7-5

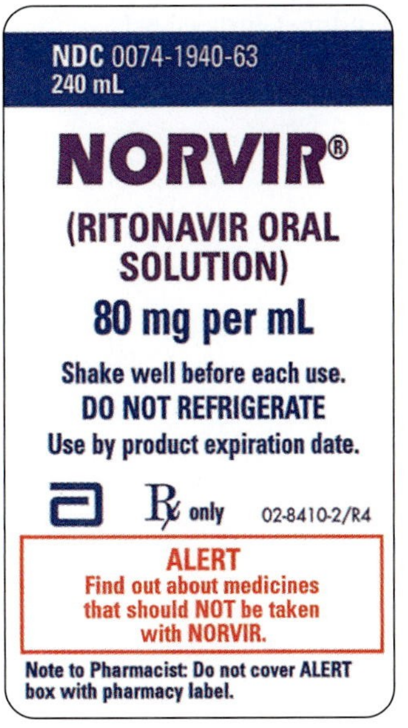

Figure 7-5

Again, the drug order and the drug label use the same units. Our desired dose is 200 mg. Reading the label tells us that the dosage unit is 1 mL and the dose on hand is 80 mg. Therefore,

$D = 200$ mg

$Q = 1$ mL

$H = 80$ mg

Follow the Procedure Checklist 7-1.

1. Fill in the proportion. (Think: *If 1 mL has 80 mg of medication, how many milliliters has 200 mg?*)

$$\frac{1\text{ mL}}{80\text{ mg}} = \frac{A}{200\text{ mg}}$$

2. Cancel units.

$$\frac{1\text{ mL}}{80\not{\text{mg}}} = \frac{A}{200\not{\text{mg}}}$$

3. Solve for the unknown.

$$80 \times A = 1\text{ mL} \times 200$$

$$A = 2.5\text{ mL} = \text{amount to administer}$$

Example 3 Find the amount to administer.

Ordered: 250 mg Ampicillin IM now

On hand: Ampicillin 0.5 g/mL

In this case, the order is written in milligrams, and the drug is labeled in grams. Before we can determine the amount to administer, we must calculate a desired dose that is in grams.

Follow the Procedure Checklist 3-1.

1. Recall that 1 gr = 1000 mg. Since we are converting to grams, our conversion factor is:

$$\frac{1\text{ g}}{1000\text{ mg}}.$$

2. The other fraction for our proportion is $\frac{D}{250\text{ mg}}$.

3. Setting the two fractions into a proportion gives us the following equation:

$$\frac{D}{250\text{ mg}} = \frac{1\text{ g}}{1000\text{ mg}}$$

4. Cancel units.

$$\frac{D}{250\,\cancel{\text{mg}}} = \frac{1\text{ g}}{1000\,\cancel{\text{mg}}}$$

5. Solve for the unknown.

$$1000 \times D = 250 \times 1\text{ g}$$
$$D = 0.25\text{ g} = \text{desired dose}$$

We now have the three necessary pieces of information: The desired dose is 0.25 g, the dosage unit is 1 mL, and the dose on hand is 0.5 g.

$D = 0.25\text{ g}$ $\quad Q = 1\text{ mL}$ $\quad H = 0.5\text{ mg}$

Follow the Procedure Checklist 7-1.

1. Fill in the proportion. (Think: *If 1 mL has 0.5 g, then how many milliliters will have 0.25 g?*)

$$\frac{1\text{ mL}}{0.5\text{ g}} = \frac{A}{0.25\text{ g}}$$

2. Cancel units.

$$\frac{1\text{ mL}}{0.5\,\cancel{\text{g}}} = \frac{A}{0.25\,\cancel{\text{g}}}$$

3. Cross-multiply and solve for the unknown.

$$0.5 \times A = 1\text{ mL} \times 0.25$$
$$A = 0.5\text{ mL} = \text{amount to administer}$$

Example 4 Find the amount to administer.

Ordered: Metformin 2 g PO daily

On hand: See label in Figure 7-6.

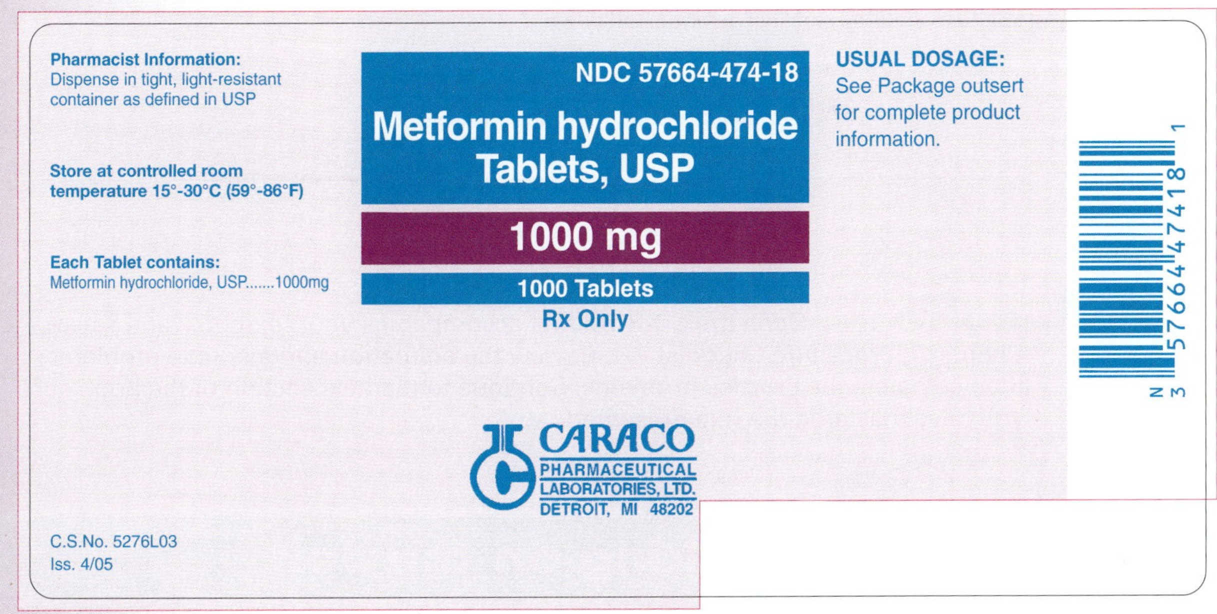

Pharmacist Information:
Dispense in tight, light-resistant container as defined in USP
Store at controlled room temperature 15°-30°C (59°-86°F)
Each Tablet contains:
Metformin hydrochloride, USP.......1000mg
NDC 57664-474-18
Metformin hydrochloride Tablets, USP
1000 mg
1000 Tablets
Rx Only
USUAL DOSAGE:
See Package outsert for complete product information.
N 3 57664 47418 1
CARACO
PHARMACEUTICAL LABORATORIES, LTD.
DETROIT, MI 48202
C.S.No. 5276L03
Iss. 4/05

Figure 7-6

In this case, the order is written in grams, and the drug is labeled in milligrams. Before we can determine the amount to administer, we must calculate a desired dose that is in milligrams.

Follow the Procedure Checklist 3-1.

1. Recall that 1 g = 1000 mg. Since we are converting to milligrams, our conversion factor is $\frac{1000\text{ mg}}{1\text{ g}}$.

2. The other fraction for our proportion is $\frac{D}{2\text{ g}}$.

3. Set up the fraction proportion equation:

$$\frac{D}{2\text{ g}} = \frac{1000\text{ mg}}{1\text{ g}}$$

4. Cancel units.

$$\frac{D}{2\cancel{\text{g}}} = \frac{1000\text{ mg}}{1\cancel{\text{g}}}$$

5. Solve for the unknown.

$$1 \times D = 1000\text{ mg} \times 2$$
$$D = 2000\text{ mg} = \text{desired dose}$$

We now have the three necessary pieces of information: The desired dose is 2000 mg, the dosage unit is 1 tablet, and the dose on hand is 1000 mg.

$D = 2000$ mg $\quad Q = 1$ tablet $\quad H = 1000$ mg

Follow the Procedure Checklist 7-1.

1. Fill in the proportion. (Think: *If 1 tablet has 1000 mg, then how many tablets will have 2000 mg?*)

$$\frac{1\text{ tablet}}{1000\text{ mg}} = \frac{A}{2000\text{ mg}}$$

2. Cancel units.

$$\frac{1\text{ tablet}}{1000\cancel{\text{mg}}} = \frac{A}{2000\cancel{\text{mg}}}$$

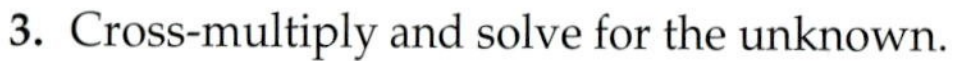

3. Cross-multiply and solve for the unknown.

$$1000 \times A = 1 \text{ tablet} \times 2000 \quad \text{(divide both sides by 1000)}$$

$$A = \frac{2000}{1000} = 2 \text{ tablets} = \text{amount to administer}$$

GO TO . . . Open the CD-ROM that accompanies your textbook, and select Chapter 7, Procedure Checklist 7-1. Review the animation and example problems, then complete the practice problems. Continue to the next section of the book once you have mastered the rule presented. ■

Critical Thinking on the Job

Use Your Critical Thinking Skills

Always use critical thinking skills to evaluate your answer before you administer a drug. For example, in Example 4 the medication ordered is 2 g, and the medication comes in bottles of 1000-mg tablets. If you set up the problem incorrectly, you may get the answer $\frac{1}{2}$ Here is how. First you determine that the desired dose is 2000 mg, and then you set up the problem, reversing the dose on hand with the desired dose.

$$\frac{1 \text{ tablet}}{2000 \text{ mg}} = \frac{A}{1000 \text{ mg}}$$

$$\frac{1 \text{ tablet}}{2000 \cancel{\text{mg}}} = \frac{A}{1000 \cancel{\text{mg}}} \quad \text{(Canceling the units makes the problem appear correct.)}$$

When you cross-multiply, you come up with the following.

$$2000 \times A = 1 \text{ tablet} \times 1000 \quad \text{(Divide both sides by 2000.)}$$

$$A = \frac{1000}{2000} = \frac{1}{2} \text{ tablet}$$

Question?

Using critical thinking, how would you know that your answer is incorrect?

Procedure Checklist 7-2

Calculating Amount to Administer by Ratio Proportion Method

1. The proportion will be set up as follows:

   Dosage unit : dose on hand : : amount to administer : desired dose or $Q : H :: A : D$
2. Cancel units.
3. Multiply the means and extremes and then solve for the missing value.

Example 1 Ordered: Famvir® 500 mg PO q8h

On hand: Famvir® 250 mg per/tablet. See Figure 7-4 on page 222.

Find the amount to administer.

The drug is ordered in milligrams, which is the same unit used on the label. Therefore, the desired dose is 500 mg. Reading the label tells us that the dosage unit is 1 tablet and the dose on hand is 250 mg. Therefore,

$D = 500$ mg

$Q = 1$ tablet

$H = 250$ mg

Follow Procedure Checklist 7-2.

1. Fill in the proportion. (Think: *If 1 tablet is 250 mg, then how many tablets make 500 mg?*)

   1 tablet : 250 mg : : A : 500 mg

2. Cancel units.

   1 tablet : 250 ~~mg~~ : : A : 500 ~~mg~~

3. Multiply the means and extremes, and then solve for the missing value.

   $250 \times A = 1 \text{ tablet} \times 500$

   $\frac{250}{250} \times A = 1 \text{ tablet} \times \frac{500}{250}$

   $A = 2$ tablets = amount to administer

Example 2 Ordered: Norvir® Liquid 40 mg PO qd

On hand: Norvir® 80 mg/mL. See Figure 7-5 page 223.

Calculate the amount to administer.

Again, the drug order and the drug label use the same units. Our desired dose is 40 mg. Reading the label, we find that the dosage unit is 1 mL and the dose on hand is 80 mg. Therefore,

$D = 40$ mg

$Q = 1$ mL

$H = 80$ mg

Follow Procedure Checklist 7-2.

1. Fill in the proportion. (Think: *If 1 mL has 80 mg, then how many milliliters has 40 mg?*)

   1 mL : 80 mg : : A : 40 mg

2. Cancel units.

   1 mL : 80 ~~mg~~ : : A : 40 ~~mg~~

3. Multiply the means and extremes, and then solve for the missing value.

   $80 \times A = 1 \text{ mL} \times 40$

   $\frac{80}{80} \times A = \frac{40 \text{ mL}}{80}$

   $A = 0.5$ mL = amount to administer

Example 3 Ordered: 250 mg Ampicillin IM now

On hand: Ampicillin 0.5 g/mL

Calculate the amount to administer.

In this case, the order is written in milligrams, and the drug is labeled in grams. Before we can determine the amount to administer, we must calculate a desired dose, converting the dose ordered, 250 mg, to grams.

Follow Procedure Checklist 3-2.

1. Recall that 1 g = 1000 mg. Since we are converting to grams, our conversion ratio is:

   $1\ \text{g} : 1000\ \text{mg}$

2. The other ratio for our proportion is the unknown *D* to the dosage ordered 250 mg.

   $D : 250\ \text{mg}$

3. Setting the two ratios into a proportion gives us the following equation:

   $1\ \text{g} : 1000\ \text{mg} :: D : 250\ \text{mg}$

4. Cancel units.

   $1\ \text{g} : 1000\ \cancel{\text{mg}} :: D : 250\ \cancel{\text{mg}}$

5. Multiply the means and extremes, and then solve for the unknown.

   $1000 \times D = 250 \times 1\ \text{g}$

   $D = 0.25\ \text{g} = \text{desired dose}$

We now have the three necessary pieces of information: The desired dose is 0.25 g, the dosage unit is 1 mL, and the dose on hand is 0.5 g.

$D = 0.25\ \text{g} \qquad Q = 1\ \text{mL} \qquad H = 0.5\ \text{g}$

Follow Procedure Checklist 7-2.

1. Fill in the proportion. (Think: *If 1 mL equals 0.5 g, then how many milliliters will equal 0.25 g*?)

   $1\ \text{mL} : 0.5\ \text{g} :: A : 0.25\ \text{g}$

2. Canceling units gives us

   $1\ \text{mL} : 0.5\ \cancel{\text{g}} :: A : 0.25\ \cancel{\text{g}}$

3. Multiply the means and extremes, and then solve for the missing value.

   $0.5 \times A = 1\ \text{mL} \times 0.25$

   $\frac{0.5}{0.5} \times A = \frac{0.25}{0.5}\ \text{mL}$

   $A = 0.5\ \text{mL} = \text{amount to administer}$

Example 4 Ordered: Metformin 2 g PO daily.

On hand: Metformin hydrochloride 1000 mg. See Figure 7-6 on page 225.

Calculate the amount to administer.

In this case, the order is written in grams, and the drug is labeled in milligrams. Before we can determine the amount to administer, we must calculate a desired dose that is in milligrams.

Follow Procedure Checklist 3-2.

1. Recall that 1 g = 1000 mg. Since we are converting to milligrams, our conversion factor is:

   1000 mg : 1 g

2. The other ratio in our proportion is:

   $D : 2$ g

3. Set up the ratio proportion equation:

   1000 mg : 1 g : : D : 2 g

4. Cancel units.

   1000 mg : 1 $\cancel{g}$: : D : 2 $\cancel{g}$

5. Multiply the means and extremes, and solve the equation.

   $1 \times D = 1000 \text{ mg} \times 2$

   $D = 2000 \text{ mg} = \text{desired dose}$

We now have the three necessary pieces of information: The desired dose is 2000 mg, the dosage unit is 1 tablet, and the dose on hand is 1000 mg.

$D = 2000$ mg $\quad Q = 1$ tablet $\quad H = 1000$ mg

Follow Procedure Checklist 7-2.

1. Fill in the ratio proportion. (Think: *If 1 tablet equals 1000 mg, then how many tablets equal 2000 mg?*)

   1 tablet : 1000 mg : : A : 2000 mg

2. Cancel units.

   1 tablet : 1000 $\cancel{mg}$: : A : 2000 $\cancel{mg}$

3. Multiply the means and extremes, and solve for the unknown.

   $1000 \times A = 2000 \times 1$ tablet (Divide both sides by 1000.)

   $A = 2$ tablets = amount to administer

GO TO . . . Open the CD-ROM that accompanies your textbook, and select Chapter 7, Procedure Checklist 7-2. Review the animation and example problems, then complete the practice problems. Continue to the next section of the book once you have mastered the rule presented. ■

Procedure Checklist 7-3

Calculating the Amount to Administer by Dimensional Analysis Method

With dimensional analysis you will not need to calculate the desired dose and amount to administer separately. You will place your unknown (amount to administer) on one side of the equation and then multiply a series of factors on the right side of the equation. Canceling units will help you determine that the equation has been set up correctly.

1. Determine the unit of measure for the answer, and place it as the unknown on one side of the equation. (In this case the answer would be the amount to administer. The unit of measure will be the same unit of measure as that of the dosage unit.)

2. On the right side of the equation, write a conversion factor with the unit of measure for the desired dose on top and the unit of measure for the dosage ordered on the bottom. *(This is necessary if the dose ordered is in a different unit of measurement than the dose on hand.)*
3. Multiply the conversion factor by a second factor—the dosage unit over the dose on hand.
4. Multiply by a third factor—dose ordered over the number 1
5. Cancel units on the right side of the equation. The remaining unit of measure on the right side of the equation should match the unknown unit of measure on the left side of the equation.
6. Solve the equation.

Example 1 Ordered: Famvir® 500 mg PO q8h

On hand: Famvir® 250 mg per tablet. See Figure 7-4 on page 222.

Find the amount to administer.

Follow Procedure Checklist 7-3.

1. The unit of measure for the amount to administer will be tablets. This is the dosage unit.

   A tablets =

2. Since the unit of measurement for the dosage ordered is the same as the dose on hand, no conversion factor is necessary.
3. The dosage unit is 1 tablet. The dosage strength is 250 mg. This is our first factor.

   $$\frac{1\text{ tablet}}{250\text{ mg}}$$

4. The dose ordered is 500 mg. Place this quantity over the number 1 for the next factor.

   $$A\text{ tablets} = \frac{1\text{ tablet}}{250\text{ mg}} \times \frac{500\text{ mg}}{1}$$

5. Cancel the units.

   $$A\text{ tablets} = \frac{1\text{ tab}}{250\text{ }\cancel{\text{mg}}} \times \frac{500\text{ }\cancel{\text{mg}}}{1}$$

6. Solve the equation.

   $$A\text{ tablets} = \frac{1\text{ tablet}}{250\text{ }\cancel{\text{mg}}} \times \frac{500\text{ }\cancel{\text{mg}}}{1}$$

   A = 2 tablets = amount to administer

Example 2 Ordered: Norvir® liquid 40 mg PO daily

On hand: Norvir® 80 mg/1 mL. See Figure 7-5 page 223.

Find the amount to administer.

Follow Procedure Checklist 7-3.

1. The amount to administer will be in milliliters. This is the dosage unit or how the medication is supplied.

   A mL =

2. Since the unit of measure for the dosage ordered is the same as that for the dose on hand, no conversion factor is necessary.

3. The dosage unit is 1 mL. The dose on hand is 80 mg. This is our first factor.

$$\frac{1 \text{ mL}}{80 \text{ mg}}$$

4. The dose ordered is 40 mg. Place this over 1

$$A \text{ mL} = \frac{1 \text{ mL}}{80 \text{ mg}} \times \frac{40 \text{ mg}}{1}$$

5. Cancel units on the right side of the equation.

$$A \text{ mL} = \frac{1 \text{ mL}}{80 \cancel{\text{mg}}} \times \frac{40 \cancel{\text{mg}}}{1}$$

6. Solve the equation.

$$A = \frac{40 \text{ mL}}{80} = 0.5 \text{ mL} = \text{amount to administer}$$

Example 3 Ordered: 250 mg Ampicillin IM now

On hand: Ampicillin 0.5 g/mL

Find the amount to administer.

Follow Procedure Checklist 7-3.

1. The unit of measure for the amount to administer will be milliliters.

$$A \text{ mL} =$$

2. The unit of measure for the dosage ordered is milligrams. The unit of measure for the dose on hand is grams. Recall the conversion factor 1 g = 1000 mg. We will be converting the dosage ordered to grams, to obtain the desired dose. Put this unit on top.

$$\frac{1 \text{ g}}{1000 \text{ mg}}$$

3. The dosage unit is 1 mL, and the dose on hand is 0.5 g. This is our second factor.

$$\frac{1 \text{ g}}{1000 \text{ mg}} \times \frac{1 \text{ mL}}{0.5 \text{ g}}$$

4. The dose ordered is 250 mg. Place this over 1

$$A \text{ mL} = \frac{1 \text{ g}}{1000 \text{ mg}} \times \frac{1 \text{ mL}}{0.5 \text{ g}} \times \frac{250 \text{ mg}}{1}$$

5. Cancel units to check your equation.

$$A \text{ mL} = \frac{1 \cancel{\text{g}}}{1000 \cancel{\text{mg}}} \times \frac{1 \text{ mL}}{0.5 \cancel{\text{g}}} \times \frac{250 \cancel{\text{mg}}}{1}$$

6. Solve the equation.

$$A \text{ mL} = \frac{250 \text{ mL}}{500} \quad \text{(Reduce the fraction to its lowest terms.)}$$

$$A = 0.5 \text{ mL} = \text{amount to administer}$$

Example 4 Ordered: Metformin 2 g PO daily

On hand: Metformin hydrochloride 1000 mg. See Figure 7-6 on page 225.

Find the amount to administer.

Follow Procedure Checklist 7-3.

1. The unit of measure for the amount to administer will be tablets.

   A tablets =

2. The unit of measure for the dosage ordered is grams. The unit of measure for the desired dose is milligrams. Recall the conversion factor 1 g = 1000 mg. We will be converting to milligrams.

   $$\frac{1000 \text{ mg}}{1 \text{ g}}$$

3. The dosage unit is 1 tablet, and the dose on hand is 1000 mg. This is our second factor.

   $$\frac{1000 \text{ mg}}{1 \text{ g}} \times \frac{1 \text{ tablet}}{1000 \text{ mg}}$$

4. The dose ordered is 2 g. Place this over 1

   $$A \text{ tablets} = \frac{1000 \text{ mg}}{1 \text{ g}} \times \frac{1 \text{ tablet}}{1000 \text{ mg}} \times \frac{2 \text{ g}}{1}$$

5. Cancel units.

   $$A \text{ tablets} = \frac{1000 \not{\text{mg}}}{1 \not{\text{g}}} \times \frac{1 \text{ tablet}}{1000 \not{\text{mg}}} \times \frac{2 \not{\text{g}}}{1}$$

6. Solve the equation.

   $$A \text{ tablets} = \frac{2000}{1000} \quad \text{(Reduce the fraction to its lowest terms.)}$$

   A = 2 tablets = amount to administer

GO TO . . . Open the CD-ROM that accompanies your textbook, and select Chapter 7, Procedure Checklist 7-3. Review the animation and example problems, then complete the practice problems. Continue to the next section of the book once you have mastered the rule presented. ■

Procedure Checklist 7-4

Calculating the Amount to Administer by the Formula Method

1. Determine the desired dose. Calculate it using the fraction proportion, ratio proportion, or dimensional analysis method. Determine the dose on hand H and dosage unit Q.
2. Fill in the formula:

   $$\frac{D}{H} \times Q = A$$

   where D = desired dose (This is the dose ordered changed to the same unit of measure as the dose on hand.)

   H = dose on hand—the amount of drug contained in each unit.

   Q = dosage unit—how the drug will be administered, such as tablets or milliliters.

   A = unknown or amount to administer.
3. Cancel the units.
4. Solve for the unknown.

Example 1 Ordered: Famvir® 500 mg PO q8h

On hand: Famvir® 250 mg per tablet (Figure 7-4)

Find the amount to administer.

Follow Procedure Checklist 7-4.

1. The drug is ordered in milligrams, which is the same unit used on the label. Therefore, the desired dose is 500 mg. Reading the label tells us that the dosage unit is 1 tablet and the dose on hand is 250 mg. Therefore,

   $D = 500 \text{ mg}$

   $Q = 1 \text{ tablet}$

   $H = 250 \text{ mg}$

2. Fill in the formula.

   $$\frac{500 \text{ mg} \times 1 \text{ tablet}}{250 \text{ mg}} = A$$

3. Cancel units.

   $$\frac{500 \cancel{\text{mg}} \times 1 \text{ tablet}}{250 \cancel{\text{mg}}} = A$$

4. Solve for the unknown.

   $A = 2 \text{ tablets} = \text{amount to administer}$

Example 2 Ordered: Norvir® liquid 40 mg PO qd

On hand: Norvir® 8 mg/1 mL. See Figure 7-5 on page 223

Calculate the amount to administer.

Follow Procedure Checklist 7-4.

1. Again, the drug order and the drug label use the same units. Our desired dose is 40 mg. Reading the label, we find that the dosage unit is 1 mL and the dose on hand is 80 mg. Therefore,

   $D = 40 \text{ mg}$

   $Q = 1 \text{ mL}$

   $H = 80 \text{ mg}$

2. Insert the numbers and units into the formula.

   $$\frac{40 \text{ mg} \times 1 \text{ mL}}{80 \text{ mg}} = A$$

3. Cancel units.

   $$\frac{40 \cancel{\text{mg}}}{80 \cancel{\text{mg}}} \times 1 \text{ mL} = A$$

4. Solve for the unknown.

   $0.5 \text{ mL} = A = \text{amount to administer}$

Example 3 Ordered: Ampicillin 250 mg IM now

On hand: Ampicillin 0.5 g/mL

Find the amount to administer.

In this case, the order is written in milligrams, and the drug is labeled in grams. Before we can determine the amount to administer, we must calculate a desired dose that is in grams. (In this example we will use fraction proportion method Procedure Checklist 3-1.)

1. Recall that 1 g = 1000 mg. Since we are converting to grams, our conversion factor is:

   $$\frac{1\text{ g}}{1000\text{ mg}}.$$

2. The other fraction for our proportion is $\frac{D}{250\text{ mg}}$.

3. Setting the two fractions into a proportion gives us the following equation:

   $$\frac{D}{250\text{ mg}} = \frac{1\text{ g}}{1000\text{ mg}}$$

4. Cancel units.

   $$\frac{D}{250\,\cancel{\text{mg}}} = \frac{1\text{ g}}{1000\,\cancel{\text{mg}}}$$

5. Solve for the unknown.

   $$1000 \times D = 250 \times 1\text{ g}$$
   $$D = 0.25\text{ g} = \text{desired dose}$$

Follow Procedure Checklist 7-4.

1. The three necessary pieces of information to complete the formula are that the desired dose is 0.25 g, the dosage unit is 1 mL, and the dose on hand is 0.5 g.

   $D = 0.25\text{ g}$ $\quad Q = 1\text{ mL}$ $\quad H = 0.5\text{ g}$

2. Insert the numbers and units into the formula.

   $$\frac{0.25\text{ g}}{0.5\text{ g}} \times 1\text{ mL} = A$$

3. Cancel units.

   $$\frac{0.25\,\cancel{\text{g}}}{0.5\,\cancel{\text{g}}} \times 1\text{ mL} = A$$

4. Solve for the unknown.

   $0.5\text{ mL} = A = \text{amount to administer}$

Example 4 Ordered: Metformin 2 g PO daily

On hand: Metformin hydrochloride 1000 mg. See Figure 7-6 on page 225

Calculate the amount to administer.

In this case, the order is written in grams, and the drug is labeled in milligrams. Before we can determine the amount to administer, we must calculate a desired dose that is in milligrams. (In this example we will use the ratio proportion method Procedure Checklist 3-2.)

1. Recall that 1 g = 1000 mg. Since we are converting to milligrams, our conversion factor is:

   1000 mg : 1 g

2. The other ratio in our proportion is:

   D : 2 g

3. Set up the ratio proportion equation.

   $1000 \text{ mg} : 1 \text{ g} :: D : 2 \text{ g}$

4. Cancel units.

   $1000 \text{ mg} : 1 \cancel{\text{g}} :: D : 2 \cancel{\text{g}}$

5. Multiply the means and extremes and solve the equation.

   $1 \times D = 1000 \text{ mg} \times 2$

   $D = 2000 \text{ mg} = \text{desired dose}$

Follow Procedure Checklist 7-4.

1. The three necessary pieces to complete the formula are the desired dose of 2000 mg, the dosage unit of 1 tablet, and the dose on hand of 1000 g.

   $D = 2000 \text{ mg} \qquad Q = 1 \text{ tablet} \qquad H = 1000 \text{ mg}$

2. Insert the numbers and units into the formula.

   $$\frac{2000 \text{ g}}{1000 \text{ g}} \times 1 \text{ tablet} = A$$

3. Cancel units.

   $$\frac{2000 \cancel{\text{g}}}{1000 \cancel{\text{g}}} \times 1 \text{ tablet} = A$$

4. Solve for the unknown.

   $2 \text{ tablets} = A = \text{amount to administer}$

GO TO . . . Open the CD-ROM that accompanies your textbook, and select Chapter 7, Procedure Checklist 7-4. Review the animation and example problems, then complete the practice problems. Continue to the next section of the book once you have mastered the rule presented. ■

Critical Thinking on the Job

When in Doubt, Check

Jorge was working in the emergency room when a patient arrived with life-threatening internal bleeding. The physician in charge told Jorge, "Aminocaproic acid 5 grams STAT! You'd better give him liquid, I don't think he's able to swallow pills." Jorge repeated the order, "Aminocaproic acid liquid 5 grams STAT."

On hand, Jorge had Amicar Syrup (aminocaproic acid) 25%, 1.25 g/5 mL (see Figure 7-7). Jorge determined the amount to administer by using the ratio proportion method.

$$5\text{ mL} : 1.25\text{ g} :: A : 5\text{ g}$$

$$5\text{ mL} : 1.25\cancel{\text{g}} :: A : 5\cancel{\text{g}}$$

$$1.25 \times A = 5\text{ mL} \times 5$$

$$A = \frac{25\text{ mL}}{1.25}$$

$$A = 20\text{ mL} = 4\text{ tsp}$$

Question?

What should Jorge do?

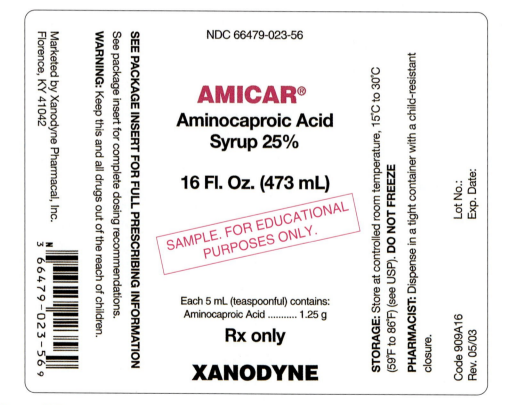

NDC 66479-023-56

AMICAR®
Aminocaproic Acid
Syrup 25%

16 Fl. Oz. (473 mL)

SAMPLE. FOR EDUCATIONAL PURPOSES ONLY.

Each 5 mL (teaspoonful) contains:
Aminocaproic Acid .... 1.25 g

Rx only

XANODYNE

STORAGE: Store at controlled room temperature, 15°C to 30°C (59°F to 86°F) (see USP). **DO NOT FREEZE**

PHARMACIST: Dispense in a tight container with a child-resistant closure.

Lot No.:
Exp. Date:

Code 909A16
Rev. 05/03

SEE PACKAGE INSERT FOR FULL PRESCRIBING INFORMATION

See package insert for complete dosing recommendations.

WARNING: Keep this and all drugs out of the reach of children.

Marketed by Xanodyne Pharmacal, Inc.
Florence, KY 41042

N 3 66479-023-56 9

Figure 7-7

REVIEW AND PRACTICE

7-2 Calculating the Amount to Administer

In Exercises 1-20, calculate the amount to administer.

1. Ordered: Thorazine® 20 mg PO tid
   On hand: Thorazine® 10 mg tablets
2. Ordered: Ranitidine hydrochloride 150 mg PO bid
   On hand: Zantac® syrup 15 mg ranitidine hydrochloride per mL
3. Ordered: Ceclor® 0.375 g PO bid
   On hand: Ceclor® Oral Suspension 187 mg per 5 mL
4. Ordered: Nitroglycerin gr $\frac{1}{100}$ SL stat
   On hand: Nitroglycerin 0.3-mg tablets
5. Ordered: Amoxicillin 250 mg PO tid
   On hand: Refer to label A.

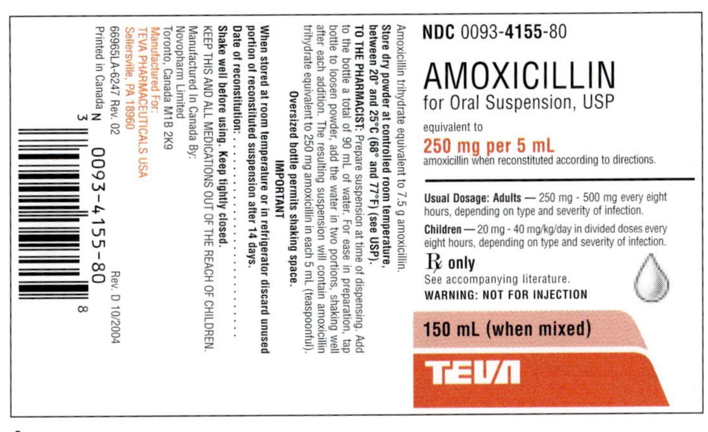

A

6. Ordered: Tricor® 108 mg PO daily
   On hand: Refer to label B.

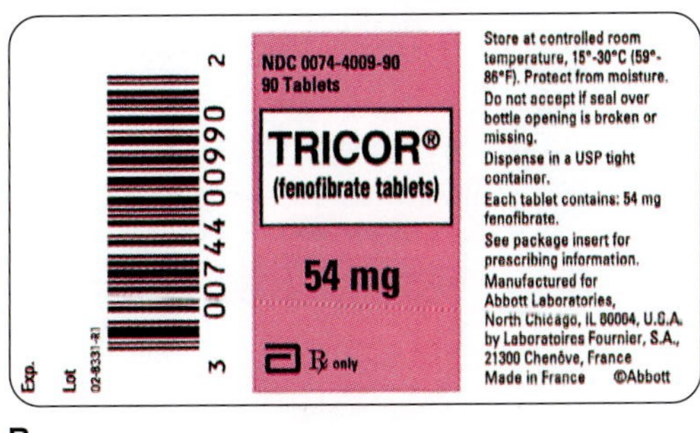

NDC 0074-4009-90
90 Tablets
TRICOR®
(fenofibrate tablets)
54 mg
℞ only
Store at controlled room temperature, 15°-30°C (59°-86°F). Protect from moisture.
Do not accept if seal over bottle opening is broken or missing.
Dispense in a USP tight container.
Each tablet contains: 54 mg fenofibrate.
See package insert for prescribing information.
Manufactured for Abbott Laboratories, North Chicago, IL 60064, U.S.A. by Laboratoires Fournier, S.A., 21300 Chenôve, France
Made in France ©Abbott

B

7. Ordered: Procardia® 20 mg PO tid
   On hand: Refer to label C.

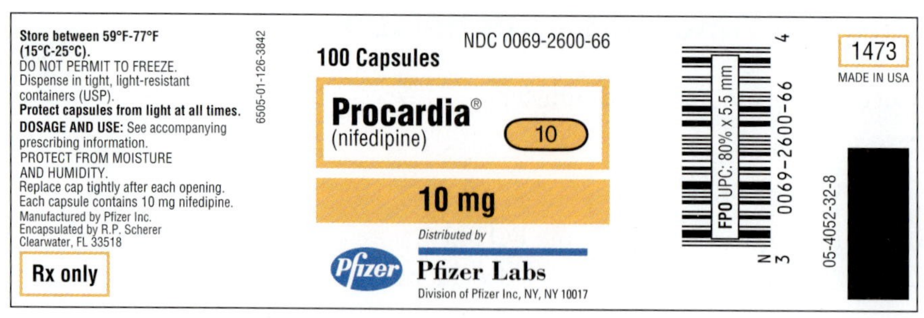

C

8. Ordered: Targretin® 150 mg PO q.d. a.c.
   On hand: Refer to label D.

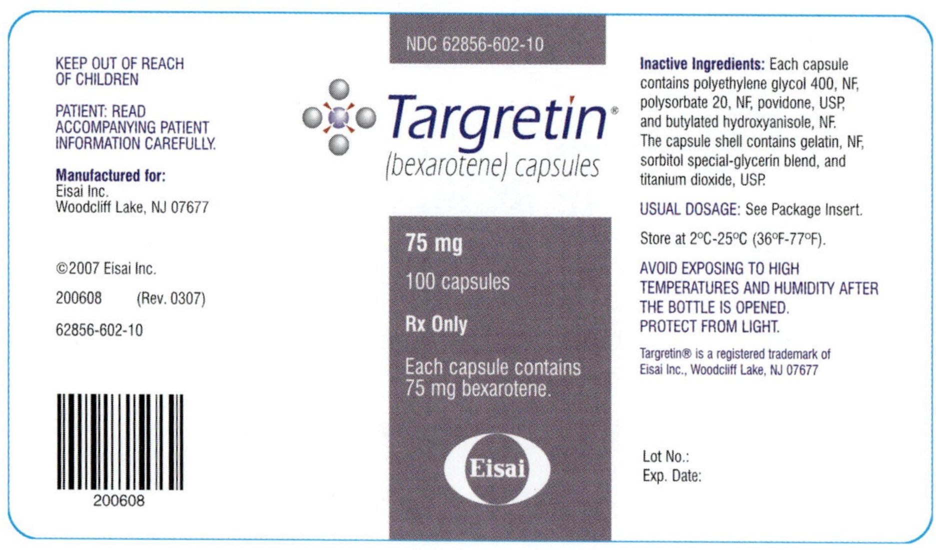

D

9. Ordered: Synthroid® 0.3 mg PO q.d.
   On hand: Refer to label E.

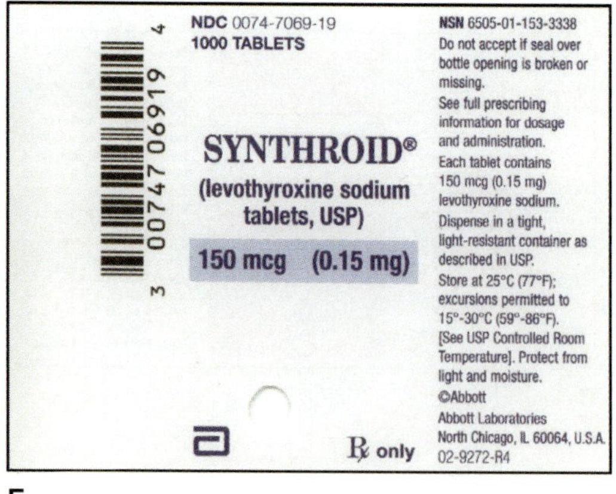

E

10. Ordered: Strattera® 0.1 g PO bid
On hand: Refer to label F.

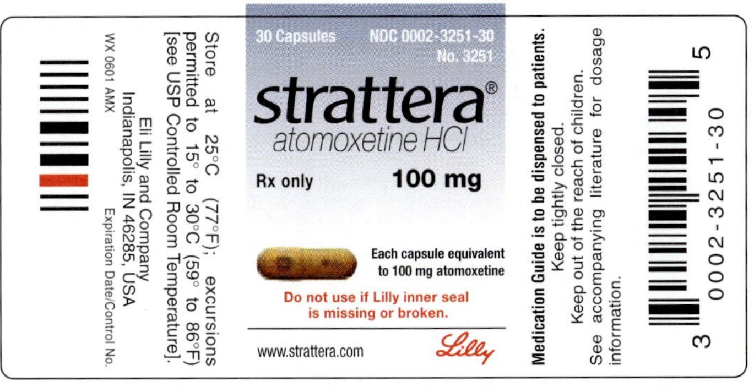

F

11. Ordered: Keflex® 500 mg PO q12h
On hand: Keflex® 250 mg per 5 mL

12. Ordered: Decadron® 6 mg IM q.i.d.
On hand: Decadron® 4 mg per mL

13. Ordered: Ketoconazole® 100 mg PO qd
On hand: Ketoconazole® 200-mg scored tablets

14. Ordered: Dilaudid® 2 mg IM prn for pain q6h
On hand: Dilaudid® for injection, 4 mg per mL

15. Ordered: Erythromycin Oral Suspension 150 mg PO bid
On hand: Refer to label G.

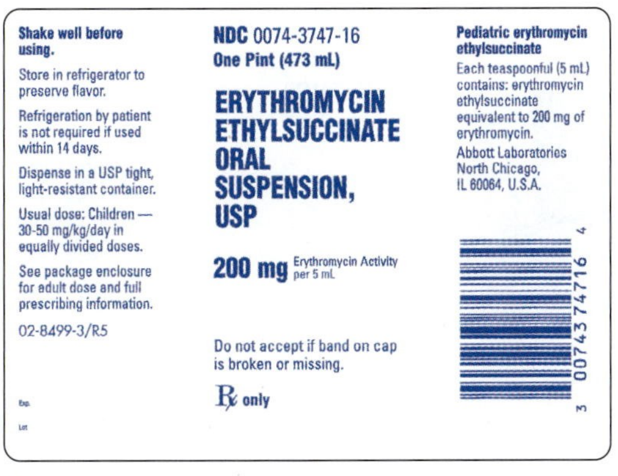

G

16. Ordered: Tranxene® 30 mg PO qhs
On hand: Refer to label H.

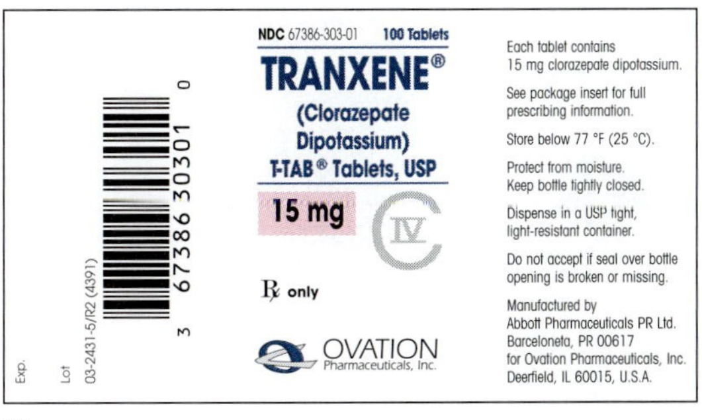

H

17. Ordered: Interferon Alfa-2b, recombinant 5 million units IM
On hand: Refer to label I.

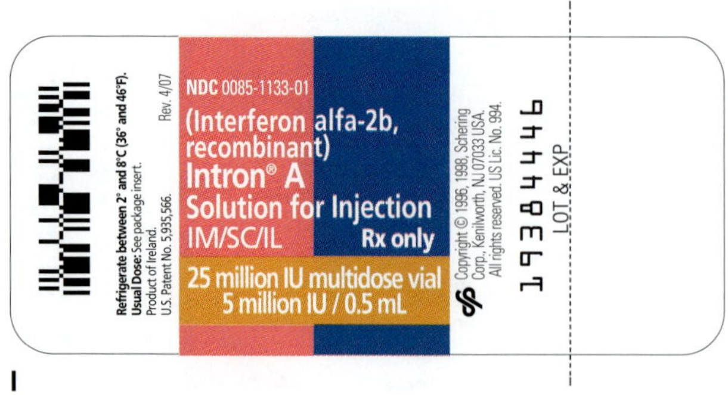

I

18. Ordered: Humulin® R 28 U sc stat
On hand: Refer to label J.

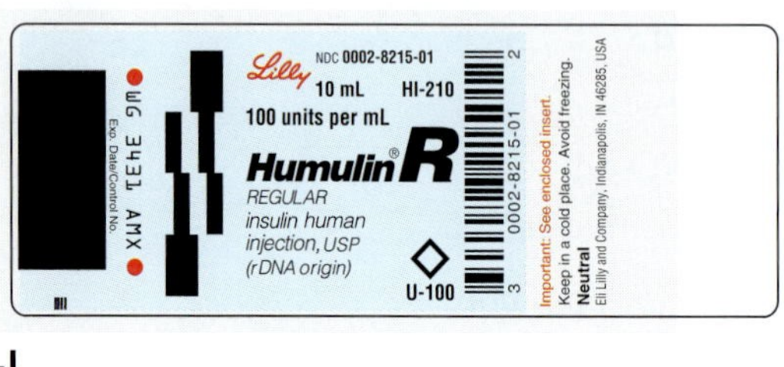

J

19. Ordered: Ritalin® 15 mg PO bid ac
On hand: Refer to label K.

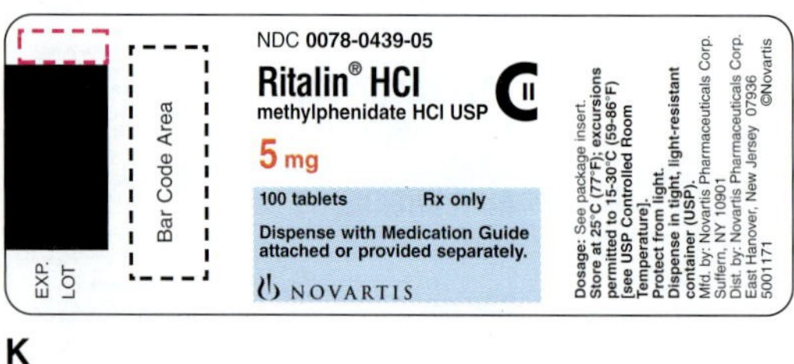

EXP.
LOT
Bar Code Area
NDC 0078-0439-05
Ritalin® HCl
methylphenidate HCl USP
CII
5 mg
100 tablets
Rx only
Dispense with Medication Guide attached or provided separately.
NOVARTIS
Dosage: See package insert.
Store at 25°C (77°F); excursions permitted to 15-30°C (59-86°F) [see USP Controlled Room Temperature].
Protect from light.
Dispense in tight, light-resistant container (USP).
Mfd. by: Novartis Pharmaceuticals Corp. Suffern, NY 10901
Dist. by: Novartis Pharmaceuticals Corp. East Hanover, New Jersey 07936
©Novartis
5001171

K

20. Ordered: Targretin® 75 mg PO q am ac
On hand: Refer to label L.

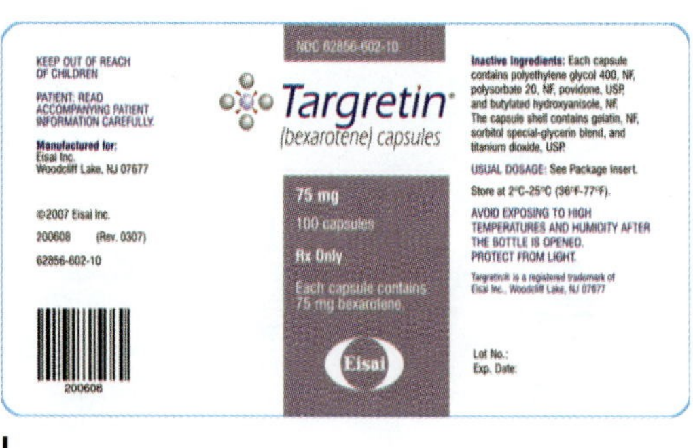

L

To check your answers, see page 520.

HOMEWORK ASSIGNMENT

Answer the following questions.

1. Define the term dosage ordered.

2. Define the term desired dose.

3. Define the term dosage unit.

4. Define the term dose on hand.

5. Define the term amount to administer.

6. Define the term dosage strength.

7. One milligram (mg) is equal to how many micrograms (mcg)?

8. One grain (gr) is equal to how many milligrams (mg)?

9. One ounce (oz) is equal to how many milliliters (mL)?

10. One kilogram (kg) is equal to how many pounds (lb)?

Select the appropriate label for the following drug orders and indicate the number of tablets/capsules/milliliters that will be required to administer the dosage ordered.

Assume that all tablets are scored. Notice that both generic and brand names are used for the orders.

11. Ordered: 2 grams amoxicillin liquid. Label ________________
Amount to administer: ________________

12. Ordered: Zemplar® 1 mcg. Label ________________
Amount to administer: ________________

13. Ordered: Meperidine 150 mg. Label ________________
Amount to administer: ________________

14. Ordered: Plendil® 5 mg. Label ________________
Amount to administer: ________________

15. Ordered: Cefprozil 500 mg liquid. Label ________________
Amount to administer: ________________

16. Ordered: Valproic acid 250 mg. Label ________________
Amount to administer: ________________

17. Ordered: Sertraline HCL 75 mg. Label ________________
Amount to administer: ________________

18. Ordered: Digoxin 0.5 mg. Label ________________
Amount to administer: ________________

19. Ordered: Omnicef® 300 mg liquid. Label ________________
Amount to administer: ________________

20. Ordered: Esomeprazole 40 mg. Label ________________
Amount to administer: ________________

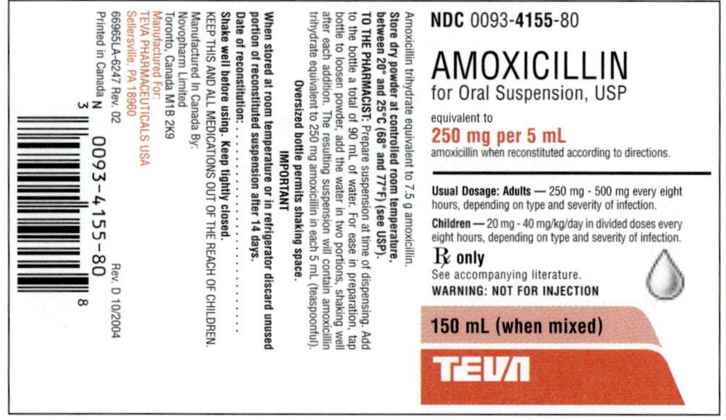

LABEL A

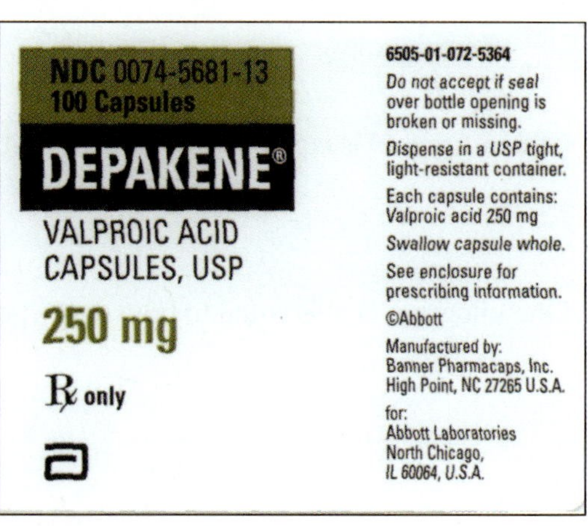

NDC 0074-5681-13
100 Capsules
DEPAKENE®
VALPROIC ACID
CAPSULES, USP
250 mg
℞ only
6505-01-072-5364
Do not accept if seal over bottle opening is broken or missing.
Dispense in a USP tight, light-resistant container.
Each capsule contains: Valproic acid 250 mg
Swallow capsule whole.
See enclosure for prescribing information.
©Abbott
Manufactured by: Banner Pharmacaps, Inc. High Point, NC 27265 U.S.A.
for: Abbott Laboratories North Chicago, IL 60064, U.S.A.

LABEL B

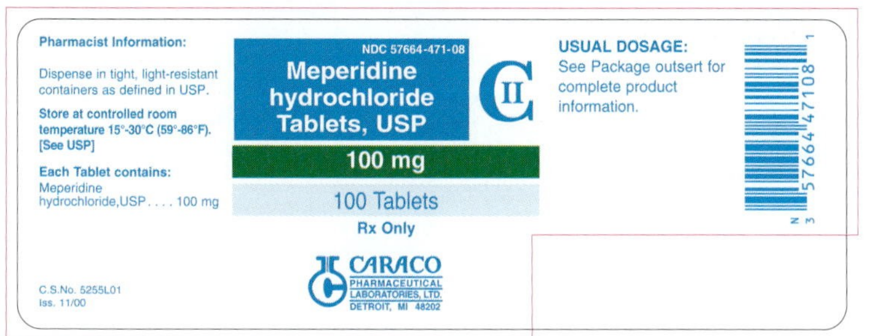

LABEL C

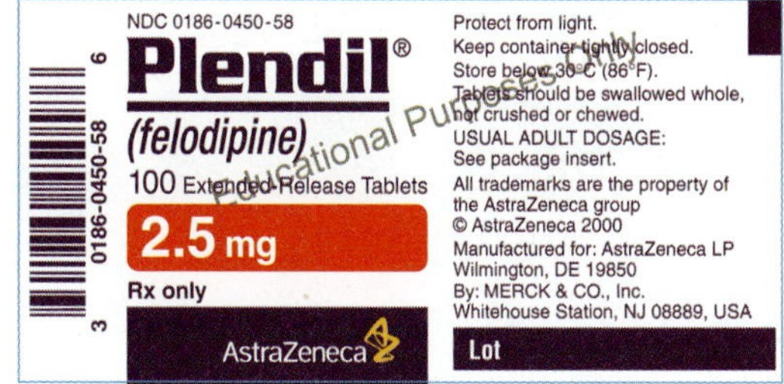

NDC 0186-0450-58
Plendil®
(felodipine)
100 Extended-Release Tablets
2.5 mg
Rx only
AstraZeneca
Protect from light.
Keep container tightly closed.
Store below 30°C (86°F).
Tablets should be swallowed whole, not crushed or chewed.
USUAL ADULT DOSAGE:
See package insert.
All trademarks are the property of the AstraZeneca group
© AstraZeneca 2000
Manufactured for: AstraZeneca LP Wilmington, DE 19850
By: MERCK & CO., Inc. Whitehouse Station, NJ 08889, USA
Lot

LABEL D

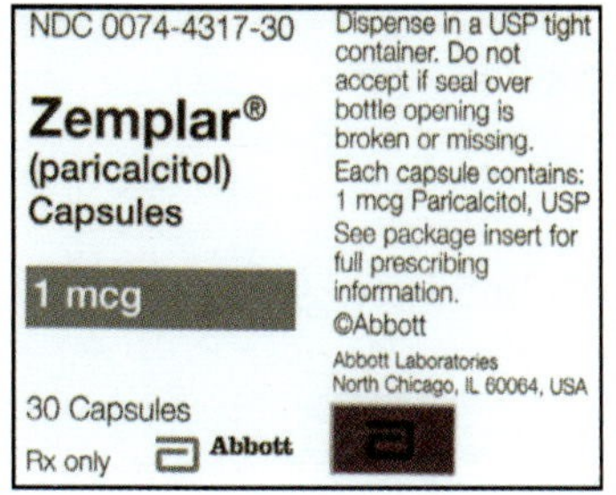

NDC 0074-4317-30
Zemplar®
(paricalcitol)
Capsules
1 mcg
30 Capsules
Rx only
Abbott
Dispense in a USP tight container. Do not accept if seal over bottle opening is broken or missing.
Each capsule contains: 1 mcg Paricalcitol, USP
See package insert for full prescribing information.
©Abbott
Abbott Laboratories North Chicago, IL 60064, USA

LABEL E

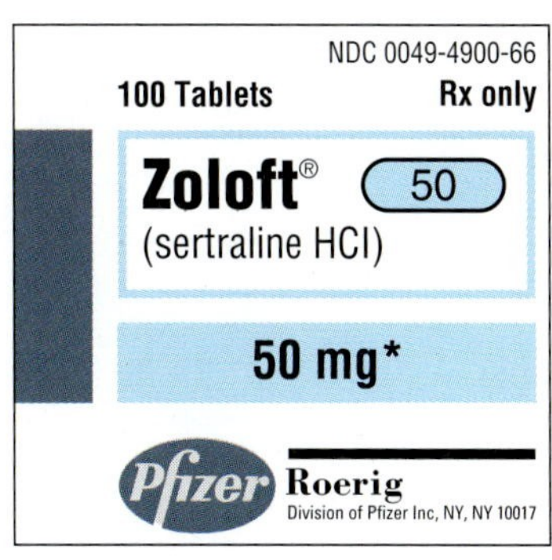

NDC 0049-4900-66
100 Tablets
Rx only
Zoloft®
50
(sertraline HCl)
50 mg*
Pfizer
Roerig
Division of Pfizer Inc, NY, NY 10017

LABEL F

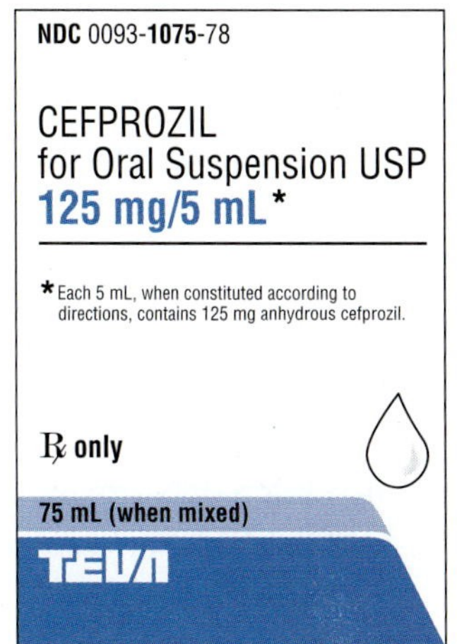

NDC 0093-1075-78

LABEL H

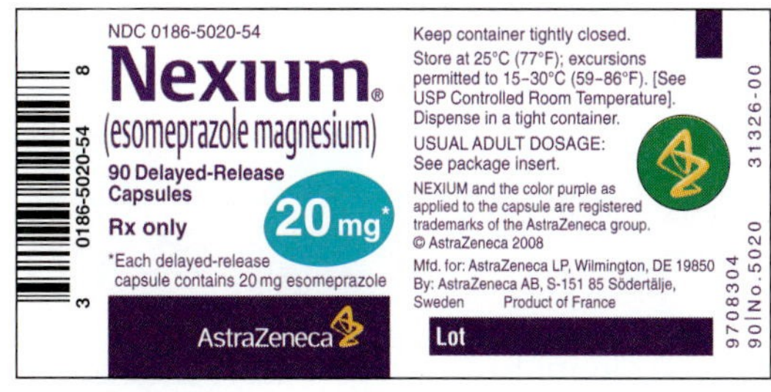

NDC 0186-5020-54
Nexium®
(esomeprazole magnesium)
90 Delayed-Release Capsules
Rx only
20 mg*
*Each delayed-release capsule contains 20 mg esomeprazole
AstraZeneca
Keep container tightly closed.
Store at 25°C (77°F); excursions permitted to 15–30°C (59–86°F). [See USP Controlled Room Temperature].
Dispense in a tight container.
USUAL ADULT DOSAGE:
See package insert.
NEXIUM and the color purple as applied to the capsule are registered trademarks of the AstraZeneca group.
© AstraZeneca 2008
Mfd. for: AstraZeneca LP, Wilmington, DE 19850
By: AstraZeneca AB, S-151 85 Södertälje, Sweden
Product of France
Lot

LABEL G

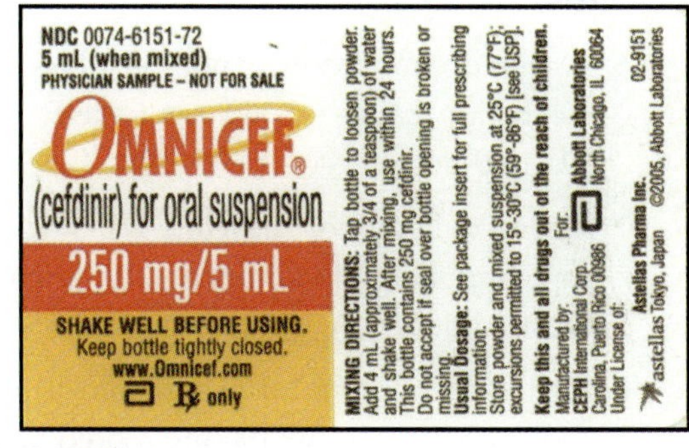

NDC 0074-6151-72
5 mL (when mixed)
PHYSICIAN SAMPLE – NOT FOR SALE
OMNICEF®
(cefdinir) for oral suspension
250 mg/5 mL
SHAKE WELL BEFORE USING.
Keep bottle tightly closed.
www.Omnicef.com
*Each 5 mL, when constituted according to directions, contains 125 mg anhydrous cefprozil.
℞ only
MIXING DIRECTIONS: Tap bottle to loosen powder. Add 4 mL (approximately 3/4 of a teaspoon) of water and shake well. After mixing, use within 24 hours. This bottle contains 250 mg cefdinir.
Do not accept if seal over bottle opening is broken or missing.
Usual Dosage: See package insert for full prescribing information.
Store powder and mixed suspension at 25°C (77°F); excursions permitted to 15–30°C (59–86°F) [see USP].
Keep this and all drugs out of the reach of children.
Abbott Laboratories North Chicago, IL 60064
Astellas Pharma Inc. Tokyo, Japan
©2005, Abbott Laboratories

LABEL J

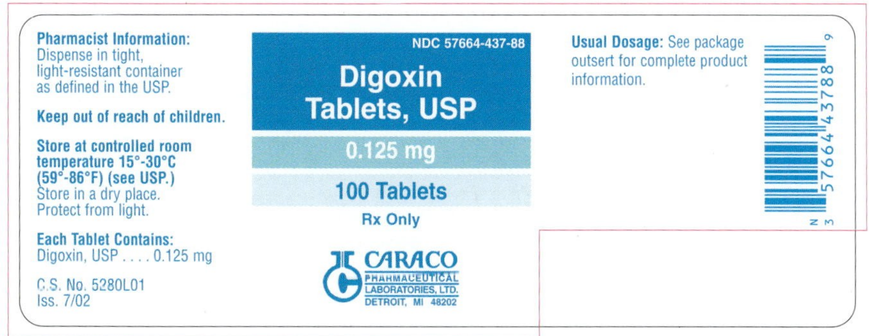

LABEL I

Chapter 7 Review

Check Up

In Exercises 1–23, calculate the desired dose. Then calculate the amount to administer.

1. Ordered: Valium® 5 mg PO tid
On hand: Valium® 2-mg scored tablets
Desired dose: __________ Amount to administer: __________

2. Ordered: Atacand® 16 mg PO bid
On hand: Atacand® 8-mg tablets
Desired dose: __________ Amount to administer: __________

3. Ordered: Cimetidine 400 mg PO qid hs
On hand: Tagamet® 200 mg tablets
Desired dose: __________ Amount to administer: __________

4. Ordered: Noroxin® 800 mg PO qd ac $\bar{c}$ H_2O
On hand: Noroxin® 400-mg tablets
Desired dose: __________ Amount to administer: __________

5. Ordered: Tenex® 2 mg PO qd hs
On hand: Tenex® 1-mg tablets
Desired dose: __________ Amount to administer: __________

6. Ordered: Tranxene® 7.5 mg PO qd hs
On hand: Tranxene® 3.75-mg tablets
Desired dose: __________ Amount to administer: __________

7. Ordered: Pergolide mesylate 100 mcg PO tid
On hand: Pergolide mesylate 0.05-mg tablets
Desired dose: __________ Amount to administer: __________

8. Ordered: Zyloprim® 0.25 g PO bid
On hand: Zyloprim® 100-mg scored tablets
Desired dose: __________ Amount to administer: __________

9. Ordered: Zaroxolyn® 7.5 mg PO daily
On hand: Zaroxolyn® 2.5-mg tablets
Desired dose: __________ Amount to administer: __________

10. Ordered: Ciprofloxacin hydrochloride 500 mg PO q12h
On hand: Refer to label A.
Desired dose: __________ Amount to administer: __________

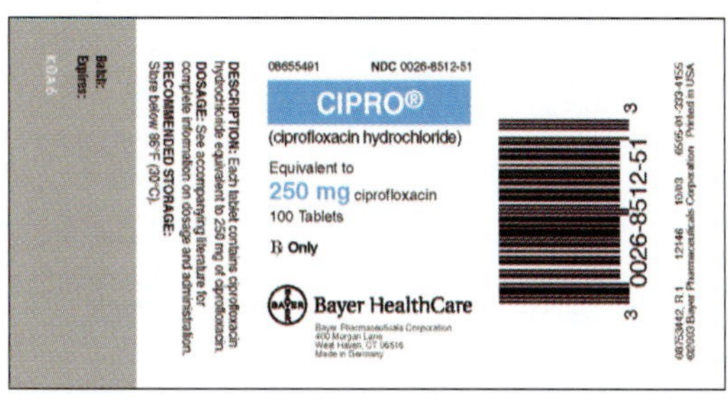

A

11. Ordered: Lexapro 20 mg PO daily
On hand: Refer to label B.

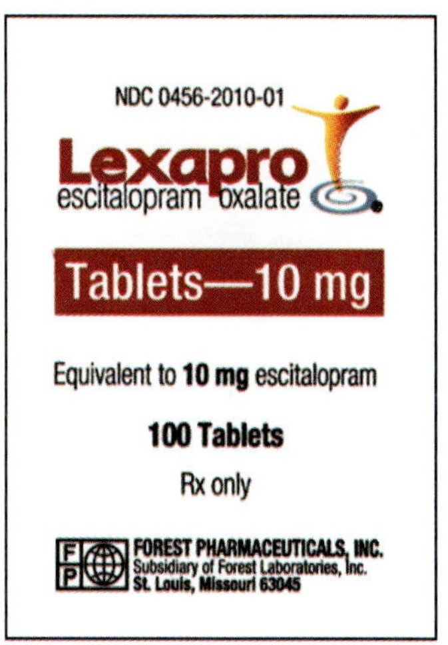

B

12. Ordered: Toprol-XL® 100 mg PO bid
On hand: Refer to label C.

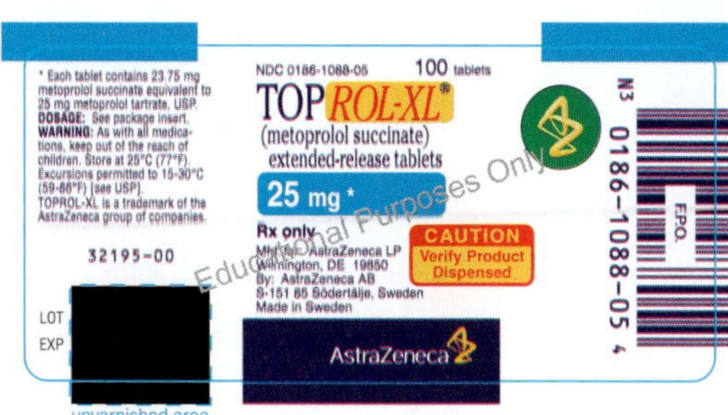

C

13. Ordered: Depakene® 250 mg PO bid
On hand: Refer to label D.

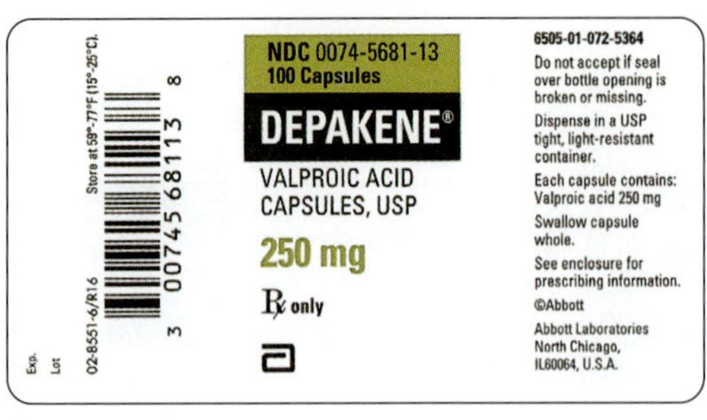

D

14. Ordered: Dilantin® 60 mg PO daily
On hand: Refer to label E.

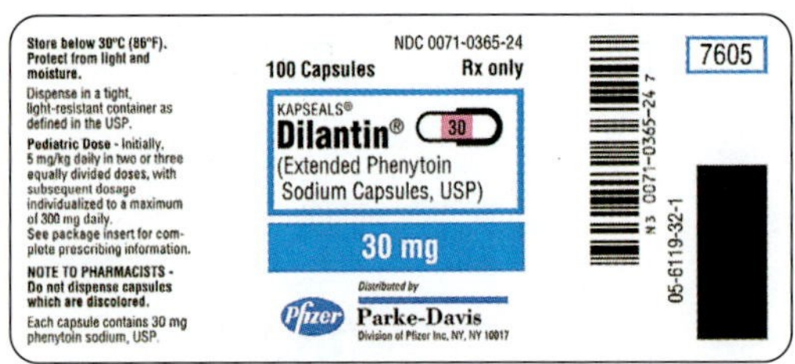

E

15. Ordered: Lisinopril 40 mg PO daily
On hand: Refer to label F.

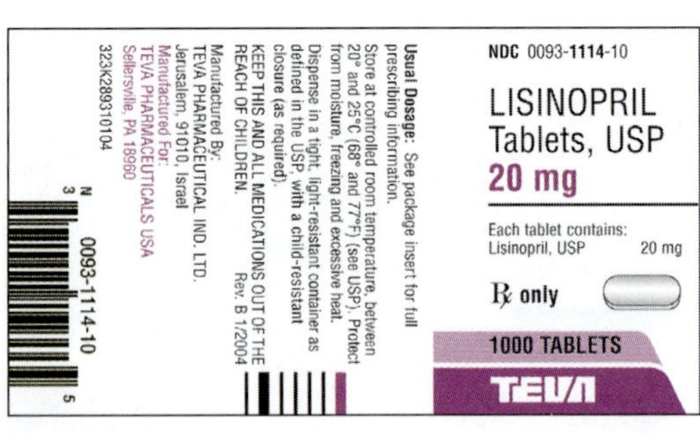

F

16. Ordered: Biaxin® 125 mg PO tid
    On hand: Refer to label G.

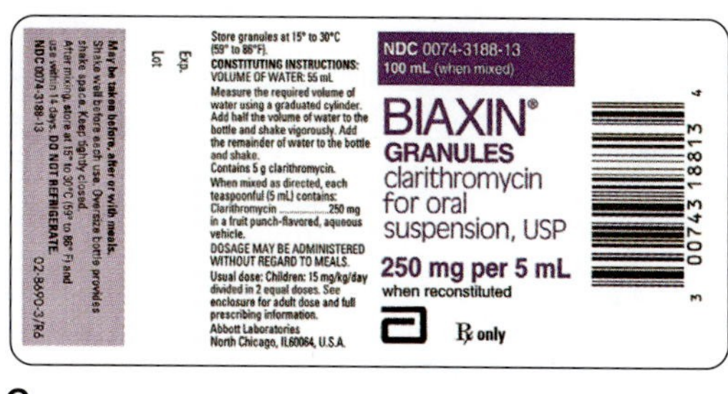

G

17. Ordered: Amoxicillin 1 gram PO bid
    On hand: Refer to label H.

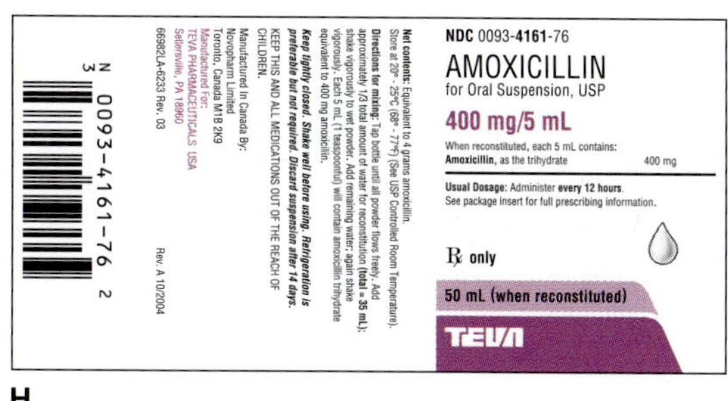

H

18. Ordered: Zyrtec® 5 mg PO daily
    On hand: Refer to label I.

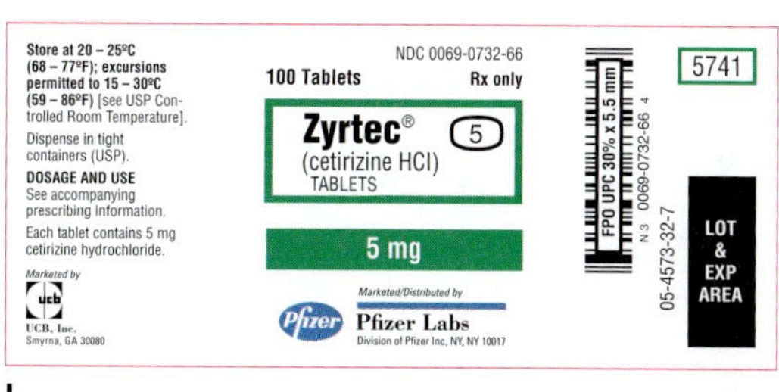

I

19. Ordered Cefprozil 200 mg PO q8h
On hand: Refer to label J.

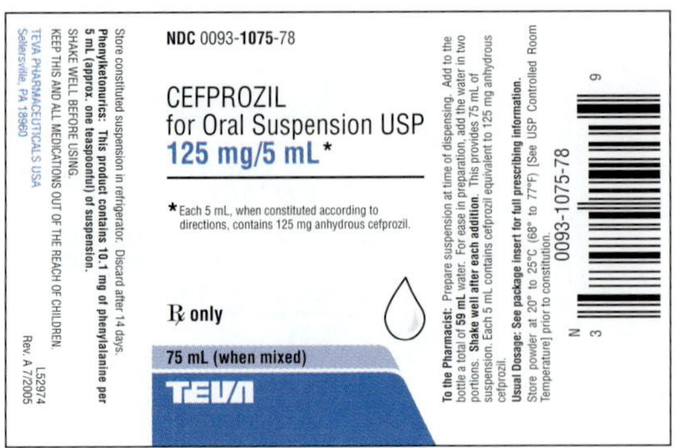

J

20. Ordered: Acyclovir 0.5 g IV
On hand: Refer to label K.

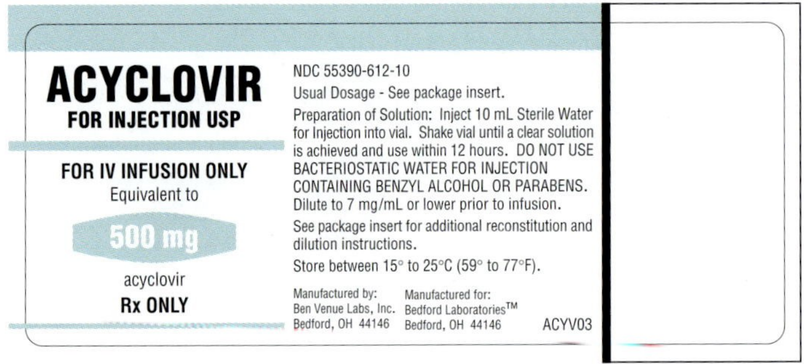

K

21. Ordered: Kytril® Oral Solution 4 mg PO tid
On hand: Refer to label L.

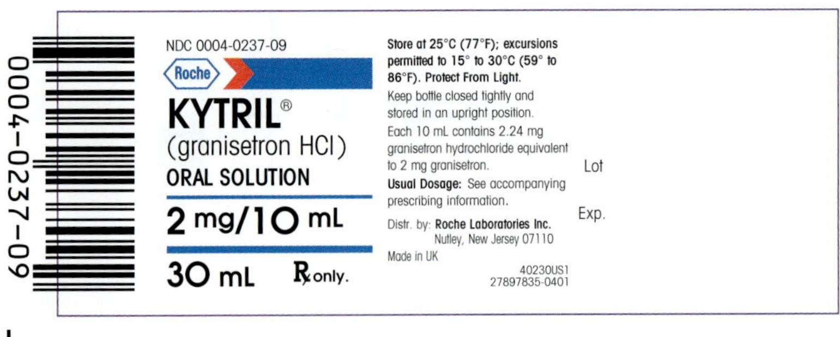

L

22. Ordered: Furosemide 100 mg IM now
    On hand: Refer to label M.

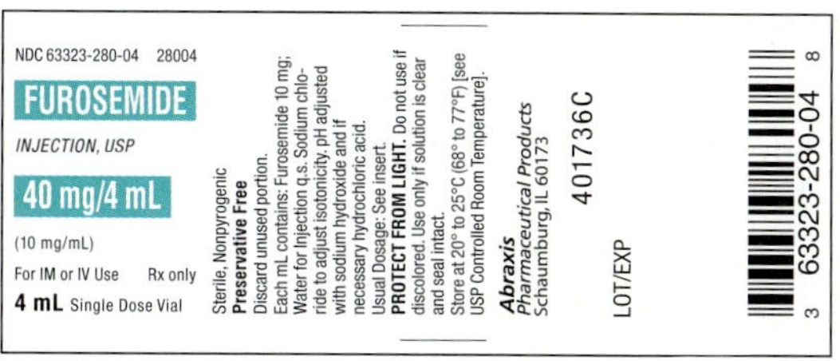

NDC 63323-280-04 28004
FUROSEMIDE
INJECTION, USP
40 mg/4 mL
(10 mg/mL)
For IM or IV Use Rx only
4 mL Single Dose Vial
Sterile, Nonpyrogenic
Preservative Free
Discard unused portion.
Each mL contains: Furosemide 10 mg; Water for Injection q.s. Sodium chloride to adjust isotonicity pH adjusted with sodium hydroxide and if necessary hydrochloric acid.
Usual Dosage: See insert.
PROTECT FROM LIGHT. Do not use if discolored. Use only if solution is clear and seal intact.
Store at 20° to 25°C (68° to 77°F) [see USP Controlled Room Temperature].
Abraxis Pharmaceutical Products Schaumburg, IL 60173
401736C
LOT/EXP
3 63323-280-04 8

M

23. Ordered: Zolpidem 20 mg PO now
    On hand: Refer to label N.

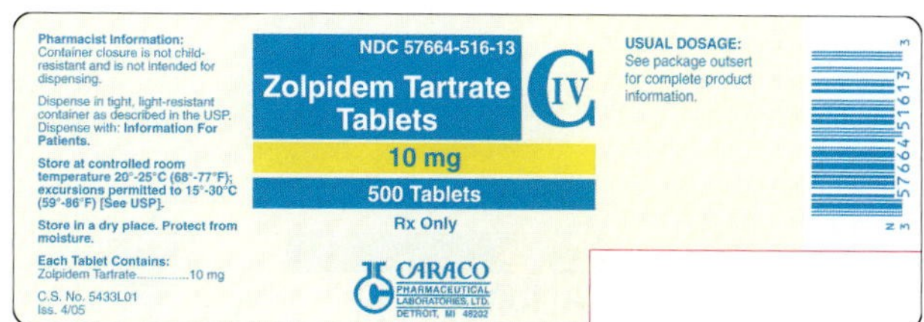

Pharmacist Information: Container closure is not child-resistant and is not intended for dispensing.
Dispense in tight, light-resistant container as described in the USP. Dispense with: Information For Patients.
Store at controlled room temperature 20°-25°C (68°-77°F); excursions permitted to 15°-30°C (59°-86°F) [See USP].
Store in a dry place. Protect from moisture.
Each Tablet Contains: Zolpidem Tartrate..............10 mg
C.S. No. 5433L01
Iss. 4/05
NDC 57664-516-13
Zolpidem Tartrate Tablets
C IV
10 mg
500 Tablets
Rx Only
CARACO PHARMACEUTICAL LABORATORIES, LTD. DETROIT, MI 48202
USUAL DOSAGE: See package outsert for complete product information.
N 3 57664 51613 3

N

Critical Thinking Applications

Use the following label to answer these questions.

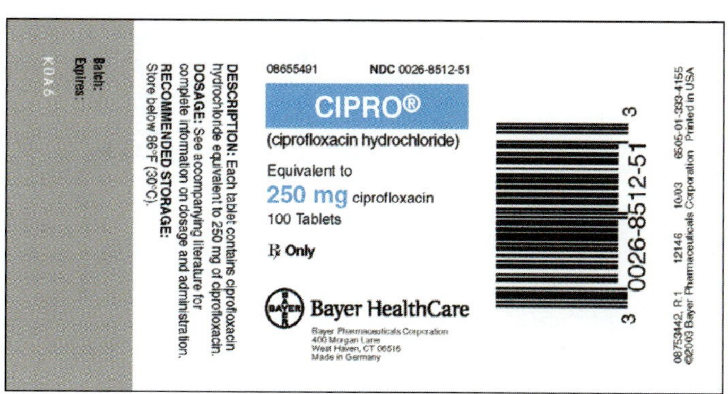

Batch:
Expires:
KDA6
DESCRIPTION: Each tablet contains ciprofloxacin hydrochloride equivalent to 250 mg of ciprofloxacin.
DOSAGE: See accompanying literature for complete information on dosage and administration.
RECOMMENDED STORAGE: Store below 86°F (30°C).
08655491
NDC 0026-8512-51
CIPRO®
(ciprofloxacin hydrochloride)
Equivalent to
250 mg ciprofloxacin
100 Tablets
℞ Only
Bayer HealthCare
Bayer Pharmaceuticals Corporation
400 Morgan Lane
West Haven, CT 06516
Made in Germany
3 0026-8512-51 3
08753442, R.1 12146 10/03 6505-01-333-4155
©2003 Bayer Pharmaceuticals Corporation Printed in USA

1. A physician's order reads Cipro® 250 mg q12 hr PO × 3 days. Calculate the amount to administer.
2. In some cases, the patient may receive 500 mg q12 hours × 10 days. Calculate the amount to administer for this dose.
3. For cutaneous anthrax a physician may order 250 mg PO bid × 60 days. Calculate the amount to administer for 1 dose.
4. In each of the cases above, determine the total number of tablets the patient will need.

Case Study

You are working in a pharmacy when the following prescription comes in: Valium® 7 mg PO tid for 7 days. The drug labels below represent what you have on hand for filling this prescription.

1. What is the amount to administer?
2. How many total tablets will the patient need?
3. What changes would you make in filling the order if the prescription read Valium 15 mg PO tid for 7 days? How many tablets will the patient need for this new order?

To check your answers, see page 521.

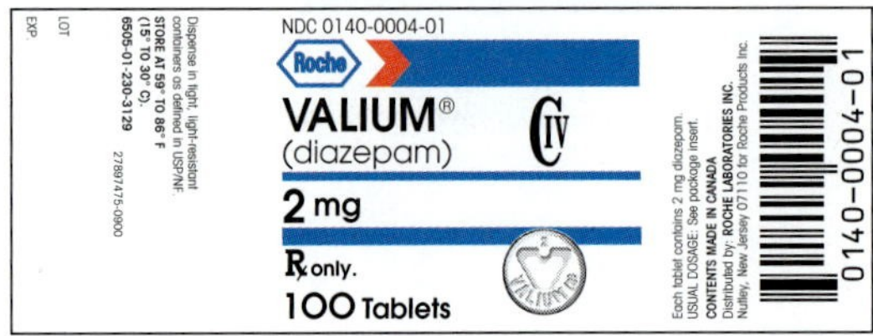

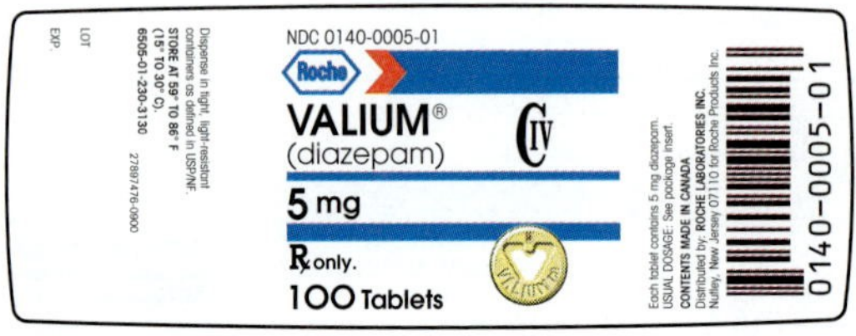

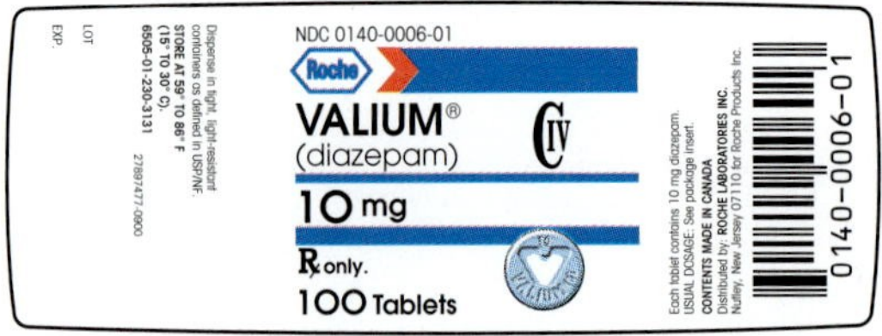

Internet Activity

Many times medications come in different dosage strengths. (Recall the dosage strength is the dose on hand per dosage unit.) If you do not look at the label carefully, you can easily select the wrong medication and/or calculate the amount to administer incorrectly. Search the Internet for at least three medications, other than ones found in this chapter, that come in different dosage strengths. List each medication and its various dosage strengths. You may want to focus on the top 200 medications at www.rxlist.com.

GO TO . . . Open the CD-ROM that accompanies your textbook, and complete a final review of the rules, practice problems, and activities presented for this chapter. For a final evaluation, take the chapter test and email or print your results for your instructor. A score of 95 percent or above indicates mastery of the chapter concepts. ■

8 Oral Dosages

If you want to achieve excellence, you can get there today. As of this second, quit doing less-than-excellent work.

—Thomas John Watson, Jr.

Learning Outcomes

When you have completed Chapter 8, you will be able to:

- Distinguish between different types of oral medications.
- Recognize the types of solid oral medications that may not be altered by crushing or opening them.
- List appropriate techniques for administering medication to patients with difficulty swallowing or with feeding tubes.
- Accurately calculate the amount of solid or liquid oral medication to administer to a patient.
- Identify common errors that occur in calculating and preparing dosages for oral administration or via a feeding tube.

Key Terms

Caplet
Capsule
Enteric-coated
Gelcap
Scored
Spansules
Sustained release
Tablet

Introduction

So far you have learned the fraction proportion, ratio proportion, formula, and dimensional analysis methods for simple calculations. In this chapter you will apply these methods to oral dosages including solids and liquids. By now you may have chosen one method with which you are most comfortable. If so, follow that method throughout this chapter, using the corresponding color coding in the examples given; then complete the practice problems, using your method of choice. While you are practicing these problems, remember that excellence is a *must* with dosage calculations.

Tables and Capsules

Solid oral medications come in several forms, including tablets, caplets, capsules, and gelcaps (see Figure 8-1).

The **tablet** is the most common form of solid oral medication. It combines an amount of drug with inactive ingredients such as starch or talc to form a solid disk or cylinder that is convenient for swallowing. Certain tablets are specially designed to be administered sublingually (under the tongue) or buccally (between the cheek and gum). Sublingual and buccal tablets release medication into an area rich in blood vessels, where it can be quickly absorbed for rapid action. Some tablets are designed to be chewed; others dissolve in water to make a liquid that the patient can drink. Always check the drug label to determine how a tablet is meant to be administered.

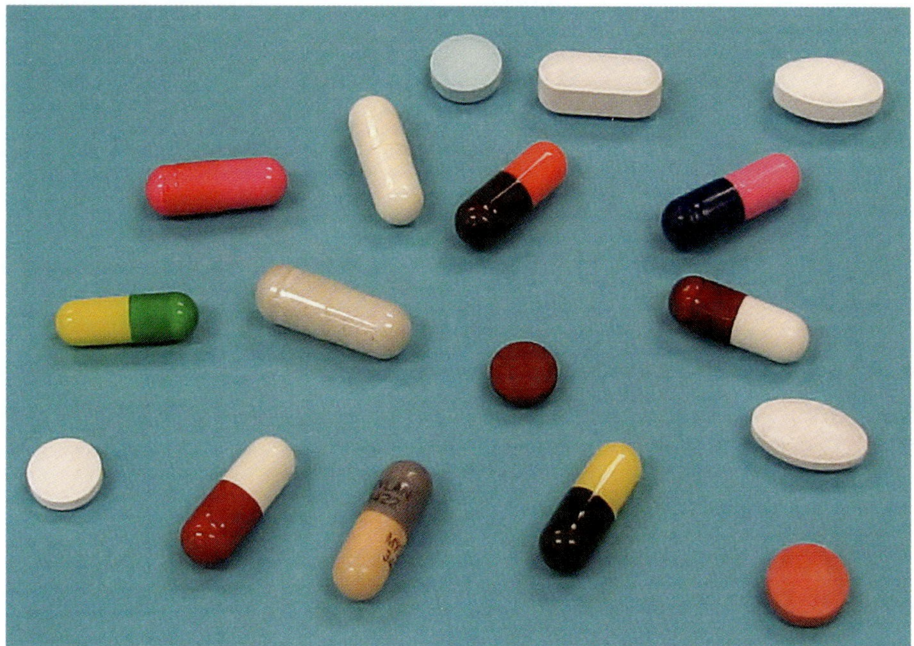

Figure 8-1 Solid oral medications including tablets, caplets, capsules, and gelcaps.

Caplets are similar to tablets. Oval-shaped, they have a special coating that makes them easy to swallow. **Capsules** are usually oval-shaped gelatin shells that contain medication in powder or granule form. The gelatin shell usually has two pieces that fit together. These pieces may sometimes be separated to remove the medication when the patient cannot swallow a pill. **Gelcaps** consist of medication, usually liquid, in gelatin shells that are not designed to be opened.

Calculating Dosages for Tablets and Capsules

Tablets are sometimes **scored** so that they can be divided when smaller doses are ordered. Most often, scored tablets divide into halves, but some are scored to divide into thirds or quarters (see Figure 8-2). The medication in scored tablets is evenly distributed throughout the tablet, allowing the dose to be divided evenly when the tablet is broken. Tablets may be broken into parts *only if they are scored,* and they must be broken only along the line of the scoring. Unscored tablets **must not** be broken into parts. Remember, breaking scored tablets to administer an ordered dosage is permitted but not optimal. Determine if another dosage strength is available before you break scored tablets.

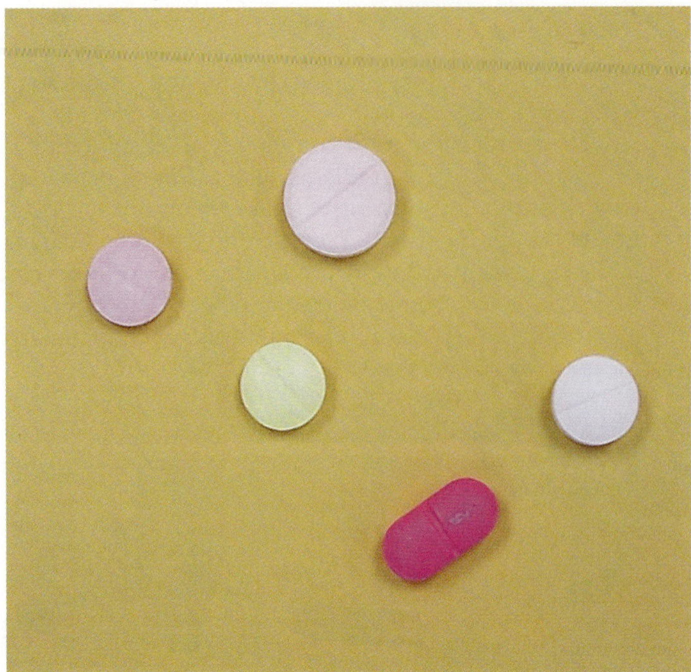

Figure 8-2 Scored tablets.

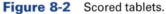

Rule 8-1 Always question and/or verify when your calculation indicates to give a portion of a tablet when the tablet is not scored.

Example 1 Do not administer $\frac{1}{2}$ of an unscored tablet.

Example 2 Do not administer $\frac{1}{3}$ or $\frac{1}{4}$ of a tablet scored for division in two.

GO TO . . . Open the CD-ROM that accompanies your textbook, and select Chapter 8, Rule 8-1. Review the animation and example problems, then complete the practice problems. Continue to the next section of the book once you have mastered the rule presented. ■

Rule 8-2 Question and recheck any calculation that indicates that you should administer more than 3 tablets or capsules.

If the situations in Rules 8-1 and 8-2 arise, recheck your calculations. If you are confident that your calculations are correct, then check with the physician or pharmacist, or your drug reference book, to be sure there is no error in the order or the dosage strength you are using.

GO TO . . . Open the CD-ROM that accompanies your textbook, and select Chapter 8, Rule 8-2. Review the animation and example problems, then complete the practice problems. Continue to the next section of the book once you have mastered the rule presented. ■

To calculate the *amount to administer,* you must know the *desired dose,* the *dose on hand,* and the *dosage unit.* The desired dose and the dose on hand must be in the same unit of measurement. Generally, the dosage unit for solid oral medications will be 1, such as 1 tablet, 1 caplet, 1 capsule, or 1 gelcap.

Rule 8-3 Follow these steps when you are determining the amount of oral medication to administer to a patient.

1. If necessary, convert the *dosage ordered O* to the *desired dose D* that has the same unit of measure as the *dose on hand H.*
2. Calculate the *amount to administer* by the method of your choice:
   - **a.** The fraction proportion method (using Procedure Checklist 7-1)
   - **b.** The ratio proportion method (using Procedure Checklist 7-2)
   - **c.** Dimensional analysis (using Procedure Checklist 7-3)
   - **d.** The formula method (using Procedure Checklist 7-4)
3. Apply critical thinking skills to determine whether the amount you have calculated is reasonable, using Rules 8-1 and 8-2. Recheck your calculation, if necessary.

Fraction Proportion Method

Example 1 The order is to give the patient 15 mg codeine PO now. You have 30 mg scored tablets available.

Solution The dosage ordered *O* is 15 mg. The dose on hand *H* is 30 mg, and the dosage unit *Q* is 1 tablet. Because the dosage ordered and the dose on hand have the same units, no conversion is needed to find the desired dose *D*, which is 15 mg.

Follow Procedure Checklist 7-1.

1. Fill in the proportion.

$$\frac{Q}{H} = \frac{A}{D} \quad \text{or} \quad \frac{\text{dosage unit}}{\text{dose on hand}} = \frac{\text{amount to administer}}{\text{desired dose}}$$

$$\frac{1 \text{ tablet}}{30 \text{ mg}} = \frac{A}{15 \text{ mg}}$$

2. Cancel units.

$$\frac{1 \text{ tablet}}{30 \cancel{\text{mg}}} = \frac{A}{15 \cancel{\text{mg}}}$$

3. Cross-multiply and solve for the unknown.

$$30 \times A = 1 \text{ tablet} \times 15$$

$$A = 1 \text{ tablet} \times \frac{15}{30}$$

$$A = 0.5 \text{ tablet}$$

Think critically about the result. Because 15 mg is one-half of 30 mg, $\frac{1}{2}$ tablet is an appropriate answer since the tablets are scored.

Example 2 The order is Inderal 80 mg PO qid. You have 40 mg tablets available.

Solution The dosage ordered *O* is 80 mg. The dose on hand *H* is 40 mg, and the dosage unit *Q* is 1 tablet. Because the dosage ordered and the dose on hand have the same units, no conversion is needed to find the desired dose *D*, which is 80 mg.

Follow Procedure Checklist 7-1.

1. Fill in the proportion.

$$\frac{1 \text{ tablet}}{40 \text{ mg}} = \frac{A}{80 \text{ mg}}$$

2. Cancel units.

$$\frac{1 \text{ tablet}}{40 \cancel{\text{mg}}} = \frac{A}{80 \cancel{\text{mg}}}$$

3. Cross-multiply and solve for the unknown.

$$40 \times A = 1 \text{ tablet} \times 80$$

$$A = 1 \text{ tablet} \times \frac{80}{40}$$

$$A = 2 \text{ tablets}$$

Thinking critically about this result, you realize 80 is twice 40 so this dose requires twice 1 tablet. The calculated dosage does not call for more than 3 tablets, so this answer seems reasonable.

Example 3 Refer to Figures 8-3 and 8-4.

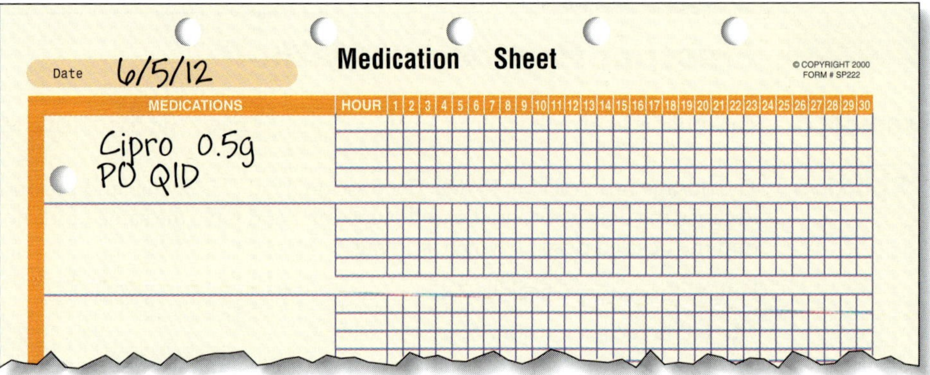

Figure 8-3 Medication ordered.

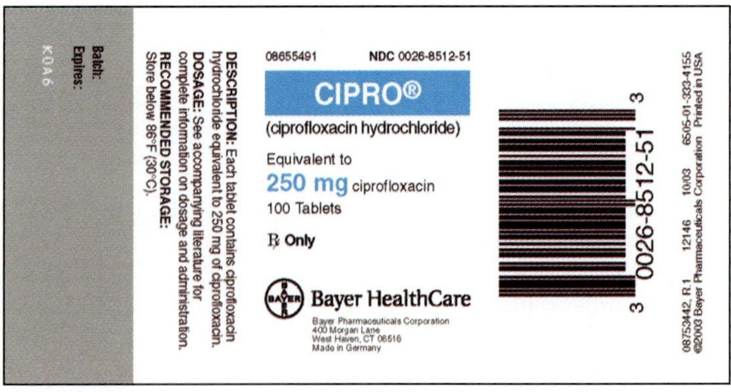

Figure 8-4 Medication on hand.

Solution The dosage ordered O is 0.5 g. The dose on hand H is 250 mg, and the dosage unit Q is 1 tablet. Because the dosage ordered and the dose on hand are in different units, you need to convert the dosage ordered to milligrams.

Follow Procedure Checklist 3-1.

1. Fill in the proportion, recalling that 1 g = 1000 mg.

$$\frac{1000 \text{ mg}}{1 \text{ g}} = \frac{D}{0.5 \text{ g}}$$

2. Cancel units.

$$\frac{1000 \text{ mg}}{1 \cancel{\text{g}}} = \frac{D}{0.5 \cancel{\text{g}}}$$

3. Cross-multiply and solve for the unknown.

$$1 \times D = 0.5 \times 1000 \text{ mg}$$
$$D = 500 \text{ mg}$$

Follow Procedure Checklist 7-1.

1. Fill in the proportion.

$$\frac{1 \text{ tablet}}{250 \text{ mg}} = \frac{A}{500 \text{ mg}}$$

2. Cancel units.

$$\frac{1 \text{ tablet}}{250 \cancel{\text{mg}}} = \frac{A}{500 \cancel{\text{mg}}}$$

3. Cross-multiply and solve for the unknown.

$$250 \times A = 1 \text{ tablet} \times 500$$
$$A = 1 \text{ tablet} \times \frac{500}{250}$$
$$A = 2 \text{ tablets}$$

Thinking critically, you recognize that 500 mg is larger than 250 mg. It is logical to give more than 1 tablet. The calculation does not call for more than 3 tablets. The answer of 2 tablets is reasonable.

Ratio Proportion Method

Example 1 The order is to give the patient 15 mg codeine PO now. You have 30 mg scored tablets available.

Solution The dosage ordered O is 15 mg. The dose on hand H is 30 mg, and the dosage unit Q is 1 tablet. Because the dosage ordered and the dose on hand already have the same units, no conversion is needed to find the desired dose D, which is 15 mg.

Follow Procedure Checklist 7-2.

1. Fill in the proportion.

$Q : H :: A : D$

1 tablet : 30 mg :: A : 15 mg

2. Cancel units.

   1 tablet : 30 ~~mg~~ :: A : 15 ~~mg~~

3. Multiply means and extremes, then solve for the missing value.

   $30 \times A = 1 \text{ tablet} \times 15$

   $A = 1 \text{ tablet} \times \frac{15}{30}$

   $A = 0.5 \text{ tablet}$

Think critically about the result. Because 15 mg is one-half of 30 mg, $\frac{1}{2}$ tablet is an appropriate answer since the tablets are scored.

Example 2 The order is Inderal® 80 mg PO qid. You have 40 mg tablets available.

Solution The dosage ordered *O* is 80 mg. The dose on hand *H* is 40 mg, and the dosage unit *Q* is 1 tablet. Because the dosage ordered and the dose on hand already have the same units, no conversion is needed to find the desired dose *D*, which is 80 mg.

Follow Procedure Checklist 7-2.

1. Fill in the proportion.

   1 tablet : 40 mg :: *A* : 80 mg

4. Cancel units.

   1 tablet : 40 ~~mg~~ :: *A* : 80 ~~mg~~

5. Cross-multiply the means and extremes, then solve for the missing value.

   $40 \times A = 1 \text{ tablet} \times 80$

   $A = 1 \text{ tablet} \times \frac{80}{40}$

   $A = 2 \text{ tablets}$

Thinking critically about this result, you realize 80 is 40 × 2, so this dose requires twice 1 tablet. The calculated dosage does not call for more than 3 tablets.

Example 3 Refer to Figures 8-3 and 8-4 on page 256.

Solution The dosage ordered *O* is 0.5 g. The dose on hand *H* is 250 mg, and the dosage unit *Q* is 1 tablet. Because the dosage ordered and the dose on hand are in different units, you need to convert the dosage ordered to milligrams since the dose on hand has a dosage strength of 250 mg/1 tablet.

Follow Procedure Checklist 3-2.

1. Fill in the proportion, recalling that 1 g = 1000 mg.

   1000 mg : 1 g :: *D* : 0.5 g

2. Cancel units.

   1000 mg : 1 ~~g~~ :: *D* : 0.5 ~~g~~

3. Cross-multiply the means and extremes, then solve for the missing value.

   $1 \times D = 1000 \text{ mg} \times 0.5$

   $D = 1000 \text{ mg} \times \frac{0.5}{1}$

   $D = 500 \text{ mg}$

Follow Procedure Checklist 7-2.

1. Fill in the proportion.

   1 tablet : 250 mg :: *A* : 500 mg

2. Cancel units.

   1 tablet : 250 ~~mg~~ :: A : 500 ~~mg~~

3. Cross-multiply the means and extremes, then solve for the missing value.

   $$250 \times A = 1 \text{ tablet} \times 500$$

   $$A = 1 \text{ tablet} \times \frac{500}{250}$$

   $$A = 2 \text{ tablets}$$

Thinking critically, you recognize that 500 mg is larger than 250 mg. It is logical to give more than 1 tablet. The calculation does not call for more than 3 tablets. The answer of 2 tablets is reasonable.

Dimensional Analysis

Example 1 The order is to give the patient 15 mg codeine PO now. You have 30 mg scored tablets available.

Solution Follow Procedure Checklist 7-3.

1. The unit of measure for the amount to administer will be tablets.

   A tablets =

2. Since the unit of measure for the dosage ordered is the same as that for the dose on hand, this step is unnecessary.

3. The dosage unit is 1 tablet. The dosage on hand is 30 mg.

   $$\frac{1 \text{ tablet}}{30 \text{ mg}}$$

4. The dosage ordered is 15 mg.

   $$A \text{ tablets} = \frac{1 \text{ tablet}}{30 \text{ mg}} \times \frac{15 \text{ mg}}{1}$$

5. Cancel units.

   $$A \text{ tablets} = \frac{1 \text{ tablet}}{30 \cancel{\text{mg}}} \times \frac{15 \cancel{\text{mg}}}{1}$$

6. Solve the equation.

   $$A \text{ tablets} = \frac{1 \text{ tablet}}{30} \times \frac{15}{1}$$

   $$A = 0.5 \text{ tablet} = \frac{1}{2} \text{ tablet}$$

Think critically about the result. Because 15 mg is one-half of 30 mg, $\frac{1}{2}$ tablet is an appropriate answer since the tablets are scored.

Example 2 The order is Inderal® 80 mg PO qid. You have 40 mg tablets available.

Solution Follow Procedure Checklist 7-3.

1. The unit of measure for the amount to administer will be tablets.

   A tablets =

2. Since the unit of measurement for the dosage ordered is the same as that for the dose on hand, this step is unnecessary.

3. The dosage unit is 1 tablet. The dose on hand is 40 mg.

$$\frac{1 \text{ tablet}}{40 \text{ mg}}$$

4. The dosage ordered is 80 mg.

$$A \text{ tablets} = \frac{1 \text{ tablet}}{40 \text{ mg}} \times \frac{80 \text{ mg}}{1}$$

5. Cancel units.

$$A \text{ tablets} = \frac{1 \text{ tablet}}{40 \cancel{\text{mg}}} \times \frac{80 \cancel{\text{mg}}}{1}$$

6. Solve the equation.

$$A \text{ tablets} = \frac{1 \text{ tablet}}{40 \text{ mg}} \times \frac{80}{1}$$

$$A = 2 \text{ tablets}$$

Thinking critically about this result, you realize 80 is 40 × 2, so this dose requires twice 1 tablet. The calculated dosage does not call for more than 3 tablets.

Example 3 Refer to Figures 8-3 and 8-4 on page 256.

Solution Follow Procedure Checklist 7-3.

1. The unit of measure for the amount to administer will be tablets.

   A tablets =

2. The unit of measure for the dosage ordered is grams. The unit of measure for the dose on hand is milligrams. Recall that 1 g = 1000 mg. Because we wish to convert to milligrams (the units of the dose on hand), our first factor must have milligrams on top.

$$\frac{1000 \text{ mg}}{1 \text{ g}}$$

3. The dosage unit is 1 tablet. The dose on hand is 250 mg. This is the second factor.

$$\frac{1000 \text{ mg}}{1 \text{ g}} \times \frac{1 \text{ tablet}}{250 \text{ mg}}$$

4. The dosage ordered is 0.5 g. Place this over 1 and set up the equation.

$$A \text{ tablets} = \frac{1000 \text{ mg}}{1 \text{ g}} \times \frac{1 \text{ tablet}}{250 \text{ mg}} \times \frac{0.5 \text{ g}}{1}$$

5. Cancel units.

$$A \text{ tablets} = \frac{1000 \cancel{\text{mg}}}{1 \cancel{\text{g}}} \times \frac{1 \text{ tablet}}{250 \cancel{\text{mg}}} \times \frac{0.5 \cancel{\text{g}}}{1}$$

6. Solve the equation.

$$A \text{ tablets} = \frac{1000}{1} \times \frac{1 \text{ tablet}}{250} \times \frac{0.5}{1}$$

$$A = 2 \text{ tablets}$$

Thinking critically, you recognize that 500 mg is larger than 250 mg. It is logical to give more than 1 tablet. The calculation does not call for more than 3 tablets. The answer of 2 tablets is reasonable.

Formula Method

Example 1 The order is to give the patient 15 mg codeine PO now. You have 30 mg scored tablets available.

Solution Follow Procedure Checklist 7-4.

1. The drug is ordered in milligrams, which is the same unit of measure as that for the dose on hand. Therefore,

   $D = 15$ mg

   $Q = 1$ tablet

   $H = 30$ mg

2. Fill in the formula.

   $$\frac{D \times Q}{H} = A \quad \text{or} \quad \frac{\text{desired dose}}{\text{dose on hand}} \times \text{dosage unit} = \text{amount to administer}$$

   $$\frac{15 \text{ mg} \times 1 \text{ tablet}}{30 \text{ mg}} = A$$

3. Cancel units.

   $$\frac{15 \cancel{\text{mg}} \times 1 \text{ tablet}}{30 \cancel{\text{mg}}} = A$$

4. Solve for the unknown.

   $$\frac{15 \times 1 \text{ tablet}}{30} = A$$

   $A = 0.5$ tablet $= \frac{1}{2}$ tablet

Think critically about the result. Because 15 mg is one-half of 30 mg, $\frac{1}{2}$ tablet is an appropriate answer since the tablets are scored.

Example 2 The order is Inderal® 80 mg PO qid. You have 40 mg tablets available.

Solution Follow Procedure Checklist 7-4.

1. The drug is ordered in milligrams, which is the same unit of measure as the dose on hand. Therefore,

   $D = 80$ mg

   $Q = 1$ tablet

   $H = 40$ mg

2. Fill in the formula.

   $$\frac{80 \text{ mg} \times 1 \text{ tablet}}{40 \text{ mg}} = A$$

3. Cancel units.

   $$\frac{80 \cancel{\text{mg}} \times 1 \text{ tablet}}{40 \cancel{\text{mg}}} = A$$

4. Solve for the unknown.

   $$\frac{80 \times 1 \text{ tablet}}{40} = A$$

   $A = 2$ tablets

Thinking critically about this result, you realize 80 is 40×2, so this dose requires twice 1 tablet. The calculated dosage does not call for more than 3 tablets.

Example 3 Refer to Figures 8-3 and 8-4 on page 256.

Solution Follow Procedure Checklist 7-4.

1. In this case the order is written in grams, and the drug is labeled in milligrams. Before we can determine the amount to administer, we must calculate a desired dose that is in milligrams. We will do this by using Procedure Checklist 3-1, the fraction proportion method.

   Recall that 1 g = 1000 mg. Since we are converting to milligrams, our conversion factor is $\frac{1000 \text{ mg}}{1 \text{ g}}$. Set up the proportion.

   $$\frac{D}{0.5 \text{ g}} = \frac{1000 \text{ mg}}{1 \text{ g}}$$

   Cancel units.

   $$\frac{D}{0.5 \not{\text{g}}} = \frac{1000 \text{ mg}}{1 \not{\text{g}}}$$

   Solve for the unknown.

   $$\frac{D}{0.5} = \frac{1000 \text{ mg}}{1}$$

   $D = 500$ mg

   We now have the three necessary pieces of information to complete the formula.

   $D = 500$ mg

   $Q = 1$ tablet

   $H = 250$ mg

2. Fill in the formula.

   $$\frac{500 \text{ mg} \times 1 \text{ tablet}}{250 \text{ mg}} = A$$

3. Cancel units.

   $$\frac{500 \cancel{\text{mg}} \times 1 \text{ tablet}}{250 \cancel{\text{mg}}} = A$$

4. Solve for the unknown.

   $$\frac{500 \times 1 \text{ tablet}}{250} = A$$

   $$2 \text{ tablets} = A$$

Thinking critically, you recognize that 500 mg is larger than 250 mg. It is logical to give more than 1 tablet. The calculation does not call for more than 3 tablets. The answer of 2 tablets is reasonable.

GO TO . . . Open the CD-ROM that accompanies your textbook, and select Chapter 8, Rule 8-3. Review the animation and example problems, then complete the practice problems. Continue to the next section of the book once you have mastered the rule presented. ■

Critical Thinking on the Job

Calculating Factors Incorrectly

A health care professional caring for a patient with chest pain is instructed to give gr $\frac{1}{200}$ of nitroglycerin PO stat. She takes the first bottle of nitroglycerin available; it contains gr $\frac{1}{100}$ tablets. (See Figure 8-5). She calculates that gr $\frac{1}{200}$ is two gr $\frac{1}{100}$ tablets.

Learning Link Recall from Chapter 1, Rule 1-15 Dividing Fractions, found on page 27.

Questions?

What is wrong with the health care professional's calculation? Using critical thinking how should the health care professional correct her mistake?

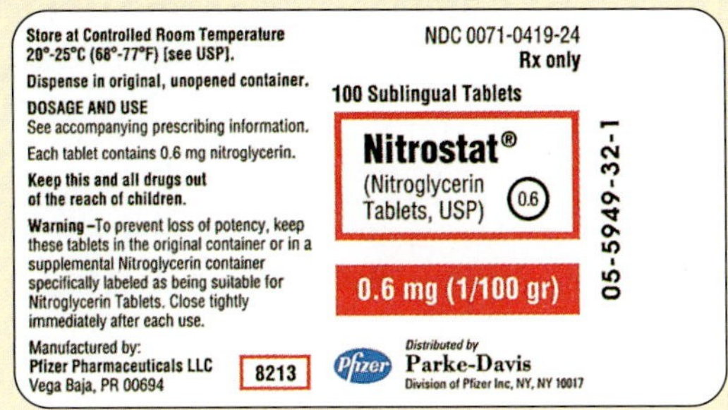
Store at Controlled Room Temperature
20°-25°C (68°-77°F) [see USP].
Dispense in original, unopened container.
DOSAGE AND USE
See accompanying prescribing information.
Each tablet contains 0.6 mg nitroglycerin.
Keep this and all drugs out of the reach of children.
Warning – To prevent loss of potency, keep these tablets in the original container or in a supplemental Nitroglycerin container specifically labeled as being suitable for Nitroglycerin Tablets. Close tightly immediately after each use.
Manufactured by:
Pfizer Pharmaceuticals LLC
Vega Baja, PR 00694
8213
NDC 0071-0419-24
Rx only
100 Sublingual Tablets
Nitrostat®
(Nitroglycerin Tablets, USP)
0.6
0.6 mg (1/100 gr)
05-5949-32-1
Pfizer
Distributed by
Parke-Davis
Division of Pfizer Inc, NY, NY 10017

Figure 8-5

Crushing Tablets or Opening Capsules

For patients who have difficulty swallowing pills, you may crush certain tablets and open certain capsules. However, in many settings such as nursing homes, you must first get a physician's order. Check with your facility about these policies before you crush tablets. See Figure 8-6.

Sometimes you can mix a crushed tablet or an opened capsule with soft food or liquid. First check for interactions between the medication and the food or fluid being mixed with it or other medications that are being administered (Table 8-1). For example, tetracycline is inactivated by milk. It must not be dissolved in foods that contain milk. In addition, it should not be given with either antacids or vitamin and mineral supplements.

Oral forms of medication are also ordered for patients with nasogastric, gastrostomy, or jejunostomy tubes. Before administering medication through the tube, you first dissolve the contents from a crushed tablet or opened capsule in a small amount of warm water.

Some medications cannot be crushed. If these medications are ordered for a patient with a feeding tube or one who cannot swallow pills, determine whether an alternative form of the medication exists. Consult a drug reference or pharmacist for information, then ask the physician if the medication could be ordered in one of these forms. Always follow the policy of the facility where you are employed regarding substituting forms of medications.

Figure 8-6 A pill crusher or mortar and pestle may be used to crush tablets when necessary.

TABLE 8-1 Examples of Food and Drug Interactions

Drug	Food	Interaction
Antipsychotics	Coffee and tea	Reduced effectiveness of drug
Bronchodilators	Caffeine	Stimulation of the nervous system
Central nervous system (CNS) depressants	Black cohosh, ginseng kava kava, St. John's Wort, valerian, ETOH	Intensified sedative effects of CNS depressant
Erythromycin	Acidic fruits or juices, carbonated beverages	Decreased antimicrobial activity
Ferrous sulfate	Tea	Decreased absorption
Haloperidol	Coffee and tea	Decreased absorption
Insulin	Coffee	Stimulated excretion
Monoamine oxidase inhibitors	Foods containing tyramine, such as hard cheeses, chocolate, red wine, and beef or chicken liver	Headache, nosebleed, chest pain, severe hypertension
Tetracyclines	Dairy products	Reduced effectiveness of drug
Antihistamines, cholesterol lowering agents, calcium channel blockers	Grapefruit and grapefruit juice	Muscle aches, fatique, fever, increase side effects

Enteric-coated tablets have a coating that dissolves only in an alkaline environment, such as the small intestine. These tablets deliver medication that would be destroyed by stomach acid or that could injure the stomach lining. Enteric-coated tablets often look like candies that have a soft center and a hard shell. Some aspirins are enteric-coated, as are certain iron tablets such as ferrous gluconate. Enteric-coated tablets must *never* be crushed, broken, or chewed. A patient must swallow them with their coating intact (see Figure 8-7).

Figure 8-7 Enteric-coated tablet. One tablet has been split for visualization only. Enteric-coated tablets should never be split when given to patients.

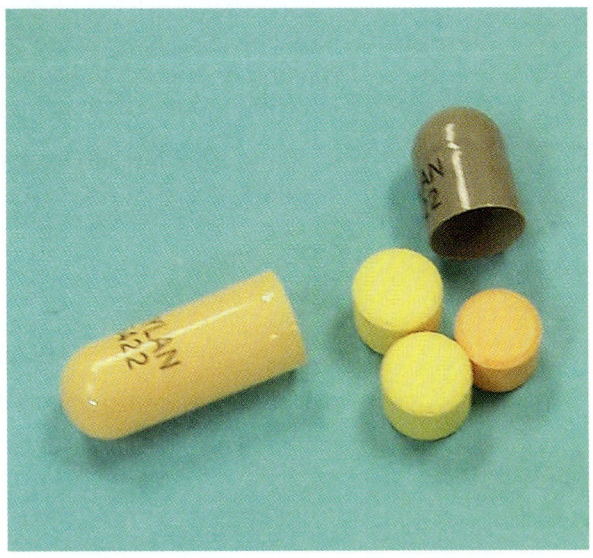

Figure 8-8 This sustained-release capsule has three smaller tablets inside that release medication into the bloodstream over a period of several hours.

Some medications are available in **sustained-release** forms. They allow the drug to be released slowly into the bloodstream over a period of several hours. If the medication is scored, you may break it at the scored line. Otherwise you must not break it. Crushing or dissolving sustained-release tablets would allow more than the intended amount of medication to be absorbed at one time, causing overdose or toxicity of the drug.

Special capsules, often called **spansules,** contain granules of medication with different coatings that delay release of some of the medication. You may open spansules and gently mix the granules in soft food, but you must not crush or dissolve the granules. (See Figure 8-8.)

Rule 8-4 To prevent an incorrect dose of medication do not crush or otherwise alter any of the following:

- Enteric-coated tablets.
- Sustained-release forms of medication.
- Any tablet with a hard shell or coating.
- Any tablet with layers or speckles of different colors.
- Tablets for sublingual or buccal use.
- Capsules with seals that prevent separating the two parts.

Example The following lists indicate some medications that should not be crushed or altered. Crushing or altering these medications could cause an inaccurate dose of the medication to be administered.

- Names that indicate sustained-release such as:

-Bid	LA	Tempule
-Dur	CR	Chronotab
Plateau cap	XL	Repetab
Span	Sequel	Tembid
SA	Spansule	
SR	Extentab	

- Names that indicate enteric-coated such as:
  EC
  Enseal

GO TO . . . Open the CD-ROM that accompanies your textbook, and select Chapter 8, Rule 8-4. Review the animation and example problems, then complete the practice problems. Continue to the next section of the book once you have mastered the rule presented. ■

Patient Education

Review with patients the following guidelines for taking tablets and capsules:

1. Perform all necessary calculations, so that you can tell patients how many pills to take.
2. Tell patients whether they need to take a medication with food. Encourage them to drink at least 8 oz of water with any medication.
3. Tell patients who need to divide tablets that pharmacists can provide this service on request. If the patients will be dividing the tablets, demonstrate and advise them as follows:
   a. Wash hands before you handle tablets.
   b. Grasp the tablet with the scored line between your fingers. Exert pressure in the same direction—downward or upward—with both hands, until the tablet breaks along the scored line.
   c. You may use a knife or pill cutter to break the tablet. Place the tablet on a clean surface, place the blade in the scored line, and press directly downward until the tablet breaks.
4. For patients who have difficulty swallowing, offer the following suggestions:
   a. Drink water before taking pills, so your mouth is moist.
   b. Place whole tablets or capsules in a small amount of food, such as applesauce or pudding. The pill will go down when the food is swallowed. *Note:* Also tell patients which foods should **not** be used.
   c. Crush tablets by placing them on a spoon and pressing another spoon down on top of them or use a pill crusher. **Note:** Warn patients not to crush any medication without first checking with the pharmacist or physician.

REVIEW AND PRACTICE

8-1 Tablets and Capsules

In Exercises 1–20, calculate the amount to administer. Unless otherwise noted, all scored tablets are scored in half.

1. Ordered: Tegretol® 400 mg PO bid
   On hand: Tegretol® 200 mg unscored tablets
   Administer: _____

2. Ordered: Seroquel® 75 mg PO tid
   On hand: Seroquel® 25 mg unscored tablets
   Administer: _____

3. Ordered: Tolectin® 300 mg PO tid
   On hand: Tolectin® 200 mg scored tablets
   Administer: _____

4. Ordered: Isordil® Titradose 15 mg PO now
   On hand: Isordil® Titradose 10 mg deep-scored tablets
   Administer: _____

5. Ordered: Felbatol® 600 mg PO qid
   On hand: Felbatol® 400 mg scored tablets
   Administer: _____

6. Ordered: Decadron® 1.5 mg PO daily
   On hand: Decadron® 0.75 mg unscored tablets
   Administer: _____

7. Ordered: Coumadin® 5 mg PO daily
   On hand: Coumadin® 2 mg scored tablets
   Administer: _____

8. Ordered: Cardizem® 90 mg PO tid
   On hand: Cardizem® 60 mg scored tablets
   Administer: _____

9. Ordered: Tambocor® 150 mg PO q12h
   On hand: Tambocor® 100 mg scored tablets
   Administer: _____

10. Ordered: Clozaril® 50 mg PO daily
    On hand: Refer to label A. Tablets are unscored.
    Administer: _____

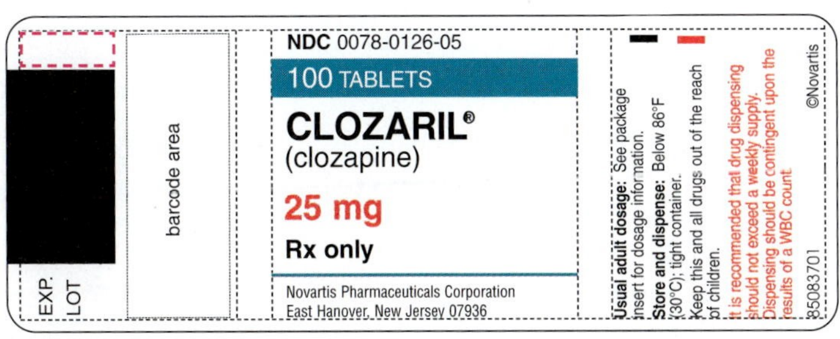

EXP.
LOT
barcode area
NDC 0078-0126-05
100 TABLETS
CLOZARIL®
(clozapine)
25 mg
Rx only
Novartis Pharmaceuticals Corporation
East Hanover, New Jersey 07936
Usual adult dosage: See package insert for dosage information.
Store and dispense: Below 86°F (30°C); tight container.
Keep this and all drugs out of the reach of children.
It is recommended that drug dispensing should not exceed a weekly supply. Dispensing should be contingent upon the results of a WBC count.
85083701
©Novartis

A

11. Ordered: Alprazolam 0.5 mg PO tid Administer: _____
On hand. Refer to label B. Tablets are scored into fourths.

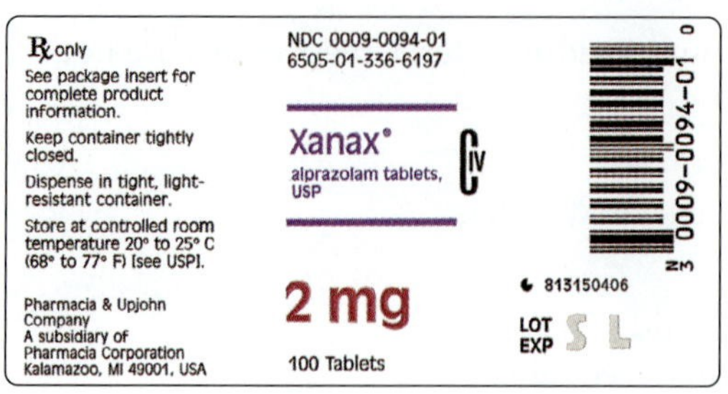

B

12. Ordered: Valium® 10 mg PO q12h Administer: _____
On hand: Refer to label C.

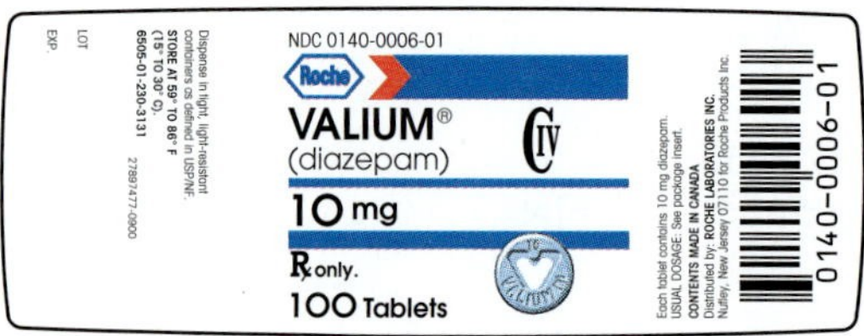

C

13. Ordered: Famvir® 250 mg PO bid Administer: _____
On hand: Refer to label D. Tablets are unscored.

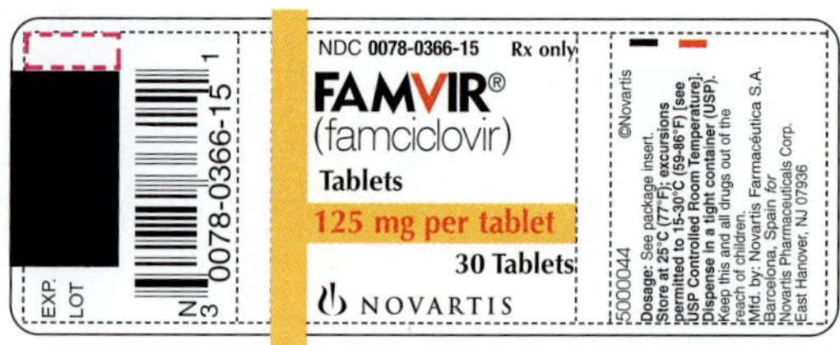

D

14. Ordered: Aricept® 10 mg PO daily Administer: _____
On hand: Refer to label E. Tablets are unscored.

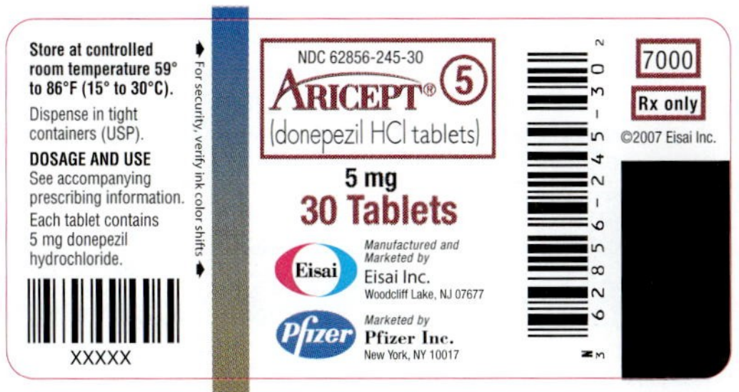

E

15. Ordered: SeroQuel® 150 mg PO bid Administer: _____
On hand: Refer to labels F. Tablets are scored. Dosage strength: _____
Determine dosage strength to use.

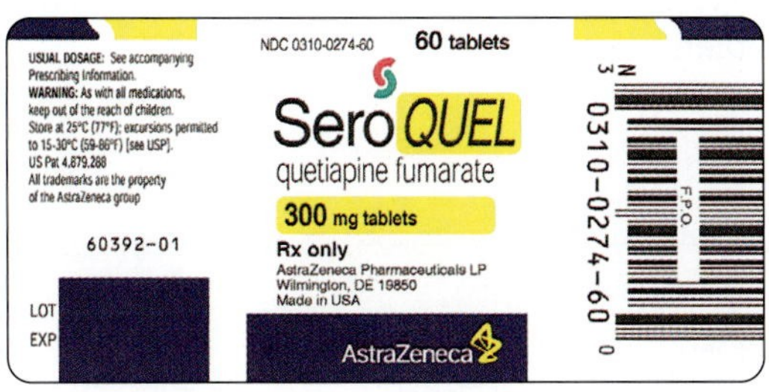

F

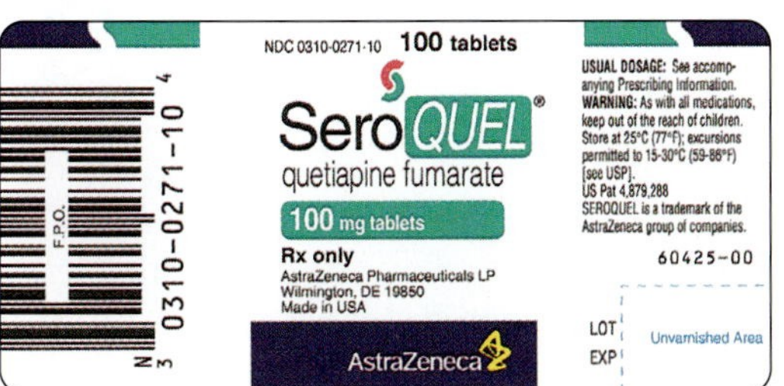

16. Ordered: Ploglitazone HCl 60 mg PO daily Administer: _____
On hand: Refer to label G. Tablets are unscored.

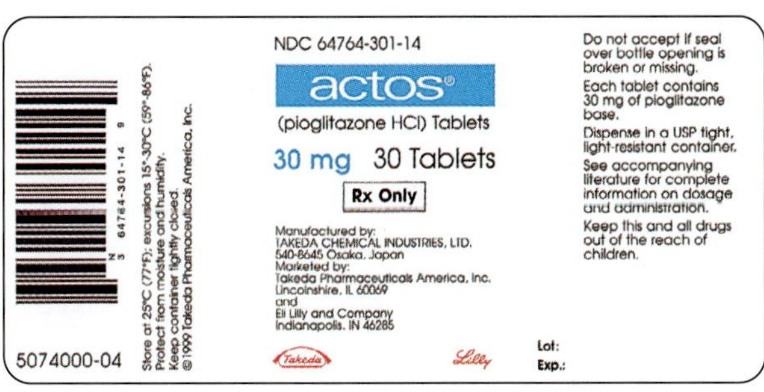

G

17. Ordered: Lipitor® 30 mg PO daily | Administer: _____
On hand: Refer to label H. Tablets are unscored

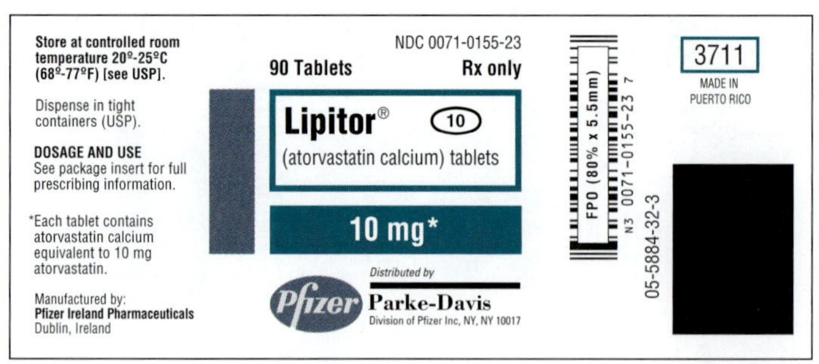

H

18. Ordered: Zoloft® 50 mg PO daily | Administer: _____
On hand: Refer to label I. Tablets are scored.

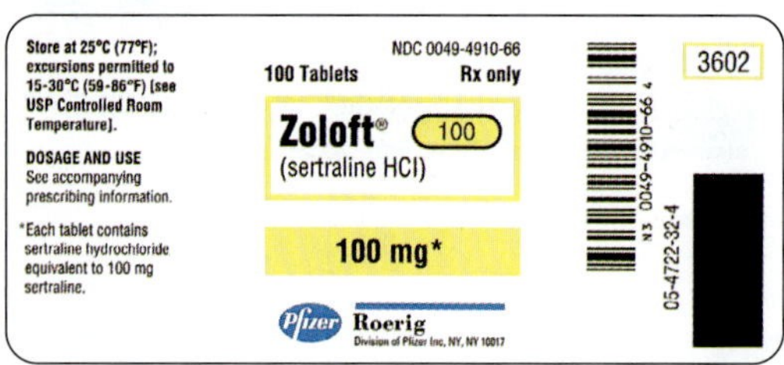

I

19. Ordered: Gleevec® 200 mg PO daily | Administer: _____
On hand: Refer to label J. Tablets are not scored.

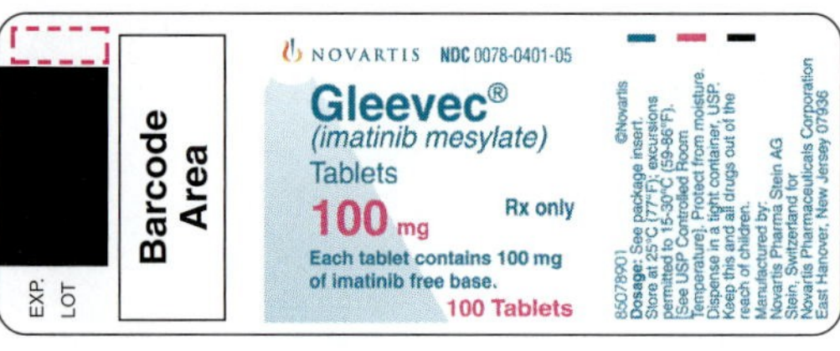

J

20. Ordered: Clonazepam 1 mg PO tid | Administer: _____
On hand: Clonazepam 2 mg scored tablets

To check your answers, see page 522.

Liquid Medications

Many medications are available in liquid form. Liquids can be measured in small units of volume; thus, a greater range of dosages can be ordered and administered. Because they are easier to swallow than tablets and capsules, they are often used for children and elderly patients. Liquids can also be administered easily through feeding tubes.

Liquids may be less stable than solid forms of drugs. Many medications that are intended to be administered as liquids are provided as powders which must be reconstituted. Many liquids, especially antibiotics, require refrigeration.

Rule 8-5 When reconstituting liquid medications:

- Use only the liquid specified on the drug label.
- Use the exact amount of liquid specified on the drug label.
- Check the label to determine whether the medication should be shaken before administering.
- Check the label to determine whether the reconstituted medication must be refrigerated.
- Write on the label the date and time you reconstitute the medication. Also, write your initials. Check the label to determine how long the reconstituted medication may be stored. Discard any medication left after this time period has passed.
- When medication can be reconstituted in different strengths, write on the label the strength that you choose.
- When medication can be reconstituted in different strengths, select the strength that will allow the desired dose to be administered in the smallest volume.
- Read the order carefully when you calculate the amount to administer. The physician usually orders the dose in units of drug, not volume of liquid. The person administering the medication calculates the volume needed to administer the desired dose.

Example 1 How would you reconstitute the following medication? Find the amount to administer.

Ordered: Kytril® 5 mg PO now

On hand: Refer to Figure 8-9.

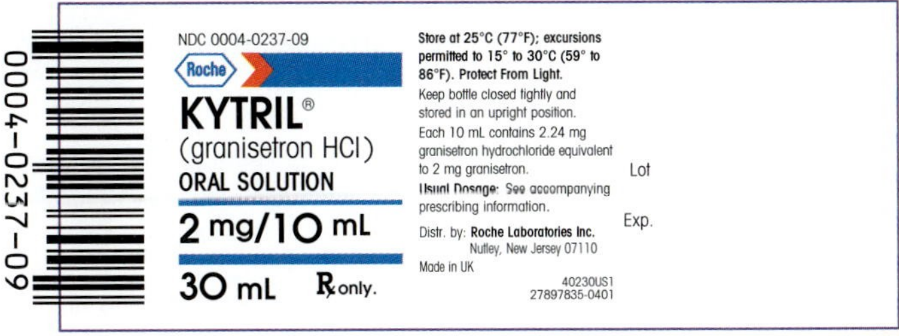

Figure 8-9

According to the label, this medication is already in liquid form as a solution, so no reconstitution is necessary.

Calculate the amount to administer. The dosage ordered is 5 mg. The dosage strength is 2 mg/10 mL, which makes the dose on hand 2 mg and the dosage unit 10 mL.

Fraction Proportion Method

Follow Procedure Checklist 7-1.

1. Fill in the proportion.

$$\frac{Q}{H} = \frac{A}{D} \quad \text{or} \quad \frac{\text{dosage unit}}{\text{dose on hand}} = \frac{\text{amount to administer}}{\text{desired dose}}$$

$$\frac{10 \text{ mL}}{2 \text{ mg}} = \frac{A}{5 \text{ mg}}$$

2. Cancel units.

$$\frac{10 \text{ mL}}{2 \cancel{\text{mg}}} = \frac{A}{5 \cancel{\text{mg}}}$$

3. Cross-multiply and solve for the unknown.

$$2 \times A = 10 \text{ mL} \times 5$$

$$A = \frac{10 \text{ mL} \times 5}{2}$$

$$A = \frac{50 \text{ mL}}{2}$$

$$A = 25 \text{ mL}$$

Ratio Proportion Method

Follow Procedure Checklist 7-2.

1. Fill in the proportion.

$Q : H :: A : D$ or dosage unit : dose on hand :: amount to administer : desired dose

10 mL : 2 mg :: A : 5 mg

2. Cancel units.

10 mL : 2 $\cancel{\text{mg}}$ :: A : 5 $\cancel{\text{mg}}$

3. Multiply the means and extremes, then solve for the missing value.

$$2 \times A = 10 \text{ mL} \times 5$$

$$A = \frac{10 \text{ mL} \times 5}{2}$$

$$A = \frac{50 \text{ mL}}{2}$$

$$A = 25 \text{ mL}$$

Dimensional Analysis

Follow Procedure Checklist 7-3.

1. The unit of measure for the amount to administer will be milliliters.

   $A \text{ mL} =$

2. Since the unit of measure for the dosage ordered is the same as the dose on hand, this step is unnecessary.
3. The dosage unit is 10 mL. The dose on hand is 2 mg. This is the first factor.

   $$\frac{10 \text{ mL}}{2 \text{ mg}}$$

4. The dosage ordered is 5 mg. Place this over 1 for the second factor.

   $$A \text{ mL} = \frac{10 \text{ mL}}{2 \text{ mg}} \times \frac{5 \text{ mg}}{1}$$

5. Cancel units.

   $$A \text{ mL} = \frac{10 \text{ mL}}{2 \cancel{\text{mg}}} \times \frac{5 \cancel{\text{mg}}}{1}$$

6. Solve the equation.

   $$A \text{ mL} = \frac{50 \text{ mL}}{2}$$

   $A = 25 \text{ mL}$

Formula Method

Follow Procedure Checklist 7-4.

1. The drug is ordered in milligrams, which is the same unit of measure as that of the dose on hand. Therefore,

   $D = 5 \text{ mg}$

   $Q = 10 \text{ mL}$

   $H = 2 \text{ mg}$

2. Fill in the formula.

   $$\frac{D \times Q}{H} = A$$

   $$\frac{5 \text{ mg} \times 10 \text{ mL}}{2 \text{ mg}} = A$$

3. Cancel units.

   $$\frac{5 \cancel{\text{mg}} \times 10 \text{ mL}}{2 \cancel{\text{mg}}} = A$$

4. Solve for the unknown.

   $$\frac{50 \text{ mL}}{2} = A$$

   $A = 25 \text{ mL}$

Example 2 How would you reconstitute the following medication? Find the amount to administer.

Ordered: E.E.S.® susp 400 mg PO q6h

On hand: Refer to Figure 8-10.

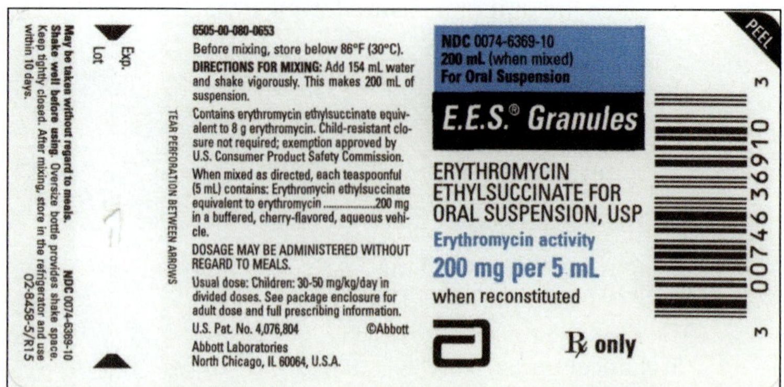

Figure 8-10

According to the label, you add 154 mL of water to the bottle of granules and shake vigorously. You then have a total of 200 mL of oral suspension that must be stored in the refrigerator for up to 10 days.

Calculate the amount to administer. The dosage ordered is 400 mg. The dosage strength is 200 mg/5 mL which makes the dose on hand 200 mg and the dosage unit 5 mL.

Fraction Proportion Method

Follow Procedure Checklist 7-1.

1. Fill in the proportion.

$$\frac{Q}{H} = \frac{A}{D} \quad \text{or} \quad \frac{\text{dosage unit}}{\text{dose on hand}} = \frac{\text{amount to administer}}{\text{desired dose}}$$

$$\frac{5 \text{ mL}}{200 \text{ mg}} = \frac{A}{400 \text{ mg}}$$

2. Cancel units

$$\frac{5 \text{ mL}}{200 \cancel{\text{mg}}} = \frac{A}{400 \cancel{\text{mg}}}$$

3. Cross-multiply and solve for the unknown.

$$200 \times A = 5 \text{ mL} \times 400$$

$$A = 5 \text{ mL} \times \frac{400}{200}$$

$$A = 5 \text{ mL} \times 2$$

$$A = 10 \text{ mL}$$

Ratio Proportion Method

Follow Procedure Checklist 7-2.

1. Fill in the proportion.

   $Q : H :: A : D$ or dosage unit : dose on hand :: amount to administer : desired dose

   5 mL : 200 ~~mg~~ :: A : 400 ~~mg~~

2. Cancel units.

   5 mL : 200 :: A : 400

3. Multiply the means and extremes, then solve for the missing value.

   $200 \times A = 5 \text{ mL} \times 400$

   $A = 5 \text{ mL} \times \frac{400}{200}$

   $A = 5 \text{ mL} \times 2$

   $A = 10 \text{ mL}$

Dimensional Analysis

Follow Procedure Checklist 7-3.

1. The unit of measure for the amount to administer will be milliliters.

   $A \text{ mL} =$

2. Since the unit of measure for the dosage ordered is the same as that for the dose on hand, this step is unnecessary.

3. The dosage unit is 5 mL. The dose on hand is 200 mg. This is your first factor.

   $\frac{5 \text{ mL}}{200 \text{ mg}}$

4. The dosage ordered is 400 mg. Place this over 1 for the second factor.

   $A \text{ mL} = \frac{5 \text{ mL}}{200 \text{ mg}} \times \frac{400 \text{ mg}}{1}$

5. Cancel units.

   $A \text{ mL} = \frac{5 \text{ mL}}{200 \cancel{\text{mg}}} \times \frac{400 \cancel{\text{mg}}}{1}$

6. Solve the equation.

   $A \text{ mL} = \frac{2000 \text{ mL}}{200}$

   $A = 10 \text{ mL}$

Formula Method

Follow Procedure Checklist 7-4.

1. The drug is ordered in milligrams, which is the same unit of measure as that for the dose on hand. Therefore,

   $D = 400 \text{ mg}$

   $Q = 5 \text{ mL}$

   $H = 200 \text{ mg}$

2. Fill in the formula.

   $$\frac{D \times Q}{H} = A$$

   $$\frac{400 \text{ mg} \times 5 \text{ mL}}{200 \text{ mg}} = A$$

3. Cancel units.

   $$\frac{400 \cancel{\text{mg}} \times 5 \text{ mL}}{200 \cancel{\text{mg}}} = A$$

4. Solve for the unknown.

   $$\frac{2000 \text{ mL}}{200} = A$$

   $A = 10 \text{ mL}$

Example 3 How would you reconstitute the following medication? Find the amount to administer.

Ordered: Omnicef® suspension 125 mg PO qid

On hand: Refer to Figure 8-11.

Tap the bottle to loosen the powder then add $\frac{1}{2}$ of the total amount of water (63 mL) for reconstitution.

$$\frac{1}{2} \times \frac{63}{1} = 31.5$$

rounded to 32 mL

Add 32 mL and shake.

Add the remaining water $63 - 32 = 31$ mL, then shake again.

When it is reconstituted, you will have total of 100 mL of suspension.

Calculate the amount to administer. The dosage ordered is 125 mg. The dosage strength is 250 mg/5 mL which makes the dose on hand 250 mg and the dosage unit 5 mL.

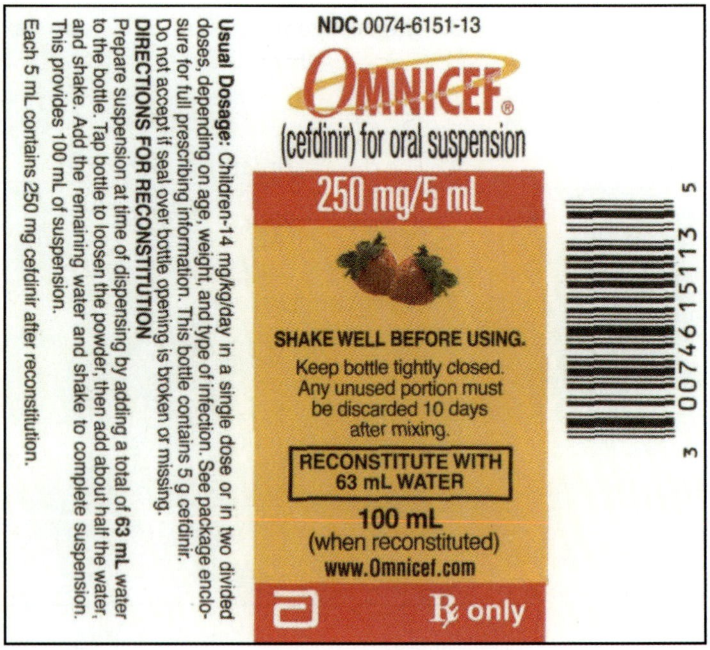

Figure 8-11

Fraction Proportion Method

Follow Procedure Checklist 7-1.

1. Fill in the proportion.

$$\frac{Q}{H} = \frac{A}{D} \quad \text{or} \quad \frac{\text{dosage unit}}{\text{dose on hand}} = \frac{\text{amount to administer}}{\text{desire dose}}$$

$$\frac{5\text{ mL}}{250\text{ mg}} = \frac{A}{125\text{ mg}}$$

2. Cancel units.

$$\frac{5\text{ mL}}{250\text{ }\cancel{\text{mg}}} = \frac{A}{125\text{ }\cancel{\text{mg}}}$$

3. Cross-multiply and solve for the unknown.

$$250 \times A = 5\text{ mL} \times 125$$

$$A = 5\text{ mL} \times \frac{125}{250}$$

$$A = 2.5\text{ mL}$$

Ratio Proportion Method

Follow Procedure Checklist 7-2.

1. Fill in the proportion.

$Q : H :: A : D$ or dosage unit : dose on hand :: amount to administer : desired dose

5 mL : 250 mg :: A : 125 mg

2. Cancel units.

5 mL : 250 $\cancel{\text{mg}}$ :: A : 125 $\cancel{\text{mg}}$

3. Multiply the means and extremes, then solve for the missing value.

$$250 \times A = 5\text{ mL} \times 125$$

$$A = 5\text{ mL} \times \frac{125}{250}$$

$$A = 2.5\text{ mL}$$

Dimensional Analysis

Follow Procedure Checklist 7-3.

1. The unit of measure for the amount to administer will be milliliters.

A mL =

2. Since the unit of measurement for the dosage ordered is the same as that for the dose on hand, this step is unnecessary.

3. The dosage unit is 5 mL. The dose on hand is 250 mg.

$$\frac{5\text{ mL}}{250\text{ mg}}$$

4. The dosage ordered is 500 mg.

$$A \text{ mL} = \frac{5 \text{ mL}}{250 \text{ mg}} \times \frac{125 \text{ mg}}{1}$$

5. Cancel units.

$$A \text{ mL} = \frac{5 \text{ mL}}{250 \cancel{\text{mg}}} \times \frac{125 \cancel{\text{mg}}}{1}$$

6. Solve the equation.

$$A \text{ mL} = \frac{625 \text{ mL}}{250}$$

$$A = 2.5 \text{ mL}$$

Formula Method

Follow Procedure Checklist 7-4.

1. The drug is ordered in milligrams, which is the same unit of measure as that for the dose on hand. Therefore,

   $D = 125$ mg

   $Q = 5$ mL

   $H = 250$ mg

2. Fill in the formula.

$$\frac{D \times Q}{H} = A$$

$$\frac{125 \text{ mg} \times 5 \text{ mL}}{250 \text{ mg}} = A$$

3. Cancel units.

$$\frac{125 \cancel{\text{mg}} \times 5}{250 \cancel{\text{mg}}} = A$$

4. Solve for the unknown.

$$\frac{625 \text{ mL}}{250} = A$$

$$A = 2.5 \text{ mL}$$

GO TO . . . Open the CD-ROM that accompanies your textbook, and select Chapter 8, Rule 8-5. Review the animation and example problems, then complete the practice problems. Continue to the next section of the book once you have mastered the rule presented. ■

Patient Education

Review with patients who are taking medications in a home environment the steps for reconstituting liquid medications. Follow Rule 8-5. If necessary, copy the rule for them, then discuss. If you are dispensing medications, give the patients the same information that the pharmacist would, if you are allowed to do so. Give patients the following information about handling liquid medication:

1. Read the label to learn how to store the medication.
2. Use the measuring device provided or a device purchased specifically to measure medications. Household teaspoons and tablespoons do not measure liquid accurately.
3. Do not store medication longer than the label indicates. Medication used after its expiration date may have lost potency, or its chemical composition may have changed.
4. Wash the measuring device with hot water and a dishwashing detergent after each use. Dry it thoroughly. Store it in a clean container such as a plastic sandwich bag.
5. Keep liquid medication in its original container. Do not transfer it to other containers.

Critical Thinking on the Job

Reconstituting Powders

A health care professional is preparing a bottle of Amoxicillin suspension for this order: Amoxicillin 500 mg PO q8h × 5. The pharmacy has available 100-mL bottles and 150-mL bottles containing 250 mg/5 mL (see Figures 8-12 and 8-13).

After calculating as follows:

$$500 \text{ mg} \times \frac{5 \text{ mL}}{250 \text{ mg}} = A$$

$$\overset{2}{\cancel{500}}\ \cancel{\text{mg}} \times \frac{5 \text{ mL}}{\underset{1}{\cancel{250}}\ \cancel{\text{mg}}} = 2 \times 5 \text{ mL} = 10 \text{ mL}$$

the health care professional determines that the patient will receive 10 mL for each dose and 3 doses each day. This will require 30 mL of suspension each day for 5 days, or a total of 150 mL. The reconstituted medication can be refrigerated for 14 days.

The health care professional selects the 150-mL bottle and adds 150 mL of water to it. However, the liquid overflows from the bottle.

Questions?

What went wrong? How will this problem affect the patient's care?

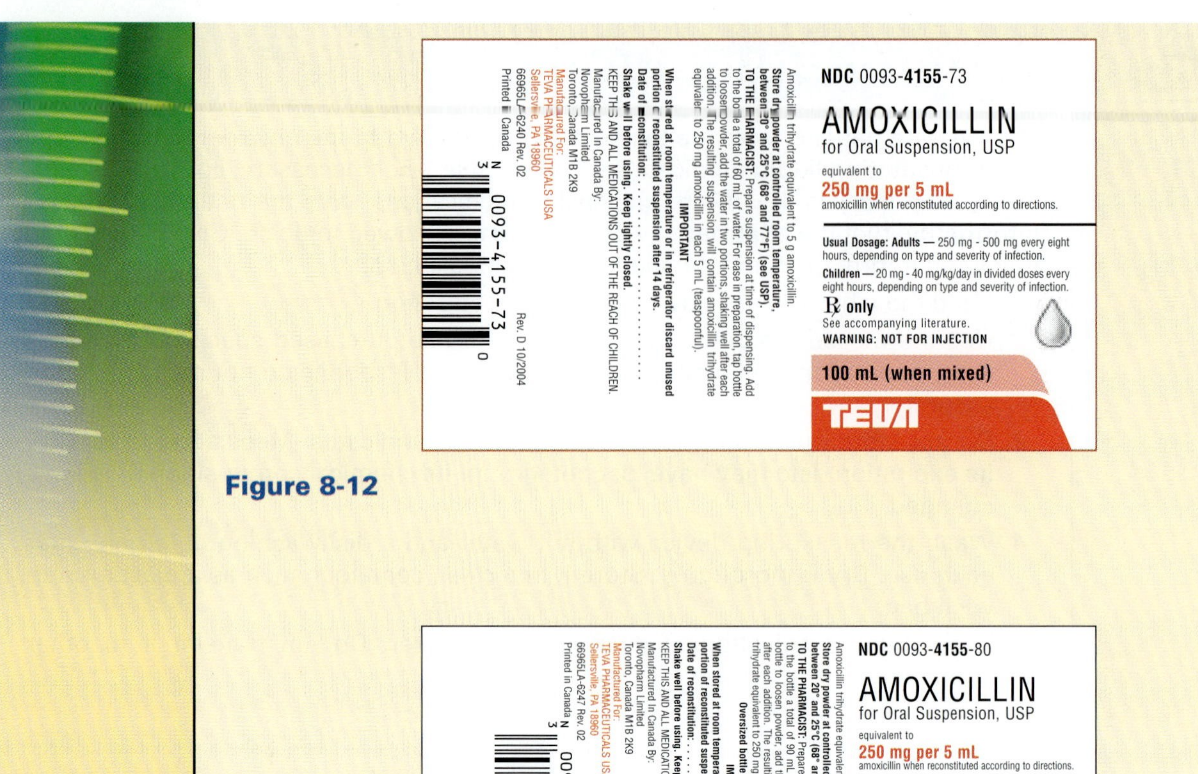

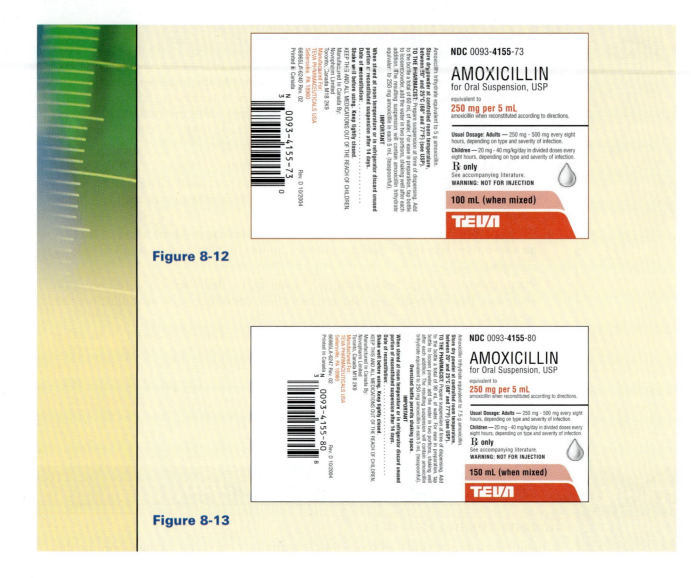

Figure 8-12

Figure 8-13

REVIEW AND PRACTICE

8-2 Liquid Medications

In Exercises 1–20, calculate the amount to administer.

1. Ordered: Trilisate® 400 mg PO tid
   On hand: Trilisate® liquid labeled 500 mg/5 mL
   Administer: _____

2. Ordered: MSIR sol 15 mg PO q4h
   On hand: MSIR solution labeled 10 mg/5 mL
   Administer: _____

3. Ordered: Megace® 200 mg PO qid
   On hand: Megace® solution labeled 40 mg/mL
   Administer: _____

4. Ordered: Norvir® 60 mg PO bid
   On hand: Norvir® solution labeled 80 mg/mL
   Administer: _____

5. Ordered: Zofran® 8 mg PO q12h
   On hand: Zofran® liquid labeled 4 mg/5 mL
   Administer: _____

6. Ordered: Motrin® 600 mg PO tid
   On hand: Motrin® liquid labeled 100 mg/5 mL
   Administer: _____

7. Ordered: E.E.S.® 500 mg PO bid
   On hand: Refer to label A.
   Administer: _____

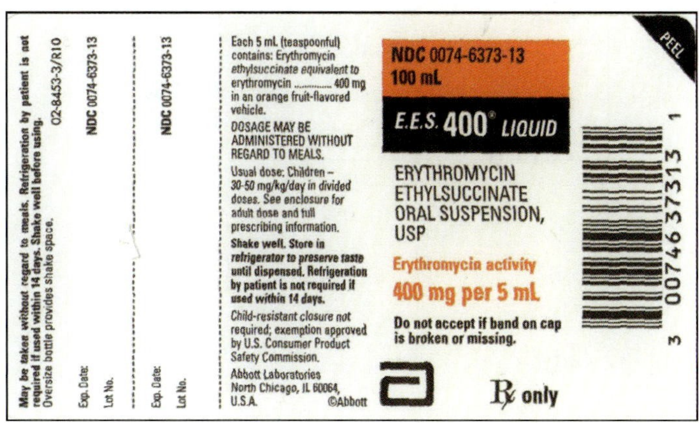

A

8. Ordered: Amoxicillin 270 mg PO q8h
   On hand: Refer to label B.
   Administer: _____

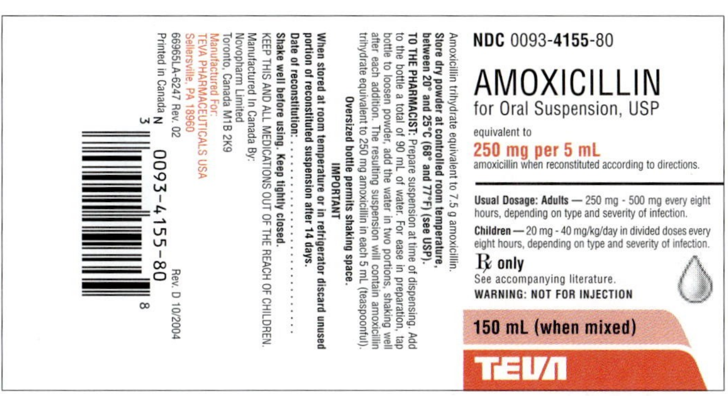

B

9. Ordered: Zithromax® 250 mg PO qd
   On hand: Refer to label C.
   Administer: _____

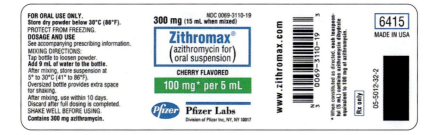

FOR ORAL USE ONLY.
Store dry powder below 30°C (86°F).
PROTECT FROM FREEZING.
DOSAGE AND USE
See accompanying prescribing information.
MIXING DIRECTIONS:
Tap bottle to loosen powder.
Add 9 mL of water to the bottle.
After mixing, store suspension at 5° to 30°C (41° to 86°F).
Oversized bottle provides extra space for shaking.
After mixing, use within 10 days.
Discard after full dosing is completed.
SHAKE WELL BEFORE USING.
Contains 300 mg azithromycin.

NDC 0069-3110-19
300 mg (15 mL when mixed)
Zithromax® (azithromycin for oral suspension)
CHERRY FLAVORED
100 mg* per 5 mL
Pfizer Labs
Division of Pfizer Inc, NY, NY 10017

www.zithromax.com

* When constituted as directed, each teaspoonful (5 mL) contains azithromycin dihydrate equivalent to 100 mg of azithromycin.
Rx only
05-5012-32-2
6415
MADE IN USA

C

10. Ordered: Depakene® 125 mg PO bid　　　　Administer: _____
On hand: Refer to label D.

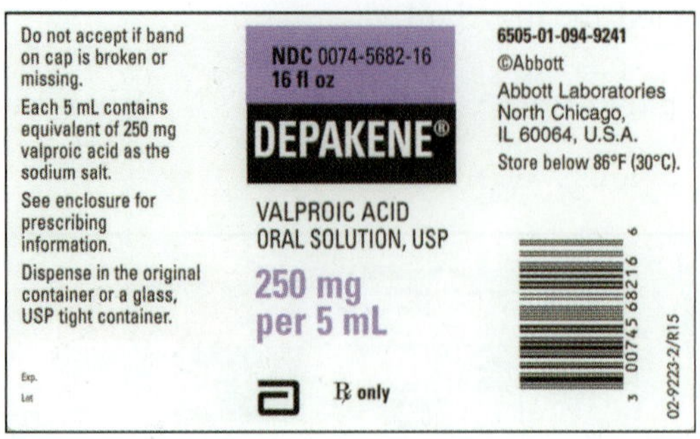

Do not accept if band on cap is broken or missing.
Each 5 mL contains equivalent of 250 mg valproic acid as the sodium salt.
See enclosure for prescribing information.
Dispense in the original container or a glass, USP tight container.
NDC 0074-5682-16
16 fl oz
DEPAKENE®
VALPROIC ACID
ORAL SOLUTION, USP
250 mg
per 5 mL
℞ only
6505-01-094-9241
©Abbott
Abbott Laboratories
North Chicago,
IL 60064, U.S.A.
Store below 86°F (30°C).
3 00745 68216 6
02-9223-2/R15

D

11. Ordered: Zyrtec® 10 mg PO daily　　　　Administer: _____
On hand: Refer to label E.

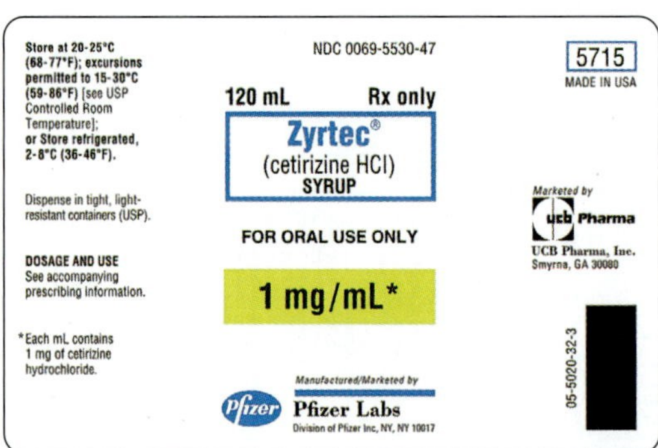

E

12. Ordered: Cellcept® 500 mg PO bid　　　　Administer: _____
On hand: Refer to label F.

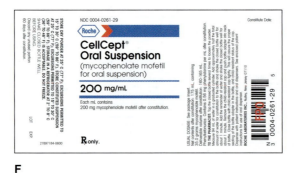

NDC 0004-0261-29
Roche
CellCept®
Oral Suspension
(mycophenolate mofetil
for oral suspension)
200 mg/mL
Each mL contains
200 mg mycophenolate mofetil after constitution.
℞only.

F

13. Ordered: Omnicef® 250 mg PO q12h Administer: _____

On hand: Refer to label G.

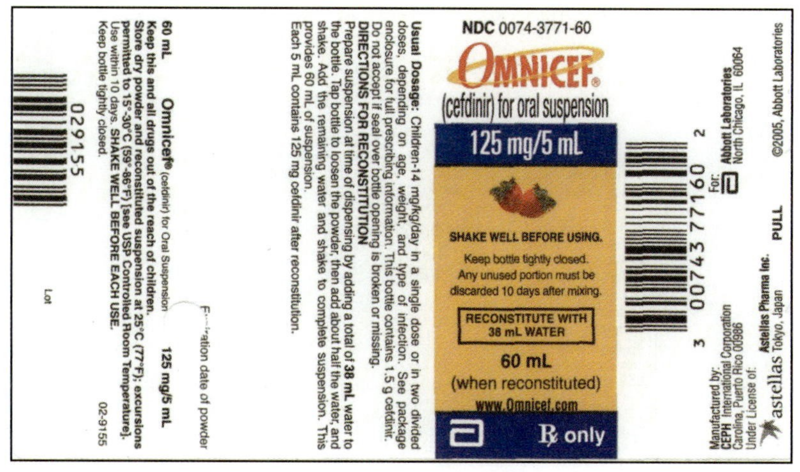

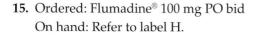

G

14. Ordered: Cefzil® 500 mg PO q24h Administer: _____

On hand: Cefzil® 125 mg/5 mL oral suspension

15. Ordered: Flumadine® 100 mg PO bid Administer: _____

On hand: Refer to label H.

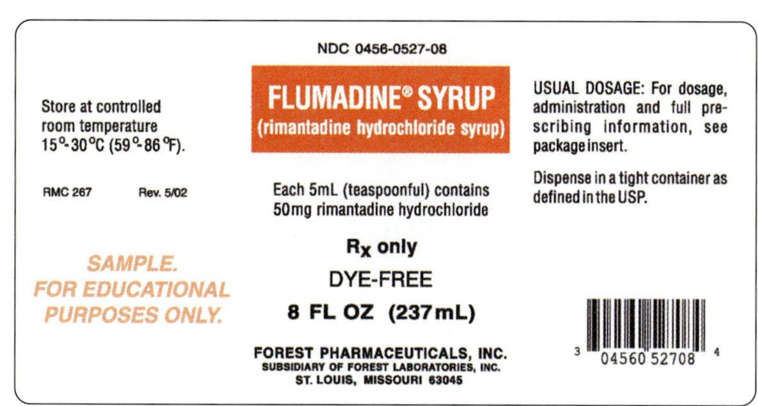

H

16. Ordered: Zithromax® 500 mg PO daily Administer: _____

On hand: Refer to label I.

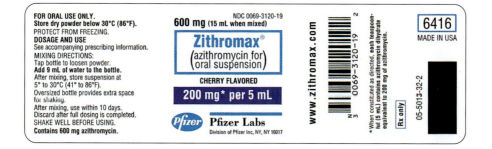

FOR ORAL USE ONLY.
Store dry powder below 30°C (86°F).
PROTECT FROM FREEZING.
DOSAGE AND USE
See accompanying prescribing information.
MIXING DIRECTIONS:
Tap bottle to loosen powder.
Add 9 mL of water to the bottle.
After mixing, store suspension at 5° to 30°C (41° to 86°F).
Oversized bottle provides extra space for shaking.
After mixing, use within 10 days.
Discard after full dosing is completed.
SHAKE WELL BEFORE USING.
Contains 600 mg azithromycin.
NDC 0069-3120-19
600 mg (15 mL when mixed)
Zithromax®
(azithromycin for oral suspension)
CHERRY FLAVORED
200 mg* per 5 mL
Pfizer Labs
Division of Pfizer Inc, NY, NY 10017
www.zithromax.com
6416
MADE IN USA
Rx only
05-5013-32-2

I

17. Ordered: Griseofulvin 500 mg PO daily Administer: _____
On hand: Griseofulvin 125 mg/5 mL suspension

18. Ordered: Depakene® 500 mg PO daily Administer: _____
On hand: Refer to label J.

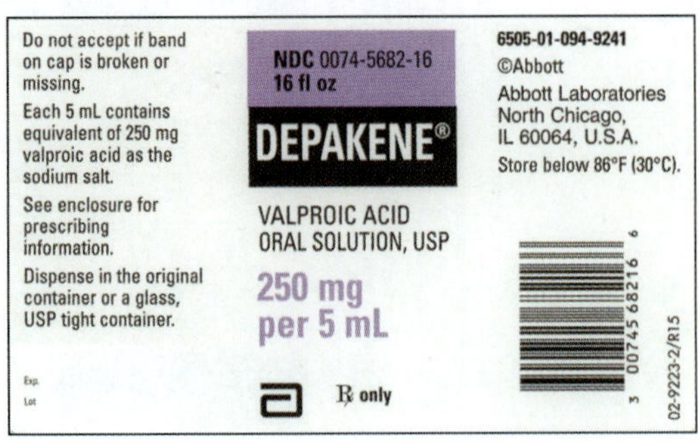

J

19. Ordered: Trileptal® 75 mg PO q6h Administer: _____
On hand: Refer to label K.

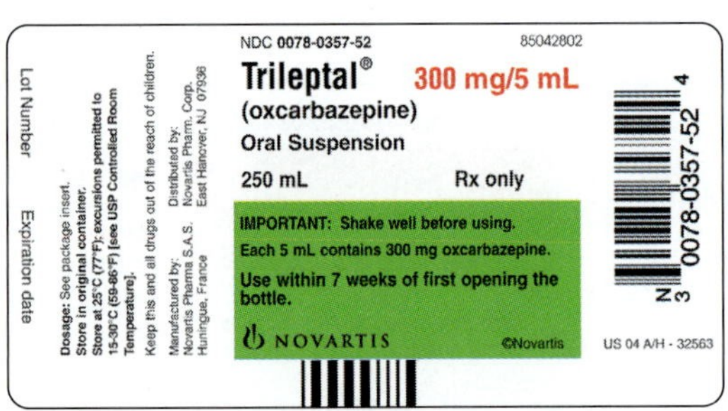

K

20. Ordered: Prozac® 60 mg PO daily Administer: _____
On hand: Prozac® 20 mg/5 mL oral solution

To check your answers, see page 523.

HOMEWORK ASSIGNMENT

Answer the following questions.

1. Define each of the following terms: scored tablet, sublingual, buccal, capsule, spansule.
2. List four solid forms of medications that may not be crushed and explain why.
3. Explain one technique for preparing a solid medication for administration through a nasogastric, gastrostomy, or jejunostomy tube.
4. List three techniques that may be suggested to a patient who complains of difficulty swallowing a tablet.
5. Children or elderly patients who require medication that is easily swallowed are often prescribed what form of medication?
6. Define the term reconstitution.
7. List four things to be written on the label of a reconstituted medication.

Use the identified drug labels to answer the following questions:

8. In Label A, what are the brand name and the form of this drug?
9. If the order was for 10 mg of this drug, what amount would you administer?

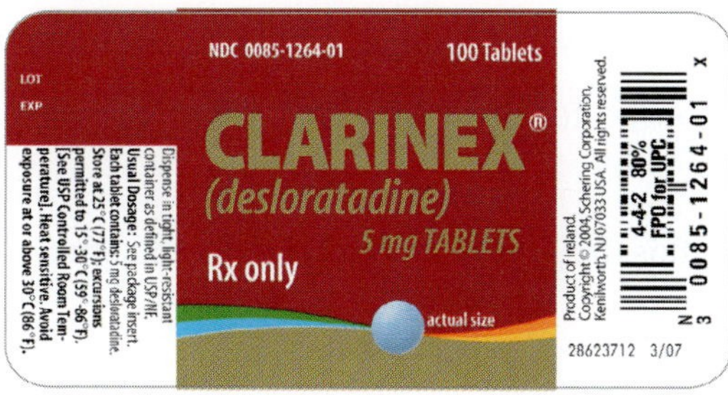

Label A

10. In Label B, what are the generic name and the form of this drug?
11. Is it acceptable to crush this drug?
12. If the order was for 40 mg of this drug, what amount would you administer?

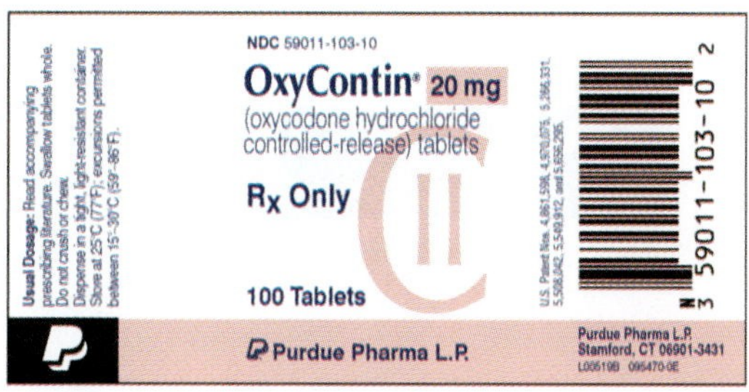

Label B

13. In Label C, what are the generic name and the form of this drug?
14. If the order was for 150 mg of this drug what action would you take?

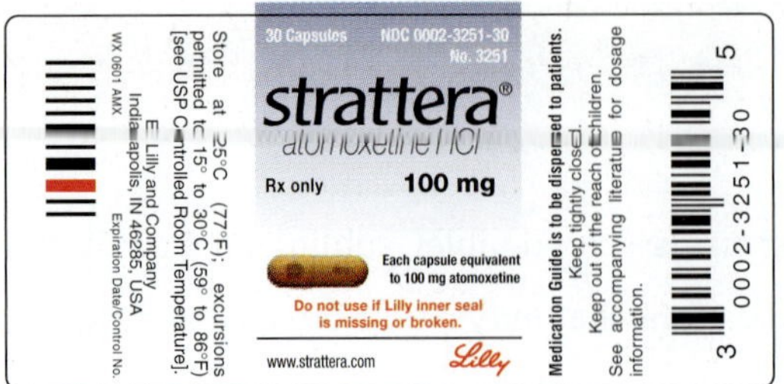

Label C

15. In Label D, what are the brand name and the form of this drug?

16. If the order was for 15 mg of this drug what amount would you administer?

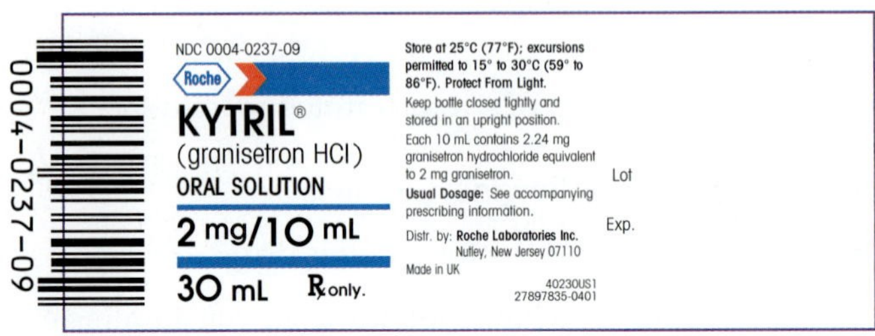

Label D

17. In Label E, how many milliliters of water would you add to reconstitute this drug to a 75 mL suspension?

18. How are you instructed to reconstitute this drug?

19. What is the warning on the label?

20. If the order was for 200 mg of this drug what amount would you administer?

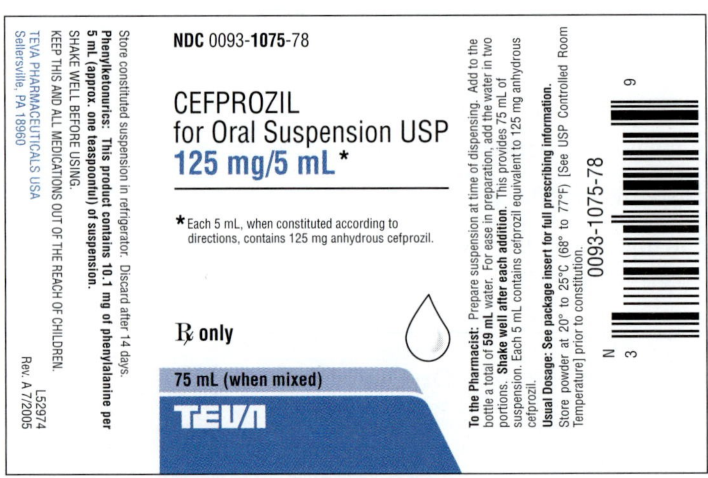

Label E

Chapter 8 Review

Check Up

In Exercises 1–17, calculate the amount to administer. Unless otherwise noted, tablets are scored in half.

1. Ordered: Dilaudid® 4 mg PO q6h — Administer: _____
   On hand: Dilaudid® 8 mg scored tablets
2. Ordered: DiaBeta® 2.5 mg PO qam ac — Administer: _____
   On hand: DiaBeta® 1.25 mg scored tablets
3. Ordered: Biltricide 450 mg PO q8h — Administer: _____
   On hand: Biltricide 600 mg tablets scored in quarters
4. Ordered: Amoxicillin 300 mg PO q12h — Administer: _____
   On hand: Amoxicillin suspension labeled 50 mg/mL
5. Ordered: Artane® 3 mg PO daily — Administer: _____
   On hand: Artane® solution labeled 2 mg/5 mL
6. Ordered: Fosamax® 10 mg PO qam 30 min ac with water — Administer: _____
   On hand: Fosamax® 5 mg unscored tablets
7. Ordered: Biaxin® liquid 62.5 mg PO q12h — Administer: _____
   On hand: Biaxin® liquid labeled 125 mg/5 mL
8. Ordered: Isoptin® 270 mg PO qam — Administer: _____
   On hand: Isoptin® 180 mg scored tablets
9. Ordered: Duricef® 0.5 g PO bid — Administer: _____
   On hand: Duricef® suspension labeled 250 mg/5 mL
10. Ordered: Levoxyl® 0.45 mg PO daily — Administer: _____
    On hand: Levoxyl® 300 mcg scored tablets
11. Ordered: Hivid 750 mcg PO q8h — Administer: _____
    On hand: Hivid 0.375 mg unscored tablets
12. Ordered: Duricef® 500 mg PO bid — Administer: _____
    On hand: Duricef® 1 g scored tablets
13. Ordered: MSIR $\frac{1}{8}$ gr PO q4h — Administer: _____
    On hand: MSIR 15 mg scored tablets
14. Ordered: Felbatol® 400 mg PO tid — Administer: _____
    On hand: Felbatol® liquid labeled 600 mg/5 mL
15. Ordered: Synthroid® 0.175 mg PO daily — Administer: _____
    On hand: Refer to label A (on page 288). Tablets are unscored.

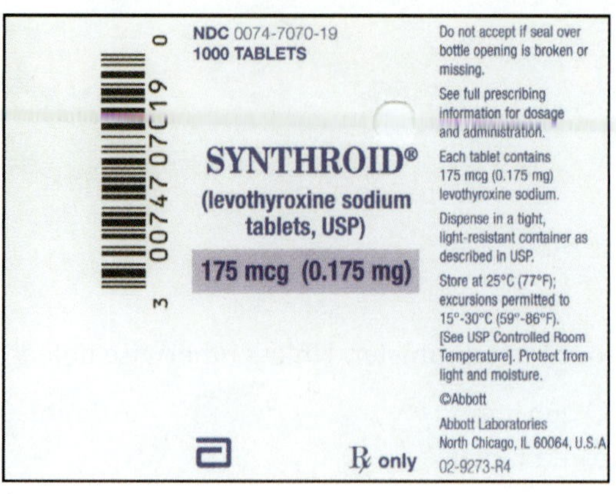

A

16. Ordered: Prilosec® 40 mg PO daily Administer: _____
    On hand: Refer to label B.

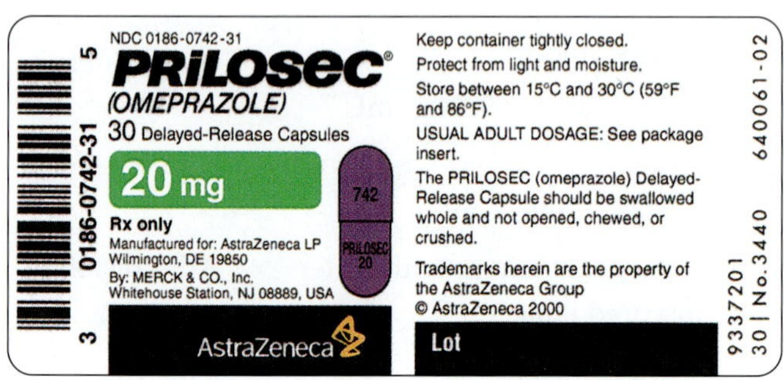

B

17. Ordered: Cefprozil 125 mg PO tid Administer: _____
    On hand: Refer to label C.

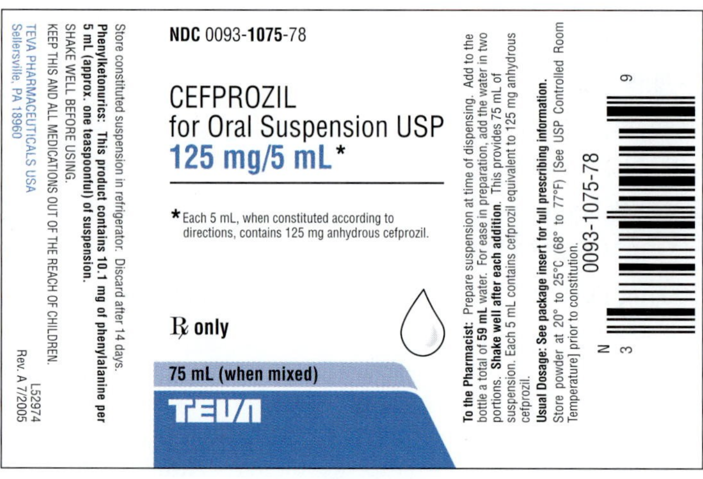

C

18. What combination will provide the desired dose with the fewest tablets?
    Ordered: Hytrin® 5 mg PO qpm
    On hand: Hytrin® 1 mg unscored tablets and Hytrin 2 mg unscored tablets
19. A patient receives 15 mL of Lortab® elixir every 6 h. Lortab® elixir contains 7.5 mg hydrocodone and 500 mg acetaminophen in each 15 mL. How much acetaminophen will this patient receive in 24 h?
20. How would you reconstitute the following medication? Find the amount to administer.
    Ordered: Cephalexin 375 mg PO bid
    On hand: Refer to label D.

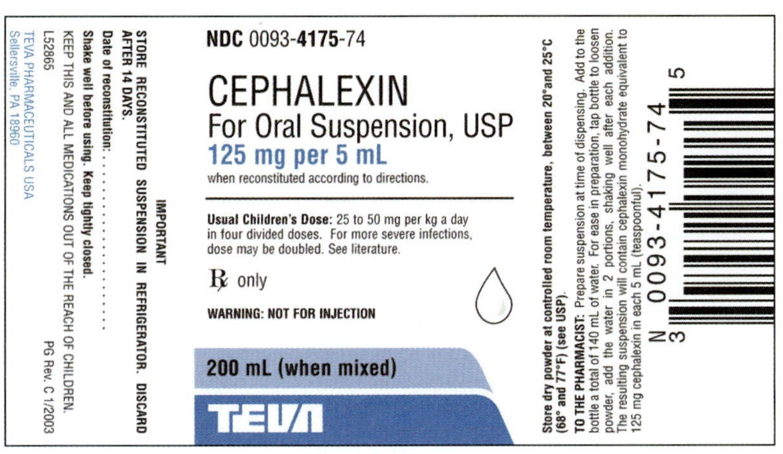

D

Each of the following sets of exercises provides a medication administration record (MAR) listing several medication orders and a variety of drug labels. Select the correct label for each medication order. Then calculate the amount to administer.

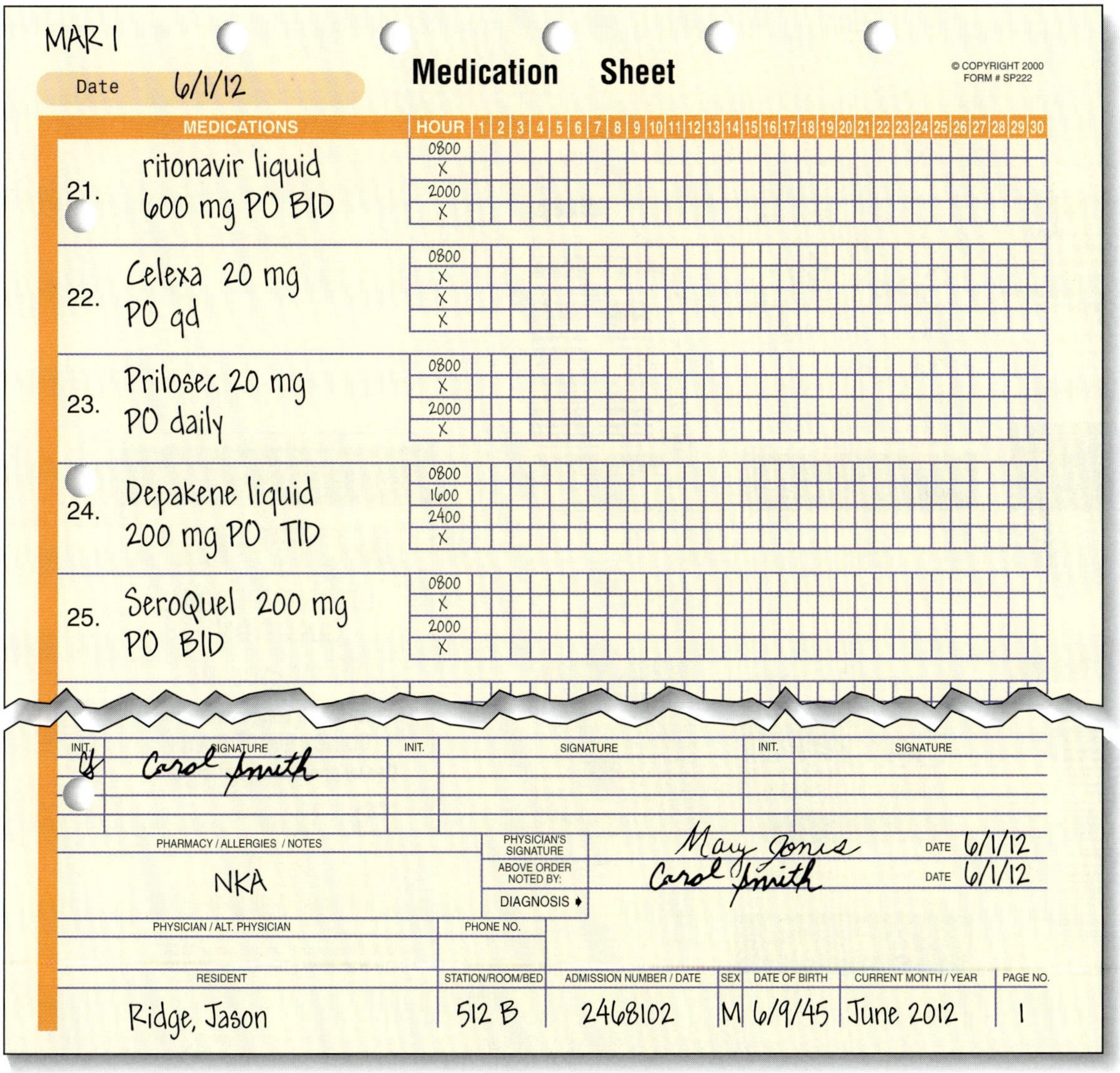

MAR 1

Date 6/1/12

Medication Sheet

© COPYRIGHT 2000
FORM # SP222

	MEDICATIONS	HOUR
21.	ritonavir liquid 600 mg PO BID	0800 / X / 2000 / X
22.	Celexa 20 mg PO qd	0800 / X / X / X
23.	Prilosec 20 mg PO daily	0800 / X / 2000 / X
24.	Depakene liquid 200 mg PO TID	0800 / 1600 / 2400 / X
25.	SeroQuel 200 mg PO BID	0800 / X / 2000 / X

INIT.	SIGNATURE	INIT.	SIGNATURE	INIT.	SIGNATURE
CS	Carol Smith				

PHARMACY / ALLERGIES / NOTES: NKA

PHYSICIAN'S SIGNATURE: Mary Jones DATE 6/1/12

ABOVE ORDER NOTED BY: Carol Smith DATE 6/1/12

DIAGNOSIS

PHYSICIAN / ALT. PHYSICIAN | PHONE NO.

RESIDENT	STATION/ROOM/BED	ADMISSION NUMBER / DATE	SEX	DATE OF BIRTH	CURRENT MONTH / YEAR	PAGE NO.
Ridge, Jason	512 B	2468102	M	6/9/45	June 2012	1

Figure 8-14

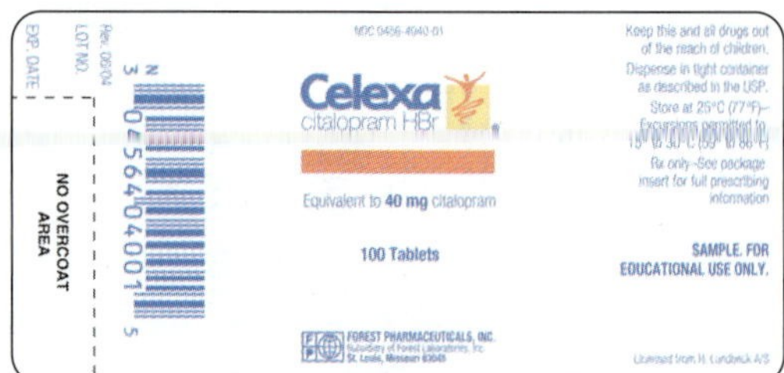

E Tablets are scored

Do not accept if band on cap is broken or missing.
Each 5 mL contains equivalent of 250 mg valproic acid as the sodium salt.
See enclosure for prescribing information.
Dispense in the original container or a glass, USP tight container.
Exp.
Lot
NDC 0074-5682-16
16 fl oz
DEPAKENE®
VALPROIC ACID
ORAL SOLUTION, USP
250 mg
per 5 mL
℞ only
6505-01-094-9241
©Abbott
Abbott Laboratories
North Chicago,
IL 60064, U.S.A.
Store below 86°F (30°C).
3 00745 68216 6
02-9223-2/R15

F

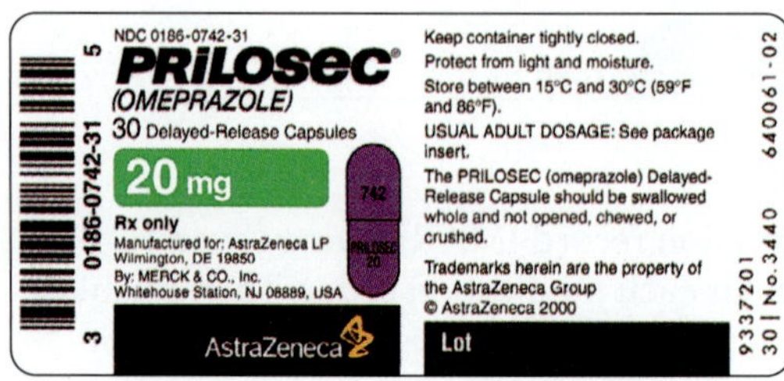

G

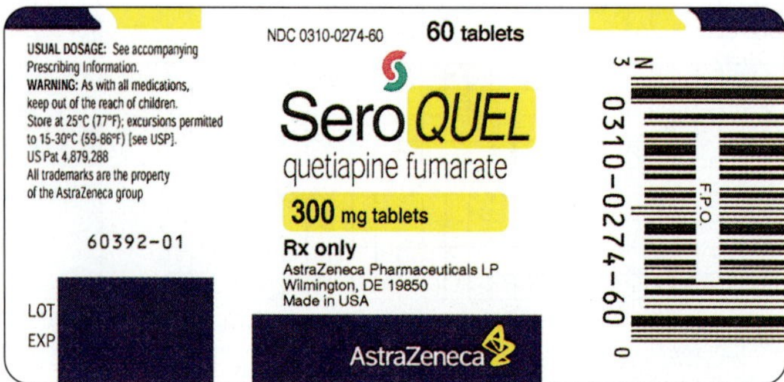

USUAL DOSAGE: See accompanying Prescribing Information.
WARNING: As with all medications, keep out of the reach of children.
Store at 25°C (77°F); excursions permitted to 15-30°C (59-86°F) [see USP].
US Pat 4,879,288
All trademarks are the property of the AstraZeneca group
60392-01
LOT
EXP
NDC 0310-0274-60
60 tablets
SeroQUEL
quetiapine fumarate
300 mg tablets
Rx only
AstraZeneca Pharmaceuticals LP
Wilmington, DE 19850
Made in USA
AstraZeneca
N 3 0310-0274-60 0
F.P.O.

H

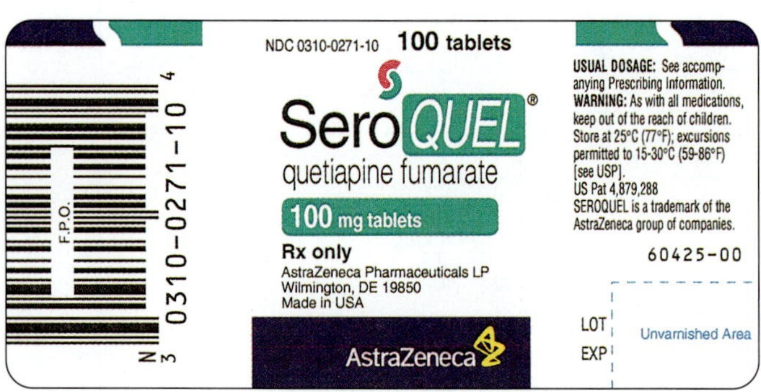

I

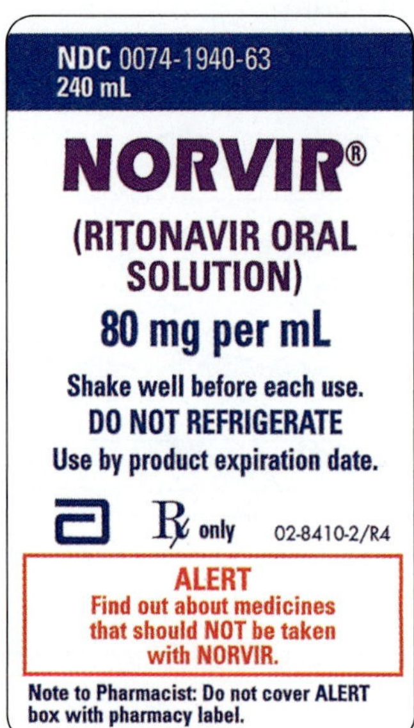

J

Refer to Figure 8-14 and labels E to J

21. Use: _____ Administer: _____

22. Use: _____ Administer: _____

23. Use: _____ Administer: _____

24. Use: _____ Administer: _____

25. Use: _____ Administer: _____

MAR II

Date 6/1/12

Medication Sheet

© COPYRIGHT 2000
FORM # SP222

MEDICATIONS	HOUR	1	2	3	4	5	6	7	8	9	10	11	12	13	14	15	16	17	18	19	20	21	22	23	24	25	26	27	28	29	30
26. Tricor 134 mg PO q AM	0800 / 1200 / 1700 / X																														
27. Prandin 1 mg PO TID with meals	0800 / 1200 / 1700 / X																														
28. Valium 6 mg PO TID	0800 / 1600 / 2400 / X																														
29. EES 500 mg PO BID	0800 / X / 2000 / X																														
30. Tranxene T-tab 15 mg PO BID	0800 / X / 2000 / X																														

INIT.	SIGNATURE	INIT.	SIGNATURE	INIT.	SIGNATURE
CS	Carol Smith				

PHARMACY / ALLERGIES / NOTES: NKA

PHYSICIAN'S SIGNATURE: Mary Jones DATE 6/1/12

ABOVE ORDER NOTED BY: Carol Smith DATE 6/1/12

DIAGNOSIS ♦

PHYSICIAN / ALT. PHYSICIAN | PHONE NO.

RESIDENT	STATION/ROOM/BED	ADMISSION NUMBER / DATE	SEX	DATE OF BIRTH	CURRENT MONTH / YEAR	PAGE NO.
Doe, Jane	112 B	1234567	F	8/10/45	May 2012	1

Figure 8-15

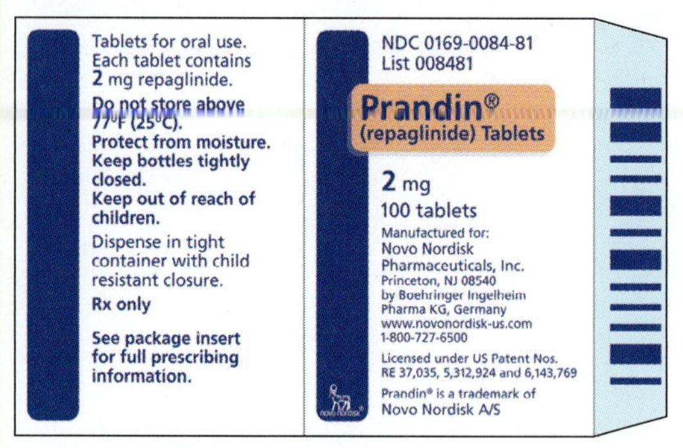
Tablets for oral use.
Each tablet contains
2 mg repaglinide.
Do not store above
77°F (25°C).
Protect from moisture.
Keep bottles tightly
closed.
Keep out of reach of
children.
Dispense in tight
container with child
resistant closure.
Rx only
See package insert
for full prescribing
information.
NDC 0169-0084-81
List 008481
Prandin®
(repaglinide) Tablets
2 mg
100 tablets
Manufactured for:
Novo Nordisk
Pharmaceuticals, Inc.
Princeton, NJ 08540
by Boehringer Ingelheim
Pharma KG, Germany
www.novonordisk-us.com
1-800-727-6500
Licensed under US Patent Nos.
RE 37,035, 5,312,924 and 6,143,769
Prandin® is a trademark of
Novo Nordisk A/S

K

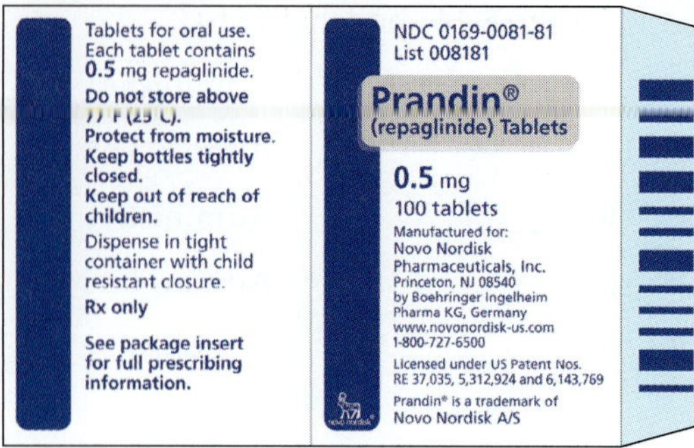
Tablets for oral use.
Each tablet contains
0.5 mg repaglinide.
Do not store above
77°F (25°C).
Protect from moisture.
Keep bottles tightly
closed.
Keep out of reach of
children.
Dispense in tight
container with child
resistant closure.
Rx only
See package insert
for full prescribing
information.
NDC 0169-0081-81
List 008181
Prandin®
(repaglinide) Tablets
0.5 mg
100 tablets
Manufactured for:
Novo Nordisk
Pharmaceuticals, Inc.
Princeton, NJ 08540
by Boehringer Ingelheim
Pharma KG, Germany
www.novonordisk-us.com
1-800-727-6500
Licensed under US Patent Nos.
RE 37,035, 5,312,924 and 6,143,769
Prandin® is a trademark of
Novo Nordisk A/S

L

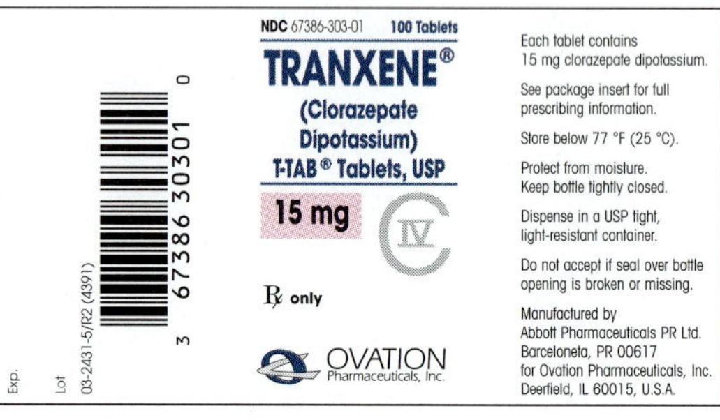
Exp.
Lot
03-2431-5/R2 (4391)
3 67386 30301 0
NDC 67386-303-01 100 Tablets
TRANXENE®
(Clorazepate
Dipotassium)
T-TAB® Tablets, USP
15 mg
C IV
℞ only
OVATION
Pharmaceuticals, Inc.
Each tablet contains
15 mg clorazepate dipotassium.
See package insert for full
prescribing information.
Store below 77 °F (25 °C).
Protect from moisture.
Keep bottle tightly closed.
Dispense in a USP tight,
light-resistant container.
Do not accept if seal over bottle
opening is broken or missing.
Manufactured by
Abbott Pharmaceuticals PR Ltd.
Barceloneta, PR 00617
for Ovation Pharmaceuticals, Inc.
Deerfield, IL 60015, U.S.A.

M

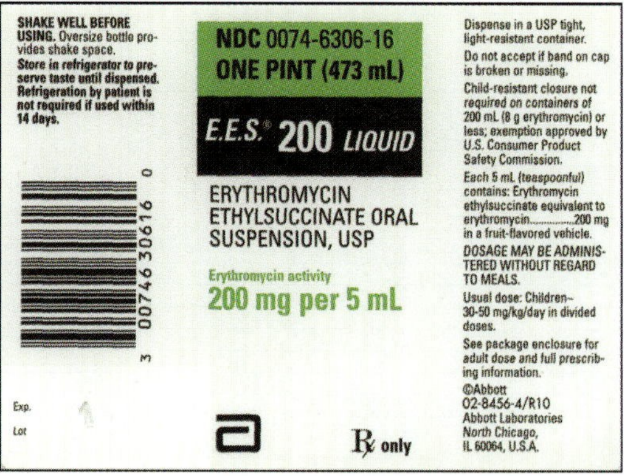
SHAKE WELL BEFORE
USING. Oversize bottle provides shake space.
Store in refrigerator to preserve taste until dispensed.
Refrigeration by patient is
not required if used within
14 days.
3 00746 30616 0
Exp.
Lot
NDC 0074-6306-16
ONE PINT (473 mL)
E.E.S.® 200 LIQUID
ERYTHROMYCIN
ETHYLSUCCINATE ORAL
SUSPENSION, USP
Erythromycin activity
200 mg per 5 mL
℞ only
Dispense in a USP tight,
light-resistant container.
Do not accept if band on cap
is broken or missing.
Child-resistant closure not
required on containers of
200 mL (8 g erythromycin) or
less; exemption approved by
U.S. Consumer Product
Safety Commission.
Each 5 mL (teaspoonful)
contains: Erythromycin
ethylsuccinate equivalent to
erythromycin..............200 mg
in a fruit-flavored vehicle.
DOSAGE MAY BE ADMINISTERED WITHOUT REGARD
TO MEALS.
Usual dose: Children–
30-50 mg/kg/day in divided
doses.
See package enclosure for
adult dose and full prescribing information.
©Abbott
02-8456-4/R10
Abbott Laboratories
North Chicago,
IL 60064, U.S.A.

N

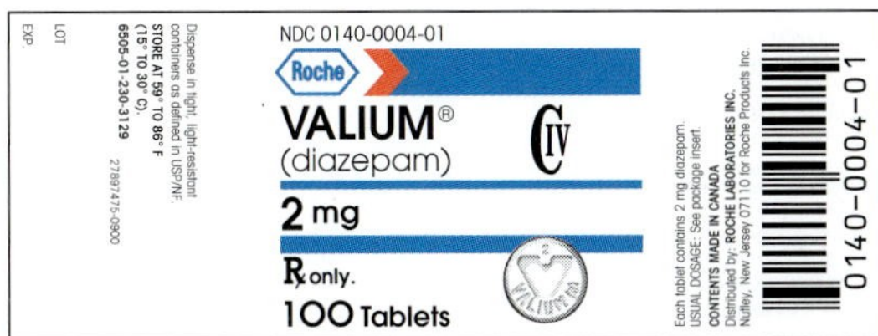

O

P

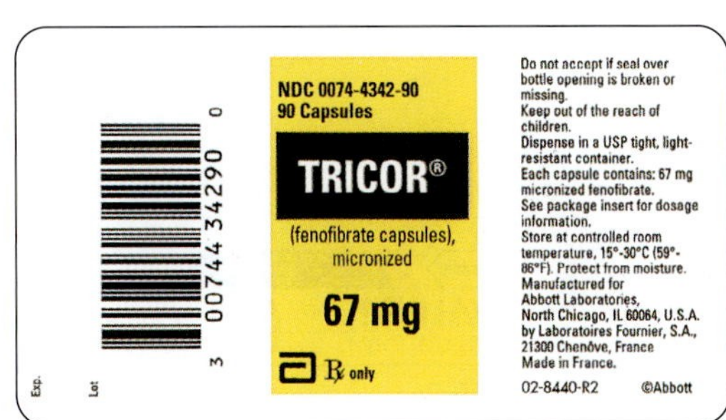

Q

Refer to Figure 8-15 and labels K to Q

26. Use: _____ Administer: _____

27. Use: _____ Administer: _____

28. Use: _____ Administer: _____

29. Use: _____ Administer: _____

30. Use: _____ Administer: _____

Critical Thinking Applications

The following medications are ordered for a patient with a gastrostomy tube who cannot swallow and must receive all medications through the tube.

a. Depakote® ER 250 mg daily

b. Valium® 4 mg qid

c. Meperidine (Demerol®) 50 mg q 6 to 8 h prn for pain

d. Cephalexin 500 mg

On hand: Refer to labels R, S, T, and U. For this exercise Meperidine and Valium® tablets are scored.

1. For each medication, calculate the amount to administer.
2. Are there any medications on the list that cannot be given as ordered?

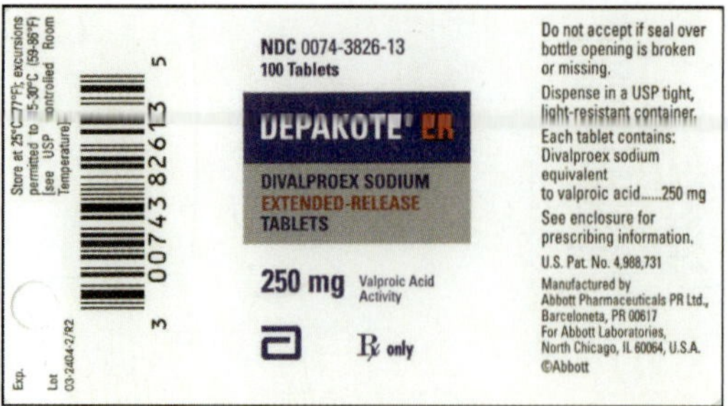
NDC 0074-3826-13
100 Tablets
DEPAKOTE® ER
DIVALPROEX SODIUM
EXTENDED-RELEASE
TABLETS
250 mg
Valproic Acid Activity
Rx only
Do not accept if seal over bottle opening is broken or missing.
Dispense in a USP tight, light-resistant container.
Each tablet contains:
Divalproex sodium equivalent to valproic acid.....250 mg
See enclosure for prescribing information.
U.S. Pat. No. 4,988,731
Manufactured by
Abbott Pharmaceuticals PR Ltd.,
Barceloneta, PR 00617
For Abbott Laboratories,
North Chicago, IL 60064, U.S.A.
©Abbott

R

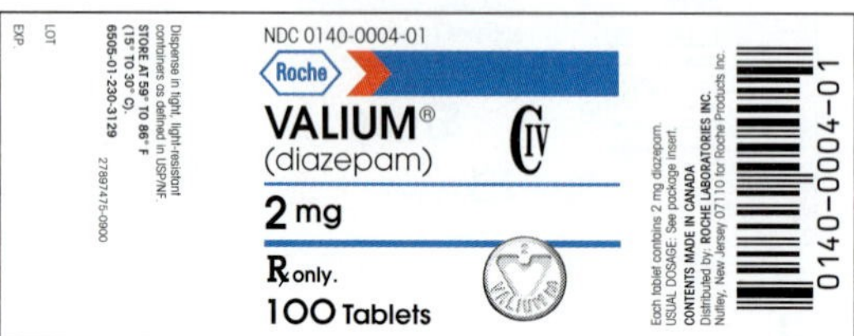
NDC 0140-0004-01
Roche
VALIUM®
(diazepam)
2 mg
Rx only.
100 Tablets
0140-0004-01

S

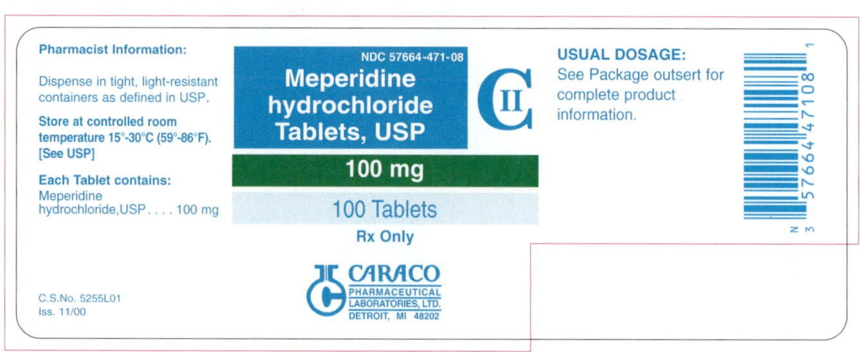
Pharmacist Information:
Dispense in tight, light-resistant containers as defined in USP.
Store at controlled room temperature 15°-30°C (59°-86°F). [See USP]
Each Tablet contains:
Meperidine hydrochloride,USP. . . . 100 mg
NDC 57664-471-08
Meperidine hydrochloride Tablets, USP
100 mg
100 Tablets
Rx Only
CII
USUAL DOSAGE:
See Package outsert for complete product information.
CARACO
PHARMACEUTICAL LABORATORIES, LTD.
DETROIT, MI 48202
C.S.No. 5255L01
Iss. 11/00

T

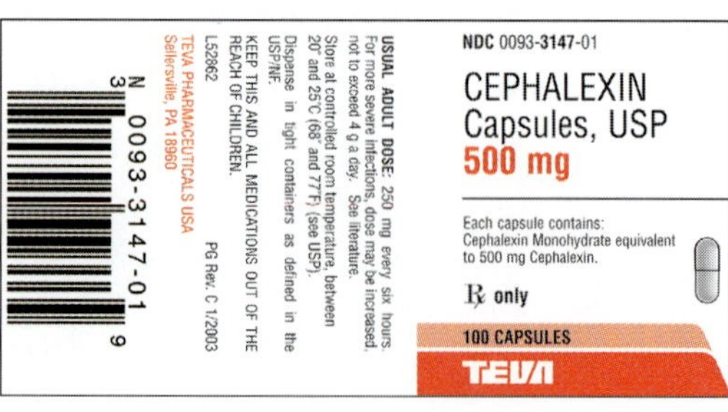
NDC 0093-3147-01
CEPHALEXIN
Capsules, USP
500 mg
Each capsule contains:
Cephalexin Monohydrate equivalent to 500 mg Cephalexin.
Rx only
100 CAPSULES
TEVA

U

3. How would you administer these medications?
4. What action would you take if you could not give a medication as ordered?

Case Study

Ordered: Biaxin® liquid 187.5 mg PO qid

On hand: Refer to label V.

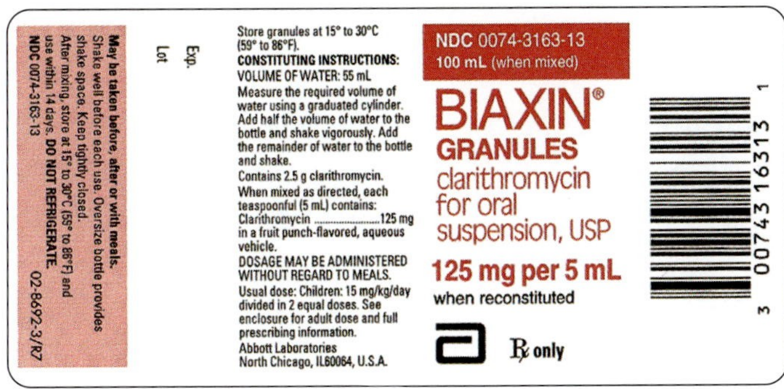

V

1. Describe how you would reconstitute this medication.
2. Calculate the amount to administer.
3. What measuring device would you use to give this dose?

To check your answers, see page 524.

Internet Activity

You are working in a physician's office where a patient with diabetes is having difficulty with blood glucose control. The physician discovers that the patient, who has trouble swallowing, has been crushing Glucotrol XL tablets. Because this sustained-release medication should not be crushed, the patient has been receiving too much medication at one time. You realize that this problem could occur with any patient who is taking a sustained-release medication.

Assignment: Search the Internet for patient education materials warning patients about crushing medications.

Suggested key words: medication + crushing; medication + swallowing; pills + swallowing

GO TO . . . Open the CD ROM that accompanies your textbook, and complete a final review of the rules, practice problems, and activities presented for this chapter. For a final evaluation, take the chapter test and email or print your results for your instructor. A score of 95 percent or above indicates mastery of the chapter concepts. ■

9 Parenteral Dosages

The best way to escape from a problem is to solve it.

—Alan Saporta

Learning Outcomes

When you have completed Chapter 9, you will be able to:

- Calculate the amount of a parenteral medication to administer.
- Select the appropriate syringe.
- Correctly reconstitute powdered medications.
- Calculate the amount of reconstituted medications to administer.
- Accurately calculate doses of inhalant, rectal, and transdermal medications.
- Identify errors that occur in calculating and preparing parenteral doses.

Key Terms

Inhalant

Intradermal

Intramuscular

Intravenous

Parenteral

Rectal

Subcutaneous

Transdermal

Introduction

Parenteral medications are *not* taken by mouth, so they bypass the digestive tract. They include medications that are administered by injection and other medications such as inhalants, rectal, and transdermal drugs. Medications delivered by injection include **intramuscular** (IM), **subcutaneous** (Sub-Q), **intradermal** (ID), and **intravenous** (IV).

- IM – into a muscle
- Sub-Q – under the skin
- IV – into a vein
- ID – between the layers of skin

This chapter will focus on dosage calculations for parenteral medications with the exception of intravenous and insulin administration, which will be covered in later chapters. So get ready to solve some problems.

Calculating Parenteral Dosages

Injections are mixtures that contain the drug dissolved in an appropriate liquid. The dosage or solution strength on an injectable medication's label indicates the amount of drug contained within a volume of solution (see Table 9-1). Dosage strength may be expressed in milligrams per milliliter, as a percent, or as a ratio.

TABLE 9-1 Sample Solution Strengths

Label Description	Interpretation
Compazine 5 mg/mL	1 mL contains 5 mg of Compazine
Epogen 3000 U/mL	1 mL contains 3000 U of Epogen
Lidocaine 1%	100 mL contains 1 g of lidocaine
Epinephrine 1:1000	1000 mL contains 1 g of epinephrine

Physicians' orders for injections will usually specify the amount of medication to be administered to the patient. Before the injection can be administered, it is necessary to calculate how many milliliters of the solution contain the desired dose of the medication.

As with oral medications, you will calculate the amount to administer, starting with the dosage ordered, dose on hand, and dosage unit. If the dosage ordered and dose on hand have different units, then you must first convert the units of the dosage ordered to the units of the desired dose. You may then calculate the amount to administer, using any of the four methods of dosage calculations, that is, fraction proportion, ratio proportion, dimensional analysis, or the formula method.

Once you have determined the amount to administer, you must select the appropriate syringe and needle. Rule 9-1 gives you the guidelines to follow for selecting the syringe. See Chapter 4 for a discussion of needles.

Rule 9-1 Select the proper syringe.

1. If the amount of injection to administer is 1 mL or more, use a standard 3-mL syringe.
2. If the amount of injection to administer is less than 1 mL but greater than or equal to 0.5 mL, use a 1-mL tuberculin syringe.
3. If the amount of injection to administer is less than 0.5 mL, use a 0.5-mL tuberculin syringe.

Example 1 The amount to administer is calculated at 2.4 mL. Since this is greater than 1 mL, a standard syringe should be used. See Figure 9-1.

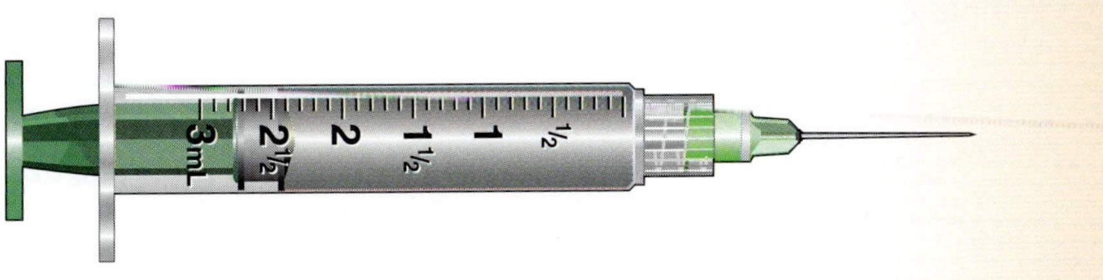

Figure 9-1 Standard syringe with 2.4 mL.

Example 2 The amount to administer is calculated at 0.6 mL. Since this is less than 1 mL and greater than 0.5 mL, a 1-mL tuberculin syringe should be used. See Figure 9-2.

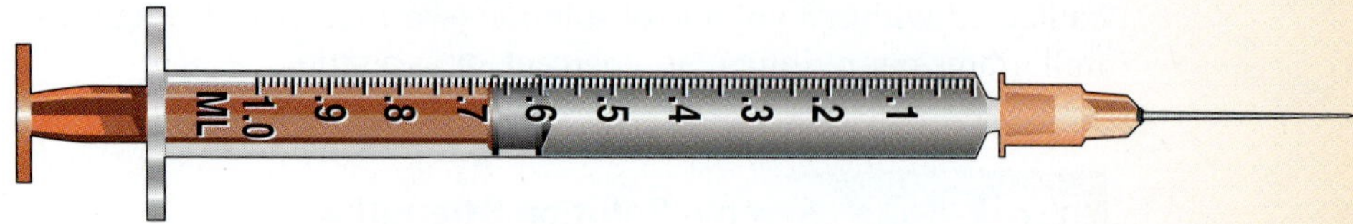

Figure 9-2 A 1-mL tuberculin syringe with 0.6 mL.

Example 3 The amount to administer is calculated at 0.34 mL. Since this is less than 1 mL, a 0.5-mL tuberculin syringe should be used when available. See Figure 9-3.

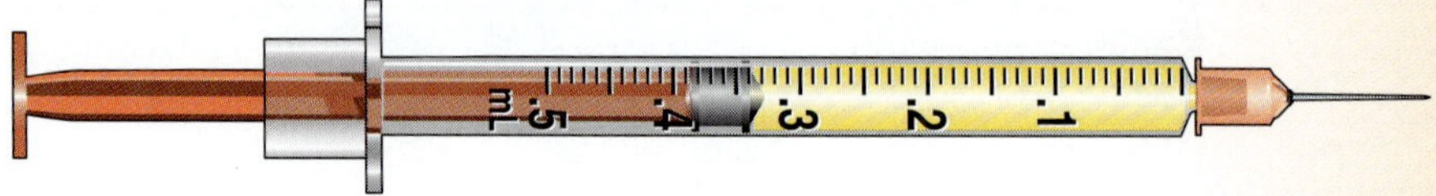

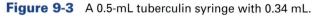

Figure 9-3 A 0.5-mL tuberculin syringe with 0.34 mL.

GO TO . . . Open the CD-ROM that accompanies your textbook, and select Chapter 9, Rule 9-1. Review the animation and example problems, then complete the practice problems. Continue to the next section of the book once you have mastered the rule presented. ■

The amount to administer will not always be in whole milliliters, and it will sometimes be necessary to round your answer.

Rule 9-2 Correctly round the amount of an injection to administer.

1. Round volumes greater than 1 mL to the nearest tenth (one decimal) because the 3 mL syringe is calibrated in tenths.
2. Round volumes less than 1 mL to the nearest hundredth (two decimals) because tuberculin syringes are calibrated in hundredths.

Example 1 The amount to administer is calculated at 1.66 mL. Since the volume is greater than 1 mL, you round 1.66 mL to the nearest tenth, which is 1.7 mL.

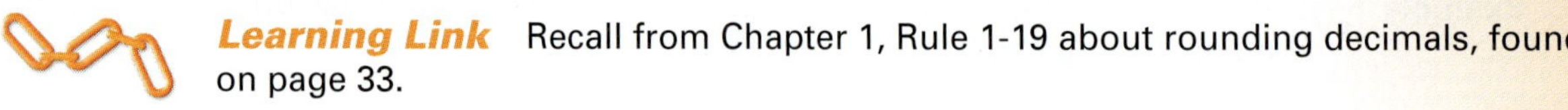

Learning Link Recall from Chapter 1, Rule 1-19 about rounding decimals, found on page 33.

Example 2 The amount to administer is calculated at 0.532 mL. Since the volume is less than 1 mL, you round 0.532 mL to the nearest hundredth, which is 0.53 mL.

Learning Link Recall from Chapter 1, Rule 1-19 about rounding decimals, found on page 33.

GO TO . . . Open the CD-ROM that accompanies your textbook, and select Chapter 9, Rule 9-2. Review the animation and example problems, then complete the practice problems. Continue to the next section of the book once you have mastered the rule presented. ■

Once you determine the proper syringe, you must also decide whether the amount to be administered can be safely injected in a single site. When the amount to administer exceeds the amount that can be safely given in one site, divide the amount into equal (or nearly equal) parts. Then administer them in separate sites.

Rule 9-3 Do not exceed maximum volume for an injection in a single site.

Intramuscular injections:

- Adult 3 mL
- Adult deltoid (arm) 2 mL
- Child 6 to 12 years 2 mL
- Child 0 to 5 years 1 mL
- Premature Infant 0.5 mL

Subcutaneous injections: 1 mL

Dosages larger than these maximum volumes are very rare and should be checked and verified.

Example The amount to administer to an adult is 4.5 mL. After checking and verifying the amount, you draw up 2 mL into one syringe and 2.5 mL into another. Inject the contents of each syringe into a different site. See Figure 9-4.

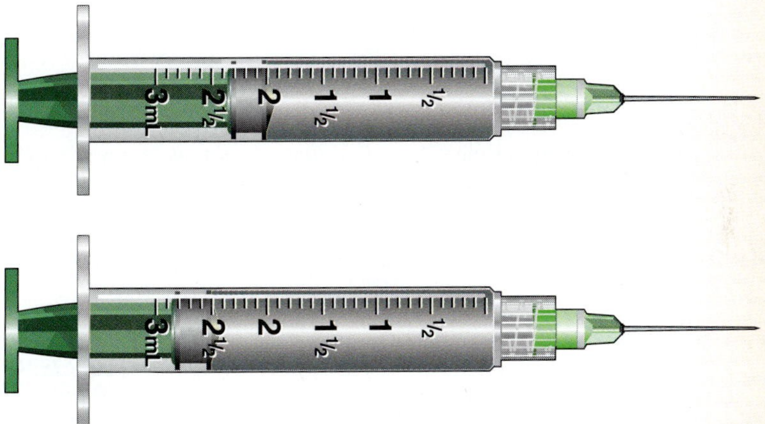

Figure 9-4 Two standard syringes would be used for a dose greater than the maximum volume for an injection such as 4.5 mL.

GO TO . . . Open the CD-ROM that accompanies your textbook, and select Chapter 9, Rule 9-3. Review the animation and example problems, then complete the practice problems. Continue to the next section of the book once you have mastered the rule presented. ■

When calculating the amount to administer for injectable medications, you must determine how many milliliters contain the desired dose of medication and then apply Rules 9-1 to 9-3.

Fraction Proportion Method

Example 1 Find the amount to administer and select the proper syringe.

Ordered: Valium® 7.5 mg IM now

On hand: Refer to Figure 9-5.

Patient: A 175-lb, 45-year-old male

Roche Laboratories Inc.
Nutley, New Jersey 07110

VALIUM® CIV
(diazepam)
5 mg/mL 10 mL Vial
Sterile. For I.M. or I.V. Use.
Each mL contains 5 mg diazepam compounded with 40% propylene glycol, 10% ethyl alcohol; 5% sodium benzoate and benzoic acid as buffers; and 1.5% benzyl alcohol as preservative.
NOTE: Solution may appear colorless to light yellow.
℞ only.
STORE AT 59° TO 86° F (15° TO 30° C).
EXPIRES
25402082-0199

Figure 9-5

Solution The dosage ordered O and the dose on hand H are already expressed in the same units, so the desired dose D is 7.5 mg. From the label we find that the dosage unit Q is 1 mL and the dose on hand H is 5 mg.

Follow Procedure Checklist 7-1.

1. Fill in the proportion.

$$\frac{Q}{H} = \frac{A}{D}$$

$$\frac{1 \text{ mL}}{5 \text{ mg}} = \frac{A}{7.5 \text{ mg}}$$

2. Cancel units.

$$\frac{1 \text{ mL}}{5 \cancel{\text{mg}}} = \frac{A}{7.5 \cancel{\text{mg}}}$$

3. Cross-multiply and solve for the unknown.

$$5 \times A = 1 \text{ mL} \times 7.5$$

$$A = 1 \text{ mL} \times \frac{7.5}{5}$$

$$A = 1.5 \text{ mL}$$

The amount to administer is 1.5 mL.

- Referring to Rule 9-1, a standard syringe would be used.
- Referring to Rule 9-2, we find that it is not necessary to round in this example.
- Referring to Rule 9-3, the total volume may be given as a single injection.

Example 2 Find the amount to administer and select the proper syringe.

Ordered: Lorazepam 800 mcg IM now

On hand: Refer to Figure 9-6.

Patient: An 85-lb, 12-year-old female

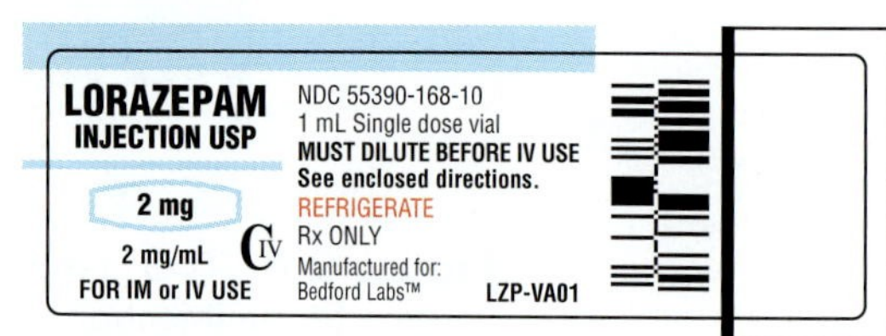

LORAZEPAM
INJECTION USP
2 mg
2 mg/mL
CIV
FOR IM or IV USE
NDC 55390-168-10
1 mL Single dose vial
MUST DILUTE BEFORE IV USE
See enclosed directions.
REFRIGERATE
Rx ONLY
Manufactured for:
Bedford Labs™
LZP-VA01

Figure 9-6

Solution The dosage ordered O is in micrograms, and the dose on hand H is in milligrams, so the desired dose D, in milligrams, must be calculated.

Follow Procedure Checklist 3-1.

1. Fill in the proportion, recalling that 1 mg = 1000 mcg.

$$\frac{1 \text{ mg}}{1000 \text{ mcg}} = \frac{D}{800 \text{ mcg}}$$

2. Cancel units.

$$\frac{1\text{ mg}}{1000\text{ }\cancel{\text{mcg}}} = \frac{D}{800\text{ }\cancel{\text{mcg}}}$$

3. Cross-multiply and solve for the unknown.

$1000 \times D = 1 \times 800\text{ mg}$

$D = 0.8\text{ mg}$

We now have the necessary pieces of information: $D = 0.8$ mg, $Q = 1$ mL, and $H = 2$ mg.

Follow Procedure Checklist 7-1.

1. Fill in the proportion.

$$\frac{1\text{ mL}}{2\text{ mg}} = \frac{A}{0.8\text{ mg}}$$

2. Cancel units.

$$\frac{1\text{ mL}}{2\text{ }\cancel{\text{mg}}} = \frac{A}{0.8\text{ }\cancel{\text{mg}}}$$

3. Cross-multiply and solve for the unknown.

$2 \times A = 1\text{ mL} \times 0.8$

$A = 1\text{ mL} \times \frac{0.8}{2}$

$A = 0.4\text{ mL}$

The amount to administer is 0.4 mL.

- Referring to Rule 9-1, a 1-mL tuberculin syringe would be used.
- Referring to Rule 9-2, we find that it is not necessary to round in this example.
- Referring to Rule 9-3, the total volume may be given as a single injection.

Ratio Proportion Method

Example 1 Find the amount to administer and select the proper syringe.

Ordered: Valium 7.5 mg IM now

On hand: Refer to Figure 9-5.

Patient: A 175-lb, 45-year-old male

Solution The dosage ordered O is 7.5 mg. The dose on hand H is 5 mg, and the dosing unit Q is 1 mL. Because the dosage ordered and the dose on hand already have the same units, no conversion is needed to find the desired dose D, which is 7.5 mg.

Follow Procedure Checklist 7-2.

1. Fill in the proportion.

$Q : H :: A : D$

$1\text{ mL} : 5\text{ mg} :: A : 7.5\text{ mg}$

2. Cancel units.

$1\text{ mL} : 5\text{ }\cancel{\text{mg}} :: A : 7.5\text{ }\cancel{\text{mg}}$

3. Multiply the means and extremes, then solve for the missing value.

$5 \times A = 1 \text{ mL} \times 7.5$

$A = 1 \text{ mL} \times \frac{7.5}{5}$

$A = 1.5 \text{ mL}$

The amount to administer is 1.5 mL.

- Referring to Rule 9-1, a standard syringe would be used.
- Referring to Rule 9-2, we find that it is not necessary to round in this example.
- Referring to Rule 9-3, the total volume may be given as a single injection.

Example 2 Find the amount to administer and select the proper syringe.

Ordered: Lorazepam 800 mcg IM now

On hand: Refer to Figure 9-6 on page 300.

Patient: An 85-lb, 12-year-old female

Solution The dosage ordered O is in micrograms, and the dose on hand H is in milligrams, so the desired dose D, in milligrams, must be calculated.

Follow Procedure Checklist 3-2.

1. Fill in the proportion, recalling that 1 mg = 1000 mcg.

   $1 \text{ mg} : 1000 \text{ mcg} :: D : 800 \text{ mcg}$

2. Cancel units.

   $1 \text{ mg} : 1000 \cancel{\text{mcg}} :: D : 800 \cancel{\text{mcg}}$

3. Multiply the means and extremes, then solve for the missing value.

   $1000 \times D = 1 \text{ mg} \times 800$

   $D = \frac{1 \text{ mg} \times 800}{1000}$

   $D = 0.8 \text{ mg}$

We now have the necessary pieces of information: $D = 0.8$ mg, $Q = 1$ mL, and $H = 2$ mg.

Follow Procedure Checklist 7-2.

1. Fill in the proportion.

   $1 \text{ mL} : 2 \text{ mg} :: A : 0.8 \text{ mg}$

2. Cancel units.

   $1 \text{ mL} : 2 \cancel{\text{mg}} :: A : 0.8 \cancel{\text{mg}}$

3. Multiply the means and extremes, then solve for the missing value.

   $2 \times A = 1 \text{ mL} \times 0.8$

   $A = \frac{1 \text{ mL} \times 0.8}{2}$

   $A = 0.4 \text{ mL}$

The amount to administer is 0.4 mL.

- Referring to Rule 9-1, a 1-mL tuberculin syringe would be used.
- Referring to Rule 9-2, we find that it is not necessary to round in this example.
- Referring to Rule 9-3, the total volume may be given as a single injection.

Dimensional Analysis

Example 1 Find the amount to administer and select the proper syringe.

Ordered: Valium 7.5 mg IM now

On hand: Refer to Figure 9-5. on page 300.

Patient: A 175-lb, 45-year-old male

Solution The dosage ordered O is 7.5 mg. The dose on hand H is 5 mg, and the dosing unit Q is 1 mL. Because the dosage ordered and the dose on hand have the same units, no conversion is needed to find the desired dose D, which is 7.5 mg.

Follow Procedure Checklist 7-3.

1. The unit of measure for the amount to administer will be milliliters.

   $A \text{ mL} =$

2. Since the unit of measure for the dosage ordered is the same as that for the dose on hand, this step is unnecessary.
3. The dosage unit is 1 mL. The dose on hand is 5 mg. This is your first factor.

   $$\frac{1 \text{ mL}}{5 \text{ mg}}$$

4. The dosage ordered is 7.5 mg. Set up the equation.

   $$A \text{ mL} = \frac{1 \text{ mL}}{5 \text{ mg}} \times \frac{7.5 \text{ mg}}{1}$$

5. Cancel units.

   $$A \text{ mL} = \frac{1 \text{ mL}}{5 \cancel{\text{mg}}} \times \frac{7.5 \cancel{\text{mg}}}{1}$$

6. Solve the equation.

   $$A \text{ mL} = \frac{1 \text{ mL}}{5} \times \frac{7.5}{1}$$

   $A = 1.5 \text{ mL}$

The amount to administer is 1.5 mL.

- Referring to Rule 9-1, a standard syringe would be used.
- Referring to Rule 9-2, we find that it is not necessary to round in this example.
- Referring to Rule 9-3, the total volume may be given as a single injection.

Example 2 Find the amount to administer and select the proper syringe.

Ordered: Lorazepam 800 mcg IM now

On hand: Refer to Figure 9-6 on page 300.

Patient: An 85-lb, 12-year-old female

Solution Follow Procedure Checklist 7-3.

1. The unit of measure for the amount to administer will be milliliters.

   A mL =

2. The unit of measure for the dosage ordered is micrograms. The unit of measure for the dose on hand is milligrams. Recall that 1 mg = 1000 mcg. Because we wish to convert to milligrams (the units of the dose on hand), our first factor must have milligrams on top.

   $$\frac{1 \text{ mg}}{1000 \text{ mcg}}$$

3. The dosage unit is 1 mL. The dose on hand is 2 mg. This is our second factor.

   $$\frac{1 \text{ mg}}{1000 \text{ mcg}} \times \frac{1 \text{ mL}}{2 \text{ mg}}$$

4. The dosage ordered is 800 mcg. Set up the equation.

   $$A \text{ mL} = \frac{1 \text{ mg}}{1000 \text{ mcg}} \times \frac{1 \text{ mL}}{2 \text{ mg}} \times \frac{800 \text{ mcg}}{1}$$

5. Cancel units.

   $$A \text{ mL} = \frac{1 \cancel{\text{mg}}}{1000 \cancel{\text{mcg}}} \times \frac{1 \text{ mL}}{2 \cancel{\text{mg}}} \times \frac{800 \cancel{\text{mcg}}}{1}$$

6. Solve the equation.

   $$A \text{ mL} = \frac{1}{1000} \times \frac{1 \text{ mL}}{2} \times \frac{800}{1}$$

   A = 0.4 mL

The amount to administer is 0.4 mL.

- Referring to Rule 9-1, a 1-mL tuberculin syringe would be used.
- Referring to Rule 9-2, we find that it is not necessary to round in this example.
- Referring to Rule 9-3, the total volume may be given as a single injection.

Formula Method

Example 1 Find the amount to administer and select the proper syringe.

Ordered: Valium 7.5 mg IM now

On hand: Refer to Figure 9-5 on page 300.

Patient: A 175-lb, 45-year-old male

Solution Follow Procedure Checklist 7-4.

1. The drug is ordered in milligrams, which is the same unit of measure as that of the dose on hand. Therefore,

   Desired dose D = 7.5 mg

   Dose on hand H = 5 mg

   Quantity to be administered Q = 1 mL

2. Fill in the formula.

$$\frac{D \times Q}{H} = A$$

$$\frac{7.5 \text{ mg} \times 1 \text{ mL}}{5 \text{ mg}} = A$$

3. Cancel units.

$$\frac{7.5 \cancel{\text{mg}} \times 1 \text{ mL}}{5 \cancel{\text{mg}}} = A$$

4. Solve for the unknown.

$$\frac{7.5 \times 1 \text{ mL}}{5} = A$$

$$1.5 \text{ mL} = A$$

The amount to administer is 1.5 mL.

- Referring to Rule 9-1, a standard syringe would be used.
- Referring to Rule 9-2, we find that it is not necessary to round in this example.
- Referring to Rule 9-3, the total volume may be given as a single injection.

Example 2 Find the amount to administer and select the proper syringe.

Ordered: Lorazepam 800 mcg IM now

On hand: Refer to Figure 9-6 on page 300.

Patient: An 85-lb, 12-year-old female

Solution The dosage ordered *Q* is in micrograms, and the dose on hand *H* is in milligrams, so the desired dose *D*, in milligrams, must be calculated.

Follow Procedure Checklist 3-1 or 3-2.

In this example, Procedure Checklist 3-1 fraction proportion method is used.

1. Fill in the proportion, recalling that 1 mg = 1000 mcg.

$$\frac{1 \text{ mg}}{1000 \text{ mcg}} = \frac{D}{800 \text{ mcg}}$$

2. Cancel units.

$$\frac{1 \text{ mg}}{1000 \cancel{\text{mcg}}} = \frac{D}{800 \cancel{\text{mcg}}}$$

3. Cross-multiply and solve for the unknown.

$$1000 \times D = 1 \times 800 \text{ mg}$$

$$D = 0.8 \text{ mg}$$

We now have the necessary pieces of information: $D = 0.8$ mg, $Q = 1$ mL, and $H = 2$ mg.

Follow Procedure Checklist 7-4.

1. Fill in the formula.

$$\frac{0.8 \text{ mg} \times 1 \text{ mL}}{2 \text{ mg}} = A$$

2. Cancel units.

$$\frac{0.8\ \cancel{mg} \times 1\text{ mL}}{2\ \cancel{mg}} = A$$

3. Solve for the unknown.

$$\frac{0.8 \times 1\text{ mL}}{2} = A$$

$$0.4\text{ mL} = A$$

The amount to administer is 0.4 mL.

- Referring to Rule 9-1, a 1-mL tuberculin syringe would be used.
- Referring to Rule 9-2, we find that it is not necessary to round in this example.
- Referring to Rule 9-3, the total volume may be given as a single injection.

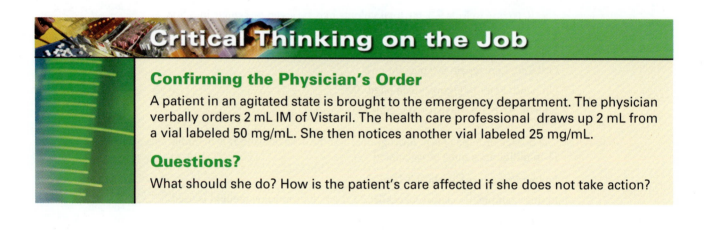

Critical Thinking on the Job

Confirming the Physician's Order

A patient in an agitated state is brought to the emergency department. The physician verbally orders 2 mL IM of Vistaril. The health care professional draws up 2 mL from a vial labeled 50 mg/mL. She then notices another vial labeled 25 mg/mL.

Questions?

What should she do? How is the patient's care affected if she does not take action?

Medications Expressed in Percent or Ratio Format

When the dosage strength is expressed as a percent or a ratio, you must convert it before calculating the amount to administer. For example, to administer 1% lidocaine, you need to rewrite the strength as 1 g (dose on hand) per 100 mL (dosage unit) before beginning your calculations. If the labeled strength were 1:2000, you would rewrite it as 1 g (dose on hand) per 2000 mL (dosage unit). Some drugs, such as heparin, are measured in units and may have their solution strength expressed in ratio format. The ratio indicates the number of units contained in 1 mL. For example, heparin sodium 1:10,000 contains 10,000 units in 1 mL.

Rule 9-4 When a solution strength is expressed as a percent or ratio:

1. Convert the percent or ratio to the dosage strength, such as grams per milliliter, milligrams per milliliter, or units per milliliter.
2. Calculate the amount to administer, then apply Rules 9-1 to 9-3.

Fraction Proportion Method

Example 1 Find the amount to administer and select the proper syringe.

Ordered: Magnesium sulfate 300 mg IM now

On hand: Magnesium sulfate 10% solution

Patient: A 75-lb, 8-year-old female

Solution Convert the dosage strength of a 10% solution to 10 g (*H*) per 100 mL (*Q*).

The dose on hand *H* is now in grams, and the dosage ordered *O* is in milligrams. To calculate the desired dose *D*, we must first convert the dosage ordered—300 mg—to the same unit of measure as that of the dose on hand, grams.

Follow Procedure Checklist 3-1.

1. Fill in the proportion, recalling that 1 g = 1000 mg.

$$\frac{1\text{ g}}{1000\text{ mg}} = \frac{D}{300\text{ mg}}$$

2. Cancel units.

$$\frac{1\text{ g}}{1000\ \cancel{\text{mg}}} = \frac{D}{300\ \cancel{\text{mg}}}$$

3. Cross-multiply and solve for the unknown.

$$1000 \times D = 300 \times 1\text{ g}$$

$$D = \frac{300 \times 1\text{ g}}{1000}$$

$$D = 0.3\text{ g}$$

We now have the necessary pieces of information: $D = 0.3$ g, $Q = 100$ mL, and $H = 10$ g.

Follow Procedure Checklist 7-1.

1. Fill in the proportion.

$$\frac{100\text{ mL}}{10\text{ g}} = \frac{A}{0.3\text{ g}}$$

2. Cancel units.

$$\frac{100\text{ mL}}{10\ \cancel{\text{g}}} = \frac{A}{0.3\ \cancel{\text{g}}}$$

3. Cross-multiply and solve for the unknown.

$$10 \times A = 100\text{ mL} \times 0.3$$

$$A = \frac{100\text{ mL} \times 0.3}{10}$$

$$A = 3\text{ mL}$$

The amount to administer is 3 mL.

- Referring to Rule 9-1, a standard syringe would be used.
- Referring to Rule 9-2, we find that it is not necessary to round in this example.
- Referring to Rule 9-3, the dose would need to be divided into two parts because of the age of the patient. Draw 1.5 mL into each of two syringes and administer them into different sites.

Example 2 Find the amount to administer and select the proper syringe.

Ordered: Epinephrine 0.2 mg Sub-Q stat

On hand: Vial of epinephrine 1:2000 solution for injection

Patient: A 150-lb, 35-year-old adult

Solution Convert the dosage strength of a 1:2000 solution to 1 g (H) per 2000 mL (Q).

The dose on hand H is now in grams, and the dosage ordered O is in milligrams. To calculate the desired dose D, we must first convert the dosage ordered 0.2 mg to the same unit of measure as that of the dose on hand, grams.

Follow Procedure Checklist 3-1.

1. Fill in the proportion, recalling that 1 g = 1000 mg.

$$\frac{1\text{ g}}{1000\text{ mg}} = \frac{D}{0.2\text{ mg}}$$

2. Cancel units.

$$\frac{1\text{ g}}{1000\text{ }\cancel{\text{mg}}} = \frac{D}{0.2\text{ }\cancel{\text{mg}}}$$

3. Cross-multiply and solve for the unknown.

$$1000 \times D = 0.2 \times 1\text{ g}$$

$$D = \frac{0.2 \times 1\text{ g}}{1000}$$

$$D = 0.0002\text{ g}$$

We now have the necessary pieces of information: $D = 0.0002$ g, $Q = 2000$ mL, and $H = 1$ g.

Follow Procedure Checklist 7-1.

1. Fill in the proportion.

$$\frac{2000\text{ mL}}{1\text{ g}} = \frac{A}{0.0002\text{ g}}$$

2. Cancel units.

$$\frac{2000\text{ mL}}{1\text{ }\cancel{\text{g}}} = \frac{A}{0.0002\text{ }\cancel{\text{g}}}$$

3. Cross-multiply and solve for the unknown.

$$1 \times A = 2000\text{ mL} \times 0.0002$$

$$A = \frac{2000\text{ mL} \times 0.0002}{1}$$

$$A = 0.4\text{ mL}$$

The amount to administer is 0.4 mL.

- Referring to Rule 9-1, a 0.5-mL tuberculin syringe would be used.
- Referring to Rule 9-2, we find that it is not necessary to round in this example.
- Referring to Rule 9-3, the dose would not need to be divided into two because the amount is less than 0.5 mL.

Example 3 Find the amount to administer and select the proper syringe.

Ordered: Heparin sodium 5000 U deep Sub-Q q8h

On hand: Heparin 1:10,000 for injection

Patient: A 145-lb, 60-year-old male

Solution Convert the dosage strength of heparin 1:10,000 to 10,000 U (*H*) per 1 mL (*Q*).

The dosage ordered *O* and the dose on hand *H* are both expressed as units, so no conversion is necessary to find the desired dose, which is 5000 U.

Follow Procedure Checklist 7-1.

1. Fill in the proportion.

$$\frac{Q}{H} = \frac{A}{D}$$

$$\frac{1 \text{ mL}}{10{,}000 \text{ U}} = \frac{A}{5000 \text{ U}}$$

2. Cancel units.

$$\frac{1 \text{ mL}}{10{,}000 \cancel{\text{U}}} = \frac{A}{5000 \cancel{\text{U}}}$$

3. Cross-multiply and solve for the unknown.

$$10{,}000 \times A = 1 \text{ mL} \times 5000$$

$$A = 1 \text{ mL} \times \frac{5000}{10{,}000}$$

$$A = 0.5 \text{ mL}$$

The amount to administer is 0.5 mL.

- Referring to Rule 9-1, a 1-mL tuberculin syringe would be used.
- Referring to Rule 9-2, we find that it is not necessary to round in this example.
- Referring to Rule 9-3, the total volume may be given as a single injection.

Ratio Proportion Method

Example 1 Find the amount to administer and select the proper syringe.

Ordered: Magnesium sulfate 300 mg IM now

On hand: Magnesium sulfate 10% solution

Patient: A 75-lb, 8-year-old female

Solution Convert the dosage strength of a 10% solution to 10 g (*H*) per 100 mL (*Q*).

The dose on hand *H* is now in grams, and the dosage ordered *O* is in milligrams. To calculate the desired dose *D*, we must first convert the dosage ordered 300 mg to the same unit of measure as that of the dose on hand, grams.

Follow Procedure Checklist 3-2.

1. Fill in the proportion, recalling that 1 g = 1000 mg.

   $1\text{ g} : 1000\text{ mg} :: D : 300\text{ mg}$

2. Cancel units.

   $1\text{ g} : 1000\,\cancel{\text{mg}} :: D : 300\,\cancel{\text{mg}}$

3. Multiply the means and extremes, then solve for the missing value.

   $1000 \times D = 1\text{ g} \times 300$

   $$D = \frac{1\text{ g} \times 300}{1000}$$

   $$D = 0.3\text{ g}$$

We now have the necessary pieces of information: $D = 0.3$ g, $Q = 100$ mL, and $H = 10$ g.

Follow Procedure Checklist 7-2.

1. Fill in the proportion.

   $100\text{ mL} : 10\text{ g} :: A : 0.3\text{ g}$

2. Cancel units.

   $100\text{ mL} : 10\,\cancel{\text{g}} :: A : 0.3\,\cancel{\text{g}}$

3. Multiply the means and extremes, then solve for the missing value.

   $100 \times A = 100\text{ mL} \times 0.3$

   $$A = \frac{100\text{ mL} \times 0.3}{10}$$

   $$A = 3\text{ mL}$$

The amount to administer is 3 mL.

- Referring to Rule 9-1, a standard syringe would be used.
- Referring to Rule 9-2, we find that it is not necessary to round in this example.
- Referring to Rule 9-3, the dose would need to be divided into two parts because of the age of the patient. Draw 1.5 mL into each of two syringes, and administer them into different sites.

Example 2 Find the amount to administer and select the proper syringe.

Ordered: Epinephrine 0.2 mg Sub-Q stat

On hand: Vial of epinephrine 1:2000 solution for injection

Patient: A 150-lb, 35-year-old adult

Solution Convert the dosage strength of a 1:2000 solution to 1 g (H) per 2000 mL (Q).

The dose on hand H is now in grams, and the dosage ordered O is in milligrams. To calculate the desired dose D, we must first convert the dosage ordered 0.2 mg to the same unit of measure as that of the dose on hand, grams.

Follow Procedure Checklist 3-2.

1. Fill in the proportion, recalling that 1 g = 1000 mg.

   $1\text{ g} : 1000\text{ mg} :: D : 0.2\text{ mg}$

2. Cancel units.

   1 g : 1000 ~~mg~~ :: D : 0.2 ~~mg~~

3. Multiply the means and extremes, then solve for the missing value.

   $100 \times D = 1 \text{ g} \times 0.2$

   $$D = \frac{1 \text{ g} \times 0.2}{1000}$$

   $$D = 0.0002 \text{ g}$$

We now have the necessary pieces of information: $D = 0.0002$ g, $Q = 2000$ mL, and $H = 1$ g.

Follow Procedure Checklist 7-2.

1. Fill in the proportion.

   2000 mL : 1 g :: A : 0.0002 g

2. Cancel units.

   2000 mL : 1 ~~g~~ :: A : 0.0002 ~~g~~

3. Multiply the means and extremes, then solve for the missing value.

   $1 \times A = 2000 \text{ mL} \times 0.0002$

   $$A = \frac{2000 \text{ mL} \times 0.0002}{1}$$

   $$A = 0.4 \text{ mL}$$

The amount to administer is 0.4 mL.

- Referring to Rule 9-1, a 0.5-mL tuberculin syringe would be used.
- Referring to Rule 9-2, we find that it is not necessary to round in this example.
- Referring to Rule 9-3, the dose would not need to be divided into two because the amount is less than 0.5 mL.

Example 3 Find the amount to administer and select the proper syringe.

Ordered: Heparin sodium 5000 U deep Sub-Q q8h

On hand: Heparin 1:10,000 for injection

Patient: A 145-lb, 60-year-old male

Solution Convert the dosage strength of heparin 1:10,000 to 10,000 U (H) per 1 mL (Q).

The dosage ordered O and the dose on hand H are both expressed as units, so no conversion is necessary to find the desired dose, which is 5000 U.

Follow Procedure Checklist 7-2.

1. Fill in the proportion.

   1 mL : 10,000 U :: A : 5000 U

2. Cancel units.

   1 mL : 10,000 ~~U~~ :: A : 5000 ~~U~~

3. Multiply the means and extremes, then solve for the missing value.

$10{,}000 \times A = 1 \text{ mL} \times 5000$

$$A = \frac{1 \text{ mL} \times 5000}{10{,}000}$$

$$A = 0.5 \text{ mL}$$

The amount to administer is 0.5 mL.

- Referring to Rule 9-1, a 1-mL tuberculin syringe would be used.
- Referring to Rule 9-2, we find that it is not necessary to round in this example.
- Referring to Rule 9-3, the total volume may be given as a single injection.

Dimensional Analysis

Example 1 Find the amount to administer and select the proper syringe.

Ordered: Magnesium sulfate 300 mg IM now

On hand: Magnesium sulfate 10% solution

Patient: A 75-lb, 8-year-old female

Solution Convert the dosage strength of a 10% solution to 10 g (*H*) per 100 mL (*Q*).

Follow Procedure Checklist 7-3.

1. The unit of measure for the amount to administer will be milliliters.

   $A \text{ mL} =$

2. The unit of measure for the dosage ordered is milligrams. The unit of measure for the dose on hand is grams. Recall that 1 g = 1000 mg. Because we wish to convert to grams (the units of the dose on hand), our first factor must have grams on top.

   $$\frac{1 \text{ g}}{1000 \text{ mg}}$$

3. The dosage unit is 100 mL. The dose on hand is 10 g. This is our second factor.

   $$\frac{1 \text{ g}}{1000 \text{ mg}} \times \frac{100 \text{ mL}}{10 \text{ g}}$$

4. The dosage ordered is 300 mg. Place this over 1 and set up the equation.

   $$A \text{ mL} = \frac{1 \text{ g}}{1000 \text{ mg}} \times \frac{100 \text{ mL}}{10 \text{ g}} \times \frac{300 \text{ mg}}{1}$$

5. Cancel units.

   $$A \text{ mL} = \frac{1 \cancel{\text{g}}}{1000 \cancel{\text{mg}}} \times \frac{100 \text{ mL}}{10 \cancel{\text{g}}} \times \frac{300 \cancel{\text{mg}}}{1}$$

6. Solve the equation.

   $$A \text{ mL} = \frac{1}{1000} \times \frac{100 \text{ mL}}{10} \times \frac{300}{1}$$

   $A = 3 \text{ mL}$

The amount to administer is 3 mL.

- Referring to Rule 9-1, a standard syringe would be used.
- Referring to Rule 9-2, we find that it is not necessary to round in this example.
- Referring to Rule 9-3, the dose would need to be divided into two parts because of the age of the patient. Draw 1.5 mL into each of two syringes, and administer them into different sites.

Example 2 Find the amount to administer and select the proper syringe.

Ordered: Epinephrine 0.2 mg Sub-Q stat

On hand: Vial of epinephrine 1:2000 solution for injection

Patient: A 150-lb, 35-year-old adult

Solution Convert the dosage strength of a 1:2000 solution to 1 g (*H*) per 2000 mL (*Q*).

The dose on hand *H* is now in grams, and the dosage ordered *O* is in milligrams. To calculate the desired dose *D*, we must first change the dosage ordered 0.2 mg to the same unit of measure as that of the dose on hand, grams.

Follow Procedure Checklist 7-3.

1. The unit of measure for the amount to administer will be milliliters.

   $A \text{ mL} =$

2. The unit of measure for the dosage ordered is milligrams. The unit of measure for the dose on hand is grams. Recall that 1 g = 1000 mg. Because we wish to convert to grams (the units of the dose on hand), our first factor must have grams on top.

   $$\frac{1 \text{ g}}{1000 \text{ mg}}$$

3. The dosage unit is 2000 mL. The dose on hand is 1 g. This is our next factor.

   $$\frac{1 \text{ g}}{1000 \text{ mg}} \times \frac{2000 \text{ mL}}{1 \text{ g}}$$

4. The dosage ordered is 0.2 mg. Place this over 1 and set up the equation.

   $$A \text{ mL} = \frac{1 \text{ g}}{1000 \text{ mg}} \times \frac{2000 \text{ mL}}{1 \text{ g}} \times \frac{0.2 \text{ mg}}{1}$$

5. Cancel units.

   $$A \text{ mL} = \frac{1 \cancel{\text{g}}}{1000 \cancel{\text{mg}}} \times \frac{2000 \text{ mL}}{1 \cancel{\text{g}}} \times \frac{0.2 \cancel{\text{mg}}}{1}$$

6. Solve the equation.

   $$A \text{ mL} = \frac{2000 \text{ mL} \times 0.2 \cancel{\text{mg}}}{1000 \cancel{\text{mg}}}$$

   $A = 0.4 \text{ mL}$

The amount to administer is 0.4 mL.

- Referring to Rule 9-1, a 0.5-mL tuberculin syringe would be used.
- Referring to Rule 9-2, we find that it is not necessary to round in this example.
- Referring to Rule 9-3, the dose would not need to be divided into two because the amount is less than 0.5 mL.

Example 3 Find the amount to administer and select the proper syringe.

Ordered: Heparin sodium 5000 U deep Sub-Q q8h

On hand: Heparin 1:10,000 for injection

Patient: A 145-lb, 60-year-old male

Solution Convert the dosage strength of heparin 1:10,000 to 10,000 U (*H*) per 1 mL (*Q*).

The dosage ordered *O* and the dose on hand *H* are both expressed as units, so no conversion is necessary to find the desired dose, which is 5000 U.

Follow Procedure Checklist 7-3.

1. The unit of measure for the amount to administer will be milliliters.

   $A \text{ mL} =$

2. Since the unit of measure for the dosage ordered is the same as that for the dose on hand, this step is unnecessary.
3. The dosage unit is 1 mL. The dose on hand is 10,000 U. This is the first factor.

   $$\frac{1 \text{ mL}}{10{,}000 \text{ U}}$$

4. The dosage ordered is 5000 U. Set up the equation.

   $$A \text{ mL} = \frac{1 \text{ mL}}{10{,}000 \text{ U}} \times \frac{5000 \text{ U}}{1}$$

5. Cancel units.

   $$A \text{ mL} = \frac{1 \text{ mL}}{10{,}000 \not{\text{U}}} \times \frac{5000 \not{\text{U}}}{1}$$

6. Solve the equation.

   $$A \text{ mL} = \frac{1 \text{ mL}}{10{,}000} \times \frac{5000}{1}$$

   $$A = 0.5 \text{ mL}$$

The amount to administer is 0.5 mL.

- Referring to Rule 9-1, a 1-mL tuberculin syringe would be used.
- Referring to Rule 9-2, we find that it is not necessary to round in this example.
- Referring to Rule 9-3, the total volume may be given as a single injection.

Formula Method

Example 1 Find the amount to administer and select the proper syringe.

Ordered: Magnesium sulfate 300 mg IM now

On hand: Magnesium sulfate 10% solution

Patient: A 75-lb, 8-year-old female

Solution Convert the dosage strength of a 10% solution to 10 g (H) per 100 mL (Q).

The dosage ordered O is in milligrams, and the dose on hand H is in grams, so the desired dose D, in grams, must be calculated.

Follow Procedure Checklist 3-1 or 3-2.

In this example, Procedure Checklist 3-1 fraction proportion method is used.

1. Fill in the proportion, recalling that 1 g = 1000 mg.

$$\frac{1\text{ g}}{1000\text{ mg}} = \frac{D}{300\text{ mg}}$$

2. Cancel units.

$$\frac{1\text{ g}}{1000\text{ }\cancel{\text{mg}}} = \frac{D}{300\text{ }\cancel{\text{mg}}}$$

3. Cross-multiply and solve for the unknown.

$$1000 \times D = 300 \times 1\text{g}$$

$$D = \frac{300 \times 1\text{ g}}{1000}$$

$$D = 0.3\text{ g}$$

We now have the necessary pieces of information: $D = 0.3$ g, $Q = 100$ mL, and $H = 10$ g.

Follow Procedure Checklist 7-4.

1. Fill in the formula.

$$\frac{0.3\text{ g} \times 100\text{ mL}}{10\text{ g}} = A$$

2. Cancel units.

$$\frac{0.3\,\cancel{\text{g}} \times 100\text{ mL}}{10\,\cancel{\text{g}}} = A$$

3. Solve for the unknown.

$$\frac{0.3 \times 100\text{ mL}}{10} = A$$

$$3\text{ mL} = A$$

The amount to administer is 3 mL.

- Referring to Rule 9-1, a standard syringe would be used.
- Referring to Rule 9-2, we find that it is not necessary to round in this example.
- Referring to Rule 9-3, the dose would need to be divided into two parts because of the age of the patient. Draw 1.5 mL into each of two syringes, and administer them into different sites.

Example 2 Find the amount to administer and select the proper syringe.

Ordered: Epinephrine 0.2 mg Sub-Q stat

On hand: Vial of epinephrine 1:2000 solution for injection

Patient: A 150-lb, 35-year-old adult

Solution Convert the dosage strength of a 1:2000 solution to 1 g (H) per 2000 mL (Q).

The dose on hand (H) is now in grams, and the dosage ordered O is in milligrams. To calculate the desired dose D, we must first convert the dosage ordered 0.2 mg to the same unit of measure as that of the dose on hand, grams.

Follow Procedure Checklist 3-1.

1. Fill in the proportion, recalling that 1 g = 1000 mg.

$$\frac{1\text{ g}}{1000\text{ mg}} = \frac{D}{0.2\text{ mg}}$$

2. Cancel units.

$$\frac{1\text{ g}}{1000\ \cancel{\text{mg}}} = \frac{D}{0.2\ \cancel{\text{mg}}}$$

3. Cross-multiply and solve for the unknown.

$$1000 \times D = 0.2 \times 1\text{g}$$

$$D = \frac{0.2 \times 1\text{ g}}{1000}$$

$$D = 0.0002\text{ g}$$

We now have the necessary pieces of information: $D = 0.0002$ g, $Q = 2000$ mL, and $H = 1$ g.

Follow Procedure Checklist 7-4.

1. Fill in the formula.

$$\frac{0.0002\text{ g} \times 2000\text{ mL}}{1\text{ g}} = A$$

2. Cancel units.

$$\frac{0.0002\ \cancel{\text{g}} \times 2000\text{ mL}}{1\ \cancel{\text{g}}} = A$$

3. Solve for the unknown.

$$\frac{0.0002 \times 2000\text{ mL}}{1} = A$$

$$0.4\text{ mL} = A$$

The amount to administer is 0.4 mL.

- Referring to Rule 9-1, a 0.5-mL tuberculin syringe would be used.
- Referring to Rule 9-2, we find that it is not necessary to round in this example.
- Referring to Rule 9-3, the dose would not need to be divided into two because the amount is less than 0.5 mL.

Example 3 Find the amount to administer and select the proper syringe.

Ordered: Heparin sodium 5000 U deep Sub-Q q8h

On hand: Heparin 1:10,000 for injection

Patient: A 145-lb, 60-year-old male

Solution Convert the dosage strength of heparin 1:10,000 to 10,000 U (H) per 1 mL (Q).

The dosage ordered O and the dose on hand H are both expressed as units, so no conversion is necessary to find the desired dose, which is 5000 U.

Follow Procedure Checklist 7-4.

1. Fill in the formula.

$$\frac{5000\text{ U} \times 1\text{ mL}}{10{,}000\text{ U}} = A$$

2. Cancel units.

$$\frac{5000\,\cancel{\text{U}} \times 1\text{ mL}}{10{,}000\,\cancel{\text{U}}} = A$$

3. Solve for the unknown.

$$\frac{5000\text{ mL}}{10{,}000} = A$$

$$0.5\text{ mL} = A$$

The amount to administer is 0.5 mL.

- Referring to Rule 9-1, a 1-mL tuberculin syringe would be used.
- Referring to Rule 9-2, we find that it is not necessary to round in this example.
- Referring to Rule 9-3, the total volume may be given as a single injection.

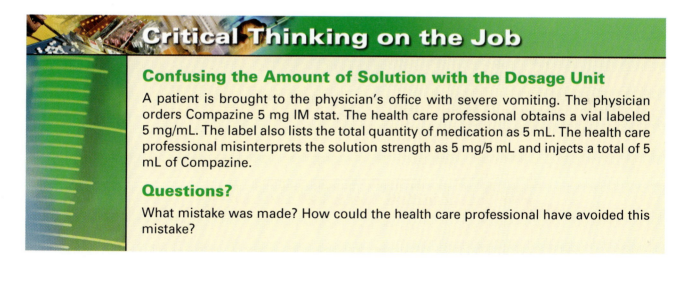

Critical Thinking on the Job

Confusing the Amount of Solution with the Dosage Unit

A patient is brought to the physician's office with severe vomiting. The physician orders Compazine 5 mg IM stat. The health care professional obtains a vial labeled 5 mg/mL. The label also lists the total quantity of medication as 5 mL. The health care professional misinterprets the solution strength as 5 mg/5 mL and injects a total of 5 mL of Compazine.

Questions?

What mistake was made? How could the health care professional have avoided this mistake?

REVIEW AND PRACTICE

9-1 Calculating Parenteral Dosages

Find the amount to administer for each of the following orders. Then mark the syringe with the correct amount to administer.

1. Ordered: Magnesium sulfate 750 mg stat
   On hand: Magnesium sulfate 20% solution

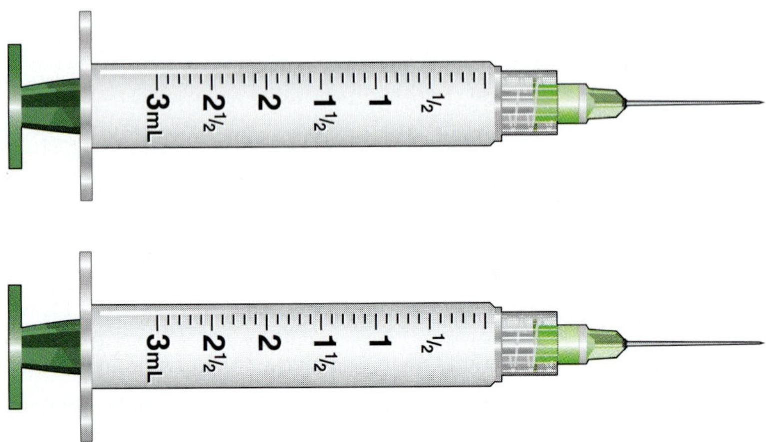

2. Ordered: Lidocaine 200 mg IM stat
   On hand: Lidocaine 10% solution

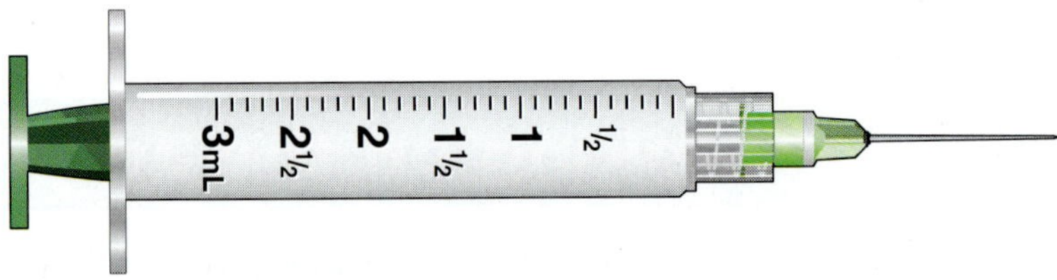

3. Ordered: Epinephrine 0.3 mg Sub-Q stat
   On hand: Epinephrine 1:1000 solution

4. Ordered: Adrenalin 0.5 mg Sub-Q stat
   On hand: Adrenalin 1:1000 solution

5. Ordered: Prostigmin 0.2 mg IM post-op q6h
   On hand: Prostigmin 1:4000 solution

6. Ordered: Prostigmin 0.5 mg IM stat
   On hand: Prostigmin 1:2000 solution

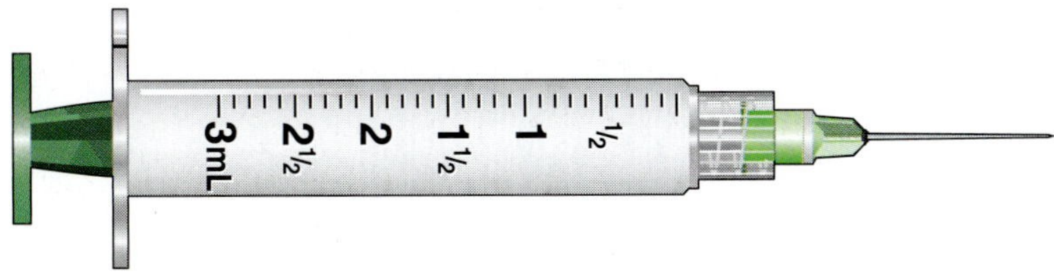

7. Ordered: Heparin sodium 8000 U deep Sub-Q q8h
   On hand: Heparin sodium 1:5000 solution

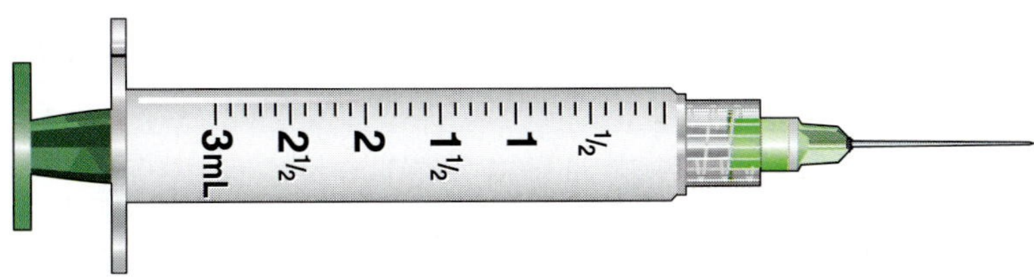

8. Ordered: Heparin sodium 5000 U Sub-Q q12h × 7
   On hand: Heparin sodium 1:10,000 solution

For Exercises 9 to 25, find the amount to administer, and then determine the proper syringe and write it in the space provided.

9. Ordered: Thiamine 100 mg IM now
   On hand: Refer to label A.
   Administer: __________ Syringe: __________

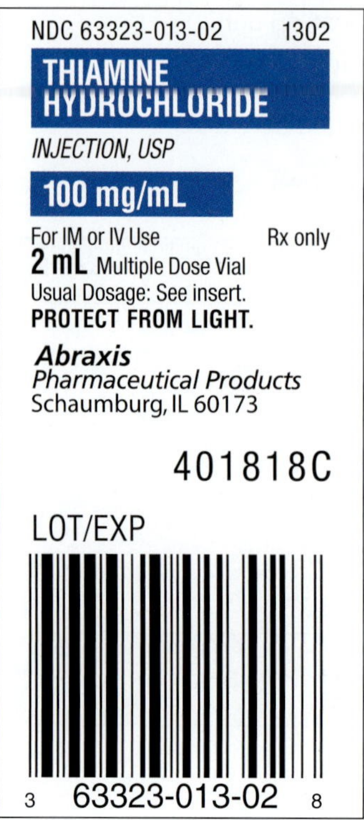
NDC 63323-013-02 1302
THIAMINE HYDROCHLORIDE
INJECTION, USP
100 mg/mL
For IM or IV Use
Rx only
2 mL Multiple Dose Vial
Usual Dosage: See insert.
PROTECT FROM LIGHT.
Abraxis
Pharmaceutical Products
Schaumburg, IL 60173
401818C
LOT/EXP
3 63323-013-02 8

A

10. Ordered: Heparin 700 units Sub-Q daily
On hand: Refer to label B.
Administer: __________ Syringe: __________

NDC 63323-540-11 504011
HEPARIN SODIUM
INJECTION, USP
1,000 USP Units/mL
(Derived from Porcine Intestinal Mucosa)
For IV or SC Use
10 mL
Multiple Dose Vial
Rx only
Sterile, Nonpyrogenic
Each mL contains: 1,000 USP Units heparin sodium; 9 mg sodium chloride; 0.15% methylparaben; 0.015% propylparaben; Water for Injection q.s. Made isotonic with sodium chloride. Hydrochloric acid and/or sodium hydroxide may have been added for pH adjustment.
Usual Dosage: See insert.
Use only if solution is clear and seal intact.
Store at 20° to 25°C (68° to 77°F) [see USP Controlled Room Temperature].
Abraxis
Pharmaceutical Products
Schaumburg, IL 60173
401586B
LOT/EXP
3 63323-540-11 1

B

11. Ordered: Tigan® 200 mg IM TID
On hand: Refer to label C.
Administer: __________ Syringe: __________

NDC 61570-541-20
100mg/mL
Tigan®
(trimethobenzamide HCl)
Injection
Rx Only
20mL Multi-Dose Vial
Monarch Pharmaceuticals®
NOT FOR USE IN CHILDREN.
For IM USE ONLY.
Store at 25° (77°F) (see insert).
Each mL of solution contains 100 mg trimethobenzamide hydrochloride compounded with 0.45% phenol as preservative, 0.5 mg sodium citrate and 0.2 mg citric acid as buffers, and sodium hydroxide to adjust pH to approximately 5.0.
Dosage: See accompanying prescribing information. For IM use only (preferably by deep IM injection).
Distributed by: Monarch Pharmaceuticals, Inc., Bristol, TN 37620
Manufactured by: King Pharmaceuticals, Inc., Bristol, TN 37620
3 61570 54120 2
540G010

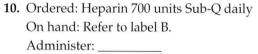

C

12. Ordered: Sandostatin® 200 mcg Sub-Q q12h
On hand: Refer to label D.
Administer: __________ Syringe: __________

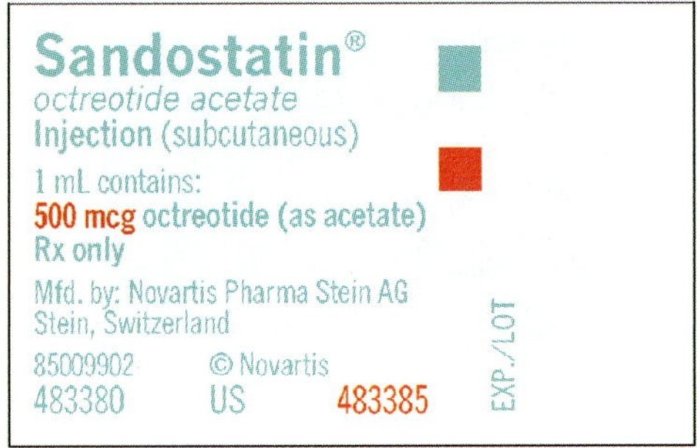

D

13. Ordered: Neupogen® 180 mcg Sub-Q qd
    On hand: Refer to label E.
    Administer: __________ Syringe: __________

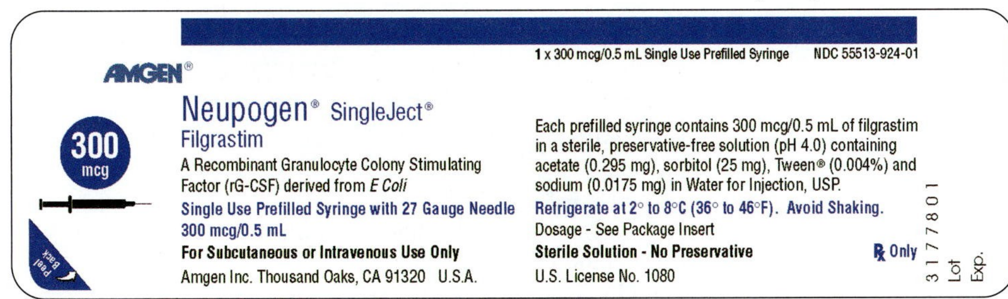

E

14. Ordered: Neupogen® 240 mcg Sub-Q qd
    On hand: Refer to label F.
    Administer: __________ Syringe: __________

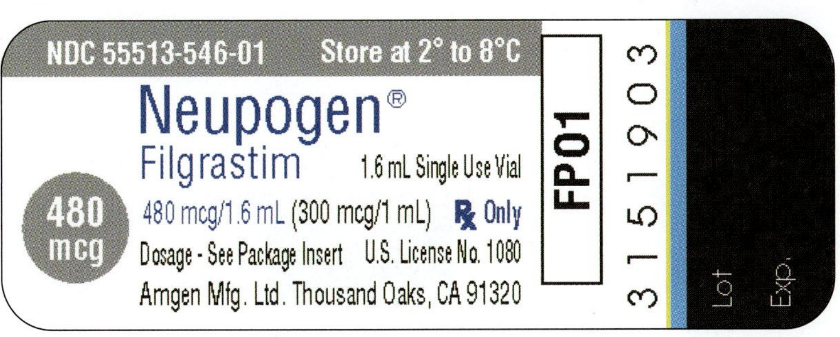

F

15. Ordered: Epogen® 3500 U Sub-Q tiw
    On hand: Refer to label G.
    Administer: __________ Syringe: __________

G

16. Ordered: Neulasta® 6 mg Sub-Q now
    On hand: Refer to label H.
    Administer: __________ Syringe: __________

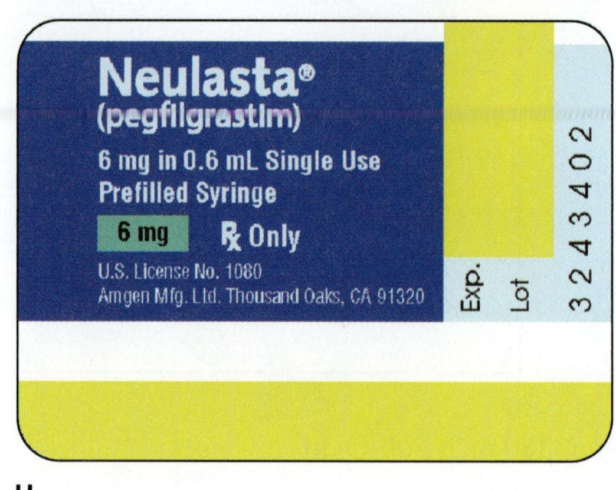

H

17. Ordered: Zemplar™ 3 mcg IM qd
    On hand: Refer to label I.
    Administer: __________ Syringe: __________

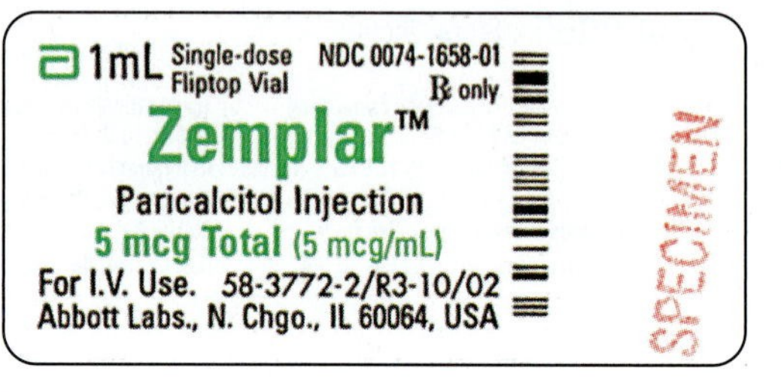

1mL Single-dose Fliptop Vial
NDC 0074-1658-01
℞ only
Zemplar™
Paricalcitol Injection
5 mcg Total (5 mcg/mL)
For I.V. Use. 58-3772-2/R3-10/02
Abbott Labs., N. Chgo., IL 60064, USA
SPECIMEN

I

18. Ordered: Fragmin® 5000 IU Sub-Q qd post-op × 5
    On hand: Refer to label J.
    Administer: __________ Syringe: __________

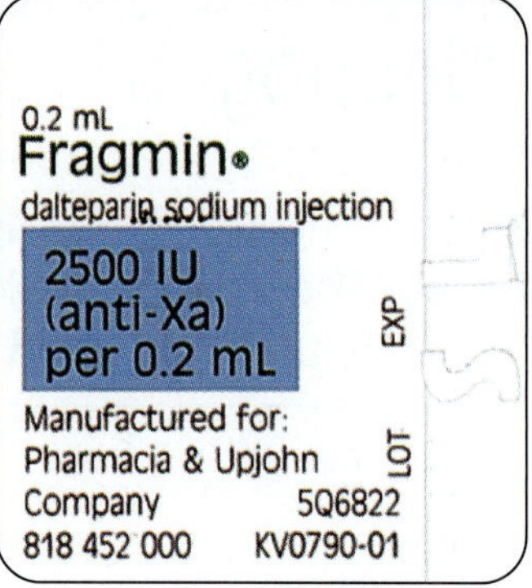

0.2 mL
Fragmin®
dalteparin sodium injection
2500 IU
(anti-Xa)
per 0.2 mL
Manufactured for:
Pharmacia & Upjohn
Company
818 452 000
5Q6822
KV0790-01
EXP
LOT

J

19. Ordered: Clindamycin 600 mg IM pre-op
    On hand: Refer to label K.
    Administer: __________ Syringe: __________

K

20. Ordered: Synagis® 35 mg IM now
On hand: Refer to label L.
Administer: __________ Syringe: __________

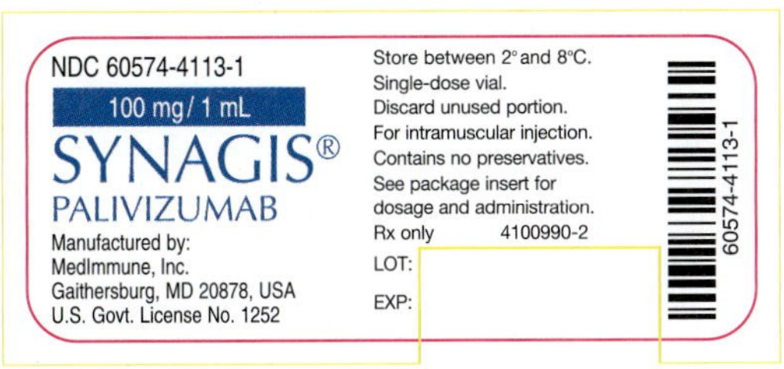

NDC 60574-4113-1
100 mg / 1 mL
SYNAGIS®
PALIVIZUMAB
Manufactured by:
MedImmune, Inc.
Gaithersburg, MD 20878, USA
U.S. Govt. License No. 1252
Store between 2° and 8°C.
Single-dose vial.
Discard unused portion.
For intramuscular injection.
Contains no preservatives.
See package insert for dosage and administration.
Rx only 4100990-2
LOT:
EXP:
60574-4113-1

L

21. Ordered: Oxytocin 15 units IM q 12h prn
On hand: Refer to label M.
Administer: __________ Syringe: __________

M

22. Ordered: Furosemide 20 mg IM at 0800
On hand: Refer to label N.
Administer: __________ Syringe: __________

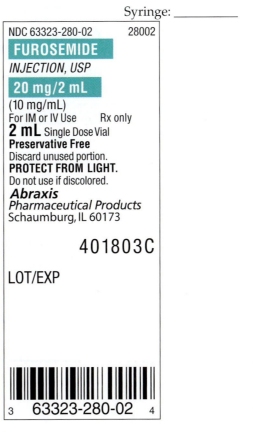

NDC 63323-280-02 28002
FUROSEMIDE
INJECTION, USP
20 mg/2 mL
(10 mg/mL)
For IM or IV Use Rx only
2 mL Single Dose Vial
Preservative Free
Discard unused portion.
PROTECT FROM LIGHT.
Do not use if discolored.
Abraxis
Pharmaceutical Products
Schaumburg, IL 60173
401803C
LOT/EXP
3 63323-280-02 4

N

23. Ordered: Aranesp® 25 mcg sub-q now
On hand: Refer to label O.
Administer: _________ Syringe: _________

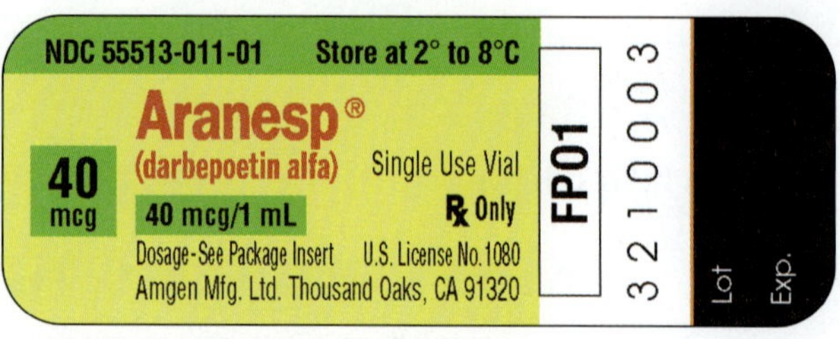

O

24. Ordered: Valium® 10 mg IM
On hand: Refer to label P.
Administer: _________ Syringe: _________

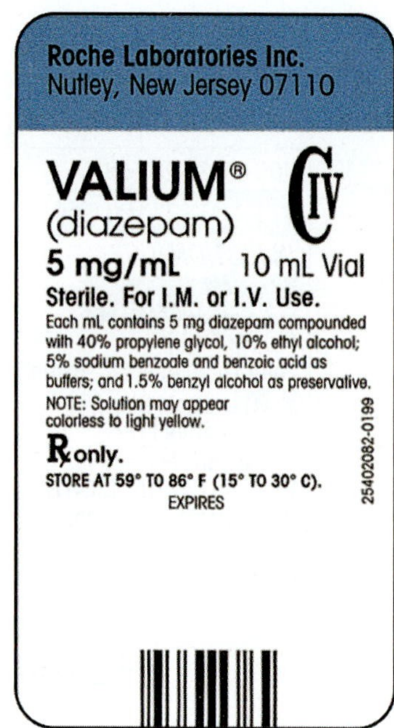

Roche Laboratories Inc.
Nutley, New Jersey 07110
VALIUM®
CIV
(diazepam)
5 mg/mL
10 mL Vial
Sterile. For I.M. or I.V. Use.
Each mL contains 5 mg diazepam compounded with 40% propylene glycol, 10% ethyl alcohol; 5% sodium benzoate and benzoic acid as buffers; and 1.5% benzyl alcohol as preservative.
NOTE: Solution may appear colorless to light yellow.
Rx only.
STORE AT 59° TO 86° F (15° TO 30° C).
EXPIRES
25402082-0199

P

25. Ordered: Lidocaine 25 mg Sub-Q
On hand: Lidocaine 5% solution
Administer: _________ Syringe: _________

To check your answers, see pages 525–526.

Reconstituting Powdered Medications

Medications that lose potency quickly in solution may be supplied in powdered form. When needed, they are reconstituted by dissolving them in an appropriate solvent (or diluent). The drug label, package insert, and PDR provide instructions for reconstituting a medication. Be sure to use the directions specific to the medication you plan to administer.

First, determine what solvent should be used to dilute the medication. Common solvents include sterile water, saline, or a bacteriostatic solution containing a preservative that prevents the growth of microorganisms. Some medications are packaged with a separate container of the appropriate solvent.

Many medications, especially antibiotics, cause severe pain when injected. They may be mixed with lidocaine, a local anesthetic, to reduce this pain. The label or package insert indicates when lidocaine can be used. *Because lidocaine is itself a medication, you need a physician's order to use it*. Therefore, check whether the physician has ordered lidocaine. Do not confuse it with the combination of lidocaine and epinephrine, because epinephrine causes vasoconstriction, a tightening of the blood vessels, which delays medication absorption.

The label or package insert lists how much solvent to combine with the medication. Read the directions carefully. Sometimes different amounts of solvent are used, based on whether the medication is for IM or IV use.

Rule 9-5 To reconstitute a powdered medication:

1. Find the directions on the medication label or package insert.
2. Use a sterile syringe and aseptic (germ-free) technique to draw up the correct amount of the appropriate diluent.
3. Inject the diluent into the medication vial.
4. Agitate the mixture by rolling, inverting, or shaking the vial. Check the directions on the label or package insert for which of these methods to use.
5. Make sure that the powdered medication is completely dissolved and that the solution is free of visible particles before you use it.

GO TO . . . Open the CD-ROM that accompanies your textbook, and select Chapter 9, Rule 9-5. Review the animation and example problems, then complete the practice problems. Continue to the next section of the book once you have mastered the rule presented. ■

You must use the exact amount of solvent indicated in the directions to produce a solution with the correct dosage strength. Powder takes up volume even when dissolved. The volume of the reconstituted medication includes the volume of the solvent and the volume of the powder.

If less than the recommended amount of solvent is used, the powder may not dissolve completely, making the solution unsafe to administer. If too much solvent is used, then the patient will not receive the desired dose. When you prepare a suspension, remember that the particles will not dissolve completely. Your goal is to distribute them evenly.

Some vials contain a single dose of medication. Many must be reconstituted immediately before administering them, because they quickly lose potency. Other such medications can be stored for a short time after reconstitution. In some facilities, medications are reconstituted in the pharmacy and delivered ready to use.

Rule 9-6 When you store a medication after reconstituting it:

1. You must record the date, the time of expiration, and your name or initials.
2. For multiple-dose medications, also record the solution strength.

Check the drug label or package insert for the length of time a reconstituted medication may be stored. Storage time may depend on whether the medication is refrigerated.

Example 1 How would you reconstitute and label the following medication?

Ordered: Glucagon 1 mg IM stat

On hand: Refer to the labels in Figure 9-7.

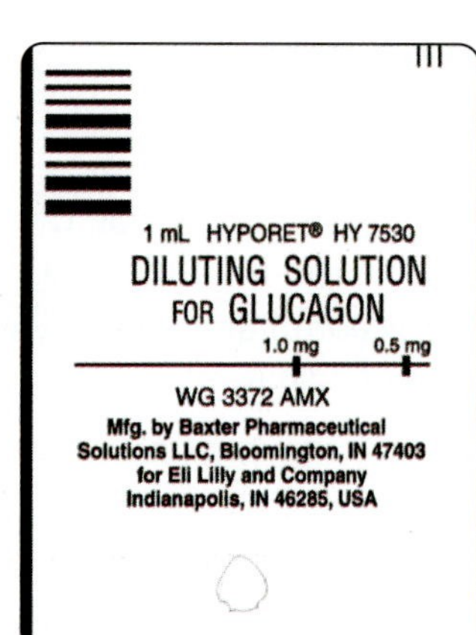

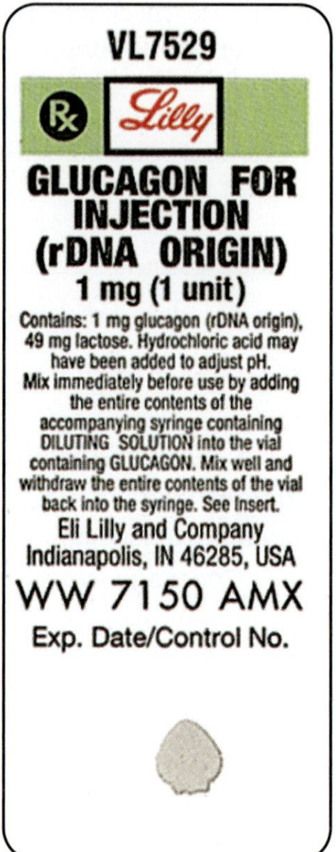

Figure 9-7

A 1-mL vial of diluting solution is provided. Once mixed, the solution must either be used immediately or be discarded. Because the mixed solution will not be stored, you do not need to label the vial.

Example 2 How would you reconstitute and label the following medication? Find the amount to administer.

Ordered: Zyprexa® 5 mg IM now

On hand: Refer to the label and portion of package insert shown in Figures 9-8 and 9-9.

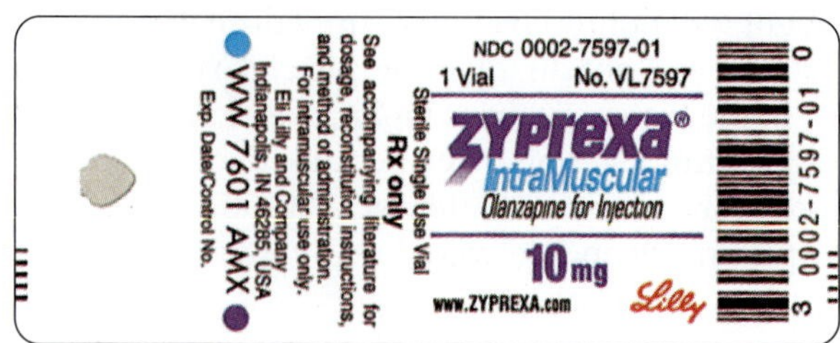

Figure 9-8

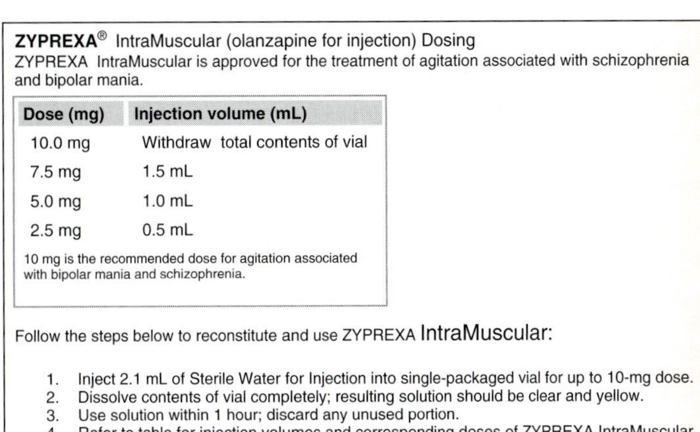

ZYPREXA® IntraMuscular (olanzapine for injection) Dosing
ZYPREXA IntraMuscular is approved for the treatment of agitation associated with schizophrenia and bipolar mania.

Dose (mg)	Injection volume (mL)
10.0 mg	Withdraw total contents of vial
7.5 mg	1.5 mL
5.0 mg	1.0 mL
2.5 mg	0.5 mL

10 mg is the recommended dose for agitation associated with bipolar mania and schizophrenia.

Follow the steps below to reconstitute and use ZYPREXA IntraMuscular:

1. Inject 2.1 mL of Sterile Water for Injection into single-packaged vial for up to 10-mg dose.
2. Dissolve contents of vial completely; resulting solution should be clear and yellow.
3. Use solution within 1 hour; discard any unused portion.
4. Refer to table for injection volumes and corresponding doses of ZYPREXA IntraMuscular.
5. Immediately after use, dispose of syringe in approved sharps box.

Figure 9-9 Package insert.

The diluent used to reconstitute this medication is 2.1 mL of sterile water for doses up to 10 mg. Contents must be dissolved completely, and fluid will be clear and yellow. The solution can only be used for 1 h. When it is prepared, you will administer 1 mL to deliver 5 mg of medication.

Example 3 How would you reconstitute and label the following medication? Find the amount to administer.

Ordered: Methylprednisolone 30 mg IM at 0930

On hand: Refer to the label (Figure 9-10) and package insert (Figure 9-11).

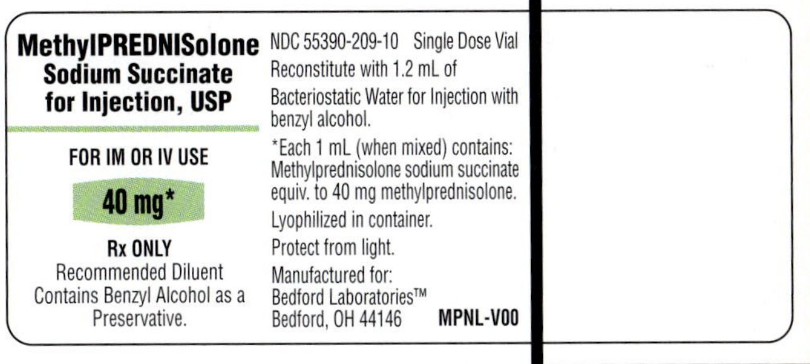

MethylPREDNISolone
Sodium Succinate
for Injection, USP
FOR IM OR IV USE
40 mg*
Rx ONLY
Recommended Diluent
Contains Benzyl Alcohol as a
Preservative.
NDC 55390-209-10 Single Dose Vial
Reconstitute with 1.2 mL of
Bacteriostatic Water for Injection with
benzyl alcohol.
*Each 1 mL (when mixed) contains:
Methylprednisolone sodium succinate
equiv. to 40 mg methylprednisolone.
Lyophilized in container.
Protect from light.
Manufactured for:
Bedford Laboratories™
Bedford, OH 44146
MPNL-V00

Figure 9-10

Methylprednisolone sodium succinate for injection may be administered by intravenous or intramuscular injection or by intravenous infusion, the preferred method for initial emergency use being intravenous injection. To administer by intravenous (or intramuscular) injection, reconstitute the product as follows:

The 40 mg single-dose vial is reconstituted by adding 1.2 mL bacteriostatic water for injection with benzyl alcohol. The 125 mg single-dose vial is reconstituted by adding 2.1 mL bacteriostatic water for injection with benzyl alcohol. The 500 mg single-dose vial is reconstituted by adding 4 mL bacteriostatic water for injection with benzyl alcohol. The 1 gram single-dose vial is reconstituted by adding 7.5 mL bacteriostatic water for injection with benzyl alcohol. The desired dose may be administered intravenously over a period of several minutes.

To prepare solutions for intravenous infusion, first prepare the solution for injection as directed. This solution may then be added to indicated amounts of 5% dextrose in water, isotonic saline solution or 5% dextrose in isotonic saline solution.

Multiple Sclerosis

In treatment of acute exacerbations of multiple sclerosis, daily doses of 200 mg of prednisolone for a week followed by 80 mg every other day for 1 month have been shown to be effective (4 mg of methylprednisolone sodium succinate for injection is equivalent to 5 mg of prednisolone).

STORAGE CONDITIONS

Protect from light.
Store unreconstituted product at 20° to 25°C (68° to 77°F). See USP controlled room temperature.
Store solution at 20° to 25°C (68° to 77°F). See USP controlled room temperature.
Use solution within 48 hours after mixing.

Figure 9-11 Package insert.

According to the label and package insert the medication should be reconstituted with 1.2 mL bacteriostatic water for injection with benzyl alcohol. Once reconstituted the vial will contain 40 mg/mL.

According the package insert the resulting solution should be used within 48 hours of mixing. If you mix the medication at 10 a.m. on 6/17/09 and store at room temperature, label the vial Exp. 6/19/09 10 a.m., 40 mg/mL. The medication should also be protected from light before and after reconstitution.

To calculate the amount to administer for example 3:

Fraction Proportion Method

Solution The dosage ordered *O* and the dose on hand *H* are already expressed in the same units, so the desired dose *D* is 30 mg. After reconstitution the dosage strength is 40 mg/mL, making the dosage unit *Q* 1 mL, and the dose on hand *H* is 40 mg.

Follow Procedure Checklist 7-1.

1. Fill in the proportion.

$$\frac{Q}{H} = \frac{A}{D}$$

$$\frac{1\text{ mL}}{40\text{ mg}} = \frac{A}{30\text{ mg}}$$

2. Cancel units.

$$\frac{1\text{ mL}}{40\,\cancel{\text{mg}}} = \frac{A}{30\,\cancel{\text{mg}}}$$

3. Cross-multiply and solve for the unknown.

$$40 \times A = 1\text{ mL} \times 30$$

$$A = 1\text{ mL} \times \frac{30}{40}$$

$$A = 0.75\text{ mL}$$

Using a 1-mL tuberculin syringe, we will administer 0.75 mL.

Ratio Proportion Method

Solution The dosage ordered *O* and the dose on hand *H* are already expressed in the same units, so the desired dose *D* is 30 mg. After reconstitution the dosage strength is 40 mg/mL, making the dosage unit *Q* 1 mL, and the dose on hand *H* is 40 mg.

Follow Procedure Checklist 7-2.

1. Fill in the proportion.

$$Q : H :: A : D$$

$$1\text{ mL} : 40\text{ mg} :: A : 30\text{ mg}$$

2. Cancel units.

$$1\text{ mL} : 40\,\cancel{\text{mg}} :: A : 30\,\cancel{\text{mg}}$$

3. Multiply the means and extremes, then solve for the missing value.

$$40 \times A = 1\text{ mL} \times 30$$

$$A = 1\text{ mL} \times \frac{30}{40}$$

$$A = 0.75\text{ mL}$$

Using a 1-mL tuberculin syringe, we will administer 0.75 mL.

Dimensional Analysis

Solution The dosage ordered O and the dose on hand H are already expressed in the same units, so the desired dose D is 30 mg. After reconstitution the dosage strength is 40 mg/mL, making the dosage unit Q 1 mL, and the dose on hand H is 40 mg.

Follow Procedure Checklist 7-3.

1. The unit of measure for the amount to administer will be milliliters.

   $A \text{ mL} =$

2. Since the unit of measure for the dosage ordered is the same as that for the dose on hand, this step is unnecessary.
3. The dosage unit is 1 mL. The dose on hand is 40 mg. This is our first factor.

   $$\frac{1 \text{ mL}}{40 \text{ mg}}$$

4. The dosage ordered is 30 mg. Place this over one and set up the equation.

   $$A \text{ mL} = \frac{1 \text{ mL}}{40 \text{ mg}} \times \frac{30 \text{ mg}}{1}$$

5. Cancel units.

   $$A \text{ mL} = \frac{1 \text{ mL}}{40 \cancel{\text{mg}}} \times \frac{30 \cancel{\text{mg}}}{1}$$

6. Solve the equation.

   $$A \text{ mL} = \frac{1 \text{ mL}}{40} \times \frac{30}{1}$$

   $$A = 0.75 \text{ mL}$$

Using a 1-mL tuberculin syringe, we will administer 0.75 mL.

Formula Method

Solution The dosage ordered O and the dose on hand H are already expressed in the same units, so the desired dose D is 30 mg. After reconstitution the dosage strength is 40 mg/mL, making the dosage unit Q 1 mL, and the dose on hand H is 40 mg.

Solution: Follow Procedure Checklist 7-4.

1. The drug is ordered in milligrams, which is the same unit of measure as that of the dose on hand. Therefore,

   Desired dose $D = 30$ mg

   Dose on hand $H = 40$ mg

   Quantity $Q = 1$ mL

2. Fill in the formula.

   $$\frac{D \times Q}{H} = A$$

   $$\frac{30 \text{ mg} \times 1 \text{ mL}}{40 \text{ mg}} = A$$

3. Cancel units.

$$\frac{30\ \cancel{mg} \times 1\ mL}{40\ \cancel{mg}} = A$$

4. Solve for the unknown.

$$\frac{30 \times 1\ mL}{40} = A$$

$$0.75\ mL = A$$

Using a 1-mL tuberculin syringe, we will administer 0.75 mL.

GO TO . . . Open the CD-ROM that accompanies your textbook, and select Chapter 9, Rule 9-6. Review the animation and example problems, then complete the practice problems. Continue to the next section of the book once you have mastered the rule presented. ■

ERROR ALERT!

Select the correct instructions for the strength and route ordered.

The package insert for a 500-mg vial of Maxipime® can be reconstituted for both IM and IV use. Suppose a nurse mistakenly reconstitutes Maxipime® 500 mg IM for 500 mg IV instead. The IV instructions indicate that the nurse should use 5 mL of diluent, producing a solution strength of 100 mg/mL. Calculate the amount to administer,

$$500\ mg \times \frac{1\ mL}{100\ mg} = A$$

$$\overset{5}{\cancel{500}}\ \cancel{mg} \times \frac{1\ mL}{\underset{1}{\cancel{100}}\ \cancel{mg}} = 5 \times 1\ mL = 5\ mL$$

The health care professional administers two injections of 2.5 mL each. The patient's discomfort increases, and the number of injection sites available for future injections is reduced. Costs increase because more diluent and syringes than necessary are used. The risk of injection complications is doubled. Correctly reconstituted for IM use, 1.3 mL of diluent will be used to produce a solution with a dosage strength of 280 mg/mL. Calculate the amount to administer.

$$500\ mg \times \frac{1\ mL}{280\ mg} = A$$

$$50\cancel{0}\ \cancel{mg} \times \frac{1\ mL}{28\cancel{0}\ \cancel{mg}} = \frac{50\ mL}{28} = 1.785 \text{ rounded to } 1.8\ mL$$

This amount 1.8 mL is the correct IM dose.

Critical Thinking on the Job

Recording Accurate Information

A health care professional receives the following order: Humatrope® 2 mg IM three times a week (see Figure 9-12).

At 0800 on 10/15/05, the health care professional prepares the medication to administer later that day. After reading the label (see Figure 9-12), she draws up all the diluent supplied with the medication (see Figure 9-13) and injects it into the vial. According to the drug label, the remaining medication may be refrigerated for 14 days if protected from light. She labels the vial "Exp: 0800 10/29/01. Refrigerate. 5 mg/mL" and signs it with her initials. The vial will not be exposed to light in the refrigerator. Otherwise, the health care professional might wrap it in foil or place it inside a paper bag.

Later that day, the health care professional calculates the amount to administer, based on the label,

$$2 \text{ mg} \times \frac{1 \text{ mL}}{5 \text{ mg}} = A$$

$$2 \cancel{\text{mg}} \times \frac{1 \text{ mL}}{5 \cancel{\text{mg}}} = 2 \times \frac{1}{5} \text{ mL} = \frac{2}{5} \text{ mL} = 0.4 \text{ mL}$$

She uses a LoDose 0.5 mL tuberculin syringe to administer the medication.

Questions?

What mistake did the health care professional make? How could she correct her mistake?

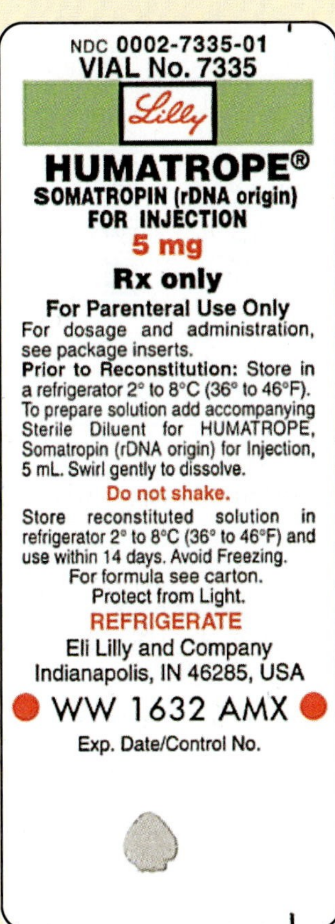

Figure 9-12

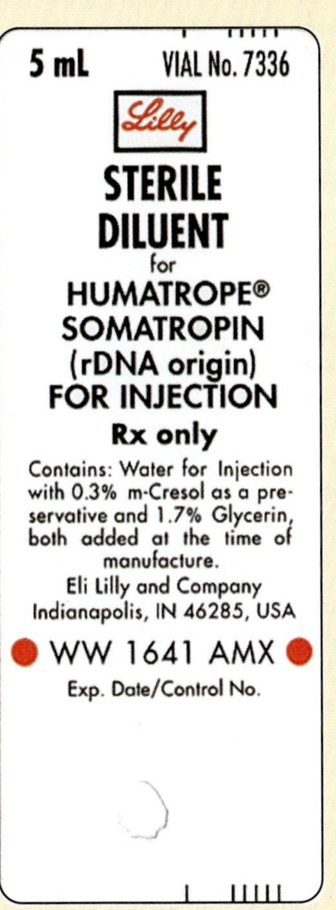

Figure 9-13

REVIEW AND PRACTICE

9-2 Reconstituting Powdered Medication

In Exercises 1–4, refer to the following label:

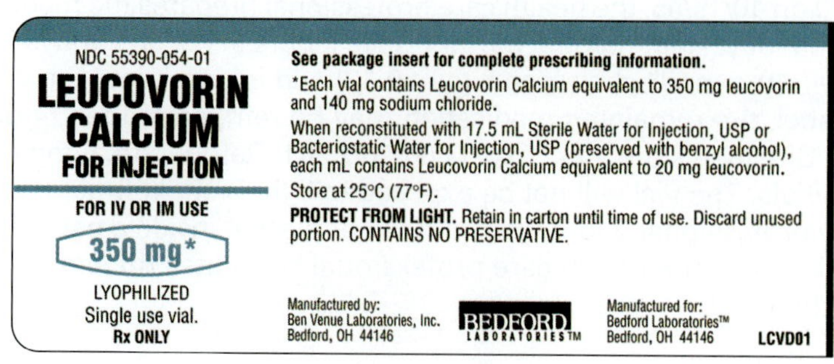

NDC 55390-054-01
LEUCOVORIN CALCIUM
FOR INJECTION
FOR IV OR IM USE
350 mg*
LYOPHILIZED
Single use vial.
Rx ONLY
See package insert for complete prescribing information.
*Each vial contains Leucovorin Calcium equivalent to 350 mg leucovorin and 140 mg sodium chloride.
When reconstituted with 17.5 mL Sterile Water for Injection, USP or Bacteriostatic Water for Injection, USP (preserved with benzyl alcohol), each mL contains Leucovorin Calcium equivalent to 20 mg leucovorin.
Store at 25°C (77°F).
PROTECT FROM LIGHT. Retain in carton until time of use. Discard unused portion. CONTAINS NO PRESERVATIVE.
Manufactured by: Ben Venue Laboratories, Inc. Bedford, OH 44146
BEDFORD LABORATORIES™
Manufactured for: Bedford Laboratories™ Bedford, OH 44146
LCVD01

1. How much diluent should you add to the 350-mg vial?
2. What solution strength should you print on the label?
3. What can be used to reconstitute the Leucovorin?
4. If the dose ordered is 15 mg IM, what would be the amount to administer?

In Exercises 5–9, refer to the following order, label, and package insert:

Ordered: Synagis® 75 mg IM q8h.

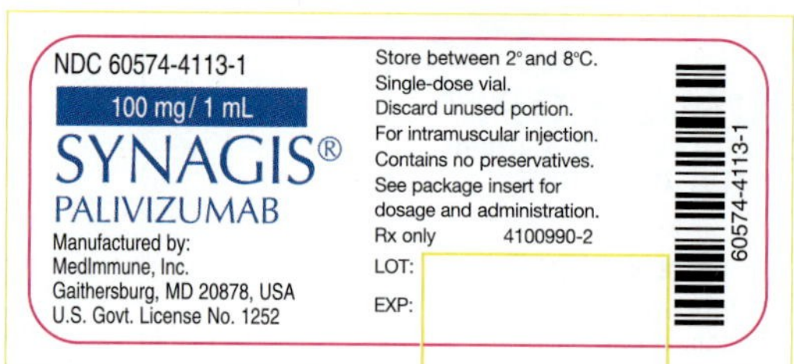

NDC 60574-4113-1
100 mg/ 1 mL
SYNAGIS®
PALIVIZUMAB
Manufactured by: MedImmune, Inc. Gaithersburg, MD 20878, USA
U.S. Govt. License No. 1252
Store between 2° and 8°C.
Single-dose vial.
Discard unused portion.
For intramuscular injection.
Contains no preservatives.
See package insert for dosage and administration.
Rx only 4100990-2
LOT:
EXP:
60574-4113-1

Preparation of Lyophilized Product for Administration:

- To reconstitute, remove the tab portion of the vial cap and clean the rubber stopper with 70% ethanol or equivalent.
- Both the 50 mg and 100 mg vials contain an overfill to allow the withdrawal of 50 mg or 100 mg Synagis® respectively when reconstituted following the directions described below.
- SLOWLY add 0.6 mL of sterile water for injection to the 50 mg vial or add 1.0 mL of sterile water for injection to the 100 mg vial. The vial should be tilted slightly and gently rotated for 30 seconds to avoid foaming. DO NOT SHAKE or VIGOROUSLY AGITATE the vial. This is a critical step to avoid prolonged foaming.
- Reconstituted Synagis® should stand undisturbed at room temperature for a minimum of 20 minutes until the solution clarifies.
- Reconstituted Synagis® should be inspected visually for particulate matter or discoloration prior to administration. The reconstituted solution should appear clear or slightly opalescent (a thin layer of micro-bubbles on the surface is normal and will not affect dosage). DO NOT use if there is particulate matter or if the solution is discolored.
- Reconstituted Synagis® does not contain a preservative and should be administered within 6 hours of reconstitution. Administer immediately after withdrawal from vial. Synagis® is supplied in single-use vials. DO NOT re-enter the vial. Discard any unused portion.

5. What diluent should you use to reconstitute Synagis®?
6. How much diluent should you add to this vial?
7. How many approximate milligrams are in 1 mL?
8. If Synagis is reconstituted at 1000 on January 3, 2009, what should you write on the label?
9. How much solution should you administer?

In Exercises 10–15, refer to the following order, label, and package insert.

Ordered: Pregnyl® 5000 USP units IM every other day

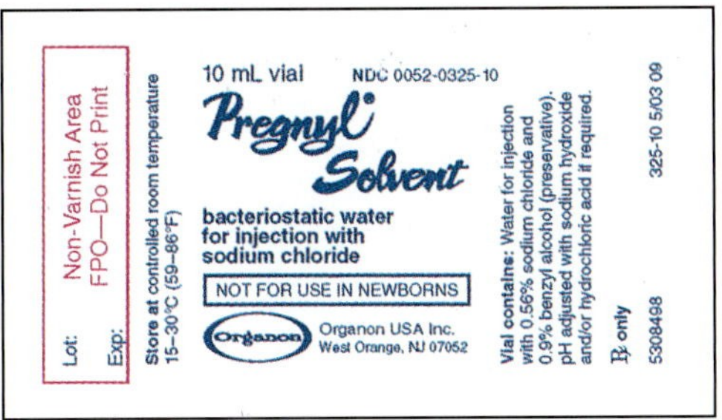

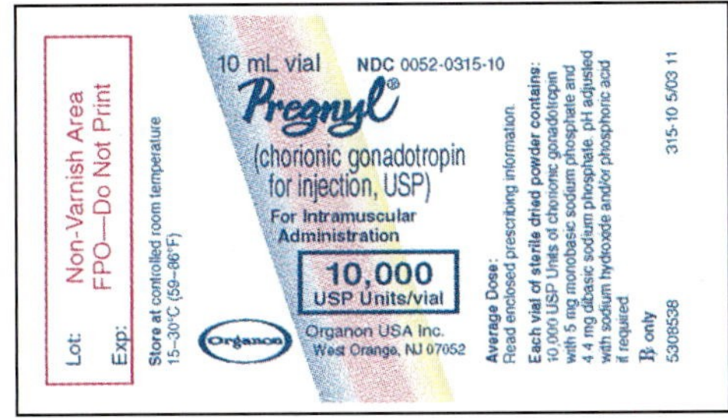

Directions for Reconstitution

Two-vial package: Withdraw sterile air from lyophilized vial and inject into diluent vial. Remove 1–10 mL from diluent and add to lyophilized vial; agitate gently until powder is completely dissolved in solution.

Parenteral drug products should be inspected visually for particulate matter and discoloration prior to administration, whenever solution and container permit.

IMPORTANT: USE COMPLETELY AFTER RECONSTITUTION. RECONSTITUTED SOLUTION IS STABLE FOR 60 DAYS WHEN REFRIGERATED.

HOW SUPPLIED

Two-vial package containing:

1-10 mL lyophilized multiple dose vial containing: 10,000 USP Units chorionic gonadotropin per vial, NDC 0052-0315-10.

1-10 mL vial of solvent containing: water for injection with sodium chloride 0.56% and benzyl alcohol 0.9%, NDC 0052-0325-10.

When reconstituted, each 10 mL vial contains:

Chorionic gonadotropin	10,000 USP Units
Monobasic sodium phosphate	5 mg
Dibasic sodium phosphate	4.4 mg
Sodium chloride	0.56%
Benzyl alcohol	0.9%

If required pH adjusted with sodium hydroxide and/or phosphoric acid.

Storage: Store at 15–30°C (59–86°F). Reconstituted solution is stable for 60 days when refrigerated.

10. What type of diluent is used for reconstituting Pregnyl®?
11. How much diluent is used to reconstitute the Pregnyl®?
12. How should the medication be mixed and then stored after reconstitution?
13. If Pregnyl® is diluted with 10 mL what is the dosage strength of the container?
14. What is the amount to administer of the Pregnyl®?
15. What should be done with the rest of the solution, and how long will it last?

In Exercises 16–21, refer to the following order, and label information.

Ordered: Gemzar® 100 mg for IV infusion

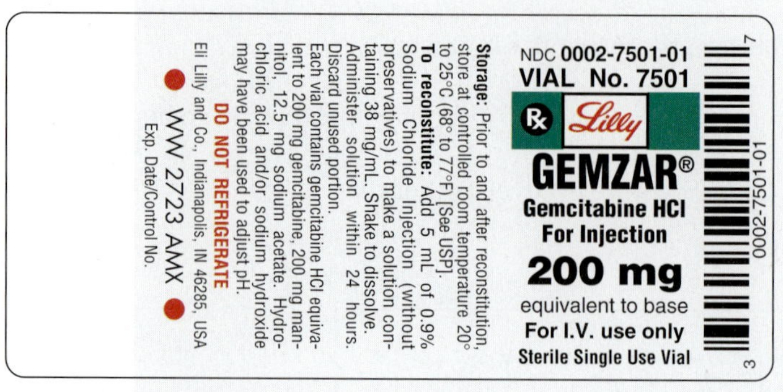

16. What diluent should you use to reconstitute this medication?
17. How much diluent should you add to the vial?
18. What solution strength should you write on the label?
19. If the Gemzar® is reconstituted at 2400 on 6/5/2012 and will be stored at room temperature, what expiration date and time should you write on the label?
20. How should the medication be stored?
21. How much solution would be used to administer 100 mg of medication?

In Exercises 22–25, refer to the following order, label, and package insert.

Ordered: Penicillin G potassium 1 million U IM q2h

22. To make a solution of 500,000 U/mL, how much diluent should you add to the vial?
23. If the medication in the vial is reconstituted with 4.8 mL of diluent, what solution strength should you write on the label?
24. If the penicillin is reconstituted at 1200 on 11/20/05 and will be stored in the refrigerator, what expiration date and time should you write on the label?
25. When reconstituted with 8.2 mL, how much solution should you administer?

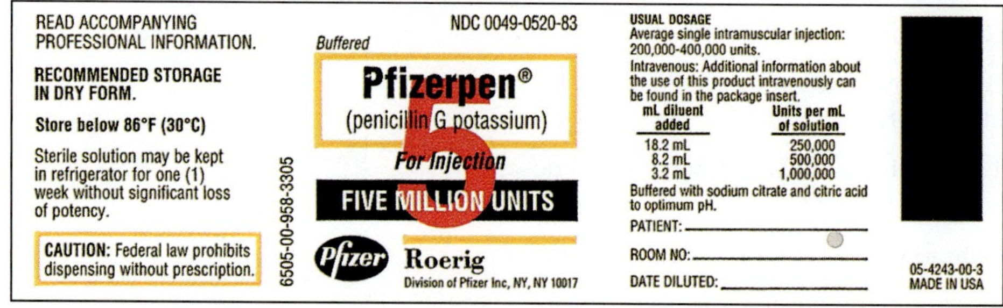

Reconstitution

The following table shows the amount of solvent required for solution of various concentrations:

Approx. Desired Concentration (units/mL)	Approx. Volume (mL) 1,000,000 units	Solvent for Vial of 5,000,000 units	Infusion Only 20,000,000 units
50,000	20.0	–	–
100,000	10.0	–	–
250,000	4.0	18.2	75.0
500,000	1.8	8.2	33.0
750,000	–	4.8	–
1,000,000	–	3.2	11.5

When the required volume of solvent is greater than the capacity of the vial, the penicillin can be dissolved by first injecting only a portion of the solvent into the vial, then withdrawing the resultant solution and combining it with the remainder of the solvent in a larger sterile container.

Buffered Pfizerpen (penicillin G potassium) for Injection is highly water soluble. It may be dissolved in small amounts of Water for Injection, or Sterile Isotonic Sodium Chloride Solution for Parenteral Use. All solutions should be stored in a refrigerator. When refrigerated, penicillin solutions may be stored for seven days without significant loss of potency.

Buffered Pfizerpen for Injection may be given intramuscularly or by continuous intravenous drip for dosages of 500,000, 1,000,000, or 5,000,000 units. It is also suitable for intrapleural, intraarticular, and other local instillations.

THE 20,000,000 UNIT DOSAGE MAY BE ADMINISTERED BY INTRAVENOUS INFUSION ONLY.

(1) Intramuscular Injection: Keep total volume of injection small. The intramuscular route is the preferred route of administration. Solutions containing up to 100,000 units of penicillin per mL of diluent may be used with a minimum of discomfort. Greater concentration of penicillin G per mL is physically possible and may be employed where therapy demands. When large dosages are required, it may be advisable to administer aqueous solutions of penicillin by means of continuous intravenous drip.

(2) Continuous Intravenous Drip: Determine the volume of fluid and rate of its administration required by the patient in a 24 hour period in the usual manner for fluid therapy, and add the appropriate daily dosage of penicillin to this fluid. For example, if an adult patient requires 2 liters of fluid in 24 hours and a daily dosage of 10 million units of penicillin, add 5 million units to 1 liter and adjust the rate of flow so the liter will be infused in 12 hours.

(3) Intrapleural or Other Local Infusion: If fluid is aspirated, give infusion in a volume equal to 1/4 or 1/2 the amount of fluid aspirated, otherwise, prepare as for intramuscular injection.

(4) Intrathecal Use: The intrathecal use of penicillin in meningitis must be highly individualized. It should be employed only with full consideration of the possible irritating effects of penicillin when used by this route. The preferred route of therapy in bacterial meningitides is intravenous, supplemented by intramuscular injection.

Parenteral drug products should be inspected visually for particulate matter and discoloration prior to administration, whenever solution and container permit.

Sterile solution may be left in refrigerator for one week without significant loss of potency.

HOW SUPPLIED

Buffered Pfizerpen® (penicillin G potassium) for Injection is available in vials containing respectively 5,000,000 units × 10's (NDC 0049-0520-83) and 20,000,000 units × 1's (NDC 0049-0530-28); buffered with sodium citrate and citric acid to an optimum pH.

Each million units contains approximately 6.8 milligrams of sodium (0.3 mEq) and 65.6 milligrams of potassium (1.68 mEq).

Store the dry powder below 86°F (30°C).

Reference

1. American Heart Association, 1977. Prevention of bacterial endocarditis. Circulation. **56**:139A-143A.

Rx only

In Exercises 26–30, refer to the following label, order, and package insert.

Ordered: Rocephin® 750 mg IM q6h
On hand: See label

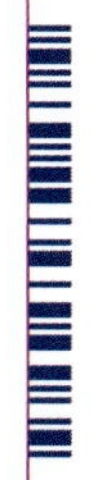

EXP

Batch

Rocephin®
(Ceftriaxone for Injection USP)
1 gram
Single-Use Vial
℞ only
Roche

For I.M. or I.V. Use
Equivalent to 1 gram ceftriaxone
For I.M. Administration:
Reconstitute with 2.1 mL 1% Lidocaine Hydrochloride Injection (USP) or Sterile Water for Injection (USP). Each 1 mL of solution contains approximately 350 mg equivalent of ceftriaxone as ceftriaxone sodium.
For I.V. Administration:
Reconstitute with 9.6 mL of an I.V. diluent specified in the accompanying package insert. Each 1 mL of solution contains approximately 100 mg equivalent of ceftriaxone as ceftriaxone sodium. **Withdraw entire** contents and dilute to the desired concentration with the appropriate I.V. diluent. USUAL DOSAGE: See package insert.
Storage Prior to Reconstitution:
Store at 20°–25°C (68°–77°F) [see USP Controlled Room Temperature].
Protect From Light.
Storage After Reconstitution:
See package insert.
Made in Switzerland
Distributed by:
Roche Laboratories Inc.
Nutley, New Jersey 07110

10079638 USA 1111

26. What diluent will be used to prepare this medication?

27. How much diluent should you add to one vial?

28. What is the dosage strength of the solution, once reconstituted?

29. If Rocephin® is reconstituted at 1600 on October 4, 2009, and will be stored in the refrigerator, what should you write on the label?

30. How much solution should you administer?

Description: Caption:

COMPATIBILITY AND STABILITY: Rocephin sterile wder should be stored at room temperature—77°F (25° C)—or below and protected from light. After reconstitution, protection from normal light is not necessary. The color of solutions ranges from light yellow to amber, depending on the length of storage, concentration and diluent used.

Rocephin *intramuscular* solutions remain stable (loss of potency less than 10%) for the following time periods:

		Storage	
Diluent	**Concentration mg/ml**	**Room Temp. (25° C)**	**Refrigerated (4°C)**
Sterile Water for Injection	100	2 days	10 days
	250,350	24 hours	3 days
0.9% Sodium Chloride	100	2 days	10 days
Solution	250,350	24 hours	3 days
5% Dextrose Solution	100	2 days	10 days
	250,350	24 hours	3 days
Bacteriostatic Water +0.9%	100	24 hours	10 days
Benzyl Alcohol	250,350	24 hours	3 days
1% Lidocaine Solution	100	24 hours	10 days
(without epinephrine)	250,350	24 hours	3 days

Rocephin *intravenous* solutions, at concentrations of 10, 20 and 40 mg/ml., remain stable (loss of potency less than 10%) for the following time periods stored in glass or PVC containers:

	Storage	
Diluent	**Room Temp. (25°C)**	**Refrigerated (4°C)**
Sterile Water	2 days	10 days
0.9% Sodium Chloride Solution	2 days	10 days
5% Dextrose Solution	2 days	10 days
10% Dextrose Solution	2 days	10 days
5% Dextrose + 0.9% Sodium Chloride Solution*	2 days	Incompatible
5% Dextrose + 0.45% Sodium Chloride Solution	2 days	Incompatible

*Data available for 10 to 40 mg/mL concentration in this diluent in PVC containers only.
*Excerpt from package insert for Rochephin by Roche.

To check your answers, see page 527.

Other Medication Routes

Medications may be given by a variety of routes besides oral and common parenteral routes. These routes are used for intradermal injections, inhalants, and rectal and transdermal medications.

Intradermal Injections

Very small doses of medication can be injected under the first layer of the skin. This route is normally used for diagnostic testing, most often screening for tuberculosis or allergies. When an **intradermal** (ID) injection is required, the physician usually orders the intended diagnostic test, such as a Mantoux (PPD) test for tuberculosis. You determine the amount of solution to use by checking the vial label or the package insert. If a dose other than the standard diagnostic dose is to be administered, the physician will order the exact amount. No calculation is required. Intradermal injections are usually 0.1 mL or less. A tuberculin syringe is always used.

Inhalants

Inhaled medications, or **inhalants,** are administered either by metered-dose inhaler (MDI), Figure 9.14, or by nebulizer. Metered-dose inhalers provide a measured dose of medication in each puff. The physician orders the number of puffs to be given. No calculation is necessary.

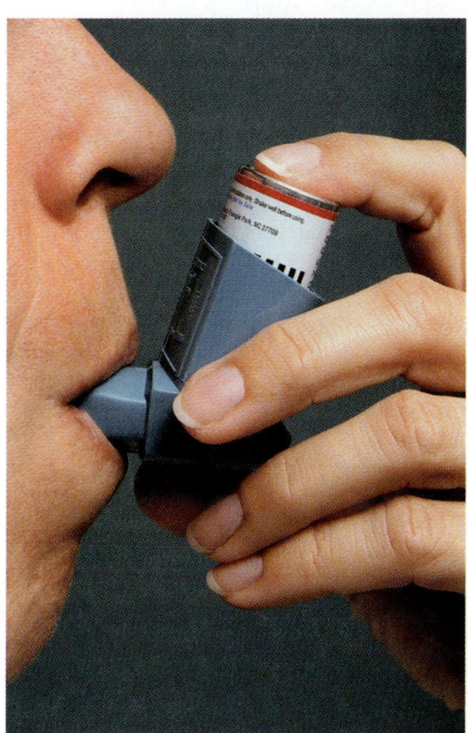

Figure 9-14 An MDI is used to deliver inhalant medications.

Medications given by nebulizer are supplied as liquids which are mixed with sterile saline solution. Single doses premixed with saline are available for most medications. A few are measured in the receptacle of the nebulizer, after which the correct amount of saline is added. Sterile saline is usually provided in 3-mL or 5-mL single-dose ampules.

Inhalant medications in multiple-dose containers are usually packaged with special droppers calibrated for the standard doses. If the dropper is not available or becomes contaminated, a sterile syringe may be used.

The physician usually specifies the solution strength and the amount of inhalant to administer. For example, the order Mucomyst 20% 3 mL via nebulizer QID is a complete order. In some cases, the physician will also order an amount of normal saline to be added to the medication. When calculations are necessary, the same methods—fraction proportion, ratio proportion, dimensional analysis, and formula—are used for inhalants as are used for parenteral medications.

Rectal Medications

Rectal medications are usually given in suppository form. Generally, suppositories cannot be accurately divided. Therefore, in most cases, only doses that are multiples of the available suppository strength may be administered.

In some cases, according to their manufacturers, suppositories can be safely divided in half. However, they are not scored. Thus, the physician's order should specify that $\frac{1}{2}$ suppository is to be given. For example, if the order reads Tigan® 50 mg p.r. t.i.d., ask the physician to clarify whether $\frac{1}{2}$ of a 100-mg suppository is acceptable. The order should then be rewritten: Tigan® 100 mg supp. Give $\frac{1}{2}$ supp.

Transdermal Systems

Transdermal medications include patches, ointments, and creams. Patches usually consist of a special membrane that releases liquid medication at a constant rate. The patch has adhesive edges to hold it in place so that the membrane rests against the skin. The dosage rate of a transdermal patch is usually expressed in milligrams or micrograms per hour. Patches cannot be divided. If a dose is larger than the amount provided by a single patch, you can use multiple patches. Before you administer a patch, be certain to remove any patches that are already in place and wipe off any residual medication. Patches should be applied to a reasonably hair-free site such as abdomen, shoulder, back, or hip.

Example 1 Find the amount to administer.

Ordered: Deponit® 0.8 mg/hr topically

On hand: Deponit® 0.2 mg/h and 0.4 mg/h patches

$$0.8 \text{ mg/h} \times \frac{1 \text{ patch}}{0.4 \text{ mg/h}} = A$$

$$\overset{2}{\cancel{0.8}\ \cancel{\text{mg/h}}} \times \frac{1 \text{ patch}}{\cancel{0.4}\ \cancel{\text{mg/h}}} = 2 \times 1 \text{ patch} = 2 \text{ patches}$$

Administer 2 patches. While you could also administer 4 of the 0.2 mg/h patches, you should use the least amount of medication possible to give the ordered dose.

Sometimes the ordered dose can be administered by using a combination of patches. In these cases, use critical thinking skills to determine which combination will work.

Example 2 Find the amount to administer.

Ordered: Vivelle® 0.125 mg/day

On hand: Vivelle® in four dosage strengths: 0.0375 mg/day, 0.05 mg/day, 0.075 mg/day, and 0.1 mg/day

Start with the patch that has the greatest dosage strength (0.1 mg/day). Because all the patches deliver more than 0.025 mg/day, none will work in combination with this patch. Next, try combinations with the 0.075 mg/day patch. Note that 0.075 + 0.05 = 0.125 Administer a combination of one 0.075 mg/day patch and one 0.05 mg/day patch.

GO TO . . . Open the CD-ROM that accompanies your textbook, and select Chapter 9, Practice 9-3. Review the animation and example problems, then complete the practice problems. Continue to the next section of the book once you have mastered the rule presented. ■

REVIEW AND PRACTICE

9-3 Other Medication Routes

In Exercises 1–8, find the amount to administer.

1. Ordered: Acetylcysteine 1 g via nebulizer
   On hand: Acetylcysteine 20% solution 10-mL vial
2. Ordered: Albuterol 2.5 mg via nebulizer
   On hand: Albuterol 5 mg/mL
3. Ordered: Atrovent® 250 mcg via nebulizer
   On hand: Atrovent® 0.02% inhalation solution 500 mcg/2.5-mL vial
4. Ordered: Numorphan 10 mg p.r. PRN (as needed)
   On hand: Numorphan 5-mg suppositories
5. Ordered: RMS morphine supp 15 mg p.r. PRN (as needed)
   On hand: RMS 5-mg, 10-mg, and 30-mg suppositories
6. Ordered: Phenergan® 12.5 mg pr PRN (as needed)
   On hand: Phenergan® 25-mg suppositories
7. Ordered: Testoderm 0.8 mg/day top
   On hand: Testoderm 0.4 mg/day patches
8. Ordered: Catapres® 0.5 mg/day top
   On hand: Catapres® TTS-1 (0.1 mg/day), TTS-2 (0.2 mg/day), TTS-3 (0.3 mg/day)
9. Ordered: Alora® 0.15 mg/day
   On hand: Alora® 0.05 mg/day, 0.075 mg/day, and 0.1 mg/day
10. Ordered: Transderm Nitro 0.3 mg/h top
    On hand: Transderm Nitro 0.1 mg/h, 0.2 mg/h, and 0.6 mg/h patches

To check your answers, see page 527.

Workspace

HOMEWORK ASSIGNMENT

Answer the following questions.

1. What is the correct syringe to select if the amount to be injected is 0.75 mL?
2. If the dosage to be administered is 1.75 mL, what syringe would you select and to what amount would the dosage be rounded?
3. List the maximum volume for an IM injection for the following patients: an adult, an adult deltoid, a child 6-12 years old.
4. If the order for a subcutaneous injection results in an amount of 2.2 mL, what action should be taken before any administration?
5. What type of injection is usually less than 0.1 mL?
6. What equipment is necessary when the order states to deliver the medication in puffs?
7. What is a common drug form for the administration of rectal medications?
8. Explain the administration of a transdermal medication.
9. List three common diluents used when reconstituting powdered medications?

Use the identified drug labels to answer the following questions:

10. Refer to Label A. What is the caution on the label?
11. If the order is for 75 mg IM, what amount would be administered?

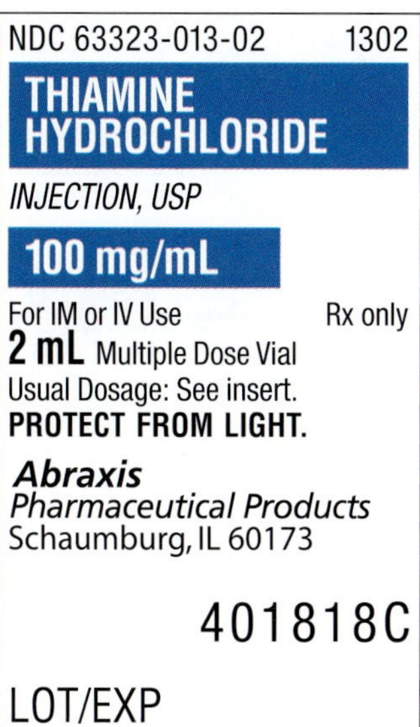

NDC 63323-013-02 1302
THIAMINE HYDROCHLORIDE
INJECTION, USP
100 mg/mL
For IM or IV Use
Rx only
2 mL Multiple Dose Vial
Usual Dosage: See insert.
PROTECT FROM LIGHT.
Abraxis
Pharmaceutical Products
Schaumburg, IL 60173
401818C
LOT/EXP

Label A

12. Refer to Label B. How many units are in each milliliter of this drug?
13. If the order is for 800 units sub-Q, what amount would be administered?
14. If the order is for 1000 units, how many times may this vial be used?

NDC 63323-540-11
504011
HEPARIN SODIUM
INJECTION, USP
(Derived from Porcine
Intestinal Mucosa)
For IV or SC Use
10 mL
Multiple Dose Vial
Rx only
Sterile, Nonpyrogenic
Use only if solution is clear
and seal intact.
Abraxis
LOT/EXP
3 63323-540-11 1

Label B

15. Refer to Label C. If the order is for 900 mg IM, what amount would be administered?

16. What extra preparations would you need to make in order to administer the amount in Question 15?

17. Other than IM, what route is acceptable for administration of this drug and what must you do before administration?

NDC 63323-282-04 28204
CLINDAMYCIN
INJECTION, USP
600 mg/4 mL
(150 mg/mL)*
For IM or IV Use
DILUTE BEFORE IV USE
4 mL Single Dose Vial
Rx only
Sterile
*Each mL contains clindamycin phosphate equivalent to 150 mg clindamycin, 0.5 mg disodium edetate and 9.45 mg benzyl alcohol as a preservative. When necessary, pH adjusted with sodium hydroxide and/or hydrochloric acid.
Usual Dosage: See package insert.
Warning: If given intravenously, must be diluted before use.
Store at 20° to 25°C (68° to 77°F) [see USP Controlled Room Temperature].
Do not refrigerate.
Vial stoppers do not contain natural rubber latex.
Abraxis
Pharmaceutical Products
Schaumburg, IL 60173
402129A
LOT/EXP
3 63323-282-04 2

Label C

18. Refer to Label D. What two diluents, and how much of each, is suggested on the label?

19. Which one of the diluents requires a physician's order?

20. If the order is 500 mg IM, what amount would be administered?

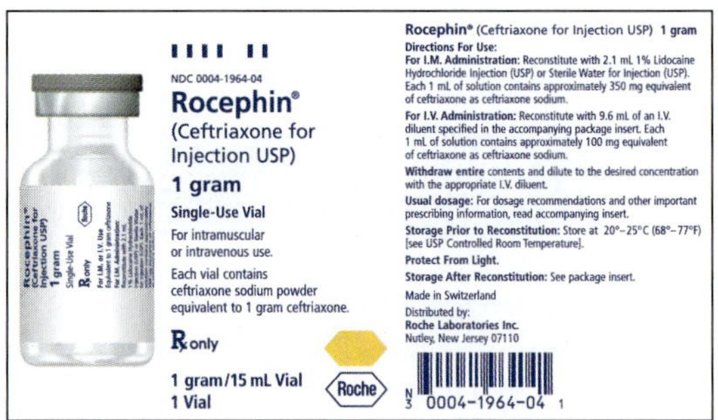
NDC 0004-1964-04
Rocephin®
(Ceftriaxone for Injection USP)
1 gram
Single-Use Vial
For intramuscular or intravenous use.
Each vial contains ceftriaxone sodium powder equivalent to 1 gram ceftriaxone.
℞ only
1 gram/15 mL Vial
1 Vial
Roche
Rocephin® (Ceftriaxone for Injection USP) 1 gram
Directions For Use:
For I.M. Administration: Reconstitute with 2.1 mL 1% Lidocaine Hydrochloride Injection (USP) or Sterile Water for Injection (USP). Each 1 mL of solution contains approximately 350 mg equivalent of ceftriaxone as ceftriaxone sodium.
For I.V. Administration: Reconstitute with 9.6 mL of an I.V. diluent specified in the accompanying package insert. Each 1 mL of solution contains approximately 100 mg equivalent of ceftriaxone as ceftriaxone sodium.
Withdraw entire contents and dilute to the desired concentration with the appropriate I.V. diluent.
Usual dosage: For dosage recommendations and other important prescribing information, read accompanying insert.
Storage Prior to Reconstitution: Store at 20°–25°C (68°–77°F) [see USP Controlled Room Temperature].
Protect From Light.
Storage After Reconstitution: See package insert.
Made in Switzerland
Distributed by:
Roche Laboratories Inc.
Nutley, New Jersey 07110
N 3 0004-1964-04 1

Label D

Chapter 9 Review

Check Up

In Exercises 1–10, find the amount to administer, then mark the syringe.

1. Ordered: INFeD® (iron dextran) 100 mg deep IM qd
   On hand: INFeD® 50 mg/mL

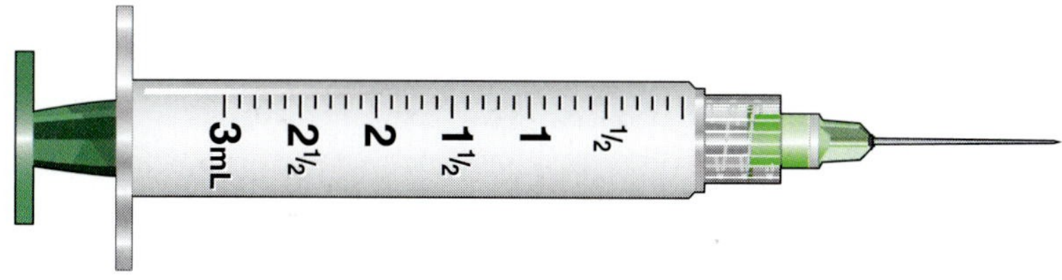

2. Ordered: Haloperidol decanoate 60 mg deep IM stat
   On hand: Haloperidol decanoate 50 mg/mL

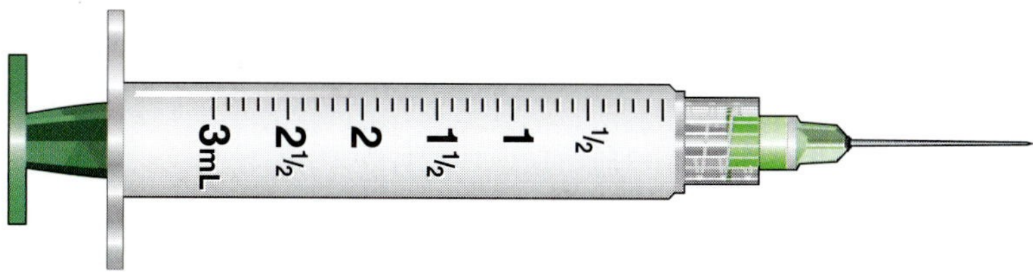

3. Ordered: Loxitane® 30 mg IM bid
   On hand: Loxitane® 50 mg/mL

4. Ordered: Epogen® 1400 U Sub-Q tiw
   On hand: Epogen® 2000 U/mL

5. Ordered: Lidocaine 300 mg IM stat
On hand: Lidocaine 20% solution

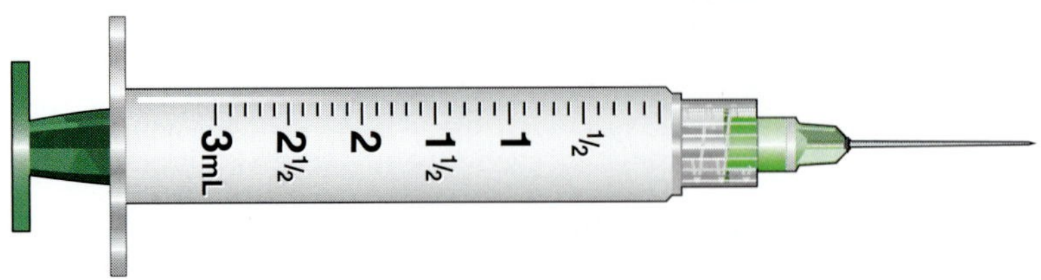

6. Ordered: Magnesium sulfate 250 mg IM qd
On hand: Magnesium sulfate 10% solution

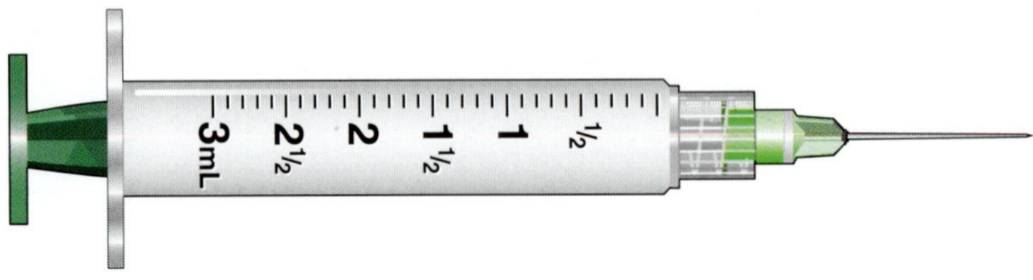

7. Ordered: Levsin® 0.4 mg IM bid
On hand: Levsin® 0.5 mg/mL

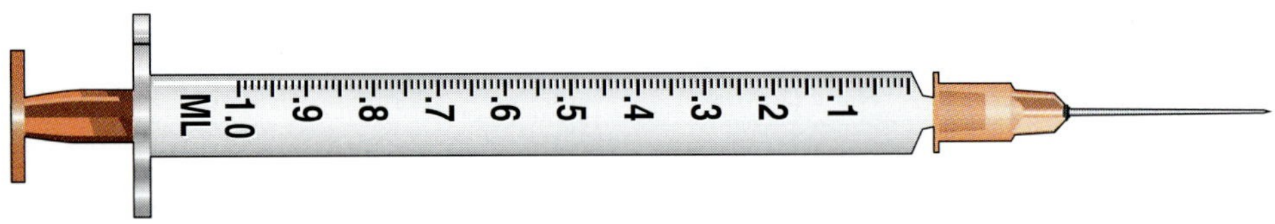

8. Ordered: Robinul® 0.15 mg IM stat
On hand: Robinul® 0.2 mg/mL

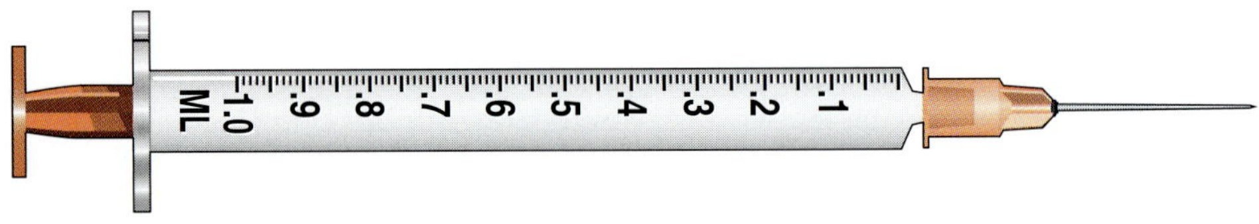

9. Order: Prostigmin 0.75 mg IM q4h
On hand: Prostigmin 1:1000 solution

10. Ordered: Epinephrine 0.5 mg Sub-Q stat
On hand: Epinephrine 1:200 solution

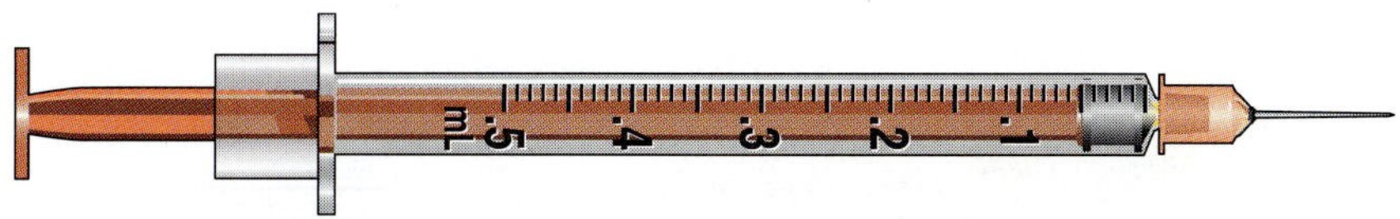

For Exercises 11–21, find the amount to administer, select the proper syringe, and write in the space provided.

11. Ordered: Adrenalin® 0.2 mg Sub-Q stat
On hand: Adrenalin® 1:2000 solution

Administer: __________ Syringe: __________________

12. Ordered: Calciferol 24,000 IU IM qd
On hand: Calciferol 500,000 IU/5 mL
Administer: __________ Syringe: __________________

13. Ordered: Heparin sodium 7500 U Sub-Q q8h
On hand: Heparin 1:20,000 solution
Administer: __________ Syringe: __________________

14. Ordered: Heparin calcium 7500 U Sub-Q q8h
On hand: Heparin calcium 5000 U/0.2 mL
Administer: __________ Syringe: __________

15. Ordered: Thiamine 200 mg IM
On hand: Refer to label A.
Administer: __________ Syringe: __________

A

16. Ordered: Heparin 400 units Sub-Q daily
On hand: Refer to label B.
Administer: __________ Syringe: __________

NDC 63323-540-11
504011
HEPARIN SODIUM
INJECTION, USP
1,000 USP Units/mL
(Derived from Porcine Intestinal Mucosa)
For IV or SC Use
10 mL
Multiple Dose Vial
Rx only
Sterile, Nonpyrogenic
Each mL contains: 1,000 USP Units heparin sodium; 9 mg sodium chloride; 0.15% methylparaben; 0.015% propylparaben; Water for Injection q.s. Made isotonic with sodium chloride. Hydrochloric acid and/or sodium hydroxide may have been added for pH adjustment.
Usual Dosage: See insert.
Use only if solution is clear and seal intact.
Store at 20° to 25°C (68° to 77°F) [see USP Controlled Room Temperature].
Abraxis
Pharmaceutical Products
Schaumburg, IL 60173
401586B
LOT/EXP
3 63323-540-11 1

B

17. Ordered: Furosemide 15 mg IM now
On hand: Refer to label C.
Administer: __________ Syringe: __________

NDC 63323-280-02
28002
FUROSEMIDE
INJECTION, USP
20 mg/2 mL
(10 mg/mL)
For IM or IV Use
Rx only
2 mL Single Dose Vial
Preservative Free
Discard unused portion.
PROTECT FROM LIGHT.
Do not use if discolored.
Abraxis
Pharmaceutical Products
Schaumburg, IL 60173
401803C
LOT/EXP
3 63323-280-02 4

C

18. Ordered: Oxytocin 20 units IM q 12h prn
On hand: Refer to label D.
Administer: __________ Syringe: __________

NDC 63323-012-10 1210
OXYTOCIN
INJECTION, USP (SYNTHETIC)
10 USP Units/mL
For IV Infusion or IM Use
10 mL Multiple Dose Vial
Rx only
Sterile
Each mL contains: Oxytocic activity equivalent to 10 USP Oxytocin Units; chlorobutanol anhydrous (chloral derivative) 0.5%; Water for Injection q.s. Acetic acid may have been added for pH adjustment.
Usual Dosage: See Insert.
Store at 20° to 25°C (68° to 77°F) [see USP Controlled Room Temperature]. Do not permit to freeze.
Abraxis
Pharmaceutical Products
Schaumburg, IL 60173
401738E
LOT/EXP
3 63323-012-10 6

D

19. Ordered: Epogen® 2500 U Sub-Q tiw
On hand: Refer to label E.
Administer: __________ Syringe: __________

NDC 55513-148-01
Store at 2° to 8°C
S4
EPOGEN®
EPOETIN ALFA
4000 Units/1 mL
1 mL Single Use Vial
Dosage - See Package Insert
℞ Only
Amgen Inc. Thousand Oaks, CA 91320
U.S. License No. 1080
3 2 5 9 6 0 0
Lot
Exp.

E

20. Ordered: Clindamycin 300 mg IM now
On hand: Refer to label F.
Administer: __________ Syringe: __________

NDC 63323-282-04 28204
CLINDAMYCIN
INJECTION, USP
600 mg/4 mL
(150 mg/mL)*
For IM or IV Use
DILUTE BEFORE IV USE
4 mL Single Dose Vial
Rx only
Sterile
*Each mL contains clindamycin phosphate equivalent to 150 mg clindamycin, 0.5 mg disodium edetate and 9.45 mg benzyl alcohol as a preservative. When necessary, pH adjusted with sodium hydroxide and/or hydrochloric acid.
Usual Dosage: See package insert.
Warning: If given intravenously, must be diluted before use.
Store at 20° to 25°C (68° to 77°F) [see USP Controlled Room Temperature].
Do not refrigerate.
Vial stoppers do not contain natural rubber latex.
Abraxis Pharmaceutical Products
Schaumburg, IL 60173
402129A
LOT/EXP
3 63323-282-04 2

F

21. Ordered: 0.25 mg Sandostatin® Sub-Q
On Hand: Refer to label G.
Administer: __________ Syringe: __________

Sandostatin®
octreotide acetate
Injection (subcutaneous)
1 mL contains:
500 mcg octreotide (as acetate)
Rx only
Mfd. by: Novartis Pharma Stein AG
Stein, Switzerland
85009902 © Novartis
483380 US 483385
EXP./LOT

G

22. Explain which of the two medications represented by labels H and I you would use for the following order.

Ordered: Furosemide 10 mg IM

NDC 63323-280-04 28004
FUROSEMIDE
INJECTION, USP
40 mg/4 mL
(10 mg/mL)
For IM or IV Use Rx only
4 mL Single Dose Vial
Sterile, Nonpyrogenic
Preservative Free
Discard unused portion.
Each mL contains: Furosemide 10 mg; Water for Injection q.s. Sodium chloride to adjust isotonicity, pH adjusted with sodium hydroxide and if necessary hydrochloric acid.
Usual Dosage: See insert.
PROTECT FROM LIGHT. Do not use if discolored. Use only if solution is clear and seal intact.
Store at 20° to 25°C (68° to 77°F) [see USP Controlled Room Temperature].
Abraxis
Pharmaceutical Products
Schaumburg, IL 60173
401736C
LOT/EXP
3 63323-280-04 8

H

NDC 63323-280-02 28002
FUROSEMIDE
INJECTION, USP
20 mg/2 mL
(10 mg/mL)
For IM or IV Use Rx only
2 mL Single Dose Vial
Preservative Free
Discard unused portion.
PROTECT FROM LIGHT.
Do not use if discolored.
Abraxis
Pharmaceutical Products
Schaumburg, IL 60173
401803C
LOT/EXP
3 63323-280-02 4

I

In Exercises 23–26, refer to label J and the package insert.

23. What diluents may be used to reconstitute Zyprexa® for IM use?

24. How much diluent should be added to the vial?

25. What is the strength of the solution when diluted for IM use?

26. How long will the solution retain its potency at room temperature?

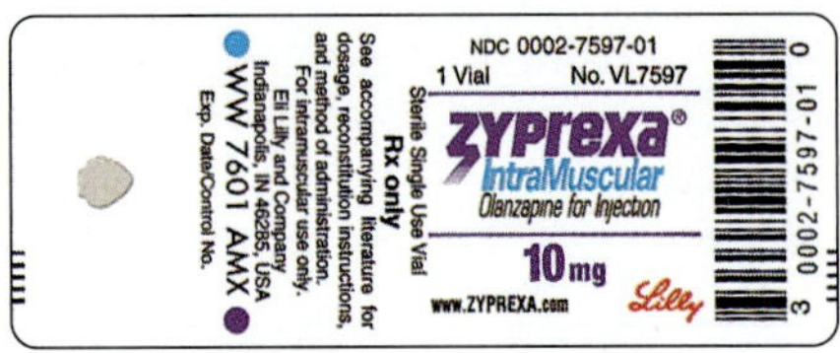

NDC 0002-7597-01
1 Vial No. VL7597
Zyprexa®
IntraMuscular
Olanzapine for Injection
10 mg
www.ZYPREXA.com
Lilly
Sterile Single Use Vial
Rx only
See accompanying literature for dosage, reconstitution instructions, and method of administration.
For intramuscular use only.
Eli Lilly and Company
Indianapolis, IN 46285, USA
WW 7601 AMX
Exp. Date/Control No.
3 0002-7597-01 0

J

ZYPREXA® IntraMuscular (olanzapine for injection) Dosing

ZYPREXA IntraMuscular is approved for the treatment of agitation associated with schizophrenia and bipolar mania.

Dose (mg)	Injection volume (mL)
10.0 mg	Withdraw total contents of vial
7.5 mg	1.5 mL
5.0 mg	1.0 mL
2.5 mg	0.5 mL

10 mg is the recommended dose for agitation associated with bipolar mania and schizophrenia.

Follow the steps below to reconstitute and use ZYPREXA IntraMuscular:

1. Inject 2.1 mL of Sterile Water for Injection into single-packaged vial for up to 10-mg dose.
2. Dissolve contents of vial completely; resulting solution should be clear and yellow.
3. Use solution within 1 hour; discard any unused portion.
4. Refer to table for injection volumes and corresponding doses of ZYPREXA IntraMuscular.
5. Immediately after use, dispose of syringe in approved sharps box.

In Exercises 27–30, refer to the package insert for Pregnyl.

Ordered: Pregnyl® 2,500 units IM every other day.

On hand: 10,000 unit vial of Pregnyl® for IM use. See Figure K and L.

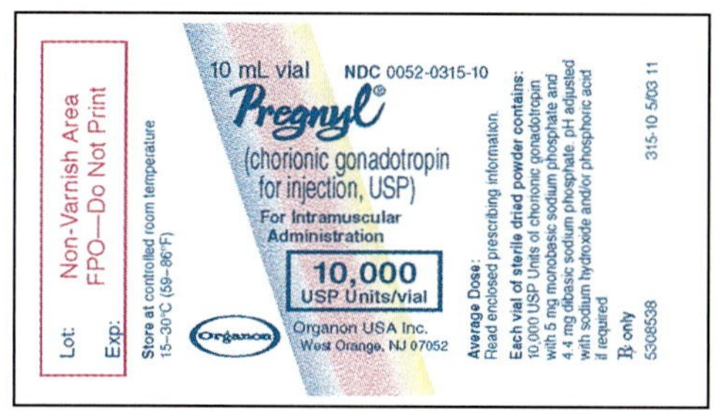

K

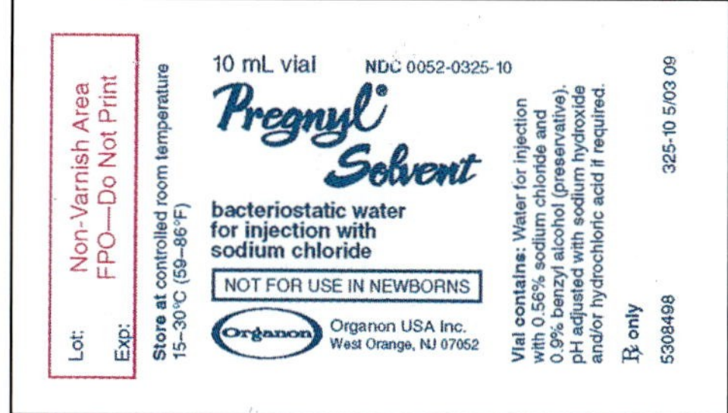

L

Directions for Reconstitution

Two-vial package: Withdraw sterile air from lyophilized vial and inject into diluent vial. Remove 1–10 mL from diluent and add to lyophilized vial; agitate gently until powder is completely dissolved in solution.

Parenteral drug products should be inspected visually for particulate matter and discoloration prior to administration, whenever solution and container permit.

IMPORTANT: USE COMPLETELY AFTER RECONSTITUTION. RECONSTITUTED SOLUTION IS STABLE FOR 60 DAYS WHEN REFRIGERATED.

HOW SUPPLIED

Two-vial package containing:

1-10 mL lyophilized multiple dose vial containing: 10,000 USP Units chorionic gonadotropin per vial, NDC 0052-0315-10.

1-10 mL vial of solvent containing: water for injection with sodium chloride 0.56% and benzyl alcohol 0.9%, NDC 0052-0325-10.

When reconstituted, each 10 mL vial contains:

Chorionic gonadotropin	10,000 USP Units
Monobasic sodium phosphate	5 mg
Dibasic sodium phosphate	4.4 mg
Sodium chloride	0.56%
Benzyl alcohol	0.9%

If required pH adjusted with sodium hydroxide and/or phosphoric acid.

Storage: Store at 15–30°C (59–86°F). Reconstituted solution is stable for 60 days when refrigerated.

27. How much diluent should be used to reconstitute Pregnyl® for IM use?

28. What solution strength should be written on the label of the vial?

29. If the Pregnyl® is reconstituted at 0200 on 8/30/2012 and will be stored in the refrigerator, what expiration date and time should be written on the label?

30. For a 2,500 unit IM dose, what is the amount to administer?

In Exercises 31–36, find the amount to administer.

31. Ordered: Acetylcysteine 800 mg via nebulizer q6h
    On hand: Acetylcysteine 10% solution

32. Ordered: Albuterol 1.25 mg via nebulizer q8h
    On hand: Albuterol 5 mg/mL

33. Ordered: Thorazine® 50 mg R as needed
    On hand: Thorazine® 25-mg and 100-mg suppositories

34. Ordered: Dilaudid® 6 mg R as needed
    On hand: Dilaudid® 3-mg suppositories

35. Ordered: Androderm® 5 mg/day top
    On hand: Androderm® 2.5 mg/day patches

36. Ordered: Nitro-Dur® 0.3 mg/h top
    On hand: Nitro-Dur® 0.1 mg/h and 0.2 mg/h patches

Critical Thinking Applications

Ordered: PegIntron™ 180 mcg sub-q weekly

On hand: Refer to the label A and package insert below.

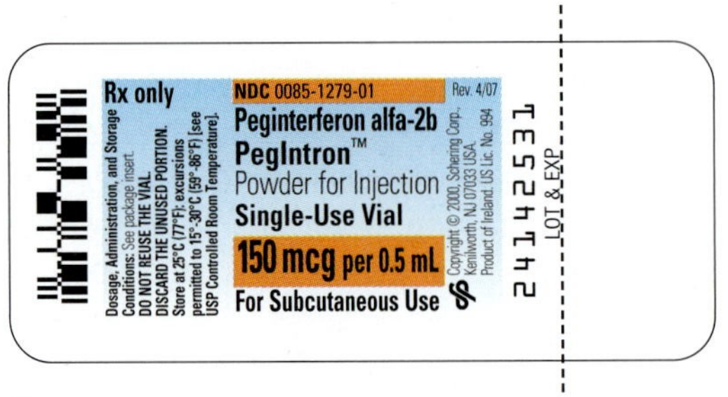

A

PegIntron™ label

Preparation and Administration

PEG-Intron Redipen™

PEG-Intron Redipen™ consists of a dual-chamber glass cartridge with sterile, lyophilized peginterferon alfa-2b in the active chamber and Sterile Water for Injection, USP in the diluent chamber. The PEG-Intron in the glass cartridge should appear as a white to off-white tablet shaped solid that is whole or in pieces, or powder. To reconstitute the lyophilized peginterferon alfa-2b in the Redipen™, hold the Redipen™ upright (dose button down) and press the two halves of the pen together until there is an audible click. Gently invert the pen to mix the solution. DO NOT SHAKE. The reconstituted solution has a concentration of either 50 µg per 0.5 mL, 80 µg per 0.5 mL, 120 µg per 0.5 mL, or 150 µg per 0.5 mL for a single subcutaneous injection. Visually inspect the solution for particulate matter and discoloration prior to administration. The reconstituted solution should be clear and colorless. Do not use if the solution is discolored or cloudy, or if particulates are present.

Keeping the pen upright, attach the supplied needle and select the appropriate PEG-Intron dose by pulling back on the dosing button until the dark bands are visible and turning the button until the dark band is aligned with the correct dose. The prepared PEG-Intron solution is to be injected subcutaneously.

The PEG-Intron Redipen™ is a single use pen and does not contain a preservative. The reconstituted solution should be used immediately and cannot be stored for more than 24 hours at 2°-8°C (see **Storage**). **DO NOT REUSE THE REDIPEN™.** The sterility of any remaining product can no longer be guaranteed. **DISCARD THE UNUSED PORTION.** Pooling of unused portions of some medications has been linked to bacterial contamination and morbidity.

PEG-Intron Vials

Two B-D Safety-Lok™ syringes are provided in the package; one syringe is for the reconstitution steps and one for the patient injection. There is a plastic safety sleeve to be pulled over the needle after use. The syringe locks with an audible click when the green stripe on the safety sleeve covers the red stripe on the needle. Instructions for the preparation and administration of PEG-Intron Powder for Injection are provided below.

Reconstitute the PEG-Intron lyophilized product with only 0.7 mL of 1.25 mL of supplied diluent (Sterile Water for Injection, USP). **The diluent vial is for single use only. The remaining diluent should be discarded.** No other medications should be added to solutions containing PEG-Intron, and PEG-Intron should not be reconstituted with other diluents. Swirl gently to hasten complete dissolution of the powder. The reconstituted solution should be clear and colorless. Visually inspect the solution for particulate matter and discoloration prior to administration. The solution should not be used if discolored or cloudy, or if particulates are present.

The appropriate PEG-Intron dose should be withdrawn and injected subcutaneously. PEG-Intron vials are for single use only and do not contain a preservative. The reconstituted solution should be used immediately and cannot be stored for more than 24 hours at 2°-8°C (see **Storage**). **DO NOT REUSE THE VIAL.** The sterility of any remaining product can no longer be guaranteed. **DISCARD THE UNUSED PORTION.** Pooling of unused portions of some medications has been linked to bacterial contamination and morbidity.

After preparation and administration of the PEG-Intron for injection, it is essential to follow the state and or local procedures for proper disposal of syringes, needles, and the Redipen™. A puncture-resistant container should be used for disposal. Patients should be instructed in how to properly dispose of used syringes, needles, or the Redipen™ and be cautioned against the reuse of these items.

Storage

PEG-Intron Redipen™

PEG-Intron Redipen™ should be stored at 2° to 8°C (36° to 46°F). After reconstitution, the solution should be used immediately, but may be stored up to 24 hours at 2° to 8°C (36° to 46°F). The reconstituted solution contains no preservative, and is clear and colorless. **DO NOT FREEZE.**

PEG-Intron Vials

PEG-Intron should be stored at 25°C (77°F); excursions permitted to 15°-30°C (59°-86°F) [see USP Controlled Room Temperature]. After reconstitution with supplied Diluent the solution should be used immediately, but may be stored up to 24 hours at 2° to 8°C (36° to 46°F). The reconstituted solution contains no preservative, is clear and colorless. **DO NOT FREEZE.**

PegIntron™ package insert

1. How should you prepare this medication?
2. How much diluent is supplied and how much should be used?
3. What should you do with the rest of the diluent?
4. How would this medication most likely be administered?
5. After reconstitution, how should this medication be stored and for how long?

Case Study

The physician orders Sandostatin® 75 mcg IM tid.

On hand: See Figure 9-15a, b, and c.

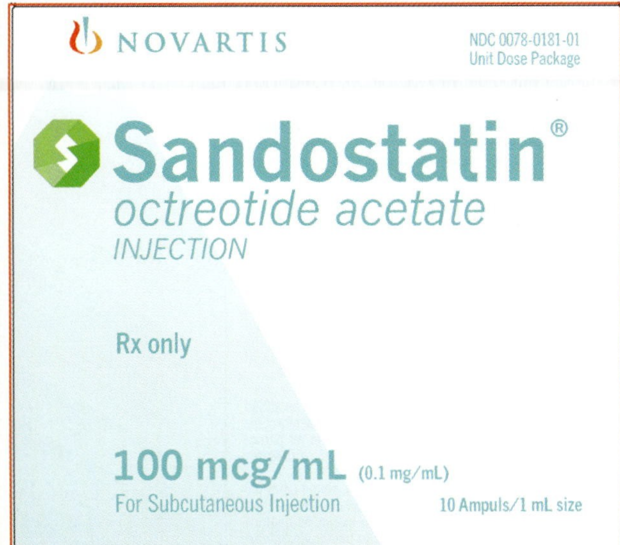

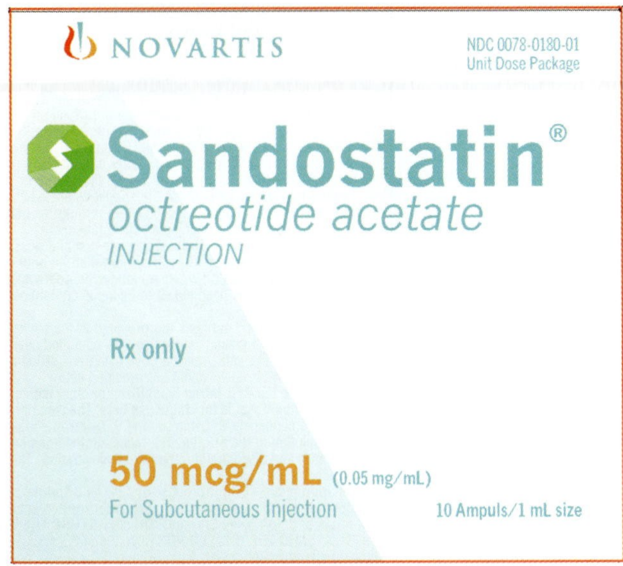

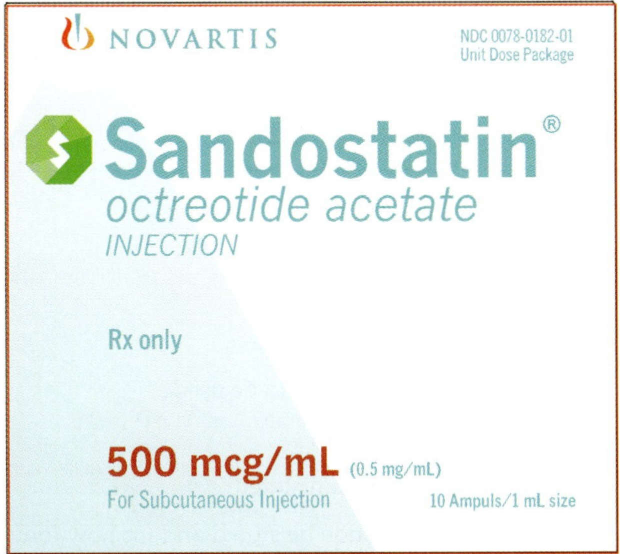

Figure 9-15

Describe what actions you should take before administering the medication. If you were going to administer the medication:

Which package would you use and why?

What would be the amount to administer?

What syringe would you use?

To check your answers, see page 530.

Internet Activity

You have a standard 3-g vial of Ticar® on hand and have been ordered to give a 36-year-old patient Ticar® 500 mg IM for a mild urinary tract infection. The package insert is not available, and you are not certain how to reconstitute the medication and calculate the dose. Search the Internet to find a package insert or other reliable source for reconstitution directions. What is the appropriate amount and solution? What is the amount to administer?

GO TO . . . Open the CD-ROM that accompanies your textbook, and complete a final review of the rules, practice problems, and activities presented for this chapter. For a final evaluation, take the chapter test and email or print your results for your instructor. A score of 95 percent or above indicates mastery of the chapter concepts. ■

10 Intravenous Dosages

One must learn by doing the thing, for though you think you know it, you have no certainty until you try.

—Aristole

Learning Outcomes

When you have completed Chapter 10, you will be able to:

- Identify the components and concentrations of IV solutions.
- Distinguish basic types of IV equipment.
- Calculate IV flow rates for both electronically controlled and manually controlled IV devices.
- Adjust the flow rate for IV infusions.
- Calculate infusion time based on volume and flow rate.
- Calculate volume based on infusion time and flow rate.
- Reconstitute and calculate medications for intermittent IV infusions.

Key Terms

Central line
Heparin lock
Hypertonic
Hypotonic
Infiltration
Infusion pumps
Intravenous
Isotonic
KVO fluids
Maintenance fluids
Patient-controlled analgesia
Phlebitis
Peripherally inserted central catheters (PICC)
Port-A-Cath
Primary line
Rate controllers
Replacement fluids
Saline lock
Secondary line
Syringe pumps
Therapeutic fluids

Introduction

Intravenous (IV) fluids are solutions, including medications, that are delivered directly into the bloodstream through a vein. Blood, a suspension, is also delivered intravenously. Fluids delivered directly into the bloodstream have a rapid effect, which is necessary during emergencies or other critical care situations when medications are needed. However, the results can be fatal if the wrong medication or dosage is given. Health care workers who administer or monitor IV solutions should know the principles discussed in this chapter.

Many IV drugs are available. Each has its own guidelines regarding its use, based on specifications developed by the manufacturers. The guidelines typically outline recommended dosages, infusion rates, compatibility, and patient monitoring. For example, some medications cannot be combined with others, or must be administered over a specific length of time.

Furthermore, most states regulate who may administer IV medications and what training is required. This chapter teaches IV calculations and theory; however, to be proficient, you must obtain the required training and learn by doing.

IV Solutions

IV solutions fall into four functional categories: **replacement fluids**, **maintenance fluids**, **KVO fluids**, and **therapeutic fluids**. *Replacement fluids* replace electrolytes and fluids lost or depleted due to hemorrhage, vomiting, or diarrhea. Examples include whole blood, nutrient solutions, or fluids administered to treat dehydration. *Maintenance fluids* help patients maintain normal electrolyte and fluid balance. They include IV fluids such as normal saline given during and after surgery. Some IVs provide access to the vascular system for emergency situations. Prescribed to keep the vein open (KVO or TKO), these *KVO fluids* include 5% dextrose in water. *Therapeutic fluids* deliver medication to the patient.

IV Labels

IV solutions are labeled with the name and exact amount of components in the solution. The label in Figure 10-1 is clearly marked as 5% dextrose and lactated Ringer's injection. Table 10-1 summarizes abbreviations often used for IV solutions.

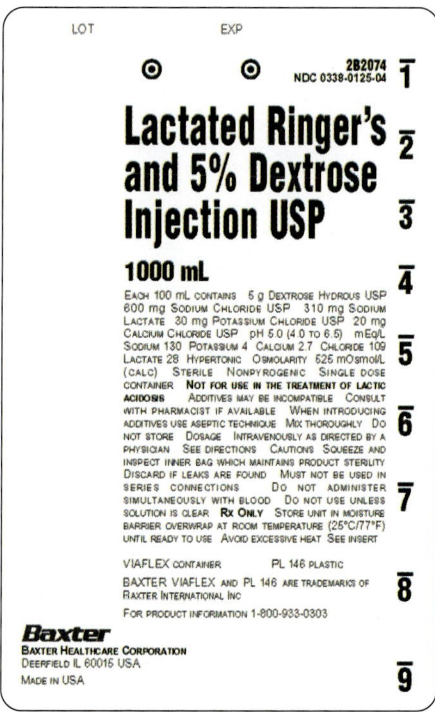
LOT
EXP
2B2074
NDC 0338-0125-04
Lactated Ringer's and 5% Dextrose Injection USP
1000 mL
EACH 100 mL CONTAINS 5 g DEXTROSE HYDROUS USP 600 mg SODIUM CHLORIDE USP 310 mg SODIUM LACTATE 30 mg POTASSIUM CHLORIDE USP 20 mg CALCIUM CHLORIDE USP pH 5.0 (4.0 TO 6.5) mEq/L SODIUM 130 POTASSIUM 4 CALCIUM 2.7 CHLORIDE 109 LACTATE 28 HYPERTONIC OSMOLARITY 525 mOsmol/L (CALC) STERILE NONPYROGENIC SINGLE DOSE CONTAINER NOT FOR USE IN THE TREATMENT OF LACTIC ACIDOSIS ADDITIVES MAY BE INCOMPATIBLE CONSULT WITH PHARMACIST IF AVAILABLE WHEN INTRODUCING ADDITIVES USE ASEPTIC TECHNIQUE MIX THOROUGHLY DO NOT STORE DOSAGE INTRAVENOUSLY AS DIRECTED BY A PHYSICIAN SEE DIRECTIONS CAUTIONS SQUEEZE AND INSPECT INNER BAG WHICH MAINTAINS PRODUCT STERILITY DISCARD IF LEAKS ARE FOUND MUST NOT BE USED IN SERIES CONNECTIONS DO NOT ADMINISTER SIMULTANEOUSLY WITH BLOOD DO NOT USE UNLESS SOLUTION IS CLEAR Rx ONLY STORE UNIT IN MOISTURE BARRIER OVERWRAP AT ROOM TEMPERATURE (25°C/77°F) UNTIL READY TO USE AVOID EXCESSIVE HEAT SEE INSERT
VIAFLEX CONTAINER PL 146 PLASTIC
BAXTER VIAFLEX AND PL 146 ARE TRADEMARKS OF BAXTER INTERNATIONAL INC
FOR PRODUCT INFORMATION 1-800-933-0303
Baxter
BAXTER HEALTHCARE CORPORATION
DEERFIELD IL 60015 USA
MADE IN USA
1
2
3
4
5
6
7
8
9

Figure 10-1 1000 mL D5LR label.

TABLE 10-1 Commonly Used Abbreviations

D10W	10% dextrose in water
D5W	5% dextrose in water
W, H_2O	Water
NS, NSS	Normal saline (0.9% NaCl)
LR	Lactated ringer's
RL	Ringer's lactate
$\frac{1}{2}$ NS, $\frac{1}{2}$ NSS	One-half normal saline solution (0.45% NaCl)
$\frac{1}{3}$ NS, $\frac{1}{3}$ NSS	One-third normal saline solution (0.3% NaCl)
$\frac{1}{4}$ NS, $\frac{1}{4}$ NSS	One-fourth normal saline solution (0.225% NaCl)

Rule 10-1 In abbreviations for IV solutions, letters identify the component and numbers identify the concentration.

Example An order for 5% dextrose in lactated Ringer's solution might be abbreviated in any of the following ways:

D5LR D_5LR 5%D/LR D5%LR

IV Concentrations

Solutions may have different concentrations of dextrose (glucose) or saline (sodium chloride, or NaCl). For example, 5% dextrose contains 5 g of dextrose per 100 mL (see Figure 10-2).

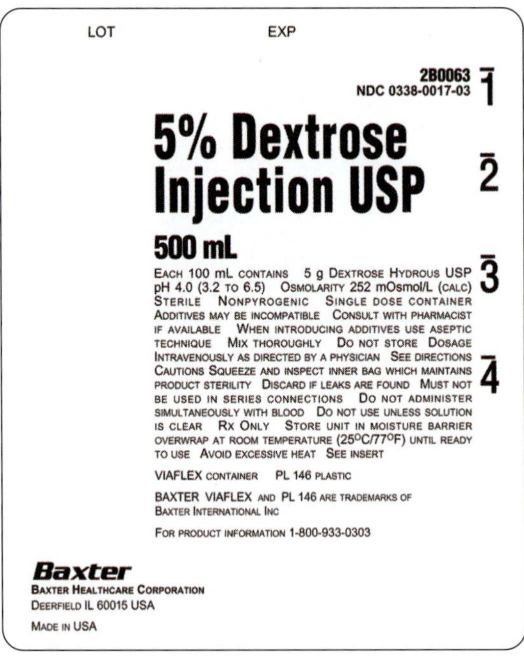
LOT
EXP
2B0063
NDC 0338-0017-03
1
5% Dextrose
Injection USP
2
500 mL
3
Each 100 mL contains 5 g Dextrose Hydrous USP pH 4.0 (3.2 to 6.5) Osmolarity 252 mOsmol/L (calc) Sterile Nonpyrogenic Single dose container Additives may be incompatible Consult with pharmacist if available When introducing additives use aseptic technique Mix thoroughly Do not store Dosage intravenously as directed by a physician See directions Cautions Squeeze and inspect inner bag which maintains product sterility Discard if leaks are found Must not be used in series connections Do not administer simultaneously with blood Do not use unless solution is clear Rx Only Store unit in moisture barrier overwrap at room temperature (25°C/77°F) until ready to use Avoid excessive heat See insert
4
VIAFLEX container PL 146 plastic
BAXTER VIAFLEX and PL 146 are trademarks of Baxter International Inc
For product information 1-800-933-0303
Baxter
Baxter Healthcare Corporation
Deerfield IL 60015 USA
Made in USA

Figure 10-2 500 mL of D5W.

Normal saline is 0.9% saline; it contains 900 mg, or 0.9 g, of sodium chloride per 100 mL (see Figure 10-3). In turn, 0.45% saline, or $\frac{1}{2}$ NS, has 450 mg of sodium chloride per 100 mL—one-half the amount of normal saline. Other saline concentrations include 0.3% saline (or $\frac{1}{3}$ NS) and 0.225% saline (or $\frac{1}{4}$ NS).

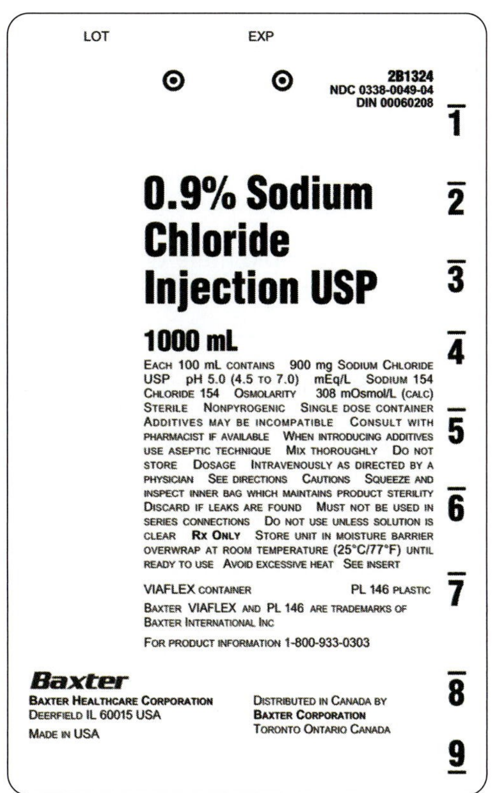

LOT
EXP
2B1324
NDC 0338-0049-04
DIN 00060208
0.9% Sodium Chloride Injection USP
1000 mL
EACH 100 mL CONTAINS 900 mg SODIUM CHLORIDE USP pH 5.0 (4.5 TO 7.0) mEq/L SODIUM 154 CHLORIDE 154 OSMOLARITY 308 mOsmol/L (CALC) STERILE NONPYROGENIC SINGLE DOSE CONTAINER ADDITIVES MAY BE INCOMPATIBLE CONSULT WITH PHARMACIST IF AVAILABLE WHEN INTRODUCING ADDITIVES USE ASEPTIC TECHNIQUE MIX THOROUGHLY DO NOT STORE DOSAGE INTRAVENOUSLY AS DIRECTED BY A PHYSICIAN SEE DIRECTIONS CAUTIONS SQUEEZE AND INSPECT INNER BAG WHICH MAINTAINS PRODUCT STERILITY DISCARD IF LEAKS ARE FOUND MUST NOT BE USED IN SERIES CONNECTIONS DO NOT USE UNLESS SOLUTION IS CLEAR RX ONLY STORE UNIT IN MOISTURE BARRIER OVERWRAP AT ROOM TEMPERATURE (25°C/77°F) UNTIL READY TO USE AVOID EXCESSIVE HEAT SEE INSERT
VIAFLEX CONTAINER PL 146 PLASTIC
BAXTER VIAFLEX AND PL 146 ARE TRADEMARKS OF BAXTER INTERNATIONAL INC
FOR PRODUCT INFORMATION 1-800-933-0303
Baxter
BAXTER HEALTHCARE CORPORATION
DEERFIELD IL 60015 USA
MADE IN USA
DISTRIBUTED IN CANADA BY
BAXTER CORPORATION
TORONTO ONTARIO CANADA
1
2
3
4
5
6
7
8
9

Figure 10-3 1000 mL of NS.

Example How much dextrose is contained in 500 mL D5W?

D5W represents 5% dextrose in water; it has 5 g of dextrose per 100 mL of water. Using ratio proportion (Procedure Checklist 3-2),

500 mL : ? g : : 100 mL : 5 g

$? \text{ g} \times 100 \text{ mL} = 500 \text{ mL} \times 5 \text{ g}$

$? = 25$

So 25 g of dextrose is contained in 500 mL D5W.

Knowledge of a patient's fluid and electrolyte balance is necessary to determine a solution's concentration. Calcium, potassium, chloride, phosphorus, and magnesium are electrolytes that can be added to an IV solution to help correct a fluid or chemical imbalance.

IV solutions are classified as **isotonic**, **hypotonic**, or **hypertonic**, depending on their effect on the fluid content of cells. Isotonic IV solutions, such as D5W, NS, and lactated Ringer's, do not affect the fluid balance of the surrounding cells or tissues. The fluid in hypotonic IV solutions such as 0.45% NS and 0.3% sodium chloride moves across the blood cell membrane into surrounding cells and tissues. This movement restores the proper fluid level in cells and tissues of patients who are dehydrated. By contrast, hypertonic IV solutions such as 3% saline draw fluids from cells and tissues across the cell membrane into the bloodstream. They are helpful for patients with severe fluid shifts, such as those caused by burns.

Rule 10-2 Patients with normal electrolyte levels are likely to receive isotonic solutions. Those with high electrolyte levels will receive hypotonic solutions. Those with low electrolyte levels will receive hypertonic solutions.

Example 1 Patient A is a 35-year-old, healthy female who will have an IV infusion during a diagnostic test. She will require an isotonic solution such as D5W, NS, or lactated Ringer's.

Example 2 Patient B is an 8-year-old female who has been vomiting and has had diarrhea for 24 h and is dehydrated. She may require a hypotonic solution like 0.45% or 0.3% NS to restore the proper fluid level in her cells and tissues.

Example 3 Patient C is a 50-year-old male with burns over 35 percent of his body. He may require hypertonic solution such as 3% NS to help draw fluids from cells and tissues.

Compatibility

Medications, electrolytes, and nutrients are additives that can be combined with IV solutions. Potassium chloride, vitamins B and C, and antibiotics are common additives. While additives are often prepackaged in the solution, you may need to mix the additive and IV solution yourself. The physician's order will tell you how much additive to administer, the amount and type of basic IV solution to use, and the length of time over which the additive/IV mixture should infuse. For example, an order may call for 20 milliequivalents (meq) of potassium chloride in 1000 mL of 5% dextrose and normal saline over 8 h, or

1000 mL D5NS c̄ 20 meq KCl IV over 8 h

Rule 10-3 Before you combine any medications, electrolytes, or nutrients with an IV solution, be sure the components are compatible.

Example Incompatible additives may cause the resulting solution to turn cloudy or crystallize, which means it hardens to crystals. If you mix an IV base solution with an additive that is incompatible (see Table 10-2), you place the patient's health at serious risk. Verify compatibility by checking with a compatibility chart, a drug reference book, the pharmacy, the Internet, or a package insert.

TABLE 10-2 Verifying Compatibility

The following are a few examples of incompatible combinations. Always check the compatibility of IV solutions and additives.

Ampicillin	5% dextrose in water
Cefotaxime sodium	Sodium bicarbonate
Diazepam	Potassium chloride
Dopamine HCl	Sodium bicarbonate
Penicillin	Heparin
Penicillin	Vitamin B complex
Sodium bicarbonate	Lactated Ringer's
Tetracycline HCl	Calcium chloride

Critical Thinking on the Job

Checking Compatibility

A patient in respiratory distress with congestive heart failure is started on D5/0.45% NaCl. The next day she is diagnosed with an upper respiratory infection. The physician orders 500 mg ampicillin IVPB q6h.

The health care professional begins to administer the ampicillin. She notices that the solution in the tubing has become cloudy.

Questions?

Why would the solution in the tubing be cloudy? What should the health care professional do?

IV Equipment

IV equipment is available in several forms. Most are electronic or have electronic components, while some are still manual.

The Primary Line

The typical IV setup consists of a bag or bottle of IV solution and tubing. IV bags come in different sizes, often containing 500 or 1000 mL of solution. You should mark them at regular time intervals to record the amount of solution that is being infused. Your facility will have specific guidelines for you to follow.

The tubing, which is the **primary line**, usually includes a drip chamber, clamp, and injection ports (see Figure 10-4). The drip chamber attaches to the IV bag. To measure the flow rate, squeeze and release the drip chamber until it is half filled with IV solution. Fluid in the chamber makes it easy to count the drops that fall into it from the bag. Use a roller clamp (Figure 10-5) or a screw clamp to set or adjust the flow rate of the IV solution. A slide clamp shuts off the IV solution flow completely without disturbing the flow rate setting at the roller or screw clamp. Injection ports allow you to inject medications or compatible fluids into the primary line or to attach a second IV line. IV bags may have ports for additives injected directly into the solution.

Tubing is available in two sizes: macrodrip or microdrip (see Figure 10-6). Macrodrip tubing allows larger drops to form before falling into the drip chamber. It is used for infusions of 80 mL/h or more and is always used for operating room infusions. Microdrip tubing allows smaller drops to enter the drip chamber. It is used for flow rates of less than 80 mL/h and is often used for KVO infusions. Microdrip tubing is especially useful for pediatric and critical care IVs, when very small volumes are used and accuracy is extremely important. Accidental increases in volume can be fatal in these situations.

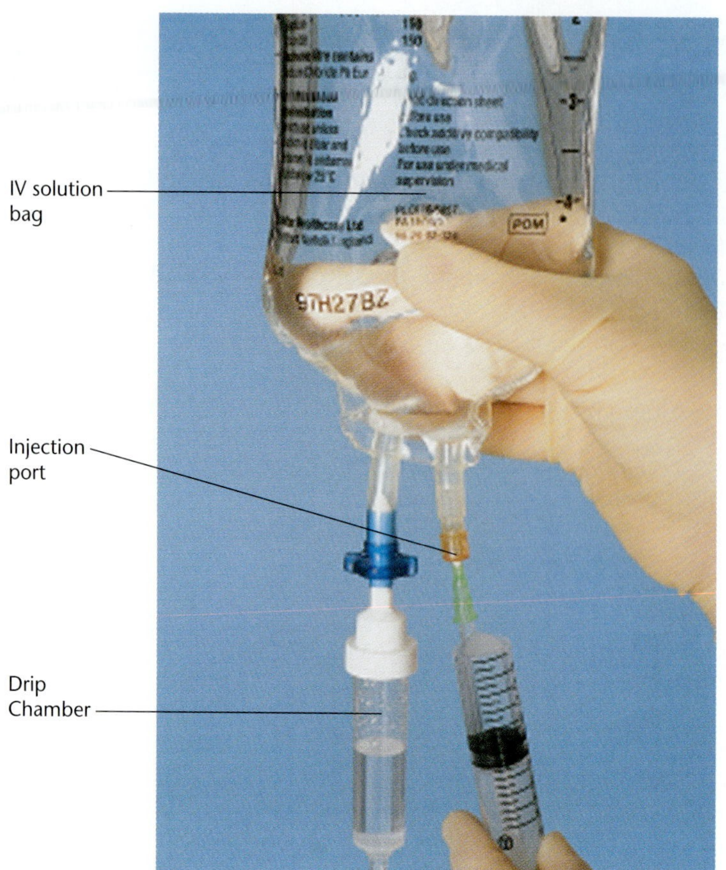

Figure 10-4 A typical IV setup.

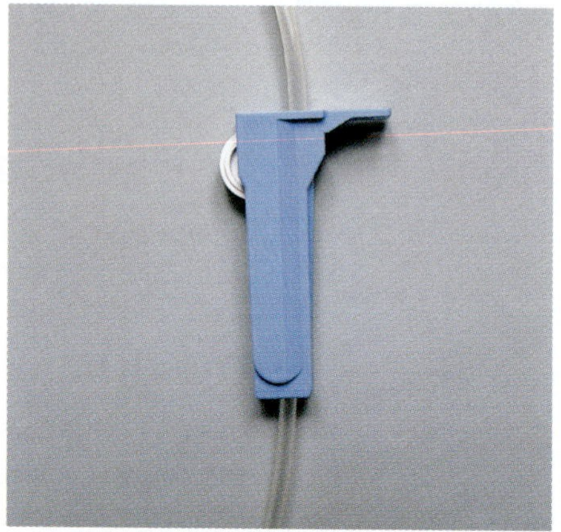

Figure 10-5 Roller clamp.

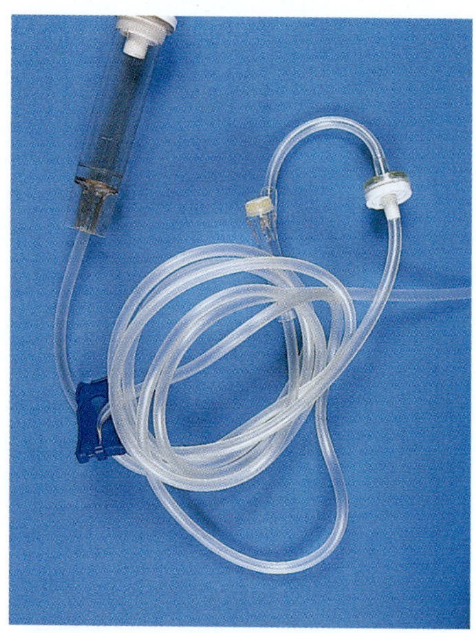

Figure 10-6 IV tubing.

Monitoring IV Equipment

IVs may be monitored manually. The bag is hung 36 inches (in.) above the patient's heart to allow gravity to draw the fluid through the tubing and into the vascular system. Whoever administers and monitors the IV adjusts the flow rate, using roller or screw clamps. Manually delivered IV fluids are usually adjusted in drops per minute (gtt/min).

Electronic devices—rate controllers, infusion pumps, syringe pumps, and *patient-controlled analgesia* (PCA) devices—can be used to regulate the flow of IV infusions. These devices often use tubing specific to the equipment. Some types of tubing may be used only with specific pumps (see Figures 10-7, 10-8, 10-9, and 10-10).

Rate controllers rely on gravity to infuse the solution, but no clamp is used to adjust the flow rate. Tubing is threaded through the controller, where a pincher maintains a preset flow rate. The controller is attached to a sensor that measures the drops or volume of solution that is delivered. An alarm sounds when the preset flow rate cannot be maintained.

Infusion pumps apply pressure sufficient to deliver a set volume of liquid every minute into the vein. They can introduce liquid into a central vein, where pressure is much higher than in peripheral veins. The desired flow rate is set on an infusion pump, either in milliliters per hour or by dosage. The unit does not rely on gravity, but forces the IV solution through the tubing. A sensor detects when the IV bag is empty or the flow is too rapid by sounding an alarm. An alarm also sounds if the flow rate cannot be maintained or if the bag is empty. A rate that is too slow may indicate too much resistance in the vein, suggesting blockage, a kink in the tubing, or that the IV catheter has come out of the vein. In some cases, the equipment will continue to pump the IV fluid, even though the catheter is out of the vein. Thus, when you use an infusion pump, you must monitor the patient's infusion site regularly for signs of infiltration (such as swelling, coolness, or discomfort).

Syringe pumps allow you to insert a syringe in the pump unit (see Figure 10-9). The syringe can deliver medication or fluids that cannot be combined with other medications or solutions. Syringe pumps are useful for pediatric medications as well as for medications that must be administered at a precisely controlled rate. Syringe pumps are often used in cases when a medication must be administered over half an hour or less; they are also used for longer time periods as well.

Patient-controlled analgesia (PCA) devices are used by patients in pain, including pain from cancer or surgery (see Figure 10-10). PCA pumps allow patients to control their own medication within limits preset according to the physician's order. By pressing a button on a handheld device, a patient administers medication. The PCA device helps monitor the effectiveness of the pain relief prescription, recording the number of times the patient uses the device.

Volume control sets such as Buretrol, Soluset, and Volutrol are used with manual IV setups and electronic rate controllers to improve accuracy, especially for small volumes of medication or fluid (see Figure 10-11). They are calibrated in 1-mL increments, with a total volume

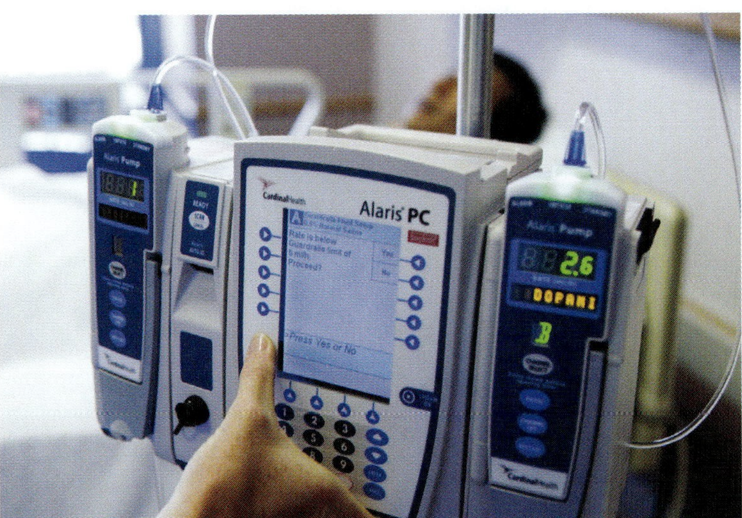

Figure 10-7 Rate controller.

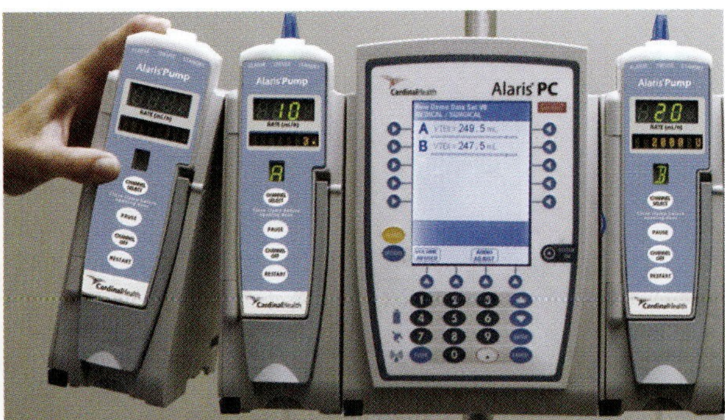

Figure 10-8 Infusion pump.

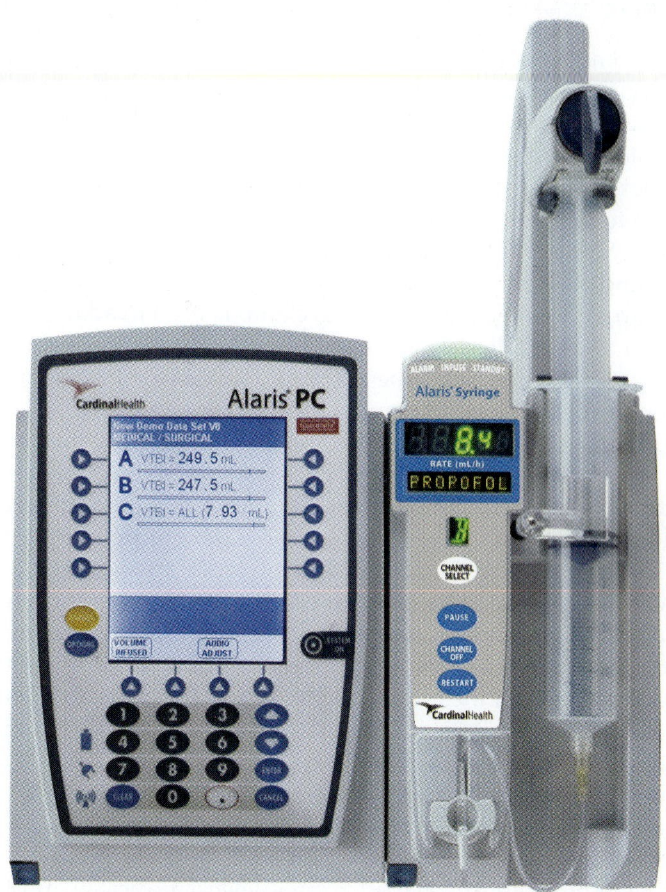

Figure 10-9 Syringe pump.

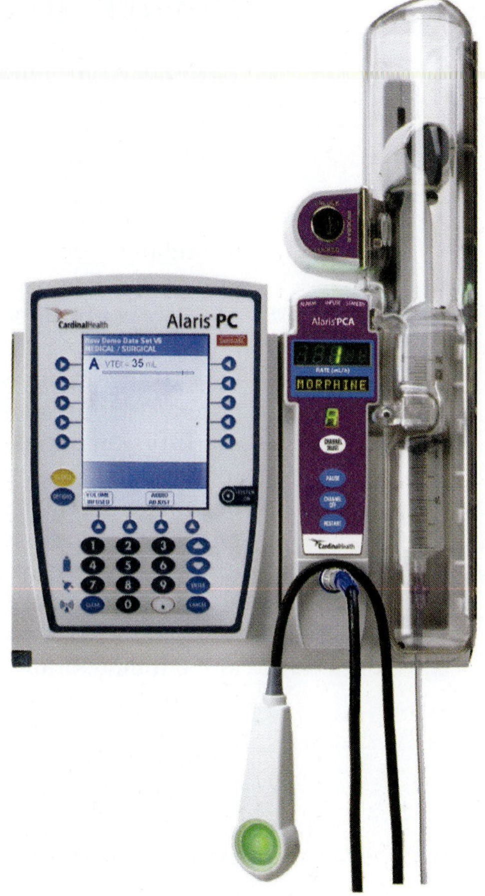

Figure 10-10 PCA pump.

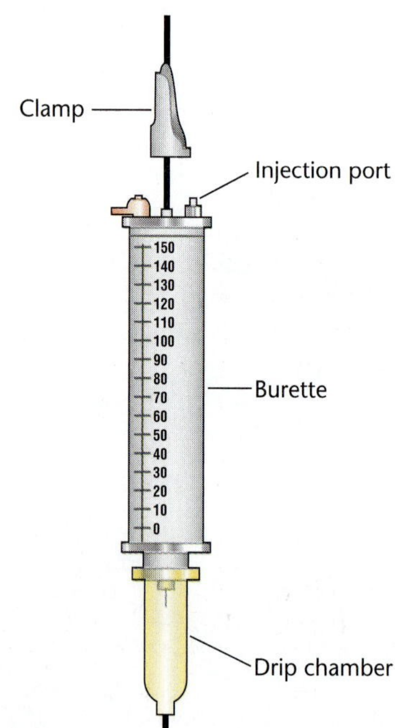

Figure 10-11 Volume control set.

capacity ranging from 100 to 150 mL. Medication is injected through an injection port into a burette—a chamber that holds a smaller controlled amount of fluid. An exact amount of IV fluid is added as a diluent to the burette chamber, where it is mixed. The fluid is delivered to the patient in microdrips. Burettes are often used in critical care or pediatric IVs because of their accuracy.

Peripheral and Central IV Therapy

Peripheral IV therapy accesses the circulatory system through a peripheral vein. Sites are usually located in the hand, forearm, foot, and leg. Because peripheral veins can be difficult to locate in small or premature infants, a peripheral IV line may be set up using a vein in the scalp.

Central IV therapy provides direct access to major veins. A **central line** is used when the patient needs large amounts of fluids, a rapid infusion of medication, infusion of highly concentrated solutions, or long-term IV therapy. Central lines can be inserted using a catheter through the chest wall or by threading a catheter through a peripheral vein. In newborn infants, a central line can be inserted into the umbilical vein or artery. These procedures are usually performed by a physician. A **peripherally inserted central catheter (PICC)** is inserted in arm veins and threaded into a central vein, often by specially trained nurses. A **Port-A-Cath** is used to deliver medication to a central vein. It is surgically placed under the skin and accessed through the skin to administer IV medication on an intermittent basis.

Rule 10-4 Never flush a sluggish IV with a syringe.

Example If you flush or irrigate an IV catheter that is clogged, you may be pushing a clot into the circulatory system. This clot, known as an *embolism*, can travel through the bloodstream and block a blood vessel. An obstruction or blockage of any blood vessel in the body is dangerous. An obstruction of a blood vessel in a vital organ such as the heart or lungs can be fatal.

Pain or swelling near an IV site may indicate an **infiltration** or phlebitis. An infiltration occurs when the needle or catheter is dislodged from the vein or penetrates the vein. Fluid is then infused into the surrounding tissue. Signs of infiltration include swelling, discomfort, and coolness at the infiltration site, as well as a sizable decrease in flow rate. **Phlebitis** is an inflammation of the vein. It can develop when the vein is irritated by IV additives, by movement of the needle or catheter, or during long-term IV therapy. In most cases of phlebitis, patients will complain of pain at or near the site. Other signs include heat, redness, and swelling of the injection site. In the case of either infiltration or phlebitis, stop the IV infusion and restart it in another limb. In the case of phlebitis, also notify the patient's physician.

REVIEW AND PRACTICE

10-1 IV Equipment

In Exercises 1–8, identify the IV equipment that you would use in each instance.

1. Add a dose of medication to an existing line.
2. Adjust the flow rate of an IV solution.

3. Stop the IV flow without disturbing the flow rate.
4. Introduce IV fluid into a central vein.
5. Set up tubing to infuse D5W at the rate of 40 mL/h.
6. Allow patient to self-administer medication for pain.
7. Administer a small amount of fluid over 15 minutes.
8. Administer an exact amount of IV fluid added as a diluent that is mixed and delivered in microdrips to a pediatric patient.

Provide a brief response to each of the following questions.

9. When would central IV therapy be used?
10. What do you look for when you monitor an IV?
11. How might you recognize infiltration?
12. What are three possible causes of phlebitis?
13. How high should the IV bag be hung?
14. What four electronic devices can be used to regulate the flow of IV infusions?
15. When is an injection port used?

To check your answers, see page 530.

Calculating Flow Rates

An order for IV fluids indicates the *amount of an IV fluid* to be administered and the *length of time* over which it is to be given. Before you can administer the IV, you must calculate a *flow rate* for the intravenous solution from these two values. For most electronic devices that regulate the flow of IV solutions, the flow rate will be expressed in milliliters per hour (mL/h).

Rule 10-5 To calculate flow rates in milliliters per hour, identify the following:

- *V* (volume) is expressed in milliliters.
- *T* (time) must be expressed in hours. (Convert the units when necessary by using fraction proportion, ratio proportion, or dimensional analysis.)
- *F* (flow rate) will be rounded to the nearest whole number.

Use the formula method with $F = \frac{V}{T}$ or dimensional analysis to determine the flow rate in milliliters per hour.

Formula Method

Example 1 Find the flow rate.

Ordered: 500 mg ampicillin in 100 mL NS to infuse over 30 minutes

In this case the volume is expressed in milliliters, and $V = 100$ mL.

Since time is expressed in minutes, you must first convert 30 minutes (min) to hours to find T. (In this example we will use the fraction proportion Procedure Checklist 3-1 to convert. You may prefer to use the ratio proportion Procedure Checklist 3-2 or dimensional analysis Procedure Checklist 3-3.)

$$\frac{1\text{ h}}{60\text{ min}} = \frac{?}{30\text{ min}}$$

$$\frac{1\text{ h}}{60\,\cancel{\text{min}}} = \frac{?}{30\,\cancel{\text{min}}}$$

$$60 \times ? = 1\text{ h} \times 30$$

$$? = \frac{1\text{ h} \times 30}{60}$$

$$? = 0.5\text{ h}$$

We now have the information needed in the proper units.

$V = 100$ mL and $T = 0.5$ h

Using the formula $F = \frac{V}{T}$, we find that

$$F = \frac{100\text{ mL}}{0.5\text{ h}}$$

$$F = 200\text{ mL/h}$$

Example 2 Find the flow rate.

Ordered: 500 mL 5% D 0.45%S over 3 h

Solution In this case the units are already expressed in milliliters and hours; $V = 500$ mL and $T = 3$ h.

Using the formula $F = \frac{V}{T}$, we have

$$F = \frac{500\text{ mL}}{3\text{ h}}$$

$$F = 166.7\text{ mL/h}$$

$F = 167$ mL/h rounded to nearest whole number

Dimensional Analysis

Example 1 Find the flow rate.

Ordered: 500 mg ampicillin in 100 mL NS to infuse over 30 min

1. Determine the units of measure for the answer (F), and place it as the unknown on one side of the equation.

   F mL/h =

2. Write a factor with the number of milliliters to be administered on top (V) and the length of time to be administered (T) on the bottom.

   $$\frac{100\text{ mL}}{30\text{ min}}$$

3. Multiply by a second factor to convert the minutes to hours and place minutes in the numerator. Set up the equation.

$$F \text{ mL/h} = \frac{100 \text{ mL}}{30 \text{ min}} \times \frac{60 \text{ min}}{1 \text{ h}}$$

4. Cancel units on the right side of the equation. The remaining unit of measure on the right side of the equation should match the unknown unit of measure on the left side of the equation.

$$F \text{ mL/h} = \frac{100 \text{ mL}}{30 \cancel{\text{min}}} \times \frac{60 \cancel{\text{min}}}{1 \text{ h}}$$

5. Solve the equation.

$$F \text{ mL/h} = \frac{100 \times 60}{30}$$

$$F = 200 \text{ mL/h}$$

Example 2 Find the flow rate.

Ordered: 500 mL 5% D 0.45%S over 3 h

Solution

1. Determine the units of measure for the answer (F) and place it as the unknown on one side of the equation.

$$F \text{ mL/h} =$$

2. On the right side of the equation, write a factor with the number of milliliters to be administered on top (V) and the length of time to be administered (T) on the bottom.

$$F \text{ mL/h} = \frac{500 \text{ mL}}{3 \text{ h}}$$

3. Since the units of measure on the left side of the equation match the units of measure on the right side of the equation, no additional conversion factors are necessary. Solve the equation.

$$F \text{ mL/h} = \frac{500 \text{ mL}}{3 \text{ h}}$$

$$F = 166.7 \text{ mL/h}$$

$F = 167$ mL/h   rounded to nearest whole number

GO TO . . . Open the CD-ROM that accompanies your textbook, and select Chapter 10, Rule 10-4. Review the animation and example problems, then complete the practice problems. Continue to the next section of the book once you have mastered the rule presented. ■

For a manually regulated IV, the flow rate needs to be calculated as the number of drops per minute (gtt/min). Before this can be calculated, you must first know how many drops are in a milliliter. IV tubing packages are labeled with a drop factor, which tells you how many drops of IV solution are equal to 1 mL when using that tubing (Figures 10-12 and 10-13). *Macrodrip* tubing has larger drops and one of three typical drop factors: 10 gtt/mL, 15 gtt/mL, or 20 gtt/mL. *Microdrip* tubing has a drop factor of 60 gtt/mL. See Figure 10-14.

In most circumstances you will determine the flow rate in milliliters per hour and then need to determine the flow rate in drops per minute. For example, you may be using an electronic device to monitor the flow rate of an IV. These devices are usually set at a flow rate of

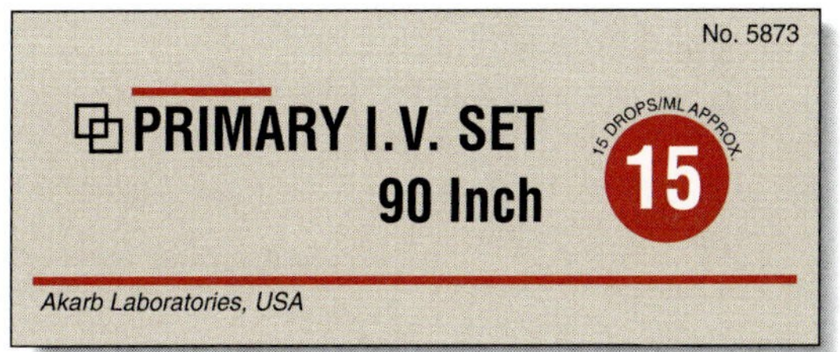

Figure 10-12 15 gtt/mL calibrated macrodrip tubing.

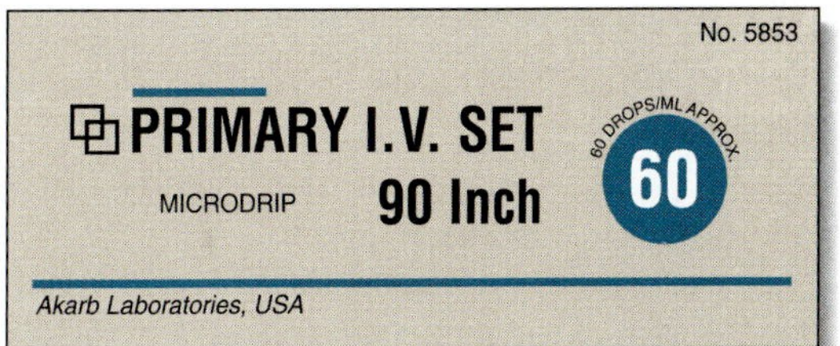

Figure 10-13 60 gtt/mL calibrated microdrip tubing.

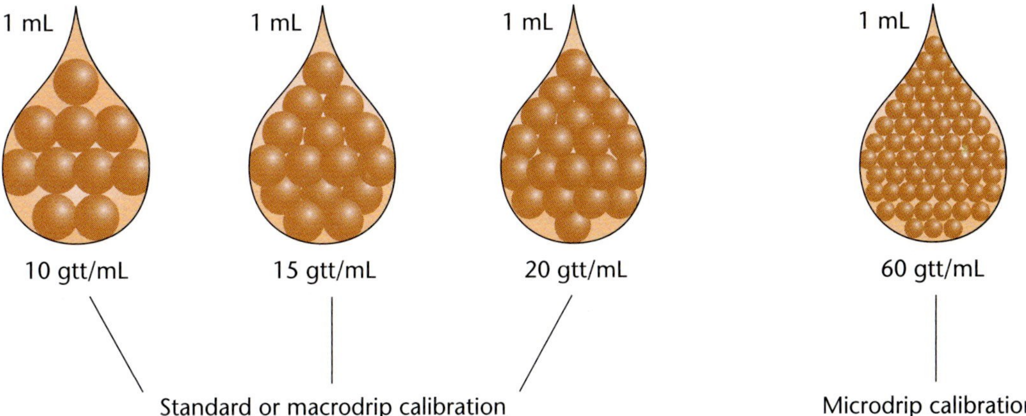

Figure 10-14 Macrodrip and microdrip calibration.

milliliters per hour. If an electronic device needs to be checked manually, then you will need to convert the flow rate of milliliters per hour to drops per minute to count the drops delivered to a patient in a minute. Once you know the desired flow rate in drops per minute and the IV is not attached to an electronic device, you adjust the roller or screw clamp so that the drops fall at the desired rate.

Rule 10-6 To determine the flow rate f in drops per minute:

Change the flow rate in milliliters per hour (F) to drops per minute (f), using the formula

$$f = \frac{F \times C}{60}$$

where F = flow rate, mL/h
C = calibration factor of tubing, gtt/mL
60 = number of minutes in 1 h

Round your answer to the nearest whole number.

Note: You can also use the dimensional analysis method for calculating the flow rate, adding the gtt factor of your tubing to the equation used for determining the flow rate F in mL/h.

Formula Method

Example 1 Find the flow rate in drops per minute that is equal to 75 mL/h when you are using 20 gtt/mL macrodrip tubing.

Solution

$F = 75 \text{ mL/h}$

$C = 20 \text{ mL/gtt}$

Substituting into the formula gives

$$f = \frac{F \times C}{60 \text{ min/h}}$$

$$f = \frac{75 \text{ mL/h} \times 20 \text{ mL/gtt}}{60 \text{ min/h}}$$

Cancel units.

$$f = \frac{75 \cancel{\text{mL}}/\cancel{\text{h}} \times 20 \text{ gtt}/\cancel{\text{mL}}}{60 \text{ min}/\cancel{\text{h}}}$$

Solve the equation.

$f = 25 \text{ gtt/min}$

Example 2 Find the flow rate in drops per minute that is equal to 35 mL/h when you are using 60 gtt/mL microdrip tubing.

Solution

$F = 35 \text{ mL/h}$

$C = 60 \text{ gtt/mL}$

Substituting into the formula gives

$$f = \frac{F \times C}{60}$$

$$f = \frac{35 \text{ mL/h} \times 60 \text{ gtt/mL}}{60 \text{ min/h}}$$

Cancel the units.

$$f = \frac{35 \cancel{\text{mL}}/\cancel{\text{h}} \times 60 \text{ gtt}/\cancel{\text{mL}}}{60 \text{ min}/\cancel{\text{h}}}$$

$f = 35 \text{ gtt/min}$

Note: The value of the flow rate is the same in drops per minute or milliliters per hour when 60 gtt/mL microdrip tubing is used. In other words, for microdrop tubing, F = f.

Dimensional Analysis

Example 1 Find the flow rate in drops per minute that is equal to 75 mL/h when you are using 20 gtt/mL macrodrip tubing.

Solution

1. Determine the unit of measure for the answer (f) and place it as the unknown on one side of the equation.

   $f \text{ gtt/min} =$

2. Determine the first factor. The number of milliliters to be administered on top (V) and the length of time to be administered (T) on the bottom.

$$\frac{75 \text{ mL}}{1 \text{ h}}$$

3. Multiply by a second factor to convert the hours to minutes, placing hours in the numerator.

$$75 \text{ mL/h} \times \frac{1 \text{ h}}{60 \text{ min}}$$

4. Multiply by the drop factor of the tubing being used. This is the third factor. Set up the equation.

$$f \text{ gtt/min} = \frac{75 \text{ mL}}{1 \text{ h}} \times \frac{1 \text{ h}}{60 \text{ min}} \times \frac{20 \text{ gtt}}{1 \text{ mL}}$$

5. Cancel units on the right side of the equation. The remaining unit of measure on the right side of the equation should match the unknown unit of measure on the left side of the equation.

$$f \text{ gtt/min} = \frac{75 \cancel{\text{mL}}}{1 \cancel{\text{h}}} \times \frac{1 \cancel{\text{h}}}{60 \text{ min}} \times \frac{20 \text{ gtt}}{1 \cancel{\text{mL}}}$$

6. Solve the equation.

$$f = \frac{75 \times 20 \text{ gtt}}{60 \text{ min}}$$

$$f = 25 \text{ gtt/min}$$

Example 2 Find the flow rate in drops per minute that is equal to 35 mL/h when you are using 60 gtt/mL microdrip tubing.

Solution

1. Determine the unit of measure for the answer (f) and place it as the unknown on the left side of the equation.

$$f \text{ gtt/min} =$$

2. Determine the first factor. The number of milliliters to be administered on top (V) and the length of time to be administered (T) on the bottom.

$$\frac{35 \text{ mL}}{\text{h}}$$

3. Multiply by a second factor to convert the hours to minutes; place hours in the numerator.

$$35 \text{ mL/h} \times \frac{1 \text{ h}}{60 \text{ min}}$$

4. Multiply by the drop factor of the tubing being used. Set up the equation.

$$f \text{ gtt/min} = \frac{35 \text{ mL}}{1 \text{ h}} \times \frac{1 \text{ h}}{60 \text{ min}} \times \frac{60 \text{ gtt}}{1 \text{ mL}}$$

5. Cancel units on the right side of the equation. The remaining unit of measure on the right side of the equation should match the unknown unit of measure on the left side of the equation.

$$f \text{ gtt/min} = \frac{35 \cancel{\text{mL}}}{1 \cancel{\text{h}}} \times \frac{1 \cancel{\text{h}}}{60 \text{ min}} \times \frac{60 \text{ gtt}}{1 \cancel{\text{mL}}}$$

6. Solve the equation.

$$f\text{ gtt/min} = 35 \times \frac{\cancel{1}}{\cancel{60}\text{ min}} \times \frac{\cancel{60}\text{ gtt}}{\cancel{1}}$$

$$f = 35\text{ gtt/min}$$

Note: The value of the flow rate is the same in drops per minute or milliliters per hour when 60 gtt/mL microdrop tubing is used. In other words, for microdrip tubing, F = f.

GO TO . . . Open the CD-ROM that accompanies your textbook, and select Chapter 10, Rule 10-6. Review the animation and example problems, then complete the practice problems. Continue to the next section of the book once you have mastered the rule presented. ■

Adjusting Flow Rates

Counting drops and timing are not always precise. What you calibrated as 25 drops per minute may actually be 25.4 drops per minute. Therefore, adjustments to flow rates sometimes need to be made. You should check at least once every hour that the IV is infusing to see if it is behind or ahead of schedule. The policy at the facility where you are employed will dictate whether you may adjust the IV flow rate or whether you should notify the physician. Always check this policy before you adjust a flow rate.

Rule 10-7 To adjust the flow rate:

- Recalculate the infusion, using the volume remaining in the IV and the time remaining in the order.
- Check the guidelines at your facility before you adjust the flow rate.

Example 1 Original order: 1500 mL NS over 12 h

The IV was infused at an original rate of 42 gtt/min using 20 gtt/mL macrodrip tubing. After 3 h, 1200 mL remains in the bag. The facility policy is that flow rate adjustments must not exceed 25 percent.

Solution

$V = 1200$ mL (volume remaining)

$T = 12$ h original $-$ 3 h elapsed $= 9$ h remaining

$C = 20$ gtt/mL

Using the formula $f = C \times \frac{V}{T}$, we must first convert 9 h to minutes. Using the ratio proportion method, we find

$$60\text{ min} : 1\text{ h} :: T : 9\text{ h}$$

$$60\text{ min} : 1\cancel{\text{h}} :: T : 9\cancel{\text{h}}$$

$$T \times 1 = 60\text{ min} \times 9$$

$$T = 540\text{ min}$$

Insert the appropriate numbers into the formula.

$$f = 20 \text{ gtt/mL} \times \frac{1200 \text{ mL}}{540 \text{ min}}$$

Cancel units.

$$f = 20 \text{ gtt/}\cancel{\text{mL}} \times \frac{1200 \cancel{\text{mL}}}{540 \text{ min}}$$

Solve for the unknown.

$$f = 20 \text{ gtt} \times \frac{1200}{540 \text{ min}}$$

$$f = 44.4 \text{ gtt/min}$$

Round to the nearest whole number.

$$f = 44 \text{ gtt/min} = \text{adjusted flow rate}$$

Determine if the adjusted rate is within 25 percent of the original.

$$25\% \text{ of original rate} = 25\% \times 42 = 10.5$$

The adjusted rate must fall within the following range:

$$\text{Minimum} = 42 - 10.5 = 31.5$$

$$\text{Maximum} = 42 + 10.5 = 52.5$$

Since 44 gtt/min falls between the minimum and maximum allowed by the policy, you may adjust the rate.

Example 2 Original order: 1500 mL NS over 12 h

Using 15 gtt/mL macrodrip tubing, the IV was infused at 30 gtt/min. After 4 h, 1100 mL remains in the bag. The facility policy is that flow rate adjustments must not exceed 25 percent.

Solution

$V = 1100$ mL (volume remaining)

$T = 12 \text{ h} - 4 \text{ h} = 8$ h remaining

$C = 15$ gtt/mL

Using the formula $f = C \times \frac{V}{T}$, we must first convert 8 h to minutes. Using the fraction proportion method, we have

$$\frac{1 \cancel{\text{h}}}{60 \text{ min}} = \frac{8 \cancel{\text{h}}}{?}$$

$$1 \times ? \text{ min} = 60 \times 8$$

$$? \text{ min} = 480$$

Insert the appropriate numbers into the formula.

$$f = 15 \text{ gtt/mL} \times \frac{1100 \text{ mL}}{480 \text{ min}}$$

Cancel units.

$$f = 15 \text{ gtt/}\cancel{\text{mL}} \times \frac{1100 \cancel{\text{mL}}}{480 \text{ min}}$$

Solve for the unknown.

$$f = 15 \text{ gtt} \times \frac{1100}{480 \text{ min}}$$

$$f = 34.4 \text{ gtt/min}$$

Round to the nearest whole number.

$$f = 34 \text{ gtt/min} = \text{adjusted flow rate}$$

Determine if the adjusted rate is within 25 percent of the original.

$$25\% \text{ of original rate} = 25\% \times 30 = 7.5$$

The adjusted rate must fall within the following range:

$$\text{Minimum} = 30 - 7.5 = 22.5$$

$$\text{Maximum} = 30 + 7.5 = 37.5$$

Since 34 gtt/min falls between the minimum and maximum allowed by the policy, you may adjust the rate.

GO TO . . . Open the CD-ROM that accompanies your textbook, and select Chapter 10, Rule 10-7. Review the animation and example problems, then complete the practice problems. Continue to the next section of the book once you have mastered the rule presented. ■

Critical Thinking on the Job

Adjusting the Flow Rate

Earlier in the day, Pat set up an IV based on the following physician's order: 750 mL D5NS to infuse over 8 h. Pat calculated that the patient should receive 94 mL of fluid per hour, with a flow rate of 16 gtt/min using 10 gtt/mL tubing.

After 4 h (one-half the time ordered for the infusion), Pat observed that 450 mL remained in the IV bag. Only one-half of the fluid, or 375 mL, should have remained in the bag. The patient had received 75 mL less fluid than expected. Pat decided to reset the flow rate for the next hour so that the patient would receive the original 94 mL/h plus the 75 mL that should have already been administered, for a total of 169 mL. After the next hour, Pat planned to reset the IV to the original flow rate of 16 gtt/min. Pat calculated the new flow rate as

$$\frac{10 \text{ gtt}}{1 \text{ mL}} \times \frac{169 \text{ mL}}{1 \text{ h}} \times \frac{1 \text{ h}}{60 \text{ min}} =$$

$$\frac{28.17 \text{ gtt}}{1 \text{ min}} = 28 \text{ gtt/min}$$

Thus, Pat adjusted the flow rate to 28 gtt/min.

Questions?

What mistake did Pat make? What should Pat have done to avoid the mistake?

REVIEW AND PRACTICE

10-2 Calculating Flow Rates

In Exercises 1–4, find the flow rate in milliliters per hour.

1. Ordered: 1000 mL LR over 6 h
2. Ordered: 300 mL NS over 2 h
3. Ordered: 3000 mL 0.45% NS q24h
4. Ordered: 40 meq KCl in 100 mL NS over 45 min

In Exercises 5–10, calculate the flow rate for IVs using electronic devices.

5. Ordered: 1500 mL RL over 12 h, using an infusion pump
6. Ordered: 1000 mL NS over 12 h, using an infusion pump
7. Ordered: 750 mL NS over 8 h, using an electronic controller set in milliliters per hour
8. Ordered: 20 meq KCl in 50 mL NS over 30 min, using an electronic rate controller set in milliliters per hour
9. Ordered: 1800 mL 0.45% S per day by infusion pump
10. Ordered: 250 mL D5W over 3 h by infusion pump

In Exercises 11–20, calculate the flow rate for manually regulated IVs.

11. Ordered: 1000 mL NS over 24 h, tubing is 20 gtt/mL
12. Ordered: 400 mL RL over 8 h, tubing is 10 gtt/mL
13. Ordered: 1500 mL 0.45% S over 12 h, tubing is 15 gtt/mL
14. Ordered: 250 mL D5W over 3 h, tubing is 10 gtt/mL
15. Ordered: 40 meq KCl in 100 mL NS over 40 min, tubing is 20 gtt/mL
16. Ordered: 500 mL NS over 8 h, tubing is 15 gtt/mL
17. Ordered: 3000 mL NS over 24 h, refer to label A.
18. Ordered: 50 mL penicillin IV over 1 h, refer to label B.
19. Ordered: 750 mL 5%D NS over 5 h, refer to label C.
20. Ordered: 100 mL gentamicin over 30 min, refer to label D.

In Exercises 21 and 22, calculate the flow rate in drops per minute.

21. Ordered: 1000 mL D5W over 9 h, using an electronic controller set in drops per minute, tubing calibration is 15 gtt/mL
22. Ordered: 750 mL RL over 8 h by electronic rate controller set in drops per minute, tubing calibration is 15 gtt/mL

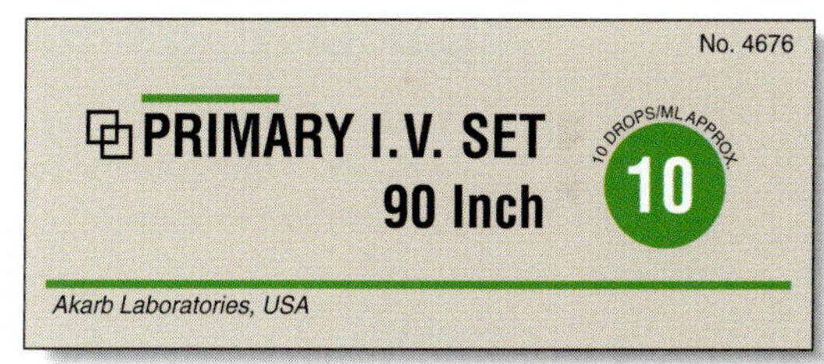

A

No. 5853
PRIMARY I.V. SET
MICRODRIP
90 Inch
60 DROPS/ML APPROX.
60
Akarb Laboratories, USA

B

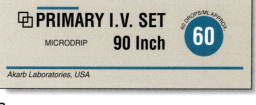

Figure 10-15 Look at the drop factor for IV tubing to calculate flow rates.

In Exercises 23–25, calculate the original flow rate. Then determine if an adjustment is necessary and calculate the adjusted flow rate. Adjustments cannot exceed 25 percent.

23. Ordered: 375 mL RL over 3 h (10 gtt/mL tubing)
    After 1 h, 175 mL has infused.

24. Ordered: 1000 mL NS over 8 h (20 gtt/mL tubing)
    With 5 h remaining, 550 mL of NS remains in the IV bag.

25. Ordered: 500 mL D5W over 4 h (15 gtt/mL tubing)
    After 2 h, 150 mL has infused.

For Exercises 26 to 30, you must manually check the IV infusion rate. Calculate the rate in drops per minute.

26. Rate controller set at 200 mL/h, tubing 10 gtt/mL

27. Infusion pump set at 125 mL/h, tubing 15 gtt/mL

28. Electronic pump set at 150 mL/h, tubing 15 gtt/mL

29. Pediatric infusion set at 75 mL/h, tubing 60 gtt/mL

30. Rate controller set at 100 mL/h, tubing 10 gtt/mL

To check your answers, see page 531.

C

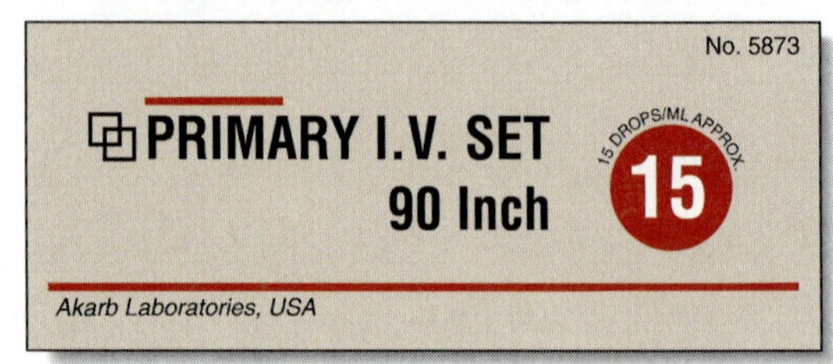

D

Figure 10-16 When calculating the gtt/min of an IV, always check the package to determine the calibration of the IV tubing you will be using.

Infusion Time and Volume

An order may call for a certain amount of fluid to infuse at a specific rate, without specifying the duration. In this case you will need to calculate the duration or amount of time the IV will take to infuse, so that you can monitor the IV properly. In other cases, you may know the duration and the flow rate, and you will need to calculate the fluid volume.

Calculating Infusion Time

In the previous section, you calculated the rate of infusion when given the volume and time. Sometimes the infusion rate and volume are given in the order, and you will need to calculate the duration or amount of time the infusion will take to be administered.

Rule 10-8 To calculate infusion time in hours T, identify:

- V (volume) expressed in milliliters.
- F (flow rate) expressed in milliliters per hour.
- Fractional hours expressed in minutes by multiplying by 60.

Use the formula $T = \frac{V}{F}$ or dimensional analysis to find T, the infusion time in hours.

Formula Method

Example 1 Find the total time to infuse.

Ordered: 1000 mL NS to infuse at a rate of 75 mL/h

Solution The volume $V = 1000$ mL. The flow rate F is expressed in milliliters per hour, $F = 75$

Substitute the values into the formula $T = \frac{V}{F}$.

$$T = \frac{1000 \text{ mL}}{75 \text{ mL/h}}$$

Cancel the units.

$$T = \frac{1000 \cancel{\text{mL}}}{75 \cancel{\text{mL}}/\text{h}}$$

$T = 13.33$ h = total time to infuse 1000 mL

Note that 0.33 h does not represent 33 minutes. Because there is 60 min in 1 h, you must multiply the fractional hours by 60 to convert it to minutes.

$0.33 \text{ h} \times 60 \text{ min/h} = 20 \text{ min}$

The total time to infuse the solution is 13 h 20 min.

Example 2 Find the total time to infuse.

Ordered: 750 mL LR to infuse at a rate of 125 mL/h started at 11 p.m.

Solution The volume $V = 750$ mL. The flow rate F is expressed in milliliters per hour, $F = 125$ mL/h. Substitute the values into the formula $T = \frac{V}{F}$.

$$T = \frac{750 \cancel{\text{mL}}}{125 \cancel{\text{mL}}/\text{h}}$$

$T = 6$ h = total time to infuse 750 mL

Dimensional Analysis

Example 1 Find the total time to infuse.

Ordered: 1000 mL NS to infuse at a rate of 75 mL/h

Solution

1. Determine the unit of measure for the answer T, and place it as the unknown on the left side of the equation.

   $T \text{ h} =$

2. The first factor is the number of milliliters to be administered on over 1. The second factor is the inverted flow rate. The flow rate is inverted in order to solve for hours (h).

$$T\text{ h} = \frac{1000\text{ mL}}{1} \times \frac{1\text{ h}}{75\text{ mL}}$$

3. Cancel units on the right side of the equation. The remaining unit of measure on the right side of the equation should match the unknown unit of measure on the left side of the equation.

$$T\text{ h} = \frac{1000\text{ }\cancel{\text{mL}}}{1} \times \frac{1\text{ h}}{75\text{ }\cancel{\text{mL}}}$$

4. Solve the equation.

$$T\text{ h} = \frac{1000\text{ h}}{75}$$

$T = 13.33\text{ h} =$ total time to infuse 1000 mL

Note that 0.33 h does not represent 33 min. Because there is 60 min in 1 h, you must multiply the fractional hours by 60 to convert it to minutes.

$0.33\text{ h} \times 60\text{ min/h} = 20\text{ min}$

The total time to infuse the solution is 13 h 20 min.

Example 2 Find the total time to infuse.

Ordered: 750 mL LR to infuse at a rate of 125 mL/h, started at 11 p.m.

Solution

1. Determine the unit of measure for the answer *T*, and place it as the unknown on the left side of the equation.

   $T\text{ h} =$

2. The first factor is the number of milliliters to be administered on top *V* over 1. The second factor is the inverted flow in order to solve for hours.

$$T\text{ h} = \frac{750\text{ mL}}{1} \times \frac{1\text{ h}}{125\text{ mL}}$$

3. Cancel units on the right side of the equation. The remaining unit of measure on the right side of the equation should match the unknown unit of measure on the left side of the equation.

$$T\text{ h} = \frac{750\text{ }\cancel{\text{mL}}}{1} \times \frac{1\text{ h}}{125\text{ }\cancel{\text{mL}}}$$

4. Solve the equation.

$$T\text{ h} = \frac{750\text{ h}}{125}$$

$T = 6\text{ h} =$ total time to infuse 750 mL

GO TO . . . Open the CD-ROM that accompanies your textbook, and select Chapter 10, Rule 10-8. Review the animation and example problems, then complete the practice problems. Continue to the next section of the book once you have mastered the rule presented.

In some cases, you will need to determine the time an infusion will be complete.

Rule 10-9 To calculate the time when an infusion will be completed, you must first know the time the infusion started in military time and the total time in hours and minutes to infuse the solution ordered. Since each day is only 24 h long, when the sum is greater than 2400 (midnight), you must start a new day by subtracting 2400 This will determine the time of completion, which will be the next calendar day.

Example 1 Determine when the following infusion will be completed.

Ordered: 1000 mL NS to infuse at a rate of 75 mL/h

You start the infusion at 7 a.m. on 6/06/12 First determine the start time in military time: 7 a.m. = 0700

Learning Link Recall from Chapter 3, Rule 3-9 using the 24-h clock, page 102.

Add the total amount of time to infuse, which was determined as 13 h 20 min, or 1320.

0700	time started in military time
+ 1320	total amount of time in hours and minutes
2020	

The infusion will be completed at 2020, which is 8:20 p.m.

Example 2 Determine when the infusion will be completed.

Ordered: 750 mL LR to infuse at a rate of 125 mL/h, started at 11 p.m. on 8/04/12

First determine the start time in military time: 11 p.m. = 2300 Add the total amount of time to infuse, which was determined as 6 h.

2300	time started in military time
+ 600	total amount of time of infusion in hours and minutes
2900	sum is greater than 2400
−2400	subtract 2400
500	

The infusion will be complete at 0500, or 5:00 a.m., on 8/05/12.

GO TO . . . Open the CD-ROM that accompanies your textbook, and select Chapter 10, Rule 10-9. Review the animation and example problems, then complete the practice problems. Continue to the next section of the book once you have mastered the rule presented.

Calculating Infusion Volume

When infusion rate and infusion time are given in the order, the volume infused over a given period of time can be calculated so that you can monitor the IV properly.

Rule 10-10 To calculate infusion volume:

Use the formula $V = T \times F$ or dimensional analysis to find the infusion volume V in milliliters, where:

- T (time) must be expressed in hours.
- F (flow rate) must be expressed in milliliters per hour.

Formula Method

Example 1 Find the total volume infused in 5 h if the infusion rate is 35 mL/h.

Solution

$T = 5\text{ h}$

$F = 35\text{ mL/h}$

Substitute the values into the formula $V = T \times F$.

$V = 5\text{ h} \times 35\text{ mL/h}$

$V = 175\text{ mL}$ = volume that will infuse over 5 h

Example 2 Find the total volume infused in 12 h if the infusion rate is 200 mL/h.

Solution Substitute the values into the formula.

$V = 12\text{ h} \times 200\text{ mL/h}$

$V = 2400\text{ mL}$ = volume that will infuse over 12 h

Dimensional Analysis

Example 1 Find the total volume infused in 5 h if the infusion rate is 35 mL/h.

Solution

1. Determine the unit of measure for the answer T, and place it as the unknown on the left side of the equation.

   $V\text{ mL} =$

2. The first factor is the length of time of the infusion over 1.

   $\frac{5\text{ h}}{1}$

3. Multiply by the flow rate of the infusion. The second factor.

   $V\text{ mL} = \frac{5\text{ h}}{1} \times \frac{35\text{ mL}}{1\text{ h}}$

4. Cancel units on the right side of the equation. The remaining unit of measure on the right side of the equation should match the unknown unit of measure on the left side of the equation.

   $V\text{ mL} = \frac{5\cancel{\text{h}}}{1} \times \frac{35\text{ mL}}{\cancel{\text{h}}}$

5. Solve the equation.

   $V\text{ mL} = 175\text{ mL}$ to be infused in 5 h

Example 2 Find the total volume infused in 12 h if the infusion rate is 200 mL/h.

Solution

1. Determine the unit of measure for the answer (V), and place it as the unknown on the left side of the equation.

   V mL =

2. The first factor is the length of time of the infusion over 1.

   $$\frac{12\text{ h}}{1}$$

3. Multiply by the flow rate of the infusion. The second factor.

   $$V\text{ mL} = \frac{12\text{ h}}{1} \times \frac{200\text{ mL}}{\text{h}}$$

4. Cancel units on the right side of the equation. The remaining unit of measure on the right side of the equation should match the unknown unit of measure on the left side of the equation.

   $$V\text{ mL} = \frac{12\cancel{\text{h}}}{1} \times \frac{200\text{ mL}}{\cancel{\text{h}}}$$

5. Solve the equation.

   V mL = 2400 mL to be infused in 12 h

REVIEW AND PRACTICE

10-3 Infusion Time and Volume

In Exercises 1–5, find the total time to infuse.

1. Ordered: 1000 mL NS at 83 mL/h using an infusion pump
2. Ordered: 500 mL LR at 125 mL/h using microdrip tubing
3. Ordered: 750 mL 0.45% NS at 31 mL/h
4. Ordered: 1000 mL NS at 200 mL/h
5. Ordered 250 mL D5W at 100 mL/h using an infusion pump

In Exercises 6–10, find when the infusion will be completed.

6. Ordered: 1500 mL D5W with 30 meq KCl/ L at a rate of 75 mL/h. You start the infusion at noon.
7. Ordered: 2000 mL NS via infusion pump at 100 mL/h. You start the infusion at 3:30 p.m.
8. Ordered: 750 mL RL at 50 mL/h. You start the IV at 1000
9. Ordered: 250 mL via a microdrip set at 40 mL/h. You start the infusion at 9:45 p.m.
10. Ordered: 500 mL $\frac{1}{2}$ NSS at 75 mL/h. The infusion started at 1615.

In Exercises 11–15, find the total volume to administer.

11. 75 mL/h 0.45% NS for 2 h 30 min using a rate controller
12. D5RL set at 100 mL/h for 8 h

13. D5W at 125 mL/h for 12 h using an infusion pump

14. An antibiotic solution infused over 2 h at 75 mL/h

To check your answers, see page 531.

Intermittent IV Infusions

IV medications are sometimes delivered on an intermittent basis. Intermittent medications can be delivered through an IV secondary line or a saline or heparin lock. Intermittent IV infusions are usually delivered through an IV secondary line when the patient is receiving continuous IV therapy. Intermittent IV infusions or IV push medications can also be delivered through a saline or heparin lock when the patient does not require continuous or replacement fluids.

Secondary Lines (Piggyback)

A **secondary line**, also known as a piggyback or IVPB, is an IV setup that attaches to a primary line. It can be used to infuse medication or other compatible fluids on an intermittent basis, such as q6h. Although shorter than primary tubing, secondary tubing has the same basic components. IVPB bags are usually smaller, often holding 50, 100, or 150 mL of fluid. (See Figure 10-17.) Some medications require a larger amount fluid as a diluent, such as 250 mL. The ADD-Vantage® system from Abbott Laboratories is a secondary system. It uses a specially designed IV bag into which you add medication directly from the vial, often in powdered form. Any mixing takes place in the bag. The solution is then infused with the medication vial remaining in place.

Intermittent Peripheral Infusion Devices

You can administer medication to a patient on a regular, though not continuous, schedule by using an *intermittent peripheral infusion device*. These devices are more commonly known as **heparin locks** or **saline locks**. To create a lock, attach an infusion port to an already inserted IV catheter. This port allows you to inject medication directly into the vein by using a syringe or to infuse medication intermittently. Physicians' orders will list IV push or bolus for medication that is injected into an IV line or through a saline or heparin lock.

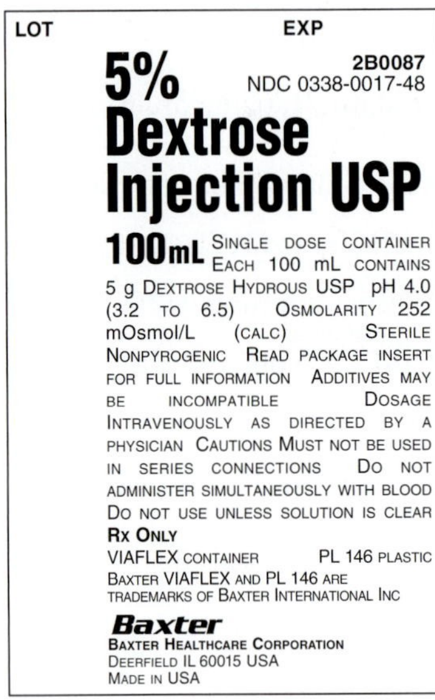

Figure 10-17 A 100 mL bag of IV fluid like this on may be used to mix and administer intermittent IV infusions.

Fluids do not flow continuously through the IV needle or catheter when a lock is used. To prevent blockage of the line, the device must be flushed 2 or 3 times a day or after administering medication. A saline lock is flushed or irrigated with saline. A heparin lock is flushed or irrigated with heparin, an anticoagulant that retards clot formation. The policy of the facility and the device will dictate the amount and concentration of solution to use. Saline or heparin fills the infusion port and IV catheter, preventing blood from entering and becoming trapped. If blood were trapped, a clot would form, blocking the catheter.

Preparing and Calculating Intermittent Infusions

Frequently intermittent medications are already reconstituted and prepared for administration by piggyback or through a heparin or saline lock. The flow rate for prepared medications is calculated in the same manner as regular IV infusions. The amount of fluid may be less and the amount of time to infuse may be less than an hour, so to calculate the flow rate you will need to change the number of minutes into hours (see Example 1 for Rule 10-5).

In some cases, you will be required to reconstitute and prepare a medication for IV infusion or calculate the amount to administer for an IV push medication.

When intermittent IV infusions are given through a saline or heparin lock, the lock should be irrigated or flushed before and/or after administration. If you meet resistance when flushing a saline or heparin lock, stop the procedure immediately so that you do not force a clot into the bloodstream.

Rule 10-11 When you prepare medication for an intermittent IV infusion:

- Reconstitute the medication, using the label and package insert.
- Calculate the amount to administer and the flow rate.

Example Ordered: Eloxatin® 75 mg in 250 mL D5W IV piggyback over 90 min

On hand: See label and package insert for Eloxatin® 100 mg (Figures 10-18a and 10-18b)

According to the package insert, Eloxatin® should be reconstituted with 20 mL of water for injection or 5% dextrose for injection. The dosage strength of the medication will be 100 mg/20 mL.

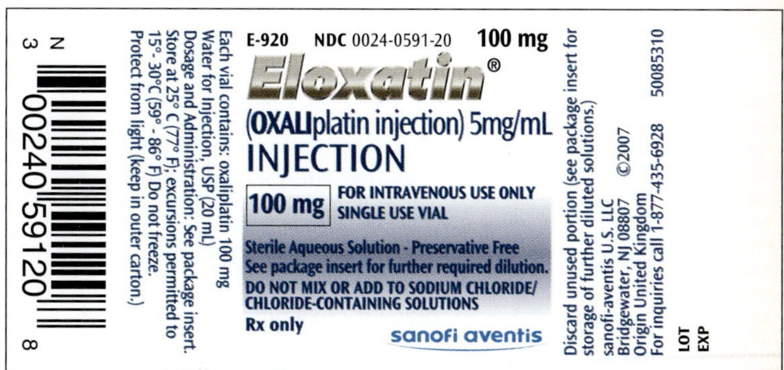

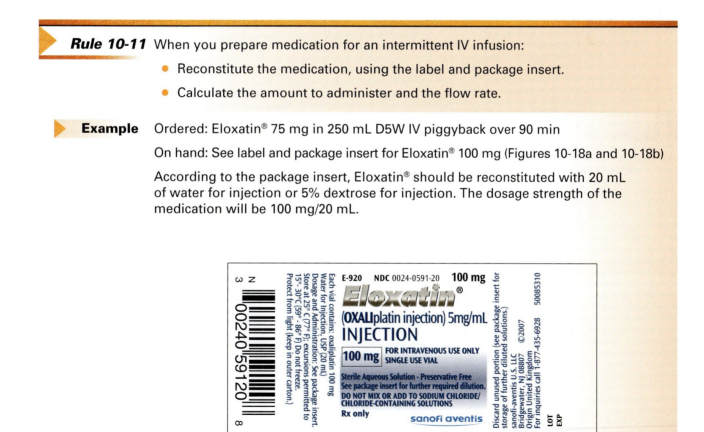

Figure 10-18a Eloxatin® label.

Dosage and Administration

Preparation of Infusion Solution

- **ELOXATIN® (oxaliplatin) IS NOT INTERCHANGEABLE WITH OTHER PLATINUMS. PLEASE VERIFY CORRECT DRUG AND DOSAGE PRIOR TO PREPARATION**
- **RECONSTITUTION OR FINAL DILUTION MUST NEVER BE PERFORMED WITH A SODIUM CHLORIDE SOLUTION OR OTHER CHLORIDE-CONTAINING SOLUTIONS**
- The lyophilized powder is reconstituted by adding 10 mL (for the 50-mg vial) or 20 mL (for the 100-mg vial) of Water for Injection, USP, or 5% Dextrose Injection, USP
- **Do not administer the reconstituted solution without further dilution.** The reconstituted solution must be further diluted in an infusion solution of 250-500 mL of 5% Dextrose Injection, USP
- After reconstitution in the original vial, the solution may be stored up to 24 hours under refrigeration (2-8°C [36-46°F])
- After final dilution with 250-500 mL of 5% Dextrose Injection, USP, the shelf life is **6 hours at room temperature (20-25°C [68-77°F]) or up to 24 hours under refrigeration (2-8°C [36-46°F])**
- **ELOXATIN® is not light sensitive**

Figure 10-18b Eloxatin® package insert.

Fraction Proportion Method

Solution Since the dosage ordered is 75 mg, you will calculate the amount to administer using the following information:

D (desired dose) = 75 mg *H* (dose on hand) = 100 mg *Q* (dosage unit) = 20 mL

Follow Procedure Checklist 7-1.

1. Fill in the proportion.

$$\frac{Q}{H} = \frac{A}{D}$$

$$\frac{20 \text{ mL}}{100 \text{ mg}} = \frac{A}{75 \text{ mg}}$$

2. Cancel units.

$$\frac{20 \text{ mL}}{100 \cancel{\text{mg}}} = \frac{A}{75 \cancel{\text{mg}}}$$

3. Cross-multiply and solve for the unknown.

$$100 \times A = 20 \text{ mL} \times 75$$

$$A = 20 \text{ mL} \times \frac{75}{100}$$

$$A = 15 \text{ mL}$$

Ratio Proportion Method

Solution Since the dosage ordered is 75 mg, you will calculate the amount to administer using the following information:

D (desired dose) = 75 mg H (dose on hand) = 100 mg Q (dosage unit) = 20 mL

Follow Procedure Checklist 7-2.

1. Fill in the proportion.

   $Q : H :: A : D$

   20 mL : 100 mg :: A : 75 mg

2. Cancel units.

   20 mL : 100 ~~mg~~ :: A : 75 ~~mg~~

3. Multiply the means and extremes, then solve for the missing value.

   $100 \times A = 20 \text{ mL} \times 75$

   $$A = 20 \text{ mL} \times \frac{75}{100}$$

   $$A = 15 \text{ mL}$$

Dimensional Analysis

Solution Since the dosage ordered is 75 mg, you will calculate the amount to administer using the following information:

D (desired dose) = 75 mg H (dose on hand) = 100 mg Q (dosage unit) = 20 mL

Follow Procedure Checklist 7-3.

1. The unit of measure for the amount to administer will be milliliters.

   A mL =

2. Since the unit of measurement for the dosage ordered is the same as that for the dose on hand, this step is unnecessary.

3. The dosage unit is 20 mL. The dose on hand is 100 mg. This is the first factor.

   $$\frac{20 \text{ mL}}{100 \text{ mg}}$$

4. The dosage ordered is 250 mg.

   $$A \text{ mL} = \frac{20 \text{ mL}}{100 \text{ mg}} \times \frac{75 \text{ mg}}{1}$$

5. Cancel units.

   $$A \text{ mL} = \frac{20 \text{ mL}}{100 \cancel{\text{mg}}} \times \frac{75 \cancel{\text{mg}}}{1}$$

6. Solve the equation.

   $$A \text{ mL} = \frac{20 \text{ mL}}{100} \times \frac{75}{1}$$

   $$A = 15 \text{ mL}$$

Formula Method

Solution Since the dosage ordered is 75 mg, you will calculate the amount to administer using the following information:

D (desired dose) = 75 mg H (dose on hand) = 100 mg Q (dosage unit) = 20 mL

Follow Procedure Checklist 7-4.

1. The drug is ordered in milligrams, which is the same unit of measure as that for the dose on hand. No conversion is necessary.
2. Fill in the formula.

$$\frac{D \times Q}{H} = A$$

$$\frac{75 \text{ mg} \times 20 \text{ mL}}{100 \text{ mg}} = A$$

3. Cancel units.

$$\frac{75 \cancel{\text{mg}} \times 20 \text{ mL}}{100 \cancel{\text{mg}}} = A$$

4. Solve for the unknown.

$$\frac{75 \times 20 \text{ mL}}{100} = A$$

$$15 \text{ mL} = A$$

Calculate the Flow Rate

From the package insert, you determine the reconstituted solutions must be further diluted with an infusion solution of 250 to 500 mL of 5% dextrose injection, USP. The order reads to use 250 mL of D5W; so using a sterile needle and proper aseptic technique, you withdraw 15 mL of the diluted medication and inject it into the 250-mL bag of D5W. Now you have a solution of 75 mg of Eloxatin® in 250 D5W, which you must deliver over 90 min. Add 15 mL medication plus 250 mL of D5W for a total volume of 265 mL. Calculate the flow rate in milliliters per hour, using Rule 10-5.

(Note: check your facility policy. In some cases the 15 mL of medication is not added to the diluent when determining the flow rate.)

In this case the volume is expressed in milliliters, and we will use the volume $V = 265$ mL.

Since time is expressed in minutes, you must first convert 90 min to hours to find T. (In this example we will use the fraction proportion method to convert; you may prefer to use the ratio proportion method.)

$$\frac{1 \text{ h}}{60 \text{ min}} = \frac{?}{90 \text{ min}}$$

$$\frac{1 \text{ h}}{60 \cancel{\text{min}}} = \frac{?}{90 \cancel{\text{min}}}$$

$$60 \times ? = 1 \text{ h} \times 90$$

$$? = \frac{1 \text{ h} \times 90}{60}$$

$$? = 1.5 \text{ h}$$

We now have the information needed in the proper units.

$V = 265$ mL and $T = 1.5$ h

Using the formula $F = \frac{V}{T}$, we find that

$$F = \frac{265 \text{ mL}}{1.5 \text{ h}}$$

$$F = 177 \text{ mL/h}$$

You would set the infusion pump to 177 mL/h.

If an infusion pump is not used, you will need to calculate the drops per minute. For this example we will use standard tubing that is 15 gtt/mL. Always check the drop factor on the tubing packaging.

Follow Rule 10-6. In this example we will use dimensional analysis to calculate the flow rate in gtt/min (f).

1. Determine the units of measure for the answer (f) and place it as the unknown on one side of the equation.

   f gtt/min =

2. The first factor is the number of mL to be administered on top (V) and the length of time to be administered (T) on the bottom.

   $$\frac{177 \text{ mL}}{\text{h}}$$

3. Multiply by a second factor to convert the hours to minutes to placing hour in the numerator.

   $$\frac{177 \text{ mL}}{\text{h}} \times \frac{1\text{h}}{60 \text{ min}}$$

4. Multiply by the drop factor of the tubing being used.

   $$f \text{ gtt/min} = \frac{177 \text{ mL}}{\text{h}} \times \frac{1 \text{ h}}{60 \text{ min}} \times \frac{15 \text{ gtt}}{1 \text{ mL}}$$

5. Cancel units on the right side of the equation. The remaining unit of measure on the right side of the equation should match the unknown unit of measure on the left side of the equation.

   $$f \text{ gtt/min} = \frac{177 \cancel{\text{mL}}}{\cancel{\text{h}}} \times \frac{1 \cancel{\text{h}}}{60 \text{ min}} \times \frac{15 \text{ gtt}}{1 \cancel{\text{mL}}}$$

6. Solve the equation.

   $$f = \frac{177 \times 15 \text{ gtt}}{60 \text{ min}}$$

   $$f = 44 \text{ gtt/min}$$

GO TO . . . Open the CD-ROM that accompanies your textbook, and select Chapter 10, Rule 10-11. Review the animation and example problems, then complete the practice problems. Continue to the next section of the book once you have mastered the rule presented.

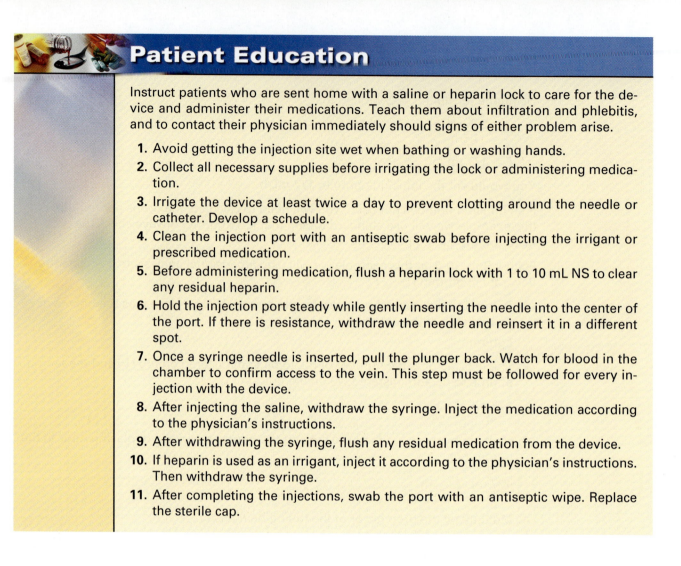

Patient Education

Instruct patients who are sent home with a saline or heparin lock to care for the device and administer their medications. Teach them about infiltration and phlebitis, and to contact their physician immediately should signs of either problem arise.

1. Avoid getting the injection site wet when bathing or washing hands.
2. Collect all necessary supplies before irrigating the lock or administering medication.
3. Irrigate the device at least twice a day to prevent clotting around the needle or catheter. Develop a schedule.
4. Clean the injection port with an antiseptic swab before injecting the irrigant or prescribed medication.
5. Before administering medication, flush a heparin lock with 1 to 10 mL NS to clear any residual heparin.
6. Hold the injection port steady while gently inserting the needle into the center of the port. If there is resistance, withdraw the needle and reinsert it in a different spot.
7. Once a syringe needle is inserted, pull the plunger back. Watch for blood in the chamber to confirm access to the vein. This step must be followed for every injection with the device.
8. After injecting the saline, withdraw the syringe. Inject the medication according to the physician's instructions.
9. After withdrawing the syringe, flush any residual medication from the device.
10. If heparin is used as an irrigant, inject it according to the physician's instructions. Then withdraw the syringe.
11. After completing the injections, swab the port with an antiseptic wipe. Replace the sterile cap.

REVIEW AND PRACTICE

10-4 Intermittent IV Infusions

For Exercises 1 to 4, refer to labels A and B.

1. Which would be used to flush a saline lock?
2. Which would be used to flush a heparin lock?
3. Standing order reads 3 mL saline lock flush q8h. Calculate the amount to administer.
4. Standing order reads Heparin lock flush 20 U q12h. Calculate the amount to administer.

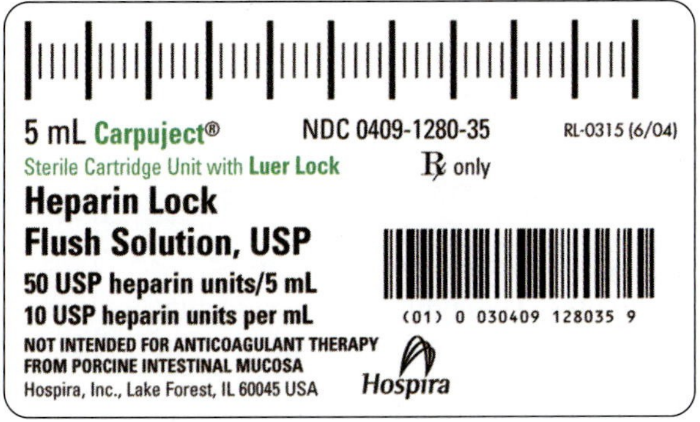

A

B

Figure 10-19 Heparin and saline lock flush.

For Exercises 5–10, determine the amount to administer and calculate the flow rate in milliliters per hour and drops per minute.

Ordered: Gemzar® 150 mg in 500 mL NSS over 2 h
On hand: Refer to IV tubing packaging C, label D, and package insert E (Figure 10-20).

5. Amount to administer: ____________________
6. Flow rate in milliliters per hour: ____________________
7. Flow rate in drops per minute: ____________________

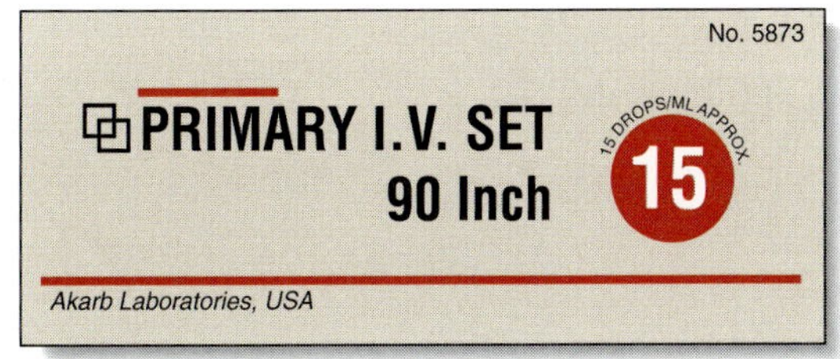

C IV tubing packaging

NDC 0002-7501-01
VIAL No. 7501
Lilly
GEMZAR®
Gemcitabine HCl
For Injection
200 mg
equivalent to base
For I.V. use only
Sterile Single Use Vial

D Gemzar® label

Figure 10-20 IV tubing packaging, Gemzar® label, and package insert.

Gemzar® may be administered on an outpatient basis.

Instructions for Use/Handling — The recommended diluent for reconstitution of Gemzar® is 0.9% Sodium Chloride Injection without preservatives. Due to solubility considerations, the maximum concentration for Gemzar® upon reconstitution is 40 mg/mL. Reconstitution at concentrations greater than 40 mg/mL may result in incomplete dissolution, and should be avoided.

To reconstitute, add 5 mL of 0.9% Sodium Chloride Injection to the 200-mg vial or 25 mL of 0.9% Sodium Chloride Injection to the 1-g vial. Shake to dissolve. These dilutions each yield a gemcitabine concentration of 38 mg/mL which includes accounting for the displacement volume of the lyophilized powder (0.26 mL for the 200-mg vial or 1.3 mL for the 1-g vial). The total volume upon reconstitution will be 5.26 mL or 26.3 mL, respectively. Complete withdrawal of the vial contents will provide 200 mg or 1 g of gemcitabine, respectively. The appropriate amount of drug may be administered as prepared or further diluted with 0.9% Sodium Chloride Injection to concentrations as low as 0.1 mg/mL.

Reconstituted Gemzar® is a clear, colorless to light straw-colored solution. After reconstitution with 0.9% Sodium Chloride Injection, the pH of the resulting solution lies in the range of 2.7 to 3.3. The solution should be inspected visually for particulate matter and discoloration, prior to administration, whenever solution or container permit. If particulate matter or discoloration is found, do not administer.

When prepared as directed, Gemzar® solutions are stable for 24 hours at controlled room temperature 20° to 25°C (68° to 77°F) [*See* USP]. Discard unused portion. Solutions of reconstituted Gemzar® should not be refrigerated, as crystallization may occur.

The compatibility of Gemzar® with other drugs has not been studied. No incompatibilities have been observed with infusion bottles or polyvinyl chloride bags and administration sets.

E Package insert

Figure 10-20 Concluded.

Ordered: Doxycycline 75 mg IVPB in 100 mL of D5W over 1 hour
On hand: Refer to label F, package insert G, and IV tubing packaging H (Figure 10-21).

8. Amount to administer: ______________________
9. Flow rate in milliliters per hour: ______________________
10. Flow rate in drops per minute: ______________________

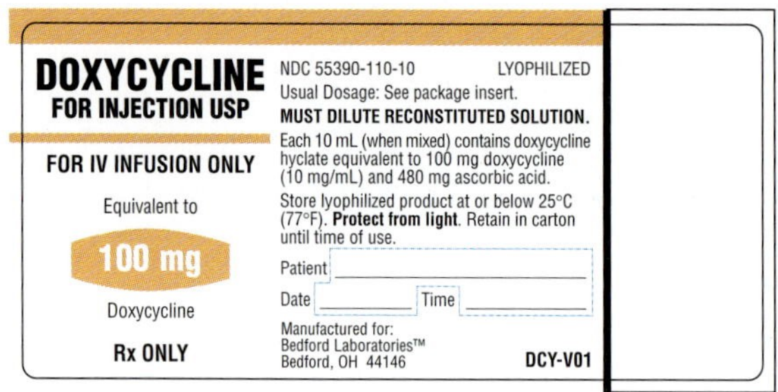

DOXYCYCLINE
FOR INJECTION USP
FOR IV INFUSION ONLY
Equivalent to
100 mg
Doxycycline
Rx ONLY
NDC 55390-110-10
LYOPHILIZED
Usual Dosage: See package insert.
MUST DILUTE RECONSTITUTED SOLUTION.
Each 10 mL (when mixed) contains doxycycline hyclate equivalent to 100 mg doxycycline (10 mg/mL) and 480 mg ascorbic acid.
Store lyophilized product at or below 25°C (77°F). Protect from light. Retain in carton until time of use.
Patient
Date
Time
Manufactured for:
Bedford Laboratories™
Bedford, OH 44146
DCY-V01

F Doxycycline Label

Figure 10-21 Dosage and Administration.
Excerpt from Doxycycline Package Insert found at www.bedfordlabs.com

Preparation of Solution: To prepare a solution containing 10 mg/mL, the contents of the vial should be reconstituted with 10 mL of Sterile Water for Injection, or any of the ten intravenous infusion solutions listed below. Each 100 mg of doxycycline (i.e., withdraw entire solution from the 100 mg vial) is further diluted with 100 to 1000 mL of the intravenous solutions listed below:

1. 0.9% Sodium Chloride Injection
2. 5% Dextrose Injection
3. Ringer's Injection
4. Invert Sugar, 10% in Water
5. Lactated Ringer's Injection
6. Dextrose 5% in Lactated Ringer's
7. Normosol-M® in D5-W (Abbott)
8. Normosol-R® in D5-W (Abbott)
9. Plasma-Lyte® 56 in 5% Dextrose (Travenol)
10. Plasma-Lyte® 148 in 5% Dextrose (Travenol)

This will result in desired concentrations of 0.1 to 1 mg/mL. Concentrations lower than 0.1 mg/mL or higher than 1 mg/mL are not recommended.

G Package Insert

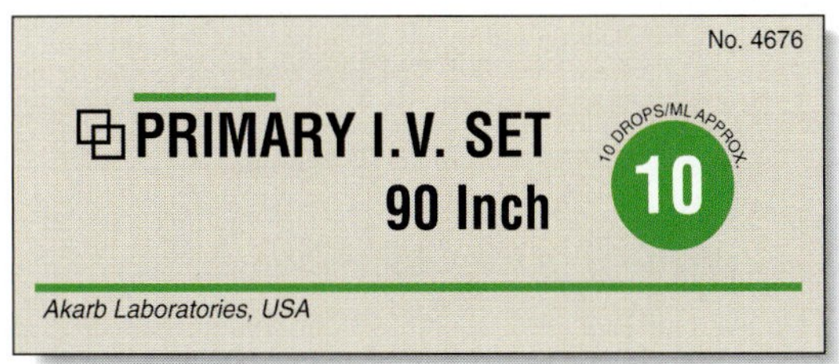

H IV Tubing Packaging

Figure 10-21 Concluded.

To check your answers, see pages 531–532.

HOMEWORK ASSIGNMENT

Answer the following questions.

1. List the four functional categories of IV solutions.
2. In the IV label abbreviation D5W, which component identifies the concentration level?
3. Define the terms isotonic, hypertonic, and hypotonic.
4. Before combining any medication with an IV solution, what must be verified first?
5. List three parts of a primary IV line.
6. List four types of electronic infusion devices.
7. Under what circumstances would a patient most likely need central line IV therapy?
8. What is the drop factor for microdrip tubing?
9. How is the flow rate measured when an infusion pump is used?
10. Name and explain two types of intermittent IV infusion methods.
11. If the order is 1000 mL to infuse over 12 hours (10 gtt/mL tubing), calculate the flow rate.
12. If the order is 50 mL to infuse over 90 minutes (microdrip tubing), calculate the flow rate.
13. If the order is 500 mL to infuse at 150 mL/hour (10 gtt/mL tubing), how long will it take to infuse?
14. The IV is infusing at 20 gtt/minute (20 gtt/mL) and the volume of the IV bag is 500 mL, how many hours will it take to infuse the IV?
15. An IV of 500 mL is begun at 11 a.m. and the flow rate is 15 gtt/minute (10 gtt/mL tubing). At what time will the infusion be complete?
16. If the order is 250 mL over 3 hours, what mL/hour setting will you useon an infusion pump?
17. If the order is 150 mL over 30 minutes, how many mL/hour will you set on the infusion pump?
18. What is the total volume to be infused for an IV infusing at 80 mL/hours for 6 hours?
19. At the end of 4 hours, what is the total volume infused for an IV infusing at 20 drops per minute (10 gtt/mL tubing)?
20. If the order is 500 mL over 4 hours (10 gtt/mL tubing), after one hour there should be how much remaining the bag?

Chapter 10 Review

Check Up

Match the following. Write the answer in the space provided.

___ **1.** Roller clamp	**a.** type of fluid given to a patient with a severe burn
___ **2.** KVO fluids	**b.** type of fluid given to a patient who is dehydrated
___ **3.** Hypertonic	**c.** the patient is in charge of delivering his or her own pain medications
___ **4.** Phlebitis	**d.** provide access to the vascular system for emergency situations
___ **5.** PCA	**e.** redness, swelling, and pain at an IV site
___ **6.** Infusion pump	**f.** applies pressure to administer an IV fluid
___ **7.** Hypotonic	**g.** used to control the rate of an IV infusion

In Exercises 8–13, calculate the flow rate for orders to be administered by an infusion pump.

8. 3000 mL D5W IV q24h

9. 500 mL LR IV q8h

10. 1200 mL 0.45% NS IV q12h

11. 250 mL NS IV q4h

12. 1 g Claforan in 100 mL D5W IV over 90 min

13. 500 mg ampicillin in 50 mL D5W IV over 30 min

In Exercises 14–19, find the flow rate for the orders, rounded to the nearest drop.

14. 2200 mL IV RL q24h (15 gtt/mL tubing)

15. 300 mL IV NS q8h (10 gtt/mL tubing)

16. 1000 mL IV D5W q6h (15 gtt/mL tubing)

17. 1800 mL IV D5/$\frac{1}{2}$ NS q12h (20 gtt/mL tubing)

18. 1500 mL IV $\frac{1}{3}$ NS q8h (10 gtt/mL tubing)

19. 300 mL D5/Ringer's IV q6h (microdrip tubing)

In Exercises 20–22, calculate the original flow rate. Then determine if an adjustment is necessary and calculate the adjusted flow rate. Determine if the rate can be adjusted safely.

20. Ordered: 1000 mL RL over 8 h (15 gtt/mL tubing)
After 2 h, 125 mL has infused.

21. Ordered: 2500 mL NS over 24 h (10 gtt/mL tubing)
After 3 h, 200 mL has infused.

22. Ordered 500 mL $\frac{1}{2}$ NSS over 8 h (60 gtt/mL tubing) After 2 h, 450 mL is remaining in the IV bag

In Exercises 23–26, find the total time to infuse.

23. Ordered: 1000 mL D5/0.45% NS at 125 mL/h via infusion pump
24. Ordered: 800 mL $\frac{1}{4}$ NS at 50 mL/h via rate controller
25. Ordered: 600 mL LR IV at 25 mL/h
26. Ordered: 1200 mL D5/NS IV at 70 mL/h

In Exercises 27–30, find when the infusion will be completed.

27. 800 mL via infusion pump at 90 mL/h, starting at 0820
28. 1000 mL at 125 mL/h, starting at 1 p.m.
29. 500 mL at 175 mL/h, starting at 2230
30. 750 mL at 35 mL/h, starting at 4 p.m.

In Exercises 31–34, find the total volume to administer.

31. $\frac{1}{4}$ NS at 125 mL/h over 5 h 30 min via infusion pump
32. RL at 25 mL/h over 12 h
33. NS at 125 mL/h over 7 h 30 min
34. D5W at 80 mL/h over 8 h 20 min

Critical Thinking Applications

The physician has ordered an adult patient with pneumonia to have clindamycin 500 mg IV q8h.

On hand: See clindamycin label and package insert (Figure 10-22)

1. How would you prepare the medication?
2. Calculate the amount to administer.
3. Is this the correct dose for treatment of pneumonia? Why or why not?
4. What fluid should not be used as a diluent for children?
5. The medication cannot be administered immediately because the patient is having a diagnostic test. What should you do?
6. Calculate the flow rate of the infusion.

NDC 63323-282-02 28202

CLINDAMYCIN

INJECTION, USP

300 mg/2 mL

(150 mg/mL)

For IM or IV Use Rx only

DILUTE BEFORE IV USE

2 mL Single Dose Vial

Do not refrigerate.

Abraxis
Pharmaceutical Products
Schaumburg, IL 60173

402128A

LOT/EXP

CLINDAMYCIN PHOSPHATE INJECTION USP

Usual adult and adolescent dose

Antibacterial
Intramuscular or intravenous, 300 to 600 mg (base) every six to eight hours; or 900 mg every eight hours.[13]

[Babesiosis (treatment)][1]
Intravenous, 300 to 600 mg clindamycin (base) four times a day with concurrent oral administration of 650 mg of quinine, three or four times a day for seven to ten days.[49] [50]

[Pneumonia, *Pneumocystis carinii* (treatment)][1]
Intravenous, 2400 to 2700 mg (base) per day in divided doses in combination with 15 to 30 mg of primaquine daily.[35] [36] [53]

[Toxoplasmosis, central nervous system (CNS) (treatment)][1]
Intravenous, 1200 to 4800 mg (base) per day in divided doses in combination with 50 to 100 mg of pyrimethamine daily.[31] [32] [33] [34] [52] [55] [56]

Usual adult prescribing limits
Up to 2.7 grams (base) daily.

Note: Doses up to 4.8 grams daily have been used. However, some medical experts recommend a maximum dose of 2.7 grams daily.

Preparation of dosage form:
To prepare initial dilution for intravenous use, each dose must be diluted as follows (it must not be administered undiluted as a bolus):

Dose (mg)	Diluent (mL)	Duration of administration (min)
300	50	10
600	100	20
900	100	30

Caution: Products containing benzyl alcohol are not recommended for use in neonates. A fatal toxic syndrome consisting of metabolic acidosis, CNS depression, respiratory problems, renal failure, hypotension, and possibly seizures and intracranial hemorrhages has been associated with this use.

Stability:
Clindamycin phosphate retains its potency for 24 hours at room temperature in intravenous infusions containing sodium chloride, dextrose, potassium, vitamin B complex, cephalothin, kanamycin, gentamicin, penicillin, or carbenicillin.[13]

Incompatibilities:
Clindamycin phosphate is physically incompatible with ampicillin, phenytoin sodium, barbiturates, aminophylline, calcium gluconate, and magnesium sulfate.[13]

*Excerpt from package insert for Clindamycin made by Abraxis

Figure 10-22 Clindamycin label and package insert.

Case Study

A patient has a PCA pump with morphine sulfate 50 mg in 500 mL D5W. Hospital policy requires you to document the dose of morphine administered during your shift. When you came on duty, the pump showed that 227 mL had infused. At the end of your shift the pump shows that 272 mL has infused. How much morphine did the patient receive during your shift?

To check you answers, see page 533.

Internet Activity

You are applying for a new job that will require you to work with IV fluids exclusively. To prepare for the position, you want to learn more about the types of fluids and when and how they are used. Additionally you would like to become familiar with IV equipment made by various companies. Research the Internet and learn more about fluids and IV equipment. You may want to make yourself a chart or table with pictures to use as a reference tool.

GO TO . . . Open the CD-ROM that accompanies your textbook, and complete a final review of the rules, practice problems, and activities presented for this chapter. For a final evaluation, take the chapter test and email or print your results for your instructor. A score of 95 percent or above indicates mastery of the chapter concepts. ■

11 Calculations for Special Populations

Perfection consists not in doing extraordinary things, but in doing ordinary things extraordinarily well.

—Angelique Arnauld

Learning Outcomes

When you have completed Chapter 11, you will be able to:

- Explain why dosages for special populations must be calculated based on the individual patient.
- Identify factors that affect the absorption, distribution, biotransformation, and elimination of drugs in special populations.
- Determine safe doses for special populations.
- Calculate patient dosages based on body weight.
- Find a patient's body surface area (BSA).
- Calculate patient dosages based on a patient's BSA.
- Describe volume and medication limitations for special populations.
- Calculate infusion rates based upon body weight.

Key Terms

Absorption
Biotransformation
Body surface area
Distribution
Daily maintenance fluid needs
Elimination
Geriatric
Pediatric
Pharmacokinetics
Polypharmacy

Introduction

There are two special populations that require extra consideration when you are calculating medication dosages. These are pediatric (children) and geriatric (mature adult) patients. Generally speaking; **pediatric** patients are under the age of 18, and **geriatric** patients are 65 and over. See Figures 11-1 and 11-2. The risk of harm to these populations is far greater because of how they break down and absorb medications. You must clarify all confusing drug orders, calculate with absolute accuracy, verifying that the dose is safe, and seek assistance from your supervisor if you have concerns. No matter how rushed you may feel, you may not take shortcuts with medication calculations for patients from special populations, but rather must calculate their dosages extraordinarily well!

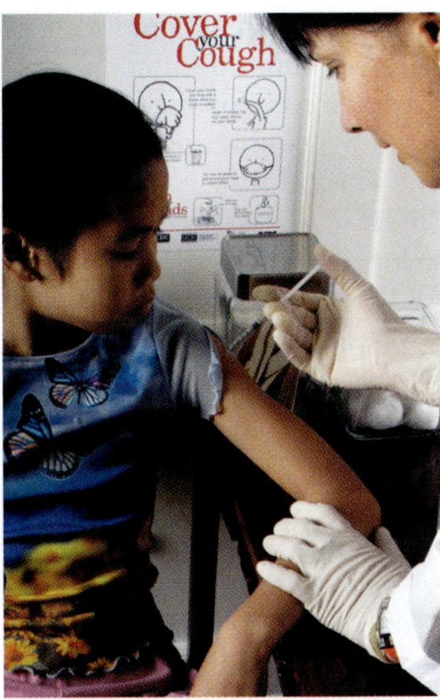

Figure 11-1 Pediatric patients, less than 18 years, require extra attention to detail when performing calculations.

Figure 11-2 Dosage calculation for geriatric patients, those 65 years and older, require extraordinary accuracy.

Individual Factors That Can Impact Dosing

For most drugs, there is a normal recommended dose standardized for an average adult weighing approximately 150 pounds. This standardized dose of a medication is based on a number of assumptions about the patient's body and age. It is assumed that the body systems are fully developed and functioning at a certain level. While these assumptions hold true for most patients, there are many situations in which the dose needs to be adjusted. If the liver or kidneys are not functioning normally, for example, the dose of many drugs needs to be decreased. If the digestive system is not functioning normally, the dose of oral medications may need to be adjusted. These changes in dosage are necessary because the drug is affecting each patient's body differently.

Pharmacokinetics—How Drugs Are Used by the Body

Pharmacokinetics is the study of what happens to a drug after it is administered to a patient. There are four processes that affect a drug after it is administered: *absorption, distribution, biotransformation*, and *elimination*. Pharmacokinetics is the study of these four processes. Understanding the processes allows adjustments to be made for patients whose body systems are not fully developed or are not functioning at a certain level.

Absorption is the process that moves a drug from the site where it is given into the bloodstream. Intravenous medications bypass the absorption process because they are administered directly into the bloodstream. Oral medications are absorbed through the digestive system, while topical medications are absorbed through the skin.

Distribution is the process that moves a drug from the bloodstream into other body tissues and fluids. The blood and each of these other areas are called *compartments*. Some compartments include blood, fat, cerebrospinal fluid, and the target site. The *target site* is the site where the drug produces its desired effect. The compartments that a drug will go to are different for different drugs, and depend on the chemical nature of the drug.

Biotransformation is the process that chemically changes the drug in the body. These changes, which occur primarily in the liver, help to protect the body from foreign chemicals, including drugs.

Elimination is the process in which the drug leaves the body. The main way of eliminating most drugs is through the urine, although the drug can also be eliminated in the air that we exhale, sweat, feces, breast milk, or any other body secretion.

The dose of a drug may need to be adjusted if one of these four processes is not functioning within certain limits. Certain conditions affect these four processes, thus affecting the dose. Some examples are found in Table 11-1. A dose adjustment is made based upon the nature and severity of the patient's condition. Thus, the dose ordered might be lower or higher than normal in some circumstances, yet still be the proper dose for the patient. These dosing considerations for various conditions are normally included in the package insert and are considered when the order for the medication is written. You are not expected to make the dosage adjustments; however, understanding these processes will make you *aware* of the many factors that need to be taken into consideration when determining the appropriate dose for an individual.

TABLE 11-1 Conditions That May Impact Dosing

Condition	Process Affected	Effect on Dosing
Stomach or intestinal disorders	Absorption	Dose of oral medications may need to be changed.
Liver disorders	Biotransformation	The dose of some drugs needs to be decreased.
Obesity	Distribution	Dose of drugs distributed to fat may need to be increased.
Kidney disease	Elimination	Dose of drugs eliminated in urine may need to be changed.

The function of many body systems changes over the life of a person. In newborns, some systems are not yet fully developed. This is especially true for premature infants. In geriatric patients, those 65 and over, the function of some body systems begins to deteriorate. Skin and veins become more fragile. The functions of the digestive and urinary systems may be affected. Table 11-2 describes some of the age-related factors that can affect dosing. This chapter will show you several methods that are used for calculating dosages for pediatric and geriatric patients.

Working with Special Populations

In addition to the variations in dosages, there are other considerations when you are working with special populations. Pediatric and geriatric patients will have a parent or caretaker who will administer or assist them with medications. These individuals will need education.You may be called upon to teach the patients or the caregivers about the medications they will be taking. Table 11-3 provides a list of what each patient or caregiver should know about medications. Geriatric patients may require extra consideration based upon their level of awareness and understanding.

Geriatric Patients

When you work with geriatric patients, show them respect. Encourage them, if they are able, to participate in planning their schedule. Listen to their concerns. Recommend that they use the same pharmacy to fill all prescriptions. Encourage them to have one doctor as their primary physician to monitor and approve all medications. Remind them to keep a list of all their medications including vitamins, herbals, and alternative medications. They should take the list or medication bottles when they see their primary physician and all their specialists.

In some cases, geriatric patients may have decreased manual dexterity that can interfere with their ability to inject medications, administer eye drops, or even open bottles. Patients may need to specifically ask their pharmacist for bottle caps that are not childproof. Patients with difficulty swallowing will need information about which medications may be crushed and mixed with applesauce or pudding. In addition, they need information about which medications may *not* be broken or crushed.

TABLE 11-2 Age-Related Factors That May Impact Dosing

Age Group	Condition	Process Affected	Effect on Dosing
Pediatric	pH of stomach is lower	Absorption	Dose of oral medications may need to be changed.
	Thinner skin	Absorption	Dose of topical medications may need to be decreased.
	Liver still developing	Biotransformation	Dose of some drugs needs to be decreased.
	Less circulaton to muscles	Absorption	Dose of IM medications may need to be increased.
Geriatric	Thinner skin	Absorption	Dose of topical medications may need to be decreased.
	Decreased liver function	Biotransformation	Dose of some drugs need to be decreased.
	Decreased kidney function	Elimination	Dose of drugs eliminated in the urine may need to be decreased.
	Poor circulation	Absorption and distribution	Dose of some drugs needs to be adjusted.

TABLE 11-3 What Patients Should Know about Medications

Name of the medication
Purpose of taking the medication
How to store the medication
How long the patient will need to take the medication
How and when to take the medication
How to know if the medication is effective
Required follow-up (lab tests, doctor appointments)
Possible side effects and what to do about them
Interactions with other drugs and foods
Symptoms to report to the doctor
What to do if a dose is missed
Keeping a list of all medications

Many patients cannot read small print. Medications may need to be labeled so that they can easily be read and will clearly describe the purpose of each medication. Do not assume patients can distinguish between colors of tablets; white and yellow may be confused, as may blue and green, or orange and pink. Make a chart for the patient with the medications, time, and description to prevent errors.

Patient Education

Family members or others who care for special population patients at home must understand and follow directions when they administer drugs. Talk to parents or caretakers about the following:

1. Explain how much medication to administer at one time, how often during the day, and for how many days. Give parents or caretakers this information in writing. Have them repeat this information, especially if their English or literacy skills are limited.
2. Discuss expected side effects. Provide a telephone number or resource for them to call in case of unexpected or serious side effects. Explain how to reach a Poison Control Center.
3. No one should ever be given someone else's medication. Patients may react differently to medications than expected.
4. Over-the-counter, herbal, and alternative remedies should never be given to a child under age 2 without first speaking to a health care provider.
5. Parents or caregivers must check dosage information on over-the-counter medications. The amount to administer changes with age and weight. Some medications may *not* be administered to children below a certain age (often 2) without first checking with a physician.
6. Demonstrate how to measure doses accurately. Calibrated equipment should be used. Droppers are not automatically interchangeable between medications.
7. The full course of prescription medications such as antibiotics must be administered, even if the patient appears to be well or resists taking the medication.
8. Never refer to medication as candy or physically force a patient to take a medication.
9. Replace childproof caps properly and keep medications out of reach of children or patients with mental confusion.

Geriatric patients often have some form of hearing loss. They may even try to hide their hearing loss from you. Have patients repeat to you the information that you give them. They may also have short-term memory loss. Determine if they need written directions and explanations. Help them work with memory tools, such as medication calendars that tell them which medications to take each day. Pharmacies often sell weekly dispensers that have a container for each day of the week; patients or family members can prepare in advance the medications for the week (see Figure 11-3).

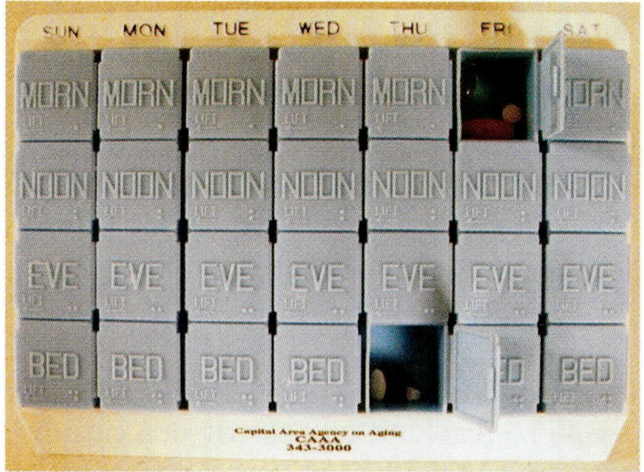

Figure 11-3 A medication container like this one would be very useful for a patient taking multiple daily medications.

Instruct patients who regularly take prescription medications not to take over-the-counter or herbal medications without first checking with their physician. They should not take more of any medication than is indicated by the label. They should avoid any medications that have expired (show them how to read the expiration date). They should never borrow medications, especially prescription ones, from anyone else.

Dosages Based on Body Weight

Many medication orders, especially pediatric and geriatric orders, are based on body weight. This is especially common for small children. An order based on body weight will often state an amount of medication per weight of the patient per unit of time. For example, the order may read 8 mg/kg/day PO q6h. This order says that, over the course of the day, the patient is to be administered 8 mg of medication for every kilogram that he or she weighs. It also says that the total daily dosage is to be divided into 4 doses given at 6-h intervals. You will calculate the amount to administer from the information on the drug order, the patient's weight, and the dose on hand.

Rule 11-1 Calculating Dosage Based on Body Weight

1. Convert the patient's weight to kilograms. For accuracy, when converting to kilograms round to the nearest hundredth (two decimal places).
2. Calculate the desired dose *D* by multiplying the dose ordered by the weight in kilograms, such as

   $\frac{\text{mg}}{\text{kg}} \times \text{kg} = \text{desired dose}$ *or* $\frac{\text{mcg}}{\text{kg}} \times \text{kg} = \text{desired dose}$

3. Confirm whether the desired dose is safe by checking the label, package insert, or product litertature. If it is unsafe, consult the physician who wrote the order.
4. Calculate the amount to administer, using the fraction proportion, ratio proportion, dimensional analysis, or formula method.

Fraction Proportion

Example Calculate the amount to administer to a 3-year-old who weighs 34 lb.

Ordered: Hyoscyamine sulfate 5 mcg/kg Sub-Q 1 h preanesthesia

On hand: Hyoscyamine sulfate 0.5 mg/mL

Solution

1. Convert the patient's weight to kilograms.

$$\frac{1 \text{ kg}}{2.2 \cancel{\text{lb}}} = \frac{?}{34 \cancel{\text{lb}}}$$

$$2.2 \times ? = 34 \times 1 \text{ kg}$$

$$? = \frac{34 \text{ kg}}{2.2}$$

$$? = 15.45 \text{ kg} = \text{patient's weight in kilograms}$$

2. Calculate the desired dose.

$$\frac{5 \text{ mcg}}{\cancel{\text{kg}}} \times 15.45 \cancel{\text{kg}} = 77.25 \text{ mcg rounded to } 77 \text{ mcg}$$

3. Confirm that the desired dose is safe.

   Checking the PDR, you find that 5 mcg/kg is the recommended dose for pediatric patients over 2 years of age.

4. Calculate the amount to administer.

   Because the unit of measure for the dose on hand is milligrams, the desired dose must also be expressed in milligrams. Convert, using the fraction proportion, ratio proportion, or dimensional analysis method. For this example we use Procedure Checklist 3-1 with the fraction proportion method.

$$\frac{1\text{ mg}}{1000\text{ }\cancel{\text{mcg}}} = \frac{?}{77\text{ }\cancel{\text{mcg}}}$$

$$1000 \times ? = 77 \times 1\text{ mg}$$

$$? = \frac{77\text{ mg}}{1000}$$

$$? = 0.077\text{ mg}$$

Using Procedure Checklist 7-1. Fill in the proportion.

$$\frac{Q}{H} = \frac{A}{D}$$

$$\frac{1\text{ mL}}{0.5\text{ mg}} = \frac{A}{0.077\text{ mg}}$$

Cancel units.

$$\frac{1\text{ mL}}{0.5\text{ }\cancel{\text{mg}}} = \frac{A}{0.077\text{ }\cancel{\text{mg}}}$$

Cross-multiply and solve for the unknown.

$$0.5 \times A = 1\text{ mL} \times 0.077$$

$$A = 1\text{ mL} \times \frac{0.077}{0.5}$$

$$A = 0.154\text{ mL}$$

Since the volume of the injection is less than 1 mL, we round to hundredth (two decimals).

$$A = 0.15\text{ mL}$$

Ratio Proportion Method

Example Calculate the amount to administer to a 3-year-old who weighs 34 lb.

Ordered: Hyoscyamine sulfate 5 mcg/kg Sub-Q 1 h preanesthesia

On hand: Hyoscyamine sulfate 0.5 mg/mL

Solution

1. Convert the patient's weight to kilograms.

$$1\text{ kg} : 2.2\text{ }\cancel{\text{lb}} :: ? : 34\text{ }\cancel{\text{lb}}$$

$$2.2 \times ? = 1\text{ kg} \times 34$$

$$? = \frac{1\text{ kg} \times 34}{2.2}$$

$$? = 15.45\text{ kg} = \text{patient's weight in kilograms}$$

Instruct patients who regularly take prescription medications not to take over-the-counter or herbal medications without first checking with their physician. They should not take more of any medication than is indicated by the label. They should avoid any medications that have expired (show them how to read the expiration date). They should never borrow medications, especially prescription ones, from anyone else.

Dosages Based on Body Weight

Many medication orders, especially pediatric and geriatric orders, are based on body weight. This is especially common for small children. An order based on body weight will often state an amount of medication per weight of the patient per unit of time. For example, the order may read 8 mg/kg/day PO q6h. This order says that, over the course of the day, the patient is to be administered 8 mg of medication for every kilogram that he or she weighs. It also says that the total daily dosage is to be divided into 4 doses given at 6-h intervals. You will calculate the amount to administer from the information on the drug order, the patient's weight, and the dose on hand.

Rule 11-1 Calculating Dosage Based on Body Weight

1. Convert the patient's weight to kilograms. For accuracy, when converting to kilograms round to the nearest hundredth (two decimal places).
2. Calculate the desired dose *D* by multiplying the dose ordered by the weight in kilograms, such as

$$\frac{\text{mg}}{\text{kg}} \times \text{kg} = \text{desired dose} \qquad \textit{or} \qquad \frac{\text{mcg}}{\text{kg}} \times \text{kg} = \text{desired dose}$$

3. Confirm whether the desired dose is safe by checking the label, package insert, or product litertature. If it is unsafe, consult the physician who wrote the order.
4. Calculate the amount to administer, using the fraction proportion, ratio proportion, dimensional analysis, or formula method.

Fraction Proportion

Example Calculate the amount to administer to a 3-year-old who weighs 34 lb.

Ordered: Hyoscyamine sulfate 5 mcg/kg Sub-Q 1 h preanesthesia

On hand: Hyoscyamine sulfate 0.5 mg/mL

Solution

1. Convert the patient's weight to kilograms.

$$\frac{1 \text{ kg}}{2.2 \cancel{\text{lb}}} = \frac{?}{34 \cancel{\text{lb}}}$$

$$2.2 \times ? = 34 \times 1 \text{ kg}$$

$$? = \frac{34 \text{ kg}}{2.2}$$

$$? = 15.45 \text{ kg} = \text{patient's weight in kilograms}$$

2. Calculate the desired dose.

$$\frac{5 \text{ mcg}}{\cancel{\text{kg}}} \times 15.45 \cancel{\text{kg}} = 77.25 \text{ mcg rounded to 77 mcg}$$

3. Confirm that the desired dose is safe.

   Checking the PDR, you find that 5 mcg/kg is the recommended dose for pediatric patients over 2 years of age.

4. Calculate the amount to administer.

   Because the unit of measure for the dose on hand is milligrams, the desired dose must also be expressed in milligrams. Convert, using the fraction proportion, ratio proportion, or dimensional analysis method. For this example we use Procedure Checklist 3-1 with the fraction proportion method.

$$\frac{1\text{ mg}}{1000\ \cancel{\text{mcg}}} = \frac{?}{77\ \cancel{\text{mcg}}}$$

$$1000 \times ? = 77 \times 1\text{ mg}$$

$$? = \frac{77\text{ mg}}{1000}$$

$$? = 0.077\text{ mg}$$

Using Procedure Checklist 7-1. Fill in the proportion.

$$\frac{Q}{H} = \frac{A}{D}$$

$$\frac{1\text{ mL}}{0.5\text{ mg}} = \frac{A}{0.077\text{ mg}}$$

Cancel units.

$$\frac{1\text{ mL}}{0.5\ \cancel{\text{mg}}} = \frac{A}{0.077\ \cancel{\text{mg}}}$$

Cross-multiply and solve for the unknown.

$$0.5 \times A = 1\text{ mL} \times 0.077$$

$$A = 1\text{ mL} \times \frac{0.077}{0.5}$$

$$A = 0.154\text{ mL}$$

Since the volume of the injection is less than 1 mL, we round to hundredth (two decimals).

$$A = 0.15\text{ mL}$$

Ratio Proportion Method

Example Calculate the amount to administer to a 3-year-old who weighs 34 lb.

Ordered: Hyoscyamine sulfate 5 mcg/kg Sub-Q 1 h preanesthesia

On hand: Hyoscyamine sulfate 0.5 mg/mL

Solution

1. Convert the patient's weight to kilograms.

$$1\text{ kg} : 2.2\ \cancel{\text{lb}} :: ? : 34\ \cancel{\text{lb}}$$

$$2.2 \times ? = 1\text{ kg} \times 34$$

$$? = \frac{1\text{ kg} \times 34}{2.2}$$

$$? = 15.45\text{ kg} = \text{patient's weight in kilograms}$$

2. Calculate the desired dose.

$$\frac{5\text{ mcg}}{\cancel{\text{kg}}} \times 15.45\ \cancel{\text{kg}} = 77.25\text{ mcg rounded to } 77\text{ mcg}$$

3. Confirm that the desired dose is safe.

   Checking the PDR, you find that 5 mcg/kg is the recommended dose for pediatric patients over 2 years of age.

4. Calculate the amount to administer.

   Because the unit of measure for the dose on hand is milligrams, the desired dose must also be expressed in milligrams.

   $1\text{ mg} : 1000\ \cancel{\text{mcg}} :: ? : 77\ \cancel{\text{mcg}}$

   $1000 \times ? = 77 \times 1\text{ mg}$

   $$? = \frac{77\text{ mg}}{1000}$$

   $? = 0.077\text{ mg}$

   Using Procedure Checklist 7-2. Fill in the proportion.

   $Q{:}H :: A{:}D$

   $1\text{ mL}{:}0.5\text{ mg} :: A{:}0.077\text{ mg}$

   Cancel units.

   $1\text{ mL}{:}0.5\ \cancel{\text{mg}} :: A{:}0.077\ \cancel{\text{mg}}$

   Multiply the means and extremes, then solve for the missing value.

   $0.5 \times A = 1\text{ mL} \times 0.077$

   $$A = 1\text{ mL} \times \frac{0.077}{0.5}$$

   $A = 0.154\text{ mL}$

   Since the volume of the injection is less than 1 mL, we round to hundredth (two decimals).

   $A = 0.15\text{ mL}$

Dimensional Analysis

Example Calculate the amount to administer to a 3-year-old who weighs 34 lb.

Ordered: Hyoscyamine sulfate 5 mcg/kg Sub-Q 1 h preanesthesia

On hand: Hyoscyamine sulfate 0.5 mg/mL

Solution 1. Convert the patient's weight to kilograms.

$$?\text{ kg} = \frac{1\text{ kg}}{2.2\ \cancel{\text{lb}}} \times 34\ \cancel{\text{lb}}$$

$? = 15.45\text{ kg}$

2. Calculate the desired dose.

$$\frac{5 \text{ mcg}}{\cancel{\text{kg}}} \times 15.45 \cancel{\text{kg}} = 77.25 \text{ mcg rounded to } 77 \text{ mcg}$$

3. Confirm that the desired dose is safe.

   Checking the PDR, you find that 5 mcg/kg is the recommended dose for pediatric patients over 2 years of age.

4. Calculate the amount to administer, using Procedure Checklist 7-3.

   The unit of measure will be milliliters.

   A mL =

   The unit of measure for the dosage ordered is mcg. The unit of measure for the dose on hand is mg. Use the conversion factor 1 mg = 1000 mcg. Since we will be converting the dosage ordered to mg place mg on top. This is the first factor.

$$\frac{1 \text{ mg}}{1000 \text{ mcg}}$$

   The dosage unit is 1 mL; the dose on hand is 0.5 mg. This is the second factor.

$$\frac{1 \text{ mg}}{1000 \text{ mcg}} \times \frac{1 \text{ mL}}{0.5 \text{ mg}}$$

   The desired dose is 0.077 mg. Place this over one and set up your equation.

$$A \text{ mL} = \frac{1 \cancel{\text{mg}}}{1000 \cancel{\text{mcg}}} \times \frac{1 \text{ mL}}{0.5 \cancel{\text{mg}}} \times \frac{77 \cancel{\text{mcg}}}{1}$$

$$A = \frac{77 \text{ mL}}{1000 \times 0.5}$$

$$A = 0.154 \text{ mL}$$

   Since the volume of the injection is less than 1 mL, we round to hundredth (two decimals).

$$A = 0.15 \text{ mL}$$

Formula Method

Example 1 Calculate the amount to administer to a 3-year-old who weighs 34 lb.

Ordered: Hyoscyamine sulfate 5 mcg/kg Sub-Q 1 h preanesthesia

On hand: Hyoscyamine sulfate 0.5 mg/mL

Solution

1. Convert the patient's weight to kilograms.

$$? \text{ kg} = \frac{1 \text{ kg}}{2.2 \cancel{\text{lb}}} \times 34 \cancel{\text{lb}}$$

$$? = 15.45 \text{ kg}$$

2. Calculate the desired dose.

$$\frac{5 \text{ mcg}}{\cancel{\text{kg}}} \times 15.45 \cancel{\text{kg}} = 77.25 \text{ mcg rounded to } 77 \text{ mcg}$$

3. Confirm that the desired dose is safe. Checking the PDR, you find that 5 mcg/kg is the recommended dose for pediatric patients over 2 years of age.

4. Calculate the amount to administer, using Procedure Checklist 7-4.

   Because the unit of measure for the dose on hand is milligrams, the desired dose must also be expressed in milligrams. Using Procedure Checklist 3-1.

$$\frac{1\text{ mg}}{1000\text{ }\cancel{\text{mcg}}} = \frac{?}{77\text{ }\cancel{\text{mcg}}}$$

$$1000 \times ? = 77 \times 1\text{ mg}$$

$$? = \frac{77\text{ mg}}{1000}$$

$$? = 0.077\text{ mg}$$

$D = 0.077$ mg

$Q = 1$ mL

$H = 0.5$ mg

Fill in the formula.

$$\frac{0.077\text{ mg} \times 1\text{ mL}}{0.5\text{ mg}} = A$$

Cancel units.

$$\frac{0.077\text{ }\cancel{\text{mg}} \times 1\text{ mL}}{0.5\text{ }\cancel{\text{mg}}} = A$$

Solve for the unknown.

$$\frac{0.077 \times 1\text{ mL}}{0.5} = A$$

$A = 0.154$ mL

Since the volume of the injection is less than 1 mL, we round to hundredth (two decimals).

$A = 0.15$ mL

Example 2 Find the amount to administer. The patient is a 6-year-old child who weighs 9 lb.

Ordered: Midazolam 0.025 mg/kg IV now

Available: Refer to the label and package insert in Figures 11-4a and 11-4b.

Figure 11-4a Midazolam label.

1. *Pediatric Patients Less Than 6 Months of Age:* Limited information is available in non-intubated pediatric patients less than 6 months of age. It is uncertain when the patient transfers from neonatal physiology to pediatric physiology, therefore the dosing recommendations are unclear. Pediatric patients less than 6 months of age are particularly vulnerable to airway obstruction and hypoventilation, therefore titration with small increments to clinical effect and careful monitoring are essential.
2. *Pediatric Patients 6 Months to 5 Years of Age:* Initial dose 0.05 to 0.1 mg/kg; total dose up to 0.6 mg/kg may be necessary to reach the desired endpoint but usually does not exceed 6 mg. Prolonged sedation and risk of hypoventilation may be associated with the higher doses.
3. *Pediatric Patients 6 to 12 Years of Age:* Initial dose 0.025 to 0.05 mg/kg; total dose up to 0.4 mg/kg may be needed to reach the desired endpoint but usually does not exceed 10 mg. Prolonged sedation and risk of hypoventilation may be associated with the higher doses.
4. *Pediatric Patients 12 to 16 Years of Age:* Should be dosed as adults. Prolonged sedation may be associated with higher doses; some patients in this age range will require higher than recommended adult doses but the total dose usually does not exceed 10 mg.

Figure 11-4b Midazolam package insert.

1. Convert the child's weight from pounds to kilograms.

   1 kg : 2.2 lb :: ? kg : 9 lb

   2.2 lb × ? kg = 1 kg × 9 lb

   ? = 22.3 kg

2. Find the daily desired dose.

   $$\frac{0.25 \text{ mg}}{1 \text{ kg}} \times 22.3 \text{ kg} = \frac{0.25 \text{ mg}}{1 \cancel{\text{kg}}} \times 22.3 \cancel{\text{kg}} = 0.5575 \text{ mg}$$

3. According to the package insert, pediatric patients from 6 to 12 may have 0.025 to 0.05 mg/kg; up to 0.4 mg/kg not to exceed 10 mg. The dosage ordered corresponds to the low end of the range and is a safe order.

4. Calculate the amount to administer. Reading the label, you determine that Midazolam has a dosage strength of 2 mg/2 mL.

We now know that

D (desired dose) = 0.5575 mg

H (dose on hand) = 2 mg

Q (dosage unit) = 2 mL

Using Procedure Checklist 7-1, the Fraction Proportion Method

1. Fill in the proportion.

   $$\frac{Q}{H} = \frac{A}{D}$$

   $$\frac{2 \text{ mL}}{2 \text{ mg}} = \frac{A}{0.5575 \text{ mg}}$$

2. Cancel units.

   $$\frac{2 \text{ mL}}{2 \cancel{\text{mg}}} = \frac{A}{0.5575 \cancel{\text{mg}}}$$

3. Cross-multiply and solve for the unknown.

   $2 \times A = 2 \text{ mL} \times 0.5575$

   $A = 2 \text{ mL} \times \frac{0.5575}{2}$

   $A = 0.5575 \text{ mL}$

4. The dose is under 1 mL, so round to the nearest hundredth.

   $A = 0.56 \text{ mL}$

Using Procedure Checklist 7-2, the Ratio Proportion Method

1. Fill in the proportion.

   $Q : H :: A : D$

   $2 \text{ mL} : 2 \text{ mg} :: A : 0.5575 \text{ mg}$

2. Cancel units.

   $2 \text{ mL} : 2 \cancel{\text{mg}} :: A : 0.5575 \cancel{\text{mg}}$

3. Multiply the means and extremes, then solve for the missing value.

   $2 \times A = 2 \text{ mL} \times 0.5575$

   $$A = 2 \text{ mL} \times \frac{0.5575}{2}$$

   $$A = 0.5575 \text{ mL}$$

4. The dose is under 1 mL, so round to the nearest hundredth.

   $A = 0.56 \text{ mL}$

Using Procedure Checklist 7-3, the Dimensional Analysis Method

1. The unit of measure will be milliliters.

   $A \text{ mL} =$

2. The dosage unit is 2 mL; the dosage strength is 2 mg.

   $$\frac{2 \text{ mL}}{2 \text{ mg}}$$

3. The desired dose is 0.5575 mg. Place this over one and set up the equation.

   $$A \text{ mL} = \frac{2 \text{ mL}}{2 \cancel{\text{mg}}} \times \frac{0.5575 \cancel{\text{mg}}}{1}$$

   $$A = 2 \text{ mL} \times \frac{0.5575}{2}$$

   $$A = 0.5575 \text{ mL}$$

4. The dose is under 1 mL, so round to the nearest hundredth.

   $A = 0.56 \text{ mL}$

Using Procedure Checklist 7-4, the Formula Method

1. We know that

   $D = 0.5575 \text{ mg}$

   $Q = 2 \text{ mL}$

   $H = 2 \text{ mg}$

2. Fill in the formula.

   $$\frac{0.5575 \text{ mg} \times 2 \text{ mL}}{2 \text{ mg}} = A$$

3. Cancel units.

$$\frac{0.5575 \cancel{mg} \times 2 \text{ mL}}{2 \cancel{mg}} = A$$

4. Solve for the unknown.

$$\frac{0.5575 \times 2 \text{ mL}}{2} = A$$

$A = 0.5575$ mL

5. The dose is under 1 mL, so round to the nearest hundredth.

$A = 0.56$ mL

GO TO . . . Open the CD-ROM that accompanies your textbook, and select Chapter 11, Rule 11-1. Review the animation and example problems, then complete the practice problems. Continue to the next section of the book once you have mastered the rule presented. ■

The total volume of a pediatric injection is limited based on the size and the age of the child. Table 11-4 summarizes the maximum volume that is appropriate for a pediatric injection. According to this table, the amount to administer in Example 2 for Rule 11-1 is within normal limits. The patient is considered a preschooler at the age of 3, and the amount to administer is 1.1 mL, which is within the maximum volume range.

The length and gauge of the needle used will also vary with the age and size of the patient as well as the location. Smaller muscles need smaller needles. The depth of an injection may also vary for geriatric patients due to their reduced muscle size. You must be aware of all these factors when administering injections to special populations. Additional details regarding these injection techniques are outside the scope of this book. Please review injection techniques before you administer any injection.

TABLE 11-4 Pediatric Injections

Stage of Development	Maximum Volume of IM Injection
Infant	0.5–1 mL
Toddler, walking for at least 1 year	1 mL
Preschooler and elementary school age	1–1.5 mL

Ensuring Safe Dosages

Drug orders may be written in several ways. If you measure or administer the medication, you have the responsibility to check whether the dose is the standard recommended dose. The recommended dose is sometimes written as a range, with a minimum and a maximum recommended dose. In this case, you will need to determine if the dose ordered is not less than the minimum or greater than the maximum recommended dose.

Rule 11-2 Ensuring Safe Dosages

When you are working with special populations, always check the package insert, drug label, or product literature to ensure the safety of the dose to be administered.

Example 1 Determine whether the following order is safe. If safe, calculate the amount to administer.

Patient: Child who weighs 14.5 kg

Ordered: Erythromycin 125 mg PO q4h

On hand: Refer to the label in Figure 11-5.

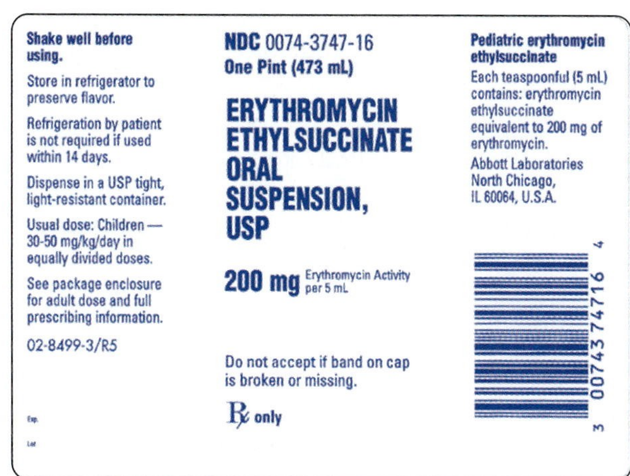

Figure 11-5

Solution The drug label lists a range of child dosages, with a minimum of 30 mg/kg/day and a maximum of 50 mg/kg/day. Multiplying the child's weight, in kilograms, by these values will give us the recommended dosage for the patient.

$$\text{Minimum recommended dosage} = \frac{30\text{ mg}}{\cancel{\text{kg}}} \times 14.5\ \cancel{\text{kg}} = 435\text{ mg/day}$$

$$\text{Maximum recommended dosage} = \frac{50\text{ mg}}{\cancel{\text{kg}}} \times 14.5\ \cancel{\text{kg}} = 725\text{ mg/day}$$

$$\text{Dosage ordered} = \frac{125\text{ mg}}{\cancel{\text{dose}}} \times \frac{6\ \cancel{\text{doses}}}{\text{day}} = 750\text{ mg/day}$$

The dosage ordered does not fall within the recommended dosage range. You should contact the physician.

Example 2 Determine whether the following order is safe. If it is safe, calculate the amount to administer.

Patient: Child who weighs 27 lb

Ordered: EryPed® drops 120 mg po q6h

On hand: Refer to the label in Figure 11-6.

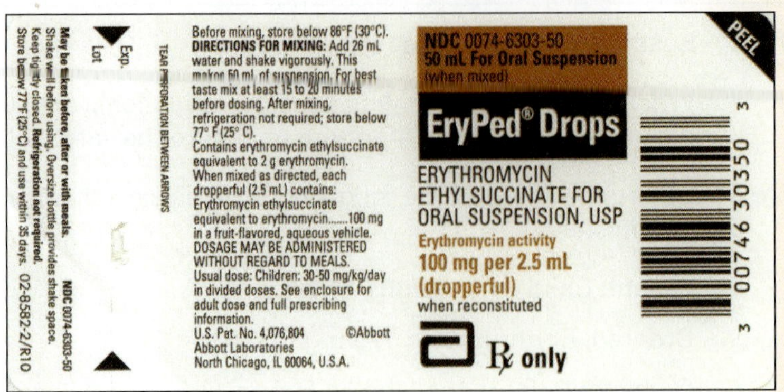

Figure 11-6

Solution The drug label lists a range of child dosages, with a minimum of 30 mg/kg/day and a maximum of 50 mg/kg/day. The patient's weight must first be converted into kilograms.

$$27 \not{lb} \times \frac{1 \text{ kg}}{2.2 \not{lb}} = 12.27 \text{ kg}$$

Multiplying the child's weight, in kilograms, by these values will give us the recommended dosage for the patient.

$$\text{Minimum recommended dosage} = \frac{30 \text{ mg}}{\not{kg}} \times 12.27 \not{kg} = 368.1 \text{ mg/day}$$

$$\text{Maximum recommended dosage} = \frac{50 \text{ mg}}{\not{kg}} \times 12.27 \not{kg} = 613.5 \text{ mg/day}$$

$$\text{Dosage ordered} = \frac{120 \text{ mg}}{\not{\text{dose}}} \times \frac{4 \not{\text{doses}}}{\text{day}} = 480 \text{ mg/day}$$

The dosage ordered is within the recommended dosage range. Calculate the amount to administer. D = 120 mg H = 100 mg Q = 2.5 mL

Using Procedure Checklist 7-1, the Fraction Proportion Method

1. Fill in the proportion.

$$\frac{Q}{H} = \frac{A}{D}$$

$$\frac{2.5 \text{ mL}}{100 \text{ mg}} = \frac{A}{120 \text{ mg}}$$

2. Cancel units.

$$\frac{2.5 \text{ mL}}{100 \not{mg}} = \frac{A}{120 \not{mg}}$$

3. Cross-multiply and solve for the unknown.

$$100 \times A = 2.5 \text{ mL} \times 120$$

$$A = 2.5 \text{ mL} \times \frac{120}{100}$$

$$A = 3 \text{ mL}$$

Using Procedure Checklist 7-2, the Ratio Proportion Method

1. Fill in the proportion.

   $Q : H :: A : D$

   2.5 mL : 100 mg :: A : 120 mg

2. Cancel units.

   2.5 mL : 100 $\cancel{\text{mg}}$ :: A : 120 $\cancel{\text{mg}}$

3. Multiply the means and extremes, then solve for the missing value.

   $$100 \times A = 2.5\text{ mL} \times 120$$

   $$A = 2.5\text{ mL} \times \frac{120}{100}$$

   $$A = 3\text{ mL}$$

Using Procedure Checklist 7-3, the Dimensional Analysis Method

1. The unit of measure will be milliliters.

   A mL =

2. The dosage unit is 2.5 mL; the dose on hand is 100 mg. This is the first factor.

   $$\frac{2.5\text{ mL}}{100\text{ mg}}$$

3. The desired dose is 120 mg. Place this over 1 and set up the equation.

   $$A\text{ mL} = \frac{2.5\text{ mL}}{100\ \cancel{\text{mg}}} \times \frac{120\ \cancel{\text{mg}}}{1}$$

   $$A = 2.5\text{ mL} \times \frac{120}{100}$$

   $$A = 3\text{ mL}$$

Using Procedure Checklist 7-4, the Formula Method

1. We know that

   D = 120 mg

   Q = 2.5 mL

   H = 100 mg

2. Fill in the formula.

   $$\frac{120\text{ mg} \times 2.5\text{ mL}}{100\text{ mg}} = A$$

3. Cancel units.

   $$\frac{120\ \cancel{\text{mg}} \times 2.5\text{ mL}}{100\ \cancel{\text{mg}}} = A$$

4. Solve for the unknown.

$$\frac{120 \times 2.5 \text{ mL}}{100} = A$$

$$A = 3 \text{ mL}$$

ERROR ALERT!

Converting ounces carefully.

Infants, especially premature infants, may be weighed in pounds and ounces or grams. On occasion you will need to convert between these two units to determine if a dose is safe. When you are doing these conversions, remember 16 oz = 1 lb. An ounce is not one-tenth of a pound. A baby whose weight is 8 lb 6 oz does not weigh 8.6 lb. Convert 6 oz to pounds, using the conversion factor of $\frac{1 \text{ lb}}{16 \text{ oz}}$.

Here, $6 \text{ oz} \times \frac{1 \text{ lb}}{16 \text{ oz}} = 0.375$ lb. Thus, 8 lb 6 oz = 8.375 lb.

Example 3 Determine whether the following order is safe for a child weighing 38 lb. If safe, calculate the amount to administer.

Ordered: Pediazole® susp 5 mL po qid

On hand: Refer to the label and package insert in Figure 11-7.

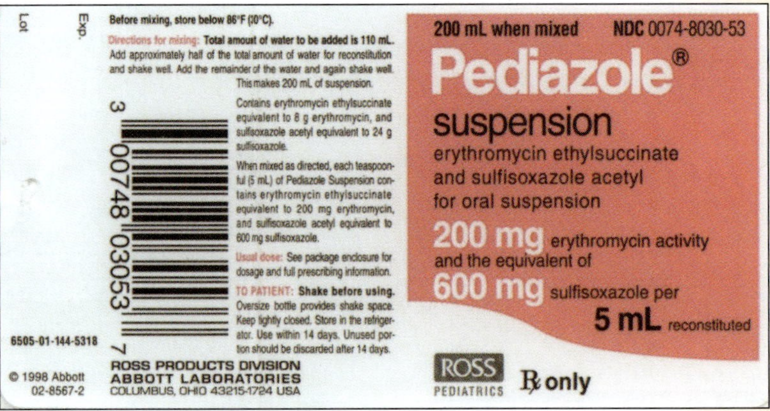

DOSAGE AND ADMINISTRATION

PEDIAZOLE SHOULD NOT BE ADMINISTERED TO INFANTS UNDER 2 MONTHS OF AGE BECAUSE OF CONTRAINDICATIONS OF SYSTEMIC SULFONAMIDES IN THIS AGE GROUP.

For Acute Otitis Media in Children: The dose of Pediazole can be calculated based on the erythromycin component (50 mg/kg/day) or the sulfisoxazole component (150 mg/kg/day to a maximum of 6 g/day). The total daily dose of Pediazole should be administered in equally divided doses three or four times a day for 10 days. Pediazole may be administered without regard to meals.

Figure 11-7 Pediazole® label and package insert.

Convert the child's weight to kilograms.

$1 \text{ kg} : 2.2 \text{ lb} :: ? \text{ kg} : 38 \text{ lb}$

$2.2 \text{ lb} \times ? \text{ kg} = 1 \text{ kg} \times 38 \text{ lb}$

$? = 17.27 \text{ kg}$

The child weighs 17.27 kg. Pediazole® is a combination drug. According to the package insert, the recommended doses of the components are 50 mg/kg/day of erythromycin ethylsuccinate and 150 mg/kg/day of sulfisoxazole acetyl.

$$\frac{50 \text{ mg}}{1 \text{ kg}} \times 17.27 \text{ kg} = 863.5 \text{ mg}$$

$$\frac{150 \text{ mg}}{1 \text{ kg}} \times 17.27 \text{ kg} = 2590.5 \text{ mg}$$

The maximum daily dosages are 863.5 mg/day of erythromycin ethylsuccinate and 2590.5 mg/day of sulfisoxazole acetyl.

The order calls for 5 mL of Pediazole® 4 times per day. According to the label, each 5 mL of reconstituted Pediazole® contains 200 mg of erythromycin and 600 mg of sulfisoxazole. The daily dosages ordered, then, are 800 mg of erythromycin and 2400 mg of sulfisoxazole. The dosage ordered of erythromycin is below the maximum safe dose of 863.5 mg/day. The dosage ordered of sulfisoxazole is below the maximum safe dose of 2590.5 mg/day. The dosage ordered of each component is within the recommended range. You may administer the medication.

The physician's order is already written as a volume of Pediazole®. No further calculation is needed. Administer 5 mL, or 1 tsp, of Pediazole® orally 4 times per day.

GO TO . . . Open the CD-ROM that accompanies your textbook, and select Chapter 11, Rule 11-2. Review the animation and example problems, then complete the practice problems. Continue to the next section of the book once you have mastered the rule presented. ■

Critical Thinking on the Job

Looking for Warnings

Elena received an order to administer 1.5 mL of Pediazole® to a 7-week-old infant who weighs 12 lb. The schedule on the package insert (see previous Example 3) has an instruction to adjust the dosage by body weight. She notices that for a 13-lb infant, the recommended dosage is 2.5 mL every 6 or 8 h. Elena administers the medication to the infant.

Questions?

Did Elena make a mistake? If so, what was it and what should she do?

REVIEW AND PRACTICE

11-1 Dosages Based on Body Weight

In Exercises 1–10, convert the following weights to kilograms. Rounding to the nearest hundredth.

1. 66 lb **2.** 77 lb **3.** 54 lb **4.** 37 lb **5.** 152 lb

6. 202 lb **7.** 16 lb 4 oz **8.** 11 lb 10 oz **9.** 9 lb 14 oz **10.** 14 lb 5 oz

In Exercises 11–22, determine if the order is safe. If it is, then determine the amount to administer.

11. The patient is a 3-day-old newborn who weighs 6 lb 5 oz.
Ordered: Nebcin® 5 mg IM q12h
On hand: Nebcin® multidose vial, 20 mg/2 mL. According to the package insert, a premature or full-term neonate up to 1 week of age may be administered up to 4 mg/kg/day in 2 equal doses every 12 h.

12. The patient is a 4-year-old child who weighs 32 lb.
Ordered: Proventil® 1 tsp syrup po tid
On hand: Proventil® Syrup, 2 mg/5 mL. According to the package insert, for children 2 to 6 years of age, dosing should be initiated at 0.1 mg/kg of body weight 3 times a day. This starting dose should not exceed 2 mg 3 times per day.

13. The patient is a 3-year old child who weighs 32 lb and has a severe infection.
Ordered: Amoxicillin 750 mg PO q8h
On hand: Refer to label A. According to the package label, the dosing regimen for children is 20 to 40 mg/kg/day q8h.

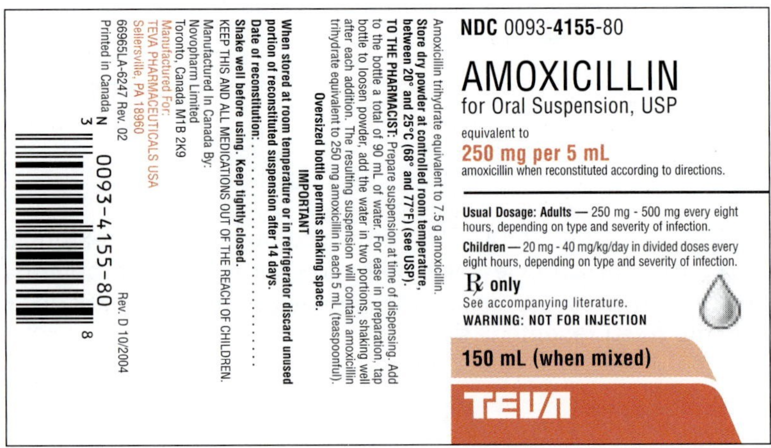

A

14. The patient is a 7-year-old child who weighs 52 lb.
Ordered: Ranitidine 30 mg IV q8h
On hand: Refer to labels B and C.

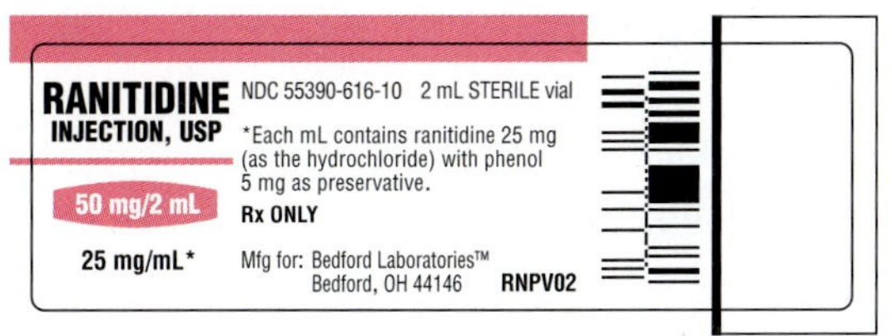

B

> **Pediatric Use:**
>
> While limited data exist on the administration of IV ranitidine to children, the recommended dose in pediatric patients is for a total daily dose of 2 to 4 mg/kg, to be divided and administered every 6 to 8 hours, up to a maximum of 50 mg given every 6 to 8 hours. This recommendation is derived from adult clinical studies and pharmacokinetic data in pediatric patients. Limited data in neonatal patients (less than one month of age) receiving ECMO have shown that a dose of 2 mg/kg is usually sufficient to increase gastric pH to >4 for at least 15 hours. Therefore, doses of 2 mg/kg given every 12 to 24 hours or as a continuous infusion should be considered.
>
> *Excerpt from package insert for Ranitidine Injection from Bedford Laboratories (www.bedfordlabs.com)

C

15. The patient is an 8-year-old child who weighs 55 lb and is being treated for streptococcal pharnygitis.
Ordered: Cephalexin Susp. 200 mg PO q6h
On hand: See label D and package insert information E below.

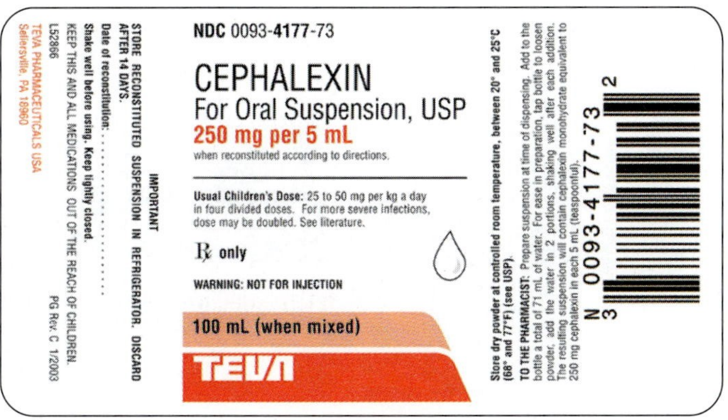

D

> *Pediatric Patients* The usual recommended daily dosage for pediatric patients is 25 to 50 mg/kg in divided doses. For streptococcal pharyngitis in patients over 1 year of age and for skin and skin structure infections, the total daily dose may be divided and administered every 12 hours.

E

16. The patient is a 44-lb child who is $5\frac{1}{2}$ years old.
Ordered: Tolectin 100 mg po qid
On hand: Tolectin 200-mg scored tablets. The package insert indicates that for children 2 years and older, the usual dose ranges from 15 to 30 mg/kg/day.

17. The patient is a 44-lb child who is $5\frac{1}{2}$ years old.
Ordered: Midazolam 1.5 mg IV
On hand: Refer to label F. The package insert indicates that for children the usual dose ranges from 0.05–0.4 mg/kg.

MIDAZOLAM
HCl INJECTION
2 mg/2 mL*
*Midazolam 1 mg/mL (as the hydrochloride) CIV
FOR IM OR IV USE Rx ONLY
NDC 55390-137-02
2 mL SINGLE USE VIAL
USUAL DOSAGE: See package insert.
DISCARD UNUSED PORTION.
Mfg for: Bedford Labs™ Bedford, OH 44146
KC-MPF-VA02

F

18. The same patient from Exercise 17 is given the following order: Midazolam 1 mg IM now.
On hand: Refer to label F. The package insert indicates that for children the usual dose ranges from 0.05–0.4 mg/kg.

19. The patient is a 4-month-old child who weighs 12 lb.
Ordered: EryPed® Drops 50 mg po q12h
On hand: Refer to label G. According to the package insert, the following are usual dosages for children over 3 months of age: For mild to moderate ear, nose, throat infections, either 25 mg/kg/day in divided doses every 12 h, or 20 mg/kg/day in divided doses every 8 h. For lower respiratory tract infections, 45 mg/kg/day in divided doses every 12 h.

20. The patient is a 1-year-old child who weighs 18 lb with severe otitis media.
Ordered: Erythromycin suspension 75 mg po q6h
On hand: Refer to label G.

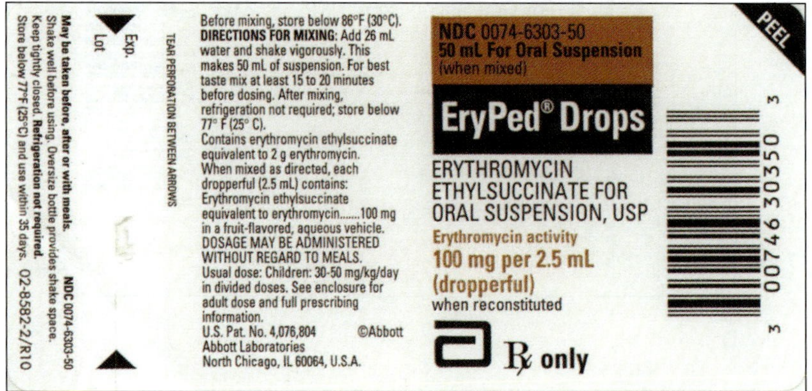

G

21. The patient weighs 47 lb.
Ordered: Antibiotic 3 mcg/kg/day IM divided in 2 equal doses
On hand: Antibiotic in 50 mcg/mL vials

22. The patient weighs 27 kg.
Ordered: Muscle relaxant 10 mg/kg/day IM daily
On hand: Muscle relaxant in 50 mg/mL suspension

To check your answers, go to page 533.

Dosages Based on Body Surface Area (BSA)

Some medications are prescribed based on a patient's body weight. Others factor in both weight and height to determine a patient's **body surface area,** or BSA. Many pediatric medications use a patient's BSA to determine the daily dosage. BSA is also important for burn victims and for patients undergoing chemotherapy, radiation treatments, and open heart surgery. BSA calculations are used to provide more accurate dosage calculations specific to the patient's size and severity of the illness.

Calculating a Patient's BSA

A patient's BSA is stated in square meters (m^2) You can calculate the BSA by using one of the two formulas listed in Rule 11-3. Your calculator should have a program or button that will help you find a square root ($\sqrt{\ }$) You can also use a special chart called a *nomogram.* Nomograms provide an estimate of BSA and are easier to use. Nomograms are available for children and adults. See Figures 11-8 and 11-9.

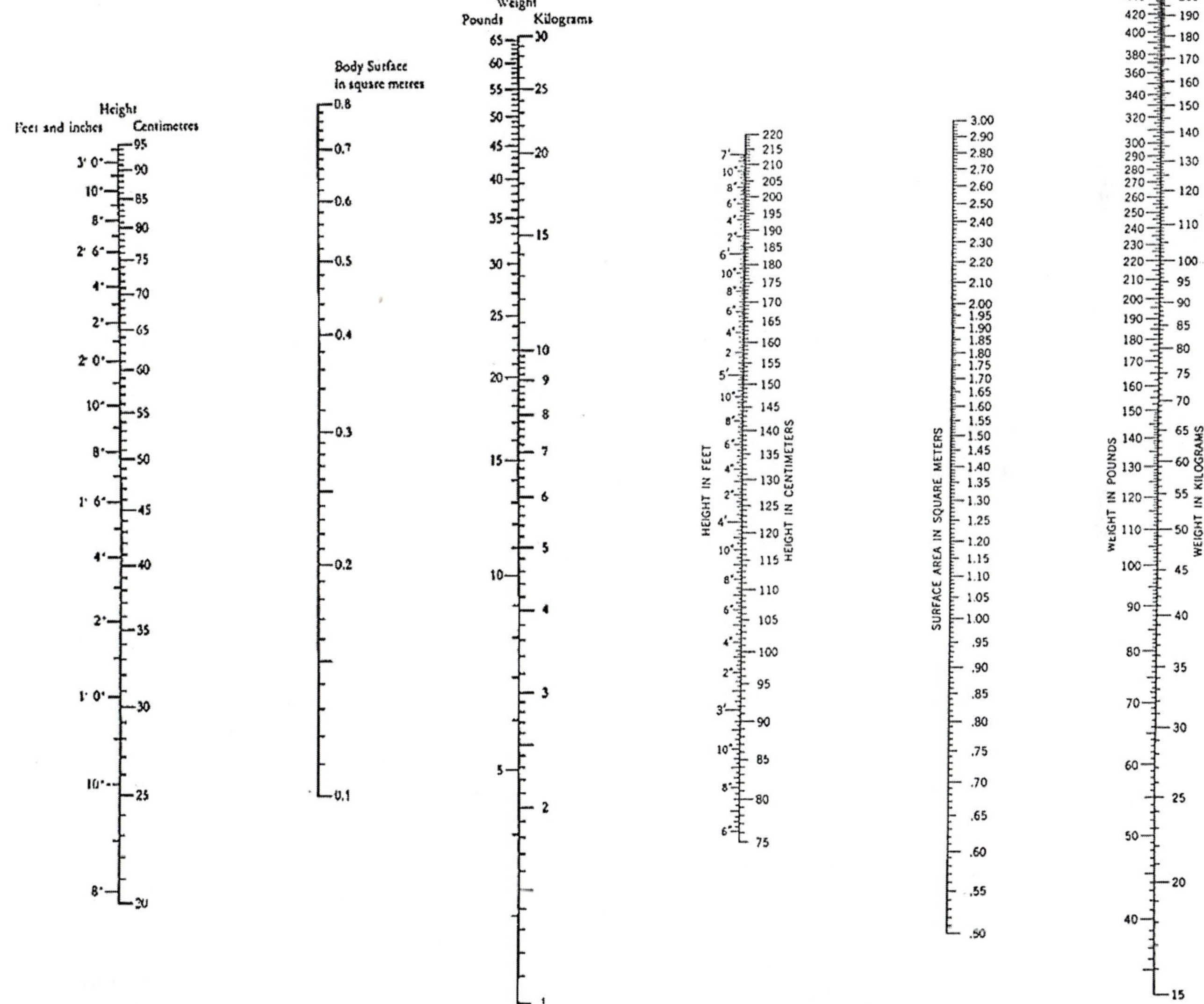

Figure 11-8 Child's nomogram.

Figure 11-9 Adult's nomogram.

Rule 11-3 Calculating the Body Surface Area Using a Formula

To determine a patient's BSA (body surface area):

1. If you know the height in centimeters and weight in kilograms, calculate

$$\text{BSA} = \sqrt{\frac{\text{height (cm)} \times \text{weight (kg)}}{3600}}\ \text{m}^2$$

2. If you know the height in inches and weight in pounds, calculate

$$\text{BSA} = \sqrt{\frac{\text{height (in)} \times \text{weight (lb)}}{3131}}\ \text{m}^2$$

Note: When using a formula to calculate BSA, if the result is less than one, round to the nearest hundredth. When the result is greater than one, round to the nearest tenth.

Example 1 Find the body surface area for a child who is 85 cm tall and weighs 13.9 kg.

Use the first of the formulas from Rule 11-3.

$$\text{BSA} = \sqrt{\frac{85 \times 13.9}{3600}}\ \text{m}^2 = \sqrt{\frac{1181.5}{3600}}\ \text{m}^2 = 0.572\ \text{m}^2 = 0.57\ \text{m}^2$$

Example 2 Find the body surface area for a baby who is 24 in. tall and who weighs 12 lb 3 oz.

Use the second of the formulas from Rule 11-3. First, convert the pounds and ounces to pounds and a fraction of a pound.

12 lb 3 oz = 12.2 lb

$$\text{BSA} = \sqrt{\frac{24 \times 12.2}{3131}}\ \text{m}^2 = \sqrt{\frac{292.8}{3131}}\ \text{m}^2 = 0.305\ \text{m}^2 = 0.31\ \text{m}^2$$

Example 3 Find the body surface area for an adult who is 5 ft 6 in. tall and who weighs 168 lb.

First, convert the height to inches. Since 1 ft equals 12 in., multiply the number of feet by 12 and then add the inches.

$5 \times 12 = 60$

$60 + 6 = 66$ in.

$$\text{BSA} = \sqrt{\frac{66 \times 168}{3131}}\ \text{m}^2 = \sqrt{\frac{11{,}088}{3131}}\ \text{m}^2 = 1.88\ \text{m}^2 = 1.9\ \text{m}^2$$

GO TO . . . Open the CD-ROM that accompanies your textbook, and select Chapter 11, Rule 11-3. Review the animation and example problems, then complete the practice problems. Continue to the next section of the book once you have mastered the rule presented.

Rule 11-4 Calculating the Body Surface Area by Using a Nomogram

Using a straightedge (such as a ruler or piece of paper), align the straightedge so that it intersects at the height and weight. Doing so will create an intersection in the BSA scale. *Note: Read the calibrations carefully, the spaces and lines vary based upon where you intersect the line.*

Example 1 Find the body surface area for a child who is 85 cm tall and weighs 13.9 kg, using the child's nomogram (Figure 11-10).

BSA = 0.57 m^2

Example 2 Find the body surface area for a baby who is 24 in. tall and weighs 12 lb 3 oz, using the child's nomogram (Figure 11-11).

BSA = 0.3 m^2

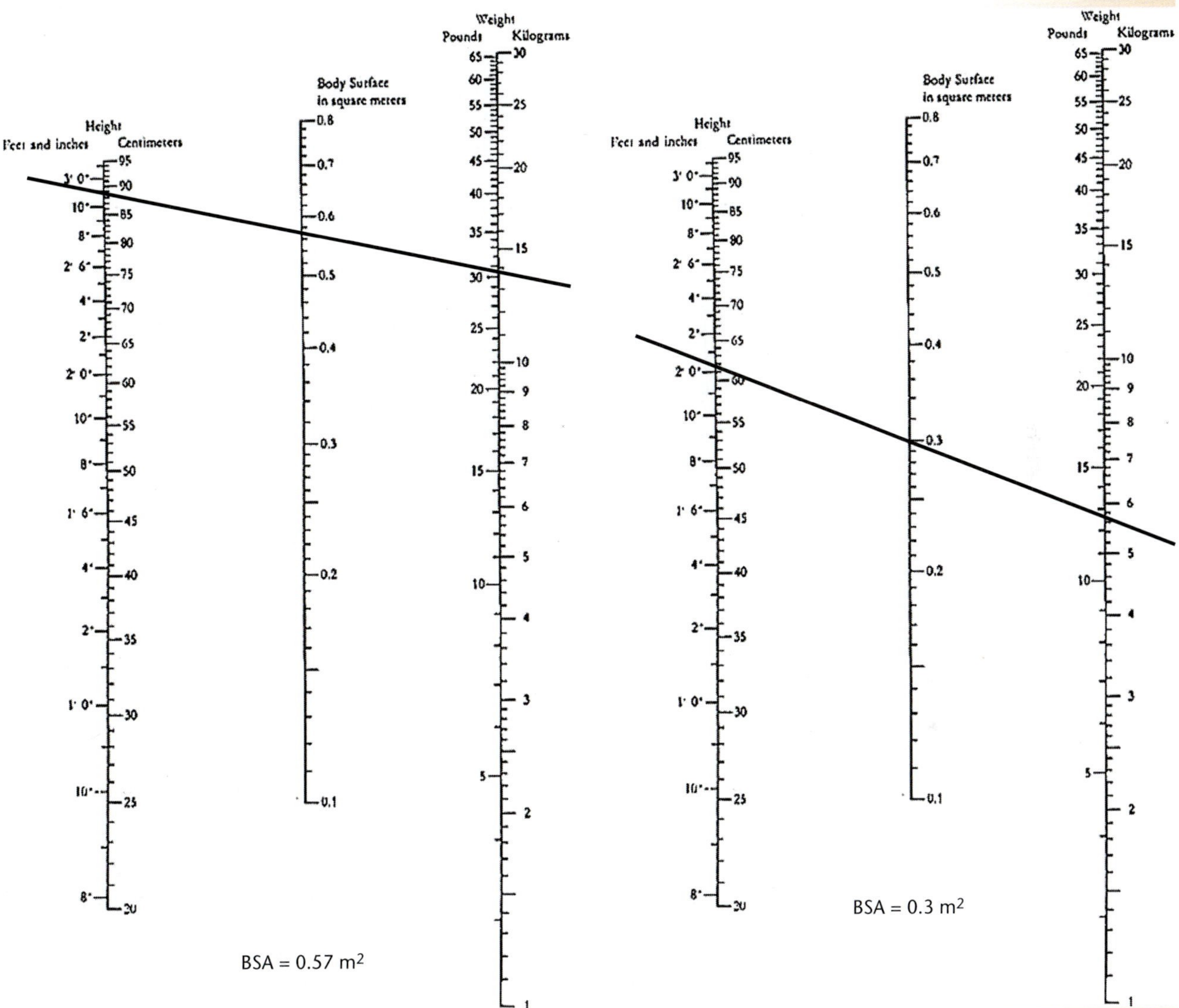

Figure 11-10 Child nomogram for example 1.

Figure 11-11 Child nomogram for example 2.

Example 3 Find the body surface area for an adult who is 5 ft 6 in. tall and who weighs 168 lb, using the adult nomogram (Figure 11-12).

BSA = 1.9 m^2

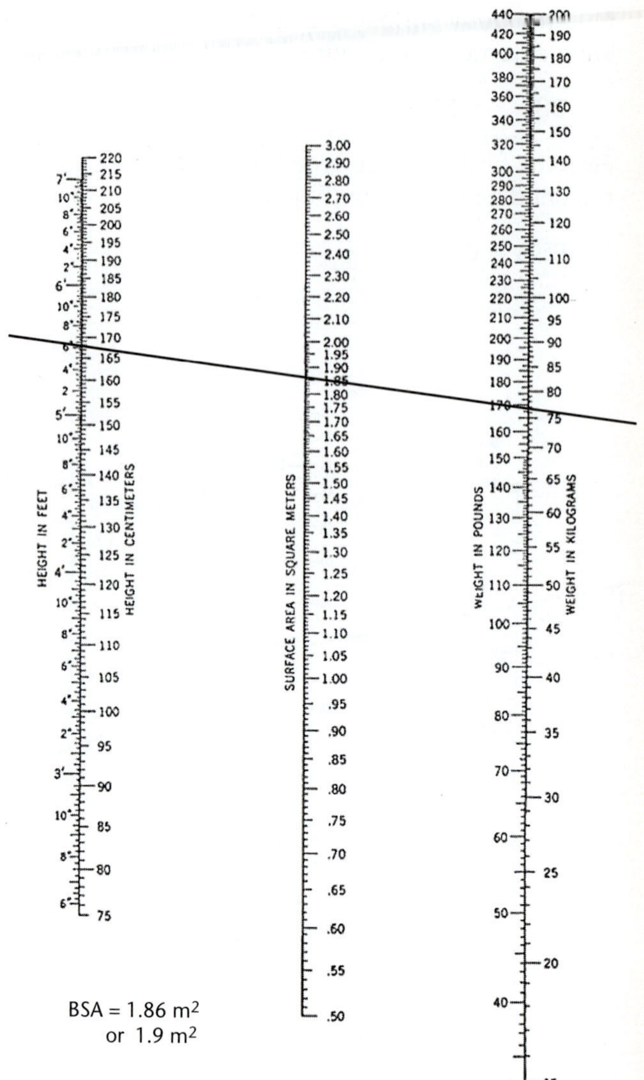

Figure 11-12 Adult nomogram for example 3.

GO TO . . . Open the CD-ROM that accompanies your textbook, and select Chapter 11, Rule 11-4. Review the animation and example problems, then complete the practice problems. Continue to the next section of the book once you have mastered the rule presented.

Rule 11-5 Calculating Dosage Based on BSA

1. Calculate the patient's BSA.
2. Calculate the desired dose:

   Dosage ordered per $m^2 \times$ BSA = desired dose

3. Confirm whether the desired dose is safe. If it is unsafe, consult the physician who wrote the order.
4. Calculate the amount to administer, using fraction proportion, ratio proportion, dimensional analysis, or the formula method.

Example 1 Ordered: CeeNU® (first dose) 140 mg now for a patient whose height is 38 in. and weight is 47 lb. According to the package insert, the first recommended dose of CeeNU® is a single oral dose providing 130 mg/m^2.

1. Because the recommended dose is per square meter (m^2), you need to find the patient's BSA. You know the patient's height and weight in inches and pounds. Use the second formula in Rule 11-3 or a nomogram.

$$BSA = \sqrt{\frac{\text{height (in.)} \times \text{weight (lb)}}{3131}}\ m^2$$

$$BSA = \sqrt{\frac{38 \times 47}{3131}}\ m^2 = \sqrt{\frac{1786}{3131}}\ m^2 = 0.76\ m^2$$

2. Calculate the desired dose.

$$\frac{130\ mg}{m^2} \times 0.76\ m^2 = 98.8\ mg$$

3. CeeNU® is available in 100-mg, 40-mg, and 10-mg capsules. The dose ordered 140 mg is above the first recommended dose of 98.8 mg. Consult the physician who wrote the order.
4. No calculation is necessary at this time.

Example 2 Ordered: CeeNU® (first dose) 150 mg now for a patient who is 99 cm tall and weighs 50 kg. According to the package insert, the first recommended dose of CeeNU® is a single oral dose providing 130 mg/m^2.

1. The recommended dose is in square meters (m^2), so you need to find the BSA. You know the patient's height and weight in centimeters and kilograms. Use the first formula in Rule 11-3.

$$BSA = \sqrt{\frac{99 \times 50}{3600}}\ m^2 = \sqrt{\frac{4950}{3600}}\ m^2 = 1.2\ m^2.$$

2. Calculate the desired dose.

$$\frac{130\ mg}{m^2} \times 1.2\ m^2 = 156\ mg$$

3. The dose ordered is safe since the dose or 150 mg is less than the calculates desired dose 152.1 mg.
4. Calculate the amount to administer. CeeNU® is available in 100-mg, 40-mg, and 10-mg capsules. Thinking critically and realizing the capsules cannot be divided you must provide the dose to the nearest whole number. This would be 150 mg. You determine to administer 1 capsule of each strength, 100 mg + 40 mg + 10 mg.

GO TO . . . Open the CD-ROM that accompanies your textbook, and select Chapter 11, Rule 11-5. Review the animation and example problems, then complete the practice problems. Continue to the next section of the book once you have mastered the rule presented. ■

REVIEW AND PRACTICE

11-2 Dosages Based on Body Surface Area (BSA)

In Exercises 1–8, use the appropriate formula to calculate the BSA for patients with the following heights and weights.

1. 88 cm and 13.2 kg
2. 58 cm and 21 kg
3. 38 cm and 6 kg
4. 48 cm and 10 kg
5. 52 in and 64 lb
6. 43 in and 35 lb
7. 22 in and 18 lb
8. 26 in and 21 lb

In Exercises 9–12, calculate the recommended dosage in the appropriate unit.

9. The child's BSA is 0.82 m^2. The recommended dosage is 175 mcg/m^2.

10. The child's BSA is 0.65 m^2. The recommended dosage is 0.4 mg/m^2.

11. The child's height is 62 cm and weight is 5 kg. The recommended dosage is 50 mcg/m^2.

12. The child's height is 41 in. and weight is 63 lb. The recommended dosage is 0.2 mg/m^2.

In Exercises 13–18, determine if the order is safe. If so, calculate the amount to administer.

13. The patient is 42 in. tall and weighs 71 lb.
Ordered: Chemotherapy medication 6 mg/m^2/day IV q12h
On hand: Chemotherapy medication 200 mcg/mL for IV use.

14. The patient is 86 cm tall and weighs 12 kg.
Ordered: Antibiotic 25 mg/m^2/day IM q6h
On hand: Antibiotic 2 mg/mL for IM use.

15. The patient is 34 cm tall and weighs 5 kg.
Ordered: Cerubidine® 25 mg/m^2 IV qw
On hand: Cerubidine® for injection. When reconstituted, each milliliter contains 5 mg of drug. The recommended pediatric dosage is 25 mg/m^2 IV the first day every week.

16. The patient, who is over 1 year old, is 42 in. tall and weighs 45 lb.
Ordered: Oncaspar® 2500 IU/m^2 IM every 14 days
On hand: Oncaspar® 5 mL/vial, 750 IU/mL. The recommended pediatric dosage is 2500 IU/m^2 for children whose BSA is greater than or equal to 0.6 m^2 and 82.5 IU/kg for children whose BSA is less than 0.6 m^2.

17. The patient is 125 cm tall and weighs 45 kg.
Ordered: Gemzar® 800 mg IV qw
On hand: Refer to label A. The usual dose is 1000 mg/m^2 over 30-min IV.

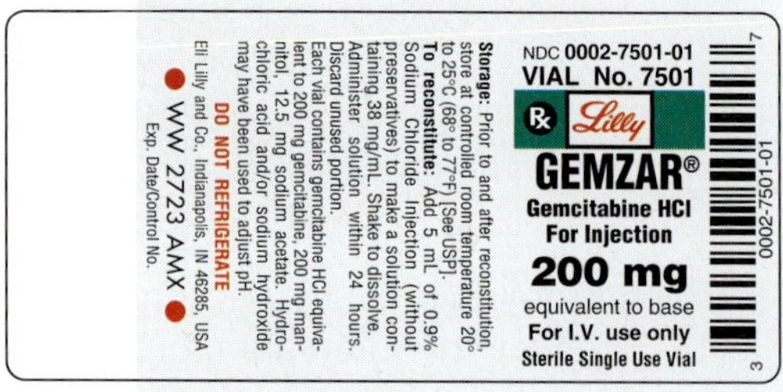

A

18. The patient is 63 in. tall and weighs 125 lb.
Ordered: Cisplatin 125 mg IV every four weeks
On hand: Refer to label B. The usual dose is 75 to 100 mg/m^2.

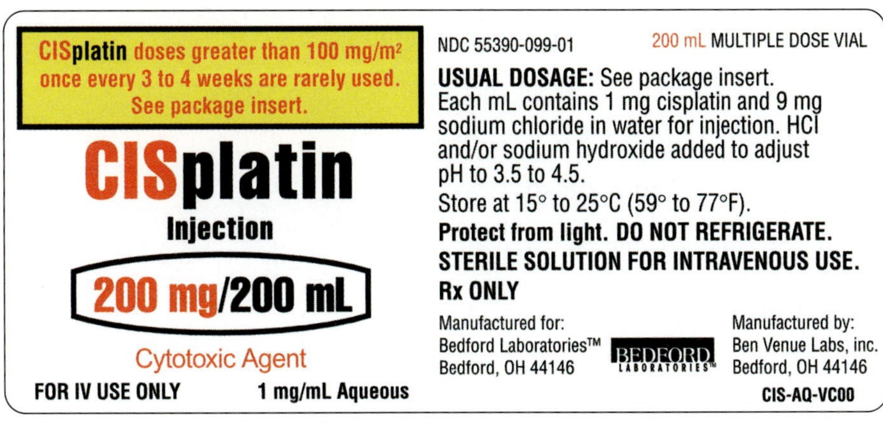

B

To check your answers, see page 534.

Age-Related Needs

Daily Maintenance Fluid Needs (DMFN)

Children's bodies contain a higher percentage of water than adults' bodies. Children, as well as critically ill patients are at greater risk for fluid overload, dehydration, or electrolyte imbalances. Therefore, you must monitor not only the amount of medication but also the amount of fluid the patient receives. Fluids may be calculated based on body weight, body surface area (BSA), metabolism, or age.

Daily maintenance fluid needs (DMFN) represent the fluid a patient needs over 24 h. It combines maintenance fluids (both orally and parenterally), medications, diluent for medications, and fluids used to flush the injection port. The amount of maintenance fluid required varies according to weight, with the smallest children requiring 100 mL/kg/day. DMFN does not include fluids needed to replace those lost to vomiting, diarrhea, or fever. These are called *replacement fluids* and are based on each patient's condition.

Rule 11-6 To calculate daily maintenance fluid needs (DMFN) based on weight:

1. If the patient weighs up to 10 kg, find

$$\frac{100 \text{ mL}}{1 \text{ kg}} \times \text{kg} = \text{DMFN mL}$$

2. If the patient weighs 10 to 20 kg, find

$$1000 \text{ mL} + \left[\frac{50 \text{ mL}}{1 \text{ kg}} \times (\text{kg} - 10)\right] = \text{DMFN mL}$$

3. If the patient weighs over 20 kg, find

$$1500 \text{ mL} + \left[\frac{20 \text{ mL}}{1 \text{ kg}} \times (\text{kg} - 20)\right] = \text{DMFN mL}$$

Example 1 Find the DMFN for a patient who weighs

a. 7 kg **b.** 16 kg **c.** 24 kg

a. The child weighs less than 10 kg.

$$\frac{100 \text{ mL}}{1 \text{ kg}} \times 7 \text{ kg} = 700 \text{ mL}$$

b. The child weighs between 10 and 20 kg.

$$1000 \text{ mL} + \left[\frac{50 \text{ mL}}{1 \cancel{\text{kg}}} \times (16 \cancel{\text{kg}} - 10)\right]$$

$$= 1000 \text{ mL} + \left[\frac{50 \text{ mL}}{1} \times 6\right]$$

$$= 1000 \text{ mL} + 300 \text{ mL} = 1300 \text{ mL}$$

c. The child weighs over 20 kg.

$$1500 \text{ mL} + \left[\frac{20 \text{ mL}}{1 \cancel{\text{kg}}} \times (24 \cancel{\text{kg}} - 20)\right]$$

$$= 1500 \text{ mL} + \left[\frac{20 \text{ mL}}{1} \times 4\right]$$

$$= 1500 \text{ mL} + 80 \text{ mL} = 1580 \text{ mL}$$

Example 2 What is the flow rate using microdrip tubing for DMFN for a child who weighs 14 kg? Find the DMFN. The patient weighs between 10 and 20 kg.

$$1000 \text{ mL} + \left[\frac{50 \text{ mL}}{1 \cancel{\text{kg}}} \times (14 \cancel{\text{kg}} - 10)\right]$$

$$= 1000 \text{ mL} + \left[\frac{50 \text{ mL}}{1} \times 4\right]$$

$$= 1000 \text{ mL} + 200 \text{ mL} = 1200 \text{ mL}$$

Next, find the microdrip tubing flow rate for 1200 mL/day.

$$\frac{1200 \cancel{\text{mL}}}{1 \cancel{\text{day}}} \times \frac{1 \cancel{\text{day}}}{24 \cancel{\text{h}}} \times \frac{1 \cancel{\text{h}}}{60 \text{ min}} \times \frac{60 \text{ gtt}}{1 \cancel{\text{mL}}}$$

$$= \frac{1200 \times 60 \text{ gtt}}{24 \times 60 \text{ min}} = \frac{50 \text{ gtt}}{1 \text{ min}}$$

The flow rate is 50 gtt/min. Recall from Chapter 10 that for microdrip tubing, $\frac{\text{mL}}{\text{h}} = \frac{\text{gtt}}{\text{min}}$. In this example, the patient should receive

$$\frac{1200 \text{ mL}}{24 \text{ h}} = \frac{50 \text{ mL}}{1 \text{ h}} = \frac{50 \text{ gtt}}{1 \text{ min}}$$

Example 3 Find the DMFN for a patient who weighs 154 lb. What would be the IV flow rate F mL/h?

First convert 154 lb to kilograms.

$$154 \cancel{\text{lb}} \times \frac{1 \text{ kg}}{2.2 \cancel{\text{lb}}} = 70 \text{ kg}$$

The patient weighs over 20 kg.

$$1500 + \left[\frac{20 \text{ mL}}{1 \not{\text{kg}}} \times (70 \not{\text{kg}} - 20)\right] = \text{DMFN mL}$$

$$2500 \text{ mL} = \text{DMFN}$$

$$\frac{2500 \text{ mL}}{24 \text{ h}} = \frac{104.1 \text{ mL}}{1 \text{ h}} = 104 \text{ mL/h} \quad \text{rounded to nearest whole number}$$

If this patient took 1300 mL of fluids by mouth, what would be the flow rate for the IV in milliliters per hour?

$$2500 - 1300 = 1200 \text{ mL IV fluids needed in 24 h}$$

$$\frac{1200 \text{ mL}}{24 \text{ h}} = \frac{50 \text{ mL}}{1 \text{ h}} = 50 \text{ mL/h}$$

GO TO . . . Open the CD-ROM that accompanies your textbook, and select Chapter 11, Rule 11-6. Review the animation and example problems, then complete the practice problems. Continue to the next section of the book once you have mastered the rule presented. ■

Rule 11-7 For pediatric and critically ill patients, the amount of solution in the IV tubing must be considered when you determine infusion times and volumes.

Five feet of standard IV tubing contains about 10 mL of solution. If this tubing is used along with a volume control chamber, a child will not begin receiving medication until the 10 mL of solution already in the tubing has infused. Low-volume (small-diameter) tubing contains only 0.3 mL of solution per 5 f and effectively eliminates this problem. Additionally most medical facilities will use electronic flow regulators or infusion pumps to ensure accuracy of medications delivered.

GO TO . . . Open the CD-ROM that accompanies your textbook, and select Chapter 11, Rule 11-7. Review the animation and example problems, then complete the practice problems. Continue to the next section of the book once you have mastered the rule presented. ■

Creatinine Clearance

Liver and kidney functions are often reduced in geriatric patients. Decreased liver function results in slower metabolism of certain drugs, delaying or prolonging the desired effect of the medication. It can also lead to a higher level of drug in the blood system, producing more intense results.

Decreased kidney function, along with decreased cardiac output, slows the excretion of medications from the body. Slower excretion (resulting from decreased kidney function) and a reduced metabolism (resulting from decreased liver function) combine their effects. Medication accumulates in the body, causing increased side effects or even toxicity.

Many chronic diseases common in the elderly can damage the kidneys. These diseases include hypertension, diabetes, and congestive heart failure. Also, some commonly used drugs, such as Lasix and aminoglycoside antibiotics, can further impair kidney function. Geriatric patients who have these diseases or are prescribed these medications must be monitored especially closely for their kidney functions.

Many package inserts discuss safe dosage levels based on creatinine clearance. Creatinine is a by-product found in the blood as a result of muscle metabolism. Creatinine clearance (CL_{CR}) is an indicator of the rate at which the kidneys filter the blood. Creatinine clearance often decreases with age because the elderly tend to have lower muscle mass, thus producing less creatinine in the blood. The filtration rate of creatinine by their kidneys is often slower. For many, this decrease is a normal part of the aging process. The creatinine clearance level of an elderly patient is often lower than the level of an average younger adult. For example, creatinine clearance is measured in mL/min and normal values usually drop by 6 mL/min for every 10 years past the age of 20.

If a patient has decreased kidney function, then the amount of creatinine excreted through the urine will decrease. At the same time, the amount of creatinine in the blood (or serum creatinine level) may increase.

The creatinine clearance level is calculated by using information that comes from an analysis of blood and urine samples. You may not know this information for every patient. The physician will usually factor in a patient's creatinine clearance when preparing a drug order. If you have any questions about administering a medication when creatinine clearance is a factor, speak with your supervisor or the physician who wrote the order.

Ideal and Actual Body Weight

A geriatric patient's body generally has a decreased proportion of lean body mass and water, along with an increased proportion of body fat. These proportions alter the distribution of drugs.

Some water-soluble drugs, such as aminoglycosides (antibiotics) and digitalis preparations (cardiac medications), are strongly bound to lean tissues. Because the elderly have less lean tissue, more of these water-soluble drugs remain in the circulating blood. Higher levels can lead to toxicity. Thus, serum drug levels (the level of drug dissolved in the blood) must be monitored.

Fat-soluble drugs are distributed to body fat. Because the elderly have a larger proportion of body fat, these drugs are distributed to more tissues. The drugs do not remain in the body fat, but are slowly released back into circulation. Thus, fat-soluble drugs have a longer duration of action, resulting in residual effects such as drowsiness.

For medications strongly bound to lean tissues (water-soluble), the dose for an overweight patient should be based on the *ideal body weight*. For patients whose weight is below ideal, the *actual* weight should be used. For medications strongly bound to body fat (fat-soluble), the dose is based on the *actual* weight.

Rule 11-8 Determining Safe Dosages for Geriatric Patients

- **Check the package insert or product literature. Double check the order to determine if it is a safe dose based upon renal function and the ideal or actual body weight of the patient.**
- **If the dose is safe, calculate the amount to administer.**

Example A 78-year-old male, 5 ft 4 in. tall, and weighing 180 lb, is given the following order. He has normal renal function and is being treated for a serious, but not life-threatening, infection.

Ordered: Garamycin® 70 mg IM q8h

On hand: Garamycin® Injectable, 40 mg/mL. According to the package insert, for patients with normal renal function, the usual dosage for serious infections is 1 mg/kg q8h. The dosage for obese patients should be based on lean body mass. For a 5 ft 4 in. patient, the ideal weight range is 122 to 157 lb.

Because 122 lb = 55 kg and 157 lb = 71 kg, after rounding, the safe dose for this patient is from 55 to 71 mg q8h. The dosage ordered, 70 mg, falls within that range and is safe. The amount to administer per dose is

$$70 \text{ mg} \times \frac{1 \text{ mL}}{40 \text{ mg}} = 1.75 \text{ mL} \quad \text{rounded to 1.8 mL}$$

GO TO . . . Open the CD-ROM that accompanies your textbook, and select Chapter 11, Rule 11-8. Review the animation and example problems, then complete the practice problems. Continue to the next section of the book once you have mastered the rule presented.

Critical Thinking on the Job

Consulting the Physician

While transcribing orders for Mrs. Bekins, who is 83 years old and weighs 118 lb, Karen notes that one of Mrs. Bekins' diagnoses is chronic renal failure. Mrs. Bekins has been given the following drug order: Tazidime 1 g IV q8h for pneumonia.

Karen knows that safe doses of antibiotics are often lower for patients with kidney disease than usual prescribed doses. From the package insert, she knows the recommended maintenance dose for Tazidime should be adjusted based on CL_{CR}. According to a table in the insert, for creatinine clearance levels of 31 to 50 mL/min, the recommended dosage is 1 g q12h; for levels of 16 to 30 mL/min, the dosage is 1 g q24h; for levels 6 to 15 mL/min, the dosage is 500 mg q24h; and for levels $<$ 5 mL/min, the dosage is 500 mg q48h.

Karen is able to determine that the patient's creatinine clearance level is 9.5 mL/min. This value is between 6 and 15 mL/min. Therefore, the safe dose of Tazidime for the patient is 500 mg q24h. This amount is considerably less than the one indicated by the drug order.

Questions?

What should Karen do? By using critical thinking skills, what patient problem was avoided?

ERROR ALERT!

For medications that are strongly bound to lean body tissue, calculate an overweight patient's dose on ideal body weight, not actual weight.

Suppose a 75-year-old female, 5 ft 1 in. tall, 190 lb. with a CL_{CR} is 30 mL/min is prescribed an initial daily dose of 0.25 mg of Lanoxin® injection.

According to the package insert, the level of Lanoxin® is based on the patient's creatinine clearance and lean body weight, not actual body weight. The patient's safe dose is 125 mcg/day (0.125 mg/day), one-half the amount prescribed. By getting too much medication, the patient could suffer digoxin toxicity. The physician who initially ordered the Lanoxin® makes the first error. Still, the health care professional who administers the Lanoxin® should check the safety of the amount and verify the order with the physician before administration.

Polypharmacy

Many geriatric patients take several medications. They often have more than one physician: an internist who is their primary physician and one or more specialists who treat very specific diseases and ailments. **Polypharmacy** refers to the practice of taking many medications at a time.

Many patients take over a dozen prescription medications each year and numerous over-the-counter medications or natural supplements. They may use medications that were initially prescribed years earlier and are past their expiration date. They may borrow medications. Because of financial pressures, they may also look for ways to limit physician costs by using older medications instead of visiting the physician. They may also order medications by mail or through the Internet, without having direct contact with the pharmacist. In addition, some medications may be prescribed without consideration of their interaction with other medications the patient is already taking.

Drug Interactions

Each additional medication a patient takes increases the likelihood of a drug interaction. These interactions can interfere with the effectiveness of one or more medications. They can also cause serious or even fatal side effects.

Rule 11-9 To identify cases of polypharmacy and reduce the risk of drug interactions, ask elderly patients about:

1. All medications they take which are prescribed by either their primary physician or specialists.
2. Any over-the-counter medications they take, including vitamins, laxatives, and allergy medications.
3. Any social drugs which they use, including alcohol, tobacco, and marijuana.
4. Medications that they borrow from family and friends.
5. Herbal and home remedies that they use, including natural supplements such as ginseng, gingko biloba, and St. John's Wort.
6. Bringing all the medications they take to be checked. This includes prescriptions, over-the-counter drugs, vitamins, minerals, and herbals.

GO TO . . . Open the CD-ROM that accompanies your textbook, and select Chapter 11, Rule 11-9. Review the animation and example problems, then complete the practice problems. Continue to the next section of the book once you have mastered the rule presented. ■

Adverse drug reactions can be caused by a variety of factors. These include advanced age, small body size, multiple illnesses (including chronic problems), multiple medications, living alone (patients with failing memories or mental capacities), and malnutrition. Elderly patients often take more than one medication to treat the same problem. Sometimes they have neglected to inform a new physician about medications prescribed by other physicians.

Sometimes multiple medications are needed to bring a problem under control. The patient may then continue to take all the medications, even though only one or two are still needed. This overuse is especially common with patients being treated for high blood pressure, constipation, or behavioral problems that occur with dementia.

Health care providers should periodically review with their elderly patients the list of medications the patients are taking. They should look especially for medications that are no longer needed as well as for multiple medications being used to treat the same condition.

Certain medications should be avoided by patients with specific diseases. Table 11-5 provides a list of some of these medications.

TABLE 11-5 Drugs to Be Avoided in Specific Diseases

These drugs are likely to cause significant adverse effects in elderly patients with the diseases noted.

Severe Risk	Drugs	Less Severe Risk	Drugs
Benign prostatic hypertrophy	Antihistamines, anti-Parkinson's drugs, GI antispasmodics, antidepressants	Benign prostatic hypertrophy	Narcotis
Cardiac dysrhythmia	Tricyclic antidepressants	Constipation	Antihistamines, anti-Parkinson's drugs, GI antispasmodics, antidepressants
Clotting disorders	Antiplatelet drugs, ASA (aspirin)	Diabetes mellitus	Steroids, beta blockers
COPD	Hypnotics, sedatives, beta blockers	GI diseases	Aspirin, potassium supplements
GI diseases	NSAIDs, ASA (aspirin)	Insomnia	Decongestants, bronchodilators, some antidepressants
Seizures	Metoclopramide (Reglan)	Seizures	Antipsychotics

REVIEW AND PRACTICE

11-3 Age-Related Needs

In Exercises 1–5, calculate the daily maintenance fluid needs, based on the following weights.

1. 8 kg **2.** 33 kg **3.** 37 lb **4.** 58 lb **5.** 121 lb

In Exercises 6–10, find the microdrip tubing flow rate for DMFN for patients, based on the following weights.

6. 21 kg **7.** 15 kg **8.** 17 lb **9.** 41 lb **10.** 165 lb

In Exercises 11 and 12, determine the recommended IV flow rate.

11. A patient who weighs 180 lb had an oral intake of 1000 mL. What should be the flow rate of his IV per hour to maintain his fluids?

12. A patient weighs 31 kg and has an oral intake of 200 mL. What would be the flow rate of his IV per hour to maintain fluids?

In Exercises 13–16, determine if the dosage ordered is safe.

13. The patient: 92-year-old female, 5 ft 6 in. tall, 130 lb, and CL_{CR} of 61 mL/min. Patient is in ideal weight range.

Ordered: Amikacin 375 mg IM q12h

According to the package insert, patients with normal renal function may be administered 7.5 mg/kg q12h or 5 mg/kg q8h. This patient has normal renal function.

14. The patient: 76-year-old female, 5 ft 2 in. tall, 126 lb, and CL_{CR} of 50 mL/min. Patient is in ideal weight range.

Ordered: Tazidime 1 g IV q12h

According to the package insert, for creatinine clearance levels of 31 to 50 mL/min, the recommended dosage is 1 g q12h; for levels of 16 to 30 mL/min, the dosage is 1 g q24h; for levels of 6 to 15 mL/min, the dosage is 500 mg q24h; and for levels < 5 mL/min, the dosage is 500 mg q48h.

15. The patient: 68-year-old male, 5 ft 7 in. tall, 188 lb, CL_{CR} of 60 mL/min, and impaired renal function. Ideal weight should be 172 lb.

Ordered: Vancocin® HCI 150 mg IV q6h

According to the package insert, the daily dosage for patients with normal renal function is 2 g divided into doses q6h or q12h. The daily dosage for patients with impaired renal function is 1545 mg/24 h for creatinine clearance of 100 mL/min; 1390 mg/24 h for 90 mL/min; 1235 mg/24 h for 80 mL/min; 1080 mg/24 h for 70 mL/min; 925 mg/24 h for 60 mL/min; 770 mg/24 h for 50 mL/min; 620 mg/24 h for 40 mL/min; 425 mg/24 h for 30 mL/min; 310 mg/24 h for 20 mL/min; and 155 mg/24 h for 10 mL/min.

16. The patient: 79-year-old female, 5 ft tall, 110 lb, CL_{CR} of 90 mL/min, and normal renal function. Patient is within ideal weight range.

Ordered: Vancocin® HCI 0.5 g IV q6h

See Exercise 15 above for information about the recommended daily dosage.

In Exercises 17 and 18, determine if the dosage ordered is safe. Then find the amount to administer.

17. The patient: 75-year-old female, 5 ft 3 in. tall, 198 lb, CL_{CR} of 56 mL/min, diagnosed with hypertension and renal impairment. Ideal weight should be 152 lb.

Ordered: Vasotec® 2.5 mg po qd

On hand: Vasotec® 5-mg scored tablets

According to the package insert, the usual dose for patients with normal renal function (over 80 mL/min creatinine clearance) is 5 mg/day; for mild impairment (over 30 and up to 80 mL/min), 5 mg/day; for moderate to severe impairment (30 or less mL/min), 2.5 mg/day.

18. The patient: 81-year-old male, 5 ft tall, 138 lb, CL_{CR} of 63 mL/min, and renal impairment. Patient is within ideal weight range.

Ordered: Ticarcillin 2 g IV q4h

On hand: Ticar 1-g vial, 200 mg/mL when reconstituted

According to the package insert, the usual dose, after the initial loading dose, for patients with infections complicated by renal insufficiency, is 3 g q4h with creatinine clearance over 60 mL/min; 2 g q4h for 30 to 60 mL/min; 2 g q8h for 10 to 30 mL/min; 2 g q12h for less than 10 mL/min; other amounts for patients with complications.

To check your answers, go to page 534.

HOMEWORK ASSIGNMENT

Answer the following questions.

1. List and explain the four processes in the body that affect a drug after it is administered.
2. Name two special populations of patients who require extra consideration when calculating medication dosages.
3. List four age-related factors that may affect the dosage of a medication for a pediatric patient.
4. List four age-related factors that may affect the dosage of a medication for a geriatric patient.
5. What is the maximum volume of an IM injection for an infant, a toddler, a preschooler, and an elementary school age child?
6. Explain the term "recommended dosage range."
7. Body surface area (BSA) uses what two body measurements to provide a more accurate dosage?
8. Explain the difference between daily maintenance fluids and replacement fluids.
9. What type of medication is strongly bound to lean tissue?
10. When would the dose of a medication be calculated on ideal body weight rather than actual weight?
11. Explain why a medication dosage may be altered based on the patient's result of a creatine clearance test.
12. Define the term polypharmacy and explain how it would increase the risk of drug interactions in a geriatric patient.

Use the identified drug labels and package insert information to answer the following questions:

13. What is the safe initial dosage range for a 4-year-old child weighing 41.8 pounds? See label A and package insert information.
14. If the order was for 850 mg of this drug, what amount would you administer? See label A and package insert information.

Depakene® package insert information:

PO (children) initial dose of 15-45 mg/kg/day

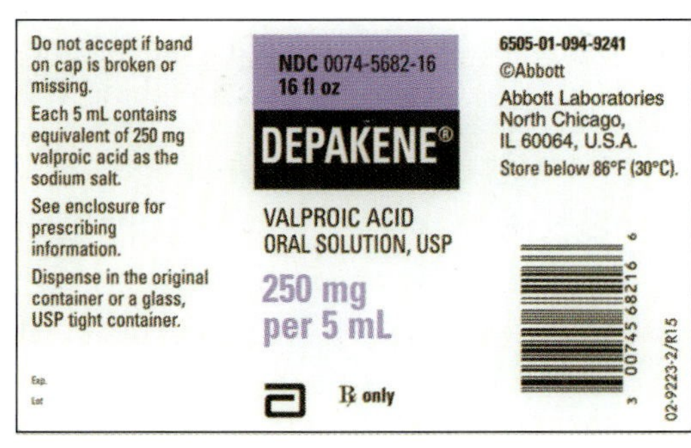

Label A

15. Calculate the correct dosage and amount to administer for a 7-year-old child weighing 61.6 pounds. See label B and package insert information.

Kytril® package insert information:

IV (adults & children 2-16 yr.) 10 mcg/kg within 30 min. prior to chemotherapy

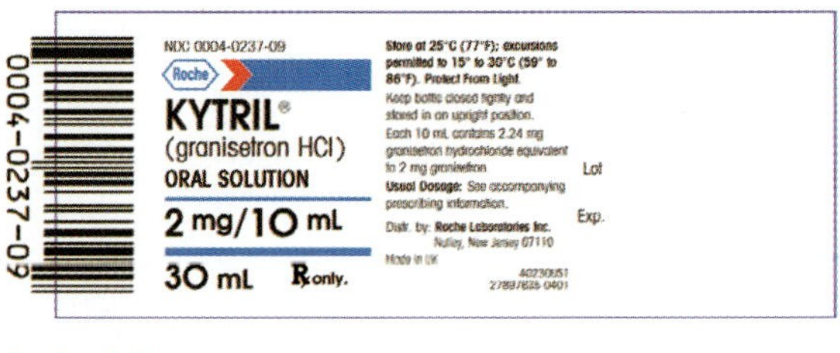

Label B

16. Calculate the safe IM dosage range for a 3-week-old infant weighing 6 pounds 12 ounces. See label C and package insert information.

17. Calculate the safe IM dosage range for a 3-year-old weighing 33 pounds. See label C and package insert information.

Clindamycin package insert information:

IM, IV (infants <1 month)
3.75-5 mg/kg every 6 hours
IM, IV (children >1 month)
5-13.3 mg/kg every 8 hours

NDC 63323-282-04 28204
CLINDAMYCIN
INJECTION, USP
600 mg/4 mL
(150 mg/mL)*
For IM or IV Use
DILUTE BEFORE IV USE
4 mL Single Dose Vial
Rx only
Abraxis Pharmaceutical Products
Schaumburg, IL 60173
LOT/EXP

Label C

18. Calculate the safe dosage range for a 66-year-old woman weighing 110 pounds. See label D and package insert information.

19. If the order was for 15 grams of this drug, what amount would you administer? See label D and package insert information.

Gammagard Liquid package insert information:

Monthly doses of approximately 300 – 600 mg/kg infused at 3 to 4 week intervals are commonly used.

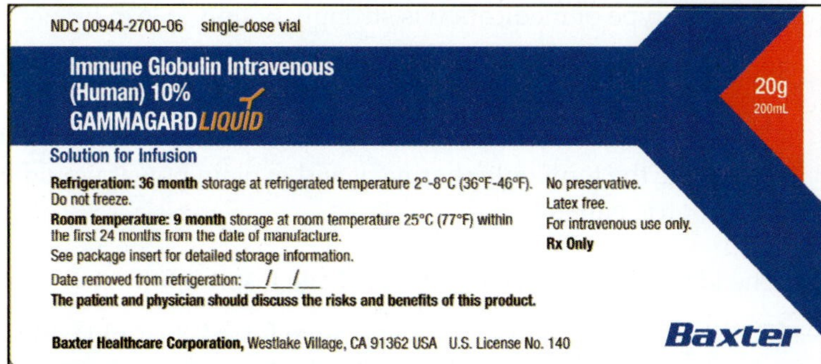

NDC 00944-2700-06 single-dose vial
Immune Globulin Intravenous
(Human) 10%
GAMMAGARD LIQUID
20g
200mL
Solution for Infusion
Refrigeration: 36 month storage at refrigerated temperature 2°-8°C (36°F-46°F). Do not freeze.
Room temperature: 9 month storage at room temperature 25°C (77°F) within the first 24 months from the date of manufacture.
See package insert for detailed storage information.
Date removed from refrigeration: __/__/__
The patient and physician should discuss the risks and benefits of this product.
No preservative.
Latex free.
For intravenous use only.
Rx Only
Baxter Healthcare Corporation, Westlake Village, CA 91362 USA U.S. License No. 140
Baxter

Label D

20. Using the blank nomogram, label E and the package insert information, calculate the correct dosage for a 5′ 8″ tall, 70-year-old man weighing 170 pounds.

NOMOGRAM

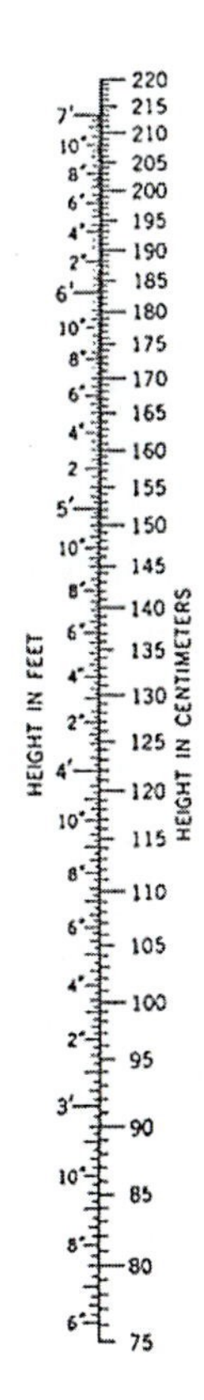

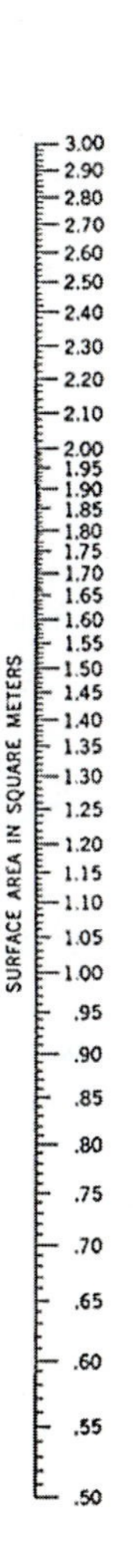

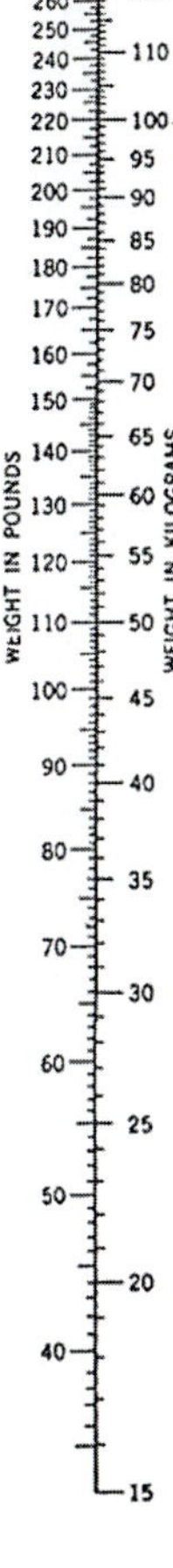

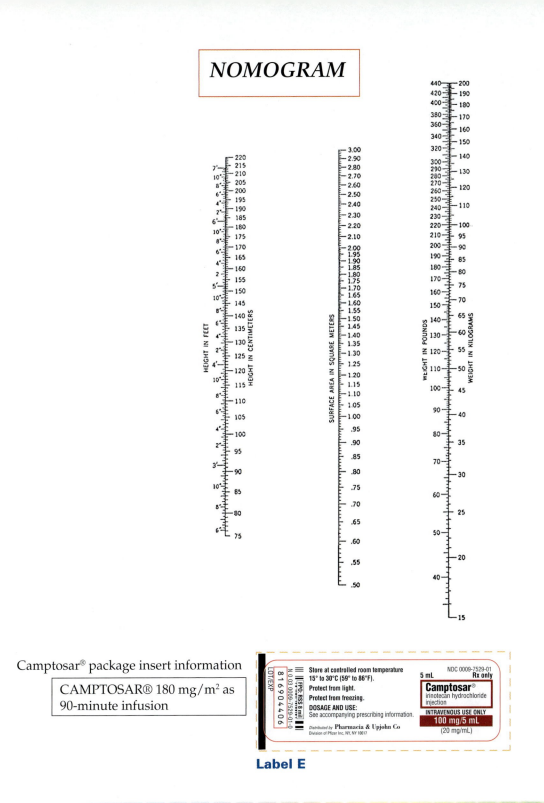

Camptosar® package insert information

CAMPTOSAR® 180 mg/m² as 90-minute infusion

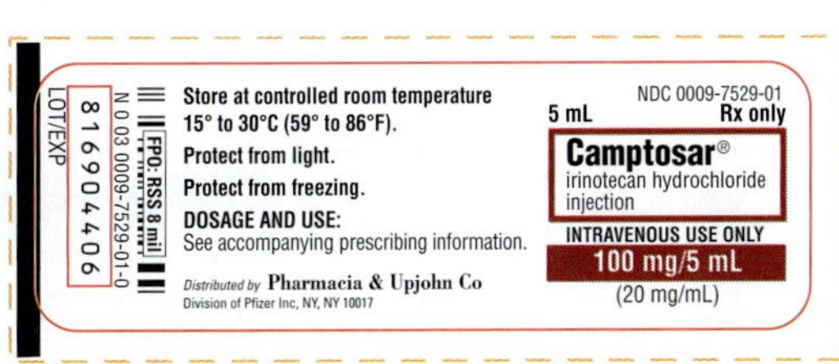

Store at controlled room temperature 15° to 30°C (59° to 86°F).
Protect from light.
Protect from freezing.
DOSAGE AND USE:
See accompanying prescribing information.
Distributed by Pharmacia & Upjohn Co
Division of Pfizer Inc, NY, NY 10017
NDC 0009-7529-01
5 mL
Rx only
Camptosar®
irinotecan hydrochloride injection
INTRAVENOUS USE ONLY
100 mg/5 mL
(20 mg/mL)
N 0 03 0009-7529-01-0
LOT/EXP

Label E

Chapter 11 Review

Check Up

In Exercises 1–4, convert the following weights to kilograms.

1. 49 lb

2. 61 lb

3. 6 lb 9 oz

4. 12 lb 13 oz

In Exercises 5–8, calculate the BSA for patients with the following heights and weights.

5. 105 cm and 19 kg

6. 74 cm and 12.1 kg

7. 41 in and 33 lb

8. 30 in and 23 lb

In Exercises 9–15, determine if the order is safe. If it is, then determine the amount to administer.

9. The child weighs 30 lb.
Ordered: Depakene® syrup 100 mg po q12h
On hand: Depakene® syrup 250 mg/5 mL. According to the package insert, the initial daily dose for pediatric patients is 15 mg/kg/day.

10. The patient is a 4-year-old child who weighs 16 kg.
Ordered: Ventolin syrup 1.6 mg po tid
On hand: Ventolin syrup 2 mg/5 mL. According to the package insert, for children from 2 to 6 years of age, dosing should be initiated at 0.1 mg/kg of body weight 3 times a day. This starting dosage should not exceed 2 mg 3 times a day.

11. The patient is 72 cm tall and weighs 16 kg.
Ordered: Oncaspar® 1300 IU IM every 14 days
On hand: Oncaspar® 5 mL/vial, 750 IU/mL. The recommended pediatric dosage is 2500 IU/m^2 for children whose BSA is greater than or equal to 0.6 m^2 and 82.5 IU/kg for children whose BSA is less than 0.6 m^2.

12. The child weighs 31 kg.
Ordered: Biaxin® susp 225 mg po q12h × 10
On hand: Refer to label A. According to the package insert, the usual recommended daily dosage for children is 15 mg/kg/day for 10 days.

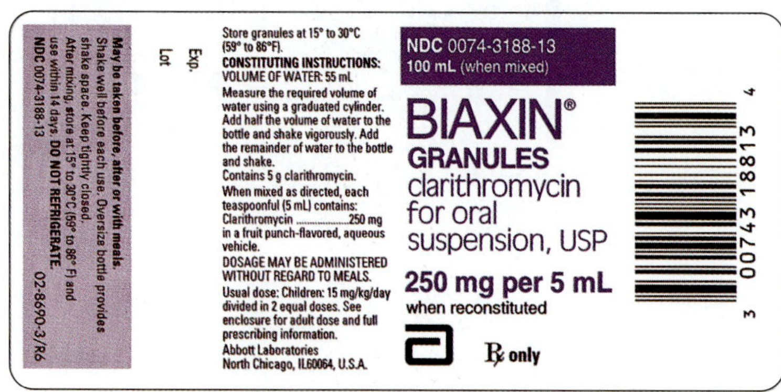

A

NOMOGRAM

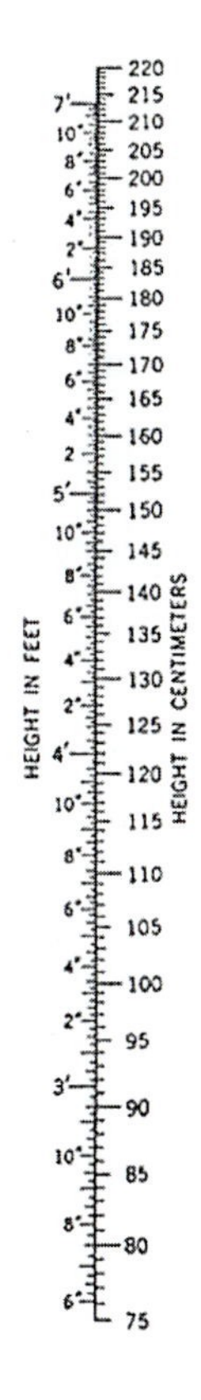

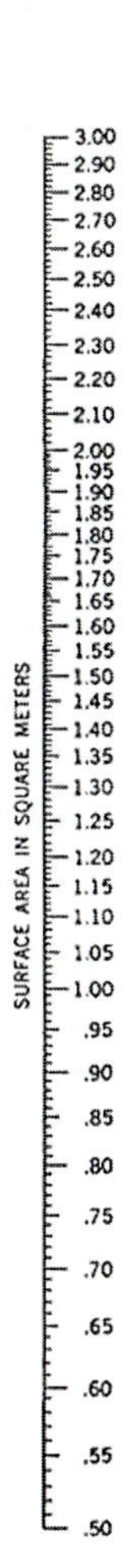

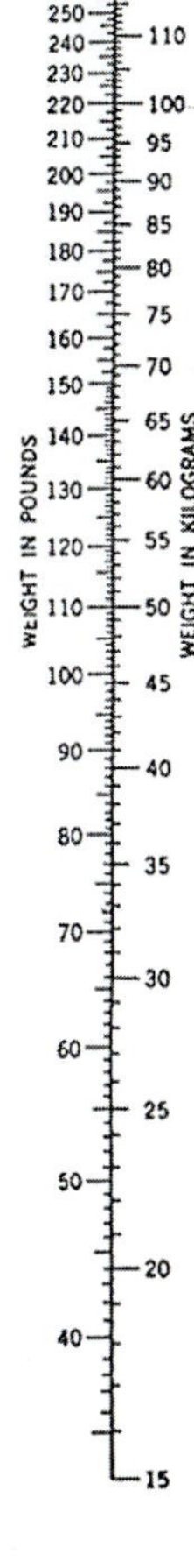

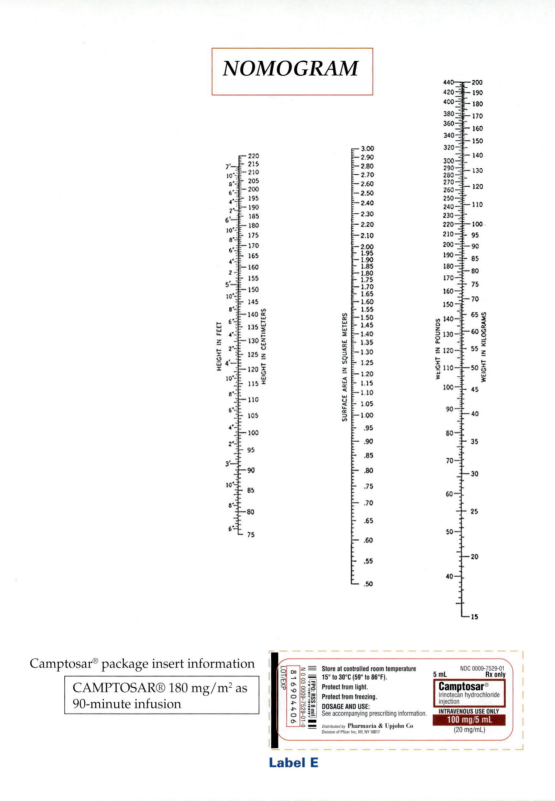

Camptosar® package insert information

CAMPTOSAR® 180 mg/m² as 90-minute infusion

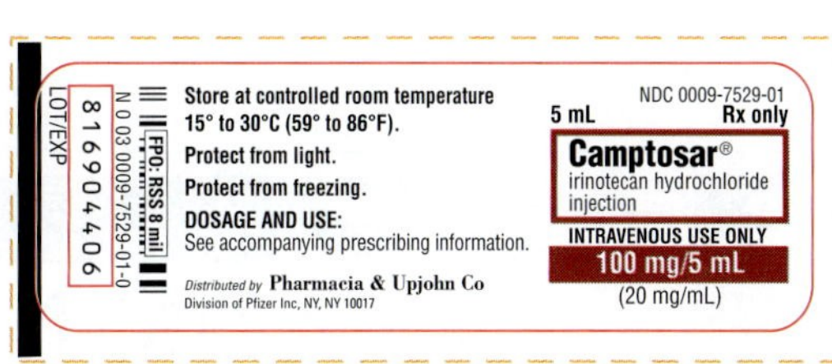

Label E

Chapter 11 Review

Check Up

In Exercises 1–4, convert the following weights to kilograms.

1. 49 lb **2.** 61 lb **3.** 6 lb 9 oz **4.** 12 lb 13 oz

In Exercises 5–8, calculate the BSA for patients with the following heights and weights.

5. 105 cm and 19 kg **6.** 74 cm and 12.1 kg **7.** 41 in and 33 lb **8.** 30 in and 23 lb

In Exercises 9–15, determine if the order is safe. If it is, then determine the amount to administer.

9. The child weighs 30 lb.
Ordered: Depakene® syrup 100 mg po q12h
On hand: Depakene® syrup 250 mg/5 mL. According to the package insert, the initial daily dose for pediatric patients is 15 mg/kg/day.

10. The patient is a 4-year-old child who weighs 16 kg.
Ordered: Ventolin syrup 1.6 mg po tid
On hand: Ventolin syrup 2 mg/5 mL. According to the package insert, for children from 2 to 6 years of age, dosing should be initiated at 0.1 mg/kg of body weight 3 times a day. This starting dosage should not exceed 2 mg 3 times a day.

11. The patient is 72 cm tall and weighs 16 kg.
Ordered: Oncaspar® 1300 IU IM every 14 days
On hand: Oncaspar® 5 mL/vial, 750 IU/mL. The recommended pediatric dosage is 2500 IU/m^2 for children whose BSA is greater than or equal to 0.6 m^2 and 82.5 IU/kg for children whose BSA is less than 0.6 m^2.

12. The child weighs 31 kg.
Ordered: Biaxin® susp 225 mg po q12h × 10
On hand: Refer to label A. According to the package insert, the usual recommended daily dosage for children is 15 mg/kg/day for 10 days.

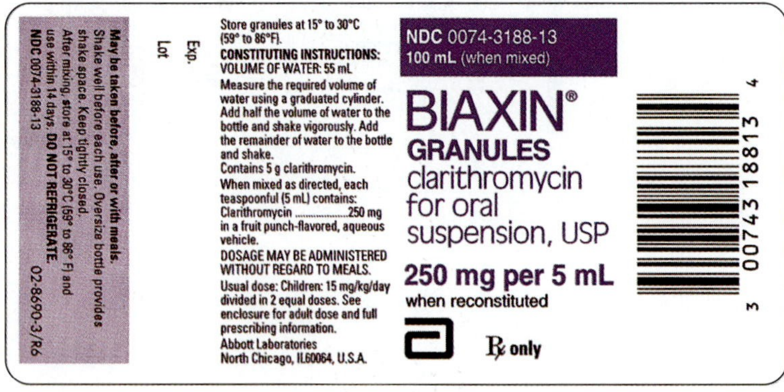

May be taken before, after or with meals.
Shake well before each use. Oversize bottle provides shake space. Keep tightly closed.
After mixing, store at 15° to 30°C (59° to 86° F) and use within 14 days. DO NOT REFRIGERATE.
NDC 0074-3188-13
02-8690-3/R6
Exp.
Lot
Store granules at 15° to 30°C (59° to 86°F).
CONSTITUTING INSTRUCTIONS:
VOLUME OF WATER: 55 mL
Measure the required volume of water using a graduated cylinder. Add half the volume of water to the bottle and shake vigorously. Add the remainder of water to the bottle and shake.
Contains 5 g clarithromycin.
When mixed as directed, each teaspoonful (5 mL) contains:
Clarithromycin ....250 mg
in a fruit punch-flavored, aqueous vehicle.
DOSAGE MAY BE ADMINISTERED WITHOUT REGARD TO MEALS.
Usual dose: Children: 15 mg/kg/day divided in 2 equal doses. See enclosure for adult dose and full prescribing information.
Abbott Laboratories
North Chicago, IL60064, U.S.A.
NDC 0074-3188-13
100 mL (when mixed)
BIAXIN®
GRANULES
clarithromycin
for oral
suspension, USP
250 mg per 5 mL
when reconstituted
℞ only
3 00743 18813 4

A

13. The child weighs 66 lb.

Ordered: Oxcarbazepine 150 mg po BID

On hand: Refer to label B. According to the package insert. Treatment should be initiated at a daily dose of 8–10 mg/kg generally not to exceed 600 mg/day, given in a BID regimen. The target maintenance dose of Trileptal® should be achieved over 2 weeks and is dependent upon patient weight, according to the following chart:

20–29 kg	900 mg/day
29.1–39 kg	1200 mg/day
.39 kg	1800 mg/day

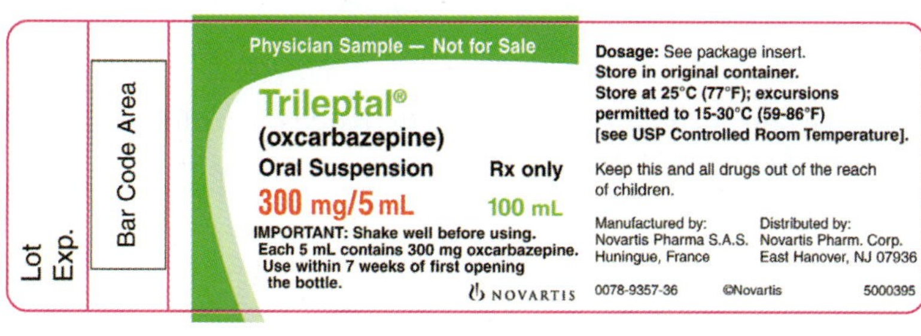

Lot
Exp.
Bar Code Area
Physician Sample — Not for Sale
Trileptal®
(oxcarbazepine)
Oral Suspension
Rx only
300 mg/5 mL
100 mL
IMPORTANT: Shake well before using.
Each 5 mL contains 300 mg oxcarbazepine.
Use within 7 weeks of first opening the bottle.
NOVARTIS
Dosage: See package insert.
Store in original container.
Store at 25°C (77°F); excursions permitted to 15-30°C (59-86°F)
[see USP Controlled Room Temperature].
Keep this and all drugs out of the reach of children.
Manufactured by:
Novartis Pharma S.A.S.
Huningue, France
Distributed by:
Novartis Pharm. Corp.
East Hanover, NJ 07936
0078-9357-36
©Novartis
5000395

B

14. The patient is a 58-lb child with severe infection.

Ordered: Erythromycin 650 mg PO tid.

On hand: Refer to label C. According to the package insert, in mild to moderate infections the usual dosage of erythromycin ethylsuccinate for children is 30 to 50 mg/kg/day in equally divided doses every 6 hours. For more severe infections this dosage may be doubled. If twice-a-day dosage is desired, one-half of the total daily dose may be given every 12 hours. Doses may also be given three times daily by administering one-third of the total daily dose every 8 hours.

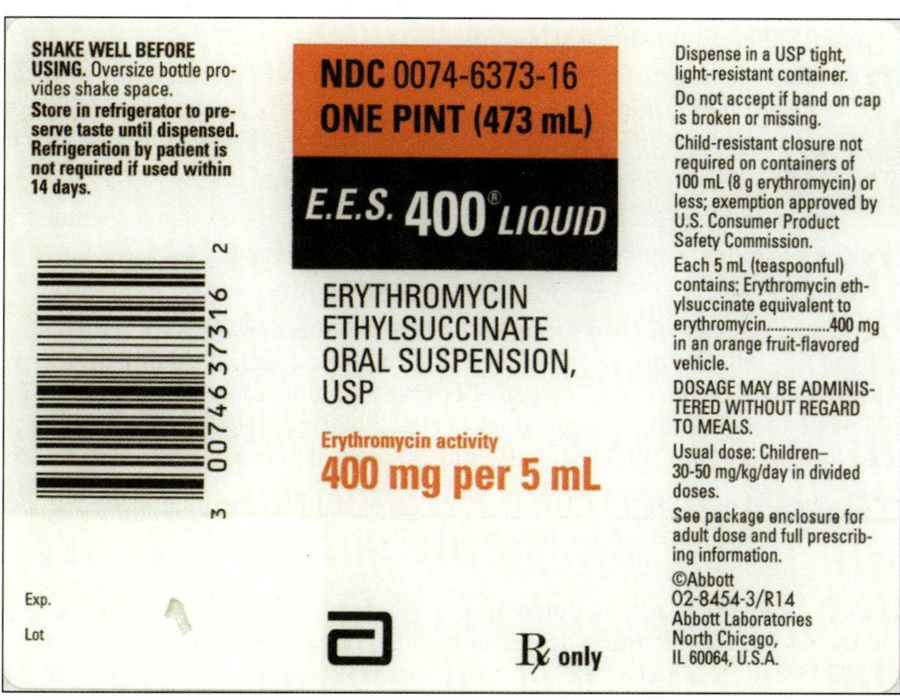

C

15. The patient is a 25-kg child receiving myelosuppressive chemotherapy.

Ordered: Neupogen® 125 mcg IVPB over 30 min

On hand: Refer to label D. According to the package insert, for patients receiving myelosuppressive chemotherapy, the recommended daily starting dose is 5 mcg/kg/day, administered as a single daily injection by Sub-Q bolus injection, by short IV infusion (15 to 30 min), or by continuous Sub-Q or continuous IV infusion.

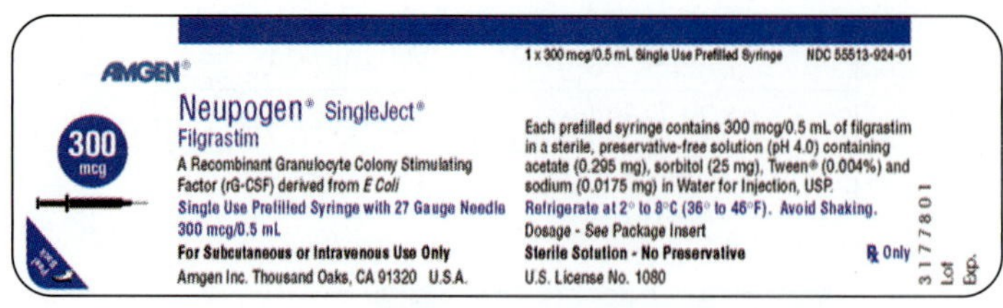

D

In Exercises 16–18, calculate the daily maintenance fluid needs, based on the following weights. Then find the flow rate in milliliters per hour (F) for the DMFN.

16. 24 kg

17. 39 lb

18. 110 lb (The patient has taken 800 mL fluid orally.)

In Exercises 19–24, determine if the dosage ordered is safe. If the order is safe, then find the amount to administer. Assume that the patients have impaired renal functions.

19. The patient: 85-year-old male, 6 ft 1 in. tall, 210 lb, CL_{CR} of 64 mL/min. Ideal weight should be 195 lb.

Ordered: Cartrol 2.5 mg po q24h

On hand: Cartrol 2.5 mg/tablet

According to the package insert, the usual dosage interval for 2.5 mg is as follows: for patients with creatinine clearance above 60 mL/min, 24 h; for 20 to 60 mL/min, 48 h; and for less than 20 mL/min, 72 h.

20. The patient: 68-year-old female, 5 ft 5 in. tall, 166 lb, CL_{CR} of 60 mL/min. Ideal weight should be 162 lb.

Ordered: Capastat 500 mg IM qd

On hand: Capastat sulfate, diluted to 300 mg/mL

According to the package insert, the estimated daily dosage required to maintain a steady level of drug is 1.29 mg/kg for creatinine clearance of 0 mL/min; 2.43 mg/kg for 10 mL/min; 3.58 mg/kg for 20 mL/min; 4.72 mg/kg for 30 mL/min; 5.87 mg/kg for 40 mL/min; 7.01 mg/kg for 50 mL/min; and 8.16 mg/kg for 60 mL/min.

21. The patient: 82-year-old female, 4 ft 10 in. tall, 102 lb, CL_{CR} of 26 mL/min. Patient is within ideal weight range.

Ordered: Acyclovir sodium (Zovirax®) 450 mg IV q12h infused over 1 h

On hand: Zovirax® for injection, 50 mg/mL when reconstituted

According to the package insert, the recommended dose for this diagnosis for patients with normal renal function is 10 mg/kg q8h. The dose is adjusted as follows for patients with impaired renal function: for creatinine clearance over 50 mL/min, 100 percent of the recommended dose every 8 h; from 25 to 50 mL/min, 100 percent of the recommended dose every 12 h; from 10 to 25 mL/min, 100 percent of the recommended dose every 24 h; for 0 to 10 mL/min, 50 percent of the recommended dose every 24 h.

22. The patient: 73-year-old male, 5 ft 8 in. tall, 154 lb, CL_{CR} of 49 mL/min, diagnosed with a complicated urinary tract infection. Patient is within ideal weight range.

Ordered: Fortaz® 1 g IV qd

On hand: Fortaz® for injection, reconstituted at 10 mg/mL

According to the package insert, the usual recommended dosage for patients with complicated urinary tract infections is 500 mg to 1 g given q8–12h. For patients with renal insufficiency, the following maintenance dosages are recommended (however, if the usual dosage is less, administer the lower amount): for creatinine clearance of 31 to 50 mL/min, 1 g q12h; for 16 to 30 mL/min, 1 g q24h; for 6 to 15 mL/min, 500 mg q24h; for less than 5 mL/min, 500 mg q48h.

23. The patient: 79-year-old male, 5 ft 9 in. tall, 149 lb, CL_{CR} of 55 mL/min. The patient does not have a life-threatening infection. Patient is within ideal weight range.

Ordered: Mandol 2 g IV q6h

On hand: Mandol reconstituted to 1 g/10 mL

According to the package insert, for patients with renal impairment and less severe infections, the following maintenance dosages are recommended: for creatinine clearance of over 80 mL/min, 1 to 2 g q6h; for 50 to 80 mL/min, 0.75 to 1.5 g q6h; for 25 to 50 mL/min, 0.75 to 1.5 g q8h; for 10 to 25 mL/min, 0.5 to 1 g q8h; for 2 to 10 mL/min, 0.5 to 0.75 g q12h; for less than 2 mL/min, 0.25 to 0.5 g q12h.

24. The patient: 92-year-old female, 5 ft 1 in. tall, 112 lb, CL_{CR} of 32 mL/min. Patient is within ideal weight range.

Ordered: Timentin® 2 g IV q4h

On hand: Timentin® reconstituted to 20 mg/mL

According to the package insert, for patients with renal impairment, the following maintenance dosages are recommended: for creatinine clearance over 60 mL/min, 3.1 g q4h; for 30 to 60 mL/min, 2 g q4h; for 10 to 30 mL/min, 2 g q8h; for less than 10 mL/min, 2 g q12h. For patients with more advanced impairments, lower dosages are recommended.

Critical Thinking Applications

You are preparing medication for two patients. Patient A has nosocomial pneumonia and a Creatinine Clearance of 37 mL/min. He has been ordered Zosyn® 4.5 gram q6h IV. Patient B has pelvic inflammatory disease and a normal creatinine clearance. She has been ordered 3.375 gram IV q6h. You have the following Zosyn® medication vials (see labels E, F, and G.) and package insert information.

1. Are the dosages for Patient A and B safe?
2. If safe, what medication will you use and how will you prepare the medication.
3. If not safe, what action should you take?
4. According to the package insert, you further dilute each medication with 150 mL of sterile water for injection. What would the IV flow rate be in mL per hour for administration of the medication to Patient A and B?

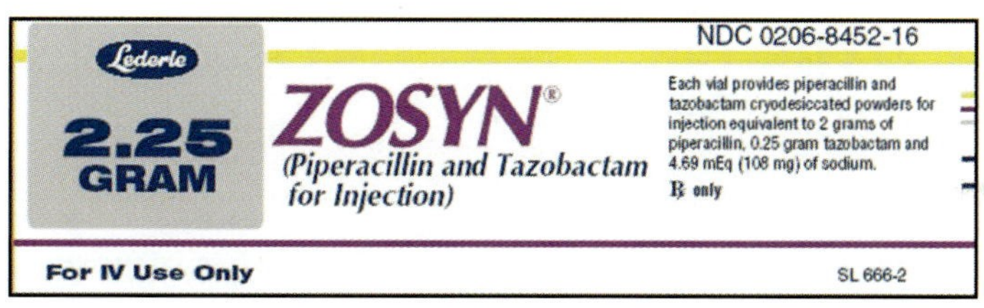

E

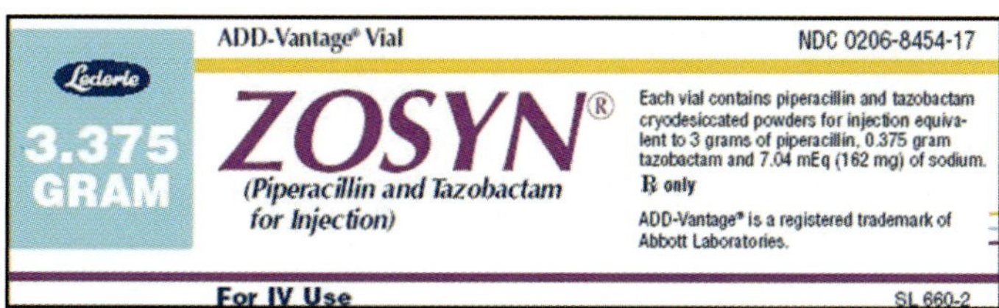

F

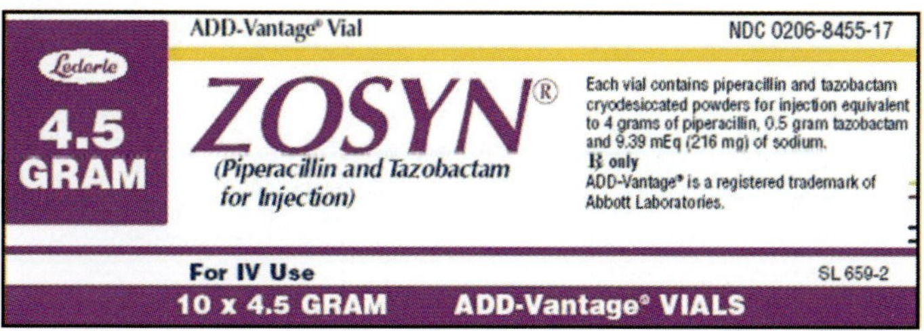

G

Directions for Reconstitution and Dilution for Use
Intravenous Administration

For conventional vials, reconstitute Zosyn® per gram of piperacillin with 5 mL of a compatible reconstitution diluent from the list provided below.

2.25 g, 3.375 g, and 4.5 g Zosyn® should be reconstituted with 10 mL, 15 mL, and 20 mL, respectively. Swirl until dissolved.

Pharmacy vials should be used immediately after reconstitution. Discard any unused portion after 24 hours if stored at room temperature [20°C to 25°C (68°F to 77°F)], or after 48 hours if stored at refrigerated temperature [2°C to 8°C (36°F to 46°F)].

Compatible Reconstitution Diluents

0.9% sodium chloride for injection
Sterile water for injection
Dextrose %
Bacteriostatic saline/parabens
Bacteriostatic water/parabens
Bacteriostatic saline/benzyl alcohol
Bacteriostatic water/benzyl alcohol
Reconstituted Zosyn® solution should be further diluted (recommended volume per dose of 50 mL to 150 mL) in a compatible intravenous diluent solution listed below. Administer by infusion over a period of at least 30 minutes. During the infusion it is desirable to discontinue the primary infusion solution.

Recommended Dosing of Zosyn in Patients with Normal Renal Function and Renal Insufficiency (As total grams piperacillin/tazobactam)

Renal Function (Creatinine Clearance, mL/min)	All Indications (except nosocomial pneumonia)	Nosocomial Pneumonia
> 40 mL/min	3.375 q 6 h	4.5 q 6 h
20-40 mL/min*	2.25 q 6 h	3.375 q 6 h
< 20 mL/min*	2.25 q 8 h	2.25 q 6 h
Hemodialysis**	2.25 q 12 h	2.25 q 8 h
CAPD	2.25 q 12 h	2.25 q 8 h

*Creatinine clearance for patients not receiving hemodialysis

** 0.75 g should be administered following each hemodialysis session on hemodialysis days

Case Study

You are working in a pediatric ICU where your patient is a 3-year-old boy who has a staphylococcal skin infection in the area surrounding his incision. His attending physician has prescribed the following treatment. You know from his chart that he weighs 40 lb.

Ordered: Rocephin® 600 mg IV over 30 min bid

On hand: Refer to label H. According to the package insert, for pediatric patients with skin infections, the recommended total daily dose is 50 to 75 mg/kg/day given once a day (or in equally divided doses twice a day). The total daily dose should not exceed 2 g. The Rocephin® is reconstituted to a dosage strength of 40 mg/mL.

1. What is the recommended range of doses for this patient?
2. Calculate the flow rate for the infusion pump used to deliver this order.
3. Is this order a safe dose for your patient? If it is not a safe dose, what steps should you take?

To check your answers, see page 536.

ROCHE LABORATORIES INC.
Nutley, New Jersey 07110

2 grams Single Use Vial
ROCEPHIN®
(ceftriaxone sodium)
For I.M. or I.V. Use
equivalent to 2 grams ceftriaxone
℞ only.
For I.M. Administration: Reconstitute with 4.2 mL 1% Lidocaine Hydrochloride Injection (USP) or Sterile Water for Injection (USP). Each 1 mL of solution contains approximately 350 mg equivalent of ceftriaxone.
For I.V. Administration: Reconstitute with 19.2 mL of an I.V. diluent specified in the accompanying package insert. Each 1 mL of solution contains approximately 100 mg equivalent of ceftriaxone. Withdraw entire contents and dilute to the desired concentration with the appropriate I.V. diluent.
USUAL DOSAGE: See package insert.
Storage Prior to Reconstitution: Store powder at room temperature 77° F (25° C) or below.
Protect From Light.
Storage After Reconstitution: See package insert.
LOT EXP.
26098278-0698

H

Figure 11-13

Internet Activity

When you are calculating dosage for special populations, you need to know the patient's actual weight. In some circumstances, it is also important to know the ideal weight for a patient based upon the patient's height. Various charts are available to determine if a patient is of ideal weight. Search the Internet and find a current and reliable height-to-weight ratio chart. Keep this chart handy when you are calculating for special populations.

GO TO . . . Open the CD-ROM that accompanies your textbook, and make a final review of the rules, practice problems, and activities presented for this chapter. For a final evaluation, take the chapter test and email or print your results for your instructor. A score of 95 percent or above indicates mastery of the chapter concepts.

12 Specialized Calculations

When on the brink of complete discouragement, success is discerning that the line between failure and success is so fine that often a single extra effort is all that is needed to bring victory out of defeat.

—Elbert Green Hubbard

Learning Outcomes

When you have completed Chapter 12, you will be able to:

- Measure insulin doses accurately.
- Measure heparin dosage and flow rates, verifying that they fall within the normal daily range.
- Calculate the hourly flow rate for IV infusions ordered in milligrams per minute.
- Calculate IV flow rates for medications ordered in mg/kg doses.
- Calculate IV flow rates for titrated medications.
- Determine the percentages of solutions, dilutions, and solids.
- Prepare solutions from a concentrate.

Key Terms

Alligation method
Anticoagulant
Dilution
Duration
Heparin
Insulin
Onset
Peak
qsad
Solute
Solvent
Titrated medications
U-100

Introduction

Depending upon your position in the health care field, you will be required to have additional knowledge related to dosage calculations. This chapter will introduce you to specialized calculations required of various health care professionals. Information relating to the use of insulin, heparin, critical care intravenous fluids, preparation of solutions, and alligations has been included. These calculations may take a little extra effort, but completing them successfully will help you step into your new career as a health care professional.

Insulin

Insulin is a pancreatic hormone that stimulates glucose metabolism. People who have low or no insulin production may have insulin-dependent diabetes. They often need regular injections of insulin to keep their glucose (blood sugar) from rising to levels that could be life-threatening. These regular injections must be rotated to various sites of the body to prevent scarring of the tissue at a single injection site.

Types of Insulin

Many types of injectable insulin exist. The oldest types are beef insulin and pork insulin, extracted from the pancreas of a cow or a pig. In the United States, animal-based insulin is being replaced by human insulin, including those types produced by using genetically engineered bacteria. A synthetic form, insulin lispro, is also available. Regular insulin or semi-lente is clear. Lente, NPH, and Ultralente insulins are cloudy.

Timing of Action

Insulins are classified by the timing of their action. Rapid-acting insulins include regular (R) and Semilente (S) insulins. Insulin lispro is an even more rapid-acting insulin that is generally administered 15 min before a meal. Other rapid-acting insulins are usually administered 30 to 60 min before eating. Intermediate-acting insulins include Lente (L), NPH (N), and Protamine Zinc (P). Ultralente (U) and Lantus® are long-acting insulins. Lantus® is a newer type of insulin that is known for maintaining a constant blood level, and having no peak.

For insulin to be effective, it must be given at a specific time relative to food intake. Therefore, before you administer insulin, you must know the onset, peak, and duration of action of each type. The **onset** is the time when the insulin begins to lower the glucose level. The **peak** is the time at which the insulin's effect is strongest. Both onset and peak are measured from the time the insulin is administered. The **duration** is the length of time the effect of the insulin lasts. It is measured from the time of onset.

For example, a regular dose of insulin administered at 0700 will begin to take effect after 30 min, at 0730 (the onset). Its peak will be 2.5 to 5 h after it is administered, between 0930 and 1200. Its effect will last until 1530 (the duration), about 8 h after the onset. Table 12-1 summarizes the action times of many types of insulin, including mixed insulins which are described later.

TABLE 12-1 Timing of Insulin Action

Insulin	Example	Onset	Peak	Duration
Rapid-acting	Humulin® R	30 min	2.5–5 h	8 h
	Novolin® R	30 min	2.5–5 h	8 h
	Velosulin® BR	30 min	1–3 h	8 h
	Humalog®	15 min	30–90 min	6–8 h
Intermediate	Novolin® L	2.5 h	7–15 h	22 h
	Humulin® N	1.5 h	4–12 h	24 h
	Novolin® N	1.5 h	4–12 h	24 h
Long-acting	Humulin® U	4–8 h	10–30 h	28 h
	Lantus®	1.5 h	None pronounced	24 h
Mixed	Humulin® 70/30	30 min	2.5–12 h	24 h
	Humulin® 50/50	30 min	2.5–12 h	24 h
	Novolin® 70/30	30 min	2.5–12 h	24 h

Insulin Labels

Like other drug labels, insulin labels identify the manufacturer, the brand name, storage information, and the expiration date (Figures 12-1 and 12-2). The concentration is usually listed twice, as the traditional dosage strength (e.g., 100 units/mL) and as the concentration. In most cases, the concentration is **U-100**, meaning that 100 units of insulin is contained in 1 mL of solution. Occasionally, the concentration is U-500, with 500 units/mL. Insulin labels also list the type (e.g., L or Lente) and the origin (beef, pork, human, or other).

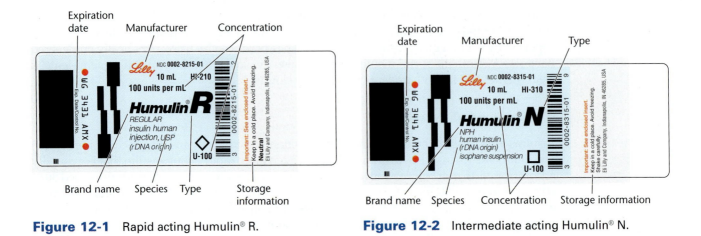

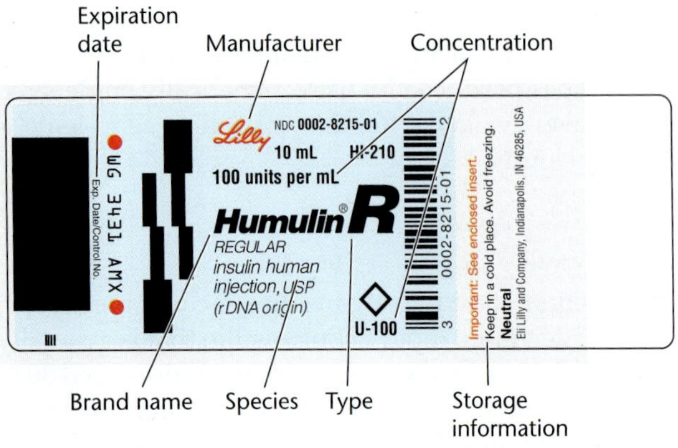

Figure 12-1 Rapid acting Humulin® R.

Expiration date

Manufacturer

Type

Lilly NDC 0002-8315-01 10 mL HI-310 100 units per mL Humulin® N NPH human insulin (rDNA origin) isophane suspension U-100

Brand name

Species

Concentration

Storage information

Figure 12-2 Intermediate acting Humulin® N.

Critical Thinking on the Job

Clarifying the Order

An order reads Iletin II 20 units Sub-Q stat. The health care professional was unaware that Iletin II is available in Lente and regular forms. He administered Iletin II regular insulin. When the patient experienced symptoms of hypoglycemia, she was given 4 oz of orange juice and 6 graham crackers.

Regular insulin has a shorter duration of action than Lente insulin. As a result, the patient's morning glucose level was elevated. Based on this elevated reading, the physician increased the dose. In turn, the patient again experienced hypoglycemia the next day and required another dosage adjustment.

Questions?

What problem occurred? How could the health care professional have avoided the problem?

Insulin Syringes

Insulin is administered with special insulin syringes marked in units.

Learning Link Recall from Chapter 4, Rule 4-4, found on page 117.

A standard U-100 insulin syringe holds up to 100 units or 1 mL of solution. Most of these syringes are calibrated for every 2 units, though some are marked for each unit. Insulin administration is different from the administration of most other injectable medications because the syringe measures the amount of insulin rather than a volume of solution.

Smaller insulin syringes, holding up to 50 units (0.5 mL of solution) or 30 units (0.3 mL), are usually calibrated for each unit. Their larger numbers make them easier to use for visually impaired patients.

Rule 12-1 For more accurate measurements use a 50-unit capacity syringe for insulin doses of less than 50 units, and a 30-unit capacity insulin syringe for insulin doses less than 30 units if these syringes are available.

Example 1 Ordered: Novolin® N 66 units

Because this order is for more than 50 units, use a U-100 insulin syringe. Find the mark for 66 units and fill the syringe to that calibration (Figure 12-3).

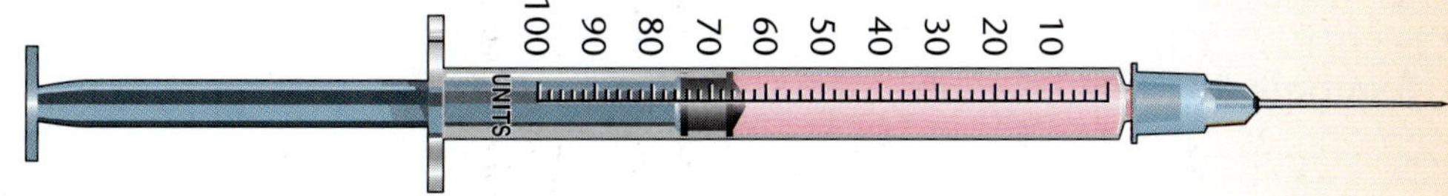

Figure 12-3 U-100 insulin syringe.

Example 2 Ordered: Humulin® R 55 units

Because this order is for more than 50 units, you will need a U-100 syringe. Your best choice would be a syringe calibrated for each unit (Figure 12-4). If you use a syringe calibrated for every 2 units, then fill it to the imaginary line between 54 and 56 units.

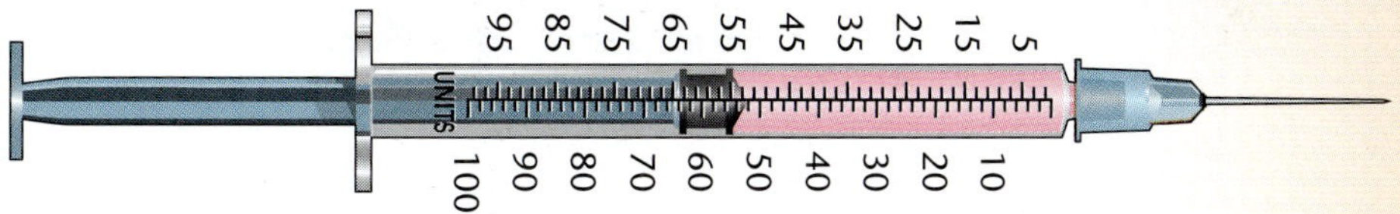

Figure 12-4 U-100 insulin syringe.

Example 3 Ordered: Humulin® R 35 units

Because this order is for less than 50 units, you may use a smaller syringe in which each unit is calibrated (Figure 12-5).

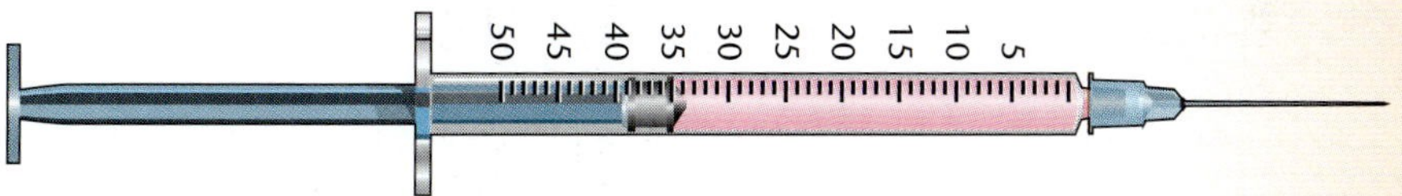

Figure 12-5 50-unit insulin syringe.

Example 4 Ordered: Novolin® R 8 units

Because this order is for less than 30 units, you may use either a 30-unit or 50-unit insulin syringe in which each unit is calibrated (Figure 12-6).

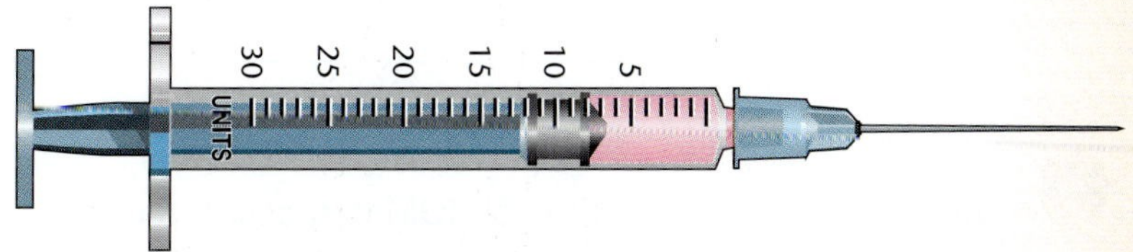

Figure 12-6 30-unit insulin syringe.

GO TO . . . Open the CD-ROM that accompanies your textbook, and select Chapter 12, Rule 12-1. Review the animation and example problems, then complete the practice problems. Continue to the next section of the book once you have mastered the rule presented. ■

U-500 insulin is used for patients with highly elevated blood sugars. Insulin may also be given intravenously for elevated blood sugars. On the occasion that U-500 is ordered or an insulin dose is over 100 units, a tuberculin or standard syringe may be necessary. U-500 insulin and U-100 insulin over 100 units will not fit in a U-100 syringe and must be measured in milliliters.

Rule 12-2 If the order is for U-500 insulin (which contains 500 units in each milliliter), use a tuberculin syringe. Calculate the amount to administer in milliliters.

Example 1 Ordered: Humulin® R U-500 insulin 80 units

$$80\text{ U} \times \frac{1\text{ mL}}{500\text{ U}} = A$$

$$\overset{4}{\cancel{80}}\ \cancel{\text{U}} \times \frac{1\text{ mL}}{\underset{25}{\cancel{500}\ \cancel{\text{U}}}} = 4 \times \frac{1}{25}\text{ mL} = \frac{4}{25}\text{ mL} = 0.16\text{ mL}$$

Administer 0.16 mL drawn up in a tuberculin syringe (see Figure 12-7).

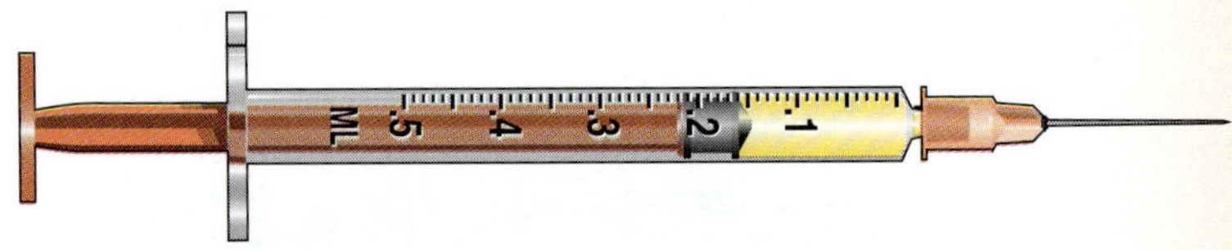

Figure 12-7 0.5 mL tuberculin syringe.

Example 2 Ordered: 150 units Humulin® R IV Stat

On hand: Humulin® R U-100 (Refer to Figure 12-8.)

$$15\cancel{0}\ \cancel{\text{units}} \times \frac{1\text{ mL}}{10\cancel{0}\ \cancel{\text{units}}} = A = 15 \times \frac{1\text{ mL}}{10} = \frac{15\text{ mL}}{10} = 1.5\text{ mL}$$

Administer 1.5 mL in a standard syringe (see Figure 12-8).

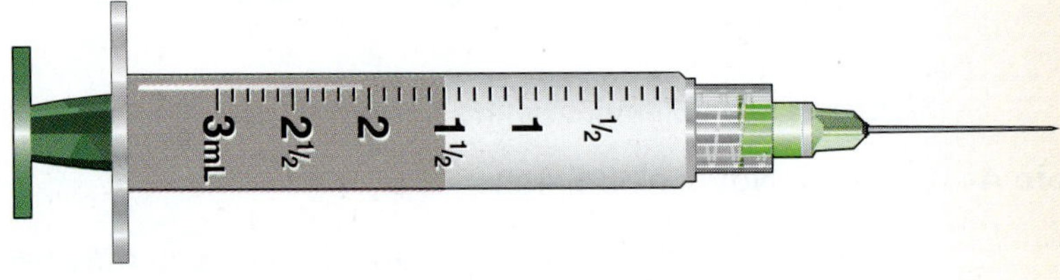

Figure 12-8 3 mL standard syringe.

GO TO . . . Open the CD-ROM that accompanies your textbook, and select Chapter 12, Rule 12-2. Review the animation and example problems, then complete the practice problems. Continue to the next section of the book once you have mastered the rule presented. ■

Patient Education

Measuring a Single Insulin Dose

Give the following information to patients:

1. Always wash your hands before handling insulin and syringes.
2. If you are using an intermediate- or long-acting insulin (Lente, NPH, UltraLente, 70/30, or 50/50), roll the vial between your palms to mix the insulin, until all the insulin looks cloudy.
3. Cleanse the rubber stopper of the vial with an alcohol wipe, using a circular motion. Start at the center of the circle and work outward.
4. Draw up an amount of air equal to your insulin dose in the syringe. Pull back the plunger until the leading ring is aligned with the correct marking on the syringe (Figure 12-9a).
5. Inject the air into the insulin vial (Figure 12-9b).
6. Keeping the needle inserted through the stopper, turn the vial upside down. Draw up your ordered dose of insulin (Figure 12-9c).
7. Avoid touching the needle during the procedure.

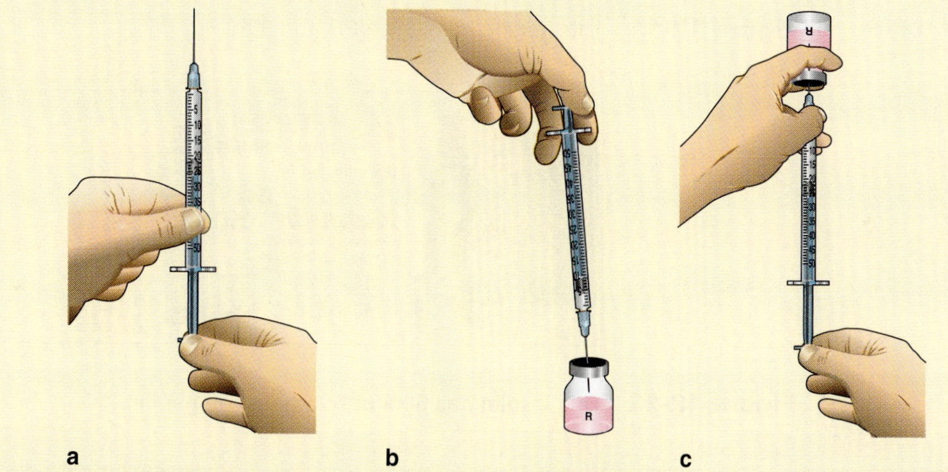

Figure 12-9 Measuring insulin. **a.** Draw up air. **b.** Inject air into insulin. **c.** Draw up dose of insulin.

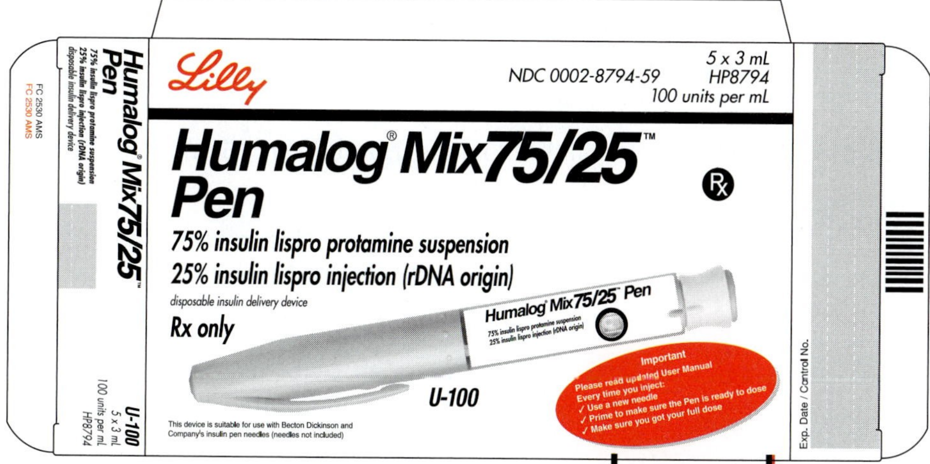

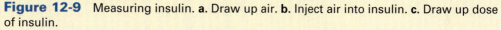

Figure 12-10 Insulin is sometimes self-administered with a pen device. A new needle must be attached for each dose.

Insulin Combinations

In some cases, in order to have greater control over glucose levels, the physician will prescribe two types of insulin for a patient. For example, the combination of a rapid-acting insulin and an intermediate-acting insulin provides the patient with the rapid onset of the first and the lengthy duration of the second. The two types of insulin can be combined in one syringe so that the patient only needs to receive one injection. Two types of insulin may be combined by the drug manufacturer. For example, Novolin® 70/30 is 70 percent intermediate-acting NPH insulin and 30 percent rapid-acting regular insulin (Figure 12-11). Humulin® 50/50 has 50 percent intermediate-acting NPH (or isophane) insulin and 50 percent rapid-acting regular insulin (Figure 12-12). In some cases, you will need to prepare the insulin combination yourself.

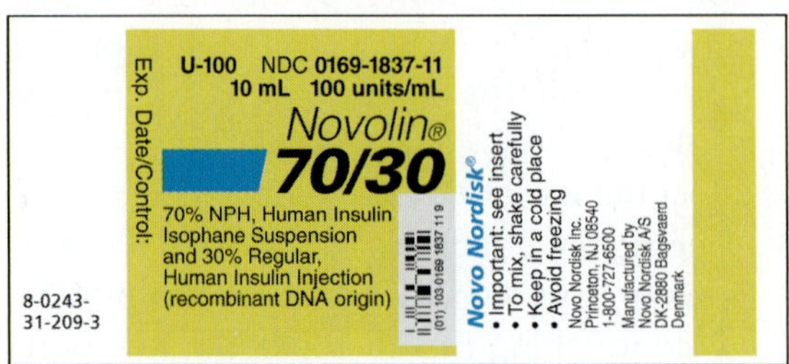

Figure 12-11 This insulin has 70% NPH and 30% R.

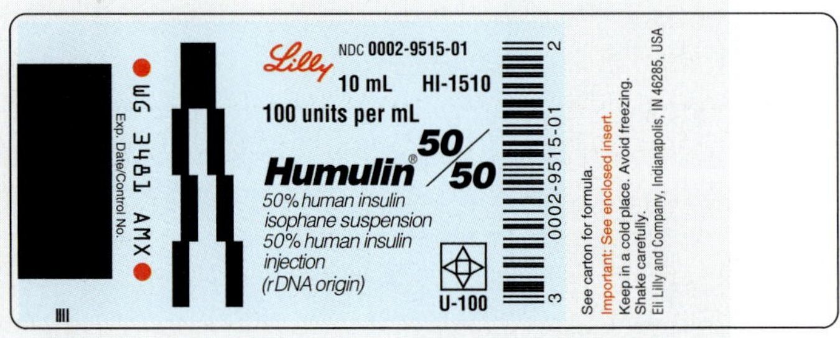

Figure 12-12 This insulin has 50% of two types of insulin.

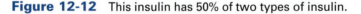

Rule 12-3 When you are preparing a combined insulin dose, always draw up the rapid-acting insulin first. *Remember:* The insulin that will act first is drawn up first.

Another way to remember Rule 12-3 is to draw up the clear insulin (rapid-acting) before the cloudy insulin (intermediate-acting).

Example Ordered: Novolin® R 20 units, Humulin® N 15 units Sub-Q now

Novolin® R is rapid-acting. Humulin® N is intermediate-acting. The rapid-acting insulin (Novolin® R) will be drawn into the syringe first.

GO TO . . . Open the CD-ROM that accompanies your textbook, and select Chapter 12, Rule 12-3. Review the animation and example problems, then complete the practice problems. Continue to the next section of the book once you have mastered the rule presented. ■

Rule 12-4 To prepare a combined insulin dose:

1. Calculate the total dose of insulin.

   Dose of rapid-acting insulin + dose of intermediate-acting insulin

   = total dose insulin

2. Draw up an amount of air equal to the dose of intermediate-acting insulin. Inject it into the intermediate insulin vial, but do not draw up the dose. Withdraw the needle from this vial.
3. Draw up an amount of air equal to the dose of rapid-acting insulin. Inject it into the rapid-acting insulin vial.
4. Without withdrawing the needle from the stopper, invert the vial. Draw up the dose of rapid-acting insulin.
5. Carefully insert the needle through the stopper of the intermediate-acting insulin vial. Invert the vial, without injecting any of the rapid-acting insulin into the vial.
6. Draw up intermediate-acting insulin until the leading ring reaches the calibration indicating the total dose. The clear insulin is drawn into the syringe first to prevent the cloudy, longer-acting insulin from entering the clear, shorter-acting insulin bottle.

Example Ordered: Humulin® N 42 units and Humulin® R 10 units Sub-Q qd

First calculate the total dose of insulin:

10 units of Humulin® R + 42 units Humulin® N = 52 units total

Next draw up 42 U of air and inject them into the vial of Humulin® N. Withdraw the needle from Humulin® N without drawing up the dose. Then draw up 10 U of air and inject them into the vial of Humulin® R. Without withdrawing the needle, invert the vial of Humulin® R and draw up 10 U of insulin (Figure 12-13a). Finally, insert the needle into the vial of Humulin® N and invert the vial. Withdraw 42 U of Humulin® N, until the leading ring of the syringe is at the calibration of 52 U, the total dose (Figure 12-13b).

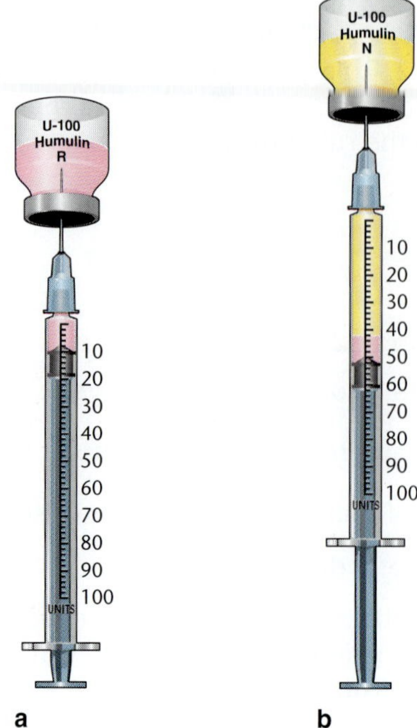

Figure 12-13 **a.** Draw up the rapid acting (clear) insulin first. **b.** Be careful when drawing up the second type of insulin.

GO TO . . . Open the CD-ROM that accompanies your textbook, and select Chapter 12, Rule 12-4. Review the animations and example problems, then complete the practice problems. Continue to the next section of the book once you have mastered the rule presented. ■

ERROR ALERT!

When two types of insulins are combined, measure the correct amount of each.

An order reads Novolin® N 37 units and Novolin® R 5 units Sub-Q stat. Suppose you draw up 37 units from the Novolin® R vial and 5 units from the Novolin® N vial. Although the patient receives 42 units of insulin, he receives a much larger dose of regular (rapid-acting) insulin than was ordered—37 units rather than 5 units. The insulin metabolizes the patient's glucose too quickly; he becomes hypoglycemic and loses consciousness. Glucagon and 50% dextrose are administered, and the patient recovers. This error can be avoided if you carefully check the order against the labels 3 times.

REVIEW AND PRACTICE

12-1 Insulin

In Exercises 1–14, refer to labels A–G. Select the label corresponding to each order. Then mark the desired amount of insulin on the syringe.

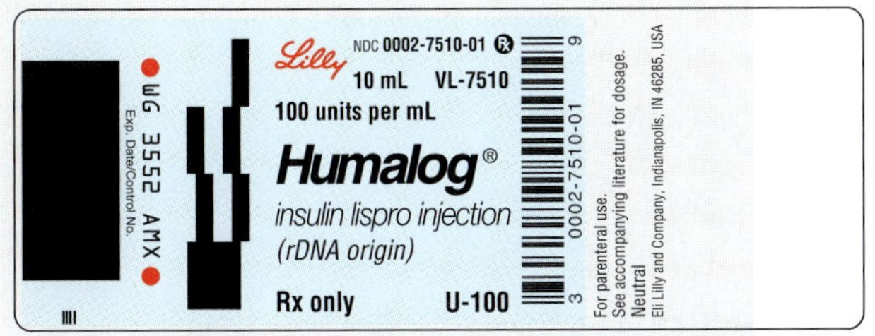

A

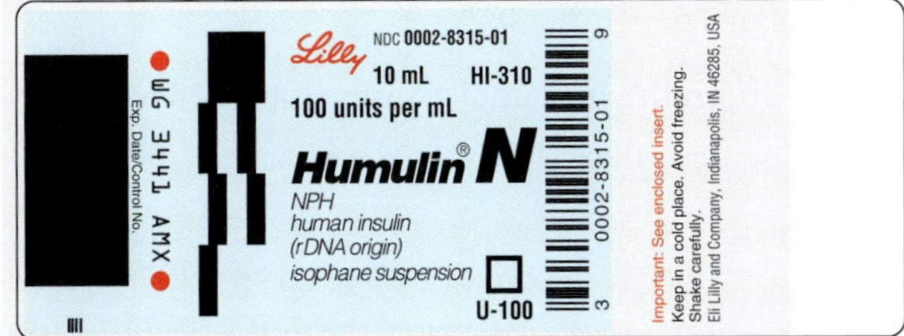

B

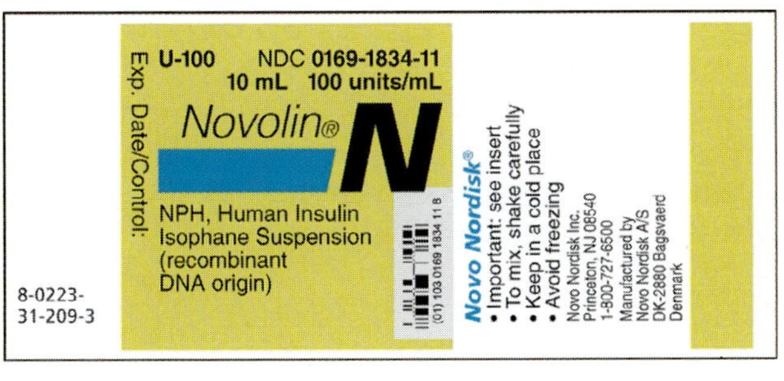

C

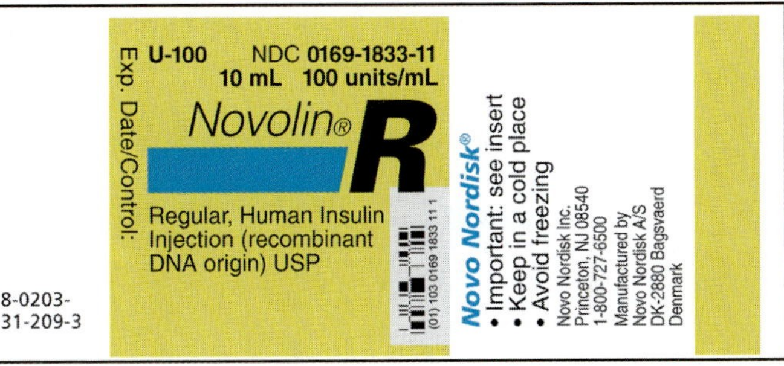

D

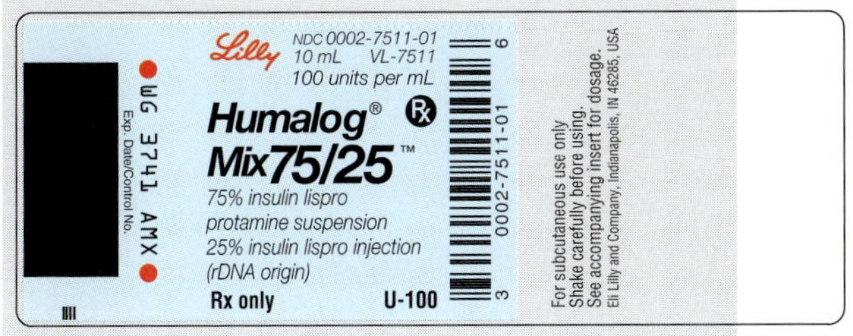

E

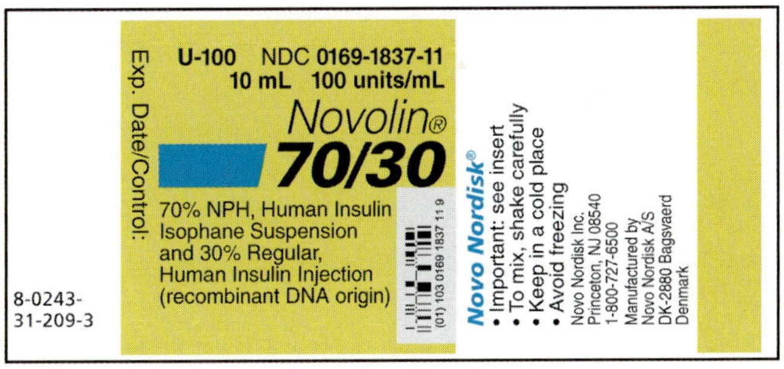

U-100 NDC 0169-1837-11
10 mL 100 units/mL
Novolin® 70/30
70% NPH, Human Insulin
Isophane Suspension
and 30% Regular,
Human Insulin Injection
(recombinant DNA origin)
Exp. Date/Control:
8-0243-
31-209-3

F

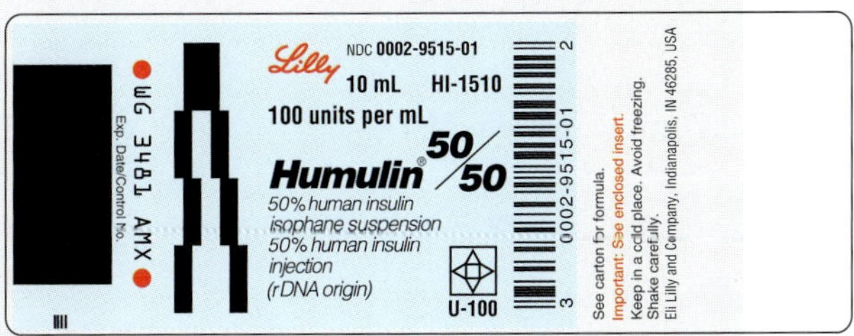

G

1. Ordered: Novolin® R 12 units Sub-Q ac breakfast
   Select vial: ____

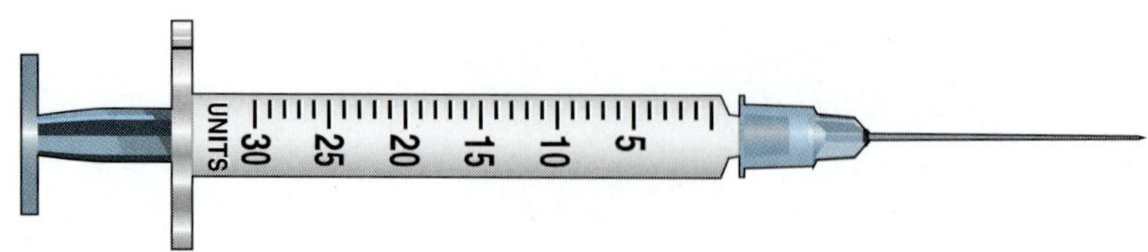

2. Ordered: Humalog® 5 units Sub-Q 15 min before lunch
   Select vial: _____

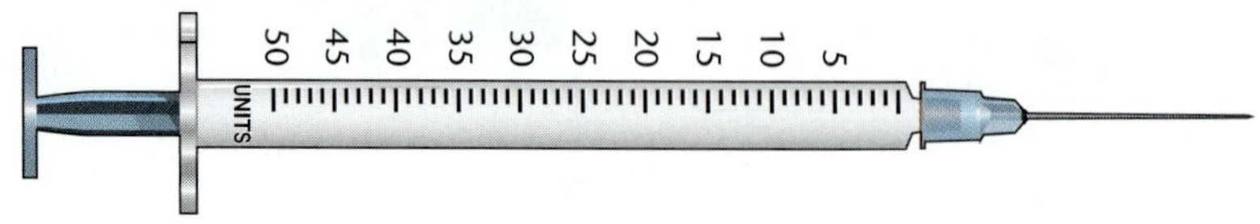

3. Ordered: Novolin® N 35 units Sub-Q qd
   Select vial: _____

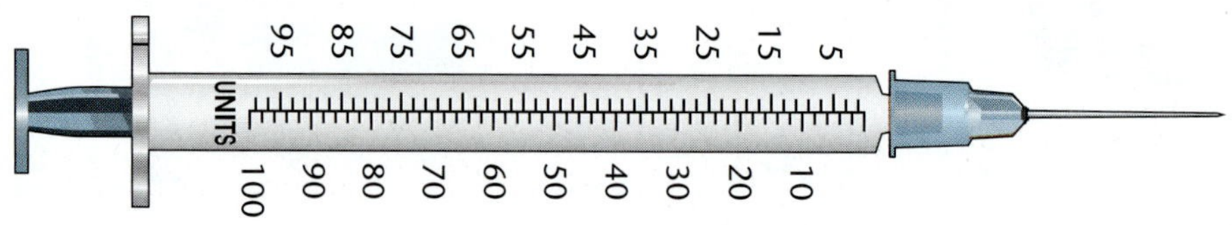

4. Ordered: Humulin® N 72 units Sub-Q qd
   Select vial: _____

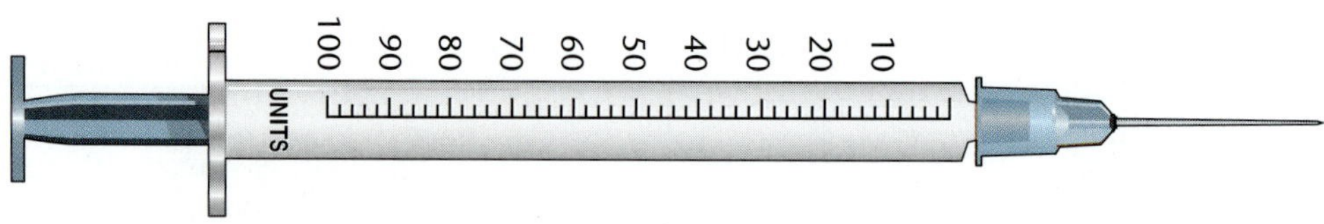

5. Ordered: Humulin® 50/50 42 units Sub-Q ac breakfast
   Select vial: _____

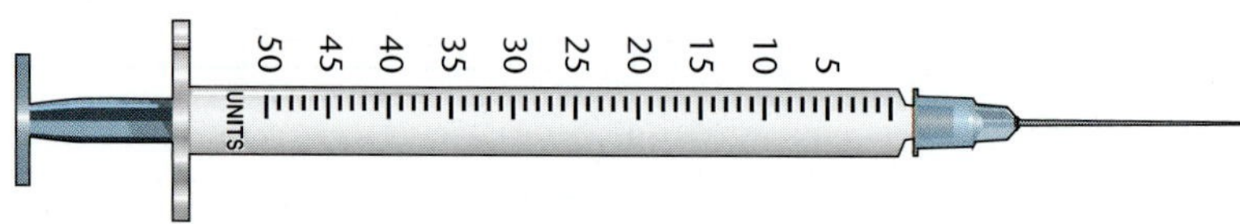

6. Ordered: Humalog® 75/25 BR 17 units ac breakfast
   Select vial: _____

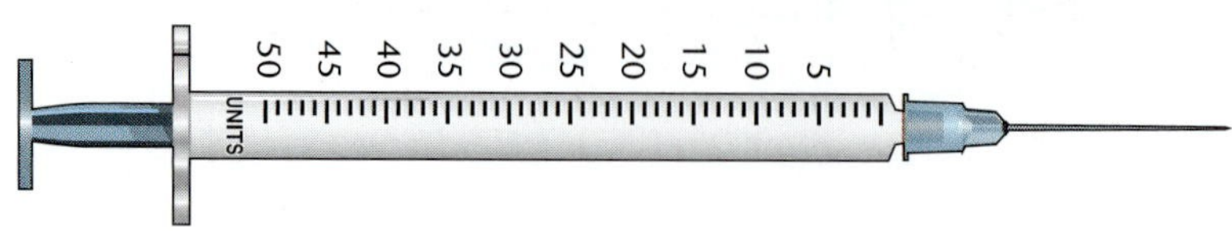

7. Ordered: Novolin® 70/30 53 units Sub-Q ac dinner
Select vial: _____

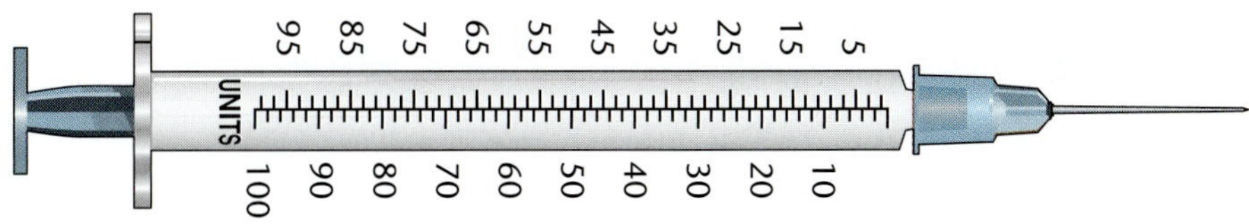

8. Ordered: Novolin® 70/30 R 26 units Sub-Q ac breakfast
Select vial: _____

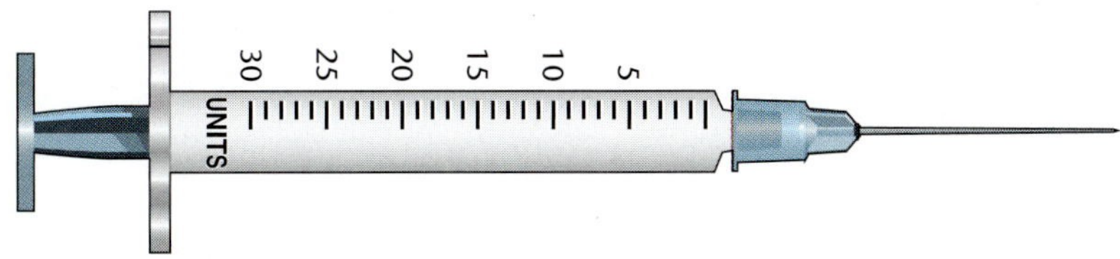

9. Ordered: Humulin® N 44 units Sub-Q ac dinner
Select vial: _____

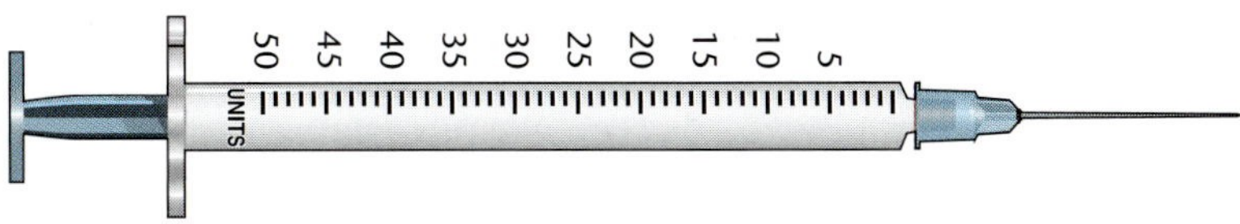

10. Ordered: Humalog® 15 units Sub-Q ac breakfast
Select vial: _____

11. Ordered: Novolin® N 64 units Sub-Q qd
Select vial: _____

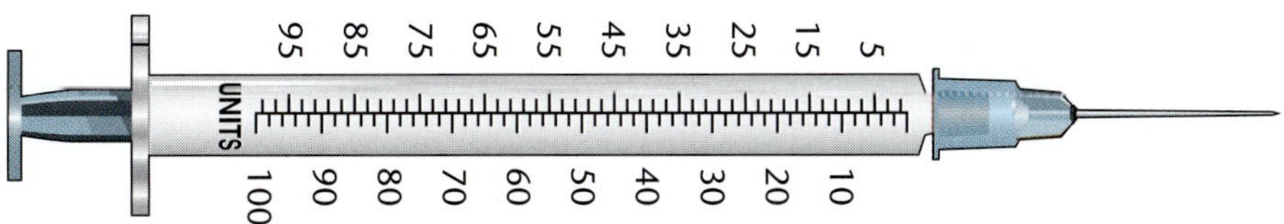

12. Ordered: Humulin® 50/50 injection 36 units Sub-Q ac dinner
Select vial: _____

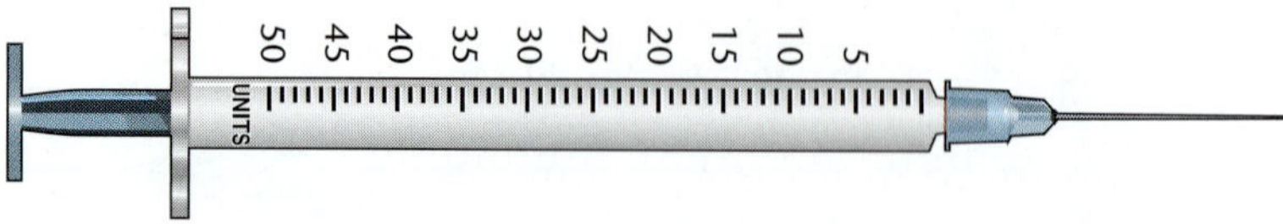

13. Ordered: Humalog® 75/25 BR 7 units Sub-Q stat
Select vial: _____

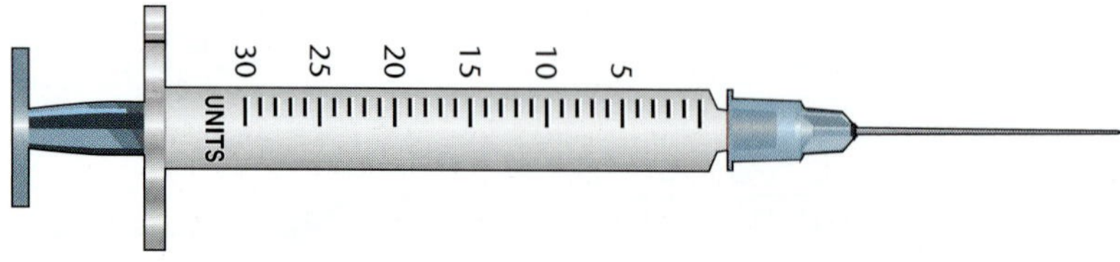

14. Ordered: Novolin® R 14 units Sub-Q ac breakfast
Select vial: _____

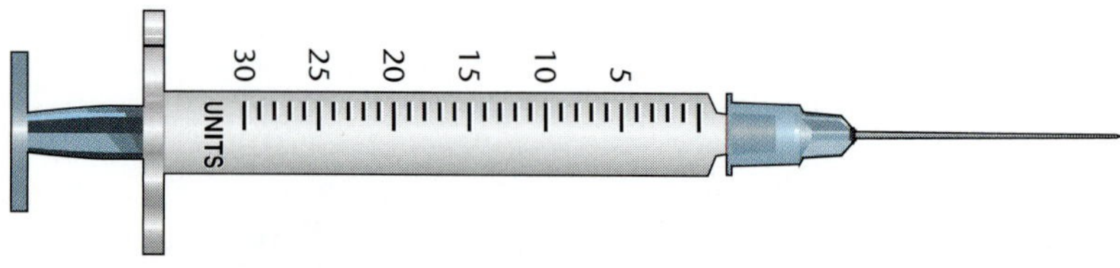

In Exercises 15 and 16, first mark on the syringe the dose of rapid-acting insulin ordered. Then mark where the leading ring will be after you draw up the intermediate-acting insulin into the same syringe.

15. Novolin® N 65 units and Novolin® R 12 units Sub-Q qam

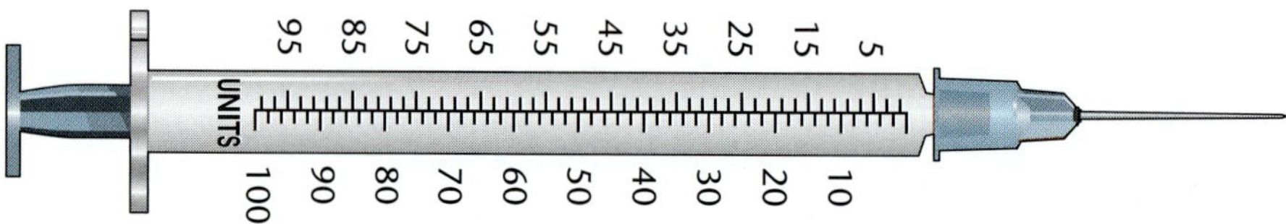

16. Humulin® N 53 units and Humulin® R 4 units Sub-Q qam

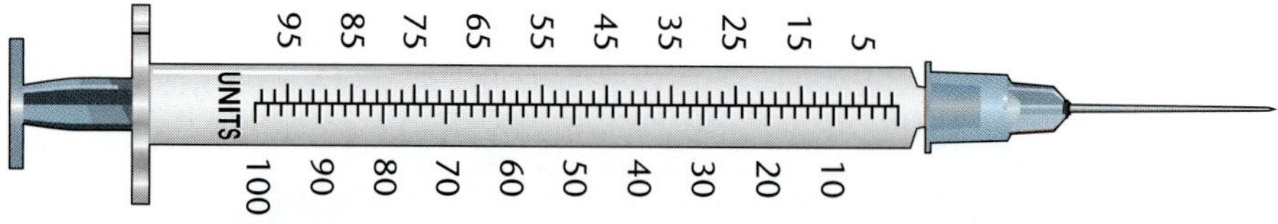

To check your answers, see pages 536–538.

Heparin

Heparin Calculations

Heparin is an **anticoagulant** used in USP units. It is given to patients to reduce or prevent the blood from clotting. As discussed in Chapter 10, heparin may be used as an irrigant to keep the blood from clotting in a heparin lock. For this purpose it is packaged in cartridges or vials of 10 to 100 units (see Figure 12-14). Heparin may also be administered intermittently Sub-Q or IV in larger dosages (see Figure 12-15). When multiple injections of heparin are administered Sub-Q, the sites of the injection should be rotated to prevent bruising. Bruising and bleeding are a great concern for patients receiving heparin, so the dosage calculations must be accurate.

When used for anticoagulant therapy, heparin may be administered safely to adults at a dosage of 20,000 to 40,000 units per 24 h. Before you administer heparin, verify that the dosage ordered falls within this range. In most cases when heparin is administered through an IV, an electronic device is used to help ensure the accuracy of the dose.

You have learned to calculate how many milliliters of solution were needed to administer a dose of heparin. Calculating the flow rate is very similar. Before, the desired dose *D* represented only a quantity of units. Here, the desired dose *D* represents a flow rate, a quantity of units per time period.

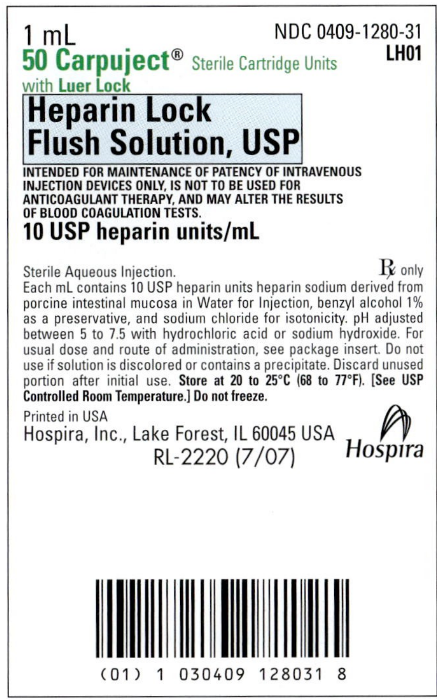

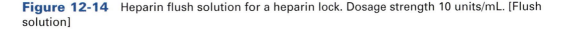

Figure 12-14 Heparin flush solution for a heparin lock. Dosage strength 10 units/mL. [Flush solution]

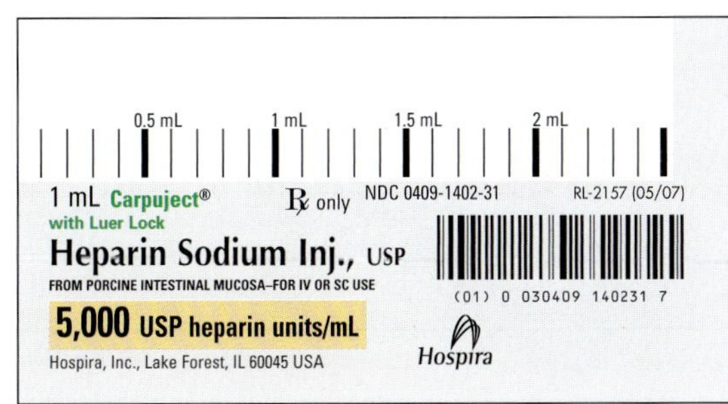

Figure 12-15 Heparin given Sub-Q or IV dosage strength 5000 units/1 mL. [Injection solution]

Rule 12-5 To determine the rate at which to administer a solution containing heparin with an electronic device that measures the infusion in milliliters per hour, find A, where:

D = rate of desired dose

Q = dosage unit

H = dose on hand

A = amount to administer

Use the fraction proportion, ratio proportion, dimensional analysis, or formula method.

Fraction Proportion Method

Example Find the hourly rate at which to administer IV heparin.

Ordered: 1000 U/h IV heparin using an infusion pump

On hand: 50,000 U heparin in 1000 mL of D5W

Solution The dosage ordered O is 1000 U/h mg. The dose on hand H is 50,000 U, and the dosage unit Q is 1000 mL. Because the dosage ordered and the dose on hand have the same units, no conversion is needed to find the desired dose D, which is 1000 U/h.

Follow Procedure Checklist 7-1.

1. Fill in the proportion.

$$\frac{Q}{H} = \frac{A}{D} \quad \text{or} \quad \frac{\text{dosage unit}}{\text{dose on hand}} = \frac{\text{amount to administer}}{\text{desired dose}}$$

$$\frac{1000 \text{ mL}}{50{,}000 \text{ units}} = \frac{A}{1000 \text{ U/h}}$$

2. Cancel units leaving mL/h for the answer.

$$\frac{1000 \text{ mL}}{50{,}000 \cancel{\text{U}}} = \frac{A}{1000 \cancel{\text{U}}/\text{h}}$$

3. Cross-multiply and solve for the answer.

$$50{,}000 \times A = 1000 \text{ mL} \times 1000 \cdot \frac{1}{\text{h}}$$

$$A = \frac{1000 \text{ mL} \times 1000 \cdot \frac{1}{\text{h}}}{50{,}000}$$

$$A = \frac{1000 \text{ mL} \times 1\cancel{000} \cdot \frac{1}{\text{h}}}{50{,}\cancel{000}}$$

$$A = 20 \text{ mL/h}$$

Ratio Proportion Method

Example Find the hourly rate at which to administer IV heparin.

Ordered: 1000 U/h IV heparin using an infusion pump

On hand: 50,000 U heparin in 1000 mL of D5W

Solution The dosage ordered O is 1000 U/h mg. The dose on hand H is 50,000 U, and the dosage unit Q is 1000 mL. Because the dosage ordered and the dose on hand have the same units, no conversion is needed to find the desired dose D, which is 1000 U/h.

Follow Procedure Checklist 7-2.

1. Fill in the proportion.

   $Q : H :: A : D$

   $1000 \text{ ML} : 50{,}000 \text{ U} :: A : 1000 \text{ U/h}$

2. Cancel units leaving mL/h for the answer.

   $1000 \text{ mL} : 50{,}000 \cancel{\text{U}} :: A : 1000 \cancel{\text{U}}/\text{h}$

3. Multiply the means and extremes, then solve for the missing value.

   $50{,}000 \times A = 1000 \text{ mL} \times 1000 \cdot \frac{1}{\text{h}}$

   $$A = \frac{1000 \text{ mL} \times 1\cancel{000} \cdot \frac{1}{\text{h}}}{50{,}\cancel{000}}$$

   $A = 20 \text{ mL/h}$

Dimensional Analysis

Example Find the hourly rate at which to administer IV heparin.

Ordered: 1000 U/h IV heparin using an infusion pump

On hand: 50,000 U heparin in 1000 mL of D5W

Solution Follow Procedure Checklist 7.3.

1. The unit of measure will be milliliters per hour.

   $A \text{ mL/h} =$

2. The dosage unit is 1000 mL; the dose on hand is 50,000 U. This is your first factor.

   $$\frac{1000 \text{ mL}}{50{,}000 \text{ U}}$$

3. The desired dose is 1000 U/h. Set up the equation.

   $$A \text{ mL/h} = \frac{1000 \text{ mL}}{50{,}000 \text{ U}} \times \frac{1000 \text{ U}}{1 \text{ h}}$$

   $$A \text{ mL/h} = \frac{1000 \text{ mL}}{50{,}\cancel{000} \cancel{\text{U}}} \times \frac{1\cancel{000} \cancel{\text{U}}}{1 \text{ h}}$$

   $A = 20 \text{ mL/h}$

Although electronic devices are usually used for IV heparin, in some cases you may need to determine the flow rate in drops per minute. This may be done to verify the flow rate of the pump or if the IV tubing is removed temporarily from the pump. To convert the flow rate from milliliters per hour to drops per minute, use the methods shown in Rule 10-6.

Learning Link Recall from Chapter 10, Rule 10-6, calculating the flow rate in drops per minute, found on page 367.

In some cases you will need to determine the hourly dose of a heparin infusion. This calculation is done to ensure the maximum and/or safe dose is not exceeded.

Rule 12-6 To calculate the hourly dose of heparin (desired dose D). Determine the following:

H = dose on hand or total amount to administer

Q = dosage unit for total amount

A = amount to administer or flow rate of infusion

Calculate, using the fraction proportion, ratio proportion, or dimensional analysis method.

Fraction Proportion Method

Example What is the hourly dose?

Ordered: 30,000 U of IV heparin in 500 mL of D5W to infuse at 25 mL/h

Solution

H = 30,000 U (dose on hand or total amount to administer)

Q = 500 mL (dosage unit for total amount)

A = 25 mL/h (amount to administer or flow rate of infusion)

D = ? (hourly or desired dose)

Thinking critically, we realize we already know the amount to administer A or flow rate of the infusion and we need to determine the hourly dose D or desired dose.

Follow Procedure Checklist 7-1.

1. Fill in the proportion.

$$\frac{Q}{H} = \frac{A}{D} \quad \text{or} \quad \frac{\text{dosage unit}}{\text{dose on hand}} = \frac{\text{amount to administer}}{\text{desired dose}}$$

$$\frac{500 \text{ mL}}{30{,}000 \text{ U}} = \frac{25 \text{ mL/h}}{D}$$

2. Cancel units.

$$\frac{500 \cancel{\text{mL}}}{30{,}000 \text{ U}} = \frac{25 \cancel{\text{mL}}\text{/h}}{D}$$

3. Cross-multiply and solve for the unknown.

$$500 \times D = 30{,}000 \text{ U} \times 25 \cdot \frac{1}{\text{h}}$$

$$D = \frac{30{,}000 \text{ U}}{500} \times 25 \cdot \frac{1}{\text{h}}$$

$$D = \frac{30{,}0\cancel{00} \text{ U}}{5\cancel{00}} \times 25 \cdot \frac{1}{\text{h}}$$

$$D = 1500 \text{ U/h}$$

Ratio Proportion Method

Example What is the hourly dose?

Ordered: 30,000 U of IV heparin in 500 mL of D5W to infuse at 25 mL/h

Solution

H = 30,000 U (dose on hand or total amount to administer)

Q = 500 mL (dosage unit for total amount)

A = 25 mL/h (amount to administer or flow rate of infusion)

D = ? (hourly or desired dose)

Thinking critically, we realize we already know the amount to administer A or flow rate of the infusion, and we need to determine the hourly dose D or desired dose.

Follow Procedure Checklist 7-2.

1. Fill in the proportion.

   $Q : H :: A : D$

   500 mL : 30,000 U : : 25 mL/h : D

2. Cancel units.

   500 $\cancel{\text{mL}}$: 30,000 U : : 25 $\cancel{\text{mL}}$/h : D

3. Multiply the means and extremes, then solve for the missing value.

   $$30{,}000 \times 25/\text{h} = 500 \times D$$

   $$\frac{30{,}0\cancel{00} \text{ U} \times 25\text{h}}{5\cancel{00}} = D$$

   $$D = 1500 \text{ U/h}$$

Dimensional Analysis

Example What is the hourly dose?

Ordered: 30,000 U of IV heparin in 500 mL of D5W to infuse at 25 mL/h

Solution

H = 30,000 U (dose on hand or total amount to administer)

Q = 500 mL (dosage unit for total amount)

A = 25 mL/h (amount to administer or flow rate of infusion)

D = ? (hourly or desired dose)

Thinking critically, we realize we already know the amount to administer A or flow rate of the infusion, and we need to determine the hourly dose D or desired dose.

Follow Procedure Checklist 7-3.

1. The hourly rate for the unknown D will be in units per hour. Place this on the left side of the equation.

   $$D\ \text{U/h} =$$

2. The dose on hand is 30,000 U. The dosage unit is 500 mL. This is your first factor.

   $$\frac{30{,}000\ \text{U}}{500\ \text{mL}}$$

3. The flow rate of the infusion is 25 mL/h. Use this as your second factor. Set up the equation.

   $$D\ \text{U/h} = \frac{30{,}000\ \text{U}}{500\ \text{mL}} \times \frac{25\ \text{mL}}{1\ \text{h}}$$

4. Cancel units. The remaining units on the right side of the equation must match those on the left side of the equation.

   $$D\ \text{U/h} = \frac{30{,}000\ \text{U}}{500\ \cancel{\text{mL}}} \times \frac{25\ \cancel{\text{mL}}}{1\ \text{h}}$$

5. Solve the equation.

   $$D\ \text{U/h} = \frac{30{,}0\cancel{00}\ \text{U}}{5\cancel{00}} \times \frac{25}{1\ \text{h}}$$

   $$D = 1500\ \text{U/h}$$

After determining the units per hour, you must confirm this is a safe rate. As discussed earlier, the safe dose for 24 h is 20,000 to 40,000 U. Before you administer the heparin, at a rate of 1500 U/h, we must multiply by 24 h/day to determine the 24-h dose.

$$\frac{1500\ \text{U}}{1\ \text{h}} \times \frac{24\ \text{h}}{1\ \text{day}} = \frac{36{,}000\ \text{U}}{1\ \text{day}}$$

So 36,000 U per 24 h is within the safe dose range for heparin.

GO TO . . . Open the CD-ROM that accompanies your textbook, and select Chapter 12, Rule 12-6. Review the animation and example problems, then complete the practice problems. Continue to the next section of the book once you have mastered the rule presented. ■

REVIEW AND PRACTICE

12-2 Heparin

In Exercises 1–5, find the flow rate.

1. Ordered: Heparin 1000 U/h IV using an infusion pump
   On hand: 20,000 U in 1500 mL of 5% DW
2. Ordered: Heparin 1000 U/h IV
   On hand: 20,000 U in 500 mL of D5W, microdrip tubing
3. Ordered: Heparin 850 U/h IV
   On hand: 40,000 U in 1500 mL of 5% DW, 15 gtt/mL tubing
4. Ordered: Heparin 1500 U/h IV
   On hand: 30,000 U in 500 mL of 5% D 0.45% NS, 10 gtt/mL tubing
5. Ordered: Heparin 750 U/h IV
   On hand: 30,000 U in 1000 mL of 5% DW using 20 gtt/mL tubing

In Exercises 6–10, find the hourly dosage for the heparin orders. Then determine whether the dosage is safe.

6. An IV with 60,000 U in 1500 mL of 5% DW infusing at 25 mL/h
7. An IV setup delivering 45 gtt/min from 25,000 U in 2500 mL of D5NS using microdrip tubing
8. 40,000 U in 1800 mL of 5% DW delivered at 25 gtt/min using 10 gtt/mL tubing
9. 30,000 U in 1500 mL of 5% D 0.45% NS delivered at 20 gtt/min using 10 gtt/mL tubing
10. 50,000 U in 850 mL NS infusing at 25 gtt/min using 20 gtt/mL

To check your answers, see page 539.

Critical Care IV

By necessity, IV medications used in critical care settings are fast-acting and potent. Because they enter the bloodstream directly, only a small amount is required to have an effect. IV medications are given to alter or maintain life-sustaining functions such as heart rate, cardiac output, blood pressure, or respiration. Because of their narrow margin of safety, you must have the right medication for the right patient in the correct concentration. Be equally certain that you are administering the proper dosage at the proper rate.

Critical care drugs are administered by IV push or bolus. However, they can be delivered via continuous IV. Using a volume-controlled burette, electronic infusion device, or microdrip tubing will improve the precision of the dosage.

Per Minute Orders

You have learned to calculate flow rates for heparin units to measure the volume to administer. You will use the same methods to calculate critical care medication. In most cases, the amount to administer will be measured in milligrams or micrograms.

In critical care, IV medication orders may express an amount of medication to be delivered per minute. You will need to convert this rate to milliliters per hour.

Rule 12-7 To convert a per minute order to an hourly flow rate:

1. Convert the order to milliliters per minute. Determine the following:

   D = rate of desired dose (mg or mcg/min)

   Q = dosage unit (mL)

   H = dose on hand (total number of mg or mcg)

   A = amount to administer (mL/min)

   Use the fraction proportion, ratio proportion, dimensional analysis, or formula method.

2. Convert milliliters per minute to milliliters per hour. When you use the fraction proportion, ratio proportion, and formula methods, convert milliliters per minute to milliliters per hour by multiplying by 60 When you are using dimensional analysis, use the factor $\frac{60}{1}$ as part of your equation. In both cases the hourly flow rate or the rate of which you will set the infusion pump is determined in milliliters per hour.

Fraction Proportion Method

Example Find the hourly flow rate.

Ordered: 5000 mg Esmolol in 500 mL of D5W at 8 mg/min via infusion pump

Solution The rate of the desired dose D is 8 mg/min. The dosage unit Q is 500 mL. The dose on hand H is 5000 mg.

Follow Procedure Checklist 7-1.

1. Fill in the proportion.

   $$\frac{Q}{H} = \frac{A}{D} \quad \text{or} \quad \frac{\text{dosage unit}}{\text{dose on hand}} = \frac{\text{amount to administer}}{\text{desired dose}}$$

   $$\frac{500 \text{ mL}}{5000 \text{ mg}} = \frac{A}{8 \text{ mg/min}}$$

2. Cancel units leaving mL/min for the answer.

   $$\frac{500 \text{ mL}}{5000 \cancel{\text{mg}}} = \frac{A}{8 \cancel{\text{mg}}/\text{min}}$$

3. Cross-multiply and solve for the unknown.

   $$500 \times A = 500 \text{ mL} \times 8/\text{min}$$

   $$A = \frac{500 \text{ mL} \times 8 \text{ /min}}{500\cancel{0}}$$

   $$A = 0.8 \text{ mL/min}$$

Multiply your answer of 0.8 mL/min × 60 to determine the milliliters per hour or the rate at which you should set your infusion pump.

$$0.8 \text{ mL/min} \times 60 = 48 \text{ mL/h}$$

Ratio Proportion Method

Example Find the hourly flow rate.

Ordered: 5000 mg Esmolol in 500 mL of D5W at 8 mg/min via infusion pump

Solution The rate of the desired dose D is 8 mg/min. The dosage unit Q is 500 mL. The dose on hand H is 5000 mg.

Follow Procedure Checklist 7-2.

1. Fill in the proportion.

   $Q : H :: A : D$

   500 mL : 5000 mg : : A : 8 mg/min

2. Cancel units leaving mL/min for the answer.

   500 mL : 5000 ~~mg~~ : : A : 8 ~~mg~~/min

3. Multiply the means and extremes, then solve for the missing value.

   $5000 \times A = 500 \text{ mL} \times 8/\text{min}$

   $$A = \frac{500 \text{ mL} \times 8/\text{min}}{5000}$$

   $A = 0.8 \text{ mL/min}$

Multiply your answer of 0.8 mL/min × 60 to determine the milliliters per hour or the rate at which you should set your infusion pump.

$0.8 \text{ mL/min} \times 60 = 48 \text{ mL/h}$

Dimensional Analysis

Example Find the hourly flow rate.

Ordered: 5000 mg Esmolol in 500 mL of D5W at 8 mg/min via infusion pump

Solution The rate of the desired dose D is 8 mg/min. The dosage unit Q is 500 mL. The dose on hand H is 5000 mg.

1. The unit of measure will be milliliters per hour.

   $A \text{ mL/h} =$

2. The dosage unit is 500 mL; the dose on hand is 5000 mg. This is the first factor.

   $$\frac{500 \text{ mL}}{5000 \text{ mg}}$$

3. The desire dose is 8 mg/min. This is the second factor.

   $$A \text{ mL/h} = \frac{500 \text{ mL}}{5000 \text{ mg}} \times \frac{8 \text{ mg}}{1 \text{ min}}$$

4. Use the third factor $\frac{60\text{ min}}{1\text{ h}}$ to convert the answer from minutes to hours.

$$A\text{ mL/h} = \frac{500\text{ mL}}{5000\text{ mg}} \times \frac{8\text{ mg}}{1\text{ min}} \times \frac{60\text{ min}}{1\text{ h}}$$

5. Cancel units. The remaining units on the right side of the equation should be milliliters per hour.

$$A\text{ mL/h} = \frac{500\text{ mL}}{5000\text{ }\cancel{\text{mg}}} \times \frac{8\text{ }\cancel{\text{mg}}}{1\text{ }\cancel{\text{min}}} \times \frac{60\text{ }\cancel{\text{min}}}{1\text{ h}}$$

6. Solve the equation.

$$A\text{ mL/h} = \frac{5\cancel{00}\text{ mL}}{50\cancel{00}} \times \frac{8}{1} \times \frac{60}{1\text{ h}}$$

$$A = 48\text{ mL/h}$$

Orders Based on Body Weight

As discussed in Chapter 11, many medication orders, especially pediatric and geriatric orders, are based on body weight. Body weight calculations are especially relevant to IV medications for critically-ill patients.

An order based on body weight will often state an amount of medication per weight of the patient per unit of time. For example, a recommended daily dosage of a medication may be 5 mg/kg/day. If the patient weighs 50 kg, then you can use a proportion to calculate the daily dosage.

$$\frac{5\text{ mg}}{1\text{ kg}} = \frac{?\text{ mg}}{50\text{ kg}}$$

$$5\text{ mg} \times 50\text{ kg} = 1\text{ kg} \times ?\text{ mg}$$

$$250 = ?$$

The daily dose in this case is 250 mg of medication.

If you know the patient's weight in pounds, then convert the weight to kilograms.

Learning Link Recall from Chapter 3, Table 3-8, 1 kg = 2.2 lb, found on page 92.

The recommended dosage may use a different time period. For example, you may need an hourly flow rate, but the order refers to a daily dosage. You would convert from days to hours. If the order lists an hourly flow rate and the recommended dosage is a daily amount convert the dosage first to hours and then to minutes.

Most drug orders already factor in the patient's body weight; therefore, in most cases you will not have to perform this calculation. However, you have already learned the techniques needed to factor in a patient's weight when calculating the amount to administer.

Rule 12-8 To find the IV flow rate based upon weight:

1. Convert the weight to kilograms.
2. Determine the desired dose.
3. Calculate the amount to administer.
4. Calculate the flow rate.

Example Find the flow rate for an adult who weighs 187 lb.

Ordered: Acyclovir 5 mg/kg IV over 1 hour, q 8 hours × 7 days

On hand: Refer to the label in Figure 12-16 below.

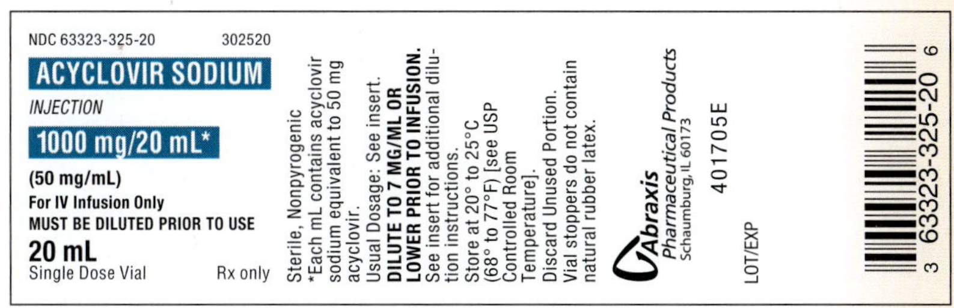

Figure 12-16 Acyclovir label.

Use macrodrip tubing calibrated for 20 gtt/mL. The recommended diluent is 100 mL of 5% DW.

1. Convert the patient's weight to kilograms.

   $$\frac{2.2 \text{ lb}}{1 \text{ kg}} = \frac{187 \text{ lb}}{? \text{ kg}}$$

   $2.2 \text{ lb} \times ? \text{ kg} = 1 \text{ kg} \times 187 \text{ lb}$

   $? = 85 \text{ kg}$

2. The patient weighs 85 kg. Determine how many milligrams of acyclovir have been ordered.

   $$\frac{5 \text{ mg}}{1 \text{ kg}} = \frac{? \text{ mg}}{85 \text{ kg}}$$

   $5 \text{ mg} \times 85 \text{ kg} = 1 \text{ kg} \times ? \text{ mg}$

   $425 \text{ mg} = ?$

3. You want to administer 425 mg of acyclovir over 1 hour (60 minutes). Now calculate the amount of milliliters to administer. The vial contains 1000 mg/20 mL.

   $$\frac{425 \text{ mg}}{? \text{ mL}} = \frac{1000 \text{ mg}}{20 \text{ mL}}$$

   $425 \text{ mg} \times 20 \text{ mL} = ? \text{ mL} \times 1000 \text{ mg}$

   $8.5 \text{ mL} = ?$

4. According to the package insert, the injection must be diluted before IV administration to a dosage strength of 7 mg/mL or less. Adding the 8.5 mL of medication to the 100 mL bag of 5%DW, you would create a dosage strength of 425 mg/108.5 mL. Dividing 425 by 108.5, you determine that the concentration is 3.9 mg/mL, which is below 7 mg/mL. You must administer 108.5 mL over 60 minutes. The tubing is calibrated for 20 gtt/mL. Determine the drops per minute flow rate.

Dimensional Analysis

Solution Follow Procedure Checklist 7-3.

1. Determine the unit of measure for the answer *f*, and place it as the unknown on the left side of the equation.

   $f \text{ gtt/min} =$

2. Write a factor with the number of milliliters to be administered on top (V) and the length of time to be administered (T) on the bottom.

   $\frac{108.5 \text{ mL}}{60 \text{ min}}$

3. Multiply by the drop factor of the tubing being used.

   $f \text{ gtt/min} = \frac{108.5 \text{ mL}}{60 \text{ min}} \times \frac{20 \text{ gtt}}{\text{mL}}$

4. Cancel units on the right side of the equation. The remaining unit of measure on the right side of the equation should match the unknown unit of measure on the left side of the equation.

   $f \text{ gtt/min} = \frac{108.5 \cancel{\text{mL}}}{60 \text{ min}} \times \frac{20 \text{ gtt}}{\cancel{\text{mL}}}$

5. Solve the equation.

   $f = \frac{108.5 \times 20 \text{ gtt}}{60 \text{ min}}$

   $f = 36.16 \text{ gtt/min}$

Here, the flow rate for administering the acyclovir is 36 gtt/min.

GO TO . . . Open the CD-ROM that accompanies your textbook, and select Chapter 12, Rule 12-8. Review the animation and example problems, then complete the practice problems. Continue to the next section of the book once you have mastered the rule presented. ■

Titrated Medications

Some critical care medications are administered at varying rates, depending on their effect on bodily functions such as heart rate and blood pressure. For example, one effect of a medication may be to raise the patient's heart rate. If the heart rate rises past a certain range, then the amount of medication must be lowered. Another example occurs when a woman is pregnant and the physician is inducing labor. The amount of medication is titrated to regulate the rate and strength of the uterine contractions of labor. The rates of such **titrated** (or regulated) **medications** have an upper and lower range. Since titrated medications are adjusted based upon their effect you must carefully monitor and adjust the rate of these medications.

When you administer titrated medications, calculate the flow rates for the lowest and highest dosages. Frequently you start administering at the lower end and increase incrementally over time until the desired effect, such as a particular heart rate, is achieved. When medications are titrated for a certain effect, you must calculate the total amount the patient has received. Start by measuring the total milliliters of infused solution. Infusion pumps have indicators that can measure the total milliliters of solution.

Rule 12-9 If you know the total amount of medication in the total volume of solution as well as the volume of solution that the patient has received, then you can use a proportion to calculate the amount of medication the patient has received (the dose).

$$\frac{\text{Total amount of medication}}{\text{Total volume of solution}} = \frac{\text{amount of medication received}}{\text{volume of solution received}}$$

Example 1 A pregnant patient has been given increasing rates of Pitocin® to induce labor. Since her arrival at the hospital, she has received 50 mL of a solution of Pitocin® that contains 20 U in 1000 mL LR. How much Pitocin® has she received?

The total amount of medication is 20 U, the total volume of solution is 1000 mL, and the volume of solution received is 50 mL.

$$\frac{20\text{ U}}{1000\text{ mL}} = \frac{?\text{ U}}{50\text{ mL}}$$

$$20\text{ U} \times 50 = 1000 \times ?$$

$$1000\text{ U} = 1000 \times ?$$

$$1\text{ U} = ?$$

The patient has received 1U of Pitocin®.

Example 2 Your patient is receiving dopamine titrated to maintain his blood pressure. His infusion started with dopamine 800 mg/D5W 250 mL at a rate of 5 mL/h. Over the last 3 h you have titrated the dopamine up to 12 mL/h to maintain the blood pressure. He has received 112 mL of the solution. How much dopamine has the patient received?

The total amount of medication is 800 mg. The total volume of solution is 250 mL. The volume of solution received is 112 mL.

$$\frac{800\text{ mg}}{250\text{ }\cancel{\text{mL}}} = \frac{?\text{ mg}}{112\text{ }\cancel{\text{mL}}}$$

$$800\text{ mg} \times 112 = 250 \times ?\text{ mg}$$

$$\frac{800\text{ mg} \times 112}{250} = ?\text{ mg}$$

$$358.4\text{ mg} = ?$$

The patient has received 358.4 mg of dopamine.

GO TO . . . Open the CD-ROM that accompanies your textbook, and select Chapter 12, Rule 12-9. Review the animation and example problems, then complete the practice problems. Continue to the next section of the book once you have mastered the rule presented. ■

REVIEW AND PRACTICE

12-3 Critical Care IV

In Exercises 1–6, find the flow rates appropriate to the IV equipment used. Round per minute rates to the nearest drop.

1. Ordered: 1.5 g Unasyn® in 50 mL of 5% DW infusing at 0.05 g/min
   a. Infusion pump
   b. Macrodrip tubing at 20 gtt/mL
2. Ordered: 250 mg dobutamine HCl in 50 mL LR infusing at 1.5 mg/min
   a. Syringe pump
   b. Microdrip tubing
3. Ordered: 2000 mg lidocaine in 500 mL NS infusing at 2 mg/min
   a. Infusion pump
   b. Macrodrip tubing at 10 gtt/mL
4. Ordered: 3000 mg lidocaine in 750 mL D5W infusing at 3 mg/min
   a. Infusion pump
   b. Macrodrip tubing at 15 gtt/mL
5. Ordered: 500 mg dobutamine HCl in 100 mL of D5W infusing at 2.4 mg/min
   a. Infusion pump
   b. Macrodrip tubing at 10 gtt/mL
6. Ordered: Demadex® 200 mg in 100 mL NS infusing at 5 mg/min
   a. Syringe pump
   b. Microdrip tubing

In Exercises 7–11, find the appropriate flow rate.

7. Ordered: Garamycin® 3 mg/kg/day IVPB q8h in 30 mL of D5W over 30 min
   On hand: Garamycin® injectable 2-mL vial with 40 mg/mL. The patient weighs 44 lb. Use microdrip tubing. The injection should be diluted with 28 mL of 5% DW.
8. Ordered: Cytoxan® 5 mg/kg IV b.i.w. in 100 mL LR over 1 h
   On hand: Cytoxan® 500-mg vial which should be dissolved in 25 mL of sterile water. The patient weighs 126 lb. Use macrodrip tubing, 15 gtt/mL. Dilute the dissolved Cytoxan in 100 mL of LR.
9. Ordered: Retrovir® 5 mg/kg/day IV q4h in 120 mL of D5W over 1 h
   On hand: Retrovir® IV, 20-mL vial with 10 mg/mL. The patient weighs 135 lb. Use macrodrip tubing, 10 gtt/mL. Dilute the desired dose in 120 mL of D5W.
10. Ordered: Mezlin 250 mg/kg/day IV q6h in 80 mL NS over 1 h
    On hand: Mezlin 20-g vial which should be reconstituted with 10 mL of sterile water for each gram. The patient weighs 152 lb. Use macrodrip tubing, 15 gtt/mL. Dilute the desired dose in 80 mL NS.
11. Ordered: Zinacef® 10 mg/ kg IVPB q8h in 100 mL of D5W over 60 min
    On hand: Zinacef® 750 mg vial 8 mL with 90 mg/mL. The patient weighs 165 lb. Use microdrip tubing. Dilute the desired dose in 92 mL of D5W.

In Exercises 12–15, find the amount of medication that has already been administered to the patient.

12. Ordered: Lidocaine 2 g in 1000 mL of D5W
    The patient has received 400 mL.
13. Ordered: Remicade® 300 mg in 250 mL of NaCl
    The patient has received 150 mL.
14. Ordered: Dobutrex® 500 mg in 500 mL of D5W
    The patient has received 120 mL.
15. Ordered: Gentamicin 80 mg in 150 mL of D5W
    The patient has received 105 mL.

For Exercises 16 to 20, the following applies:

Ordered: Dopamine: Begin infusion of 2 mcg/kg/min for bradycardia and systolic BP < 90 mmHg. Increase infusion up to 10 mcg/kg/min as needed. You have a premixed bag of dopamine with 200 mg in 500 mL of D5W. The patient weighs 165 lb.

16. At what flow rate in milliliters per hour should you start the IV?

17. The blood pressure remains low, and you increase the flow rate to 35 mL/h. What is the dosage that the patient is receiving at this time?

18. After another hour the blood pressure remains low, and you increase the rate to 50 mL/h. What is the dosage the patient is receiving at this time?

19. At the end of your shift the patient has received 425 mL of the premixed bag of 200 mg in 500 mL of D5W. What is the total amount of dopamine the patient has received on your shift?

20. What is the maximum flow rate at which the IV should run?

To check your answers, see page 539.

Preparation of Solutions, Dilutions, and Solids

Solutions are liquid mixtures containing two or more different chemicals. The liquid that is used to dissolve the other chemicals is called the **solvent,** while the chemicals dissolved in the solvent are called **solutes**. The most commonly used solvent is water, which is sometimes referred to as the *universal solvent*. An example of a solution is normal saline, which contains 0.9 g of sodium chloride (table salt) in every 100 mL of solution. In this example, sodium chloride is the solute, and water is used as the solvent.

The manufacturer prepares most of the solutions used in health care. Some common examples are injections, eyedrops, and cough syrups. It is occasionally necessary to prepare a solution "from scratch" or to dilute a solution that is more concentrated than what is needed. To do this, you will need to know how concentrations of solutions are expressed.

Percent Concentration

One of the most common ways of expressing the concentration is as a percent. Remember that *percent* means *per hundred*. When a concentration is expressed as a percent, it tells you how much of the solute is found in every 100 mL of the solution.

When the solute is a solid, the percent concentration tells you how many grams of the solute are contained in every 100 mL of the solution. For example, a 2% lidocaine solution contains 2 g of lidocaine in every 100 mL of solution. In other words, 100 mL of 2% lidocaine solution = 2 g lidocaine mixed with enough solvent to make a total of 100 mL.

When the solute is a liquid, the percent concentration tells you how many milliliters of solute are contained in every 100 mL of the solution. A 70% isopropyl alcohol solution has 70 mL of isopropyl alcohol in every 100 mL of solution. In this case, 100 mL of 70% isopropyl alcoho = 70 mL isopropyl alcohol mixed with enough solvent to make a total of 100 mL.

When the solute is a solid and the solvent is a solid, the percent concentration tells you how many grams of the solute are contained in 100 g of the product. For example, 2% hydrocortisone ointment means that every 100 g of ointment will contain 2 g of hydrocortisone. If the preparation were being compounded in the pharmacy, 2 g of hydrocortisone would be incorporated in 98 g of petroleum jelly.

Preparing Percent Solutions and Solids

Note that in these two examples we stated that the solution contained *enough solvent to make a total of 100 mL*. In the first example, the 2% lidocaine solution, you may think that you would need 100 mL of solvent. This, however, would not take into account the volume occupied by the lidocaine. To prepare a solution, you would first measure the solute and then add a *sufficient quantity of solvent to bring the total to the desired volume.*

Example 1 A "recipe" for preparing 100 mL of 2% lidocaine solution would look like this:

2% Lidocaine Solution	
Lidocaine	2 g
Water	qsad 100 mL

(The abbreviation **qsad** is taken from a Latin phrase meaning "a sufficient quantity to adjust the dimensions to . . . ." In this case, you are adding enough water to make the final volume 100 mL. When you are preparing a liquid solution, the diluent should be added up to the desired volume.

In this example you were asked how to prepare 100 mL of the solutions, and no calculations were needed. For a percent solution prepared from a solid solute, the percent strength is equal to the number of grams of solute contained in 100 mL of the solution.

Example 2 Write a recipe for preparing 100 g of 10% zinc oxide from zinc oxide powder and petroleum jelly.

10% Zinc Oxide	
Zinc oxide	10 g
Petroleum jelly	90 g

In this example a solid (zinc oxide) is being added to another solid (an ointment base). Once you determine the number of grams for the solute, you must subtract from the total number of grams to create the recipe to compound this ointment. You do not use the expression *qsad* when preparing solid mixtures.

Example 3 Write out a recipe for preparing 250 mL of 0.9% sodium chloride.

Up to this point, you have always been asked to make 100 mL of solution. Here you are asked to make a larger volume, which means that you will need more than 0.9 g of NaCl. You can calculate the quantity needed by using the methods introduced earlier in this text—the fraction proportion method, the ratio proportion method, or dimensional analysis.

Procedure 3-1

Converting by the Fraction Proportion Method

1. Write a conversion factor with the units needed in the answer in the numerator and the units you are converting from in the denominator.
2. Write a fraction with the unknown? in the numerator and the number that you need to convert in the denominator.
3. Set up the two fractions as a proportion.
4. Cancel units.
5. Cross-multiply, then solve for the unknown value.

Example 3

1. The percent concentration tells us that 100 mL solution contains 0.9 g NaCl. We need to calculate how many grams of NaCl are needed to prepare 250 mL of solution. Since we are calculating grams, our conversion factor will be written $\frac{0.9 \text{ g NaCl}}{100 \text{ mL solution}}$.
2. The other fraction for our proportion has the unknown? for a numerator and 250 mL solution as the denominator: $\frac{?}{250 \text{ mL solution}}$.
3. Setting up the two fractions as a proportion gives us the following equation:

$$\frac{?}{250 \text{ mL solution}} = \frac{0.9 \text{ g NaCl}}{100 \text{ mL solution}}$$

4. Cancel units.

$$\frac{?}{250 \cancel{\text{mL solution}}} = \frac{0.9 \text{ g NaCl}}{100 \cancel{\text{mL solution}}}$$

5. Cross-multiply to solve for the unknown.

$$100 \times ? = 0.9 \text{ g NaCl} \times 250$$

$$\frac{100 \times ?}{100} = \frac{0.9 \text{ g NaCl} \times 250}{100}$$

$$? = 2.25 \text{ g NaCl}$$

Procedure 3-2

Converting by the Ratio Proportion Method

1. Write the conversion factor as a ratio $A : B$ so that A has the units needed in the answer.
2. Write a second ratio $C : D$ so that C is the missing value and D is the number that is being converted.
3. Write the proportion in the form $A : B :: C : D$.
4. Cancel units.
5. Solve the proportion by multiplying the means and extremes.

Example 3

1. Since we are converting to grams, our conversion ratio will have grams as the first part. In a 0.9% NaCl solution, 100 mL solution contains 0.9 g NaCl. Our first ratio is 0.9 g NaCl : 100 mL.
2. The second ratio is ? : 250 mL.
3. Our proportion is:

   0.9 g NaCl : 100 mL : : ? : 250 mL
4. Cancel units.

   $0.9 \text{ g NaCl} : 100 \cancel{\text{mL}} :: ? : 250 \cancel{\text{mL}}$
5. Solve for the missing value.

$$100 \times ? = 0.9 \text{ g NaCl} \times 250$$

$$\frac{100 \times ?}{100} = \frac{0.9 \text{ g NaCl} \times 250}{100}$$

$$? = 2.25 \text{ g NaCl}$$

Procedure 3-3

Converting Using the Dimensional Analysis Method

1. Write the unknown? alone on one side of an equation.
2. On the other side of the equation, write a conversion factor with the units of measure for the answer in the numerator and the units you are converting from in the denominator.
3. Multiply the conversion factor by the number that is being converted over 1
4. Cancel units.
5. Solve the equation.

Example 3

1. ? =
2. The percent concentration tells us that 100 mL solution contains 0.9 g NaCl. We need to calculate how many grams of NaCl are needed to prepare 250 mL of solution. Since we are calculating grams, our conversion factor will be written $\frac{0.9 \text{ g NaCl}}{100 \text{ mL solution}}$.

   $? = \frac{0.9 \text{ g NaCl}}{100 \text{ mL solution}}$
3. $? = \frac{0.9 \text{ g NaCl}}{100 \text{ mL solution}} \times \frac{250 \text{ mL solution}}{1}$
4. $? = \frac{0.9 \text{ g NaCl}}{100 \cancel{\text{ mL solution}}} \times \frac{250 \cancel{\text{ mL solution}}}{1}$
5. ? = 2.25 g NaCl

Regardless of the method used, we find that 2.25 g of NaCl is needed to prepare 250 mL of a 0.9% solution. Our "recipe" looks like this:

0.9% Sodium Chloride	
NaCl	2.25 g
Water	qsad 250 mL

Preparing a Dilution from a Concentrate

In the preceding section you were asked to calculate how to prepare a solution using a solid solute. Sometimes, however, you need to prepare a solution by mixing a solution that is more concentrated than needed with one that is less concentrated than needed. (The less concentrated solution often is pure water, which would have a concentration of zero.) For example, you may need a 50% ethanol solution but find that you have only a 90% solution. In this case, you need to prepare a **dilution.**

Two methods will be presented for calculating how to prepare dilutions. One method is called the **alligation method,** and the other uses a formula. As before, you have the option of choosing a method—both will give the same answer.

Rule 12-10 Preparing Dilutions from a Concentration

To prepare a dilution from a concentration, determine:

- The volume needed V_n
- The concentration needed C_n
- The concentration(s) available C_a
- If water is being used, one of the concentrations is zero

Then use the alligation method or the formula method to obtain your answer.

Procedure Checklist 12-1

The Alligation Method

1. Write out a tic-tac-toe grid, and fill in the following values.

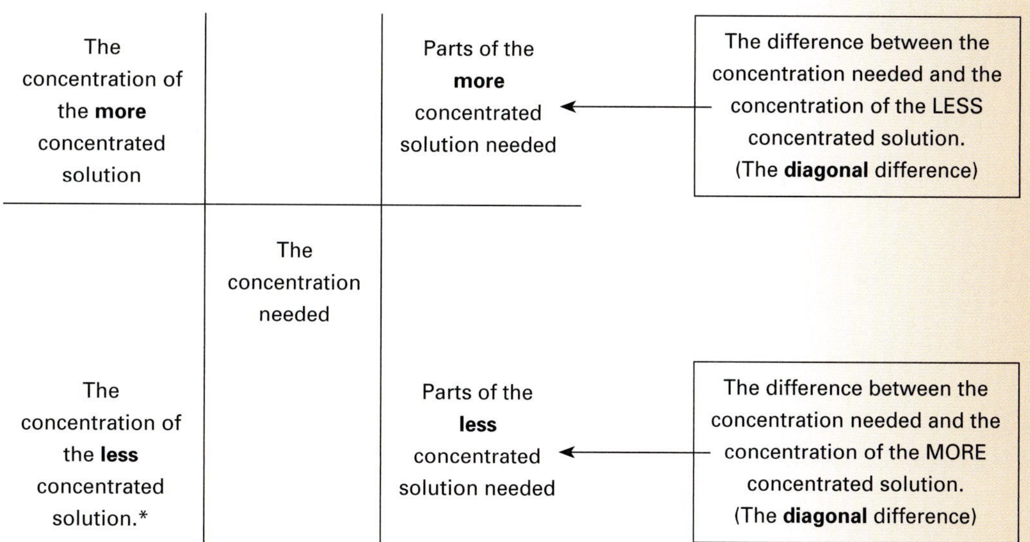

*When you are diluting with water, the less concentrated solution has a concentration of zero.

2. Find the total number of parts in the solution by adding the two values in the right column.
3. Find the volume of one part by dividing the total number of parts into the volume needed.
4. Multiply the volume of one part (answer from step 3) by the number in the top right of the grid. The result is the amount of the more concentrated solution needed.
5. Add a sufficient quantity of the less concentrated solution to bring the final volume up to the desired volume.

Example How would you prepare 500 mL of 50% ethanol from 90% ethanol? (You will use water as a diluent.)

Before you do the calculation, break down the problem to find the information that is needed.

- The volume needed is 500 mL.
- The concentration needed is 50%.
- The concentrations available are 90% and 0%.

Solution **Using Procedure Checklist 12-1, the Alligation Method**

1. Write the concentration of the more concentrated solution in the upper left, the concentration of the less concentrated solution in lower left, and the

concentration needed in the center of a tic-tac-toe grid, and then take the differences diagonally.

90		50
	50	
0		40

2. Add the numbers in the right column to find the total number of parts.

   $50 \times 40 = 90$ parts total

3. Determine the volume of 1 part by dividing the volume needed by the total number of parts.

   $\frac{500 \text{ mL}}{90 \text{ parts}} = 5.56 \text{ mL/part}$

4. Determine how much of the more concentrated solution is needed.

   5.56 mL/part $\times$ 50 parts of concentrate $=$ 278 mL

   So 278 mL of the 90% solution is needed to prepare 500 mL of a 50% solution.

5. Since the desired volume is 500 mL, you would dilute the 90% solution by adding water up to a final volume of 500 mL.

GO TO . . . Open the CD-ROM that accompanies your textbook, and select Chapter 12, Procedure Checklist 12-1. Review the animation and example problems, then complete the practice problems. Continue to the next section of the book once you have mastered the rule presented. ■

Procedure Checklist 12-2

The Formula Method

1. Identify the following information in the problem.
   - **a.** The volume of solution needed (This is V_n.)
   - **b.** The concentration of the solution needed (This is C_n.)
   - **c.** The concentration of the solution that is available (C_a.)
2. Plug the values into the following formula, in which V_a is the volume of the available solution needed to prepare the dilution.

   $$V_a = \frac{V_n \times C_n}{C_a}$$

3. Cancel units.
4. Solve the equation for V_a.

Note: The formula method for dilutions can be used only when one of the solutions being mixed has a concentration of zero.

Exampe 1 How would you prepare 500 mL of 50% ethanol from 90% ethanol? (You will use water as a diluent.)

Before you do any calculations, first you must break down the problem to find the information that is needed.

- The volume needed is 500 mL.
- The concentrations available are 90% and 0%.
- The concentrations available are 90% and 0%.

Solution **Using Procedure Checklist 12-2, the Formula Method**

1. Identify the information in the problem.
   a. The volume of solution needed: $V_n = 500$ mL
   b. The concentration of the solution needed: $C_n = 50\%$
   c. The concentration of the solution that is available: $C_a = 90\%$
2. Plug the values into the formula.

$$V_a = \frac{500 \text{ mL} \times 50\%}{90\%}$$

3. Cancel units.

$$V_a = \frac{500 \text{ mL} \times 50\cancel{\%}}{90\cancel{\%}}$$

4. Solve the equation.

$$V_a = 278 \text{ mL}$$

Regardless of the method used, we find that 278 mL of the 90% ethanol solution is needed to prepare 500 mL of a 50% solution. Our "recipe" looks like this:

50% Ethanol	
90% Ethanol	278 mL
Water	qsad 500 mL

Exampe 2 How would you prepare 100 mL of 5% iodine from 10% and 2% iodine solutions?

In this case, because neither of the solutions has a concentration of zero, we must use the alligation method.

1. Write the concentration of the more concentrated solution in the upper left, the concentration of the less concentrated solution in the lower left, and the concentration needed in the center of a tic-tac-toe grid, and then take the differences diagonally.

10		3
	5	
2		5

2. Add the numbers in the right column to find the total number of parts.

$3 + 5 = 8$ parts total

3. Determine the volume of 1 part by dividing the volume needed by the total number of parts.

$$\frac{100 \text{ mL}}{8 \text{ parts}} = 12.5 \text{ mL/part}$$

4. Determine how much of the more concentrated solution is needed.

   12.5 mL/part × 3 parts of 10% solution = 37.5 mL

   So 37.5 mL of the 10% solution is needed to prepare 100 mL of a 10% solution.

5. Since the desired volume is 100 mL, you would dilute the 10% solution by adding 2% iodine up to a final volume of 100 mL.

Our "recipe" looks like this:

5% Iodine	
10% Iodine	37.5 mL
2% Iodine	qsad 100 mL

GO TO . . . Open the CD-ROM that accompanies your textbook, and select Chapter 12, Procedure Checklist 12-2. Review the animation and example problems, then complete the practice problems. Continue to the next section of the book once you have mastered the rule presented. ■

REVIEW AND PRACTICE

12-4 Preparation of Solutions, Dilutions, and Solids

In Exercises 1 to 5, write a recipe for creating a percent solution or solid.

1. Write a recipe for preparing 500 mL of 0.9% sodium chloride.
2. Write a recipe for preparing 50 mL of a 2% lidocaine solution.
3. Write a recipe for preparing 100 g of a 3% hydrocortisone ointment from hydrocortisone powder and petroleum jelly.
4. Write a recipe for preparing 250 mL of a $\frac{1}{2}$ NS solution.
5. Write a recipe for preparing 75 g of a 20% zinc oxide ointment from zinc oxide powder and petroleum jelly.

In Exercises 6 to 10, write a recipe for preparing a dilution from a concentrate.

6. How much D5W must be mixed with D20W to make 1 L of a D12W solution?
7. How many milliliters of 95% ethyl alcohol should be mixed with water to make 1.5 L of 30% ethyl alcohol solution?
8. How would you prepare 100 mL of 20% iodine from 25% and 10% iodine solutions?
9. Using water as a diluent, how would you prepare 2 L of 50% ethanol from 90% ethanol?
10. How many grams of 10% ointment should be mixed with a 2% ointment to make a $\frac{1}{2}$ lb of a 5% ointment?

To check your answers, see pages 539–540.

HOMEWORK ASSIGNMENT

Answer the following questions.

1. What is the onset, peak, and duration for Novolin® R, a rapid acting insulin?

2. What is the onset, peak, and duration for Humulin® N, an intermediate acting insulin.

3. What is the onset, peak, and duration for Lantus®, a long acting insulin.

4. Explain what U100 means when referring to insulin.

5. List the steps in preparing a combined insulin dose.

6. Calculate the total units of insulin if the order is for 16 units Novolin® R and 30 units of Humulin® N. What type and size of syringe would you select?

7. When administering heparin Sub-Q, what action should be taken to prevent bruising?

8. Explain the term titrated.

9. A liquid solution labeled 10% hydrogen peroxide would have how many mL of hydrogen peroxide in each 100 mL of solution?

10. Using a tic-tac-toe grid to calculate a dilution is called what?

Use the identified drug labels and package inserts to answer the following questions:

11. Refer to Label A. If the order is for 1500 units heparin Sub-Q, what amount would your administer?

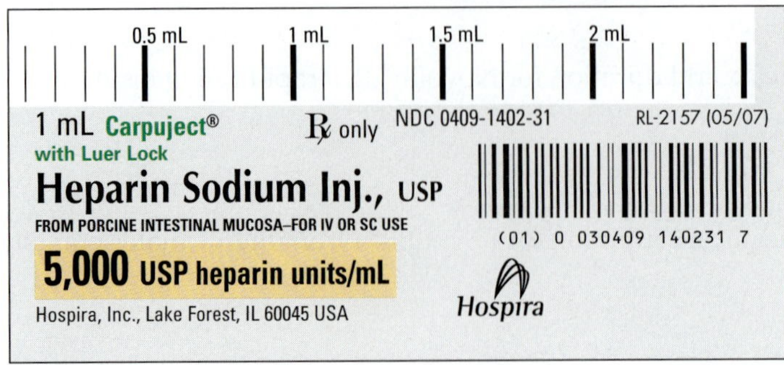

Label A

12. Refer to Label B. If the entire vial is mixed in 250 mL of 0.9% NaCl, how many units of insulin will be in each one mL?

13. Refer to Label B and Novolin® package insert information. If the patient weighs 198 pounds, what is the recommended dose per hour?

Novolin® R package insert information

> IV (adults): 0.1 unit/kg/hour as a continuous infusion

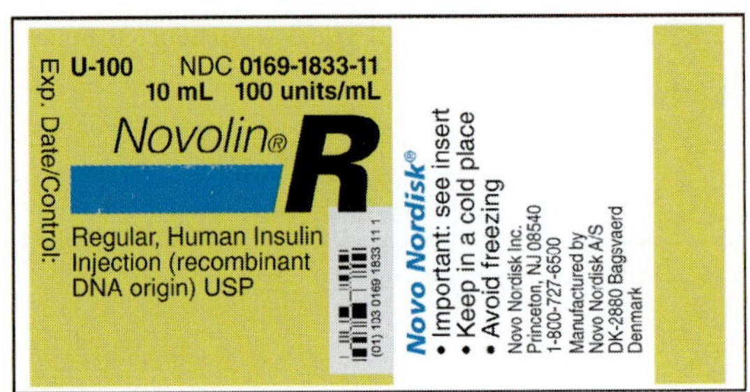

Label B

14. Refer to Label C and Ontak® package insert information. In how many milliliters of normal saline would you mix the entire vial?

15. Refer to Label C and package insert information. If your patient weighs 149.6 pounds, what is the recommended minimum dose?

16. Refer to Label C and Ontak® package insert information. If your patient weighs 50 kg, what is the recommended maximum dose?

Ontak® package insert information

> IV (adults): 9-18 mcg/kg/day over 30 minutes. For each 1 mL of ONTAK® from the vial, no more than 9 mL of sterile saline without preservative should then be added to the IV bag.

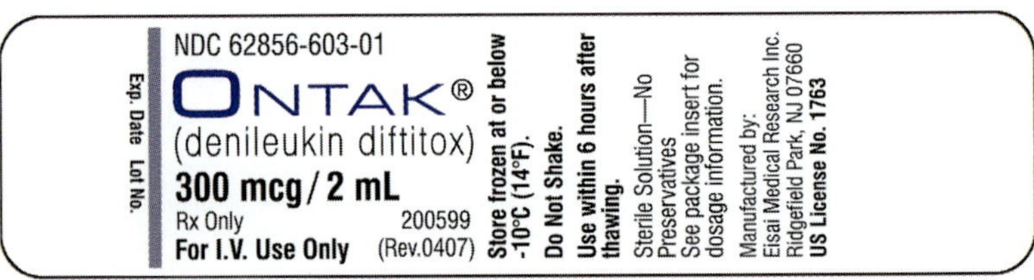

Label C

17. Refer to Label D and amiodarone HCl package insert information. If the order is for 1.8 mg/hour, what is the mL/hour when using an infusion pump?

18. Refer to Label D and amiodarone HCl package insert information. If the order is for 360 mg over 6 hours, what is the mL/hour when using an infusion pump?

Amiodarone package insert information

> IV (adults): Add 900 mg of Amiodarone to 500 mL of D5W. Infuse 150 mg over 10 minutes followed by 360 mg over the next 6 hours and then 540 mg over the next 18 hours. Continuous infusion at 0.5 mg/minute until oral therapy is begun.

NDC 63323-616-03 601603
AMIODARONE HCl
INJECTION
150 mg (50 mg/mL)
3 mL
Single Dose Vial
MUST BE DILUTED
FOR IV USE ONLY
Rx only
Sterile
Each mL contains: 50 mg amiodarone HCl, 100 mg polysorbate 80, and 20.2 mg benzyl alcohol in water for injection.
Usual Dosage: See package insert.
Protect from light and excessive heat.
Use carton to protect contents from light until used.
Store at 20° to 25°C (68° to 77°F) [see USP Controlled Room Temperature].
Vial stoppers do not contain natural rubber latex.
Abraxis
Pharmaceutical Products
Schaumburg, IL 60173
401956C
LOT/EXP
3 63323-616-03 8

Label D

19. Refer to Label E and oxytocin package insert information.. If the IV started with 0.5 milliunits/min at 3 pm and the maximum increased titration rate was used, how many milliunits/min would the patient be receiving at 4:30 pm?

Oxytocin package insert information

> To create a concentration of milliunits/mL dilute 10 units in 1 L of 0.9% NaCl. Infuse 0.5–2 milliunits/minute; increase by 1–2 milliunits/minute every 15 minutes.

NDC 63323-012-10 1210
OXYTOCIN
INJECTION, USP (SYNTHETIC)
10 USP Units/mL
For IV Infusion or IM Use
10 mL Multiple Dose Vial
Rx only
Sterile
Each mL contains: Oxytocic activity equivalent to 10 USP Oxytocin Units; chlorobutanol anhydrous (chloral derivative) 0.5%; Water for Injection q.s. Acetic acid may have been added for pH adjustment.
Usual Dosage: See Insert.
Store at 20° to 25°C (68° to 77°F) [see USP Controlled Room Temperature]. Do not permit to freeze.
Abraxis
Pharmaceutical Products
Schaumburg, IL 60173
401738E
LOT/EXP
3 63323-012-10 6

Label E

20. Refer to Label F. If the order is for 50 mL/hour, how many mEq of potassium chloride will the patient receive each hour?

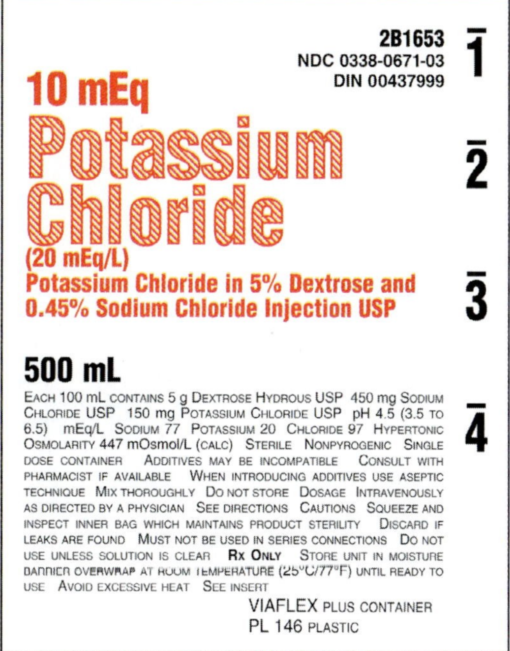

2B1653
NDC 0338-0671-03
DIN 00437999
10 mEq
Potassium Chloride
(20 mEq/L)
Potassium Chloride in 5% Dextrose and 0.45% Sodium Chloride Injection USP
500 mL
EACH 100 mL CONTAINS 5 g DEXTROSE HYDROUS USP 450 mg SODIUM CHLORIDE USP 150 mg POTASSIUM CHLORIDE USP pH 4.5 (3.5 TO 6.5) mEq/L SODIUM 77 POTASSIUM 20 CHLORIDE 97 HYPERTONIC OSMOLARITY 447 mOsmol/L (CALC) STERILE NONPYROGENIC SINGLE DOSE CONTAINER ADDITIVES MAY BE INCOMPATIBLE CONSULT WITH PHARMACIST IF AVAILABLE WHEN INTRODUCING ADDITIVES USE ASEPTIC TECHNIQUE MIX THOROUGHLY DO NOT STORE DOSAGE INTRAVENOUSLY AS DIRECTED BY A PHYSICIAN SEE DIRECTIONS CAUTIONS SQUEEZE AND INSPECT INNER BAG WHICH MAINTAINS PRODUCT STERILITY DISCARD IF LEAKS ARE FOUND MUST NOT BE USED IN SERIES CONNECTIONS DO NOT USE UNLESS SOLUTION IS CLEAR RX ONLY STORE UNIT IN MOISTURE BARRIER OVERWRAP AT ROOM TEMPERATURE (25°C/77°F) UNTIL READY TO USE AVOID EXCESSIVE HEAT SEE INSERT
VIAFLEX PLUS CONTAINER
PL 146 PLASTIC
1
2
3
4

Label F

Chapter 12 Review

Check Up

In Exercises 1–6, refer to labels A–F. Select the label corresponding to each exercise. Then mark the desired amount of insulin on the syringe.

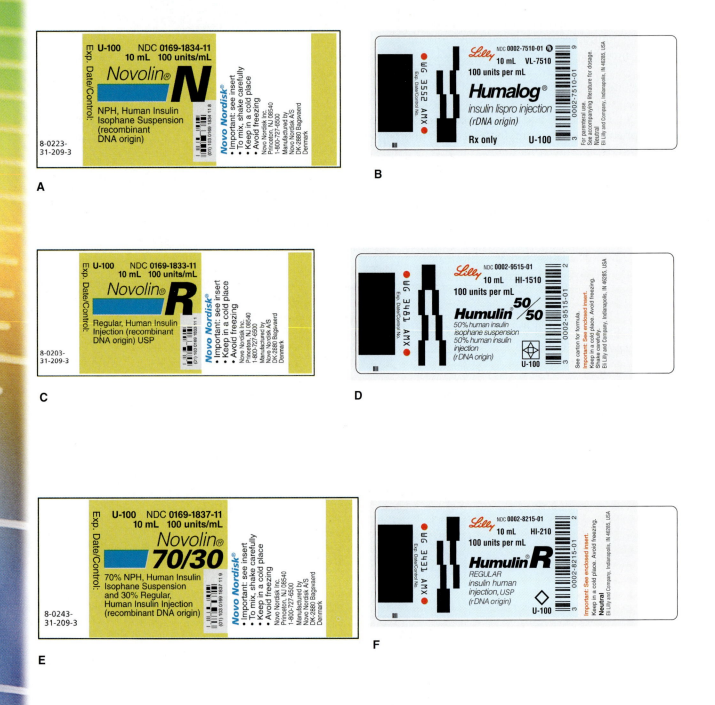

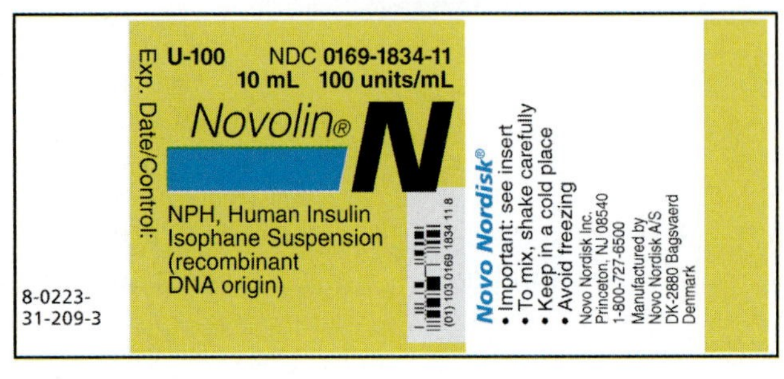

A

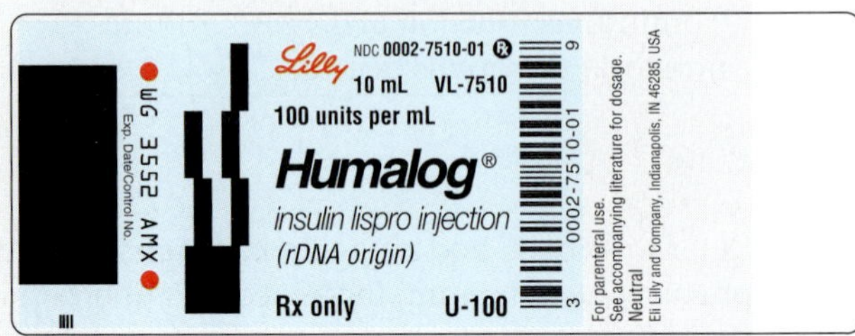

Lilly NDC 0002-7510-01
10 mL VL-7510
100 units per mL
Humalog®
insulin lispro injection
(rDNA origin)
Rx only U-100
For parenteral use.
See accompanying literature for dosage.
Neutral
Eli Lilly and Company, Indianapolis, IN 46285, USA

B

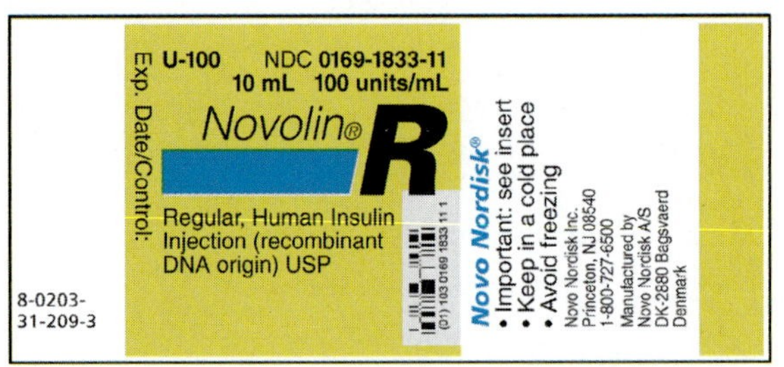

C

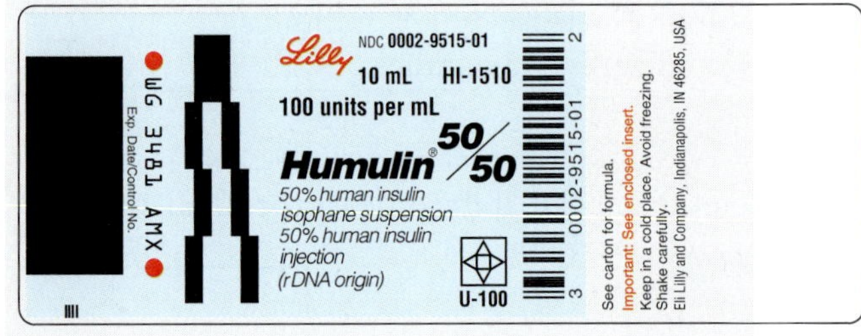

Lilly NDC 0002-9515-01
10 mL HI-1510
100 units per mL
Humulin® 50/50
50% human insulin isophane suspension
50% human insulin injection
(rDNA origin)
U-100
See carton for formula.
Important: See enclosed insert.
Keep in a cold place. Avoid freezing.
Shake carefully.
Eli Lilly and Company, Indianapolis, IN 46285, USA

D

U-100 NDC 0169-1837-11
10 mL 100 units/mL
Novolin® 70/30
70% NPH, Human Insulin Isophane Suspension and 30% Regular, Human Insulin Injection (recombinant DNA origin)
Exp. Date/Control:
Novo Nordisk®
• Important: see insert
• To mix, shake carefully
• Keep in a cold place
• Avoid freezing
Novo Nordisk Inc.
Princeton, NJ 08540
1-800-727-6500
Manufactured by
Novo Nordisk A/S
DK-2880 Bagsvaerd
Denmark
8-0243-
31-209-3

E

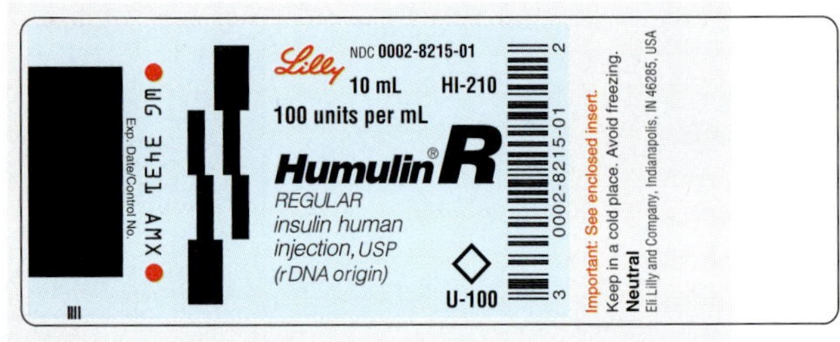

Lilly NDC 0002-8215-01
10 mL HI-210
100 units per mL
Humulin® R
REGULAR
insulin human injection, USP
(rDNA origin)
U-100
Important: See enclosed insert.
Keep in a cold place. Avoid freezing.
Neutral
Eli Lilly and Company, Indianapolis, IN 46285, USA

F

1. Ordered: Humulin® R 11 units Sub-Q ac breakfast
   Select vial: _____

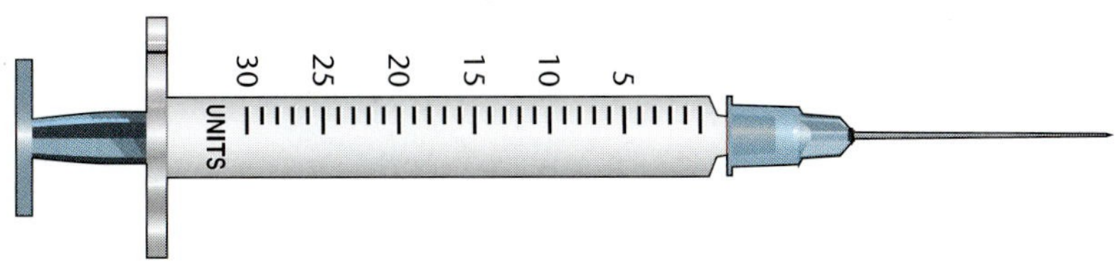

2. Ordered: Humulin® 50/50 48 units Sub-Q ac dinner
   Select vial: _____

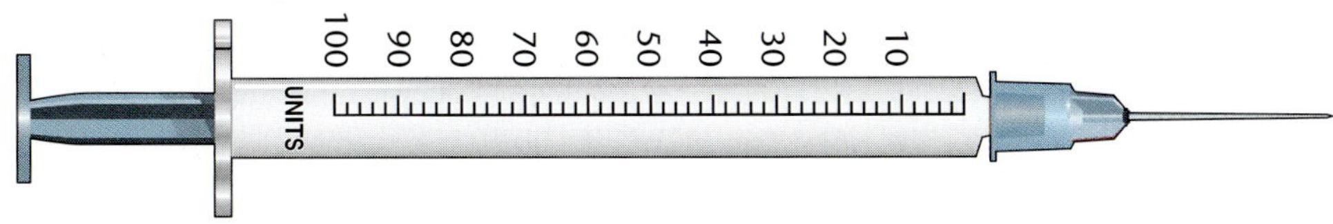

3. Ordered: Novolin® 70/30 57 units Sub-Q ac breakfast
   Select vial: _____

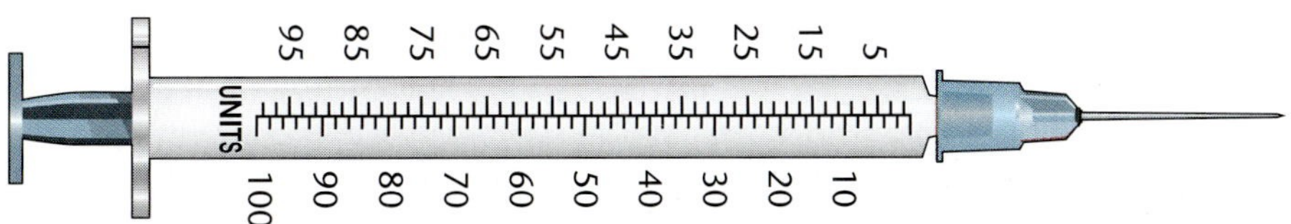

4. Ordered: Humalog® 24 units Sub-Q qd
   Select vial: _____

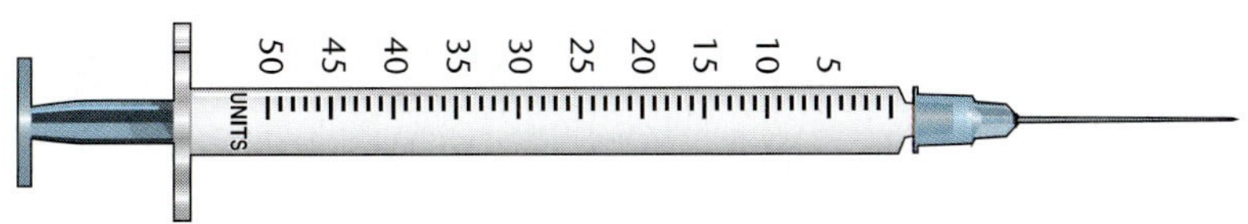

5. Ordered: Novolin® N 65 units Sub-Q ac dinner
   Select vial: _____

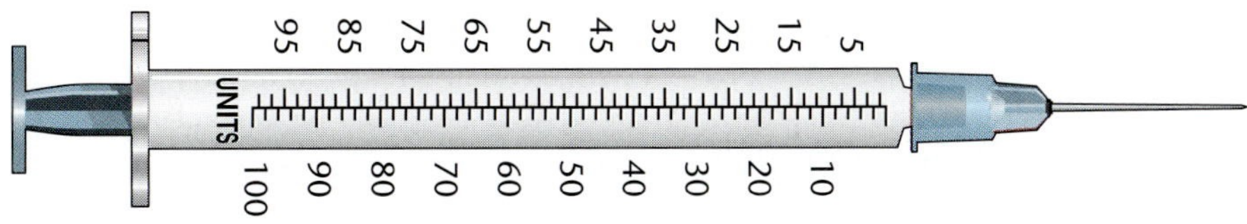

6. Ordered: Novolin® R insulin 21 units Sub-Q ac dinner
   Select vial: _____

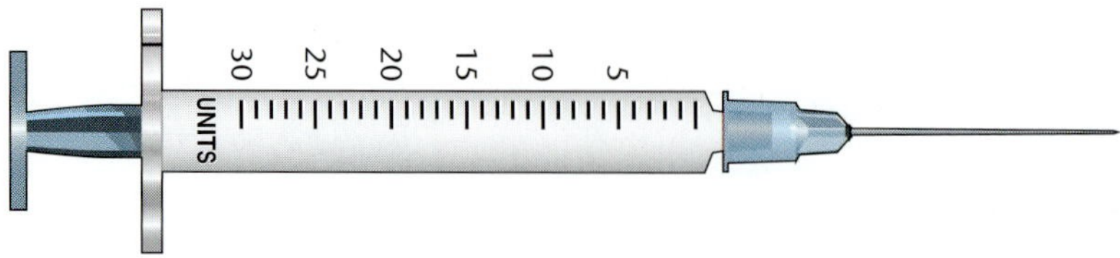

In Exercises 7 and 8, first mark on the syringe the dose of rapid-acting insulin ordered. Then mark where the leading ring will be after you draw up the intermediate-acting insulin into the same syringe.

7. Humulin® N 27 units and Humulin® R 8 units Sub-Q qam

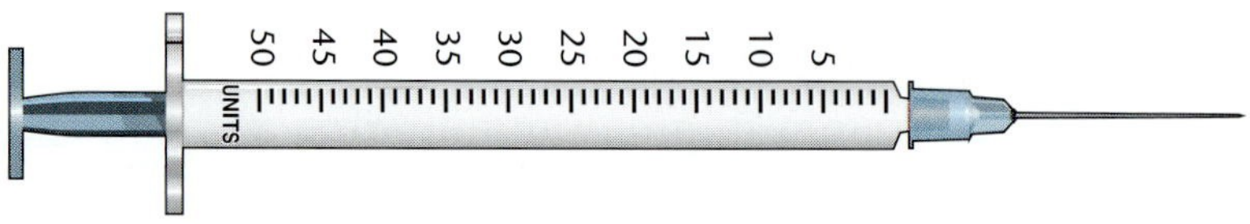

8. Novolin® R 13 units and Novolin® N 57 units Sub-Q qam

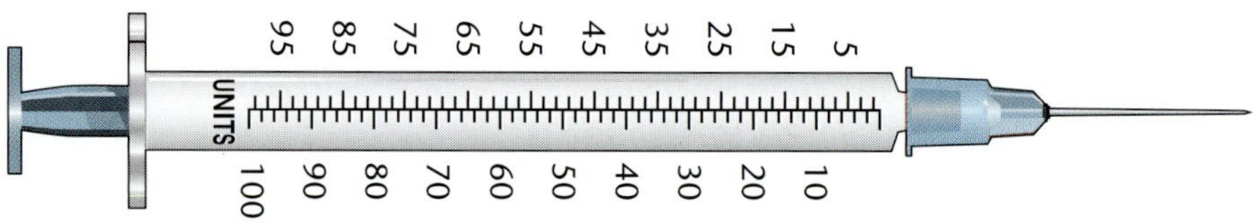

In Exercises 9–12, find the flow rate.

9. Ordered: Heparin 1500 U/h IV
   On hand: 50,000 U in 1000 mL D5W, 15 gtt/mL tubing
10. Ordered: Heparin 1500 U/h IV
    On hand: 100,000 U in 1000 mL NS, 20 gtt/mL tubing
11. Ordered: Heparin 1200 U/h IV
    On hand: 40,000 U in 500 mL D5W via infusion pump
12. Ordered: Heparin 800 U/h IV
    On hand: 20,000 units in 1000 mL NS using microdrip tubing

In Exercises 13–16, find the hourly dosage for the heparin orders. Determine whether the dosage is within the safe daily range for adults.

13. 20,000 U in 1000 mL D5W infusing at 30 gtt/min using 20 gtt/mL tubing
14. 30,000 U in 1000 mL NS infusing at 7 gtt/min using 10 gtt/mL tubing
15. 40,000 U in 1000 mL NS infusing at 40 gtt/min via microdrip tubing
16. 50,000 U in 500 mL D5W infusing at 10 mL/h via infusion pump

In Exercises 17–20, find the flow rates appropriate to the IV equipment used.

17. Ordered: Nitroprusside sodium 50 mg in 500 mL D5W at 0.35 mg/min
a. Infusion pump **b.** Macrodrip tubing at 15 gtt/mL

18. Ordered: Dopamine 800 mg in 500 mL NS at 0.9 mg/min
a. Infusion pump **b.** Microdrip tubing

19. Ordered: Isuprel 1 mg in 250 mL D5W at 3 mcg/min
a. Infusion pump **b.** Macrodrip tubing at 10 gtt/mL

20. Ordered: 4 g lidocaine in 1000 mL NS at 3 mg/min
a. Infusion pump **b.** Macrodrip tubing at 20 gtt/mL

In Exercises 21 and 22, find the appropriate flow rate.

21. Ordered: Ticar 200 mg/kg/day IV q4h in 60 mL LR over 90 min
On hand: Ticar 3-g vial. Each gram is reconstituted with 4 mL of diluent. The patient weighs 182 lb. Use macrodrip tubing, 20 gtt/mL. Dilute the reconstituted Ticar in 48 mL of LR.

22. Ordered: Septra NaCl 15 mg/kg/day IV q8h in D5W over 1 h
On hand: Septra NaCl 5-mL vial, with 80 mg/mL. The patient weighs 108 lb. Use macrodrip tubing, 15 gtt/mL. Each 5 mL must be further diluted with 75 mL of D5W.

In Exercises 23 and 24, find the amount of medication that has already been administered to the patient.

23. Ordered: Magnesium sulfate 20 g in 500 mL LR
The patient has received 200 mL.

24. Ordered: Nitroprusside 50 mg in 500 mL D5W
The patient has received 30 mL.

For Exercises 25 and 26, the following applies:

Ordered: Isuprel 4 mg IV per 500 mL D5W at 6 mcg/min. Titrate up to 10 mcg/min for systolic BP less than 90.

25. Set the infusion pump rate at ____ mL/h to start.

26. Maximum infusion pump rate to reach desired effect is ____ mL/h.

In Exercises 27–30, write a recipe for creating percent solutions or solids.

27. How can you prepare 50 g of 1% hydrocortisone cream using a 2.5% hydrocortisone cream and a cream base?

28. How would you prepare 250 mL of 40% dextrose from 80% and 25% dextrose solutions?

29. How would you prepare 1000 mL of 35% iodine from 70% and 20% iodine solutions?

30. Using distilled water as a diluent, how would you prepare 500 mL of 6% Betadine solution, using 10% Betadine solution?

Critical Thinking Applications

A patient with malignant hypertension is being treated in the critical care unit. The physician writes the following order: Nipride 50 mg in 500 mL D5W to start at 1 mcg/kg/min, and titrate to maintain the systolic BP under 140. (When you measure a patient's blood pressure, the first number represents the systolic blood pressure.) The patient weighs 176 lb. According to the product insert, the maximum safe dose of Nipride is 10 mcg/kg/min.

1. At what rate should you initially set the infusion?
2. What is the maximum safe rate for the infusion?
3. At 1600, the patient's BP is 210/105. The Nipride infusion is running at 360 mL/h. What should you do?
4. At 2000, the patient's BP is 170/90. The Nipride infusion is running at 480 mL/h. What should you do?

Case Study

A patient has a PCA pump with morphine sulfate 50 mg in 500 mL in D5W. Hospital policy requires you to document the dose of morphine administered during your shift. When you came on duty, the pump showed that 227 mL had infused. At the end of your shift the pump shows that 272 mL has infused. How much morphine did the patient receive during your shift?

To check your answers, see page 542.

Internet Activity

You are working in an intensive care unit that uses many IV critical care drugs. Checking the safe ranges of these drugs during emergency situations is both inconvenient and time-consuming. You and your coworkers decide to search the Internet for guides to dosages of commonly used drugs. Try to find answers to the following questions online.

1. What is the usual adult loading dose of amiodarone IV?
2. You are administering nitroglycerine IV and titrating the dose.
   a. By how many micrograms can you increase the dose at a time?
   b. How often can you increase the dose?
3. What treatment must be in place before tubocurarine is given?

Go To . . . Open the CD-ROM that accompanies your textbook, and complete a final review of the rules, practice problems, and activities presented for this chapter. For a final evaluation, take the chapter test and email or print your results for your instructor. A score of 95 percent or above indicates mastery of the chapter concepts. ■

Comprehensive Evaluation

The following test will help you check your dosage calculation skills. Throughout the book you have learned a variety of methods for calculating dosages. In these questions, you should use the methods with which you are most comfortable and competent. Since medication errors are serious, mastery is considered when all problems are calculated correctly.

In Exercises 1–8, refer to MAR 1.

1. What dose of Neurontin® should be administered?
2. By what route should Desyrel® be administered?
3. When should Reglan® be administered?
4. Why are no times listed for Ativan®?
5. Are any of the orders incomplete? If so, what information is missing?
6. What is the likely reason that Reglan® is administered half an hour before Neurontin®?
7. If the order for Neurontin® read q8h instead of TID, when would the second and third doses be administered?
8. If Ativan® is administered at 1330, when can the patient receive another dose?

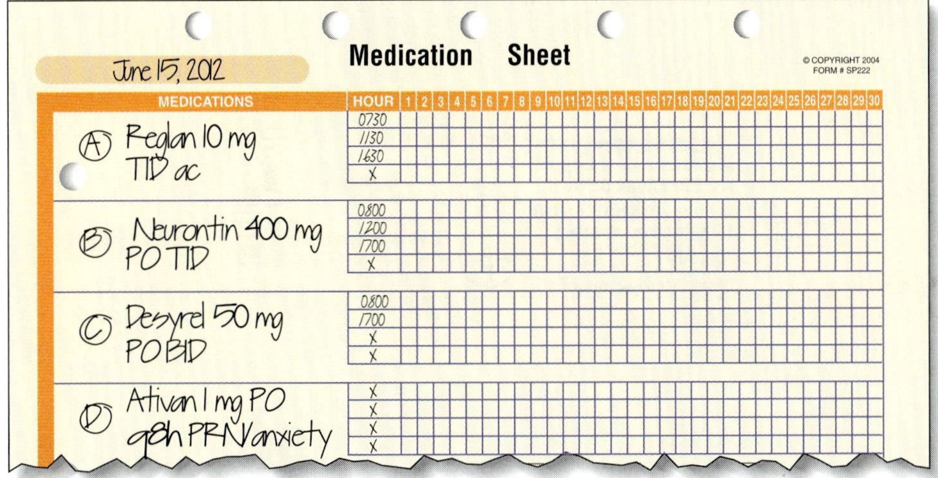

June 15, 2012 — Medication Sheet — © COPYRIGHT 2004 FORM # SP222

MEDICATIONS	HOUR
(A) Reglan 10 mg TID ac	0730, 1130, 1630, X
(B) Neurontin 400 mg PO TID	0800, 1200, 1700, X
(C) Desyrel 50 mg PO BID	0800, 1700, X, X
(D) Ativan 1 mg PO q8h PRN/anxiety	X, X, X, X

In Exercises 9–12, refer to label A.

9. What is the generic name of the drug?
10. At what temperature should the drug be stored?
11. What is the dosage strength?
12. If an adult took twice the usual adult dose, how long would the container last?

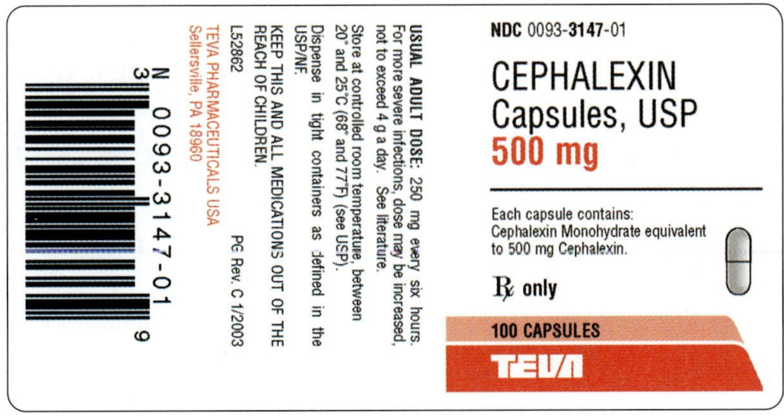
NDC 0093-3147-01
CEPHALEXIN Capsules, USP
500 mg
Each capsule contains: Cephalexin Monohydrate equivalent to 500 mg Cephalexin.
℞ only
100 CAPSULES
TEVA
USUAL ADULT DOSE: 250 mg every six hours. For more severe infections, dose may be increased, not to exceed 4 g a day. See literature.
Store at controlled room temperature, between 20° and 25°C (68° and 77°F) (see USP).
Dispense in tight containers as defined in the USP/NF.
KEEP THIS AND ALL MEDICATIONS OUT OF THE REACH OF CHILDREN.
L52862
PC Rev. C 1/2003
TEVA PHARMACEUTICALS USA
Sellersville, PA 18960
N 3 0093-3147-01 9

A

In Exercises 13 and 14, refer to label B.

13. How much fluid is used to reconstitute the entire container of suspension?

14. If the dosage prescribed for a child is 250 mg, how many doses are in the container?

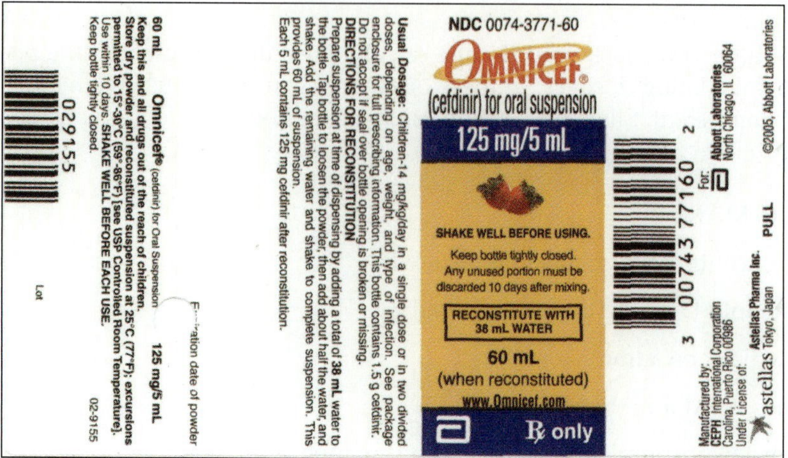

B

In Exercises 15 and 16, refer to labels C and D.

15. Using the diluent provided how much fluid is used to reconstitute the entire vial for Sub-Q use?

16. When it is reconstituted, how many international units are in 0.2 mg?

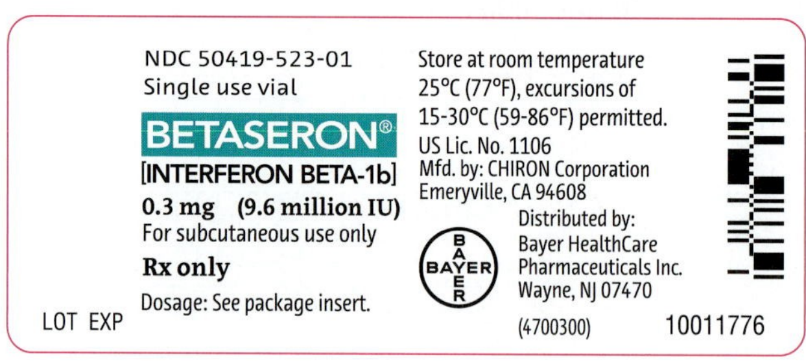

NDC 50419-523-01
Single use vial
BETASERON®
[INTERFERON BETA-1b]
0.3 mg (9.6 million IU)
For subcutaneous use only
Rx only
Dosage: See package insert.
LOT EXP
Store at room temperature 25°C (77°F), excursions of 15-30°C (59-86°F) permitted.
US Lic. No. 1106
Mfd. by: CHIRON Corporation Emeryville, CA 94608
Distributed by: Bayer HealthCare Pharmaceuticals Inc. Wayne, NJ 07470
BAYER
(4700300)
10011776

C

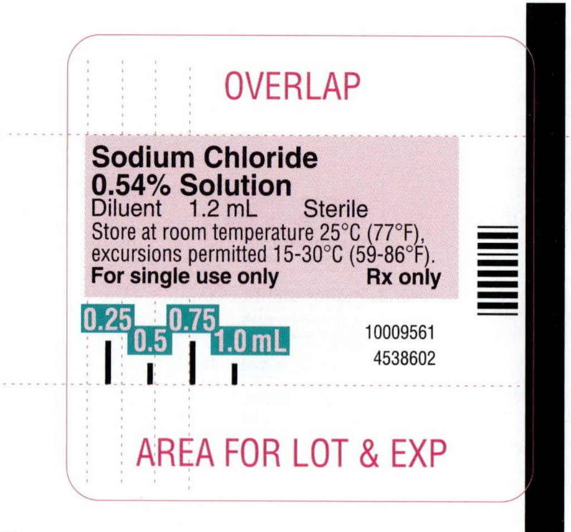

OVERLAP
Sodium Chloride
0.54% Solution
Diluent 1.2 mL Sterile
Store at room temperature 25°C (77°F), excursions permitted 15-30°C (59-86°F).
For single use only Rx only
0.25
0.5
0.75
1.0 mL
10009561
4538602
AREA FOR LOT & EXP

D

In Exercises 17–32, calculate the amount to administer.

17. Ordered: Zoloft® 75 mg PO qd
 On hand: Zoloft® 50-mg scored tablets

18. Ordered: Zovirax® 0.2 g PO q4h 5x/day
 On hand: Zovirax® suspension 200 mg/5 mL

19. Ordered: Nitroglycerin gr $\frac{1}{200}$ SL stat
 On hand: Nitroglycerin 0.3-mg tablets

20. Ordered: Morphine sulfate gr $\frac{1}{4}$ Sub-Q q4h prn/pain
 On hand: Morphine sulfate 10 mg/mL vial

21. Ordered: Claforan® 0.6 g IM 30 min pre-op
 On hand: Claforan® 300 mg/mL when reconstituted

22. Ordered: Sandostatin® 0.3 mg Sub-Q tid
 On hand: Sandostatin® 200 mcg/mL multidose vial

23. The patient is 14 years old and weighs 97 lb.
 Ordered: Agenerase® sol 17 mg/kg PO tid
 On hand: Agenerase® Oral Solution, 15 mg/mL

24. The patient is 10 years old and weighs 62 lb.
 Ordered: Vancocin® 10 mg/kg IV q6h
 On hand: Vancocin® 500 mg/100 mL

25. Ordered: Follistim® 200 IU Sub-Q qd
 On hand: Follistim® reconstituted to 225 IU/mL

26. The patient is 7 years old and weighs 49 lb.
 Ordered: Zinacef® 20 mg/kg IM q6h
 On hand: Zinacef® 220 mg/mL when reconstituted

27. Ordered: Prilosec® 40 mg PO daily
 On hand: Refer to label E.

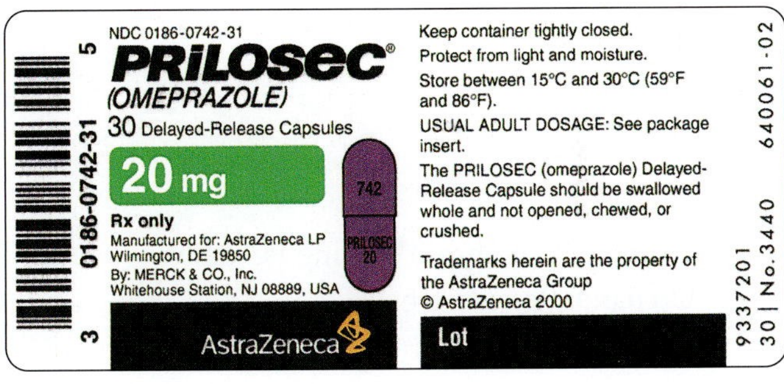

E

28. Ordered: Phenytoin extended
 300 mg PO daily
 On hand: Refer to label F.

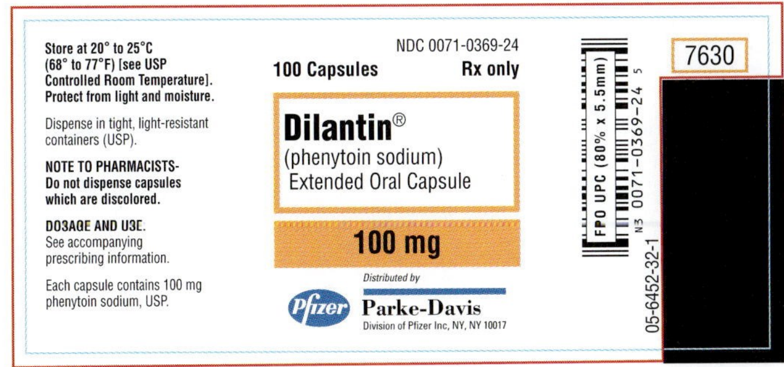

F

29. The patient is 7 years old and weighs 55 lb.
Ordered: Trileptal® 10 mg/kg PO bid
On hand: Refer to label G.

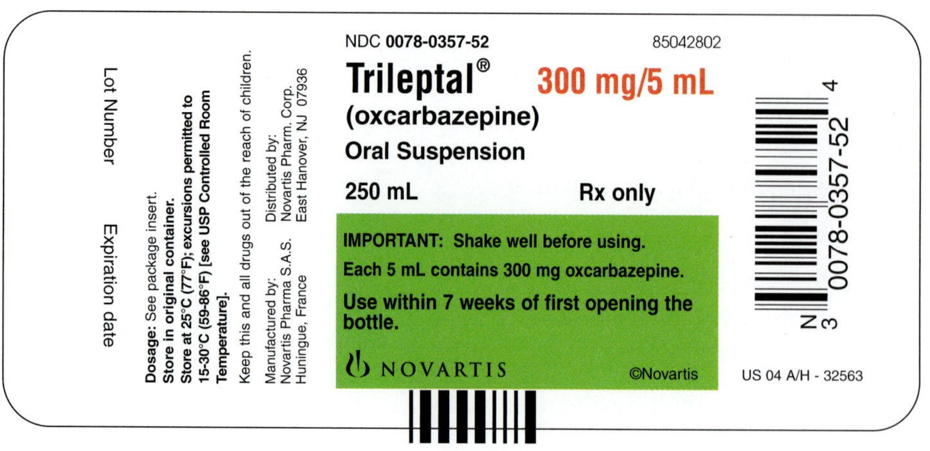

G

30. Ordered: Biaxin® XL 1 g PO q12h
On hand: Refer to label H.

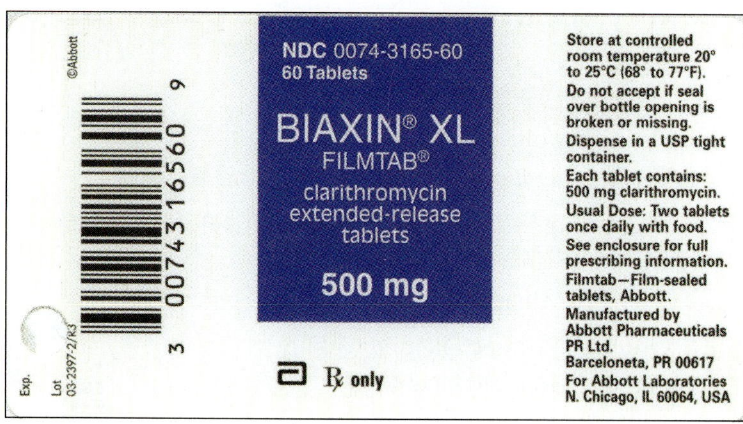

H

31. Ordered: Levothroid® 0.1 mg PO daily
On hand: Refer to label I.

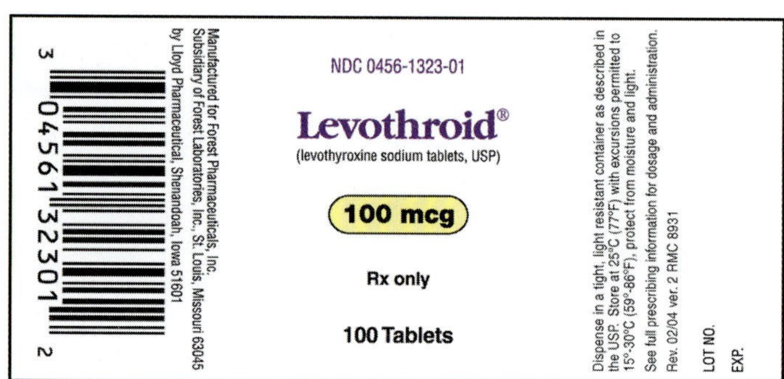

I

32. Ordered: Furosemide 75 mg IV now
On hand: Refer to label J.

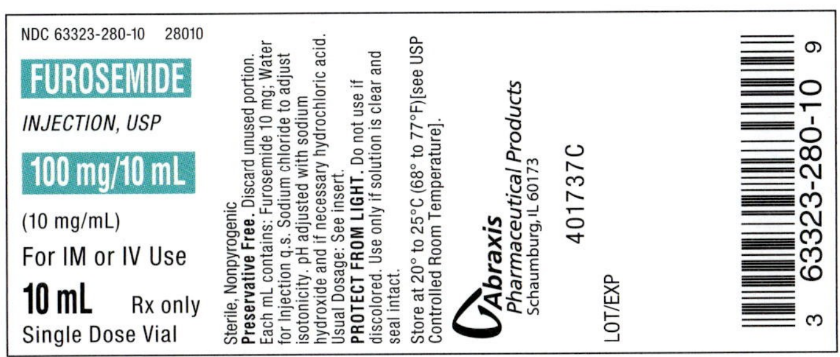

NDC 63323-280-10 28010
FUROSEMIDE
INJECTION, USP
100 mg/10 mL
(10 mg/mL)
For IM or IV Use
10 mL Rx only
Single Dose Vial
Sterile, Nonpyrogenic
Preservative Free. Discard unused portion.
Each mL contains: Furosemide 10 mg; Water for Injection q.s. Sodium chloride to adjust isotonicity, pH adjusted with sodium hydroxide and if necessary hydrochloric acid.
Usual Dosage: See insert.
PROTECT FROM LIGHT. Do not use if discolored. Use only if solution is clear and seal intact.
Store at 20° to 25°C (68° to 77°F)[see USP Controlled Room Temperature].
Abraxis Pharmaceutical Products
Schaumburg, IL 60173
401737C
LOT/EXP
3 63323-280-10 9

J

33. Which of the following insulins is rapid-acting?
Refer to labels K and L.

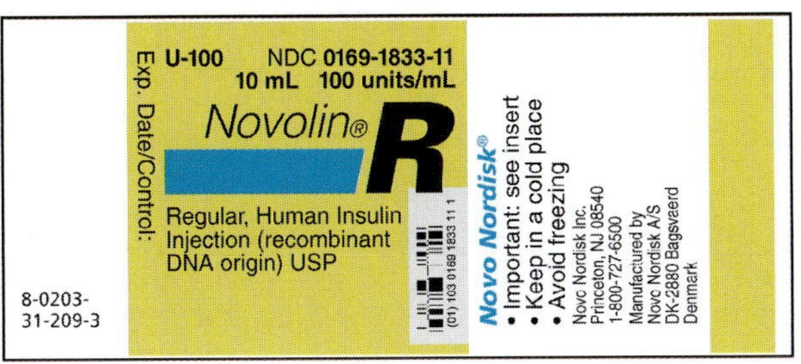

U-100 NDC 0169-1833-11
10 mL 100 units/mL
Novolin® R
Regular, Human Insulin Injection (recombinant DNA origin) USP
Exp. Date/Control:
8-0203-31-209-3
Novo Nordisk®
• Important: see insert
• Keep in a cold place
• Avoid freezing
Novo Nordisk Inc.
Princeton, NJ 08540
1-800-727-6500
Manufactured by
Novo Nordisk A/S
DK-2880 Bagsvaerd
Denmark

K

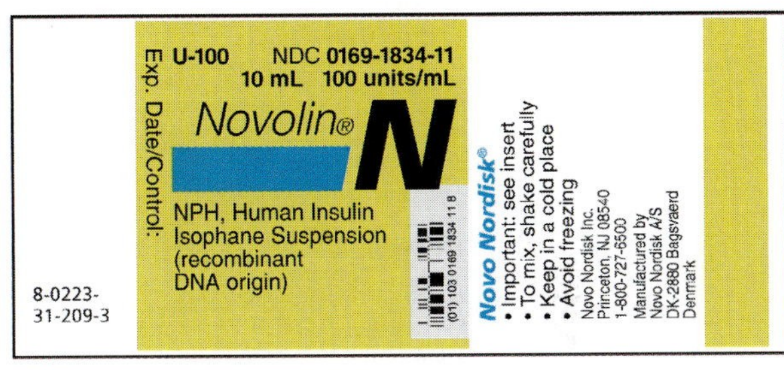

U-100 NDC 0169-1834-11
10 mL 100 units/mL
Novolin® N
NPH, Human Insulin Isophane Suspension (recombinant DNA origin)
Exp. Date/Control:
8-0223-31-209-3
Novo Nordisk®
• Important: see insert
• To mix, shake carefully
• Keep in a cold place
• Avoid freezing
Novo Nordisk Inc.
Princeton, NJ 08540
1-800-727-6500
Manufactured by
Novo Nordisk A/S
DK-2880 Bagsvaerd
Denmark

L

34. In what order will you draw the insulin into the syringe for the following order?
Ordered: Humulin® N 46 units and Humulin® R 8 units Sub-Q ac breakfast

In Exercises 35–42, find the flow rate.

35. Ordered: 1000 mL RL over 8 hours, using an infusion pump

36. Ordered: 600 mL 5%D NS over 4 h, 15 gtt/mL tubing

37. Find the flow rate for an adult who weighs 147 lb.
Ordered: Garamycin® 1.2 mg/kg IV q8h over 45 min
On hand: Garamycin® injectable, 2-mL vial with 40 mg/mL, and 15 gtt/mL tubing; the injection is diluted with 68 mL of D5W.

38. Find the flow rate for a child who weighs 68 lb.
Ordered: Zofran® 0.1 mg/kg IV over 4 min
On hand: Zofran®, premixed with 32 mg in 5% dextrose, 50 mL, and 10 gtt/mg tubing

39. Find the flow rate for an adult who weighs 134 lb.
Ordered: Ciprofloxacin 10 mg/kg in 150 mL 5% DW over 60 min. q12h
On hand: Refer to label M. Tubing is 20 gtt/mL.

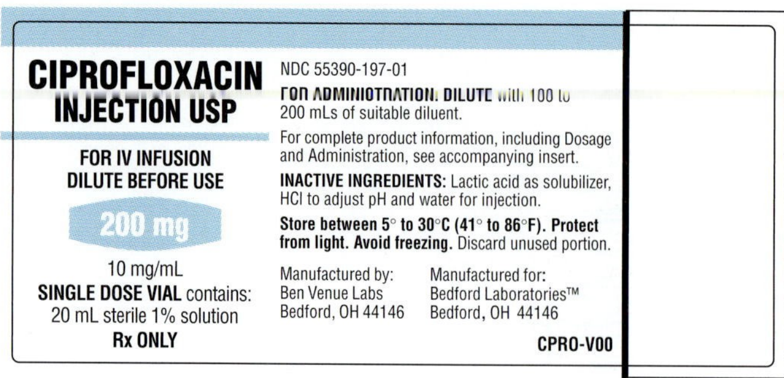

M

40. Ordered: Rocephin® 750 mg in 100 mL NS IVPB over 30 min q8h via infusion pump.
On hand: Refer to label N.

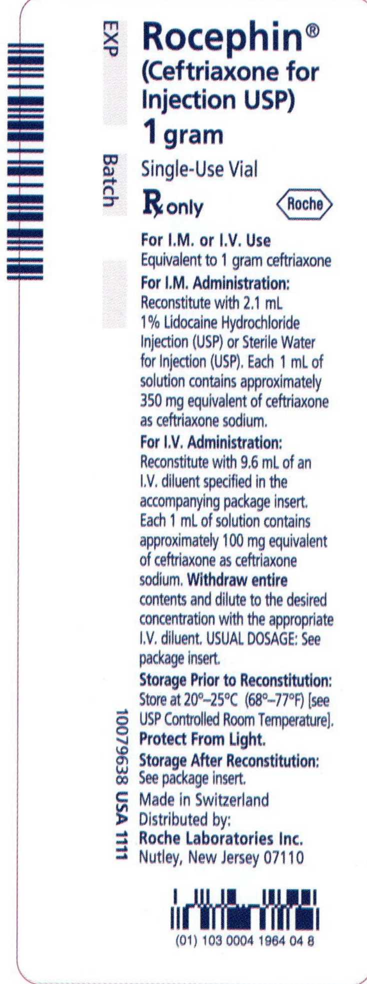

N

41. Ordered: Heparin 1300 U/h IV
On hand: 50,000 U in 1000 mL D5W, 20 gtt/mL tubing

42. Ordered: Heparin 850 U/h IV
On hand: 40,000 U in 500 mL D5W via infusion pump

43. Find the amount of medication that has already been administered to the patient.
Ordered: Nitroprusside 40 mg in 500 mL D5W
The patient has received 175 mL.

44. Calculate the original flow rate for the following order. Then determine if an adjustment is necessary and calculate the adjusted flow rate.
Ordered: 650 mL NS over 8 h (15 gtt/mL tubing)
With 5 h remaining, 490 mL of NS remains in the IV bag.

45. The patient's height is 59 in. and weight is 93 lb. What is the patient's BSA?

46. The patient's height is 150 cm and weight is 61 kg. What is the patient's BSA?

47. The adult patient's height is 66 in. and weight is 125 lb.
Ordered: BiCNU mg/m^2 IV over 2 h
How many milligrams of BiCNU should be administered?

48. The adult patient's height is 60 in. and weight is 103 lb. What is the flow rate?
Ordered: Leucovorin Calcium 200 mg/m^2 IV over 6 min
On hand: Refer to label O. Tubing is 10 gtt/mL.

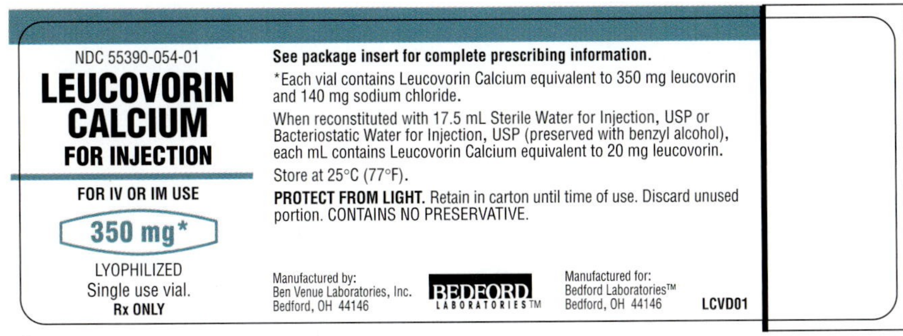

NDC 55390-054-01
LEUCOVORIN CALCIUM
FOR INJECTION
FOR IV OR IM USE
350 mg*
LYOPHILIZED
Single use vial.
Rx ONLY
See package insert for complete prescribing information.
*Each vial contains Leucovorin Calcium equivalent to 350 mg leucovorin and 140 mg sodium chloride.
When reconstituted with 17.5 mL Sterile Water for Injection, USP or Bacteriostatic Water for Injection, USP (preserved with benzyl alcohol), each mL contains Leucovorin Calcium equivalent to 20 mg leucovorin.
Store at 25°C (77°F).
PROTECT FROM LIGHT. Retain in carton until time of use. Discard unused portion. CONTAINS NO PRESERVATIVE.
Manufactured by: Ben Venue Laboratories, Inc. Bedford, OH 44146
BEDFORD LABORATORIES™
Manufactured for: Bedford Laboratories™ Bedford, OH 44146
LCVD01

O

49. Find the daily maintenance fluid needs for a child who weighs 18 kg. Then find the microdrip tubing flow rate for DMFN.

50. The patient: 78-year-old male, 5 ft 7 in. tall, 148 lb, CL_{CR} of 48 mL/min. Determine if the order is safe. If it is, then find the amount to administer.
Ordered: Timentin® 1.7 g IV q4h
On hand: Timentin® reconstituted to 20 mg/mL
According to the package insert, for patients with renal impairment, the following maintenance dosages are recommended: for creatinine clearance over 60 mL/min, 3.1 g q4h; for 30 to 60 mL/min, 2 g q4h; for 10 to 30 mL/min, 2 g q8h; for less than 10 mL/min, 2 g q12h. For patients with more advanced impairments, lower dosages are recommended.

Score 2 points for each exercise you have answered correctly. To check your answers, see pages 542–545.

Exercise Work Space

Exercise Work Space

Exercise Work Space

Exercise Work Space

Answers

PRETEST ANSWERS

1. $4\frac{2}{3}$
2. $\frac{31}{8}$
3. 6
4. 18
5. $\frac{2}{5}$
6. $\frac{3}{8}$
7. $1\frac{7}{40}$
8. $2\frac{1}{21}$
9. $\frac{9}{20}$
10. $5\frac{11}{12}$
11. $\frac{1}{15}$
12. 14
13. $\frac{5}{6}$
14. 2
15. 1.009
16. 14
17. 6.1
18. 19.20
19. 3.8
20. $\frac{9}{200}$
21. 15.3
22. 5.112
23. 14.7
24. 99.43
25. 0.224
26. 20
27. 0.008
28. 99 percent
29. $2\frac{3}{5}$
30. 112.5 percent
31. $7{:}12 = \frac{7}{12}$
32. 1 : 5
33. 0.08
34. 2 : 5
35. 37.5 percent rounded to 38 percent
36. 1 : 200
37. ? = 4
38. ? = 15
39. ? = 6
40. ? = 8
41. The health care professional gives 6 teaspoons each day.
42. The desired supply is short by $2\frac{1}{2}$ bottles. The health care professional will order 3 bottles.
43. 58.7 milliliters (mL) remains in the bottle.
44. The patient receives 1.875 milligrams (mg) over 5 days.
45. 200 mL/h
46. 3 mg
47. 8.32 mcg
48. 5 mg : 500 mL = 1 mg : 100 mL
49. 25 mg : 1 tablet
50. 125 mg/3 mL

CHAPTER 1 ANSWERS

Review and Practice

1-1 Working with Fractions and Mixed Numbers

1. 17
2. 8
3. 100
4. 1
5. a. $\frac{4}{12}$ b. $\frac{8}{12}$
6. a. $\frac{6}{20}$ b. $\frac{14}{20}$
7. $\frac{16}{3}$
8. $\frac{4}{15}$
9. $\frac{3}{4}$
10. a. = b. < c. >
11. a. > b. < c. =
12. $7\frac{1}{6}$
13. $5\frac{2}{3}$
14. 5
15. 1
16. $1\frac{3}{5}$
17. $6\frac{17}{25}$
18. $1\frac{4}{12} = 1\frac{1}{3}$
19. $\frac{50}{17}$
20. $\frac{80}{9}$
21. $\frac{11}{10}$
22. $\frac{33}{8}$
23. $\frac{311}{3}$
24. $\frac{55}{8}$
25. $\frac{41}{5}$

Review and Practice

1-2 Equivalent Fractions

For Exercises 1–8, answers may vary.

1. $\frac{8}{10}, \frac{16}{20}, \frac{20}{25}$
2. $\frac{2}{20}, \frac{3}{30}, \frac{10}{100}$
3. $\frac{2}{1}, \frac{8}{4}, \frac{16}{8}$
4. $\frac{5}{3}, \frac{30}{18}, \frac{45}{27}$
5. $\frac{18}{2}, \frac{27}{3}, \frac{81}{9}$
6. $\frac{48}{2}, \frac{72}{3}, \frac{96}{4}$
7. $\frac{14}{6}, \frac{21}{9}, \frac{28}{12}$
8. $\frac{11}{3}, \frac{66}{18}, \frac{99}{27}$
9. $\frac{14}{24}, \frac{21}{36}, \frac{28}{48}, \frac{42}{72}, \frac{70}{120}$
10. $\frac{34}{8}, \frac{51}{12}, \frac{68}{16}, \frac{85}{20}, \frac{119}{28}$
11. 6
12. 9
13. 4
14. 6
15. 12
16. 60
17. 210
18. 17
19. 16
20. 8

Review and Practice

1-3 Simplifying Fractions to Lowest Terms

1. $\frac{5}{6}$
2. $\frac{1}{2}$
3. $\frac{1}{3}$
4. $\frac{1}{2}$
5. $\frac{1}{10}$
6. $\frac{11}{20}$
7. $\frac{4}{5}$
8. $\frac{6}{17}$
9. $\frac{7}{9}$
10. $\frac{7}{10}$
11. $\frac{8}{15}$
12. $\frac{7}{12}$
13. $\frac{5}{14}$
14. $\frac{5}{8}$
15. $\frac{1}{3}$
16. $\frac{5}{7}$
17. $\frac{3}{8}$
18. $\frac{7}{11}$
19. $\frac{1}{4}$
20. $\frac{6}{11}$

Review and Practice

1-4 Finding Common Denominators

1. LCD: 21
   $\frac{7}{21}$ and $\frac{3}{21}$
2. LCD: 40
   $\frac{8}{40}$ and $\frac{5}{40}$
3. LCD: 200
   $\frac{8}{200}$ and $\frac{5}{200}$
4. LCD: 72
   $\frac{3}{72}$ and $\frac{2}{72}$
5. LCD: 12
   $\frac{6}{12}$ and $\frac{1}{12}$
6. LCD: 18
   $\frac{3}{18}$ and $\frac{1}{18}$
7. LCD: 42
   $\frac{35}{42}$ and $\frac{24}{42}$
8. LCD: 8
   $\frac{6}{8}$ and $\frac{5}{8}$
9. LCD: 36
   $\frac{4}{36}$ and $\frac{1}{36}$
10. LCD: 96
    $\frac{20}{96}$ and $\frac{9}{96}$
11. LCD: 55
    $\frac{44}{55}$ and $\frac{45}{55}$
12. LCD: 12
    $\frac{10}{12}$ and $\frac{7}{12}$
13. LCD: 240
    $\frac{88}{240}$ and $\frac{63}{240}$
14. LCD: 144
    $\frac{15}{144}$ and $\frac{14}{144}$
15. LCD: 12
    $\frac{6}{12}, \frac{4}{12}$, and $\frac{3}{12}$
16. LCD: 72
    $\frac{12}{72}, \frac{32}{72}$, and $\frac{39}{72}$
17. LCD: 45
    $\frac{30}{45}, \frac{20}{45}$, and $\frac{21}{45}$
18. LCD: 48
    $\frac{12}{48}, \frac{40}{48}$, and $\frac{21}{48}$
19. LCD: 20
    $\frac{4}{20}, \frac{6}{20}$, and $\frac{7}{20}$
20. LCD: 120
    $\frac{20}{120}, \frac{25}{120}$, and $\frac{27}{120}$

Review and Practice

1-5 Comparing Fractions

1. $<$
2. $>$
3. $=$
4. $>$
5. $=$
6. $>$
7. $>$
8. $>$

9. <
10. <
11. >
12. <
13. <
14. >
15. =
16. $\frac{5}{6}, \frac{3}{4}, \frac{4}{7}, \frac{2}{5}$
17. $\frac{4}{7}, \frac{5}{9}, \frac{1}{2}, \frac{1}{3}$
18. $\frac{5}{2}, 2\frac{1}{10}, 1\frac{3}{16}, \frac{9}{8}$
19. $2\frac{1}{2}, 1\frac{5}{6}, \frac{5}{3}, \frac{12}{9}$
20. $\frac{6}{7}, \frac{5}{8}, \frac{3}{5}, \frac{1}{2}$
21. Too Little
22. Yes, $\frac{25}{125} = \frac{1}{5}$
23. North wing
24. Martha
25. You have more time since $1\frac{3}{4}$ equals $1\frac{9}{12}$ hours

Review and Practice

1-6 Adding Fractions

1. $\frac{1}{2}$
2. $\frac{4}{7}$
3. $\frac{2}{7}$
4. $\frac{2}{3}$
5. $\frac{13}{24}$
6. $\frac{12}{25}$
7. $1\frac{5}{24}$
8. $1\frac{11}{18}$
9. $2\frac{4}{5}$
10. $3\frac{8}{11}$
11. $1\frac{5}{6}$
12. $2\frac{23}{40}$
13. $1\frac{19}{36}$
14. $5\frac{29}{35}$
15. $3\frac{5}{8}$
16. $\frac{33}{40}$
17. $\frac{47}{60}$
18. $1\frac{17}{24}$
19. $1\frac{23}{30}$
20. $\frac{19}{20}$
21. $160\frac{1}{4}$ pounds
22. $11\frac{1}{4}$ ounces
23. $2\frac{2}{3}$ cups
24. 2 tablespoons
25. 12 hours

Review and Practice

1-7 Subtracting Fractions

1. $3\frac{1}{5}$
2. $\frac{1}{5}$
3. $3\frac{1}{3}$
4. $\frac{3}{7}$
5. $\frac{7}{18}$
6. $\frac{7}{12}$
7. $1\frac{5}{8}$
8. $2\frac{1}{8}$
9. $5\frac{1}{2}$
10. $3\frac{3}{4}$
11. $11\frac{17}{30}$
12. $4\frac{9}{35}$
13. $20\frac{15}{16}$
14. $4\frac{19}{20}$
15. $5\frac{1}{3}$
16. $6\frac{4}{7}$
17. $4\frac{2}{5}$
18. $\frac{1}{6}$
19. $9\frac{11}{15}$
20. $7\frac{1}{12}$
21. $\frac{3}{8}$ cup
22. $\frac{5}{8}$ cup
23. $3\frac{3}{4}$ bottles
24. $143\frac{3}{4}$ pounds
25. $2\frac{3}{4}$ degrees

Review and Practice

1-8 Multiplying Fractions

1. $\frac{1}{48}$
2. $\frac{6}{35}$
3. $\frac{3}{8}$
4. $\frac{1}{9}$
5. $\frac{1}{6}$
6. $\frac{1}{6}$
7. 1
8. $1\frac{1}{2}$
9. $4\frac{1}{2}$
10. $2\frac{1}{2}$
11. $1\frac{1}{2}$
12. 2
13. $19\frac{7}{12}$
14. $14\frac{3}{10}$
15. $\frac{7}{24}$
16. $\frac{9}{56}$
17. $\frac{3}{4}$
18. $\frac{1}{2}$
19. $\frac{2}{5}$
20. 1

21. 234 doses

22. $1\frac{1}{8}$ grains

23. a. $1\frac{1}{3}$ teaspoons

b. $\frac{3}{4}$ teaspoon

c. $\frac{7}{12}$ teaspoon

24. 30 ounces

25. 1750 milligrams

Review and Practice

1-9 Dividing Fractions

1. $\frac{28}{45}$
2. $\frac{15}{44}$
3. $\frac{3}{4}$
4. $\frac{2}{9}$
5. $2\frac{2}{5}$
6. $1\frac{7}{15}$
7. $1\frac{1}{2}$
8. 1
9. $2\frac{5}{8}$
10. $\frac{1}{2}$
11. $1\frac{7}{9}$
12. $2\frac{4}{5}$
13. $\frac{5}{24}$
14. $9\frac{3}{5}$
15. $1\frac{1}{8}$
16. $1\frac{7}{9}$
17. $1\frac{1}{7}$
18. $1\frac{1}{6}$
19. $6\frac{1}{4}$
20. $2\frac{2}{15}$
21. 160 doses
22. 60 doses
23. 16 times
24. 48 vials
25. $12\frac{3}{4}$

Review and Practice

1-10 Working with Decimals

1. 0.2
2. 0.17
3. 6.5
4. 7.19
5. 0.003
6. 0.023
7. 5.067
8. 7.151
9. >
10. >
11. <
12. >
13. <
14. >
15. >
16. <
17. <
18. >
19. >
20. <

Critical Thinking on the Job

Rounding Errors with 9

Because the health care professional forgets to carry a unit from the tenths place into the ones place, an error results. The patient does not receive a full dose of medication. With correct rounding, the patient should receive 5.0 mL, not 4.0 mL.

Review and Practice

1-11 Rounding Decimals

1. 14.3
2. 3.5
3. 0.9
4. 0.2
5. 1
6. 0.2
7. 152.7
8. 9.29
9. 55.17
10. 4.01
11. 2.21
12. 5.52
13. 12
14. 767.46
15. 11
16. 20
17. 2
18. 51
19. 3.8 milliliters
20. 0.38 milliliters

Review and Practice

1-12 Converting Fractions into Decimals

1. 0.4
2. 0.35
3. 0.75
4. 0.5
5. 0.333
6. 0.444
7. 0.556
8. 0.583

9. 1.5
10. 2.2
11. 2.333
12. 1.125
13. 1.8
14. 2.1
15. 6.75
16. 3.5
17. 0.875
18. 0.2
19. 2.857
20. 7.667

Review and Practice

1-13 Converting Decimals into Fractions

1. $1\frac{1}{5}$
2. $98\frac{3}{5}$
3. $\frac{3}{10}$
4. $\frac{221}{500}$
5. $5\frac{3}{100}$
6. $\frac{301}{1000}$
7. $100\frac{1}{25}$
8. $206\frac{7}{100}$
9. $10\frac{17}{25}$
10. $7\frac{11}{25}$

Review and Practice

1-14 Adding and Subtracting Decimals

1. 10.82
2. 165.12
3. 13.66
4. 26.512
5. 2.51
6. 2.305
7. 0.805
8. 4.025
9. 14.25
10. 14.625
11. 5.57
12. 13.8
13. 0.82
14. 11.228
15. 7.924
16. 11.98
17. 9.6
18. 177.79
19. 65.99
20. 518.69
21. 1.9 degrees
22. $7.45
23. 2.25 grams
24. 37.4 mL
25. 74.62 kg

Review and Practice

1-15 Multiplying Decimals

1. 60.68
2. 9.031
3. 1.26
4. 0.1216
5. 0.275
6. 0.0108
7. 0.00006
8. 0.1875
9. 14.42
10. 1.08
11. 0.062
12. 0.16004
13. 16.12
14. 37.5 milliliters
15. 4.48 milliliters
16. 1.125 milligrams
17. 0.2 milligrams
18. 17.376
19. 2700.9 ounces
20. 191.4 ounces

Critical Thinking on the Job

Placing Decimals Correctly

The health care professional got confused when placing the decimal in the answer. She followed the rule for addition rather than that for multiplication of decimals. The result, had the error not been caught, would have been a disastrous overdose. The baby should have received 1.5625 g of medication, which rounds to 1.56 g.

Review and Practice

1-16 Dividing Decimals

1. 2
2. 27
3. 122.5
4. 2
5. 2.5
6. 0.4
7. 0.0125 Rounded to 0.013
8. 3.15

9. 50
10. 0.1
11. 0.2905 Rounded to 0.291
12. 0.00348 Rounded to 0.004
13. 1322.2
14. 6060
15. 0.8887 Rounded to 0.889
16. 1.231
17. 80 doses
18. 80 doses
19. 0.25 grams per dose
20. 3.75 pounds per month

Chapter 1 Review

1. $\frac{19}{8}$
2. $\frac{9}{7}$
3. $\frac{99}{10}$
4. $\frac{155}{12}$
5. $\frac{1}{3}$
6. $\frac{13}{16}$
7. 5
8. $7\frac{1}{4}$
9. LCD: 10

   $\frac{3}{10}$ and $\frac{8}{10}$
10. LCD: 18

    $\frac{15}{18}$ and $\frac{8}{18}$
11. LCD: 24

    $\frac{9}{24}$, $\frac{18}{24}$, and $\frac{4}{24}$
12. LCD: 60

    $\frac{42}{60}$, $\frac{15}{60}$, and $\frac{40}{60}$
13. $>$
14. $<$
15. $=$
16. $<$
17. $2\frac{11}{12}$
18. 2
19. $\frac{29}{100}$
20. $7\frac{3}{8}$
21. $\frac{8}{9}$
22. $\frac{1}{20}$
23. $1\frac{3}{8}$
24. $2\frac{5}{7}$
25. $\frac{5}{9}$
26. $\frac{1}{6}$
27. 8
28. $4\frac{1}{8}$
29. $\frac{4}{21}$
30. 16
31. $15\frac{3}{4}$
32. $\frac{4}{15}$
33. $>$
34. $>$
35. $=$
36. $<$
37. 0.23
38. 7.09
39. 46
40. 9.89
41. 4.3
42. 3.7
43. 7
44. 0.1
45. 9
46. 21
47. 1
48. 12
49. 0.5
50. 0.625
51. 2.6
52. 8
53. $\frac{41}{50}$
54. $\frac{13}{20}$
55. $3\frac{1}{2}$
56. $1\frac{1}{1000}$
57. 19.61
58. 4.79
59. 1.021
60. 13.704
61. 7.11
62. 0.89
63. 0.242
64. 13.222
65. 11.27
66. 0.00066
67. 4.995
68. 12.07005
69. 18.5
70. 1.5
71. 20
72. 20
73. Type O—$\frac{4}{9}$

    Type A—$\frac{5}{18}$

    Type AB—$\frac{1}{9}$

    Type B—$\frac{1}{6}$
74. Receives $\frac{2}{3}$ cup; amount is $\frac{5}{6}$ cup less than the patient should receive
75. $8\frac{3}{4}$ cups
76. a. 64 doses

    b. 40 doses

Critical Thinking Applications

$\frac{1}{16}$, $\frac{1}{8}$, $\frac{3}{16}$, $\frac{1}{4}$, $\frac{5}{16}$, $\frac{7}{16}$, $\frac{1}{2}$

$\frac{3}{8}$-diameter instrument is missing.

Case Study

Day 2, 8:00 a.m. $+1\frac{1}{2}$ pounds

Day 3, 8:00 a.m. $+1\frac{3}{4}$ pounds

Day 3, 2:00 p.m. -1 pound

Day 4, 8:00 a.m. -2 pounds

Day 4, 4:00 p.m. $-2\frac{1}{4}$ pounds

Internet Activity

Pages on the Internet may change. Check web links listed or search for new sites to review mathematics.

CHAPTER 2 ANSWERS

Review and Practice

2-1 Working with Percents

1. 0.14
2. 0.3
3. 0.02
4. 0.09
5. 1.03
6. 3
7. 0.00021
8. 0.004
9. 0.425
10. 0.038
11. 0.045
12. 2.508
13. 0.237
14. 0.018
15. 0.145
16. 404 %
17. 230 %
18. 70 %
19. 33 %
20. 6 %
21. 1.3 %
22. 1500 %
23. 3200 %
24. 12,100 %
25. $\frac{11}{50}$
26. $\frac{1}{25}$
27. $1\frac{29}{50}$
28. 3
29. $\frac{1}{1000}$
30. $\frac{1}{125}$
31. $\frac{9}{1000}$
32. $\frac{7}{500}$
33. $\frac{3}{1000}$
34. 75 %
35. 80 %
36. 17 %
37. 56 %
38. 110 %
39. 225 %
40. 175 %
41. 40 %
42. 567 %

Critical Thinking on the Job

Confusing Percent Strength with Percent Conversions

The health care professional confused percent strength with percent conversion. The definition of the percent strength is the amount of the drug, in grams, that is dissolved in 100 mL. In this case, the health care professional did not need to convert 2% to a decimal. Instead, the health care professional should have realized that two (2) grams of drug dissolved in 100 mL of solution would equal 2%.

Review and Practice

2-2 Percent Strengths of Mixtures

1. 5 g
2. 6 g
3. 10 g
4. 10 g
5. 10 g
6. 1 g
7. 100 g
8. 25 g
9. 50% solution
10. 75 g

Review and Practice

2-3 Working with Ratios

1. $\frac{3}{4}$
2. $\frac{4}{9}$
3. $1\frac{2}{3}$
4. $\frac{10}{1}$
5. $\frac{1}{20}$
6. $\frac{1}{250}$
7. $\frac{1}{3}$
8. 2 : 3
9. 6 : 7
10. 5 : 4
11. 7 : 3
12. 15 : 8
13. 10 : 3
14. 3 : 5
15. 2 : 3
16. 1 : 50
17. 1 : 75
18. 29 : 5
19. 0.25
20. 0.13
21. 0.75
22. 0.4
23. 50
24. 12.5
25. 2.67
26. 0.83
27. 0.07
28. 9 : 10
29. 3 : 10
30. 1 : 100
31. 9 : 20
32. 6 : 1
33. 12 : 5
34. 8 : 1
35. 49 : 5
36. 25 %
37. 4 %
38. 22 %
39. 41 %
40. 2000 %
41. 750 %
42. 37.5 %
43. 7 : 50
44. 13 : 20
45. 4 : 1
46. 7 : 4
47. 3 : 500
48. 9 : 5000
49. 21 : 25
50. 57 : 10,000

Critical Thinking on the Job

Reversing Terms in Ratio Strength

The health care professional should have used 100 g of dextrose in 1000 mL. Note that 100 g : 1000 mL = 1 g : 10 mL. Thinking critically about the task, the health care professional should have realized that, in a 1 g : 10 mL solution, the number of grams should be smaller than the number of milliliters. The fact that 10,000 is larger than 1000 should have alerted the professional to the error.

Review and Practice

2-4 Ratio Strengths

1. 1 g : 20 mL
2. 1 g : 20 mL
3. 5 mg : 1 capsule
4. 40 mg : 1 tablet
5. 1 mg : 5 mL
6. 1 mg : 10 mL
7. 1 g : 5 mL
8. 500 mg : 1 capsule
9. 1 mg : 2 mL
10. 250 mg : 1 tablet
11. 1 mL : 10 mL
12. 1 mL : 5 mL
13. 5 gr : 1 tablet
14. 250 mg : 1 tablet
15. 30 mg : 1 tablet
16. 10 mg : 1 tablet
17. 1 mL : 5 mL
18. 1 mL : 2 mL
19. 5 gr : 1 tablet
20. 1 g : 4 tablets

Review and Practice

2-5 Writing Proportions

1. $\frac{4}{5} = \frac{8}{10}$
2. $\frac{5}{12} = \frac{10}{24}$
3. $\frac{1}{10} = \frac{100}{1000}$
4. $\frac{2}{3} = \frac{20}{30}$
5. $\frac{50}{25} = \frac{10}{5}$
6. $\frac{6}{4} = \frac{18}{12}$
7. $\frac{5}{24} = \frac{10}{48}$
8. $\frac{75}{100} = \frac{150}{200}$
9. $\frac{4}{16} = \frac{16}{64}$
10. $\frac{125}{100} = \frac{375}{300}$
11. 3 : 4 :: 75 : 100
12. 1 : 5 :: 3 : 15
13. 8 : 4 :: 2 : 1
14. 8 : 7 :: 24 : 21
15. 18 : 16 :: 9 : 8
16. 10 : 1 :: 40 : 4
17. 5 : 7 :: 15 : 21
18. 45 : 5 :: 9 : 1
19. 1 : 100 :: 100 : 10,000
20. 36 : 12 :: 72 : 24

Review and Practice

2-6 Means and Extremes

1. True
2. Not true
3. Not true
4. True
5. True
6. 16
7. 8
8. 100
9. 20
10. 35
11. 4
12. 9
13. 65
14. 4
15. 4
16. 150 mg
17. 40 g
18. 15 mg
19. 300 mL
20. 360 mL

Critical Thinking on the Job

Confusing Multiplying Fractions with Cross-Multiplying

The health care professional realizes that mixing 10,000 g for a solution with 500 mL is not reasonable. Looking at her work, she realizes she has treated the problem as if she were multiplying fractions rather than solving a proportion problem. She has mistakenly multiplied the two numerators and set their product equal to the product of the two denominators.

To calculate correctly, the health care professional should cross-multiply 100 × ? and 5 × 500. In turn,

$$100 \times ? = 2500$$
$$? = 25.$$

She will need 25 g, *not* 10,000 g, of dextrose. The health care professional did not double-check her answer by making sure that the answer gave a true proportion. If she had, she would have found that 5 g/100 mL is not equal to 10,000 g/500 mL. Fortunately, the answer that she calculated was not reasonable, and she was able to detect the error. Sometimes, however, even a wrong answer can be reasonable. *Always double-check your answer when performing dosage calculations.*

Review and Practice

2-7 Cross-Multiplying

1. Not true
2. True
3. True
4. Not true
5. Not true
6. 1
7. 25
8. 24
9. 150
10. 200
11. 1
12. 1
13. 9
14. 6
15. 30
16. 42 tablets
17. 300 mg
18. 24 h
19. 250 mg
20. 1250 mL

Critical Thinking on the Job

Setting Up the Correct Proportion

The health care professional realizes while solving the equation that he will not be able to cancel any units, and that his answer will contain both milligrams and milliliters. After looking at the proportion once again, he discovers that he has set up the proportion incorrectly, and that the first fraction is written upside down. He rewrites the proportion as 125 mg/mL = 250 mg/?. After canceling the milligrams in the numerator, he solves the equation and correctly determines that the dose is 10 mL. Critical thinking is especially important when you use a calculator. Had the health care professional ignored the units and simply punched in the numbers, he would have come up with the wrong dose. Always include the units when you perform drug calculations.

Chapter 2 Review

Exercise	Fraction	Decimal	Ratio	Percent
1.	$\frac{2}{3}$	0.67	2 : 3	67
2.	$\frac{5}{4}$	1.25	5 : 4	125
3.	$\frac{28}{100}$	0.28	28 : 100	28
4.	$\frac{3}{100}$	0.03	3 : 100	3
5.	$\frac{40}{8}$	5	40 : 8	500
6.	$\frac{4}{12}$	0.33	4 : 12	33
7.	$\frac{9}{27}$	0.33	9 : 27	33
8.	$\frac{7}{5}$	1.4	7 : 5	140
9.	$\frac{1}{200}$	0.005	1 : 200	0.5
10.	$\frac{3}{50}$	0.06	3 : 50	6
11.	$\frac{1}{4}$	0.25	1 : 4	25
12.	$\frac{6}{1}$	6	6 : 1	600
13.	$\frac{1}{9}$	0.11	1 : 9	11
14.	$\frac{150}{100}$	1.5	3 : 2 150 : 100	150
15.	$\frac{6}{97}$	0.06	6 : 97	6
16.	$\frac{128}{10}$	12.8	64 : 5 128 : 10	1280

17. 40
18. 1
19. 4
20. 25
21. 48
22. 1
23. 2
24. 5
25. 75 mg
26. 300 mL
27. 25 g
28. 150 mL
29. 2 mL
30. 30 g : 100 mL

Critical Thinking Applications

5 g

Case Study

1. 250 mg
2. 10 mL

Internet Activity

Answers may vary.
Solutions: (1) Auralgan otic solution, (2) Eye irrigating solution, (3) SSKI solution
Lotions: (1) Cordran® lotion, (2) Hytone® lotion, (3) Klaron lotion
Creams: (1) Bactroban® cream, (2) Benzashave cream, (3) Silvadene® 1% cream
Suspensions: (1) Carafate® suspension, (2) Betaxon opthalmic suspension, (3) Codeine suspension
Ointments: (1) Cetamide ointment, (2) Cortosporin® ointment, (3) Neosporin® opthalmic ointment

CHAPTER 3 ANSWERS

Review and Practice

3-1 Understanding Metric Notation

1. d
2. d
3. c
4. d
5. b
6. c
7. d
8. b
9. c
10. a
11. 4.5 mL
12. 0.62 g
13. 0.75 mL
14. 0.7 m
15. 12 L
16. 0.75 kg
17. 157 km
18. 7.75 cm
19. 93 mcg
20. 0.08 mg

Critical Thinking on the Job

Placing the Decimal Point Correctly

When you convert quantities from one unit of measure to another, pay close attention to the decimal point. In going from milligrams (mg) to micrograms (mcg), the quantity should be multiplied by 1000; the decimal should move three places to the right. The problem should have been calculated as follows:

$$0.05 \text{ mg} \times 1000 = 50 \text{ mcg}$$

$$0.050 \text{ mg} = \underset{\longrightarrow}{0050.} \text{ mcg}$$

Be even more careful when the patient is a child. Dosages that are perfectly safe for adults may be life-threatening for children. As you will learn in Chapter 6, you must carefully read the labels on all drugs. In the case of Lanoxin®, both the elixir and the injection have labels marked in both micrograms (mcg) and milligrams (mg). A careful look at the labels would help prevent this error.

Review and Practice

3-2 Converting Within the Metric System

1. 7000 mg
2. 1.2 g
3. 0.023 kg
4. 8000 g
5. 8010 mL
6. 0.1 L
7. 3600 mm
8. 5.233 m
9. 0.5 km
10. 3250 m
11. 250 mcg
12. 462,000 mcg
13. 0.25 mg
14. 0.075 mg
15. 60,000 mcg
16. 500,000 mcg
17. 0.008 g
18. 0.02 g
19. 56.2 cm
20. 0.0432 m

Review and Practice

3-3 Other Systems of Measurement

1. m or ♏
2. dr or ʒ
3. gr
4. oz or ℥
5. gtt
6. tsp or t
7. tbsp or T
8. pt
9. mEq
10. U
11. gr vii or gr $\overline{\text{vii}}$
12. ℥ v or ℥ $\overline{\text{v}}$
13. 3 oz or ℥ iii or ℥ $\overline{\text{iii}}$
14. 8 oz or ℥ viii or ℥ $\overline{\text{viii}}$
15. gr 14 or gr $\overline{\text{xiv}}$
16. gr 17 or gr $\overline{\text{xvii}}$
17. $\frac{1}{2}$ tsp
18. $\frac{1}{2}$ tbsp
19. gr ss
20. $\frac{1}{2}$ oz or ℥ ss
21. $2\frac{1}{2}$ oz or ℥ iiss
22. $5\frac{1}{2}$ oz or ℥ vss
23. 2000 units
24. 40 mEq
25. 49 units

Critical Thinking on the Job

Selecting the Correct Conversion Factor

Even though Greg set up the proportion correctly, he used the wrong conversion factor. Certain equivalent measures, such as 1 tsp = 5 mL, are so commonly used that you should memorize them. When you are using a conversion chart, always double-check that you select the correct unit of measure and be sure that you read across the same line. Had Greg used the correct conversion, 1 tsp = 5 mL, he would have calculated as follows:

$$\frac{?}{30 \text{ mL}} = \frac{1 \text{ tsp}}{5 \text{ mL}}$$

$$? \times 5 = 1 \text{ tsp} \times 30$$

$$? = 6 \text{ tsp}$$

The patient would be told to take 6 tsp of medication, not 2 tsp.

Review and Practice

3-4 Converting Among Metric, Apothecary, and Household Systems

1. 4 tsp (using Table 3-7)
2. 25 tsp
3. 24 tsp
4. 8 oz
5. gr $\frac{1}{4}$
6. 900 mg
7. gr $\frac{1}{6}$
8. gr 37ss
9. 92.4 lb
10. 20 kg
11. 180 mL
12. $1\frac{1}{2}$ pt
13. 10 mL
14. 143 lb
15. 85 kg
16. $2\frac{2}{3}$ tbsp (using Table 3-7)
17. 960 mL
18. 180 mg (using gr i̇ = 60 mg)
19. gr ss
20. 22.5 mL

Review and Practice

3-5 Temperature

1. 93.2°F
2. 105.8°F
3. 35°C
4. 38.9°C
5. 7.4°C
6. 100°C
7. 77°F
8. 212°F
9. 15°C
10. 152.6°F

Review and Practice

3-6 Time

1. 0235
2. 0757
3. 0008
4. 0055
5. 1349
6. 1514
7. 2354
8. 2219
9. 1859
10. 0426
11. 12:11 a.m.
12. 12:36 a.m.
13. 3:25 a.m.
14. 8:49 a.m.
15. 1:13 p.m.
16. 3:27 p.m.
17. 9:45 p.m.
18. 11:59 p.m.
19. 8:37 p.m.
20. 6:18 p.m.

Chapter 3 Review

Check Up

1. 25.5 kg
2. 0.45 cm
3. 40 mcg
4. 0.75 L
5. 0.9 mg
6. gr iss
7. 0.375 g
8. 12 mL
9. 60 mg
10. 0.125 mg
11. 4 m
12. 7.5 mm
13. 0.965 L
14. 8 mL
15. 320 g
16. 50 mcg
17. 0.988 km
18. 0.01725 km
19. 0.368 g
20. 0.247 kg
21. gr 120
22. 150 mg
23. 6 tbsp
24. 25 mL
25. 30 mL
26. 40 oz
27. gr ix
28. 45 mg
29. 81 kg
30. 103.4 lb
31. 325 mg
32. 20 kg
33. 15 mL
34. 45 mL
35. $2\frac{1}{2}$ qt per day
36. 90 kg
37. 36.4°C
38. 22.2°C
39. 14.1°C
40. 28.2°C
41. 75.2°F
42. 110.8°F
43. 60.1°F
44. 47.8°F
45. 0321
46. 1642
47. 2247
48. 1120
49. 12:29 a.m.
50. 2:17 p.m.
51. 8:53 p.m.
52. 9:12 a.m.

Critical Thinking Applications

1. Preferably a calibrated or baking tablespoon, or a calibrated or baking teaspoon
2. 1 tbsp of medication every 8 h
3. 2 half-pint bottles

Case Study

Between $2.\overline{2}$ and 5°C

Internet Activity

Try *refdesk.com*. Click on "Convert anything" on bottom of left column

a. Answer varies based upon weight.

b. 36.7777778°C

c. 0.5 g

d. 1 oz

CHAPTER 4 ANSWERS

Critical Thinking on the Job

Use the Correct Dropper

This error could easily be avoided. During the initial visit, the father should have been instructed to use *only* the dropper that accompanies the medication. When he calls 2 days later, he should be asked if the *correct equipment* was used.

Review and Practice

4-1 Oral Administration

1. False
2. False
3. False
4. True
5. False
6. False
7. True
8. True
9. False
10. False
11. False
12. False
13. True
14. False
15. d
16. c
17. d
18. b
19. c
20. d

Critical Thinking on the Job

Finishing What You Start

This overdose could be avoided. The first health care professional should check the prescribed dose and immediately discard the extra medication from the syringe. He should personally deliver the dose or in some cases identify its contents prior to leaving the room. The second health care professional also should take extra precautions. She should confirm directly with the first health care professional the amount to deliver and check the patient's chart before administering medication. She should not have administered from a syringe unless she was certain that its contents matched the amount ordered.

Review and Practice

4-2 Hypodermic Syringes

1. Tenths of a mL (0.1 mL)
2. Units (1 unit or 0.01 mL)
3. Hundredths of a mL (0.01 mL)
4. Two-tenths of a mL (0.2 mL)
5. True
6. True
7. False
8. True
9. False
10. False
11. False
12. False
13. True
14. Type: standard, volume: 1 mL
15. Type: insulin, volume: 49 units
16. Type: tuberculin, volume: 0.05 mL
17. Type: large-capacity, volume: 4.2 mL
18. Type: standard, volume: 2.2 mL
19. Type: insulin, volume: 22 units
20. Type: tuberculin, volume: 0.8 mL
21. Type: standard, volume: 3 mL
22. Type: large-capacity, volume: 9.5 mL or 9.4 mL (closest calibration mark)
23. Type: tuberculin, volume: 0.2 mL
24. Type: large-capacity, volume: 2.8 mL
25. Type: insulin, volume: 1 unit

Review and Practice

4-3 Other Administration Forms

1. F
2. A
3. E
4. B
5. G
6. B
7. C
8. D
9. F, B, G
10. C

Critical Thinking on the Job

Document the Use of Patches

This problem can be avoided if the second health care professional is more careful in assessing the patient before placing a second patch. The second health care professional should not assume that the patch is not in place. At the same time, the first health care professional should indicate clearly on the patient's chart when the patch was placed and its exact location. Because patches are so potent, they make it easy for health care professionals to give an overdose of the drug. You must carefully document the placement of patches and ensure that no patch is already in place when you administer a new one.

Chapter 4 Review

Check Up

1. A and C
2. C
3. D
4. C and D
5. D
6. A
7. C
8. B
9. A
10. D
11. True
12. False
13. False
14. True
15. True
16. False
17. True
18. False
19. False
20. False
21.

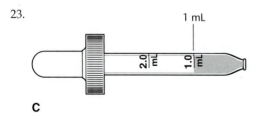

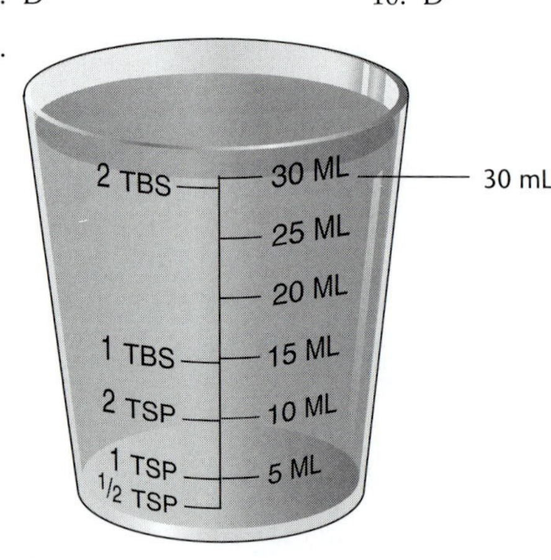

A

22.

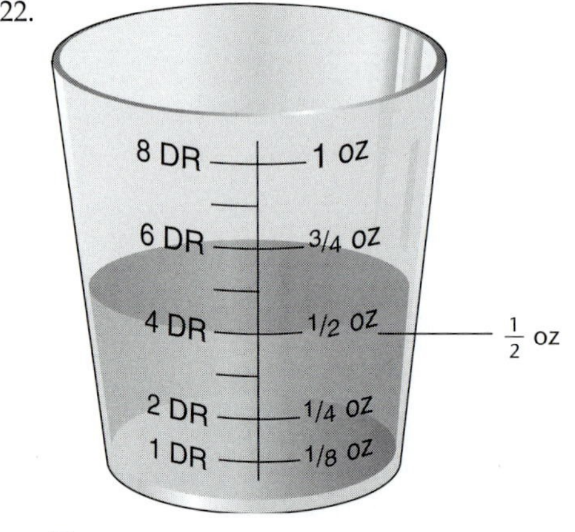

B

23.

1 mL

2.0 mL 1.0 mL

C

24.

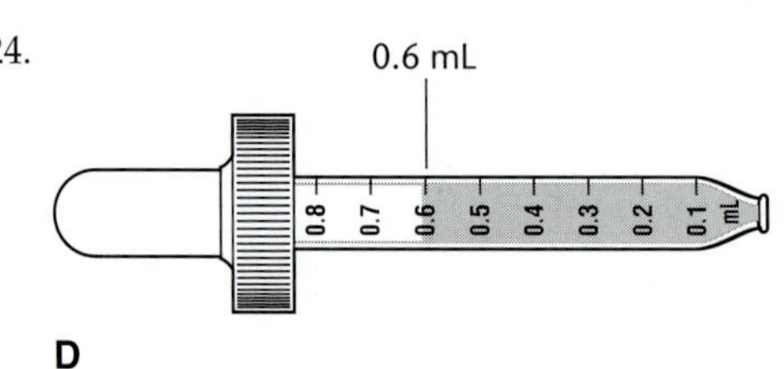

25.

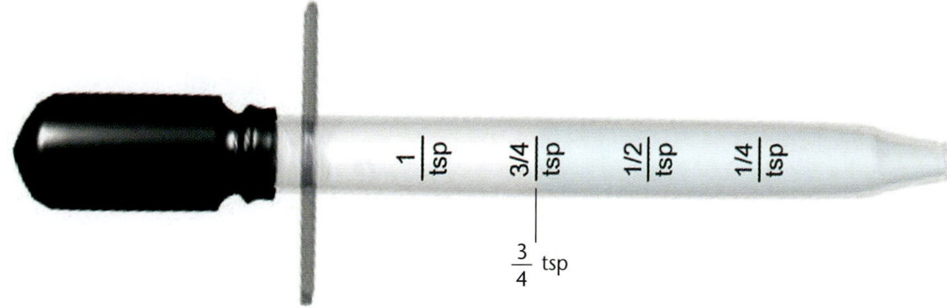
1 tsp
3/4 tsp
1/2 tsp
1/4 tsp
3/4 tsp
E

26.

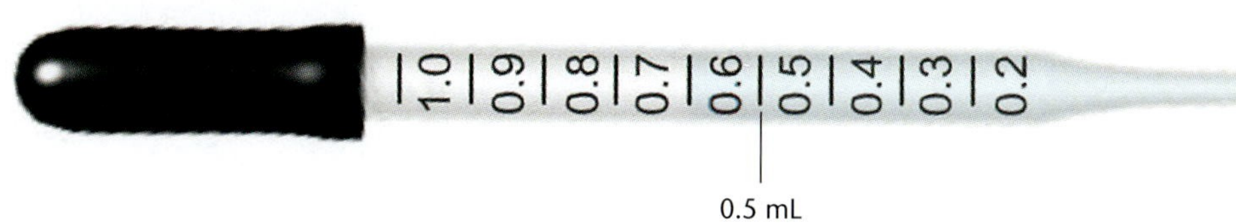
1.0
0.9
0.8
0.7
0.6
0.5
0.4
0.3
0.2
0.5 mL
F

27.

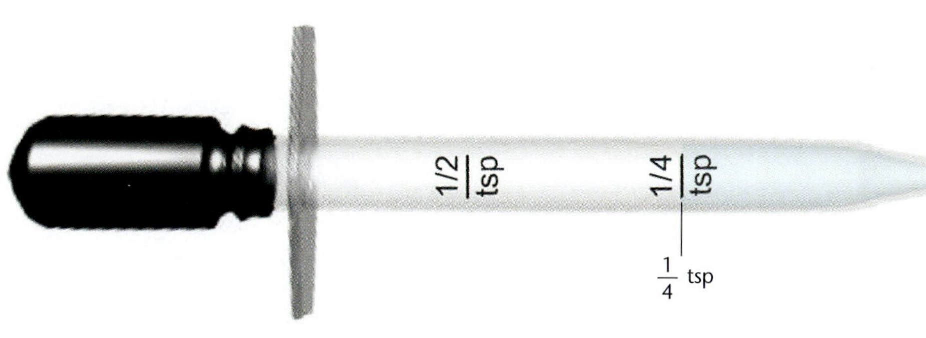
1/2 tsp
1/4 tsp
1/4 tsp
G

28.

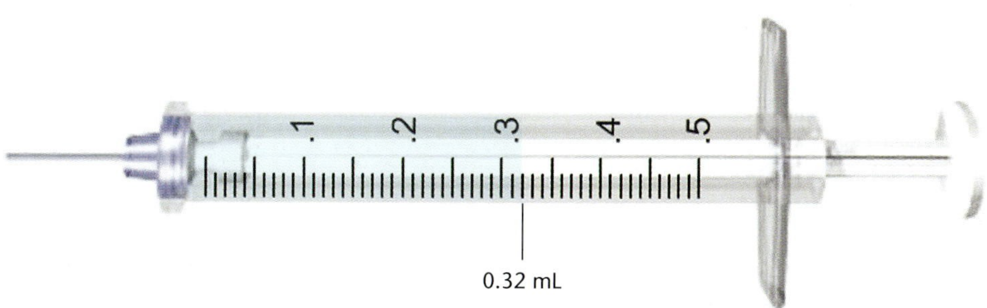
.1
.2
.3
.4
.5
0.32 mL
H

29.

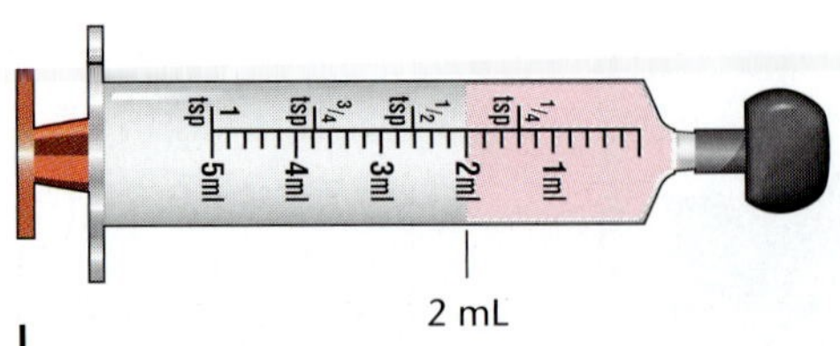

30.

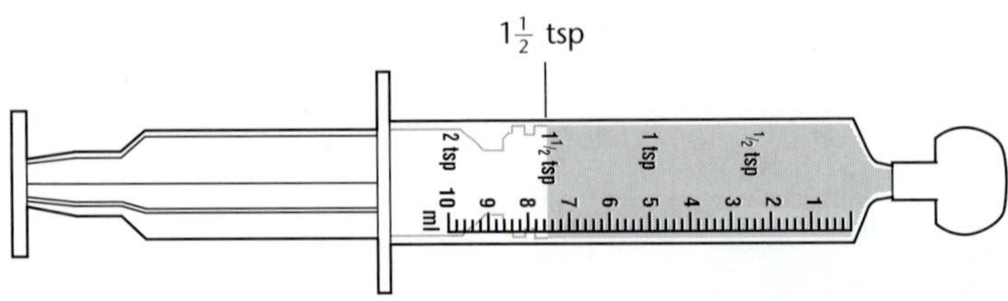

31.

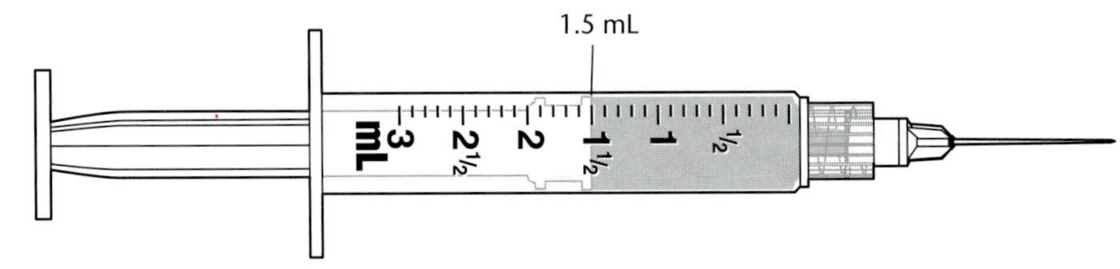

32.

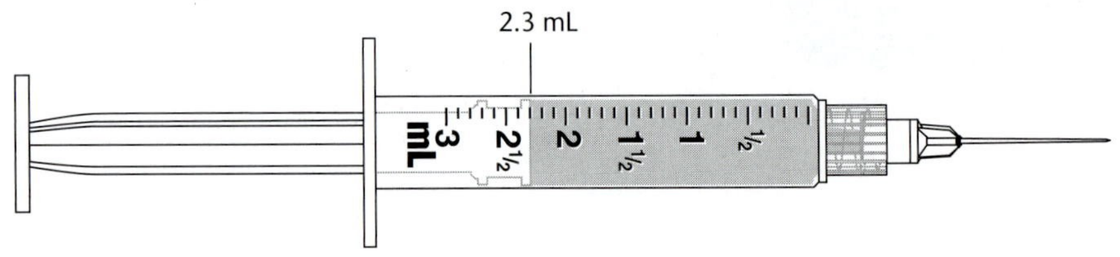

33.

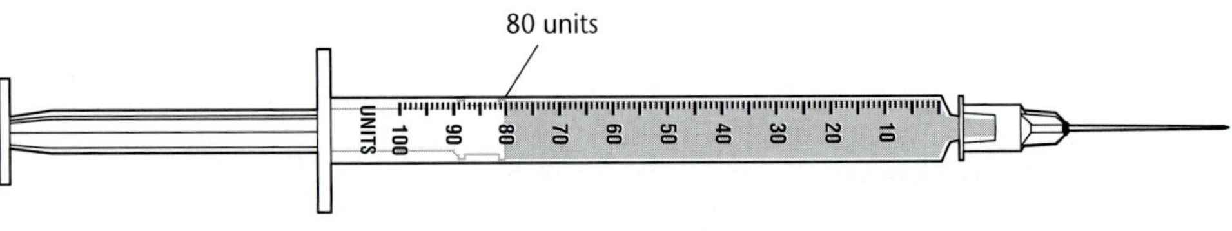

34.

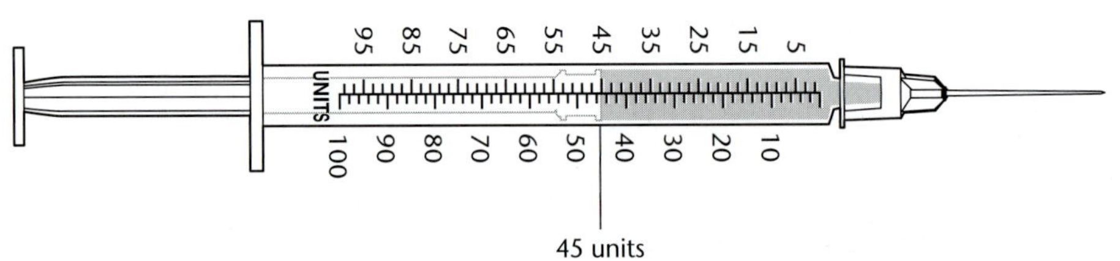

35.

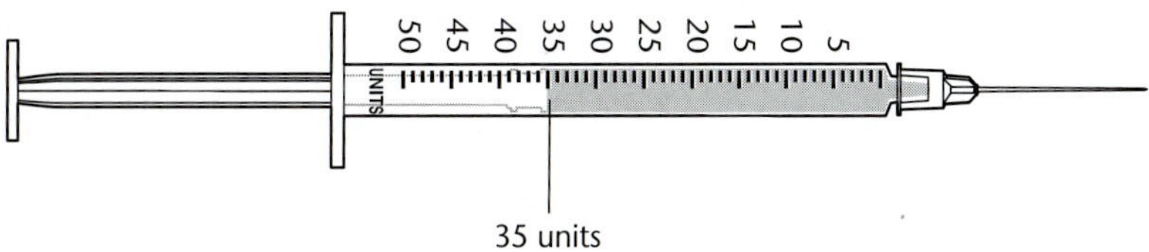

36.

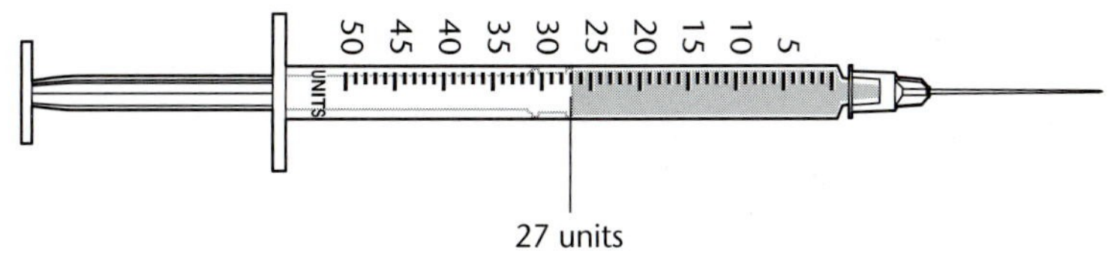

37.

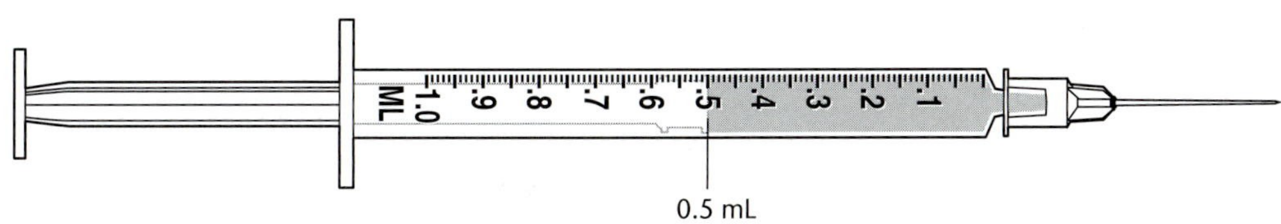

38.

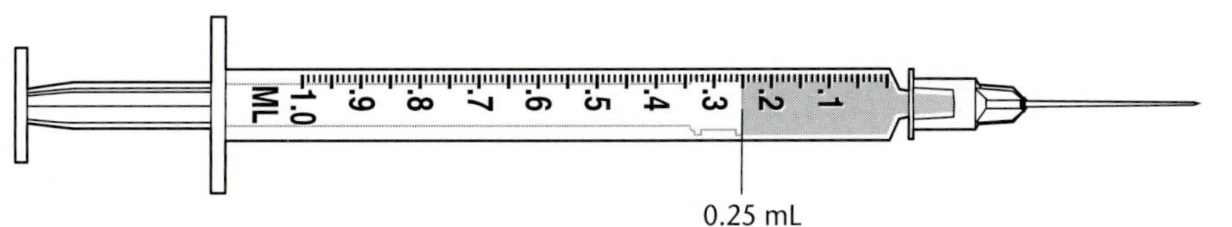

39.

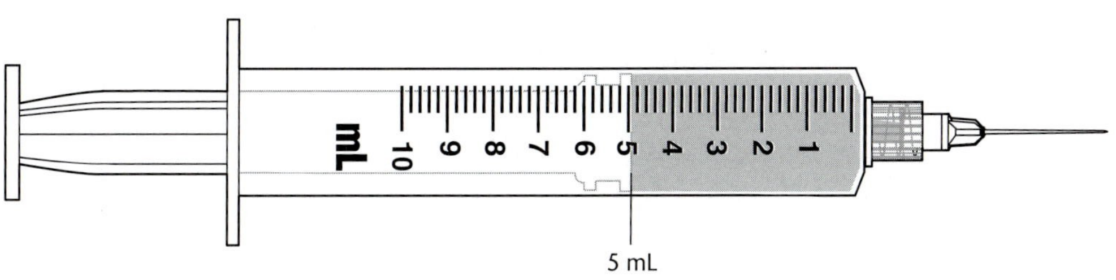

40.

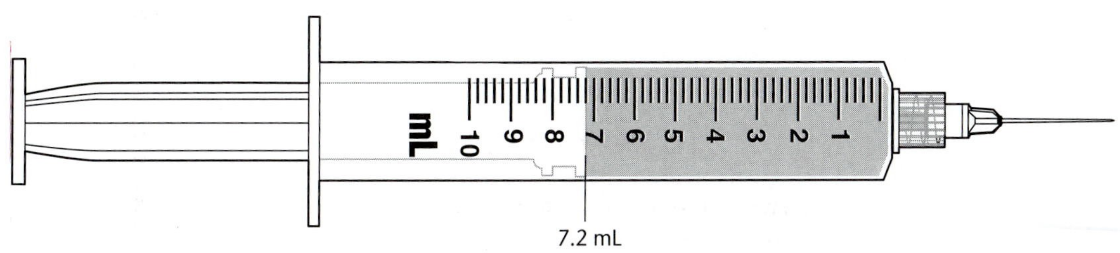

Critical Thinking Applications

1. Standard, safety, or needleless syringe
2. U-30 Insulin syringe
3. Tuberculin syringe
4. Oral syringe
5. Medicine cup, calibrated spoon, or oral syringe
6. Standard, safety, or needleless syringe
7. Medicine cup, calibrated spoon, or oral syringe
8. Large-capacity syringe
9. Transdermal patch
10. Dropper or oral syringe
11. Teaspoon dropper, calibrated spoon, or oral syringe

Case Study

The dose to be administered is 0.75 tsp. It could be measured in a calibrated spoon. Another acceptable form of administration would be 3.75 mL in an oral syringe.

Internet Activity

Your report should indicate knowledge of at least two needleless systems or types of safety syringes. Try *www.safetysyringes.com, www.bd.com,* or *www.index.com.*

CHAPTER 5 ANSWERS

Review and Practice

5-1 The Rights of Medication Administration

1. D
2. H
3. A
4. G
5. I
6. C
7. E
8. B
9. F

Critical Thinking on the Job

Understanding the Order of Roman Numerals

The numeral with the smaller value comes immediately before the one with the larger value. Instead of adding i and x, the health care professional should subtract i from x to calculate the correct dose of 9 gr.

Review and Practice

5-2 Roman Numerals

1. 6
2. 12
3. 9
4. 14
5. 24
6. 18
7. $5\frac{1}{2}$
8. $9\frac{1}{2}$
9. $11\frac{1}{2}$
10. 25
11. 19
12. $8\frac{1}{2}$

13. 21
14. 26
15. 11
16. 10
17. 16
18. 7
19. 18
20. 25

Critical Thinking on the Job

The Importance of the Right Dose

The pediatrician made the initial error by not specifying the dose. This error does not relieve the health care professional of her responsibilities. All of them should have recognized that one of the rights of medication administration—the right dose—was missed in this order. They should have called the physician to clarify the desired dose.

Critical Thinking on the Job

The Importance of the Right Drug

The health care professional should have read the label three times to be certain that he had the correct drug. If he had also shown the label to the physician, the error could have been avoided.

Review and Practice

5-3 Physicians' Orders and Prescriptions

1. The card is complete.
2. 90
3. Three times a day
4. 50 mg
5. 180 days or approximately 6 months
6. The strength of the oral suspension is not listed.
7. 100 mL; however, the pharmacy technician still needs to know the solution strength.
8. One teaspoon (i tsp)
9. It cannot be refilled.
10. Every 8 h (q8h)
11. Prescription C does not include adequate information to determine the dose. Patient Talbott should have a complete first name. Norvasc® comes in strengths of 2.5, 5, and 10 mg per tablet. It is unknown which strength to use. The frequency of dosing is missing and should read daily to ensure proper dosing. The form of the medication is also missing.
12. 30; however, the pharmacy technician still needs to know the proper dosage.
13. One; 1 po, but the frequency of dosing needs to be clarified.
14. This prescription can be refilled 2 times.
15. It is not possible to determine how long it will last since the number per day is not written.
16. Three times a day (TID) before meals (AC)
17. Instilled into the left eye (OS)
18. Once a day (qd)
19. 0.4 mg
20. Orally (po)
21. Twice a day (BID)
22. 0.25 mg
23. At bedtime (HS)

24. Sublingually (SL)
25. Seven

Critical Thinking on the Job

When in Doubt, Check

By using her critical thinking skills, the health care professional has prevented a serious error. The Lasix® order appears to be 600 mg. However, she realizes that the second zero of 600 is actually the loop of the *q* from the Digoxin order. The intended dose of Lasix® is 60 mg. If she had not checked the physician's orders, she might have given 10 times more than the physician intended. When in doubt, always verify the order written on the chart. In this case, if daily had been used instead of qd the error would not have occurred.

Review and Practice

5-4 Medication Administration Systems

1. Check if the patient's systolic blood pressure is below 110.
2. 5 mg
3. 0630 (6:30 a.m.)
4. Sub-Q or subcutaneous injection
5. The Maalox® is administered 2 h after meals and at bedtime.
6. 30 mL
7. 45 units
8. It is to be given PRN or as necessary, and the time would be charted when given.
9. Normdyne, Atrovent®, and Buspar®
10. Humulin®, Dulcolax®
11. Humulin® does not include a route, Dulcolax® does not include a reason to give PRN.
12. po—by mouth
13. 2 puffs
14. Stat—immediately
15. OU or into both eyes
16. IVPB or intravenously through a vein
17. Dilantin®, Ancef®, Viroptic
18. Adalat 3 doses, Dilantin® 3 doses, Ancef® 3 doses, Viroptic 6 doses, Furosemide 1 dose
19. Six
20. Furosemide and Ancef

Review and Practice

5-5 Medication Reference Materials

1. B
2. C
3. D
4. A

Chapter 5 Review

Check Up

1. Administer one 2-mg tablet of Dilaudid® orally when necessary for pain, as often as every 4 h.
2. Administer one 30-mg tablet of Codeine orally 4 times a day.
3. Instill 2 drops in the right eye 4 times a day.
4. Twice
5. Instilled into the right eye
6. "For ophthalmic use"
7. Every 8 h
8. By inhalation, using a metered dose inhaler
9. 1000
10. 8 tablets
11. 200 mg
12. Twice a day
13. Intravenous
14. 10 mg
15. Trental® and Rocephin®
16. Three times a day
17. 8:00 a.m.
18. Order B (the Right Time rule), order C (does not list a specified route), order D (does not list the frequency)
19. OD or in the right eye
20. 1.25 mg
21. 8 a.m., 2 p.m., 8 p.m.
22. Heparin and Proventil®
23. 0600
24. Every 8 h or 3 times in 24 h
25. New times would be marked at 8 a.m., 2 p.m., and 8 p.m.
26. Bleph-10 does not include the dose, and Premarin does not include the route.
27. Determine if the patients ate at lunchtime. Determine if any medications have to be taken on an empty stomach. Double-check your calculations to ensure that they are correct and that your answers make sense. Make sure you use the appropriate measuring device to administer medications. Finally, ensure that there are no time considerations for administering these medications after a meal.
28. Ask patients their names; verify their names with their patient identification bracelets. Do not administer drugs prescribed for the elderly to children or vice versa.

Critical Thinking Applications

Contact the physician to find the strength that should be given. Administer the tablet immediately. Be sure that the revised order is signed. Also, there is not a patient's name on the order sheet.

Case Study

1. Tell Mr. Burke to take one Xanax® tablet with breakfast, lunch, and dinner. One tablet three times a day with meals.
2. Give 120 tablets at one time. You may refill the prescription once.

Internet Activity

According to the article on cephalexin, the usual adult dose of cephalexin is 1 to 4 g daily, in divided doses. If the patient receives 500 mg q6h, the patient will receive 500 mg every 6 h or 4 times a day. Four doses of 500 mg is a total of 2000 mg, or 2 g, per day. This amount is within the acceptable range. You may want to check for more information by using *cephalexin* as your search word.

CHAPTER 6 ANSWERS

Critical Thinking on the Job

Read Labels Carefully

In this example, the health care professional made an initial mistake. He did not carefully compare the drug order with the drug label. Looking at Figures 6-19 and 6-20, you may think that the health care professional's error was reasonable given the similarity between the labels. Still, this error should have been avoided. The health care professional should have read the label three times before trying to administer the drug. This rule is especially important when you administer a drug that is available in different dosage strengths or is designed for different routes of administration.

Fortunately, the patient gave the health care professional an opportunity to catch the error. When she questioned the color, the health care professional listened to the patient and rechecked his work. However, if he had not listened, or if the patient had not alerted him, he may have administered 3 times the amount of the drug that was ordered.

Review and Practice

6-1 Locating Information on Drug Labels and Package Inserts

1. Nexium®
2. esomeprazole magnesium
3. Multiple doses, 30 capsules
4. Astra Zeneca
5. 20 mg/capsule
6. 25°C (77°F), excursions permitted to 15–30°C (59–86°F)
7. clorazepate dipotassium
8. Tranxene
9. 15 mg/tablet
10. Pink-colored, distinctive T shape, marked with a code for product identification
11. 03-2376-5/R1(4391)
12. Protected from moisture, stored below 77°F (25°C), bottle kept tightly closed, dispense in USP tight, light-resistant container
13. IntronA
14. Schering Corp.
15. 25 million IU/1 mL
16. injection
17. Refrigerate between 2 and 8°C (36 and 46°F)
18. To learn about the usual dose of the medication or how to inject the drug
19. cefdinir
20. Orally
21. Label does not state; refer to the package insert
22. 125 mg/5 mL

23. 2 tsp
24. 6 days and 2 bottles equals 120 mL divided by 20 mL per day.
25. furosemide
26. Ancef®
27. Intramuscular (IM) or intravenous (IV) injection
28. Protect from light, do not use if discolored, use only if solution is clear and seal is intact, and store at controlled room temperature 20 to 25°C (68 to 77°F).
29. 40 mg
30. Usual dose is not given on the label. However, if a usual dose is 10 mg then there are 4 doses in this vial.
31. 75 mL.
32. Tap the bottle until all the powder flows freely. Add half of the total water (59 mL) and shake vigorously. Add remaining water and shake vigorously.
33. Approximately 30 mL
34. 125 mg/5 mL
35. Stored in the refrigerator and discarded after 14 days.
36. 15 doses
37. There is no registration mark, so this is a generic drug only, it has no trade name.
38. Metformin hydrochloride is the generic name of the drug.
39. 500 mg/tablet
40. At 15 to 30°C or 59 to 86°F in a light-resistant container
41. Orally. The label states that the medication is a tablet.
42. 90 tablets in the bottle divided by 30 tablets in each prescription equals three prescriptions per bottle.
43. Caraco is the manufacturer.
44. B, G
45. A, B, D, F, G
46. F
47. F, C
48. C, E
49. E, F
50. G

Review and Practice

6-2 Oral Drugs

1. 57664-474-99
2. 1000 mg/tablet
3. Yes
4. 90
5. Levoxyl®
6. King Pharmaceuticals
7. 50 mcg/tablet
8. 100
9. morphine sulfate controlled-release
10. 63323-739-12
11. 20 mg/2 mL
12. Intravenous Use Only After Dilution
13. 60 mL
14. 250 mg/5 mL
15. 100 mL
16. 14 days
17. rimantadine hydrochloride
18. 100 mg/tablet
19. Flumadine®
20. Controlled room temperature 15°–30°C (59°–86°F)
21. B. Levoxyl®, E. Flumadine®
22. C. famotidine and D. amoxicillin because they are both liquids
23. D. amoxicillin
24. A. metformin, B. Levoxyl®, D. amoxicillin in powder form only, E. Flumadine® (C. famotidine does not indicate the storage information on the label.)
25. A. metformin, B. Levoxyl®, E. Flumadine®

Review and Practice

6-3 Parenteral Drugs

1. 10 mg/mL
2. IV administration only
3. Use strict aseptic technique. Contamination can cause fever, infection, and/or other life-threatening illness; single-patient use; contains no preservatives; supports microbial growth. Begin use promptly after opening, discard within specified time, do not use if contaminated. Shake well before using.
4. See package insert.
5. 200 mg erythromycin activity with the equivalent of 600mg sulfisoxazole per 5 mL
6. Ross Pediatrics, a division of Abbott labs
7. erythromycin ethylsuccinate and sulfisoxazole acetyl for oral suspension
8. Yes, all directions are clearly indicated on the label.
9. Store in a controlled room temperature 20 to 25 degrees C (68 to 77 degrees F)
10. 0.5 mcg/capsule
11. Orally
12. 25 doses (each dose is 2 capsules)
13. rDNA origin
14. 100 units/mL
15. insulin lispro injection
16. 100 doses
17. 100 mg/mL
18. trimethobenzamide HCl
19. Each dose is 1 mL, 20-mL multidose vial would hold 20 doses.
20. Storage should be at 25°C (77°F).

Critical Thinking on the Job

Avoid Unnecessary Risks

The health care professional should send the vial back to the pharmacy for a replacement because, when she returned and read the drug label, she noticed that you should not start the administration more than two hours after Gammagard® S/D is reconstituted. Since she reconstituted at 0830 and it is now 1100, it has been 2½ hours and the drug should not be administered. By sending the medication back to the pharmacy, the health care professional observed the special instructions on the drug label and, in turn, protected the patient's rights.

Review and Practice

6-4 Drugs Administered by Other Routes

1. albuterol USP
2. 90 mcg per actuation
3. 2 inhalations every 4 to 6 h
4. Away from heat or flame, 25°C (77°F)
5. Topically
6. Store refrigerated 2 to 8°C (36 to 46°F)
7. 0.01 percent of the gel
8. Becaplermin
9. Nasal spray (Intranasal)
10. 50 mcg/spray
11. 28 metered sprays
12. No, intranasal only
13. Transdermally
14. 100 mcg/h for 72 h

15. opioid-tolerant
16. 5
17. Calcitonin-salmon
18. 200 units/spray
19. Intranasal use only
20. 30

Chapter 6 Review

Check Up

1. The generic name is the one official name recognized by the USP and the NF. Brand or trade names can be registered with the U.S. Patent and Trademark Office. A drug may have several different trade names, but only one generic name.
2. When the order is for a combination drug and the specific components are required
3. IM means intramuscular, or into the muscle. IV means intravenously, or into the vein.
4. Unscored tablets, gelcaps, caplets, enteric-coated tablets, and controlled-, sustained-, or extended-release capsules; breaking them would change the action or absorption of the drug
5. Use a package insert, when the label does not have enough information to administer the drug correctly
6. A lot number provides a code that enables a company to know when and where the drug was manufactured. A company can use the lot number to recall a product in case of contamination.
7. 0.125 mg
8. 100 tablets
9. Caraco Laboratories
10. Orally
11. Flunisolide
12. By oral inhalation
13. 250 mcg per activation
14. 50 metered inhalations
15. Orally
16. 100 mg
17. meperidine hydrochloride
18. 100 tablets
19. rDNA origin
20. R (regular), which means short acting
21. By injection
22. 100 units/mL

Critical Thinking Applications

The tablets would probably be preferable for an adult homeless patient. The oral suspension needs refrigeration, a condition that would not be available to someone living on the streets or in a shelter. In addition, a homeless person might not have adequate facilities to wash the medicine cup. Finally, the tablets can be stored in a much smaller container than the liquid and are, therefore, more portable.

Case Study

1. Read the package insert, check any warnings on the package insert, and note any warnings on the label. Note that this medication is for pediatric patients and must be diluted for IV use.
2. The drug will need to be administered intravenously. It will need to be diluted before it is administered. You will need to check the label and the package insert for specific directions.
3. This is a single-dose vial. Dispose of the container, following the guidelines at your facility. Destroy leftover medication.

Internet Activity

Pages on the Internet may change. By doing a search using words such as *coumadin* and *interactions*, you will find a variety of sites. Sample sites include

http://www.druginfonet.com/faq/faqcouma.htm
http://www.thriveonline.com/medical
http://www.healthtouch.com

CHAPTER 7 ANSWERS

Review and Practice

7-1 Calculating the Desired Dose

1. Desired dose: 250 mg
2. Desired dose: 500 mg
3. Desired dose: 30 mg
4. Desired dose: 250 mg
5. Desired dose: 150 mcg
6. Desired dose: 1 tsp
7. Desired dose: 10 cc
8. Desired dose: 1000 mg
9. Desired dose: 15 mg
10. Desired dose: 50 mcg
11. Desired dose: 88 mcg
12. Desired dose: 500 mg
13. Desired dose: 0.25 mg
14. Desired dose: 7.5 mL
15. Desired dose: $1\frac{1}{2}$ tsp
16. Desired dose: 137 mcg
17. Desired dose: 112 mcg
18. Desired dose: 0.25 mg
19. Desired dose: 0.75 mg
20. Desired dose: 1000 mg

Critical Thinking on the Job

Use Your Critical Thinking Skills

If 1 tablet of Metformin is 1000 g and you need to give the patient 2000 g, you realize that you need to give more than one tablet. With critical thinking you would determine that the answer appears incorrect and should recalculate. *Never administer a medication if you are uncertain or uncomfortable with the answer you obtain.*

When in Doubt, Check

As he looked at his answer of 4 tsp, Jorge's initial thought was that 4 tsp was a lot of medication to give the patient. First, Jorge rechecked his calculations, which were correct. Still not sure that 4 tsp was an appropriate amount to give the patient, Jorge went to the physician in charge and confirmed the order. In this case, Jorge heard the order correctly. The usual first dose of aminocaproic acid is large: 5 g, which amounts to either 4 tsp of Amicar syrup 25% or 10 of the 500-mg Amicar capsules.

Because of the unusually large volume of liquid or number of tablets, Jorge was correct to reconfirm the order. Since the order was given orally in an emergency situation, it was important that he check directly with the ordering physician before administering the drug. If Jorge's question had arisen in another circumstance, when a physician was not readily available, Jorge could consult the pharmacist or refer to the PDR (*Physicians' Desk Reference*) or another reliable reference to check the usual dose.

Review and Practice

7-2 Calculating the Amount to Administer

1. Amount to administer: 2 tablets
2. Amount to administer: 10 mL
3. Amount to administer: 10 mL
4. Amount to administer: 2 tablets
5. Amount to administer: 5 mL
6. Amount to administer: 2 tablets
7. Amount to administer: 2 capsules
8. Amount to administer: 2 capsules
9. Amount to administer: 2 tablets
10. Amount to administer: 1 capsule
11. Amount to administer: 10 mL
12. Amount to administer: $1\frac{1}{2}$ or 1.5 mL
13. Amount to administer: $\frac{1}{2}$ tablet
14. Amount to administer: 0.5 mL
15. Amount to administer: 3.75 mL
16. Amount to administer: 2 tablets
17. Amount to administer: 0.5 mL
18. Amount to administer: $\frac{28}{100}$ mL = 0.28 mL or 28 units
19. Amount to administer: 3 tablets
20. Amount to administer: 1 capsule

Chapter 7 Review

Check Up

1. Desired dose: 5 mg

   Amount to administer: $2\frac{1}{2}$ tablets
2. Desired dose: 16 mg

   Amount to administer: 2 tablets
3. Desired dose: 400 mg

   Amount to administer: 2 tablets
4. Desired dose: 800 mg

   Amount to administer: 2 tablets
5. Desired dose *D*: 2 mg

   Amount to administer: 2 tablets
6. Desired dose *D*: 7.5 mg

   Amount to administer: 2 tablets
7. Desired dose *D*: 0.1 mg

   Amount to administer: 2 tablets
8. Desired dose *D*: 250 mg

   Amount to administer: $2\frac{1}{2}$ tablets
9. Desired dose *D*: 7.5 mg

   Amount to administer: 3 tablets
10. Desired dose *D*: 500 mg

    Amount to administer: 2 tablets
11. Desired dose *D*: 20 mg

    Amount to administer: 2 tablets
12. Desired dose *D*: 100 mg

    Amount to administer: 4 tablets

    Since > 3 tabs check before administering
13. Desired dose: *D*: 250 mg

    Amount to administer: 1 capsule
14. Desired dose: 60 mg

    Amount to administer: 2 capsules
15. Desired dose: 40 mg

    Amount to administer: 2 tablets
16. *D*: 125 mg, *A*: 2.5 mL or $\frac{1}{2}$ tsp
17. *D*: 1000 mg, *A*: 12.5 mL or $2\frac{1}{2}$ tsp
18. *D*: 5 mg, *A*: 1 tablet
19. *D*: 200 mg, *A*: 8 mL
20. *D*: 500 mg, *A*: 1 caplet
21. *D*: 4 mg, *A*: 20 mL or 4 tsp
22. *D*: 100 mg, *A*: 10 mL (This is a large amount to give IM. You should check with pharmacist or physician to verify this order.)
23. *D*: 20 mg, *A*: 2 tablets

Critical Thinking

1. Amount to administer: 1 tablet
2. Amount to administer: 2 tablets
3. Amount to administer: 1 tablet
4. #1 6 tablets

   #2 40 tablets

   #3 120 tablets

Case Study

1. Amount to administer: 1 of the 5 mg tablets and 1 of the 2 mg tablets.
2. The patient should take 2 tablets 3 times a day orally for 7 days. This would be total of 42 tablets. The patient will need 21 of the 2 mg tablets and 21 of the 5 mg tablets.
3. The order now calls for 15 mg, rather than 7 so the patient would need one 5 mg and one 10 mg tablet per dose. Each dose is given three times a day for seven days for a total of 21 of each of the 5 mg and 10 mg tablets.

Internet Activity

Answers will vary. Some examples of medications with different dosage strengths are

Avandia®	Valium®	Metformin	Nitroglycerin
Augmentin®	Aerobid®	Biaxin®	Levothroid®
Amoxil®	Dilaudid®	Univasc®	Coumadin®

CHAPTER 8 ANSWERS

Critical Thinking on the Job

Calculating Factors Incorrectly

If the health care professional administers two gr 1/100 tablets, the patient will receive gr 2/100 or gr 1/50, *4 times the ordered dose of nitroglycerin.* As a result, the patient is likely to have a severe decrease in blood pressure.

If the health care professional thinks critically, she will realize that when an object is divided into 200 parts, each part is smaller than when the object is broken into 100 parts. Therefore, gr 1/20 must be smaller than 1/100. This conclusion should alert her to rework the calculation as follows:

$$\text{gr}\,\frac{1}{200} \times \frac{1 \text{ tablet}}{\text{gr}\,\frac{1}{100}} = A$$

$$\frac{1}{200} \times 1 \text{ tablet} \div \frac{1}{100} = A$$

$$\frac{1}{\cancel{200}_{2}} \times 1 \text{ tablet} \div \frac{\overset{1}{\cancel{100}}}{1} = A$$

$$\frac{1}{2} \times 1 \text{ tablet} = \frac{1}{2} \text{ tablet} = A$$

When you divide a number by a fraction, multiply that number by the reciprocal of the fraction. Instead of dividing 1 tablet by 1/100, multiply 1 tablet by 100/1. The *amount to administer* is ½ tablet. Because nitroglycerin tablets are not scored, the health care professional must look for a bottle of gr 1/200 tablets. The health care professional **must not** administer ½ of an unscored tablet.

Review and Practice

8-1 Tablets and Capsules

1. 2 tablets
2. 3 tablets
3. $1\frac{1}{2}$ tablets
4. $1\frac{1}{2}$ tablets
5. $1\frac{1}{2}$ tablets
6. 2 tablets
7. $2\frac{1}{2}$ tablets
8. $1\frac{1}{2}$ tablets
9. $1\frac{1}{2}$ tablets
10. 2 tablets
11. $\frac{1}{4}$ tablet
12. 1 tablet
13. 2 tablets
14. 2 tablets
15. use $\frac{1}{2}$ of a 300 mg tablet
16. 2 tablets
17. 3 tablets
18. $\frac{1}{2}$ tablet
19. 2 tablets
20. $\frac{1}{2}$ tablet

Critical Thinking on the Job

Reconstituting Powders

Checking the label, the health care professional realizes that 90 mL of water, not 150 mL, should have been used. Furthermore, only 1/2 of the water, or 45 mL, should have been added at first to wet the powder, followed by the remaining 45 mL.

The powder's volume must be considered when you calculate the final volume of solution. The manufacturer has performed this calculation and listed the correct amount of water to add to the powder. The label calls for 90 mL of water to be added. When the health care professional added 150 mL of water, instead of 90 mL, a volume of 210 mL of suspension was produced, instead of 150 mL.

During the week, the patient will be administered 150 mL of liquid medication, yet 210 mL has been prepared. The patient will receive of 150/210 the total solution and, in turn, only 150/210 of the drug itself. Thus, instead of receiving 250 mg/5 mL, the patient will receive $\frac{150}{210} \times 250$, or 179 mg, in each 5 mL.

This lesser amount of medication may be ineffective in treating the patient's infection. If the patient does not improve, the physician may order a different antibiotic. The health care professional's error, if not corrected, could cause the patient to suffer symptoms for a longer time period, to be exposed to side effects from an unnecessary medication, and to spend additional money to purchase a drug that would not otherwise have been needed.

Review and Practice

8-2 Liquid Medications

1. 4 mL
2. 7.5 mL
3. 5 mL
4. 0.78125 mL = 0.8 mL
5. 10 mL
6. 30 mL
7. 6.25 mL
8. 5.4 mL
9. 12.5 mL
10. 2.5 mL
11. 10 mL
12. 2.5 mL or $\frac{1}{2}$ teaspoon
13. 10 mL
14. 20 mL
15. 10 mL
16. 12.5 mL
17. 20 mL
18. 2 teaspoons or 10 mL
19. 1.25 mL
20. 15 mL

Chapter 8 Review

Check Up

1. $\frac{1}{2}$ tablet
2. 2 tablets
3. $\frac{3}{4}$ tablet
4. 6 mL
5. 7.5 mL
6. 2 tablets
7. 2.5 mL
8. $1\frac{1}{2}$ tablets
9. 10 mL
10. $1\frac{1}{2}$ tablets
11. 2 tablets
12. $\frac{1}{2}$ tablet
13. $\frac{1}{2}$ tablet
14. 3.3 mL
15. 1 tablet
16. 2 capsules
17. 1 teaspoon or 5 mL
18. Two 2-mg tablets and one 1-mg tablet
19. 2000 mg = 2 g
20. Add 140 mL in 2 portions.
    Amount to administer: 3 teaspoons or 15 mL
21. Use: J
    Administer: 7.5 mL
22. Use: E
    Administer: $\frac{1}{2}$tablet
23. Use: G
    Administer: 1 capsule
24. Use: F
    Administer: 4 mL
25. Use: I
    Administer: 2 tablets
26. Use: Q
    Administer: 2 tablets
27. Use: L
    Administer: 2 tablets
28. Use: O
    Administer: 3 tablets
29. Use: N
    Administer: 12.5 mL
30. Use: M
    Administer: 1 tablet

Critical Thinking Applications

1. a. Depakote® ER: 1 tablet
   b. Valium®; 2 tablets
   c. meperidine tablet
   d. cephalexin: 1 capsule
2. Depakote® ER
3. Valium® and meperidine would be crushed and dissolved in water. The cephalexin capsules would be opened and the powder dissolved in water. The three would then be administered through the tube.
4. Consult either the pharmacist or a drug reference. Depakote® is now available in multiple forms. Consult the physician about how to handle the Depakote® order.

Case Study

1. First add one-half of 55 mL of water to the powdered drug, and shake. Then add the other half of the water, and shake again.
2. 7.5 mL
3. An oral syringe, calibrated spoon, or medicine cup marked in tenths of a milliliter

Internet Activity

Pages on the Internet may change. Try these sites
http://www.pharmasave.com
http://ohioline.ag.ohio-state.edu/ss-fact/0132.html
http://www.ntl.sympatico.ca/healthyway/HEALTHYWAY/com_4.html

CHAPTER 9 ANSWERS

Critical Thinking on the Job

Confirming the Physician's Order

The health care professional realizes that the physician has ordered a quantity of solution rather than a quantity of drug. She checks with the physician, who clarifies that the intended dose is 50 mg. If the health care professional administers the injection without thinking, the patient receives 100 mg, twice the intended dose. The health care professional calculates that 1 mL of 50 mg/mL solution contains the 50-mg dose. She discards 1 mL of the original dose and administers the remaining 1 mL.

Critical Thinking on the Job

Confusing the Amount of Solution with the Dosage Unit

The actual solution strength is 5 mg/1 mL, not 5 mg/5 mL. When 5 mL is administered, the patient receives 25 mg of Compazine®, 5 times the desired dose.

This error would have been avoided if the health care professional reads the label three times. The health care professional must take extra care not to confuse the total volume in the vial (5 mL) with the dosage unit (1 mL), as indicated by the dosage strength (5 mg/mL).

Review and Practice

9-1 Calculating Parenteral Dosages

1. Administer 3.8 mL in 2 syringes.

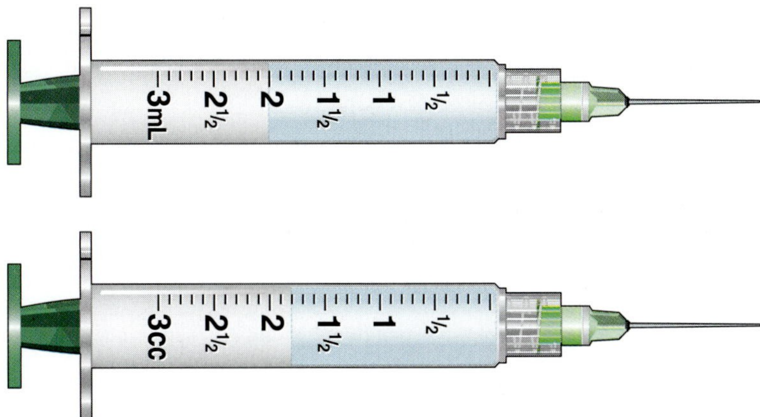

2. Administer 2 mL.

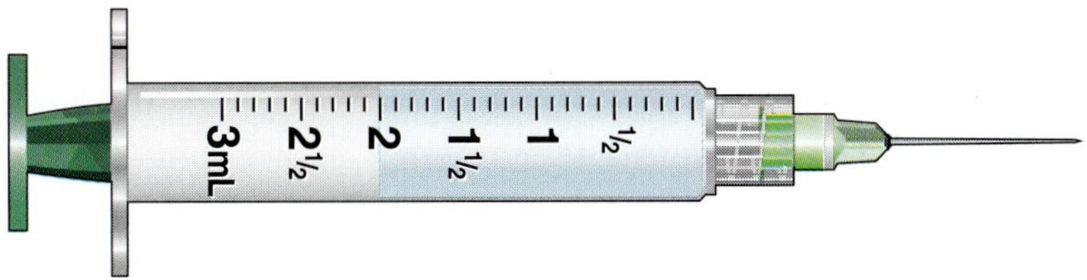

3. Administer 0.3 mL.

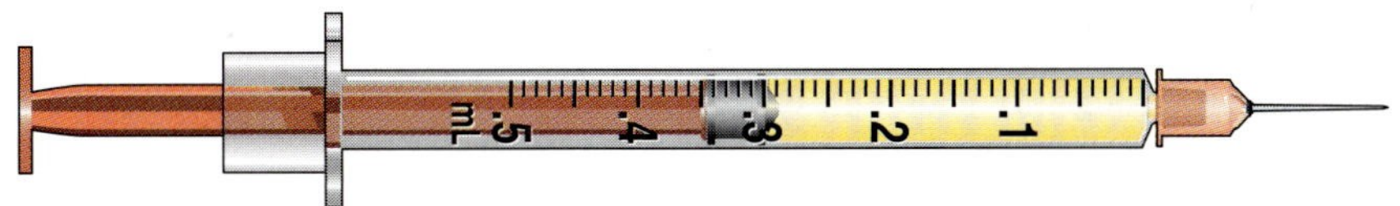

4. Administer 0.5 mL.

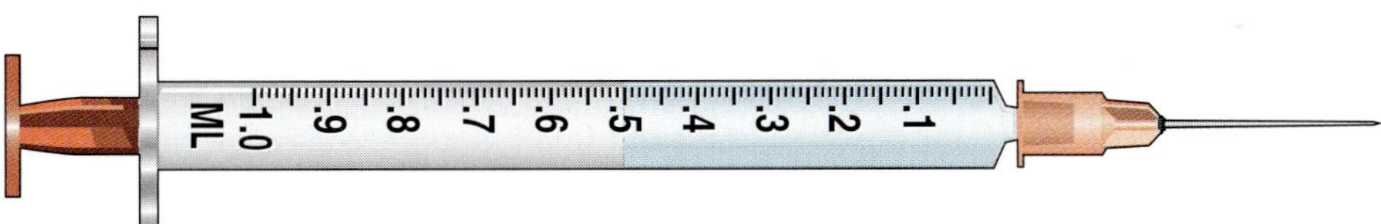

5. Administer 0.8 mL.

6. Administer 1 mL.

7. Administer 1.6 mL.

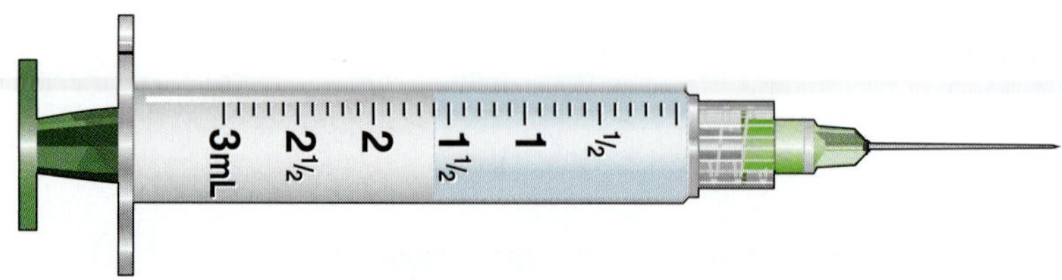

8. Administer 0.5 mL.

9. Administer: 1 mL   Syringe: standard 3 mL syringe
10. Administer: 0.7 mL   Syringe: 1 mL tuberculin syringe
11. Administer: 2 mL   Syringe: standard syringe
12. Administer: 0.4 mL   Syringe: 0.5 mL tuberculin syringe
13. Administer: 0.3 mL   Syringe: 0.5 mL tuberculin syringe
14. Administer: 0.8 mL   Syringe: 1 mL tuberculin syringe
15. Administer: 0.88 mL   Syringe: 1 mL tuberculin syringe
16. Administer: 0.6 mL   Syringe: 1 mL tuberculin syringe
17. Administer: 0.6 mL   Syringe: 1 mL tuberculin syringe
18. Administer: 0.4 mL   Syringe: 0.5 mL tuberculin syringe
19. Administer: 4 mL   Syringe: 2 standard 3 mL syringes, dose divided evenly between syringes, 2 mL each, given in two injections
20. Administer: 0.35 mL   Syringe: 0.5 mL tuberculin syringe
21. Administer: 1.5 mL   Syringe: standard 3 mL syringe
22. Administer: 2 mL   Syringe: standard 3 mL syringe
23. Administer: 0.63 mL   Syringe: 1 mL tuberculin syringe
24. Administer: 2 mL   Syringe: standard syringe
25. Administer: 0.5 mL   Syringe: 1 mL tuberculin syringe

Critical Thinking on the Job

Recording Accurate Information

While the health care professional has followed instructions carefully, she has mislabeled the vial. She used 5 mL of sterile diluent, not 1 mL. Her label should indicate 5 mg/5 mL so that her calculation is:

$$2\text{ mg} \times \frac{5\text{ mL}}{5\text{ mg}} = A$$

$$2\,\cancel{\text{mg}} \times \frac{\overset{1}{\cancel{5}}\text{mL}}{\cancel{5}\,\cancel{\text{mg}}} = 2 \times 1\text{ mL} = 2\text{ mL}$$

She would administer this amount using a standard syringe. Because of her labeling error, the patient receives only $\frac{1}{5}$ the amount of medication ordered.

Review and Practice

9-2 Reconstituting Powdered Medication

1. 17.5 mL of sterile diluent
2. 20 mg/1 mL
3. Sterile water or bacteriostatic water for injection
4. 0.75 mL
5. Sterile water for injection
6. 1 mL
7. 100 mg/mL
8. Exp. 1/3/06, 1600 plus initials; usually dose is given immediately—only stable 6 h
9. 0.75 mL
10. Bacteriostatic water for injection with sodium chloride—the Pregnyl solvent is provided.
11. 10 mL
12. Withdraw sterile air from lyophilized vial and inject into diluent. Remove 1 to 10 mL from diluent and add to lyophilized vial; agitate gently until powder is completely dissolved in solution. Store at 15 to 30°C (59 to 86°F).
13. 10,000 USP units/10 mL
14. 5 mL
15. Place in the refrigerator. It will last for 60 days.
16. 0.9% sodium chloride injection
17. 5 mL
18. 38 mg/mL
19. 2400 on 6/6/2012
20. At a controlled room temperature. Do not refrigerate.
21. 2.63 mL
22. 8.2 mL
23. 750,000 units/mL
24. Exp. 11/27/05, 1200
25. 2 mL
26. 1% lidocaine hydrochloride or sterile water injection
27. 2.1 mL
28. 350 mg/mL
29. Discard at 1600 on October 7, 2012
30. 1 mL

Review and Practice

9-3 Other Medication Routes

1. 5 mL
2. 0.5 mL
3. 1.25 mL
4. 2 suppositories
5. One 10-mg and one 5-mg suppository
6. $\frac{1}{2}$ suppository. Check manufacturer's directions to verify that suppository can be divided before administration.
7. 2 patches
8. One TTS-3 patch and one TTS-2 patch
9. One 0.1 mg/day patch with one 0.05 mg/day patch, or two 0.75 mg/day patches
10. One 0.2 mg/h patch with one 0.1 mg/h patch

Chapter 9 Review

1. Administer 2 mL.

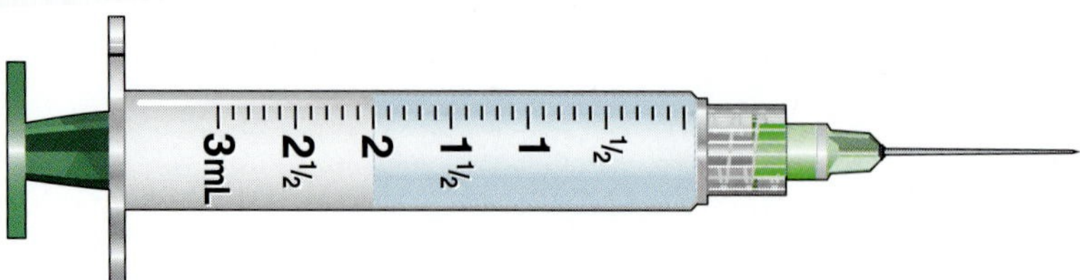

2. Administer 1.2 mL.

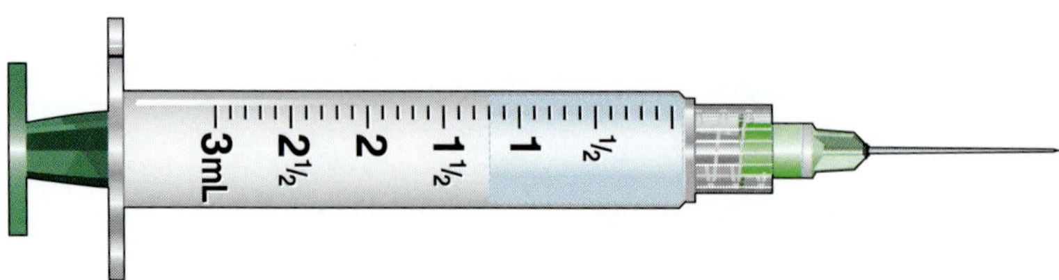

3. Administer 0.6 mL.

4. Administer 0.7 mL.

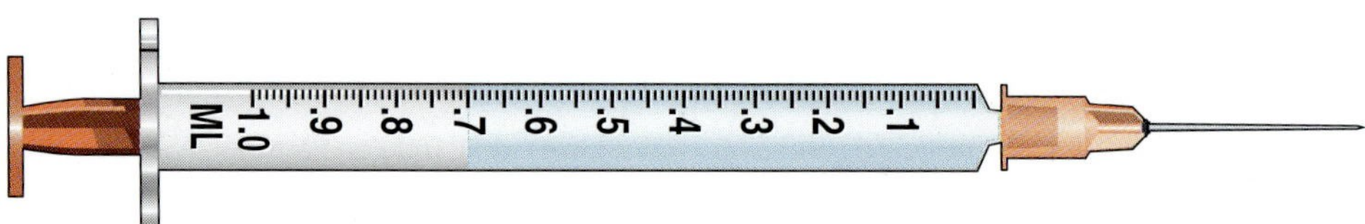

5. Administer 1.5 mL.

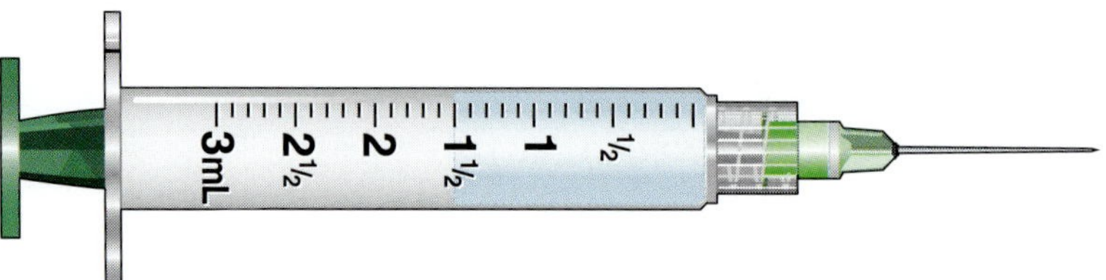

6. Administer 2.5 mL.

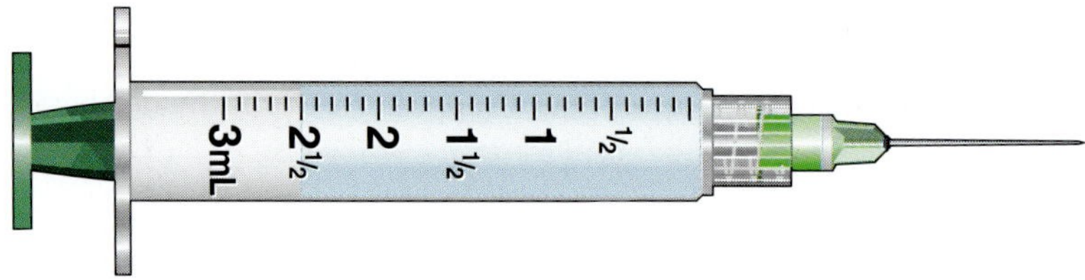

7. Administer 0.8 mL.

8. Administer 0.75 mL.

9. Administer 0.75 mL.

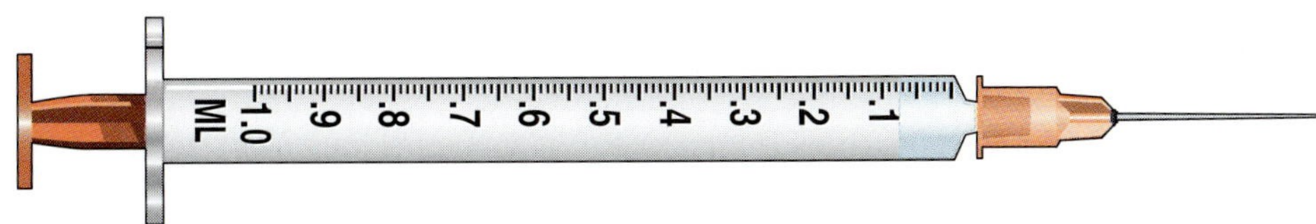

10. Administer 0.1 mL.

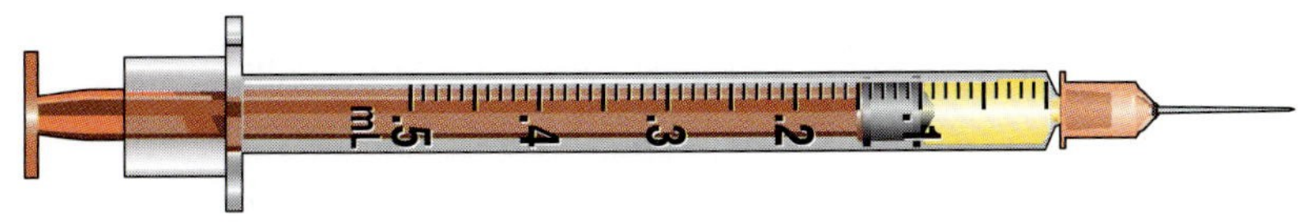

11. Administer: 0.4 mL Syringe: 0.5 mL tuberculin syringe
12. Administer: 0.24 mL Syringe: 0.5 mL tuberculin syringe
13. Administer: 0.38 mL Syringe: 0.5 mL tuberculin syringe
14. Administer: 0.3 mL Syringe: 0.5 mL tuberculin syringe
15. Administer: 2 mL Syringe: standard 3 mL syringe
16. Administer: 0.4 mL Syringe: 1 mL tuberculin syringe
17. Administer: 1.5 mL Syringe: standard 3 mL syringe
18. Administer: 2.0 mL Syringe: standard 3 mL syringe
19. Administer: 0.63 mL Syringe: 1 mL syringe
20. Administer: 2.0 mL Syringe: standard 3 mL syringe
21. Administer: 0.5 mL Syringe: 1 mL tuberculin syringe
22. one vial of label I (40 mg/4 mL) or two vials of label H (20 mg/2 mL). For either medication you would administer 4 mL of medication.
23. Sterile water
24. 2.1 mL
25. 10 mg/mL
26. 1 hour
27. Bacteriostatic water for injection with sodium chloride; the Pregnyl sovent is provided
28. 10,000 units/10 mL
29. 0200 on 10/30/2012
30. 2.5 mL
31. 8 mL
32. 0.25 mL
33. Two 25-mg suppositories
34. 2 suppositories
35. 2 patches
36. One 0.2 mg/h patch with one 0.1 mg/h patch

Critical Thinking Applications

1. Using the diluent that is provided with the vial—sterile water
2. Supplied 1.25 mL, used 0.7 mL
3. Discard
4. Subcutaneously
5. Stored at a temperature of 2 to 8°C (36 to 46°F) for up to 24 h

Case Study

First of all, the physician should be contacted because all three packages of medication indicate they are for subcutaneous use only. Assuming the physician changes the order to sub-q, you would use the package 9-14A with a dosage strength of 100 mcg/mL and administer 0.75 mL in a tuberculin syringe. Using 9-14B, 50 mcg/mL would cause the amount to administer to be over 1 mL, which is not acceptable for a sub-q injection. Using 9-14C, 500 mcg/mL would make the amount to administer only 0.15 which is very small and more difficult to measure.

Internet Activity

Answer found at manufacturer's website GlaxoKlineSmith at *www.gks.com.*

Intramuscular use (concentration of approximately 385 mg/mL): For initial reconstitution use sterile water for injection, USP, sodium chloride injection, USP, or 1% lidocaine hydrochloride solution (without epinephrine).

Each gram of ticarcillin should be reconstituted with 2 mL of sterile water for injection, USP, sodium chloride injection, USP, or 1% lidocaine hydrochloride solution (without epinephrine) and **used promptly**. Each 2.6 mL of the resulting solution will then contain 1 g of ticarcillin. Amount to administer: 0.5 mL.

CHAPTER 10 ANSWERS

Critical Thinking on the Job

Checking Compatibility

The health care professional suspects that the additive is not compatible with the IV solution. She stops the administration, then uses a compatibility chart to verify her suspicions. After that, she calls the physician to obtain a new order. In this case, the health care professional was fortunate that a change in the fluid's appearance alerted her to the problem. However, in many cases, failing to verify compatibility **before** introducing an additive can have severe consequences for the patient, including death.

Review and Practice

10-1 IV Equipment

1. Injection port
2. Roller or screw clamp
3. Slide clamp
4. Infusion pump
5. Microdrip tubing
6. PCA pump
7. Syringe pump
8. Burette
9. If a patient needed a large amount of fluids, a rapid infusion of medication, an infusion of highly concentrated solutions, or long-term IV therapy
10. Flow rate, whether an ordered medication has been infused properly, times at which an IV bag needs to be changed, and signs of infiltration or phlebitis
11. Swelling, coolness, or discomfort at the IV site
12. Irritation of the vein by IV additives, movement of the needle or catheter, and long-term IV therapy
13. 36 inches above the heart
14. Rate controller, infusion pump, syringe pump, patient-controlled analgesia device
15. To add medications or other additives to an IV

Critical Thinking on the Job

Adjusting the Flow Rate

Pat's error could have serious consequences for the patient by providing far too much solution in a limited time period. The new flow rate of 28 gtt/min is a 75 percent increase over the original flow rate of 16 gtt/min, well beyond facility guidelines. Instead of making up the entire schedule in 1 h, Pat should have adjusted the remainder of the entire schedule. The IV bag still had 450 mL of fluid with 4 h of infusion time remaining. The calculation should be

$$\frac{10\text{ gtt}}{1\text{ mL}} \times \frac{450\text{ mL}}{4\text{ h}} \times \frac{1\text{ h}}{60\text{ min}} = \frac{18.75\text{ gtt}}{1\text{ min}} = 19\text{ gtt/min}$$

The increase from 16 gtt/min to 19 gtt/min is a 19 percent increase, within the acceptable amount of change.

Review and Practice

10-2 Calculating Flow Rates

1. 167 mL/h
2. 150 mL/h
3. 125 mL/h
4. 133 mL/h
5. 125 mL/h
6. 83 mL/h
7. 94 mL/h
8. 100 mL/h
9. 75 mL/h
10. 83 mL/h
11. 14 gtt/min
12. 8 gtt/min
13. 31 gtt/min
14. 14 gtt/min
15. 50 gtt/min
16. 16 gtt/min
17. 21 gtt/min
18. 50 gtt/min
19. 50 gtt/min
20. 50 gtt/min
21. 28 gtt/min
22. 23 gtt/min
23. Original flow rate: 21 gtt/min
    Adjusted flow rate: 17 gtt/min
24. Original flow rate: 42 gtt/min
    Adjusted flow rate: 37 gtt/min
25. Original flow rate: 31 gtt/min
    Adjusted flow rate: 44 gtt/min
    The flow rate can only be increased 25%-contact physician
26. 33 gtt/min
27. 31 gtt/min
28. 38 gtt/min
29. 75 gtt/min
30. 17 gtt/min

Review and Practice

10-3 Infusion Time and Volume

1. 12 h 3 min
2. 4 h
3. 24 h, 11 min
4. 5 h
5. $2\frac{1}{2}$ h or h 30 min
6. The infusion will be finished at 0800 the next day.
7. The infusion will be finished the next day at 11:30 a.m.
8. The infusion will be finished at 0100 the next day.
9. The infusion will be finished at 4 a.m. the next day
10. The infusion will be finished at 2325, or 11:25 p.m.
11. 187.5 mL
12. 800 mL
13. 1500 mL
14. 150 mL

Review and Practice

10-4 Intermittent IV Infusions

1. B
2. A
3. 3 mL
4. 2 mL

5. 3.9 mL added to 500 mL NSS
6. 250 mL/h
7. 63 gtt/min
8. 7.5 mL
9. 107.5 rounded to 108 mL/h
10. 17 to 18 gtt/min

Chapter 10 Review

Check Up

1. g
2. d
3. a
4. e
5. c
6. f
7. b
8. Flow rate = 125 mL/h
9. Flow rate = 63 mL/h
10. Flow rate = 100 mL/h
11. Flow rate = 63 mL/h
12. Flow rate = 67 mL/h
13. Flow rate = 100 mL/h
14. 23 gtt/min
15. 6 gtt/min
16. 42 gtt/min
17. 50 gtt/min
18. 31 gtt/min
19. 50 gtt/min
20. Original flow rate: 31 gtt/min
    Adjusted flow rate: 36 gtt/min
21. Original flow rate: 17 gtt/min
    Adjusted flow rate: 18 gtt/min
22. Original flow rate: 63 gtt/min
    Adjusted flow rate: 75 gtt/min
23. 8 h
24. 16 h
25. 24 h
26. 17 h 8 min or 17 h 9 min
27. The infusion will be finished at 1713.
28. The infusion will be finished at 9 p.m.
29. The infusion will be finished the next day at 0121 or 0122
30. The infusion will be finished the next day at 1:25 to 1:26 p.m.
31. 688 mL
32. 300 mL
33. 938 mL
34. 667 mL

Critical Thinking Applications

1. You use 2 mL of medication from one vial (300 mg) and 1.3 mL (200 mg) from a second vial of clindamycin for a total of 3.3 mL. Add this to 100 mL of diluent.
2. X = 3.3 mL
3. No. For pneumonia, the package insert suggests the patient should receive 2400 to 2700 mg per day. The order is for 500 mg three times a day (1500 mg), which is less than the lowest recommended dose.
4. Any product that contains benzyl alcohol.
5. According to the package insert, the medication can be used for up to 24 h with no refrigeration. You should label the IV bag with the date and time of expiration, the solution strength, and your initials.
6. 300 mL/h

Case Study

The patient received 4.5 mg of morphine sulfate during the shift.

Internet Activity

Internet sites include various IV fluids and equipment. Try the following sites as part of your research.
http://gucfm.georgetown.edu/welchjj/netscut/fen/ivfluids.html
www.nursewise.com/courses/IV_hour.htm
www.tpub.com/content/medical/14274/css/14274_204.htm
www.worldhistory.com/wiki/I/Intravenous-therapy.htm

CHAPTER 11 ANSWERS

Critical Thinking on the Job

Looking for Warnings

Elena did make a mistake; she should have read the dosage and administration section of the package insert. She would have found a warning that Pediazole® should not be administered to patients less than 2 months of age. This patient, 7 weeks old, should not receive Pediazole®. Elena should have contacted the physician to discuss the order further.

Review and Practice

11-1 Dosages Based on Body Weight

1. 30 kg
2. 35 kg
3. 24.55 kg
4. 16.82 kg
5. 69.09 kg
6. 91.82 kg
7. 7.39 kg
8. 5.28 kg
9. 4.49 kg
10. 6.51 kg
11. Two doses of 5 mg, or 10 mg/day, are within a safe range.

    Amount to administer: 0.5 mL
12. The order is above the appropriate starting dosage for a patient of this weight. Consult the physician.
13. The order of 750 mg per dose is above the safe dose for children with a severe infection. According to the dosage regimen, the child should receive between 290 mg to 580 mg of medication. Consult the prescribing physician.
14. The ordered dose of 30 mg is within a safe range.

    Amount to administer 1.2 mL
15. The ordered dose is safe to administer to this child.

    Amount to administer: 4 mL
16. The safe range is 300 to 600 mg/day or 75 to 150 mg/dose qid. This order is within the safe range. Amount to administer: $\frac{1}{2}$ tablet
17. Based on the patient's weight of 20 kg, the safe range is 1.0 to 8.0 mg. This order is within the safe range.

    Amount to administer: 1.5 mL.
18. The order of 1 mg IM is safe. Administer 1 mL IM to the patient.
19. Dose is safe. Amount to administer: 1.25 mL
20. Dose is safe. Amount to administer: 1.9 mL

21. 0.64 mL/dose
22. 5.4 mL/dose

Review and Practice

11-2 Dosages Based on Body Surface Area (BSA)

1. 0.57 m^2
2. 0.58 m^2
3. 0.25 m^2
4. 0.37 m^2
5. 1.03 m^2
6. 0.69 m^2
7. 0.36 m^2
8. 0.42 m^2
9. 143.5 mcg
10. 0.26 mg
11. 14.65 mcg
12. 0.18 mg
13. 14.7 mL/dose
14. 1.7 mL/dose
15. safe dose

    1.1 mL
16. The order is consistent with the recommended dosage.

    Amount to administer: 2.6 mL
17. BSA: 1.3 m^2

    D: 1250 mg

    The ordered dose of 800 mg is within this range and is safe.

    Amount to administer: 21 mL (Note you will need 4 vials)
18. Upper range: $1.6\ \cancel{m^2} \times \frac{75\text{ mg}}{\cancel{m^2}} = 120\text{ mg}$

    Lower range: : $1.6\ \cancel{m^2} \times \frac{100\text{ mg}}{\cancel{m^2}} = 160\text{ mg}$

    The ordered dose of 125 mg is within this range and is safe.

    Amount to administer: 125 mL

Critical Thinking on the Job

Consulting the Physician

Karen should have immediately contacted the physician to discuss the order. As it turns out, the physician did not consider the chronic renal failure when prescribing the medication for the pneumonia. A new order is written: Tazidime 500 mg q24h.

If Karen had administered the original amount of Tazidime, Mrs. Bekins would probably have developed an accumulation of the drug, producing symptoms of toxicity such as seizures. By using her critical thinking skills, Karen helped the patient to receive the correct dosage.

Review and Practice

11-3 Age-Related Needs

1. 800 mL
2. 1760 mL
3. 1341 mL
4. 1627 mL
5. 2200 mL
6. 63 gtt/min
7. 52 gtt/min
8. 32 gtt/min
9. 60 gtt/min
10. 108 gtt/min
11. 72 mL/h
12. 63 mL/h
13. Safe
14. Safe
15. Safe
16. Safe
17. Safe; amount to administer: $\frac{1}{2}$ tablet
18. Safe; amount to administer: 10 mL q4h

Chapter 11 Review

Check Up

1. 22.27 kg
2. 27.73 kg
3. 2.98 kg
4. 5.82 kg
5. 0.74 m^2
6. 0.5 m^2
7. 0.66 m^2
8. 0.47 m^2
9. The ordered dose of 100 mg may be administered safely.

   Amount to administer: 2 mL
10. The usual dose corresponds to the dosage ordered. The order of 1.6 mg may be administered safely.

    Amount to administer: 4 mL
11. The ordered dose is 1300 IU, slightly less than the recommended dosage, and, therefore, is safe to administer.

    Amount to administer: 1.7 mL
12. The maximum safe dose is 232.5 mg/dose. The ordered dose of 225 mg may be administered safely. Amount to administer: 4.5 mL
13. 1. Convert the patient's weight to kg.
       30 kg
    2. Multiply weight in kg by suggested dose
       $8 \text{ mg} \times 30 \text{ kg} = 240 \text{ mg}$
       $10 \text{ mg} \times 30 \text{ kg} = 300 \text{ mg}$
    3. Determine ordered daily dose
       $150 \text{ mg} \times 2 = 300 \text{ mg}$
    4. Dose is safe; determine amount to administer
       2.5 mL
14. Safe dose: Amount to administer 8.125 mL rounded to 8 mL.

    Patient weight in kg equals 26.4 lb.

    Minimum safe dose: 780 mg per day for mild to moderate infection;

    1560 mg per day for severe infection;

    Maximum safe dose: 1300 mg per day for mild to moderate infection;

    2600 mg per day for severe infection.
15. The ordered dose matches the recommended dose and is safe. Amount to administer: 0.21 mL
16. DMFN: 1580, flow rate: 66 mL/h
17. DMFN: 1386, flow rate: 58 mL/h
18. DMFN: 2100 mL − 800 mL taken orally = 1300 mL, flow rate: 54 mL/h
19. Safe; amount to administer: 1 tablet
20. Safe; amount to administer: 1.7 mL
21. Safe; amount to administer: 9 mL
22. Safe; amount to administer: 100 mL
23. Not safe; consult the physician
24. Safe; amount to administer: 100 mL

Critical Thinking Applications

1. The dose for patient A is not safe. He has a creatinine clearance of 37 mL/min and nosocomial pneumonia. He should receive 3.375 gram. A dose of 4.5 gram would not be safe because of his lowered renal function. The dose for patient B is safe.
2. For patient B, reading the information of the package insert you would select the 3.375 g vial and mix it with 15 mL of any of the compatible reconstitution solutions. *Always double check the package instructions with any medication you are going to reconstitute.*

3. For patient A, you would notify the prescribing physician and inform him of the patient's creatinine clearance of 37 mL/min.

4. For patient B, you end up with a volume of 165 mL (15 mL in syringe and 150 mL of diluent) to infuse over $^1/_2$ hour. The flow rate in mL per hour would be 330 mL per hour.

Case Study

1. Upper range: 1364 mg/day

   Lower range: 909 mg/day

2. The hourly flow rate is 30 mL/h.

3. The order for 1200 mg (600 mg twice a day) is within the accepted range of 909 to 1364 mg. It may be administered safely.

Internet Activity

Some height-to-weight ratio charts can be found at
http://kidshealth.org/teen/question/dieting/weight_height.html
http://www.healthstatus.com/calculate/iwc
http://www.halls.md/ideal-weight/met.htm
http://www.thecolumn.org/ideal-body-weight-chart-view.asp?sf=wlf

CHAPTER 12 ANSWERS

Critical Thinking on the Job

Clarifying the Order

The physician's initial order was incomplete. The health care professional should have realized that a correct order would include the type of insulin. The health care professional should have checked with the physician before administering the dose.

Review and Practice

12-1 Insulin

1. Vial D

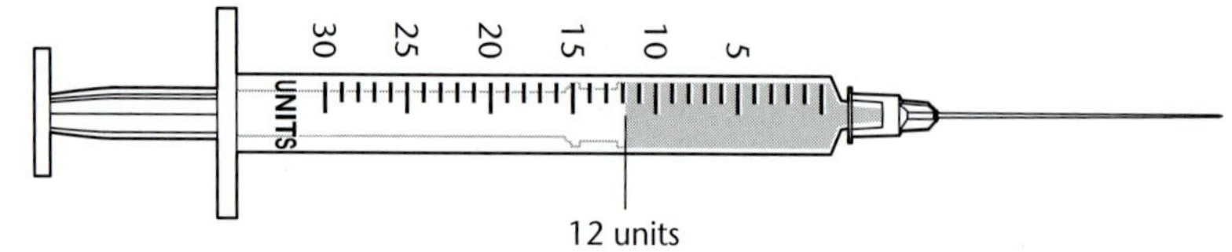

2. Vial A

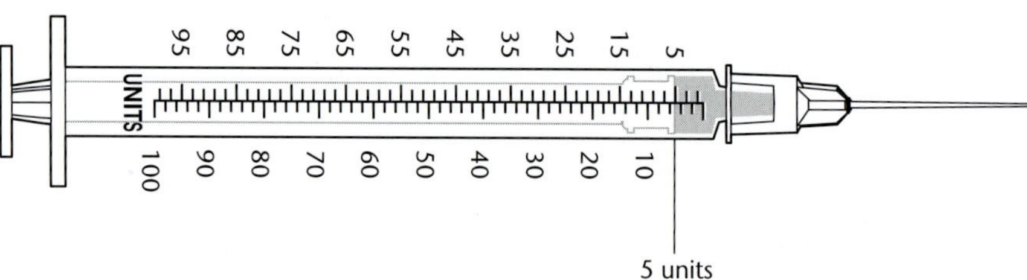

3. Vial C

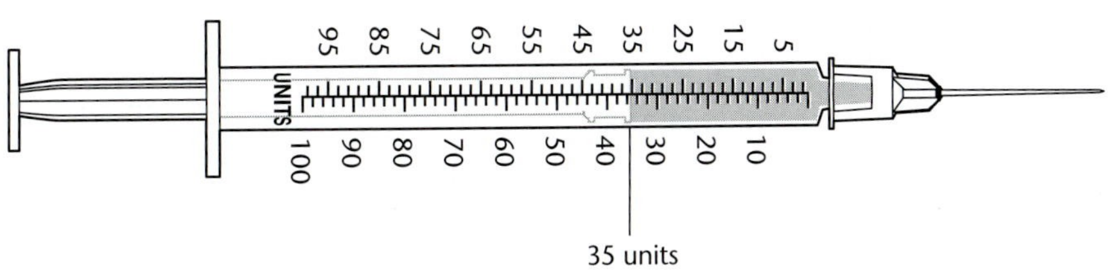

4. Vial B

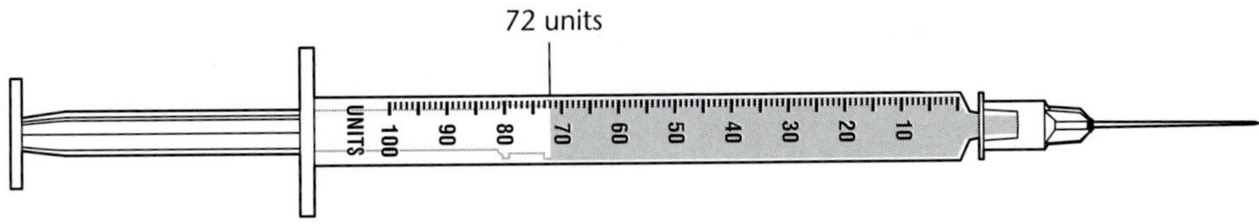

5. Vial G

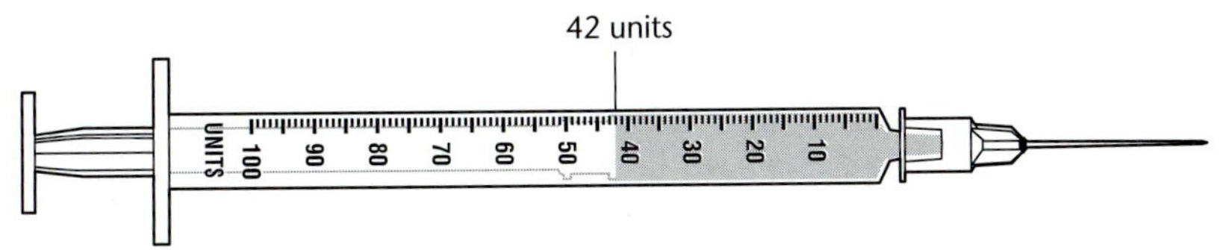

6. Vial E

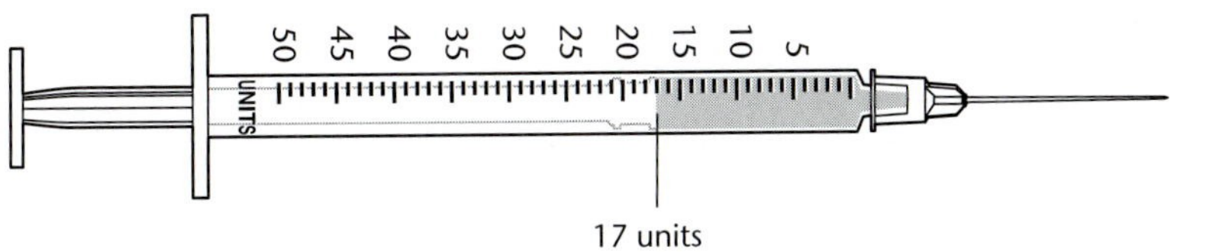

7. Vial F

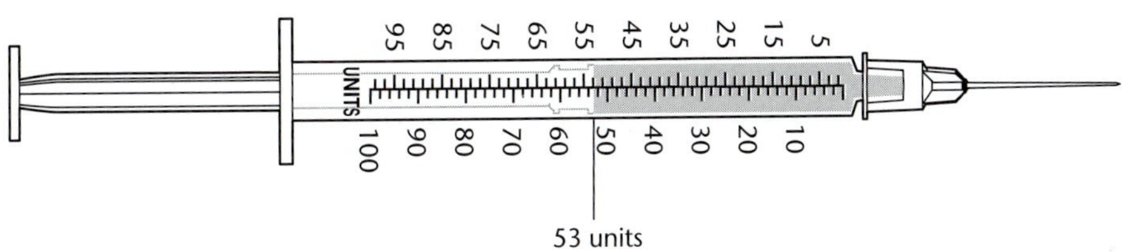

8. Vial F

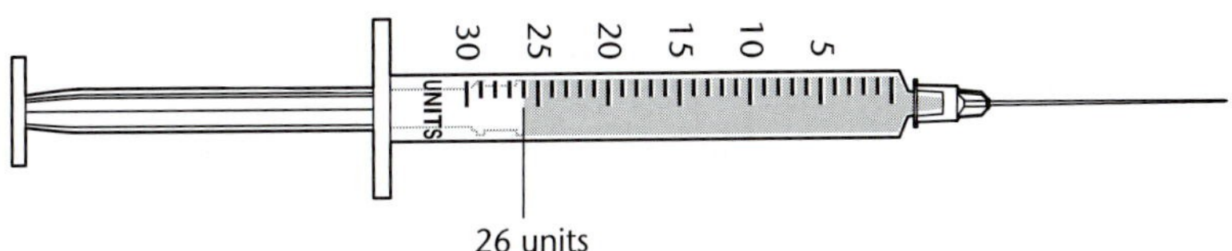

9. Vial B

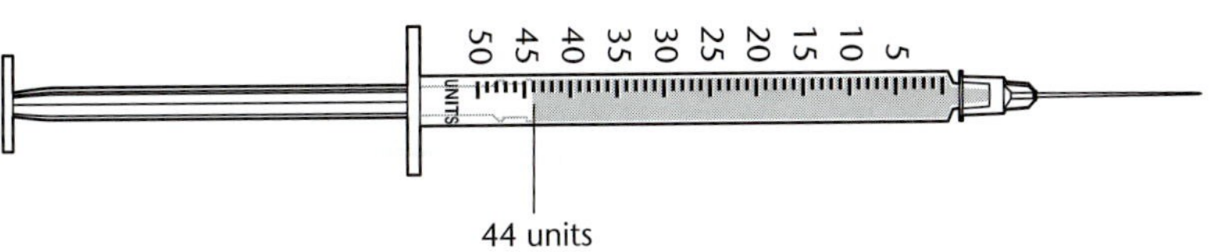

10. Vial A

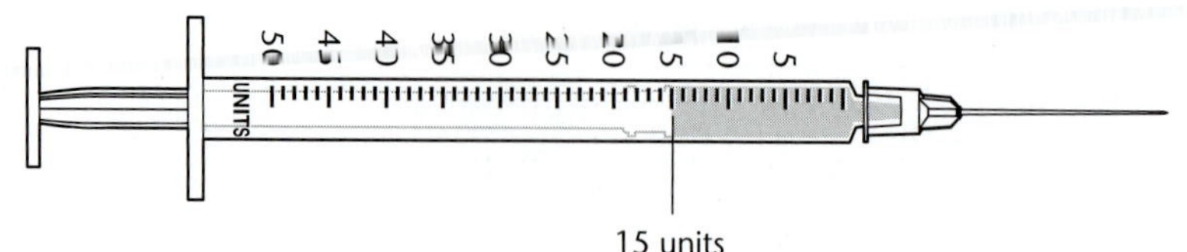

11. Vial C

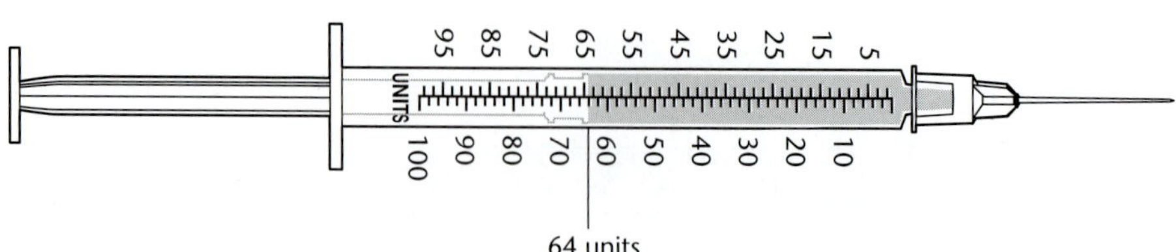

12. Vial G

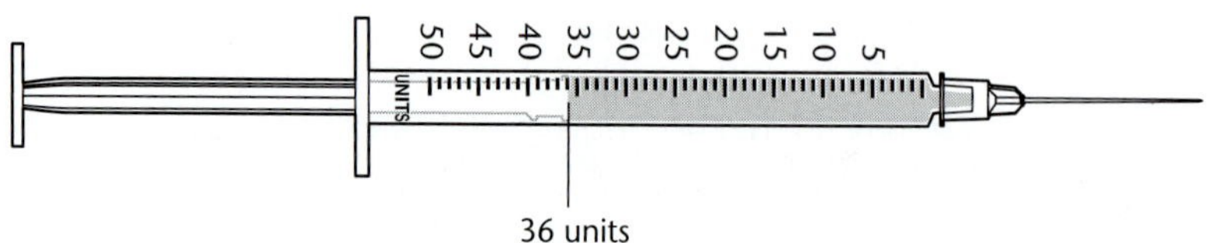

13. Vial E

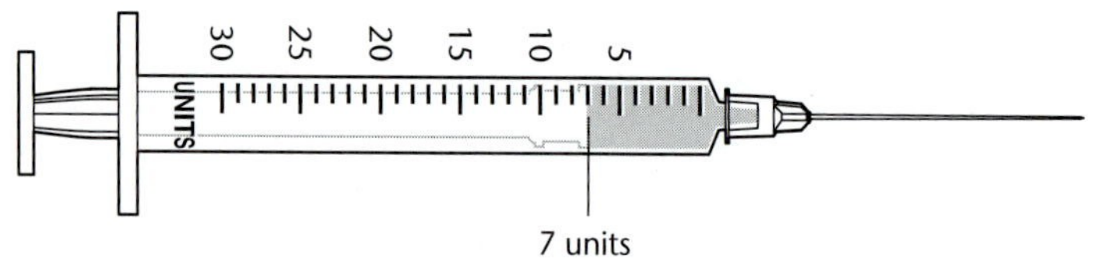

14. Vial D

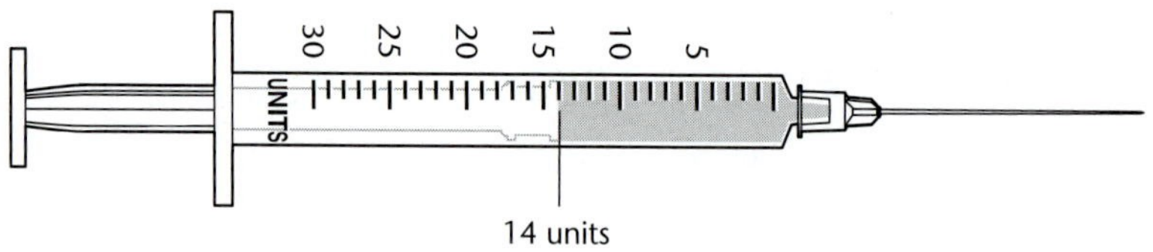

15.

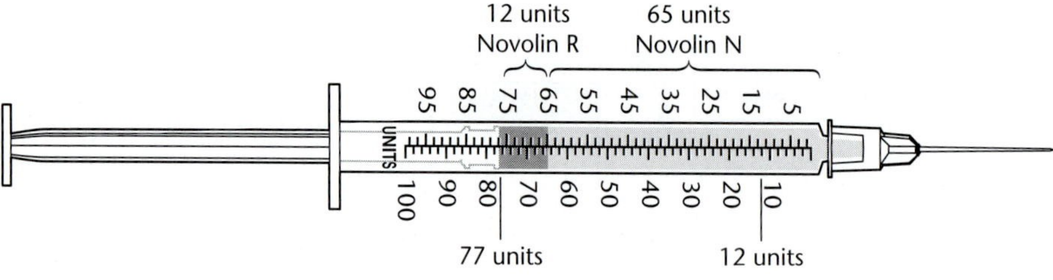

16.

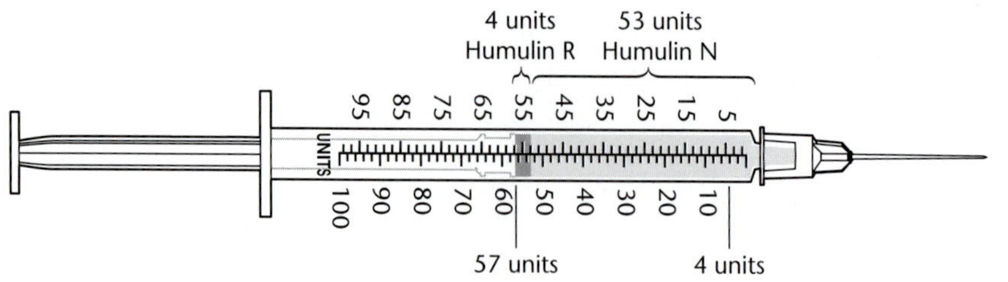

Review and Practice

12-2 Heparin

1. 75 mL/h
2. 25 gtt/min
3. 8 gtt/min
4. 4 gtt/min
5. 8 gtt/min
6. The hourly dose is 1000 units/h. This dosage is within the safe daily rate of heparin.
7. The hourly dose is 450 units/h. This dosage lies below the safe daily rate, so is safe to administer.
8. The hourly dose is 3333 units/h. This dosage is well above the safe daily rate of heparin. It would not be safe to administer.
9. The hourly dose is 2400 units/h. This dosage lies above the safe daily rate of heparin. It would not be safe to administer.
10. The hourly dose is 4412 units/h. This dosage lies above the safe daily rate of heparin. It would not be safe to administer.

Review and Practice

12-3 Critical Care IV

1. a. 100 mL/h
   b. 33 gtt/min
2. a. 18 mL/h
   b. 18 gtt/min
3. a. 30 mL/h
   b. 5 gtt/min
4. a. 45 mL/h
   b. 11 gtt/min
5. a. 29 mL/h
   b. 5 gtt/min
6. a. 150 mL/h
   b. 150 gtt/min
7. 60 gtt/min
8. 29 gtt/min
9. 21 gtt/min
10. 31 gtt/min
11. Weight in kilograms: 75
    Ordered: 750 mg
    Amount to administer 100 mL
    flow rate 100 gtt/min
12. The patient has received 0.8 g of Lidocaine.
13. The patient has received 180 mg of Remicade®.
14. The patient has received 120 mg of Dobutrex®.
15. The patient has received 56 mg of Gentamicin.
16. 23 mL/h
17. 233 mcg/min
18. 333 mcg/min
19. 170 mg dopamine has been received.
20. Flow rate 113 mL/h

Review and Practice

12-4 Preparation of Solutions, Dilutions, and Solids

1.

0.9% Sodium chloride	
NaCl	4.5 g
Water	qsad 500 mL

2.

2% Lidocaine	
Lidocaine	1 g
Water	qsad 50 mL

3.

3% Hydrocortisone ointment	
Hydrocortisone	3 g
Petroleum jelly	97 g

4.

0.45% Sodium chloride	
NaCl	1.125 g
Water	qsad 250 mL

5.

20% Zinc oxide ointment	
Zinc oxide	15 g
Petroleum jelly	60 g

6.

12% Dextrose water	
20% Dextrose	466.7 mL
5% Dextrose	qsad 1000 mL

7.

30% Ethyl alcohol	
95% Ethyl alcohol	473.7 mL
Water	qsad 1500 mL

8.

20% Iodine	
25% Iodine	67 mL
10% Iodine	qsad 100 mL

9.

50% Ethanol	
90% Ethanol	1111.1 mL
Water	qsad 2000 mL

10.

5% Ointment	
10% Ointment	85.1 g
2% Ointment	141.9 g

or

5% Ointment	
10% Ointment	0.1875 lb
2% Ointment	0.3125 lb

Chapter 12 Review

Check Up

1. Vial F

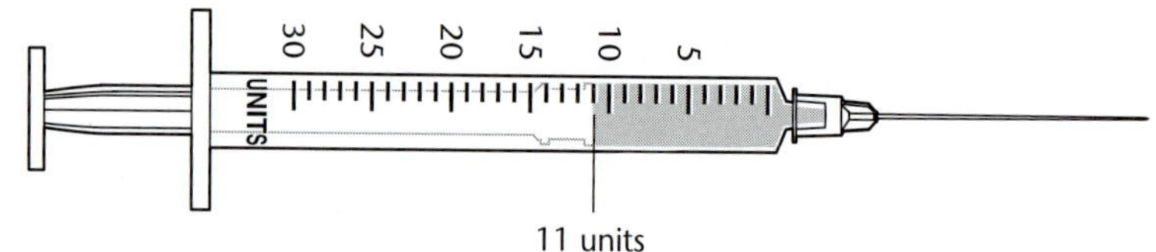

2. Vial D

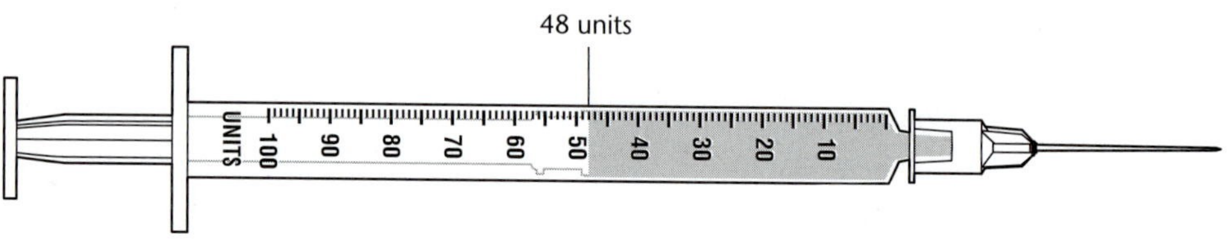

3. Vial E

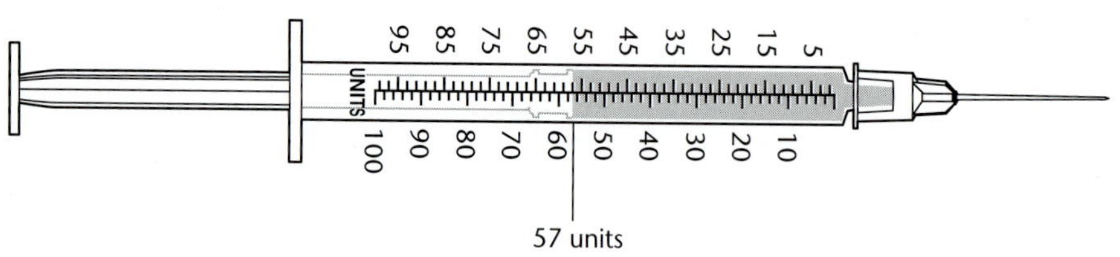

4. Vial B

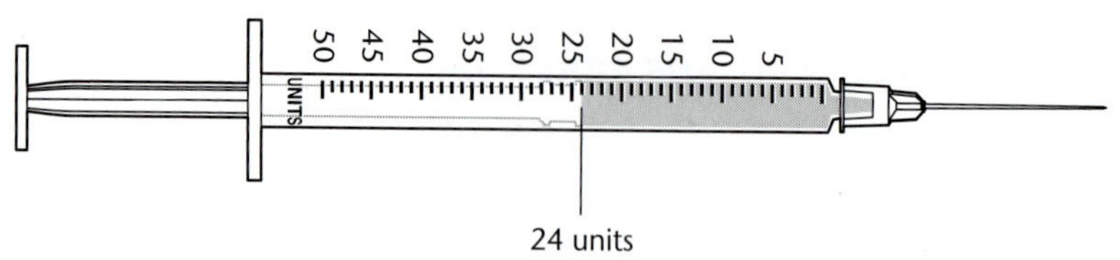

5. Vial A

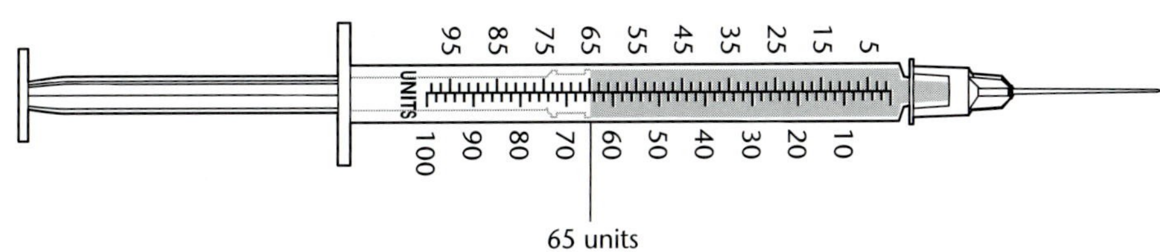

6. Vial C

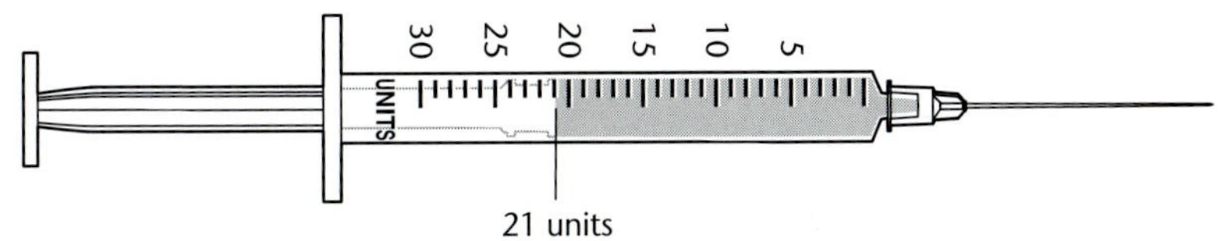

7.

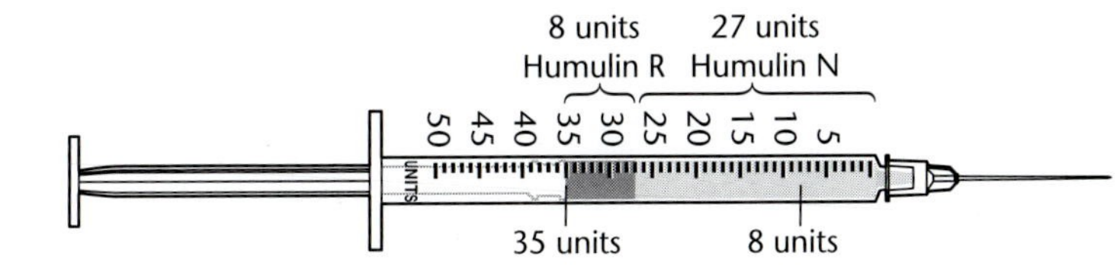

8.

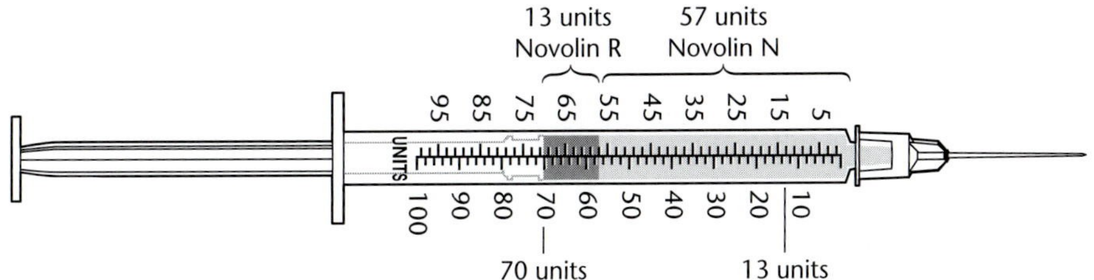

9. 8 gtt/min
10. 5 gtt/min
11. 15 mL/h
12. 40 gtt/min
13. 1800 U/h

    This dosage lies above the safe daily rate of heparin. It would not be safe to administer.
14. 1260 U/h

    This dosage is within the safe daily rate of heparin.
15. 1600 U/h

    This dosage is within the safe daily rate of heparin.
16. 1000 U/h

    This dosage is within the safe daily rate of heparin.
17. a. 210 mL/h

    b. 53 gtt/min
18. a. 34 mL/h

    b. 34 gtt/min
19. a. 45 mL/h

    b. 8 gtt/min
20. a. 45 mL/h

    b. 15 gtt/min
21. 13 gtt/min
22. 20 gtt/min
23. The patient has received 8 g of magnesium sulfate.

24. The patient has received 3 mg of nitroprusside.

25. $F\ \text{mL/h} = \frac{500\ \text{mL}}{4\ \text{mg}} \times \frac{6\ \text{mcg}}{1\ \text{min}} \times \frac{60\ \text{min}}{1\ \text{h}} \times \frac{1\ \text{mg}}{1000\ \text{mcg}}$

$F\ \text{mL/h} = \frac{500\ \text{mL}}{4\ \cancel{\text{mg}}} \times \frac{6\ \cancel{\text{mcg}}}{1\ \cancel{\text{min}}} \times \frac{60\ \cancel{\text{min}}}{1\ \text{h}} \times \frac{1\ \cancel{\text{mg}}}{1000\ \cancel{\text{mcg}}}$

$F = 45\ \text{mL/h}$

26. $F\ \text{mL/h} = \frac{500\ \text{mL}}{4\ \text{mg}} \times \frac{10\ \text{mcg}}{1\ \text{min}} \times \frac{60\ \text{min}}{1\ \text{h}} \times \frac{1\ \text{mg}}{1000\ \text{mcg}}$

$F\ \text{mL/h} = \frac{500\ \text{mL}}{4\ \cancel{\text{mg}}} \times \frac{10\ \cancel{\text{mcg}}}{1\ \cancel{\text{min}}} \times \frac{60\ \cancel{\text{min}}}{1\ \text{h}} \times \frac{1\ \cancel{\text{mg}}}{1000\ \cancel{\text{mcg}}}$

$F = 75\ \text{mL/h}$

27.

1% Hydrocortisone	
2.5% Hydrocortisone	20 g
Cream base	30 g

28.

40% Dextrose	
80% Dextrose	68.2 mL
25% Dextrose	qsad 250 mL

29.

35% Iodine	
70% Iodine	300 mL
20% Iodine	qsad 1000 mL

30.

6% Betadine	
10% Betadine	300 mL
Distilled water	qsad 500 mL

Critical Thinking Applications

1. Set the initial rate at 48 mL/h.
2. The maximum safe rate is 480 mL/h.
3. The infusion rate has not yet reached the maximum safe dose. Therefore, it may be increased in an attempt to lower the patient's systolic pressure.
4. The injection is now running at the maximum safe rate and cannot be raised further. Because the patient's systolic pressure is still above the one desired, you should contact the patient's physician for further instructions.

Case Study

The patient received 4.5 mg of morphine sulfate during the shift.

Internet Activity

1. 5 to 10 mg/kg in 5 min
2. a. 5 to 10 mcg
   b. Every 5 to 10 min
3. Mechanical ventilatory support should be in place.

COMPREHENSIVE EVALUATION ANSWERS

1. 400 mg
2. Orally (po)
3. 0730, 1130, and 1630
4. The order indicates that Ativan® should be administered when necessary for pain, rather than according to a set schedule.
5. The order for Reglan® is incomplete. It does not list the route.
6. Reglan® is administered before meals (ac).
7. 1600 and 2400. If the order was for q8h, Neurontin would be administered every 8 h.
8. 2130. The order indicates that Ativan® may be administered every 8 h. If Ativan® is administered at 1330, then the patient may not receive another dose for another 8 h, at 2130.
9. Cephalexin
10. 20 to 25°C (68 to 77°F)
11. 500 mg/capsule

12. The usual adult dose is 250 mg every 6 h. Therefore, the usual adult dose is 2 capsules a day, since each capsule is 500 mg. Twice the usual dose, then, is 4 capsules a day. The container has 100 capsules. At the rate of 4 capsules a day, the container would last 100 ÷ 4 or 25 days.

13. 38 mL

14. A 250 mg dosage is 10 mL of solution. The bottle contains 60 mL, so it will last for 6 doses.

15. 1.2 mL. The diluent provided with the medication.

16. 6.4 million international units

17. Desired dose *D*: 75 mg

Dose on hand *H*: 50 mg

Dosage unit *Q*: 1 tablet

$75 \text{ mg} \times \frac{1 \text{ tablet}}{50 \text{ mg}} = \overset{3}{\cancel{75}} \cancel{\text{mg}} \times \frac{1 \text{ tablet}}{\underset{2}{\cancel{50}} \cancel{\text{mg}}}$

$= 3 \text{ t} \times \frac{1}{2} \text{ tablet}$

$= \frac{3}{2} \text{ tablets}$

Amount to administer: $1\frac{1}{2}$ tablets

18. $0.2 \text{ g} \times \frac{1000 \text{ mg}}{1 \text{ g}} \times \frac{5 \text{ mL}}{200 \text{ mg}}$

$= 0.2 \cancel{\text{g}} \times \frac{\overset{5}{\cancel{1000}} \cancel{\text{mg}}}{1 \cancel{\text{g}}} \times \frac{5 \text{ mL}}{\underset{1}{\cancel{200}} \cancel{\text{mg}}}$

$= 0.2 \times 5 \times 5 \text{ mL} = 5 \text{ mL}$

Amount to administer: 5 mL

19. First convert gr $\frac{1}{200}$ to milligrams.

gr $\frac{1}{200}$: ? mg :: 1 gr : 60 mg

$? \times 1 = \frac{1}{200} \times 60$

$? = \frac{60}{200} = 0.3$

Desired dose *D*: 0.3 mg

Dose on hand *H*: 0.3 mg

Dosage unit *Q*: 1 tablet

$0.3 \text{ mg} \times \frac{1 \text{ tablet}}{0.3 \text{ mg}} = \overset{1}{\cancel{0.3}} \cancel{\text{mg}} \times \frac{1 \text{ tablet}}{\underset{1}{\cancel{0.3}} \cancel{\text{mg}}}$

$= 1 \times 1$ tablet

$= 1$ tablet

Amount to administer: 1 tablet

20. gr $\frac{1}{4} \times \frac{60 \text{ mg}}{1 \text{ gr}} \times \frac{1 \text{ mL}}{10 \text{ mg}}$

$= \cancel{\text{gr}} \frac{1}{4} \times \frac{\overset{6}{\cancel{60}} \cancel{\text{mg}}}{1 \cancel{\text{gr}}} \times \frac{1 \text{ mL}}{\underset{1}{\cancel{10}} \cancel{\text{mg}}}$

$= \frac{1}{4} \times 6 \times 1 \text{ mL} = \frac{6}{4} \text{ mL}$

Amount to administer: 1.5 mL

21. $0.6 \text{ g} \times \frac{1000 \text{ mg}}{1 \text{ g}} \times \frac{1 \text{ mL}}{300 \text{ mg}}$

$= 0.6 \cancel{\text{g}} \times \frac{\overset{10}{\cancel{1000}} \cancel{\text{mg}}}{1 \cancel{\text{g}}} \times \frac{1 \text{ mL}}{\underset{3}{\cancel{300}} \cancel{\text{mg}}}$

$= 0.6 \times 10 \times \frac{1 \text{ mL}}{3} = 2 \text{ mL}$

Amount to administer: 2 mL

22. $0.3 \text{ mg} \times \frac{1000 \text{ mcg}}{1 \text{ mg}} \times \frac{1 \text{ mL}}{200 \text{ mcg}}$

$= 0.3 \cancel{\text{mg}} \times \frac{\overset{5}{\cancel{1000}} \cancel{\text{mcg}}}{1 \cancel{\text{mg}}} \times \frac{1 \text{ mL}}{\underset{1}{\cancel{200}} \cancel{\text{mcg}}}$

$= 0.3 \times 5 \times 1 \text{ mL} = 1.5 \text{ mL}$

Amount to administer: 1.5 mL

23. $97 \cancel{\text{lb}} \times = \frac{1 \text{ kg}}{2.2 \cancel{\text{lb}}} = 44.1 \text{ kg}$

$44 \cancel{\text{kg}} \times \frac{17 \text{ mg}}{1 \cancel{\text{kg}}} = 748 \text{ mg}$

$748 \cancel{\text{mg}} \times \frac{1 \text{ mL}}{15 \cancel{\text{mg}}} = \frac{748}{15} \text{ mL} = 50 \text{ mL}$

The amount to administer is 50 mL.

24. $62 \cancel{\text{lb}} \times \frac{1 \text{ kg}}{2.2 \cancel{\text{lb}}} = 28.2 \text{ kg}$

$28.2 \cancel{\text{kg}} \times \frac{10 \text{ mg}}{1 \cancel{\text{kg}}} = 282 \text{ mg}$

$282 \cancel{\text{mg}} \times \frac{\cancel{100} \text{ mL}}{\cancel{500} \cancel{\text{mg}}} = \frac{282}{5} = 56.4 \text{ mL}$

The amount to administer is 56 mL.

25. Desired dose *D*: 200 IU

Dose on hand *H*: 225 IU

Dosage unit *Q*: 1 mL

$200 \text{ IU} \times \frac{1 \text{ mL}}{225 \text{ IU}} = \overset{8}{\cancel{200}} \cancel{\text{IU}} \times \frac{1 \text{ mL}}{\underset{9}{\cancel{225}} \cancel{\text{IU}}}$

$= 8 \times \frac{1}{9} \text{ mL} = \frac{8}{9} \text{ mL}$

Amount to administer: 0.89 mL

26. $49 \cancel{\text{lb}} \times \frac{1 \text{ kg}}{2.2 \text{ lb}} = 22.3 \text{ kg}$

$23.3 \cancel{\text{kg}} \times \frac{20 \text{ mg}}{1 \cancel{\text{kg}}} = 446 \text{ mg}$

$446 \cancel{\text{mg}} \times \frac{1 \text{ mL}}{220 \cancel{\text{mg}}} = 446 \times \frac{1}{220} \text{ mL}$

$= \frac{446}{220} \text{ mL}$

The amount to administer is 2 mL.

27. Desired dose *D*: 40 mg

Dose on hand *H*: 20 mg

Dosage unit Q: 1 capsule

$40 \text{ mg} \times \frac{1 \text{ capsule}}{20 \text{ mg}} = \overset{2}{\cancel{40}} \cancel{\text{mg}} \times \frac{1 \text{ capsule}}{\underset{1}{\cancel{20}} \cancel{\text{mg}}}$

$= 2 \times 1$ capsule

$= 2$ capsule

Amount to administer: 7 capsules

28. Desired dose *D*: 300 mg

Dose on hand *H*: 100 mg

Dosage unit Q: 1 tablet

$$300\text{ mg} \times \frac{1\text{ tablet}}{100\text{ mg}} = \overset{3}{\cancel{300}}\ \cancel{\text{mg}} \times \frac{1\text{ tablet}}{\underset{1}{\cancel{100}}\ \cancel{\text{mg}}}$$

$= 3 \times 1$ tablet

$= 3$ tablets

Amount to administer: $\frac{1}{2}$ tablet

29. $55\ \cancel{\text{lb}} \times \frac{1\ \cancel{\text{kg}}}{2.2\ \cancel{\text{lb}}} = 25\ \cancel{\text{kg}}$

$$25\ \cancel{\text{kg}} \times \frac{10\text{ mg}}{1\ \cancel{\text{kg}}} = 250\text{ kg}$$

$$25\cancel{0}\ \cancel{\text{mg}} \times \frac{5\text{ mL}}{30\cancel{0}\ \cancel{\text{mg}}} = \frac{125\text{ mL}}{30} = 4.2\text{ mL}$$

Amount to administer: 4.2 mL

30. The amount to administer is 2 tablets.
31. Amount to administer: 1 tablet
32. Amount to administer: 7.5 mL
33. Novolin® R is a rapid-acting insulin. Novolin® N is an intermediate-acting insulin.
34. Draw the rapid-acting insulin into the syringe first, then the intermediate-acting insulin. In this case, draw Humulin® R first, then Humulin® N.
35. $\frac{1000\text{ mL}}{8\text{ h}} = \frac{125\text{ mL}}{1\text{ h}}$

Administer: 125 mL/h

36. $\frac{15\text{ gtt}}{1\text{ mL}} \times \frac{600\text{ mL}}{4\text{ h}} \times \frac{1\text{ h}}{60\text{ min}}$

$$\frac{\overset{1}{\cancel{15}}\text{ gtt}}{1\ \cancel{\text{mL}}} \times \frac{\overset{150}{\cancel{600}}\ \cancel{\text{mL}}}{\underset{1}{\cancel{4}}\ \cancel{\text{h}}} \times \frac{1\ \cancel{\text{h}}}{\underset{4}{\cancel{60}}\text{ min}}$$

$= 1\text{ gtt} \times 150 \times \frac{1}{4}\text{ min}$

$= \frac{150\text{ gtt}}{4\text{ min}} = 37.5\text{ gtt/min}$

Flow rate: 38 gtt/min

37. $147\ \cancel{\text{lb}} \times \frac{1\text{ kg}}{2.2\ \cancel{\text{lb}}} = 66.8\text{ kg}$

$$66.8\ \cancel{\text{kg}} \times \frac{1.2\text{ mg}}{1\ \cancel{\text{kg}}} = 80\text{ mg}$$

You want to administer 80 mg of Zofran® over 45 min. The vial had 2 mL of 40 mg/mL injection, or 80 mg/2 mL. It is mixed with 68 mL of D5W, leading to a solution with 80 mg/70 mL.

$$\frac{80\text{ mg}}{?\text{ mL}} = \frac{80\text{ mg}}{70\text{ ML}}$$

No further calculation is needed to see that you will need to administer 70 mL over 45 min. The tubing is calibrated for 15 gtt/mL. Therefore,

$$\frac{15\text{ gtt}}{1\text{ mL}} \times \frac{70\text{ mL}}{45\text{ min}}$$

$$\frac{\overset{1}{\cancel{15}}\text{ gtt}}{1\ \cancel{\text{mL}}} \times \frac{70\ \cancel{\text{mL}}}{\underset{3}{\cancel{45}}\text{ min}} = 1\text{ gtt} \times \frac{70}{3\text{ min}} = \frac{23\text{ gtt}}{1\text{ min}}$$

Flow rate: 23 gtt/min

38. $68\ \cancel{\text{lb}} \times \frac{1\text{ kg}}{2.2\ \cancel{\text{lb}}} = 31\text{ kg}$

$$31\ \cancel{\text{kg}} \times \frac{0.1\text{ mg}}{1\ \cancel{\text{kg}}} = 3.1\text{ mg}$$

You want to administer 3.1 mg of Zofran®.

$$\frac{3.1\text{ mg}}{?\text{ mL}} = \frac{32\text{ mg}}{50\text{ mL}}$$

$3.1\text{ mg} \times 50\text{ mL} = ?\text{ mL} \times 32\text{ mg}$

$4.8 = ?$

You want to administer 4.8 mL of Zofran® over 4 min.

The tubing is calibrated for 10 gtt/mL. Therefore,

$$\frac{10\text{ gtt}}{1\text{ mL}} \times \frac{4.8\text{ mL}}{4\text{ min}}$$

$$= \frac{\overset{5}{\cancel{10}}\text{ gtt}}{1\ \cancel{\text{mL}}} \times \frac{4.8\ \cancel{\text{mL}}}{\underset{2}{\cancel{4}}\text{ min}} = 5\text{ gtt} \times \frac{4.8}{2\text{ min}} = \frac{12\text{ gtt}}{1\text{ min}}$$

Flow rate: 12 gtt/min

39. $134\ \cancel{\text{lb}} \times \frac{1\text{ kg}}{2.2\ \cancel{\text{lb}}} = 61\text{ kg}$

$$61\ \cancel{\text{kg}} \times \frac{10\text{ mg}}{1\ \cancel{\text{kg}}} = 610\text{ mg}$$

You want to administer 610 ciprofloxacin.

$$\frac{610\text{ mg}}{?\text{ mL}} = \frac{10\text{ mg}}{1\text{ mL}}$$

$610\text{ mg} \times 1\text{ mL} = ?\text{ mL} \times 10\text{ mg}$

$61\text{ mL} = ?$

You add 61 mL of ciprofloxacin to 150 mL of 5% DW for a total volume of 211 mL to be administered over 60 minutes.

The tubing is calibrated for 20 gtt/mL. Therefore,

$$\frac{20\text{ gtt}}{1\text{ mL}} \times \frac{211\text{ mL}}{60\text{ min}}$$

$= 61$

Flow rate: 70 gtt/min

40. You want to administer 750 mg Rocephin® in 100 mL of NS over 30 minutes. According to the drug label you must add 9.6 mL of diluent; then each 1 mL contains 100 mg. Adding 7.5 mL to 100 mL of NS, you will have 107.5 mL to infuse over 30 minutes. Your flow rate will be 215 mL/h.

41. $\frac{20\text{ gtt}}{1\ \cancel{\text{mL}}} \times \frac{1300\ \cancel{\text{U}}}{1\ \cancel{\text{h}}} \times \frac{1000\ \cancel{\text{mL}}}{50{,}000\ \cancel{\text{U}}} \times \frac{1\ \cancel{\text{h}}}{60\text{ min}}$

$$= \frac{9\text{ gtt}}{1\text{ min}}$$

42. $\frac{850\ \cancel{\text{U}}}{1\text{ h}} \times \frac{500\text{ mL}}{40{,}000\ \cancel{\text{U}}} = 11\text{ mL/h}$

43. $\frac{40\text{ mg}}{500\text{ mL}} = \frac{?\text{ mg}}{175\text{ mL}}$

$40\text{ mg} \times 175\text{ mL} = 500\text{ mL} \times ?\text{ mg}$

$14 = ?$

The patient has received 14 mg of nitroprusside.

44. $\frac{15 \text{ gtt}}{1 \text{ mL}} \times \frac{650 \text{ mL}}{8 \text{ h}} \times \frac{1 \text{ h}}{60 \text{ min}} = F$

$$\frac{\overset{1}{\cancel{15}} \text{ gtt}}{1 \cancel{\text{mL}}} \times \frac{\overset{325}{\cancel{650}} \cancel{\text{mL}}}{\underset{4}{\cancel{8}} \cancel{\text{h}}} \times \frac{1 \cancel{\text{h}}}{\underset{4}{\cancel{60}} \text{ min}} = 20.31 \text{ gtt/min}$$

The original flow rate is 20 gtt/min. The amount that should have infused in the first 3 h is

$$3 \cancel{\text{h}} \times \frac{60 \cancel{\text{min}}}{1 \cancel{\text{h}}} \times \frac{20 \cancel{\text{gtt}}}{1 \cancel{\text{min}}} \times \frac{1 \text{ mL}}{15 \cancel{\text{gtt}}} = 240 \text{ mL}$$

The amount that has infused is 650 mL − 490 mL = 160 mL. The solution has been infusing too slowly and needs to be adjusted.

The amount of solution remaining is 490 mL.

Recalculate the flow rate for 490 mL over 5 h.

$$\frac{15 \text{ gtt}}{\text{mL}} \times \frac{490 \text{ mL}}{5 \text{ h}} \times \frac{1 \text{ h}}{60 \text{ min}} = F$$

$$\frac{\overset{1}{\cancel{15}} \text{ gtt}}{1 \cancel{\text{mL}}} \times \frac{\overset{98}{\cancel{490}} \cancel{\text{mL}}}{\underset{1}{\cancel{5}} \cancel{\text{h}}} \times \frac{1 \cancel{\text{h}}}{\underset{4}{\cancel{60}} \text{ min}} = 24.5 \text{ gtt/min}$$

The adjusted flow rate should be 25 gtt/min. Check if it is within 25 percent of the original flow rate:

$25\% \times 20 \text{ gtt/min} = 5$

Because 25 gtt/min is 5 gtt/min more than the original flow rate, 25 gtt/min is within the acceptable range.

45. $\text{BSA} = \sqrt{\frac{59 \times 93}{3131}} \text{ m}^2 = \sqrt{\frac{5487}{3131}} \text{ m}^2 = 1.324 \text{ m}^2$ rounded to 1.3 m

46. $\text{BSA} = \sqrt{\frac{150 \times 61}{3600}} \text{ m}^2 = \sqrt{\frac{9150}{3600}} \text{ m}^2 = 1.594 \text{ m}^2$ rounded to 1.6 m^2

47. $\text{BSA} = \sqrt{\frac{66 \times 125}{3131}} \text{ m}^2 = \sqrt{\frac{8250}{3131}} \text{ m}^2 = 1.623 \text{ m}^2$ rounded to 1.6 m^2

$$1.623 \text{ m}^2 \times \frac{200 \text{ mg}}{\text{m}^2} = 324.7 \text{ mg}$$

Administer: 324.7 mg

48. $\text{BSA} = \sqrt{\frac{60 \times 103}{3131}} \text{ m}^2 = \sqrt{\frac{6180}{3131}} \text{ m}^2 = 1.405 \text{ m}^2$ rounded to 1.4 m^2

$$1.4 \cancel{\text{m}^2} \times \frac{200 \text{ mg}}{\cancel{\text{m}^2}} = 280 \text{ mg}$$

You want to administer 280 mg of Leucovorin Calcium. According to the label, the reconstituted Leucovorin will have a dosage strength of 20 mg/mL.

$$\frac{280 \text{ mg}}{? \text{ mL}} = \frac{20 \text{ mg}}{1 \text{ mL}}$$

$280 \text{ mg} \times 1 \text{ mL} = ? \text{ mL} \times 20 \text{ mg}$

$14 = ?$

You want to administer 14 mL of Leucovorin Calcium over 6 min. The tubing is calibrated for 10 gtt/mL. Therefore,

$$\frac{10 \text{ gtt}}{1 \text{ mL}} \times \frac{14 \text{ mL}}{6 \text{ min}}$$

$$= \frac{\overset{5}{\cancel{10}} \text{ gtt}}{1 \cancel{\text{mL}}} \times \frac{14 \cancel{\text{mL}}}{\underset{3}{\cancel{6}} \text{ min}} = 5 \text{ gtt} \times \frac{14}{3 \text{ min}} = \frac{23 \text{ gtt}}{1 \text{ min}}$$

Flow rate: 23 gtt/min

49. $1000 \text{ mL} + \left[\frac{50 \text{ mL}}{1 \cancel{\text{kg}}} \times (18 \cancel{\text{kg}} - 10)\right]$

$$= 1000 \text{ mL} + \left(\frac{50 \text{ mL}}{1} \times 8\right)$$

$$= 1000 \text{ mL} + 400 \text{ mL}$$
$$= 1400 \text{ mL}$$

DMFN = 1400 mL; for microdrip tubing,

$$\frac{1400 \text{ mL}}{1 \cancel{\text{day}}} \times \frac{1 \cancel{\text{day}}}{24 \text{ h}} = \frac{1400 \text{ mL}}{24 \text{ h}} = \frac{58 \text{ gtt}}{1 \text{ min}} = 58 \text{ gtt/min}$$

50. The patient's CL_{CR} is 48 mL/min. The recommended maintenance dosage is 2 g q4h. This amount is more than the order, which is safe to administer.

$$1.7 \cancel{\text{g}} \times \frac{1000 \cancel{\text{mg}}}{1 \cancel{\text{g}}} \times \frac{1 \text{ mL}}{20 \cancel{\text{g}}} = 85 \text{ mL}$$

Glossary

A

Absorption Movement of a drug from the site where it is given into the bloodstream.

Alligation One method for calculating dilutions.

Amount to administer The volume of liquid or number of solid dosage units that contains the desired dose.

Ampule Sealed container that usually holds 1 dose of liquid medication.

Anticoagulant A class of medication that reduces the blood's ability to clot.

Apothecary system An older system of measurement based upon a grain of wheat; other common units are the ounce, minum, and dram.

B

Biotransformation Chemical changes of a drug in the body.

Body surface area (BSA) Surface area of a patient's body, factoring in both height and weight, stated in square meters, or m^2.

C

Calibrated spoons Specially marked spoons used to administer oral medications with accuracy.

Calibrations Markings on medication equipment at various intervals.

Caplet Oval-shaped pill similar to a tablet but having a coating for easy swallowing.

Capsule Oval-shaped gelatin shell, usually in two pieces, that contains powder or granules.

Cartridge Prefilled container shaped like a syringe barrel, generally used with a reusable syringe.

Celsius A temperature scale with 0 degrees as freezing and 100 degrees as boiling; also called centigrade.

Centi (c) Metric prefix that indicates $\frac{1}{100}$ of the basic unit.

Centigrade A temperature scale with 0 degrees as freezing and 100 degrees as boiling; Celsius.

Central line An IV line that administers large amounts of medications to major veins.

Complex fraction A fraction in which the numerator and the denominator are themselves fractions.

Critical care Area of a medical facility in which patients are more seriously ill and fast-acting, potent medications are given.

Cross-multiplying Multiplying the numerator of the first fraction by the denominator of the second fraction and the denominator of the first fraction by the numerator of the second fraction, for example, $\frac{A}{B} = \frac{C}{D}$ or $A \times D = B \times C$

Cubic centimeter Measure of volume that is the same as a milliliter (mL).

D

Daily maintenance fluid need (DMFN) Amount of fluids a patient needs over 24 h both oral and parenteral.

D5W solution Intravenous solution of 5% dextrose in water.

Denominator The bottom number of a fraction; represents the whole.

Desired dose Amount of drug to be given at a single time.

Dilution A solution created from an already prepared concentrated solution.

Dimensional analysis A method of dosage calculations that utilizes a series of factors to calculate dosages.

Distribution Movement of a drug from the bloodstream into other body tissues and fluids.

Dosage ordered Amount of drug to give and how often it is to be given.

Dosage strength Dose on hand per dosage unit; the amount of drug over the form of the drug, for example, 500 mg/tablet or 250 mg/5 mL; ratio strength.

Dosage unit The unit by which the drug will be measured when administered.

Dose on hand Amount of drug contained in each dosage unit.

Dram ʒ Common unit of volume in the apothecary system.

Drip chamber An area on the IV equipment where the drop of fluid is visualized during an infusion.

Duration The length of time that the effect of the insulin lasts.

E

Eccentric Off-center.

Electronic medication administration record (eMAR) A method of utilizing bar code reading technology to monitor the bedside administration of medications.

Elimination How a drug leaves the body.

Embolism A traveling blood clot.

Enema Means of delivering medication or fluids into the rectum.

Enteric-coated Medications that only dissolve in the alkaline environment of the small intestines.

Equivalent fractions Two fractions, written differently but having the same value.

F

Fahrenheit A temperature scale in which 32 degrees is freezing and 212 degrees is boiling; average human body temperature is 98.6 degrees.

Formula A method of dosage calculation that utilizes a set equation (formula) to calculate the amount to administer; $\frac{D}{H}XQ = A$

Fraction proportion Mathematical statement that indicates two fractions are equal.

G

Gelcap Medication, usually liquid in a gelatin shell; not designed to be opened.

Generic name A drug's official name.

Geriatric Typically considered anyone over the age of 65.

Grain Basic unit of measurement in the apothecary system.

Gram Basic unit for measurement of weight in the metric system.

H

Heparin An anticoagulant or medication that reduces the blood's ability to clot.

Heparin lock An infusion port attached to an already inserted catheter for IV access; flushed with heparin.

Household Common system of measurement that utilizes the teaspoon, ounce, cup, pint, quart, and gallon.

Hypertonic Describes fluids that draw fluids from cells and tissues across the cell membrane into the bloodstream, such as 3% saline.

Hypodermic syringe Syringe used to deliver medication under the skin such as intravenously, intramuscularly, intradermally, and subcutaneously.

Hypotonic Describes fluids that move across the cell membrane into surrounding cells and tissues, such as 0.45% NS and 0.3% NS.

I

Infiltration Event of an IV infusion delivering fluid outside of a blood vessel into the surrounding tissue.

Infusion pump Device that applies pressure to maintain the rate of an IV infusion, using a sensor to monitor both the rate and when the bag is empty.

Inhalant Medication administered directly to the lungs, usually through a metered-dose inhaler or nebulizer.

Instillations Also known as drops; usually placed in the eyes or ears.

Insulin A pancreatic hormone that stimulates glucose metabolism.

International unit (IU) Amount of medication needed to produce a certain effect; standardized by an international agreement.

Intradermal (ID) Describes medication administered between the layers of skin.

Intramuscular (IM) Describes medication administered into a muscle by injection.

Intravenous (IV) Describes medication delivered directly to the bloodstream through a vein.

Isotonic Describes fluids that do not affect the fluid balance of the surrounding cells or tissues such as D5W, NS, and lactated Ringer's.

J

Jejunostomy tube Tube that delivers medication and nutrients directly into the small intestines.

K

Kilo (k) Metric prefix that indicates the basic unit times 1000.

KVO fluids Fluids that provide access to the vascular system for emergency situations.

L

Leading ring The wide ring on the tip of the plunger of a syringe that is closest to the needle; the medication is measured here.

Least common denominator The smallest number that is a common multiple of all the denominators in a group of fractions.

Liter (L) The basic unit for measurement of volume in the metric system.

M

Macrodrip tubing Type of IV tubing that delivers 10, 15, or 20 drops of fluid per milliliter.

Maintenance fluids Fluids that maintain the fluid and electrolyte balance for patients.

Means and extremes For the equation $A : B :: C : D$, A and D are the extremes (ends) and B and C are the means (middle).

Medication administration record (MAR) Record that contains a list of medications ordered for a patient and a space to document the administration of those medications.

Medicine cup A calibrated cup used to measure and deliver medications that usually holds 30 mL or 1 oz.

Meniscus A slight curve in the surface of a liquid.

Meter (m) Basic unit for measurement of length in the metric system.

Metered-dose inhaler (MDI) Device to deliver medication into the lungs.

Metric A widely used system of measurements based upon the meter for length, gram for weight, and liter for volume.

Micro (mc) Metric prefix that indicates $\frac{1}{1,000,000}$ of the basic unit.

Microdrip tubing Type of IV tubing that delivers 60 drops of fluid per milliliter.

Milli (m) Metric prefix that indicates $\frac{1}{1000}$ of the basic unit.

Milliequivalents (mEq) A unit of measure based upon the chemical combining power of the substance; defined as $\frac{1}{1000}$ of an equivalent of a chemical.

Minim ♏ Common unit of volume in the apothecary system.

Mixed number Fractions that are greater than one that combine a whole number with a fraction.

N

Nasogastric Type of tube that carries medication through the nose to the stomach.

Nomogram A special chart used to determine a patient's body surface area (BSA).

Numerator The top number of a fraction; represents parts of the whole.

O

Onset Moment when insulin begins to lower the glucose (blood sugar) level.

Ounce ℥ Generally implies a fluid ounce volume when discussing medications.

P

Package insert Paper insert that provides complete and authoritative information about a medication.

Parenteral Route of administration other than oral; medications that are delivered outside of the digestive tract; most often refers to injections.

Patient-controlled analgesia (PCA) Technique that allows the patient to control the amount of pain medication delivered through an IV.

Peak The time when insulin has its strongest effect upon the glucose level.

Pediatric Describes patients under the age of 18 years.

PEG tube Tube that delivers medication directly into the stomach.

Percent Means per 100 or divided by 100

Percent strength Represents the number of grams or milliliters of medication contained in 100 mL of a mixture.

Peripherally inserted central catheter (PICC) IV line that is inserted in arm veins and threaded into a central vein, often by a specially trained nurse.

Pharmacokinetics The study of what happens to a drug after it is administered to a patient.

Phlebitis Inflammation of a vein, which can be caused by an irritated IV site.

Physician order form Written or computerized form for medication orders used in an inpatient facility; can list multiple medications.

***Physician's Desk Reference* (PDR)** A compilation of information from package inserts of medications; reprinted every year.

Polypharmacy The practice of taking many medications at one time.

Port-A-Cath A device placed surgically under the skin in the chest in order to deliver drugs into a large vein.

Prefilled syringe Syringe that comes from the manufacturer with the medication already inside; usually marked in milliliters (mL) and milligrams (mg).

Prescription Written or computerized form for medication orders; used in outpatient settings

Primary line The main tubing that delivers an IV infusion, usually consisting of a drip chamber, clamp, and injection port(s).

Prime number Number other than 1 that can be evenly divided by only itself and 1, such as 2, 3, 5, 7, 11, 13, 17, 19, 23, and 29

PRN drug A medication or drug that is given only when necessary.

Proportion A mathematical statement that two ratios are equal.

Q

QSAD Abbreviation of a Latin phrase meaning "a sufficient quantity to adjust the dimensions to ..."; used when preparing solutions.

R

Rate controller Device that controls the rate of an IV infusion by using a pincher and sensor; the infusion relies on gravity.

Ratio Expression of the relationship of a part to the whole.

Ratio proportion Mathematical statement that indicates two ratios are equal.

Ratio strength The amount of drug in a solution or the amount of drug in a solid dosage such as a tablet or capsule; dosage strength.

Reconstitute Add liquid to a powder medication; must be done before administering.

Rectal Describes medication administered through the rectum, usually a suppository.

Replacement fluids Fluids that replace electrolytes or fluids lost from dehydration, hemorrhage, vomiting, or diarrhea.

Roman numerals A numeral system in which letters indicate numbers; I = 1, V = 5, and X = 10

Route Method by which a medication is to be delivered to the patient.

S

Saline lock An infusion port attached to an already inserted catheter for IV access; flushed with saline.

Scored Describes medications having indented lines indicating where they may be broken or divided.

Secondary line Also known as *piggyback*; line used to add medications or other additives to an existing IV or infusion port.

Sig Indicates the instructions for the container; found on a prescription.

Solute Chemicals dissolved in a solvent, making a solution; drug or substance being dissolved in a solution.

Solution Combined mixture of solute and solvent or diluent.

Solution strength The amount of dry drug in grams per 100 mL of solution.

Solvent (diluent) liquid used to dissolve other chemicals, making a solution.

Spansule Special capsule that contains coated granules to delay the release of medication.

Subcutaneous (sub-Q) Describes medication administered under the skin by an injection.

Sustained release Describes medication that releases slowly into the bloodstream over several hours.

Syringe Device used to deliver parenteral medications that includes a barrel, plunger, hub, leading ring, and needle.

Syringe pump Pump that provides precise control of IV infusions via a syringe inside of a pump.

T

Tablet A solid disk or cylinder that contains a drug plus inactive ingredients.

Therapeutic fluids IV fluids that deliver medication to patients.

Titrated medication Medication that is adjusted or regulated based upon its effect.

Topical medication Medication applied to the skin.

Trade name Name of a drug owned by a specific company; also called *brand name.*

Trailing ring The ring on the plunger of the syringe farthest from the needle. Do *not* measure medication from this ring.

Transdermal Describes medication administered through the skin, typically via a patch.

Tuberculin syringe A small syringe used for delivering 1 mL of medication or less parenterally.

U-100 Common concentration of insulin meaning that 100 units of insulin are contained in 1 mL of solution.

Unit Also known as USP unit; amount of a medication required to produce an effect.

United States Pharmacopeia A medication guide or reference for health care professionals.

Vaporizer Device that uses boiling water to create a mist from liquid medications; also known as steam inhaler.

Verbal order A medication spoken by the physician to a qualified health care employee who records it.

Vial Container that holds 1 or more doses of medication and is closed by a rubber stopper.

Warnings Statements found on the medication label that help the health care worker to deliver medications safely.

Credits

Reprinted with permission of Abraxis BioScience, Inc. Amiodarone HCI®, Furosemide®, Famotidine®, Gentamicin®, Arcyclover Sodium®, Oxytocin®, Thiamine Hydrochloride®, Heparin Sodium®, Clindamycin®

Courtesy of Abbott Laboratories and Knoll Labs. E.E.S. 200 Liquid®, Synthroid®, Tranxene®, Isoptin® SR, Dilaudid-HP®, Depakote® ER, Norvir®, Erythromycin®, E.E.S. Granules®, Omnicef®, Pediazole®, E.E.S. 400 Liquid®, Depakene®, Tricor®, Biaxin®, Zemplar®, EryPed®, Zosyn®

Published or replayed with permission of Amgen Inc. Neupogen®, Epogen®, Neulasta®, Aranesp®

Reprinted with permission of AstraZeneca Pharmaceuticals LP. Prilosec®, Diprivan®, Nexium®, Toprol XL®, Sero-QUEL®

Images courtesy of Baxter Healthcare Corporation. All rights reserved. Gammagard liquid®, Cipro®, Lacatated Ringer's and 5% Dextrose®, 5% Dextrose®, 0.9% Sodium chloride®

Baxter and Viaflex are registered trademarks of Baxter International Inc. Diluting Solution for Glucagon®

Courtesy of Bedford Laboratories. Ciprofloxacin®, Lorazepam®, MethylPREDNISolutions Sodium Succinate®, Leucovorin®, Doxycycline®, Midazolam®, Ranitidine®, CISplatin®

Courtesy of Caraco Pharmaceutical Laboratories. Metformin®, Digoxin®, Meperidine®

Copyright 2007, Eisai Inc. All rights reserved. Used with permission. Ontak®

© Coyright Eli Lily and Company. All rights reserved. Used with permission. Humulin®, Humalog®, Actos®, Strattera®, Glucagon®, Zyprexa®, Humatrope®, Sterile Diluent®, Gemzar®, **are registered trademarks of Eli Lily and Company**

Courtesy of Forest Laboratories, Inc. Aerobid-M®, Aerobid®, Flumadine®, Celexa®

Courtesy of Genzyme, Inc. Hectorol®

Hospira Inc., Lake Forrest, IL. USA. Heparin Sodium®, Heparin Lock Flush Solution, USP®, 0.9% Sodium Chloride®

Courtesy of Janssen Pharmaceuticals. Regranex®, Duragesic®

Courtesy of King Pharmaceuticals. Tigan®, Levoxyl®

Courtesy of Medimmune, Inc. Synagis®

Copyright © Novartis Pharmaceutical Corp. used by permission. Clozaril®, Miacalcin®, Famvir®, Gleevec®, Trileptal®, Sandostatin®

Courtesy of Novo Nordisk Pharmaceuticals, Inc. Novolin®, Prandin®

Reproduced with permission of Organon USA Inc. Pregnyl® **is a trademark of N.V. Organon.**

Copyright permission granted by Ovation Pharmaceuticals, Inc." Tranxene® **is a Registered Trademark of Sanofi-Aventis liscensed to Ovation Pharmaceuticals, Inc.**

Used with permission from Pfizer Inc. Provera®, Nitrostat®, Vantin®, Aricept®, Vistaril®, Camptosar®, Xanax®, Lipitor®, Zoloft®, Zithromax®, Zyrtec®, Fragmin®, Pfizerpen®

© Purdue Pharma L.P., used with permission. OxyContin®

Reprinted with the permission of Roche Laboratories Inc. All rights reserved. Kytril®, Rocephin®, Valium®, CellCept®

Coutresy of Sanofi-Aventis. Eloxatin®

Reproduced with permission of Schering Corporation. All rights reserved. Intron® A, Diprolene®, Nasonex®, Proventil®, Rebetol®, Clarinex®, PegIntron®

Courtesy of Teva Pharmaceuticals USA. Amoxicillin and Clavulanate Potassium®, Azithromycin®, Amoxicillin®, Cefprozil®, Cephalexin®

Courtesy of Xanodyne Pharmaceuticals. Amicar®

Index

E

F

G

H

I

J

K

L

M

N

O

P

Q

R

S

T

Converting by the Fraction Proportion Method

Procedure Checklist 3-1:

1. Write a conversion factor with the units needed in the answer in the numerator and the units you are converting from in the denominator.
2. Write a fraction with the unknown, "?", in the numerator and the number that you need to convert in the denominator. *(For most calculations, the unknown is the desired dose (D). The number you need to convert is the dosage ordered (O).)*
3. Set the two fractions up as a proportion.
4. Cancel units.
5. Cross multiply, then solve for the unknown value.

Converting by the Ratio Proportion Method

Procedure Checklist 3-2:

1. Write the conversion factor as a ratio A:B so that A has the units of the value that you are converting *(the dosage ordered)* and B has the unit of value of the *dose on hand.*
2. Write a second C:D so that C is the missing value (*desired dose*) and D is the number that is being converted *(the dosage ordered).*
3. Write the proportion in the form A:B: :C:D. *Note: When using the ratio proportion method to calculate the desired dose, C indicates the unknown value (desired dose).*
4. Cancel units.
5. Solve the proportion by multiplying means and extremes.

Converting using the Dimensional Analysis Method

Procedure Checklist 3-3:

1. Determine the units of measure for the answer and place it as the unknown on one side of the equation.
2. On the other side of the equation, write a conversion factor with the units of measure for the answer on top and the units you are converting from on the bottom.
3. Multiply the conversion factor by the number that is being converted over 1.
4. Cancel units on the right side of the equation. The remaining unit of measure on the right side of the equation should match the unknown unit of measure on the left side of the equation.
5. Solve the equation.

IV Flow Rates: F = mL/h and f = gtt/min

$$F = \frac{V}{T}, \text{ where V = volume, T = time}$$

$$f = \frac{F \times C}{60}, \text{ where C = tubing gtt/mL, 60 minutes}$$

Infusion time and volume:

$$T = \frac{V}{F} \text{ and V = T} \times \text{F, where V = volume in mL, F = mL/h, T = time}$$

BSA:

$$\text{BSA} = \sqrt{\frac{\text{height (cm)} \times \text{weight (kg)}}{3600}}\ \text{m}^2$$

$$\text{BSA} = \sqrt{\frac{\text{height (in.)} \times \text{weight (lb)}}{3131}}\ \text{m}^2$$

DMFN:

- Weight up to 10 kg, find $\frac{100 \text{ mL}}{1 \text{ kg}} \times \text{kg} = \text{DMFN mL}$
- Weight 10 to 20 kg, find $1000 \text{ mL} + \left[\frac{50 \text{ mL}}{1 \text{ kg}} \times (\text{kg} - 10)\right] = \text{DMFN mL}$
- Weight over 20 kg, find $1500 \text{ mL} + \left[\frac{20 \text{ mL}}{1 \text{ kg}} \times (\text{kg} - 20)\right] = \text{DMFN mL}$

Calculating Amount to Administer by Ratio Proportion

Procedure Checklist 7-2:

1. Calculate the desired dose using the ratio proportion 3-2 if the unit measure of the dose ordered is different than the unit of measure of the dose on hand.
2. The proportion will be set up as follows:

dosage unit : dose on hand : : amount to administer : desired dose

or

$$Q:H::A:D$$

3. Cancel units.
4. Multiply the means and extremes then solve for the missing value.

Calculating Amount to Administer by Fraction Proportion

Procedure Checklist 7-1:

1. Calculate the desired dose using fraction proportion (3-1) if the unit of measure of the dose ordered is different than the unit of measure of the dose on hand.
2. The proportion will be set up as follows:

$$\frac{\text{dosage unit}}{\text{dose on hand}} = \frac{\text{amount to administer}}{\text{desired dose}}$$

or

$$\frac{Q}{H} = \frac{A}{D}$$

3. Cancel units.
4. Cross multiply, then solve for the unknown value.

Calculating Amount to Administer by the Formula Method

Procedure Checklist 7-4:

1. Determine the desired dose. Calculate using the fraction proportion 3-1, ratio proportion 3-2, or dimensional analysis 3-3 if the unit measure of the dose ordered is different than the unit of measure of the dose on hand. Determine the dose on hand (H) and dosage unit (Q).

2. Fill in the formula: $\frac{D}{H} \times Q = A$

   D = desired dose (This is the dose ordered changed to the same unit of measure as the dose on hand.)
   H = dose on hand - the amount of drug contained in each unit.
   Q = dosage unit - how the drug will be administered such as tablets or mL.
   A = the unknown or the amount to administer.

3. Cancel the units.
4. Solve for the unknown.

Calculating Amount to Administer by Dimensional Analysis

Procedure Checklist 7-3:

Do not calculate the desired dose and amount to administer separately. Place the unknown (amount to administer) on one side of the equation then multiply a series of factors on the right side of the equation. Cancel units to determine if the equation has been setup correctly.

1. Determine the units of measure for the answer and place it as the unknown on one side of the equation. (In most cases this will be the *amount to administer.* The unit of measure will be the same unit of measure as the *dosage unit.*)
2. On the right side of the equation, write a conversion factor with the units of measurement for the desired dose on top and the units of measurement for the dose on hand. *(This is necessary if the dose ordered is a different unit of measurement than the dose on hand.)*
3. Multiply the conversion factor by a second factor: the dosage unit over the dose on hand.
4. Multiply by a third factor, the dose ordered over the number one.
5. Cancel units on the right side of the equation. The remaining unit of measure on the right side of the equation should match the unknown unit of measure on the left side of the equation.
6. Solve the equation.